Food Engineering Series

Springer's *Food Engineering Series* is essential to the Food Engineering profession, providing exceptional texts in areas that are necessary for the understanding and development of this constantly evolving discipline. The titles are primarily reference-oriented, targeted to a wide audience including food, mechanical, chemical, and electrical engineers, as well as food scientists and technologists working in the food industry, academia, regulatory industry, or in the design of food manufacturing plants or specialized equipment.

More information about this series at http://www.springer.com/series/5996

Javier Raso • Volker Heinz
Ignacio Alvarez • Stefan Toepfl
Editors

Pulsed Electric Fields Technology for the Food Industry

Fundamentals and Applications

Second Edition

Editors
Javier Raso
Tecnología de los Alimentos
Facultad de Veterinaria
Instituto Agroalimentario de Aragón-IA2
Universidad de Zaragoza
Zaragoza, Spain

Ignacio Alvarez
Tecnología de los Alimentos
Facultad de Veterinaria
Instituto Agroalimentario de Aragón-IA2
Universidad de Zaragoza
Zaragoza, Spain

Volker Heinz
German Institute of Food Technologies
Quakenbrück, Niedersachsen, Germany

Stefan Toepfl
Elea Technology GmbH
Quakenbrück, Germany

ISSN 1571-0297
Food Engineering Series
ISBN 978-3-030-70588-6 ISBN 978-3-030-70586-2 (eBook)
https://doi.org/10.1007/978-3-030-70586-2

This Springer imprint is published by the registered company Springer Nature Switzerland AG
The registered company address is: Gewerbestrasse 11, 6330 Cham, Switzerland

Preface

Since the publication of the first edition of the book Pulsed Electric Fields for the Food Industry: fundamentals and applications in 2006, the new research and breakthroughs that have occurred in the field, certainly justify the need to publish a second edition of the book. In the last years, it has been observed a growing interest in the multidisciplinary subject of PEF which has result in a major increase in the research activity, publications, training schools, conferences and applications in the food industry.

The concern of the food industry for making their activities more energy-efficient and sustainable has been an important driving force for conducting new research on PEF. There has been also an increased interest in the application of PEF as a pretreatment for the recovery of high-added value compounds from microbial cells for different applications in the food, cosmetic, pharmaceutical or bio-fuel industries. PEF for the recovery of valuable compounds from food wastes and by-products generated during food processing and biorefinery applications have also attracted considerable attention.

The Cost Action TD1104 European network for the development of electroporation-based technologies and treatments that started in 2012 contributed to the development of new applications of PEF by integrating multidisciplinary research teams. During the 4 years of duration of the action, almost 600 researchers and industrial technicians from 42 countries working in different scientific domains such as Biology, Food Science, Medicine, Biotechnology, or Environmental Science were involved in this action. In the framework of this COST action, two training Schools on Pulsed Electric Field Applications in Food and Biotechnology were organized and the event has continued to be held annually. With the aim of keeping the networking established between researchers working in different fields during the COST Action, the International Society of Electroporation based Technologies and treatments (ISEBTT) was created in 2016. This society has among its objectives to organize every two years the World Congress on Electroporation and Pulsed Electric Fields in Biology, Medicine, and Food & Environmental Technologies. Three editions of this congress have been held so far with an average of around 400 attendants.

The development in the last 15 years of pulse power systems responding to the food industry requirements has allowed the successful transfer of the technology for industrial applications. Currently, different industrial scale PEF units for different applications are manufactured for different companies and more than 150 of these units have been installed in the food industry all around the world.

This second edition tries to provide comprehensive, updated information and a view of the current state of the art in fundamentals and applications of PEF for the food industry. All chapters have been updated or revised and 13 new chapters have been added to provide the latest information on electrochemical reactions, process validation, environmental impacts, performance analysis, legal regulations and application in the food industry. The chapters have been written by recognized authors from academia, industry and R&D laboratories from many different countries of the world. The book aims to have an interdisciplinary character addressing students, academia, consultants and industrial readership.

This second edition is the result of the effort enthusiasm and interest of many contributors. We would like to express our sincerest gratitude to all of them for their efforts in completing this project.

Contents

Part I
Fundamentals of Application of Pulsed Electric Fields in the Food Industry

History of Pulsed Electric Fields in Food Processing

Werner Sitzmann, Eugene Vorobiev, Javier Raso, Ignacio Álvarez, and Nikolai Lebovka

Introduction

The first industrial applications of electricity in foods are dated about 100 years ago. Continuous and alternating electric currents were applied to heat the media and to kill germs in milk. This procedure is known today as "ohmic heating."

At the end of the 1940s Ukrainian researches, Zagorul'ko (Mil'kov and Zagorul'ko 1949) and Flaumenbaum (Flaumenbaum and Yablochnik 1950) showed that vegetable and fruit cells could be cracked by electric AC fields. This process was named in the former USSR as "electroplasmolysis." Later on, Zagorul'ko (Zagorul'ko 1958b) used pulsed electric fields (PEF) in order to enhance extraction in sugar beet processing. Already in the 1950s and 1960s numerous studies were conducted by Ukrainian and Moldavian researchers on the application of electric fields for the treatment of food plants (Sitzmann et al. 2016a, b). However, due to the technical problems and the lack of financial means, these studies never led to industrial implementation. Soviet works were practically unknown in Europe and the USA.

W. Sitzmann
Solids Process Engineering and Particle Technology, Technical University of Hamburg, Hamburg, Germany

E. Vorobiev
Laboratoire de Transformations Intégrées de la Matière Renouvelable, EA 4297, Centre de Recherches de Royallieu, Université de Technologie de Compiègne, BP, Compiègne Cedex, France
e-mail: eugene.vorobiev@utc.fr

J. Raso (✉) · I. Álvarez
Tecnología de los Alimentos, Facultad de Veterinaria, Instituto Agroalimentario de Aragón-IA2, Universidad de Zaragoza, Zaragoza, Spain
e-mail: jraso@unizar.es

N. Lebovka
Institute of Biocolloidal Chemistry named after F.D. Ovcharenko, NAS of Ukraine, Kyiv, Ukraine

J. Raso et al. (eds.), *Pulsed Electric Fields Technology for the Food Industry*, Food Engineering Series, https://doi.org/10.1007/978-3-030-70586-2_1

Very high electric field strengths were applied by Früngel (1960) in Germany using capacitor discharges which create electric arcs and a quickly expanding plasma. Mainly, a shock wave of up to 5000 bar was identified responsible for the bactericidal effects and the cracking of composite materials. This process, then titled "electrohydraulic treatment," is nowadays used, for example, by the company ELEA GmbH in Quakenbrück, Germany, to improve the structure of ham.

Generation, application, and impact of PEF on the cell membranes of food plants and fish were reported and patented in the 1960s–1980s by the German engineer Heinz Doevenspeck. The PEF process in its current applications goes back to the works of Heinz Doevenspeck, who cooperated closely with Krupp company in the 1980s.

In the early 1990s, leading university professors – in particular, Professor Knorr from the Technical University of Berlin and Professor Barbosa-Cánovas from the Washington State University – quickly took up the results, which had primarily produced at Krupp. Since this moment it was initiated a profound research work that constitutes the basis of the current industrially implemented technology under the term PEF (pulsed electric fields).

Early Works Prior to the Middle of the Twentieth Century

Bactericidal Effects

The bactericidal effects of electricity can include chemical, thermal, mechanical, and other effects. The attempts to apply electricity for killing bacteria and sterilization purposes were applied at the end of the nineteenth century. Table 1 presents the collection of the representative pioneering inventions related to the bactericidal effects of electricity. In early studies, the electricity was applied to damage the microbes in meat (Jones et al. 1886), purification of water and other liquids (Schroeder 1891; Roeske 1893), sterilization of beer (Wagner and Marr 1895), milk (Goucher 1909; Shelmerdine 1915; Mittelmann 1951), manufacturing fermented beverages (Kuhn 1909), juices (Schweizer 1924), and sterilization of other liquids (Bartley 1909; Bloom 1915; Rudd 1920; Louder 1933; Wright 1937). The mechanisms of the microbial killing effects were attributed to the formation of toxic compounds by electrolysis (Cohn and Mendelsohn 1879; Thornton 1912), a specific manifestation of the effect of electricity (Spilker and Gottstein 1891), and thermal or electrochemical effects (Heller 1897; Thiele and Wolf 1899). For example, the killing of yeast suspended in beer wort or grape juice was attributed to the formation of toxic compounds (Kleiber 1924). The lethal effects of AC treatment on yeast cells in grape juice were explained by the formation of toxic substances like free chlorine. However, the stimulation effects of electricity on the growth of bacteria and yeast were also observed (Stone 1909).

Very active studies devoted to the application of AC treatment for milk pasteurization were conducted (Beattie and Lewis 1913; Lewis 1914). The disinfection

Table 1 Representative inventions related to the preservation of foodstuffs by electric field treatments

Patent nos.	Title and comments	References
US337334A	Applying of electricity for destroying living organisms in the bodies of slaughtered animals	Jones et al. (1886)
US447585A	Liquid purifier	Schroeder (1891)
US501732	Method of and apparatus for purifying water	Roeske (1893)
US535267A	Electrolytic conduit for beer or other liquids.	Wagner and Marr (1895)
US645569A	Receptacle for sterilized perishable substances	Roberts (1900)
US922134A	Method of treating milk	Goucher (1909)
US930023	Process sterilizing fluids	Bartley (1909)
US936328A	Process of manufacturing fermented beverages	Kuhn (1909)
US1044201A	Process of preserving	Lincoln (1912)
US1147558A	Apparatus for the sterilization of milk	Shelmerdine (1915)
US1162213A	Process and apparatus for the electrical treatment of liquids and fluids and the products resulting therefrom	Bloom (1915)
US1360447A	Apparatus for electrically treating liquids	Rudd (1920)
DE415090C	A method for preserving feedstuffs juicy by the action of the electric current	Schweizer (1924)
US1479725A	Electric fluid sterilizer	Wright (1924)
US1900509A	Process for sterilization of liquids	Louder (1933)
US1934703A US1934704A	Electrical sterilizing apparatus	Golden (1933a, b)
US2046467A	Sterilization of liquids by means of oligodynamy	Krause (1936)
US2081243A	Apparatus for pasteurizing liquids	Wright (1937)
US2132708A	Method for treating materials and electrical treating apparatus, especially sterilization	Smith (1938)
US2365924A	Process of preserving fresh food products	Willard (1944)
US2550584A	Milk pasteurization method and apparatus	Mittelmann (1951)
US2576862A	Method and apparatus for preserving foodstuffs	Smith and Grinnell (1951)

effects were explained by heat produced by the electric current (Anderson and Finkelstein 1919), whereas the importance of the nonthermal effects of electricity was also stated (Beattie and Lewis 1925). The “electropure process” with heating the milk up to 70 °C allowed effective inactivation of different microorganisms (Fetterman 1928; Gelpi Jr and Devereux 1930; Getchell 1935). In the 1930s, the “electropure process” was accepted as a safe milk pasteurization technology in Europe and the USA (Moses 1938). This milk pasteurization technology was widely used up to 1950, but later on, it was replaced by milder thermal technology (Reitler 1990; Lewis and Neil 2000).

Assistance of Food Processing Operations

The effects of treatment by electric currents in different food processing operations were also intensively studied. More than one century ago, electricity was applied for the assistance of sugar extraction from sugar beet (Schwerin 1901, 1903). The effects were explained by the processes of electrolysis and electroosmosis.

The first application of electricity for cooking of foods was done at the end of the nineteenth century (Capek 1890; Crompton 1894; Oneill 1895). It was noted that "the use of electricity for cooking was more economical and efficient than the use of coal" (RAF 1893). The different fruits, vegetables, and meat were used in cooking experiments with the application of alternating electric currents (Sater 1935). The coking assisted by electricity allowed a significant reduction of the cooking time and improving the quality of products with retention of the natural flavor of the fruits and vegetables. The representative inventions related to the assistance of typical food processing operation by electricity are presented in Table 2. Examples of application of electricity for cleaning the sugar juice (Schwerin 1929), pasteurizing of cheese (Parsons and Richardson 1930), cooking (Berkeley 1935; Griffith 1942), heating (Courtright 1934; Chester and Rexford 1943; Simpson and Stirling 1984; Reznik 1997), extraction (Reld 1946), osmotic impregnation (Berthold 1940), and other processes (Schade 1951; Miyahara 1986, 1988) have been described.

Table 2 Some inventions related to applications of passing an electric current directly through the food products for different food processing operations

Patent nos.	Title and comments	References
SU7981A1	The method of cleaning sugar juice by electric current	Schwerin (1929)
US1774610A	Electrothermal treatment of cheese. The pasteurizing of cheese under pressure and heating by electric current	
US1970360A	Electric food heating apparatus. The apparatus for cooking or heating food products (sausages, frankfurters)	Courtright (1934)
US2025085A	Electric cooking apparatus. The apparatus for cooking or heating by passing an electric current directly through the food product	Berkeley (1935)
US2219772A	Method for treating food. The treating of foods by an electric current for better impregnation with sugar or salt	Berthold (1940)
US2299088A	Apparatus for electric conductance cooking. The cooking of food (e.g., pieces of meat) by electrical conductance	Griffith (1942)
US2324837A	Electric heater. The heating by passing a current through the liquid products, such as milk, vegetable and fruit juices, beer, and the like in order to pasteurize the liquid	Chester and Rexford (1943)
US2400951A	Manufacture and treatment of animal and vegetable materials. Application of electric treatment for the extraction process	Reld (1946)
US2569075A	Prevention of enzymatic discoloration of potatoes by passing a high-frequency alternating electric current through them	Schade (1951)

Plasmolysis and "Electrocution" of Foods

The plasmolysis of plant cells reflects the water loss from cells and possible loss of their vitality under the influence of external factors such as irradiation by visible, ultraviolet or infrared light, X-rays, ultrasonication, mechanical pressing, heating, freezing, magnetic and electric fields, or chemical additives (ionic substances, solvent, polymers, and surfactants) (Nasonov and Aleksandrov 1940; Stadelmann 1956). The simple osmotic concept to explain the water transfers in plants was first proposed by Dutrochet (1837). The pioneering studies revealed the effects of cell permeability on transfers of different cell substances (various neutral salts, saccharides, coloring matters) (Nägeli 1853; Pringsheim 1854; De Vries 1871; Martin 1905). It was demonstrated that exchange between the living matter (protoplasm) and outside medium is delimited by some sort of integumentary layer called plasmatic membrane (Pfeffer 1877; Tswett 1896). A regulatory role was attributed to this boundary layer in the metabolism of cell. At the beginning of the twentieth century, intensive studies of plant cells' molecular structure started. It was established that the impermeability of the plasmatic membrane is a peculiarity of the living cell and it disappears after death (Lillie 1909). The thickness of the membrane was estimated as ≈6 nm (Devaux 1903; Zangger 1908; Weis 1926). The significantly different plasmolysis effects were revealed for different plant materials (Martin 1905). The electrical measurements of the impedance of suspensions of erythrocytes allowed to conclude that cells are composed of low conducting membranes surrounding highly conducting cytoplasm (Höber 1910). The early studies of cells evidenced the low conductivity of membranes and the presence of capacities of the intracellular compartment (plasma membrane, tonoplast, vacuole, cytoplasm) (Philippson 1921; Fricke and Morse 1925; Cole 1940).

However, for many years the nature of the plasmatic membranes was not well understood, and many discussions on this subject were initiated until the mid of the twentieth century. Experimental methods used at that time for the study of plasmolysis were based on staining cells with different dyes, and significant effects of dyes on the observed results were not excluded. It was recognized that dyeing can impair the vitality of the cells (Küster 1939). Moreover, some investigators have denied the existence of the plasma membrane and proposed different mechanisms for explanation of staining (Nasonov and Aleksandrov 1940) and electric properties of the cells (Lepeschkin 1930).

For the first time, the effects of electrical treatment (direct electric current) on plasmolysis of plant cells were visualized using the optical microscopy for observation of onion epidermis (*Allium cepa* L.) (Fig. 1) (Becker and Beckerowa 1934). The effect of electricity on the osmotic movement of water into and out of plant cells and tissues was taken into account (Bennet-Clark and Bexon 1943; Kramer and Currier 1950). It was speculated that there is an electroosmotic component in the uptake of water by bulk storage tissue (Teoreli 1949).

In the line with these ideas, Poland sugar technologist Jaroslav Dedek (1890–1962) conducted in 1947 some experiments studying the effects of DC on

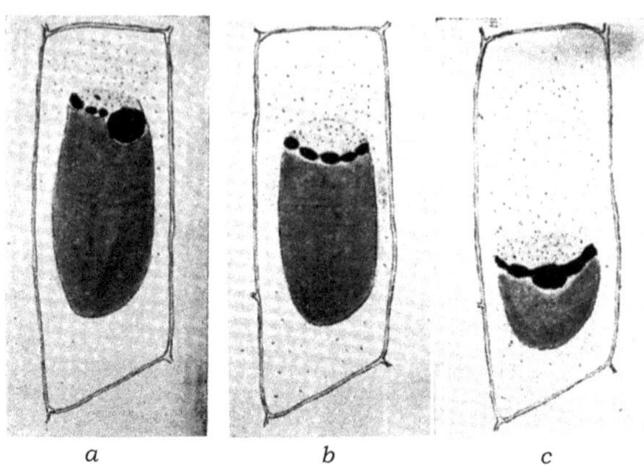

Fig. 1 Damage of an isolated tonoplast when direct current flows through the onion epidermis (*Allium cepa* L.). Neutral red was used to stain cells (Adapted from (Becker and Beckerowa 1934))

the sugar beet tissue. Short biography of Dedek is presented by Fronek (2012). In his experiments, Dedek treated the sugar beet samples at an electric field strength of 40 V/cm (Zagorul'ko 1958b). A significant increase of the electric current during the treatment was observed, and it was speculated that it reflects the damage of cells. Dedek named this process as "electrocution." This word is derived from "electro" and "execution," and it is also used for accidental death caused by electric shock.

Electroplasmolysis

Early Studies in the 1950s

The Role of Anatolii Zagorul'ko

Dedek's work on "electrocution" highly stimulated Anatolii Zagorul'ko to apply electricity for the assistance of sugar beet plasmolysis and the facilitation of sugar extraction. Anatolii Ya. Zagorul'ko (1920–1983) (Fig. 2) was born in the village of Globin, Poltava Oblast (Ukraine), in the family of a teacher (Chernyavskaya 2019).

He graduated from the Chemical Department of Kharkiv University (Ukraine) in 1941. From 1943, he started to work as a chemical engineer at different Ukrainian sugar factories. In 1955–1983, he worked at the Research Institute of Sugar Industry (Kyiv) as a head of the laboratory. He defended his PhD thesis in 1958 (Zagorul'ko 1958b) and doctor science habilitation thesis in 1974 (Zagorul'ko 1974). His PhD thesis "Obtaining of diffusion juice with the help of electroplasmolysis" was completely devoted to the problem of plasmolysis of sugar beet under the action of electricity.

In the 1949–1953, Zagorul'ko in cooperation with Mil'kov patented different constructions of continuous electrotreatment apparatuses (so-called electroplasmolyzers) for the treatment of sugar beets (Mil'kov and Zagorul'ko 1949, Zagorul'ko and Myl'kov 1953). Figure 3 shows the principal scheme of electrotreatment

Fig. 2 Anatolii Ya. Zagorul'ko (1920–1983)

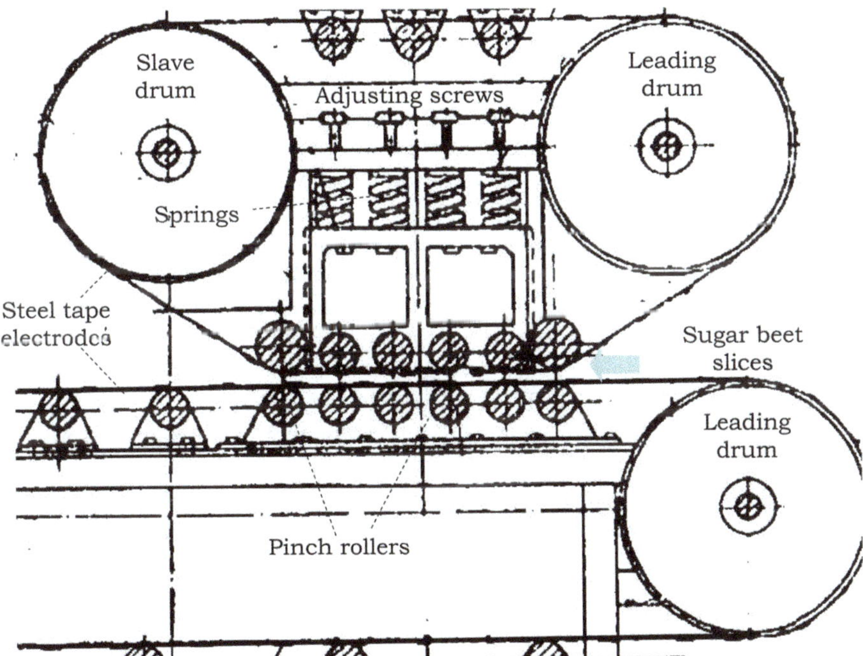

Fig. 3 Continuous electrotreatment apparatus (electroplasmolyzer) of the belt type (Mil'kov and Zagorul'ko 1950)

apparatus of belt type. The electricity can be applied between two rotating steel tapes (electrodes).

To explain the effects of electric treatment on the sugar beet tissue, it was speculated that electricity provokes the selective damage of plasmatic shells with insignificant heating of surrounding medium (Zagorul'ko 1958b, Zagorul'ko 1957b). This type of plasmolysis was called as electroplasmolysis. Zagorul'ko explained electroplasmolysis by exceptional electroosmotic jerking of plasmatic shells. He noted that the electroplasmolysis has an insignificant effect on damage of cell walls, whereas the traditional thermal plasmolysis causes disruption of cell walls and resulted in release of pectin substances into the juice. He concluded that the use of electroplasmolysis allows obtaining cleaner sugar beet juice.

The reversible (or unfinished) electroplasmolysis has been also observed at relatively small electric fields (with strength below 400 V/cm). Such type of electroplasmolysis allows effective extraction of juice using pressing or diffusion techniques at moderate temperatures (below 60 °C). He estimated energy consumptions of ≈4–5 kJ/kg and the electrical potential on the plasmatic shells ($u_m \approx 0.953$ V) required for the electroplasmolysis of the sugar beet (Zagorul'ko and Myl'kov 1953). Moreover, Zagorul'ko estimated the electrical conductivity of the membrane ($\sigma \approx 10^{-6}$ S/cm), so it behaves as an insulator layer between intra- and extracellular media.

For characterization of efficiency of electrical damage, Zagorul'ko introduced the degree of plasmolysis index defined as (Zagorul'ko 1958b)

$$Z_\rho = (\rho - \rho_i)/(\rho_d - \rho_i), \quad (1)$$

where ρ, is the electrical resistivity of the cell tissue and the subscripts "*i*" and "*d*" correspond to the resistivity of untreated (initial) and completely damaged tissues, respectively.

Zagorul'ko investigated the time dependencies of Z_ρ for sugar beet treated at different electric field strengths, E, and temperatures, T (Zagorul'ko 1958a). For better regulation of treatment efficiency, he proposed the treatment by pulsed electric fields (PEF). His first PEF generator produced exponential pulses with a duration of 20μs and with electric field strength up to E = 20 kV/cm using the small distance between electrodes (1–2 mm). Figure 4 presents the dependence $Z_\rho(t)$ obtained for the PEF treatment of sugar beet at the room temperature. Inset shows the principal scheme of the pulse generator developed by Zagorul'ko in 1955–1957 (Zagorul'ko 1958a). The PEF treatment allowed obtaining a high level of electroplasmolysis for the treatment time above several microseconds. All these discoveries were revolutionary for that time.

The Role of Boris Flaumenbaum

Nearly at the same starting period, similar studies on juice extraction from different fruits and vegetables assisted by electricity were started in Odessa (Ukraine) by Flaumenbaum. Boris L. Flaumenbaum (1910–1996) (Fig. 5) was born in the family of a journalist and lived in Odessa (Bezusov et al. 2010). He finished 2-year chemical school, and in 1931 he graduated from the Odessa Food Institute with the qualification of engineer and food technologist. During 1931–1935, he worked at different food factories of Ukraine and Moldova, and in 1936–1941 he was a senior researcher at the Odessa All-Union Research Institute of the Canning Industry (VNIIKP).

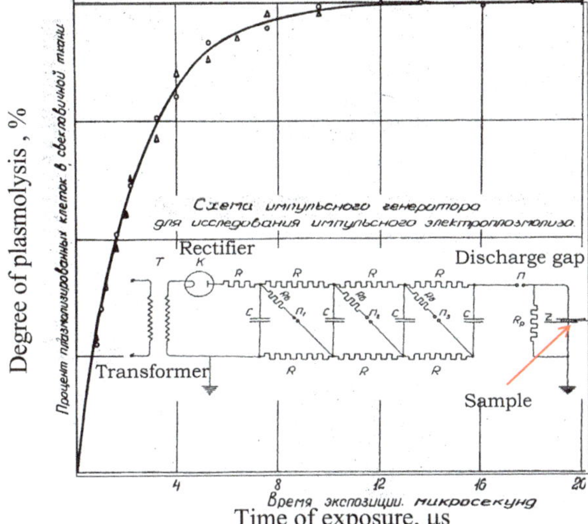

Fig. 4 Electroplasmolysis degree Z_ρ versus the time of treatment t dependences for sugar beet tissue at different values of electric field strength E (Zagorul'ko 1958b)

Fig. 5 Boris L. Flaumenbaum (1910–1996)

During this period, Flaumenbaum published about two dozen articles, and in 1941 he defended his PhD thesis "Extraction of juice from vegetable raw material" (Flaumenbaum 1941). Demobilized after the War in December 1945, Flaumenbaum returned to Odessa, and in January 1946 he started to work as a senior lecturer at the Department of Conservation Technology in Odessa Technological Institute of Food and Refrigeration Industry (now known as Odessa State Academy of Food Technologies). His scientific activity was related to the extraction of juices from fruits, thermal sterilization, and canning of foods. He is author of well-known textbooks (more than 20) published in Russian, such as *Technology for preservation of fruits and vegetables* (Fan-Jung et al. 1956), *Theoretical foundations of sterilization of canned food* (Flaumenbaum 1960) (1960), and many others. In 1969, he defended his doctor science habilitat thesis "Problems of intensification of technological processes of preservation of food products" (Flaumenbaum 1969).

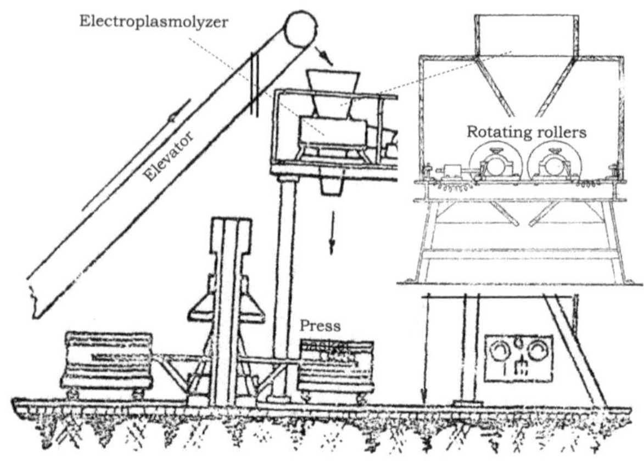

Fig. 6 Industrial-scale setup for continuous electroplasmolysis based on the electroplasmolyzer with rotating electrodes (Flaumenbaum 1953)

In 1949, Flaumenbaum published his first study related to the electrical treatment of grapes, apples, and carrots (Flaumenbaum 1949). At the same time, he patented and constructed (in cooperation with engineer Yablochnik) a roller-type electroplasmolyzer for the treatment of vegetable and fruit tissues by low-frequency electrical current (Flaumenbaum and Yablochnik 1950) (Fig. 6). The industrial-scale prototype for AC electroplasmolysis was manufactured, installed, and operated on 1949–1953 on Tiraspol fruit factory (grapes, apricots, plums, apples), Chisinau and some other Moldavian wineries (grapes), and Odessa juice plant (carrots) (Flaumenbaum 1953) (Fig. 6). The calculations of the rotation rate required for the effective plasmolysis were performed. In 1951, the similar testing of electroplasmolyzer was also performed at the Chisinau Glavvino plant and the Chisinau branch of the Magarach Institute (Prichko and Kudryavtseva 1951). The authors observed an increase in the yield of wort for grapes treated by electroplasmolyzer.

In all performed tests, a strong dependence of the efficiency of electroplasmolysis on the type of treated material was observed. The significant heterogeneity of electrical treatment was observed for different materials (Grudinovker 1954a). Flaumenbaum discussed the problem of the construction of the universal device suitable for all types of raw materials (Flaumenbaum 1953). He has proposed to use a device with finned electrodes that allow the simultaneous capture and grinding of the raw materials.

Later on in the 1960s–1980s, Flaumenbaum with coworkers formulated the concept of the electrostability of fruits and berries (Flaumenbaum and Kazandzhii 1966; Kazandzhii 1966; Kazandzhii and Flaumenbaum 1977). Electrical resistances of fruit tissues steeped in sugar syrups were measured to determine the degree of plasmolysis (Flaumenbaum and Shin 1990). The studies for different fruits and berries allowed concluding that high level of electroplasmolysis required some optimum values of treatment time and electric field strength, E. For characterization of the efficiency of electroplasmolysis, Flaumenbaum introduced the value of electrostability defined as

$$K = tE^2, \quad (2)$$

where t is the time of electroplasmolysis and E is the electric field strength.

It was demonstrated that the value of K depends upon the protocol of treatment (AC or DC), degree of tissue grinding, and type of the material. For example, the highest electrostability was observed for the apples and the lowest for the strawberries. Applications of electroplasmolysis for the production of fruit juices (Flaumenbaum and Al-Saadi 1966; Flaumenbaum 1968; Shengeliia 1974) and red wines (Flaumenbaum et al. 1991) were also discussed. The electrohydraulic effects in the processing of grape pulp were also studied (Yatsko and Zhuravleva 1970; Yatsko et al. 1971). The high voltage pulse device was included in the continuous production line of the Odessa cannery with a productivity of 3.7 tons per hour. The processing allowed increasing the juice yield by 7.7% as compared with control. The increase in the iron content of juice was also reported.

Other Studies of Electroplamolysis in the 1950s

Table 3 presents a collection of the representative inventions related to early studies of electroplasmolysis performed in the 1950s. To ensure the uniformity of electric treatment, some devices were proposed based on knife electrodes. Figure 7 presents the device with circular cutting knives that serve as electrodes (a) (Grudinovker 1954b) and the device with an obliquely mounted channel and a rectangular cross-section (b) (Rozhdestvensky 1955).

After fundamental works of Zagorul'ko and Flaumenbaum, the idea of electrical treatment of food products became very popular in the 1950s in the former USSR. For example, it was demonstrated that the electrical treatment between roller electrodes allows increasing the degree of plasmolysis of sugar beet slices up to 95–96% (Kartashov and Koval 1955, 1956).

The special conference on the electrical methods of treatment of food products was organized on 7–13 September 1958 in Kiev Technological Institute of the Food Industry, Ukraine. It included about 350 participants, and more than 50 reports were presented (Vologdin 1959). The applications of electrostatic fields; direct, low-frequency, and high-frequency currents; ultrasound; and infrared, ultraviolet radiation, X-rays, and gamma irradiations for processing of food raw materials were discussed. The problems of electroplasmolysis, electrical breakdown, and burns of food products were also discussed (Pavlov 1959). The feasibility of using electroplasmolysis in the food industry was noted. The scientific activity on the electrical methods of food product treatment in the 1950s was presented in the annotated bibliographic index of domestic (former USSR) and foreign literature (Anonymous 1962).

Table 3 Inventions related to applications of electroplasmolysis performed in the 1950s

Patent nos.	Title and comments	References
SU89009 A1	Electroplasmolyzer for obtaining and purification of diffusion juice from sugar beet. The sugar beet slices are treated by industrial frequency electric current between the rotating roller electrodes	Mil'kov and Zagorul'ko (1949)
SU92191 A1	Method of electroplasmolysis of sugar beet slices and similar slices. The device consists of two converging metal (steel) tapes for electrical treatment of slices under the pressure (Fig. 3)	Mil'kov and Zagorul'ko (1950)
SU83502 A1	Electroplasmolyzer for processing vegetables and fruits. The device of the roller type for AC treatment of the crushed raw materials (apples, cherries, grapes, carrots, etc.) (Fig. 6). After treatment the processed materials is pressed	Flaumenbaum and Yablochnik (1950)
SU96911 A1	Electroplasmolyzer for the processing of sugar beets and other vegetables. The device includes the circular cutting knives that serve as electrodes (Fig. 7a)	Grudinovker (1954b)
SU100094 A1	Electroplasmolyzer for plant materials. The device represents an obliquely mounted channel with a rectangular cross-section (Fig. 7b). The cylindrical (graphite or carbon) electrodes are located perpendicularly to the axis of the channel. The device ensures the uniformity of electric treatment	Rozhdestvensky (1955)
SU100624 A1	Device for the electroplasmolysis of sugar beet slices. The cutting of sugar beet is carried out by adjacent knifes simultaneously with electric current treatment	Sichevoi (1955)
SU104718 A1 SU112554	Electroplasmolyzer for plant materials. The device is constructed as the centrifuge rotor. The walls are formed by vertically mounted electrodes with knives. Inside the rotor four diametrically located electrode blades are installed. The material cutting and treatment with electric current are done simultaneously	Kogan (1955) Kogan (1958)
SU109588 A1	Method of electroplasmolysis of sugar beets during its cutting and a device for implementing the method. The centrifugal cutting device includes the electroplasmolyzer between diffusion knives	Zagorul'ko (1957a)

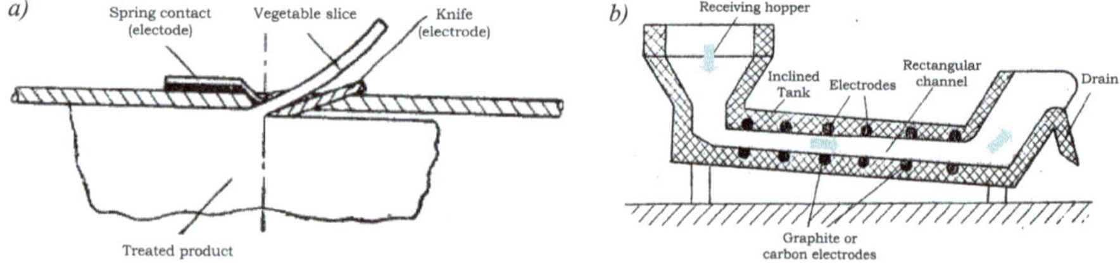

Fig. 7 The device with the circular cutting knives that serve as electrodes (**a**) (Grudinovker 1954b) and the device with the obliquely mounted channel and a rectangular cross-section (**b**) (Rozhdestvensky 1955)

Food-Oriented Applications of Electroplasmolysis in the 1960s–1990s

Experimental Studies and Development of Methods

Starting from 1960s the interest to studies of the DC and AC electroplasmolysis was renewed in a former USSR, mainly in Moldova and Ukraine. Table 4 presents some inventions based on these studies.

In the 1960s, different pilot experiments on DC and AC electroplasmolysis were realized on the base of the Institute of Applied Physics, Chisinau, Moldova. The pilot-scale machine with a productivity of 3000 kg/hour was tested at Bendery canning factory in 1966 (Kogan 1968). The electrical treatment (E = 150–500 V/cm, t = 1–50 ms) allowed enhancement in juice yield from whole apples up to ≈11–20%. Particularly, the effects of electricity on juice yield, the regimes of electroplasmolysis assisted by vacuum (Lazarenko et al. 1967), and electroplasmolysis of grinded raw plant material (Lazarenko et al. 1968) were studied. Different regimes of

Table 4 Some inventions related to methods of electroplasmolysis issued in the 1960s–1990s

Patent nos.	Title and comments	References
SU194016	Method of electroplasmolysis of cells of raw plant material. Vacuum-assisted electroplasmolysis	Lazarenko et al. (1967)
SU212147	Method of electroplasmolysis of grinded raw plant material. The electric treatment by AC of industrial frequency with strength of 50–350 V/cm	Lazarenko et al. (1968)
SU446266 A1	Method of electroplasmolysis of plant raw materials. The method uses the separate flows with increase of the electric field strength from 30 to 150 V/cm	Scheglov and Chebanu (1974)
SU799711 A1	Method for producing of juice from fruits. The raw materials (grinded to a size of 0.1–1 cm3) are treated by the electric field with strength of 100–150 V/cm and material is moved at velocity of 0.1–2 m/s	Papchenko (1981)
SU854984 A1	Method for producing of diffusion juice. The sugar beet slices are pre-scalded and moved to the diffusion chamber. For de-sugaring of the slices, they are moved between carbon (+) and steel (−) electrodes in a DC with an electric field strength of E = 30–40 V/cm. The counterflow by the preheated water (40–50 °C) is used to produce the diffusion juice. The total time of de-sugaring is 60–70 min, and the diffusion temperature is controlled by changes of E	Bazhal et al. (1981)
SU912755 A1	Method for producing of diffusion juice. The method uses de-sugaring of sugar beet slices using the treatment by DC with electric field strength of 30–40 V/cm. The de-sugaring is accompanied by electrodialysis of the juice through bipolar and anion exchange membranes	Bazhal et al. (1982)
SU1789563 A1	Method for the production of diffusion juice. The sugar beet slices are subjected to electroplasmolysis and washed. Then de-sugaring of slices in a counter current extractor is performed	Katroha et al. (1993)

electroplasmolysis of grinded plant tissues were tested at electric field strengths of E = 50–350 V/cm (Shcheglov et al. 1967). The effects of current frequency (50–2400 Hz) on the efficiency of electroplasmolysis of plant tissues were investigated (Fursov and Bordiyan 1974). An increase in frequency reduced the nonuniformity of ohmic heating. The maximum juice output was observed at some optimum temperature during electric treatment. Tests were done for carrots, plums, apricots, grapes, and apples on the industrial scale of Odessa juice plant (Lazarenko et al. 1977).

The changes in electrical conductivity of tissues during application of electric field were analyzed for various biological materials (Papchenko et al. 1992; Botoshan et al. 1994). Some relationships were proposed to calculate the volume of the extracting agent needed to provide solid-liquid extraction during electroplasmolysis (Botoshan and Berzoj 1994a). The application of electroflotation was discussed to separate electrically charged suspended substances from the juice obtained by electroplasmolysis (Botoshan and Berzoj 1994b).

Electroplasmolysis was also applied to treat sea products. The electrotreatment of fish meal allowed decreasing the fat and moisture contents. Final products contained less than 10% fat (Kolpakov et al. 1991). The application of electrotreatment allowed better oil extraction and drying the fish meal. The application of electroplasmolysis for fish raw material treatment was discussed (Skimbov and Berzoj 1992). The treatment allowed enhancing the output of the fat and vitamin A. Different regimes of electroplasmolysis and centrifugation were discussed (Berzoj et al. 1993b). The effects of electroplasmolysis on the Antarctic shrimps were also studied (Berzoj et al. 1993a). The electrotreatment of fish stock prior to heat treatment allowed producing of fodder meals with better qualitative and quantitative characteristics (Skimbov and Berzoj 1994).

Many works were devoted to the intensification of sugar extraction from electrotreated sugar beet slices. The action of electric field on cells of sugar beet was discussed (Karpovich et al. 1981). The aqueous extraction at moderate temperatures (50–60 °C) assisted by AC and DC electric fields improved extraction of sugar and quality of extracted juice (Kupchik et al. 1982, 1984; Bazhal et al. 1984; Kupchik 1986). For example, the treatment of sugar beet slices was done by DC at the electric field strength of E = 0–20 V/cm and temperature of 50 °C (Kupchik et al. 1984). In other studies, the DC electric field strength of 30–40 V/cm was applied for the extraction and electrodialysis of juice through bipolar and anion exchange membranes (Bazhal et al. 1981, 1982, 1983). Changes in the ultrastructure of sugar beet cells during diffusion under the temperature and electric field were discussed (Kupchik et al. 1987). Empirical relations were proposed for the disintegration degree of sugar beet in function of electric field strength and treatment time (Shulika et al. 1987). Application of the combined sequential AC and DC treatments allowed an increase of sugar diffusivity (Shulika 1988). This combined AC + DC method was tested at the pilot scale at the Yagotin sugar factory (Ukraine) in 1986–1987. Besides the indicated studies, during the 1980s and 1990s the effects of the electric field on sugar extraction and pH values of juice-cossette mixture (Katroha et al. 1984; Katrokha et al. 1987), the mechanisms of the thermal and electroplasmolysis

(Dolinskaya et al. 1992; Dankevich 1995), and the effects of electric treatment on sucrose diffusivity (Matvienko 1996a) have been studied, including also the development of numerical models (Bazhal 1995). Application of electroplasmolysis for better recovery of phenolic compounds into the grape juice was discussed (Kalmykova 1992). The anthocyanin extraction from the grape skin was done at the treatment conditions: electric field strength of $E = 650–750$ V/cm and time of treatment $t = 0.4–0.6$ s.

Several books summarizing obtained results related to electroplasmolysis of foods were published in this period in Russian language (Kogan 1968; Rogov and Gorbatov 1974; Lazarenko et al. 1977; Bologa and Litinsky 1988; Rogov 1988). During this period many PhD and doctor of science (habil) dissertations were also defended in Krasnodar (Reshetko 1970; Scheglov 1971), Kiev (Papchenko 1979; Cebanu 1988; Shulika 1988; Dolinskaya 1993; Bazhal 1995; Dankevich 1995; Matvienko 1996b) and Odessa (Kalmykova 1993; Ilyeva 1996) and Moscow (Kupchik 1991).

Construction of Electroplasmolyzers and Electroextractors

Different constructions of the electroplasmolyzers were patented in the period between the 1960s and mid-1990s. Most of them were patented in USSR and some of them were also patented in the USA, Germany, and France (Table 5). Cumulative number of SU patents (former USSR inventor's certificate) versus the year of publication is shown in Fig. 8. After the initial stagnation period in 1959–1965, the number of patents gradually increased and reached up to about 90 in 1995.

Figure 9 presents the drum-type device (Scheglov et al. 1969). In this device, three rows of movable paddles are placed on the surface of the drum. The gaps between the drum and sieve gradually decrease downward. The device is powered by a three-phase source.

Figure 10 presents a device with rotating disk electrodes (Biryukova and Preobrazhensky 1970). The device was tested for treatment by AC at electric field strength of $E = 628–733$ V/cm and time of treatment of $t = 0.1–0.3$ s. The elevation of temperature during the treatment was up to 32–38 °C.

Figure 11 presents electroplasmolyzers of cylindrical type for the treatment of crushed plant material (pulp) (Lazarenko et al. 1976, 1979). The pulp is treated by electrical pulses in the cylindrical chambers. The crushed material is treated in the cylindrical chamber between outer annular electrodes and the central grounded cylindrical electrode. The treated material can be preliminary densified in the perforated chamber (Fig. 26a). For the construction presented in Fig. 26b, the external electrodes are divided by dielectric rings. It allows multistep treatment (Lazarenko et al. 1976).

Figure 12 presents different constructions of devices with two pairs of roll electrodes (Gatin et al. 1975, 1978). The device presented in Fig. 12a provides a more uniform loading of the power source (Gatin et al. 1975). The device presented in Fig. 12b is constructed for the treatment of lumpy raw materials. The upper pair of

Table 5 Some inventions on devices for food electroplasmolysis issued in the 1960–1990s

Patent nos.	Title and comments	References
SU244880 A1	Electroplasmolyzer for extracting juice from pulp. The drum type device (Fig. 9)	Scheglov et al. (1969)
SU286491 A1	Electroplasmolyzer for extracting juice from plant materials. A device with rotating disk electrodes (Fig. 10)	Biryukova and Preobrazhensky (1970)
DE2402947A1	Vorrichtung zur durchfuehrung des plasmolyseverfahrens zerkleinerter pflanzlicher rohstoffe (device for carrying out the plasmolysis process of vegetable raw materials)	Lasarenko et al. (1975)
SU407457 A1 SU429601 A1	Electroplasmolyzers for treatment of crushed plant materials The pulp is treated by electrical pulses in the cylindrical chambers (Fig. 11a, b)	Lazarenko et al. (1976) Lazarenko et al. (1979)
SU459211 A1 SU627813 A1	Electroplasmolyzers for treatment of plant materials. The devices include two pairs of roll electrodes (Fig. 12a, b)	Gatin et al. (1975) Gatin et al. (1978)
SU592406 A1	Electroplasmolyzer for treatment of plant materials. The device includes roll electrodes of various sizes (one large and three smaller) (Fig. 13)	Reshetko (1978)
SU745427 A1	Device for the electrical treatment of plant materials before drying (Fig. 14)	Martynenko et al. (1980)
SU786966 A1 SU888921 A1	Electroplasmolyzer for plant materials. The device consists of a rotating drum with conical dielectric fingers (Fig. 15)	Papchenko et al. (1980) Papchenko and Kuharuk (1981)
SU808047 A1	Electroplasmolyzer for treatment of plant material. The device with two rollers covered with flexible electrodes (Fig. 16)	Bejenar (1981)
SU1588363 A1	Device for electric treatment of food raw materials. The cylindrical screw type device (Fig. 17)	Skimbov and Berzoy (1990)
US4608920A	Electroplasmolyzer for processing vegetable stock	Scheglov et al. (1986)
FR2621456A1	Procédé de traitement de matières premières végétales et dispositive pour sa réalisation (method for treating vegetable raw materials and device for carrying it out)	Papchenko et al. (1990)
US5031521A	Electroplasmolyzer for processing plant raw material	Grishko et al. (1991)

rolls is equipped with radically fixed knife electrodes (Gatin et al. 1978). The devices provide more uniform electrical treatment of the material.

Figure 13 presents electroplasmolyzer with one large and three small roll electrodes (Reshetko 1978). The device provides more uniform multistep treatment.

Figure 14 shows a device for the electrical treatment of plant materials before drying (Martynenko et al. 1980). The device includes roller electrodes, a conveyor, and an additional roller electrode located above the conveyor.

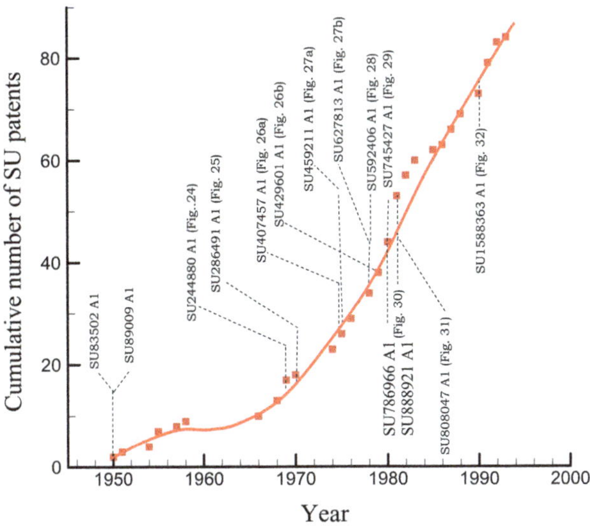

Fig. 8 Cumulative number of SU patents on electroplasmolyzers (former USSR inventor's certificate) versus the year of publication. Constructions of some devices are presented on Figs. 9–17

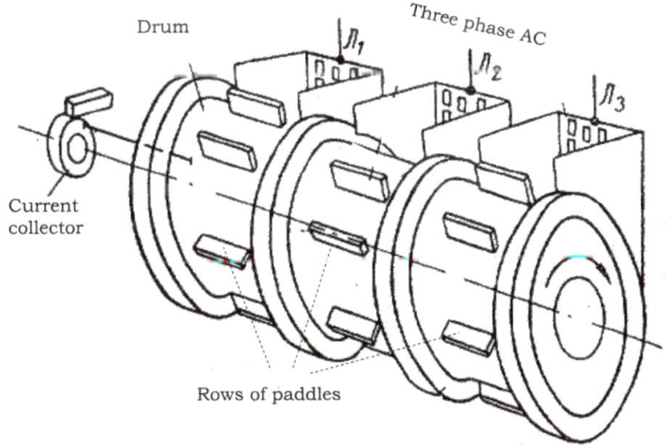

Fig. 9 Electroplasmolyzer for extracting juice from pulp. The drum-type device (Scheglov et al. 1969)

Figure 15 presents a drum-type electroplasmolyzer for treatment of plant materials (Papchenko et al. 1980; Papchenko and Kuharuk 1981). The device consists of a rotating drum with conical dielectric fingers. The drum is divided into three compartments. The drum in the middle compartment is formed by metallic plates (electrodes).

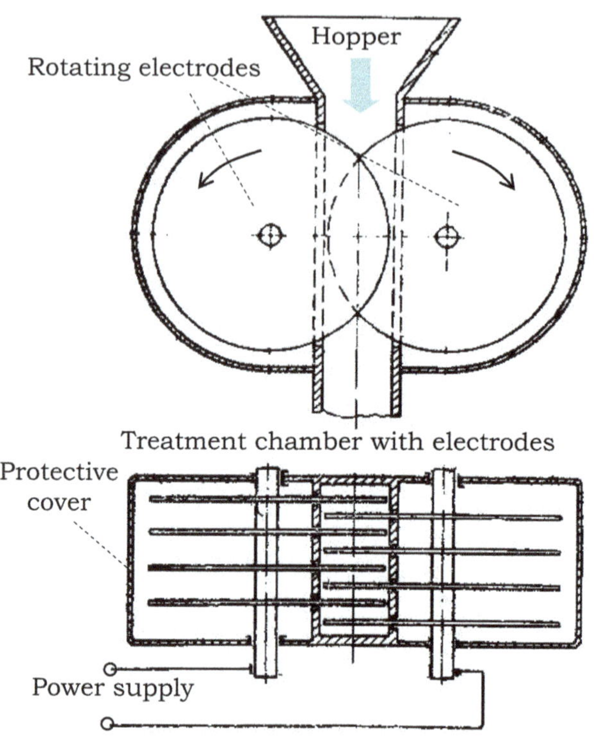

Fig. 10 Electroplasmolyzer for extracting juice from plant materials. A device with rotating disk electrodes (Biryukova and Preobrazhensky 1970)

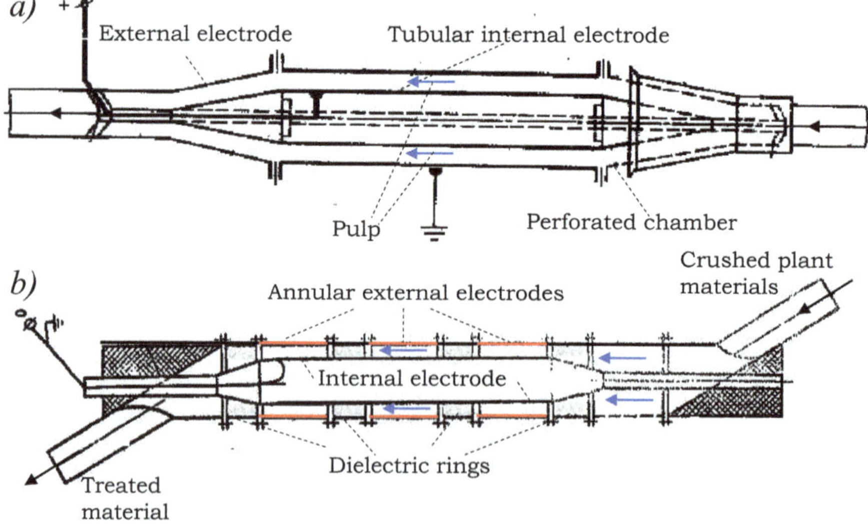

Fig. 11 Electroplasmolyzers for the treatment of crushed plant materials. The pulp is treated by electrical pulses in the cylindrical chamber between outer annular electrodes and central grounded cylindrical electrode. (**a**) The treated material was preliminary densified in the perforated chamber (Lazarenko et al. 1979). (**b**) The external electrodes are divided by dielectric rings. It allows multistep treatment (Lazarenko et al. 1976)

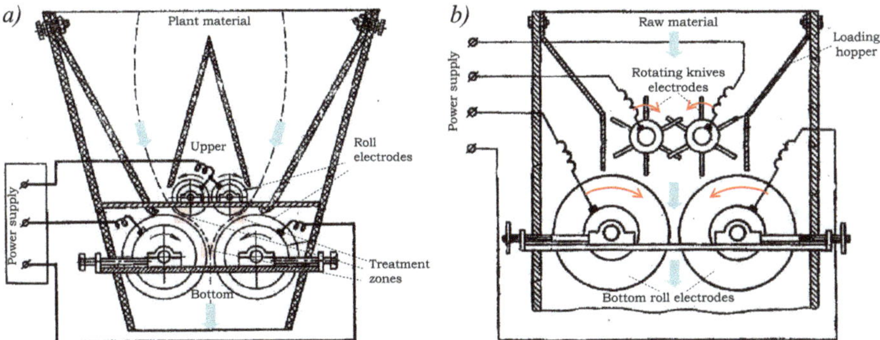

Fig. 12 Electroplasmolyzers for the treatment of plant materials. The devices include two pairs of roll electrodes. (**a**) The device provides a more uniform loading of the power source (Gatin et al. 1975). (**b**) The device for the treatment of lumpy raw materials. The upper pair of rolls is equipped with radically fixed knife electrodes (Gatin et al. 1978)

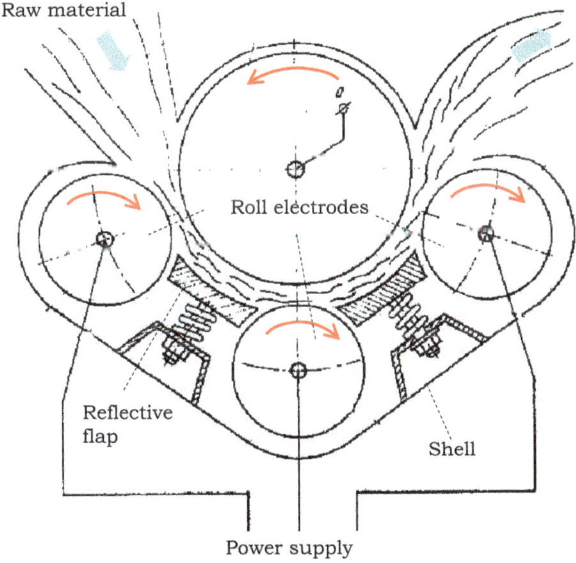

Fig. 13 Electroplasmolyzer for the treatment of plant materials. The device includes roll electrodes of various sizes (one large and three smaller) (Reshetko 1978)

Figure 16 presents the roller-type electroplasmolyzer for the treatment of plant material. The device includes two rollers covered with flexible electrodes. It provides a good uniformity of treatment (Bejenar 1981).

Figure 17 shows the screw-type device for the electric treatment of raw materials (Skimbov and Berzoy 1990). It is a cylindrical device with plate electrodes mounted along the wall of the cylinder.

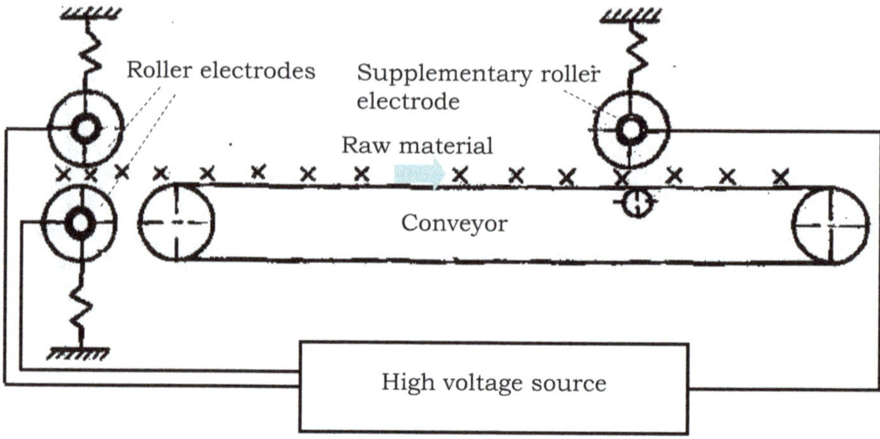

Fig. 14 A conveyor device for the electrical treatment of plant materials before drying (Martynenko et al. 1980)

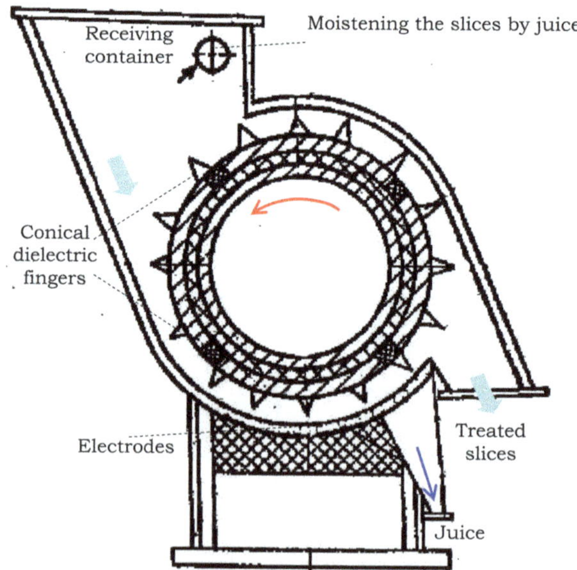

Fig. 15 Electroplasmolyzer for plant materials. The device consists of a rotating drum with conical dielectric fingers (Papchenko et al. 1980; Papchenko and Kuharuk 1981)

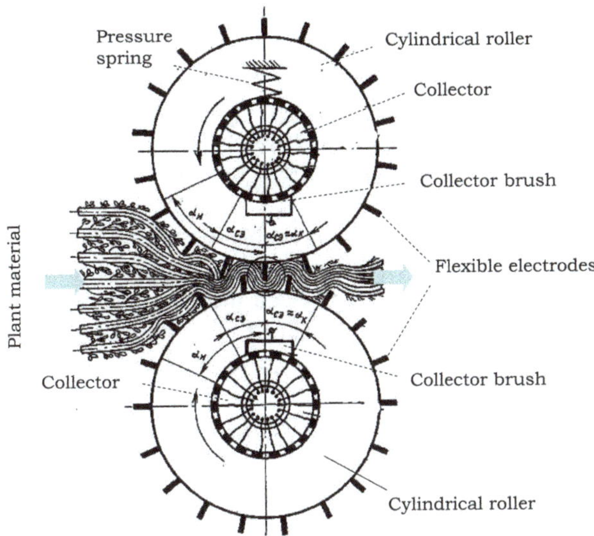

Fig. 16 The roller-type electroplasmolyzer for the treatment of plant material (Bejenar 1981)

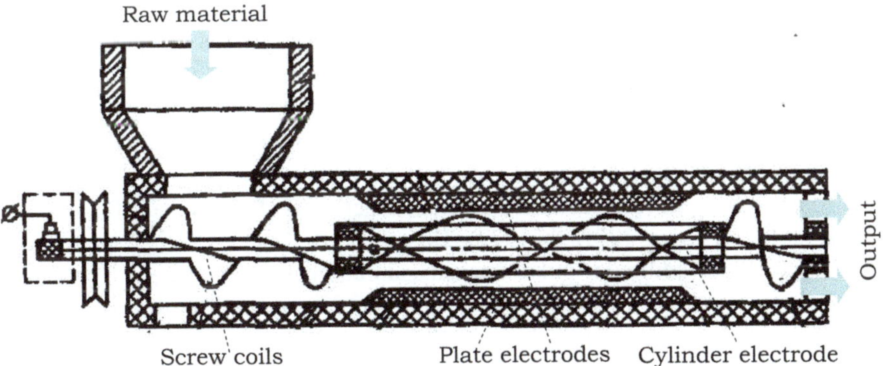

Fig. 17 The cylindrical screw-type device for the electric treatment of food raw materials (Skimbov and Berzoy 1990)

Pulsed Electric Fields

Early Concepts of Electroporation and Treatment by PEF

Concept of Electroporation

The idea about the formation of pores in plasma membrane was speculated in the 1930s. It was assumed that membrane behaves as a porous sieve for the molecules and the permeability of the plasma membrane might depend upon the solubility of the molecules in the compounds of the membrane (Höber 1930, 1933). However, the first evidence about the stimulation of pores in membranes in the external electric field probably appeared in 1958 after the publication of the experimental report on reversible electrical breakdown of the excitable membrane of a Ranvier node (Stämpfli 1958). This study provoked a series of experimental works devoted to the mechanisms of pore formation under the action of pulsed electric fields in the 1960s–1980s.

The formation of large pores and irreversible loss of the membrane's function as a semipermeable barrier between the bacterial cell and its environment may result in cell death. The first systematic studies on the effects of PEF on the killing of bacteria and yeasts have been started in the 1967 (Hamilton and Sale 1967; Sale and Hamilton 1967, 1968). The lethal effects for vegetative bacteria and yeasts were observed in absence of electrolysis and heating with the application of 20μs high voltage pulses ($E = 25$ kV/cm) (Sale and Hamilton 1967). The dependencies of the killing efficiency on the field strength and total time of the treatment were observed (Hamilton and Sale 1967), and the critical transmembrane potential required for the breakdown of the membrane was estimated at level $u_m \approx 1$ V (Sale and Hamilton 1968). These conclusions were confirmed in the series of works performed in the 1970s (Zimmermann et al. 1974, 1976a, b, c, 1980). For example, the breakdown of the membranes was observed at $u_m \approx 0.85$ V for *Valonia utricularis* (cells of the seawater alga) (Neumann and Rosenheck 1972) and at $u_m \approx 1.6$ V for red blood cells and bacteria (Zimmermann et al. 1974). A polarization mechanism was suggested for explaining the observed effects.

The important conclusion about the possibility of reversible changes in the membrane permeability in human and bovine red blood cells under the application of moderate PEF treatment (10^3–10^4 V/cm) was made (Riemann et al. 1975). It was concluded that resealing of membranes can be used for incorporation of the foreign molecules inside the resealed, but otherwise intact, cells (Kinosita and Tsong 1977; Kinosita Jr and Tsong 1977a, b, 1979). The lethal efficiency of high-voltage pulses as a function of additives, pulse intensity, duration, temperature, and pulse frequency have been also discussed (Hülsheger and Niemann 1980; Hülsheger et al. 1981, 1983). The different sensibility of the Gram-positive and Gram-negative species has been observed (Hülsheger et al. 1983). Based on basic experimental works, the electroporation concept was theoretically grounded (Weaver and Chizmadzhev 1996).

Early Practical Application of PEF: The Role of Heinz Doevenspeck

Heinz Helmut Doevenspeck was born on 15 August 1917 in Bremen, Germany, where he died on 1 July 1993 (Fig. 18). He served his apprenticeship as a locksmith at AG Weser in Bremen, and during the years of World War II he did an apprenticeship as a machine constructor and finished his education as a mechanical engineer. Due to missing documents, it is not clear at which institution he obtained his engineering degree. During World War II, he worked for the "Organisation Todt" in

Fig. 18 Heinz Doevenspeck (≈ 1963)

different countries which were occupied by the German Wehrmacht in those years. His ingenuity as an engineer resulted in the first patent applications in the 1940s. During the turmoil of the post-war era, Doevenspeck had several jobs, including work for the British military government in the City of Bremen. Early after the war, his smartness and his aspiration for independence led to his self-employment as a consulting engineer.

The existing documents owned by the author do not show when and how exactly Doevenspeck came up with the idea of using pulsed electric fields for different applications. It is not even clear without any doubt that it was his invention. Only in one of his progress reports from the 1970s he mentioned that in 1958, he was inspired by a publication of Sven Carlsson on the influence of electric current on the growth of microorganisms. Based on Carlson's publications, he started his own experiments in 1958. Doevenspeck argued that Carlson's mistake was to change the physical properties of the treated materials in a way that particularly a targeted electric field strength could not be adjusted during the whole duration of treatment. According to Doevenspeck, defined portions of electric charges determined by the size of a capacitor had to be used to overcome this problem. He was obsessed with the idea of influencing substance systems by pulsed electric fields. Doevenspeck was an autodidact and not a scientifically educated person.

As Fig. 19 shows (original document), he was a nonprofessional interested in physical, chemical, and biological details. He wanted to understand how the material systems he was working with were constructed and how PEF could modify these materials. He was so convinced of his ideas that with his private means he converted the basement of his one-family house in Minden (Germany) into an almost professional laboratory. Over the years, he invested in different small-scale pulse generators, experimental equipment made of glass, microscopes, centrifuges, filters, etc. Despite his limited financial and scientific possibilities and his very empirical approach, he astonishingly came to still valid conclusions, which he described in 1960 in his first patent application (Doevenspeck 1967) on a "process and device for gaining different phases from dispersed systems" (Fig. 20).

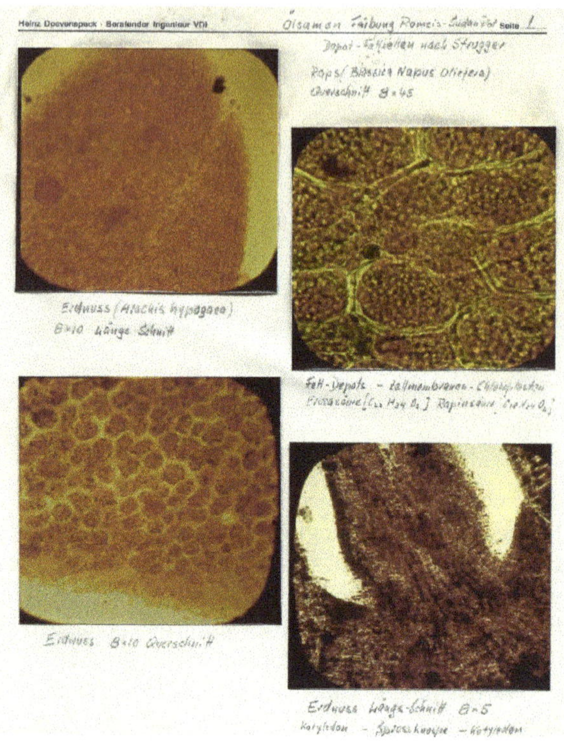

Fig. 19 Doevenspeck's studies on oilseeds (peanut and rapeseed cell materials)

In his patent application, he identified and described in detail that PEF had the following advantages over existing technologies using electricity: extensive suppression of electrolysis; almost no temperature increase; low energy consumption; gentle treatment of raw materials and preservation of biological activity; and killing of pathogenic germs.

As early as in 1960, Doevenspeck postulated that surface charges of inorganic and organic substances could be influenced by PEF. Thus, undesired ingredients in wastewater can be flocculated and separated and oil/water emulsions can be separated into their components. Doevenspeck was an enthusiastic sailor, and he contacted fish processing companies in North Germany. In the early 1960s, he carried out several trials and found out that – when processing fish material to produce fish oil and fish meal – protein-containing wastewater ("stickwater") as by-products could be cleaned by electroflocculation using PEF.

His observations concerning cell cracking were undoubtedly pioneering. When conducting PEF experiments, he noted that it was possible to release oils or fats from animal raw materials under low temperatures. In 1960, he described that when applying PEF, muscle fascicles contracted and cell-containing liquids were set free.

BUNDESREPUBLIK DEUTSCHLAND

DEUTSCHES PATENTAMT

PATENTSCHRIFT

1 237 541

Int. Cl.: B 01 j

Deutsche Kl.: 12 g - 5/01

Nummer: 1 237 541
Aktenzeichen: D 32565 IV a/12 g
Anmeldetag: 8. Februar 1960
Auslegetag: 30. März 1967
Ausgabetag: 12. Oktober 1967

Patentschrift stimmt mit der Auslegeschrift überein

1

An den Grenzflächen von dispersen Phasen sind Kräfte wirksam, die man als Oberflächenspannung bezeichnet. An den Oberflächen bilden sich oft elektrische Potentiale aus, die z. B. durch Dissoziation, Ionenadsorbtion und Reibungselektrizität verursacht werden.

Die Veränderung der Ladungen der dispersen Phasen eines Systems kann auf verschiedene Weise geschehen, z. B. durch physikalische und/oder chemische Reaktionen, elektrokinetische Prozesse, wie Elektrophorese, Elektroosmose, Elektrodialyse, sowie Aufladung durch Elektrolyse oder Strahlung. Zu erwähnen ist noch die Veränderung der Ladung dieser Systeme durch photochemische Prozesse.

Verfahren und Vorrichtung zur Gewinnung der einzelnen Phasen aus dispersen Systemen

Patentiert für:
Gesellschaft für Getreidehandel Aktiengesellschaft,
Düsseldorf, Kavalleriestr. 2

Als Erfinder benannt:
Heinz Doevenspeck, Bremen

Fig. 20 Doevenspeck's main patent DE1237541B on PEF from 1960 (Doevenspeck 1967)

The same happened to oil-containing deposit cells of fish. Those effects could be proven and visualized for the first time by Sitzmann in the 1980s.

In cooperation with the North German company Baumgarten Fischindustrie in Bremerhaven in the early 1960s, Doevenspeck produced fish meal and fish oil from malodorous herring offal by cracking the cells with pulsed electric fields before separating solids and liquids by using a screw press. The original correspondence between Doevenspeck and his partners shows that the experts on site especially noticed that contrary to expectations the whole process has very weak odor.

Together with the German fish meal-producing factory Lohmann & Co. in Cuxhaven, he experimented on fish processing. Figure 21 displays an original flow diagram from 1964 drawn by Doevenspeck personally. It shows a typical fish meal production line as it is still used nowadays. The only difference lies in the fact that a PEF system is used instead of a conventional cooking system.

Figure 22 shows Sankey diagrams drawn by Doevenspeck in 1960 demonstrating that "his" new process design worked on a much lower energy level compared to the conventional process.

Still, existing original reports and the correspondence between Lohmann and a participating feed-producing company show that Doevenspeck's fish meal was of high quality and had a long shelf life. The vitamin A content did not change

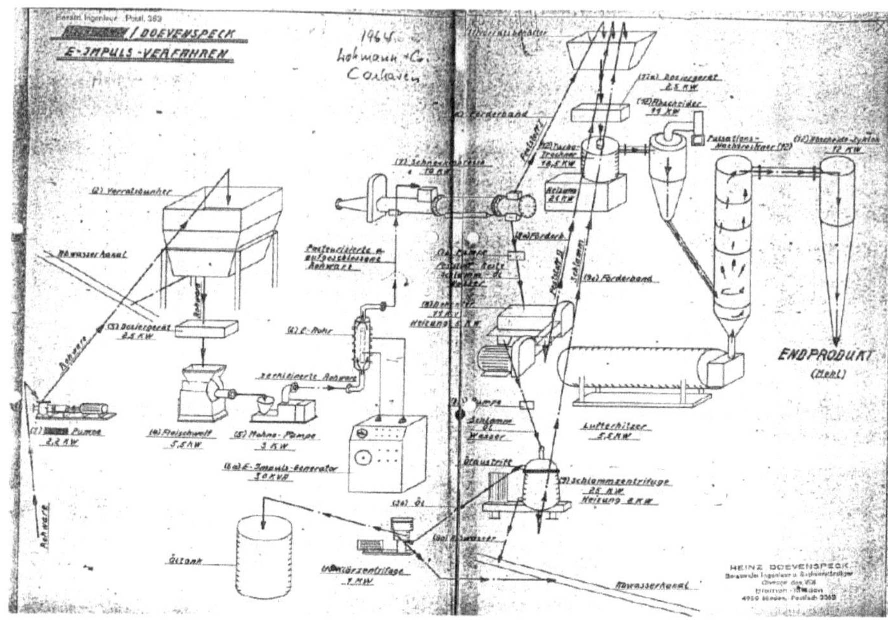

Fig. 21 Production line for fish meal and oil from herring. Original flow diagram drawn by Doevenspeck in 1964, owned by the author (W. Sitzmann)

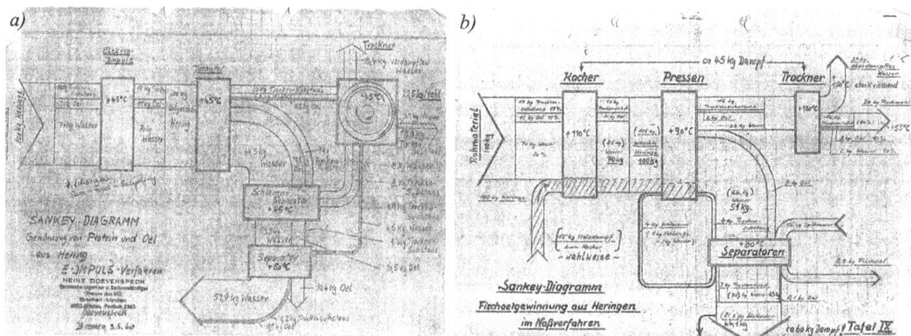

Fig. 22 Production of fish meal and fish oil using PEF treatment “E-Impuls-Verfahren” (a) and conventional process (b). Original Sankey diagram drawn by Doevenspeck, 1960, owned by the author (W. Sitzmann)

essentially during a storage period of 6 months. There was almost no oxidation of the residual oil in the final fish meal. Feeding experiments with laying hens showed that “Doevenspeck” fish meal produced from poor raw materials low in protein was much better than a conventional fish meal from Peru concerning the feed conversion rate. Independent experts certified that the eggs produced by the abovementioned laying hens tasted distinctively better than eggs from hens fed with conventional Peru meal.

Doevenspeck realized very early that the impact of electric pulses on biological matter strongly correlates to the electric field strengths being applied. During experiments with *Escherichia coli* bacteria in his own lab in 1962/1963, he discovered that cultures being treated with field strengths below 3 kV/cm seemed to increase their growth rates, whereas those being treated with field strengths above this value showed reduced growth velocities. Very high field strengths killed microorganisms. Encouraged by these observations, he extracted corn mash containing microorganisms from a technical scale fermenter and applied electric pulses with a field strength of about 2.4 kV/cm. As a result, he could measure a distinct acceleration of the fermentation process. In cooperation with the sewage plant operators of the German city of Nienburg, Doevenspeck achieved a 20% increase in the methane yield by using PEF. In 1967, he applied for a patent together with a well-known big German beer brewery in Dortmund. The subject of their common invention was a low temperature (25 °C) conservation method for beer with the positive side effect of almost no changes in color and taste during the subsequent storage phase. Despite all his efforts and achievements, he had to realize that without any scientific background his work was far from being approved by the public and by possible investors. Consequently, at the end of the 1970s and the beginning of the 1980s, Doevenspeck contacted the Institute of Biophysics of the University of Hannover (Glubrecht, Niemann) and the Institute of Microbiology of the Hannover Medical School (Hülsheger, Potel) where his working hypothesis on germ-killing was confirmed and scientifically examined for the first time (Hülsheger et al. 1981, 1983).

Until 1983, Doevenspeck conducted many trials with interested companies, communities, and research institutes. He took part in symposia and was able to inspire people. But in the end, he failed to achieve a remarkable financial success with his process, despite 20 years of intensive work and investing the money he had earned from his job as a consulting engineer.

Doevenspeck and Krupp Company

Doevenspeck's chance came in the early 1980s due to the ban of perchlorethylene as a solvent for the extraction of oils and fats from animal raw materials in rendering plants. Krupp Industrietechnik GmbH, a German company for plant construction and engineering located in Hamburg, was very successful in selling this type of extraction plants for many years. In this critical phase, Ernst Wilhelm Münch (Fig. 23), the area manager of the abovementioned extraction plants at Krupp Company, became acquainted with Doevenspeck.

Münch realized the high potentials of Doevenspeck's new technology which could be an alternative for the old (banned) solvent extraction process. In 1985, he engaged him as a consultant. In the R&D center of Krupp company in Hamburg, a PEF pilot plant was erected and taken into operation in 1985 (Fig. 24).

Fig. 23 Münch (CEO, left) and Sitzmann (Head of R&D, right) in the later 1980s

Fig. 24 Pilot plant at Krupp Company, Hamburg, Germany

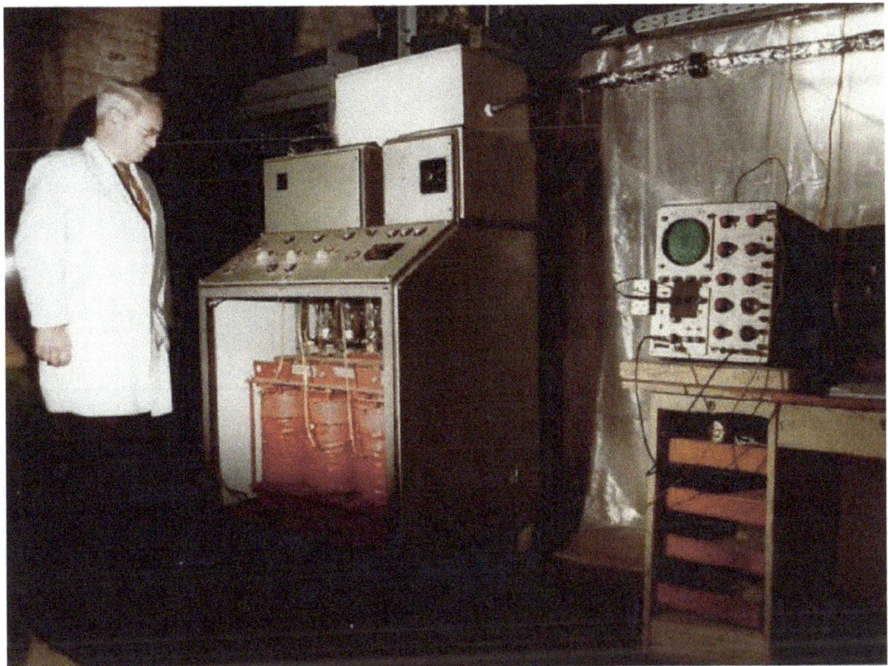

Fig. 25 Doevenspeck and his pulse generator (Krupp pilot plant, 1985)

The main component of this plant was Doevenspeck's 80 kW pulse generator (Fig. 25). Frequencies ranging from 1 to 16.7 Hz and capacities of 5, 10, 15, 20, 25, 30, and 35μF could be adjusted; the maximal charging voltage was 8 kV. The generator could only be used for "trial and error" experiments. It was nearly impossible to set repeatable electric parameters and thus to reveal the interaction between cause (e.g., the field strength) and impact (for instance, the cell cracking or the killing of germs) of the process.

Krupp was mainly interested in gentle cell cracking of animal and herbal materials. From 1986 to 1990, Sitzmann's (Fig. 23) task as Head of R&D department was to investigate different products with respect to their processability, to research the operating principle of PEF, and to make the results available to a scientific public (Sitzmann and Münch 1987, 1988a, b).

In order to protect Doevenspeck's ideas, several patents were applied, and Krupp company registered the trademarks ELCRACK© and ELSTERIL© (Sitzmann and Münch 1988a, c). Sitzmann was the first to visualize Doevenspeck's ideas regarding low-temperature cell cracking of fish material (Fig. 26). Irreversible damage was obviously the consequence of sufficiently high field strengths. Surprisingly it also turned out that PEF caused fish proteins to coagulate already at low temperatures of about 30 °C.

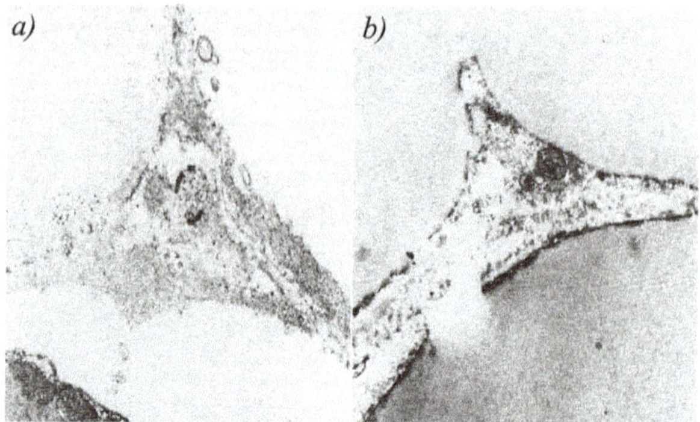

Fig. 26 Herring cells before (**a**) and after (**b**) PEF treatment (Sitzmann 1995). 4900-fold magnification

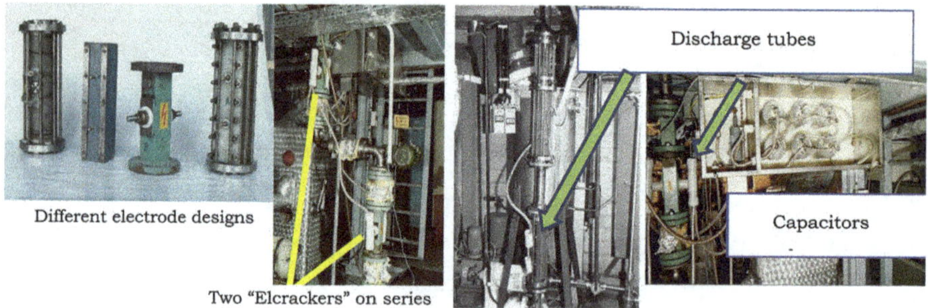

Fig. 27 Krupp's pilot plant equipment (1985–1990)

Fig. 28 Krupp's trade-fair appearance at the ACHEMA fair, Frankfurt, Germany, 1985

In Krupp's continuously operating pilot plant, the material throughput could be varied between 100 and about 250 kg/h. The distances of the discharge electrodes were 50 and 70 mm, respectively. Several tests regarding different electrode geometries and raw materials were carried out in the pilot facility, as displayed in Fig. 27.

The positive results in the pilot plant in the 1980s encouraged Krupp company to present their new technology to the international public. This was first done at the ACHEMA fair in Frankfurt in 1985 (Fig. 28). The left picture shows Doevenspeck

Table 6 PEF treated different materials in Krupp's pilot plant (throughput, 100–250 kg/h; discharge voltage, <6 kV; distances of electrodes, 50 and 70 mm; capacities, 5–40μF)

Material	Experimental purpose
Sugar beet	Enhancement of sugar yield
Krill	Dewatering
Bones from cattle	Defatting of residual meat
Hide shaving	Production of glue
Back rind from pigs	Defatting
Gelatine bones	Production of gelatine
Chicken by-product	Defatting
Rendering plant materials	Defatting
Cod liver	Extraction of cod liver oil
Animal organs	Alternative to chemical cell cracking
Herring	Fish oil production
Cocoa nuts	Defatting
Palm fruits	Defatting
Copra expeller	Aflatoxin reduction
Olives	Deoiling
Fruits	Improvement of juice yield
Milk, juices, protein suspensions	Killing of microorganisms, process modelling (ELSTERIL[R])[a]
Sewage sludge	Cell cracking, protein agglomeration

[a]In cooperation with Technical University of Hamburg (Märkl, Grahl)

together with Ziemann and one of his employees. Ziemann company designed and manufactured the electrodes for Doevenspeck's equipment in those years. On the right-hand side next to Doevenspeck is Dr. Münch from Krupp company.

From 1986 to 1994, in his quality as Head of R&D at Krupp company and later on, between 1994 and 1996, as founder and CEO of the consulting company NaFuTec, Sitzmann's main tasks were to investigate different products (Table 6) concerning their processability, to research the operating principle of PEF, and to make the results available to a scientific public.

Figure 29 shows the treatment of gelatine bones for the production of gelatine (Sitzmann and Münch 1988c) as an example for the processing of animal raw materials with PEF. Leading figures in food engineering like Barbosa-Cánovas, Washington State University, and Knorr, Technical University of Berlin, were very interested in Sitzmann's work and the lessons learned up to then in PEF technology.

They visited Krupp's pilot plant in the late 1980s and the first common investigations were realized (Geulen et al. 1992). In the following years, their main merit was to initiate research projects and to spread the knowledge about this new emerging technology worldwide. Based on the work done at Krupp Company, the PEF technology was made available not only for use in science but also for use in industry. The main results of Krupp's R&D work included mathematical models for describing the discharge curves and the efficiency in germ elimination depending on

Fig. 29 PEF treatment of gelatine bones (pilot plant Krupp company)

electrical parameters of the materials to be treated, a better understanding of the operating principle of pulsed electric fields, and a way to quantify cell cracking effects by using the impedance of the materials as a measure.

From 1988 to 1990, Krupp, together with the Technical University of Hamburg, Germany, initiated an EU-funded project on continuous sterilization of juices and milk in a pilot plant. In Grahl's doctoral thesis (Grahl 1994), the influence of PEF on food components like vitamins, enzymes, and flavor carriers was investigated for the first time (Grahl et al. 1992).

Figure 30 shows Grahl's modern generator that can be used in scientific work (in contrast to Doevenspeck's homemade generator (Fig. 25)). Two industrial-scale plants (a 10 t/h plant for producing rendered fats from slaughterhouse by-products in Duisburg, Germany, and a 22 t/h plant in Egersund, Norway, for low temperature production of high-grade fish meal and fish oil from herring) were engineered and commissioned in those years.

Figure 31 gives some impressions of the Norwegian plant for fish processing. This depiction was part of a brochure published in 1988 after commissioning. Another two plants for processing slaughterhouse offal and bones were designed at the end of the 1980s but never implemented. Not only the ELCRACK© system was brand new in the Norwegian plant and caused many problems especially regarding the material the discharge vessels were made of. Also, the double screw presses for liquid-solid separation, the fluidized bed drier for reduction of moisture in the press cakes of fish material, and the ultrafiltration plant for wastewater cleaning and protein recycling were newly designed.

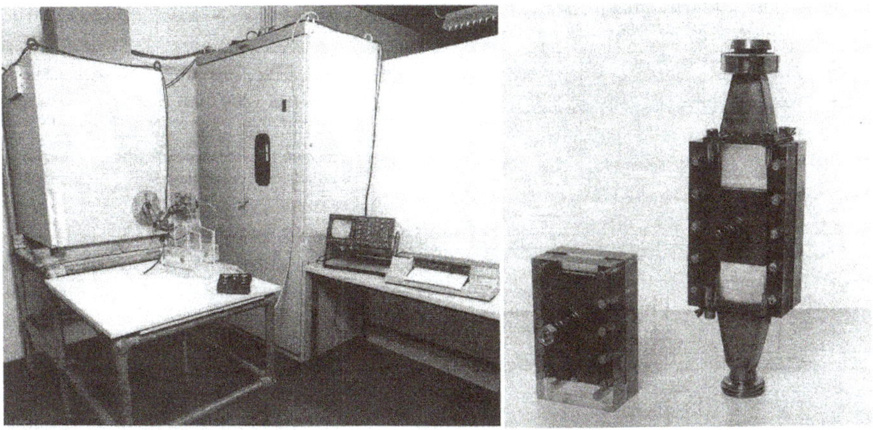

Fig. 30 Cooperation Krupp – Technical University of Hamburg (Grahl 1994)

Fig. 31 Fish processing by the ELCRACK© process

There were many difficulties during the startup phase, which could not be solved within an appropriate time. The lack of short-term economic success induced the clients to stop the commissioning. The plants had to be dismantled and replaced by conventional technology. From then on, Krupp stopped all activities concerning the PEF technology.

For the summary, Table 7 presents main patents issued by Doevenspeck on activities related to PEF treatment of different materials.

Table 7 Doevenspeck's main patents

Patent nos.	Title and comments	References
DE1237541B	Method and device for obtaining the individual phases from disperse systems	Doevenspeck (1967)
AT283245B	Method and apparatus for the treatment of disperse systems, preferably liquids, such as beer	Doevenspeck (1970)
US3625843A	Method for treating beer. Electrical method of and a device for treating disperse systems, preferably liquids, particularly beer. The disperse systems are subjected to impulse-discharged direct current fields of gradually increased voltage	Doevenspeck (1971)
DE3413583A1	Electric pulse method for the treatment of materials and apparatus for performing the method	Doevenspeck (1985)
DE3481977D1	Electric impulse method and device for treating substances	Doevenspeck (1990)

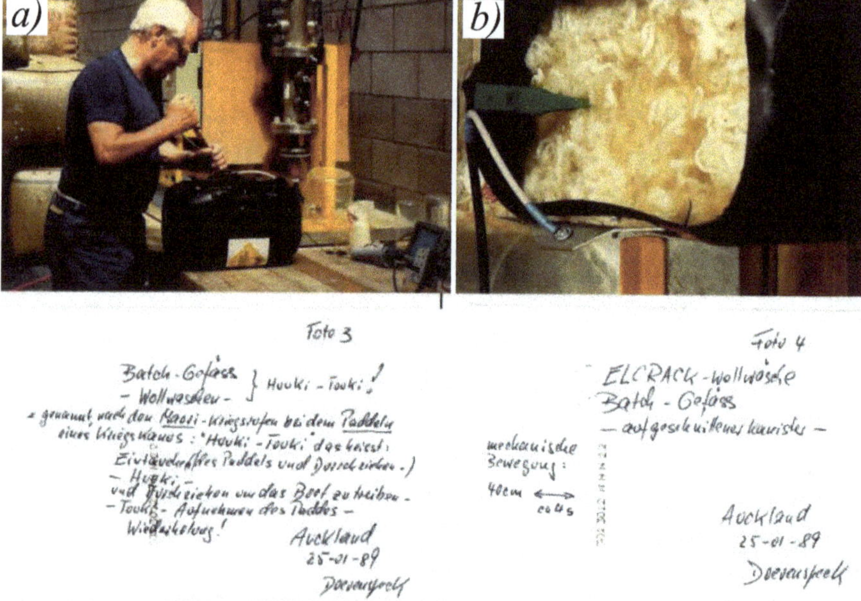

Fig. 32 Heinz Doevenspeck converting a canister into a discharge vessel (**a**) and the discharge vessel used for batch-wise PEF treatment of sheep wool (**b**) (personal communication to the W. Sitzmann)

Doevenspeck and Sitzmann

Therefore, the cooperation contract between Doevenspeck and Krupp Company was terminated at the end of the 1980s. But Doevenspeck – still believing in "his" technology – continued his work (e.g., extraction of wool fat from sheep wool and cleaning of wastewater with PEF) on a private basis in New Zealand (Fig. 32b).

Heinz Doevenspeck died on 1 July 1993, in a hospital in Bremen, Germany. In his final days, he contacted Sitzmann and encouraged him to continue working on PEF and to finalize their common ideas in his stead. Sitzmann – after years of close cooperation with Doevenspeck – was convinced that PEF had many potential applications.

After Krupp had stopped all activities in the field of PEF at the end of the 1990s, Sitzmann took over the rights to both the ELSTERIL© and ELCRACK© brands from Krupp. Besides, he took over one of the two generator systems worldwide unique at that time, suitable for the operation of the PEF process on a pilot scale. He signed a cooperation agreement with Prof. Knorr aimed at the joint development of the PEF process. For this purpose – and because the entire environment for the operation of such a plant, such as analytics, technical and scientific personnel, and above all the know-how in the field of food process technology, was available at the TU Berlin – Sitzmann decided to place his equipment at the TU Berlin.

With the foundation of NaFuTec GmbH, Sitzmann intensified his experimental work at the TU Berlin. In cooperation with TU Berlin, he investigated until 1996 several possible applications. For example, cell cracking of fruits (Knorr et al. 1994; Sitzmann and Heinz 1996) in order to increase the yield of juices. At the end of 1995, shortly before Sitzmann had to give up his PEF activities for the time being due to a move to Bahlsen, Germany, in mid-1996, he initiated a joint, publicly subsidized project for the electric disintegration of grated potatoes and the decontamination of the resulting potato fruit water using the PEF process. Apart from a large, German, starch-producing company, the participation of the University of Oldenburg, Germany, was planned.

The first successful tests in January 1996 at the TU Berlin were the basis of this project. In addition to the head of research at the starch factory, Knorr's doctoral candidate Volker Heinz, who later became director of the German Institute of Food Technology (DIL) in Quakenbrück, Germany, was also actively involved. Due to the promising results at the TU Berlin and above all the competence of the institute in the field of food technology, the circle of project participants was extended to include TU Berlin, Institute.

Without doubt, these tests and the project based on them, which was publicly subsidized by the German Ministry of Research and in which Sitzmann as initiator was no longer able to participate due to his changing career, formed the basis for the meanwhile impressive success of the PEF technology, especially in the potato processing industry.

PEF: From the Lab to the First Industrial Applications (1980–2010)

Despite the studies conducted on the effects of PEF since the early second half of the twentieth century, the feasibility of this technology for food processing began to be widely considered by researchers and food companies in the late 1980s and early

1990s. In this period, different technologies able to inactivate microbial cells and enzymes without increasing the food temperature started to be deeply investigated in response to consumer demands for fresh-tasting, convenient, and safe refrigerated foods. PEF was included in a group of techniques called nonthermal methods of food preservation, emerging preservation technologies, or minimal processing technologies aiming to prevent or greatly reduce detrimental changes in the sensory and physical properties of foods caused by thermal processes.

Early research and development for commercialization of liquid foods preserved by pulse energy were conducted in the 1980s by the company PurePulse Technologies, a subsidiary of Maxwell Laboratories located in San Diego, California. The basic science behind this application can be found in the paper published by Sale and Hamilton at the end of the 1960s (Sale and Hamilton 1967, 1968; Hamilton and Sale 1967). These researchers observed that the electric field strength, pulse duration, and size and shape of microbes were relevant factors affecting microbial inactivation and demonstrated that the permanent loss of permeability properties of the bacterial membrane caused by PEF led to cell death.

PurePulse Technologies developed two new processes for killing microorganisms, PureBright® (pulse light) and CoolPure™ (pulsed electric fields), based on storing electrical energy in a capacitor and releasing it in short, high-intensity pulses used to create intense pulses of light or pulses of high electric field. A 200 L/h continuous flow CoolPure™ pilot system that included heat exchangers and temperature controls to regulate product inlet temperatures and to cool the product after the treatment was developed to evaluate the benefits of PEF for preserving different fluid foods such as dairy products, fruit juices, and fluid eggs. This company owns several US patents related to PEF technology (Dunn and Pearlman 1987; Dunn et al. 1989, Bushnell et al. 1993, 1995a, 1995b, 1996). On 7 July 1995, the Food and Drug Administration (FDA) expressed no objection to the CoolPure™ pulsed electric fields process developed by PurePulse Technologies for antimicrobial treatment of liquids and pumpable foods, representing the first regulatory effort to implement PEF technology at an industrial level (Barbosa-Canovas and Sepulveda 2005).

In the early 1990s, Prof. Barbosa-Cánovas and Prof. Knorr decided to explore the use of PEF technology for food processing at Washington State University (WSU) and the Technical University of Berlin (TUB), respectively.

The laboratory of Prof. Barbosa at WSU established a comprehensive program aiming for food pasteurization by PEF. The first reported results were obtained using a modified version of an electroporator (International Biotech, Inc., New Haven, CT) that only permitted to treat a few volumes of liquid foods. Later on, a PEF system including a power supply capable of delivering a peak voltage of 40 kV at frequencies up to 10 Hz was designed and constructed. Numerical simulation techniques were used by the WSU group to support parallel plate and coaxial continuous treatment chamber's designs and system effectiveness. Studies conducted were mainly focused on PEF preservation of liquid whole eggs (LWE), milk, and fruit juices.

In the middle of 1990s, Prof. Barbosa established at WSU the **Center for Nonthermal Processing of Food (CNPF)** for researching the latest methods in nonthermal food processing including high hydrostatic pressure, oscillating magnetic fields, ultrasound, and ultraviolet irradiation in addition to PEF. Since its creation, Prof. Barbosa offered the opportunity of conducting research with the facilities available in the **CNPF** center to many visiting scholars that did not have access to these emerging technologies in their laboratories. These research stays have certainly contributed to spread out research on nonthermal technologies around the world because many of these researchers initiated research programs on the topic after this training period. The first book on PEF technology for the food industry was published in 1999 by Prof. Barbosa-Cánovas's group (Table 8).

Initial research on PEF in Prof. Knorr's laboratory was focused on the application of the technology for improving the processing of plant tissues. Studies on the application of the technology for the extraction of potato starch and sugar beet and drying enhancement were initiated. A technique to determine cell permeability and define a cell disintegration index was developed (Angersbach et al. 1999), and since 1997 the design of their own pulse generator and treatment chambers was initiated. This generator designed by researches of TUB was installed in the laboratories of some of the partners of the project "High electric field pulses: food safety, quality, and critical process parameters." This project that involved research institutions (TU Berlin, KU Leuven, the Universities of Montpellier and Zaragoza, SIK, Icetek) and industrial partners (Tetra Pak, Unilever, Best Foods, and Pernod Ricard) represented the first project funded by the European Community on this topic. Results obtained in this project demonstrated that this technology was of special interest to the food industry because it provided an attractive alternative to conventional thermal processing, required minimal energy, was highly flexible and applicable to a wide range of products, and could also be applied as effective preprocessing technology (i.e., to improve heat and mass transfers). After founding this first project, several projects based total or partially on the PEF technology have been founded in different EU Research and Innovation programs (Table 9).

In 1994, H. Zhang after completing his Ph.D. and postdoc in Prof. Barbosa-Cánovas's lab at WSU joined Ohio State University (OSU). Prof. Zhang initiated a compressive program to demonstrate the commercial feasibility of PEF as a pasteurization process. OSU in cooperation with Diversified Technologies, Inc. (DTI) developed bench and pilot-scale PEF systems. The system called OSU-4 was installed in different laboratories in the USA as well as in Europe. In 2000, DTI installed a commercial-scale pulsed electric field (PEF) system at Ohio State University's Department of Food Technology. This unit capable of treating 1000–5000 liters/hour was installed in the framework of a consortium of governmental, academic, and industrial participants (Natick, Tetra Pak, Diversified Technologies, Kraft, AEP, General Mills, Hirzel, and AmeriQual) with Prof. Zhang as principal investigator (Dunne 2011). In 2005, 2 years later of the official conclusion of the project, the PEF technology was industrially adopted for the first time by an organic fruit juice processor. Genesis Juice Company licensed the patented PEF technology from OSU and started commercial production of organic juices including apple, strawberry, and other flavors that were pasteurized by PEF at a flow of 200 L/h (Clark 2006). The PEF pasteurized products reach a small market sector in

Table 8 Edited books on pulsed electric field technology for the food industry

Title	Publication date	Editors/authors	Editorial
Preservation of Foods with Pulsed Electric Fields	1999	G.V. Barbosa-Cánovas, M. M. Góngora-Nieto, Usha R. Pothakamury, B.G. Swanson	Academic Press
Pulsed Electric Fields in Food Processing: Fundamental Aspects and Applications	2001	G.V. Barbosa-Cánovas, Q.H. Zhang	CRC Press
Pulsed Electric Fields Technology for the Food Industry: Fundamentals and Applications	2006	J. Raso, V. Heinz	Springer US
Food Preservation by Pulsed Electric Fields: From Research to Application	2007	H.L.M. Lelieveld, S. Notermans and S.W.H. de Haan	Woodhead
Electrotechnologies for Extraction from Food Plants and Biomaterials	2008	E. Vorobiev, N. Lebovka	Springer
Pulsed Electric Fields (PEF): Technology, Role in Food Science and Emerging Applications	2016	S. Lynch	Nova
Pulsed Electric Fields to Obtain Healthier and Sustainable Food for Tomorrow	2020	F. Barba, O. Parniakov, A. Wiktor	Academic Press
Processing of Foods and Biomass Feedstocks by Pulsed Electric Energy	2020	E. Vorobiev, N. Lebovka	Springer
Pulsed Electric Fields Technology for the Food Industry: Fundamentals and Applications (second edition)	2021	J. Raso, V. Heinz, I. Álvarez, S. Toepfl	Springer

Eugene and Portland (Oregon). As a result, Genesis and DTI were awarded by the Institute of Food with the Food Technology Industrial Achievement Award in 2007.

The coverage given to PEF technology by pioneering groups and the increasing availability of PEF units around the late twentieth century and the beginning of the twenty-first century stimulated research groups all around the world to investigate the potential of PEF for food processing. It is estimated that there are more than 30 research groups researching PEF at these times. The growth of research on PEF around the world provided the basic science needed for the implementation of the technology for different applications in the food industry. A large amount of research describing the inactivation kinetics of a wide range of microorganisms and enzymes and the effect of PEF on food quality parameters have permitted the optimization of the current processing parameters used for the preservation of fruit juices and smoothies. Simultaneously, much research on the fundamentals of improving mass transfer by PEF in different operations of the food industry, including solid-liquid extraction, pressing, drying, or osmotic dehydration, were conducted. Finally,

Table 9 Examples of projects based total or partially on the PEF technology founded in different EU Research and Innovation programs

Acronym	Title	Program	Coordinator	Start/end date
HELP	High electric field pulses: food safety, quality, and critical process parameters	FP4-FAIR	Technische Universitat Berlin	1/11/1997–31/10/2000
	Industrial process of food preservation by pulsed electric fields	FP5-LIFE QUALITY	Enertronic Sa	3/07/2000–2/07/2001
NOVELQ	Novel processing methods for the production and distribution of high-quality and safe foods	FP6-FOOD	Stichting Dienst Landbouwkundig Onderzoek	1/03/2006–28/02/2011
HIGHQ RTE	Innovative nonthermal processing technologies to improve the quality and safety of ready-to-eat (RTE) meals	FP6-FOOD	University of Bologna	1/10/2006–30/09/2009
OILPULSE	Increased virgin olive oil yield and quality by complementing existing and new olive oil mills with pulsed electric field (PEF) technology	FP7-SME	Technische Universitat Berlin	1/04/2010–28/02/2013
SMARTMILK	A novel system for the treatment of milk based on the combination of ultrasounds and pulsed electric field technologies	FP7-SME	Iris Technology Solutions	1/10/2010–31/12/2012
PHENOLIVE	Revalorization of wet olive pomace through polyphenol extraction and steam gasification	FP7-SME	Technische Universitaet Wien	1/10/2013–30/06/2016
FieldFOOD	Integration of PEF in food processing for improving food quality, safety, and competitiveness	H2020	Universidad de Zaragoza	1/04/2015–31/03/2018
i^3-food	Process integration for rapid implementation of sustainable innovative food processing	H2020	Elea	1/04/2015–28/02/2018
EcoPROLIVE	Ecofriendly processing system for the full exploitation of the olive health potential in products of added value	H2020	Contactica S.L.	1/09/2015–31/12/2017

(continued)

Table 9 (continued)

Acronym	Title	Program	Coordinator	Start/end date
AQUABIOPROFIT	Aquaculture and agriculture biomass side stream proteins and bioactives for feed, fitness, and health-promoting nutritional supplements	H2020	Nofima As	1/04/2018–31/03/2022
SHEALTHY	Nonthermal physical technologies to preserve fresh and minimally processed fruit and vegetables	H2020	Enco Srl	1/03/2019–30/04/2023
FOX	Innovative downscaled food processing in a box	H2020	DIL -	1/06/2019–30/11/2023

Source: https://cordis.europa.eu/es

efforts of multidisciplinary groups, formed by food scientists and electrical engineers, contributed to the developed of continuous systems for processing liquid and solid foods.

In 2006, Volker Heinz, assistant professor at TU Berlin, was appointed Director General at the German Institute of Food Technologies (DIL). Then, he started a program for commercialization of pulse modulators for pilot- and industrial-scale treatments. The same year, DIL was the owner of the ELCRACK brand (Toepfl et al. 2007). The first commercial application in Europe started in 2009, with the installation of a 1500 L/h juice preservation line. In parallel, DIL initiated a project with the potato processor company Wernsing Feinkost GmbH with the purpose of developing an industrial-scale unit able to process up to 50 tons/h of potatoes. The unit was installed in the company in 2012. The electroporation of potatoes reduced turgor pressure of cells. As consequence, a significant tissue softening occurs that facilitates cutting of potatoes and replacing conventional preheating in French fries processing.

PEF: *The Consolidation of the Technology and the Development of Future Applications (2010–2020)*

The concern of the food industry for making their activities more energy-efficient and sustainable has been an important driving force for conducting new research on PEF applications in the last 10 years (Vorobiev and Lebovka 2020). In order to achieve the full potential of PEF for industrial implementation and commercial exploitation, issues related to improving productivity and reducing energy inputs and environmental impact are increasingly attracting the attention of the food processors since they can represent significant reductions of the processing costs (Barba et al. 2015). Reducing malaxation time and temperature in olive oil production, maceration time in wine production, energy consumption in tomato peeling, energetic requirements for mashing of fruits for fruit juice production, or drying time or temperature are some examples that may be achieved when different raw materials are pretreated by PEF. There has been also an increased interest on the application of PEF as a pretreatment for the recovery of high-added value compounds (proteins, lipids, carotenoids, chlorophylls, sterols, etc.) from microbial cells (yeast and microalgae) for different applications in the food, cosmetic, pharmaceutical, and biofuel industries. Finally, over the last years, PEF for recovery of valuable compounds from food wastes and by-products generated during food processing and biorefinery application has also attracted considerable attention.

The Cost Action TD1104 European network for development of electroporation-based technologies and treatments that started in 2012 certainly contributed to the development of these new applications of PEF by integrating multidisciplinary research teams. The main objective of this action was the coordination and interdisciplinary exchange of knowledge between researchers working in the electroporation and PEF in different scientific domains such as biology, food science, medicine, biotechnology, or environmental science. During the 4 years of duration of the action, almost 600 researchers and industrial technicians from 42 countries working in the field were involved in this action. In the framework of the COST action, two training schools on pulsed electric field applications in food and biotechnology were organized in 2014 and 2015 at the University of Zaragoza (Spain) and the University of Salerno (Italy), respectively. Due to the success of these two first schools, the event has continued to be held annually at University College of Dublin (Ireland), University of Natural Resources and Life Sciences (Austria), Osnabrück University of Applied Sciences (Austria), and University of Bologna (Italy). The course that offers the opportunity of promoting networking and communication between young scientists, experts, and industrial partners aims to provide an overview of knowledge and understanding of the basic principles involved in processing by pulsed electric fields and practice in the use of these principles for food processing. Other remarkable inputs of this action were the edition of the Handbook of Electroporation (a work of 4 volumes and 157 chapters) (Miklavcic 2017) and the preparation of guidelines on the key information to be reported in studies of the application of PEF technology in food and biotechnological processes with the purpose that other researchers will be able to repeat, judge, and evaluate experiments and data obtained in different laboratories (Raso et al. 2016).

With the aim of keeping the networking established between researchers working in different fields during the COST Action, the International Society of Electroporation-Based Technologies and Treatments (ISEBTT) was created in 2016. The first president of this society was Prof. Damijan Miklavčič from the Faculty of Electrical Engineering of the University of Ljubljana. This society with more than 200 members has among its objectives to organize every 2 years the World Congress on Electroporation and Pulsed Electric Fields in Biology, Medicine,

Table 10 Manufacturing companies of PEF units

Company name	Country	Website
Arc Aroma	Sweden	www.arcaroma.com
BASIS Électronique de Puissance	France	www.basis-ep.com
Diversified Technologies	USA	www.divtecs.com
ELEA	Germany	www.elea-technology.de
Energy Pulse Systems	Portugal	www.energypulsesystems.pt/eps
PEF Technologies	Netherlands	www.pef-technologies.nl/
PurePulse	Netherlands	www.purepulse.eu
Pulsemaster	Netherlands	www.pulsemaster.us
ScandiNova	Sweden	www.scandinovasystems.com
Nutri-Pulse	Netherlands	www.e-cooking-technology.com
SureView	India	www.sureview.xyz
Vitave	Germany/Czech Republic	www.vitave.eu

and Food & Environmental Technologies. Three editions of this congress have been held in Portorož (Slovenia), Norfolk (VA, USA), and Toulouse (France) so far with an average of around 400 attendants.

During many years, the lack of reliable industrial-scale equipment limited the commercial exploitation of PEF in the food industry. However, the development in the last 15 years of pulse power systems responding to the food industry requirements (high processing capacity, low energetic requirements, and easy implementation in the existing processing lines) has allowed the successful transfer of the technology for industrial applications. Currently, different industrial-scale PEF units for different applications are manufactured for different companies (Table 10), and more than 100 of these units have been installed in the food industry all around the world.

Acknowledgment WS thanks Torsten Kochems, Ernst-Wilhelm Münch, H. and W. Ziemann, Sebastian Biedermann. EV and NL thank L. I. Cherniavska.

References

Anderson AK, Finkelstein R (1919) A study of the electropure process of treating milk. J Dairy Sci 2:374–406

Angersbach A, Heinz V, Knorr D (1999) Electrophysiological model of intact and processed plant tissues: cell disintegration criteria. Biotechnol Prog 15:753–762

Anonymous (1962) Electrical methods of food treatment: annotated bibliographic index of domestic and foreign literature for 1950–1961. Central Institute of Scientific and Technical Information of the Food Industry of the USSR, Moscow

Barba FJ, Parniakov O, Pereira SA, Wiktor A, Grimi N, Boussetta N, Saraiva JA, Raso J, Martin-Bellos O, Witrowa-Rajchert D, Lebovka N, Vorobiev E (2015) Current applications and new

opportunities for the use of pulsed electric fields in food science and industry. Food Res Int 77:773–798
Barbosa-Cánovas GV, Sepulveda D (2005) Present status and the future of PEF technology. In: Barbosa-Cánovas GV, Tapia MS, Cano MP (eds) Novel food processing technologies. CRC Press, New York, pp 1–44
Bartley CB (1909) Process sterilizing fluids, US930023 (Patent)
Bazhal MI (1995) Mathematical modeling of electrical treatments of sugar beet slices to accelerate the process of extraction of sucrose. PhD Thesis (Candidate of technical sciences), Ukrainian Technological University of Food Industry, Kiev, Ukraine
Bazhal IG, Gulyi IS, Vorona LG, et al (1981) Method for producing of diffusion juice. SU854984 A1 (Inventor's certificate)
Bazhal IG, Gulyy IS, Bobrovnik LD, et al (1982) Method for producing of diffusion juice. SU912755 A1 (Inventor's certificate)
Bazhal IG, Kupchik MP, Guly IS (1983) Desugaring of sugar beet slices in an electric field. Sakharnaya Promyshlennost (Sugar Ind) 3:28–30
Bazhal IG, Kupchik MP, Vorona LG et al (1984) Extraction of sugar from sugar beet in electrical field. Electron Treat Mater (Elektronnaya Obrab Mater J Inst Appl Physics, Chisinau, Repub Mold) N1:79–82. (in Russian)
Beattie JM, Lewis FC (1913) The utilisation of electricity in the continuous sterilization of milk. J Pathol Bacteriol 18:120–122
Beattie JM, Lewis FC (1925) The electric current (apart from the heat generated). A bacteriologibal agent in the sterilization of milk and other fluids. Epidemiol Infect 24:123–137
Becker WA, Beckerowa Z (1934) Über einige Desintegrationserscheinungen des Protoplasten (Some disintegration phenomena of the protoplasts). Acta Soc Bot Pol 11:367–392
Bejenar GS (1981) Electroplasmolyzer for treatment of plant material. SU808047 A1 (Inventor's certificate)
Bennet-Clark TA, Bexon D (1943) Water relations of plant cells. III. The respiration of plasmolysed tissues. New Phytol 42:65–92
Berkeley LJ (1935) Electric cooking apparatus. US2025085A (Patent)
Berthold G (1940) Method for treating food. US2219772A (Patent)
Berzoj SE, Botoshan NI, Bologa MK et al (1993a) Thermophysical characteristics and humidity keeping capability of the electric plasmolysed crevette stuffing. Elektron Obrab Mater (Surf Eng Appl Electrochem) N5:58–64
Berzoj SE, Kolpakov EK, Botomak NI, Tsyrdya ID (1993b) Technological features of fish food meal production of raw material subjected to preliminary electroplasmolysis. Elektron Obrab Mater (Surf Eng Appl Electrochem) N2:61–63
Bezusov AT, Verkhivker YG, Storozhuk VN, Belyavskaya NP (eds) (2010) Flaumenbaum Boris Lvovich: on the occasion of the 100th birthday. Information-methodical collection. Odessa National Academy of Food Technologies, Odessa
Biryukova SN, Preobrazhensky AA (1970) Electroplasmolyzer for extracting juice from plant materials. SU 286491 A1 (Inventor's certificate)
Bloom JE (1915) Process and apparatus for the electrical treatment of liquids and fluids, and the products resulting therefrom. US1162213A (Patent)
Bologa MK, Litinsky GA (1988) Electric antiseptic effects in the food industry. Ştiinţe, Chişinău
Botoshan NI, Berzoj SE (1994a) Diffusion process and its specific features during electroplasmolysis. Elektron Obrab Mater (Surf Eng Appl Electrochem) N7:64–66
Botoshan NI, Berzoj SE (1994b) The electroplasmolysis phenomenon in the biological media and the perspective of its application in fool processing industry. Elektron Obrab Mater (Surf Eng Appl Electrochem) N3:73–76
Botoshan NI, Berzoj SE, Papchenko AY (1994) Effect of dielectric current stability in biological media on electroplasmolysis. Elektron Obrab Mater (Surf Eng Appl Electrochem) N11:47–51
Bushnell A H, Dunn JE, Clark RW (1993). High pulsed voltage systems for extending the shelf life of pumpable food products. U. S. Patent 5,235,905

Bushnell AH, Clark RW, Dunn JE, Lloyd SW (1995a) Prevention of electrode fouling in high electric field systems for killing microorganisms in food products. US5393541A (Patent)

Bushnell AH, Clark RW, Dunn JE, Lloyd SW (1995b) Prevention of electrochemical and electrophoretic effects in high-strength-electric-field pumpable-food-product treatment systems. US5447733A (Patent)

Bushnell, AH, Clark, RW, Dunn JE, Lloyd, SW (1996). Process for reducing levels of microorganisms in pumpable food products using a high pulsed voltage system. U. S. Patent 5,514,391

Capek J V (1890) Electrical cooking-stove. US424922 (Patent)

Cebanu VG (1988) Improving the processing efficiency of plant materials by electroplasmolysis. PhD Thesis (Candidate of technical sciences), Kiev Technological Institute of Food Industry, Ukraine

Chernyavskaya LI (2019) Zagoruilko Anatoliy Yakovlevich. In: Encyclopedia of modern Ukraine. Institute of Encyclopedic Research, NAS of Ukraine, Kyiv

Chester IH, Rexford NY (1943) Electric heater. US2324837A (Patent)

Clark P (2006) Pulsed electric field processing. Food Technol 60:66–67

Cohn F, Mendelsohn B (1879) Ueber Einwirkung des electrischen Stromes auf die Vermehrung von Bacterien (About the effect of electric current on the growth of bacteria). Beitrage Biol der Pflauzen 3:141–162

Cole KS (1940) Permeability and impermeability of cell membranes for ions. In: Cold Spring Harbor Symposia on quantitative biology, pp 110–122

Courtright A V (1934) Electric food heating apparatus. US1970360A (Patent)

Crompton RE (1894) The use of electricity for cooking and heating. J Soc Arts Lond 43:511

Dankevich GN (1995) Intensification of sugar extraction process by thermal and electrical treatment of sugar beet. PhD Thesis (Candidate of technical sciences), Kiev Technological Institute of Food Industry, Kiev, Ukraine (in Russian)

De Vries H (1871) Sur la permeabilite du protoplasme de betteraves rouges (On the permeability of the red beet protoplasm). Arch Neerl Sci Exactes Nat 6:117–126

Devaux H (1903) Sur une réaction nouvelle et générale des tissus vivants. Essai de détermination directe des dimensions de la micelle albuminoïde (On a new and general reaction of living tissue. Essay for direct determination of the dimensions of the albuminoid micelle). Extr des comptes rendus des séances la Société linnéenne Bordeaux (publiés seuls ou avec les Actes la Société linnéenne Bordeaux), Procès-verbaux des séances la Société des Sci Phys Nat Bordeaux 3–7 (séance du 19 novembre 1903) and 259–264 (séance du 2 décembre 1903)

Doevenspeck H (1967) Verfahren und Vorrichtung zur Gewinnung der einzelnen Phasen aus dispersen Systemen (Method and device for obtaining the individual phases from disperse systems). DE1237541B (Patent)

Doevenspeck H (1970) Method and apparatus for the treatment of disperse systems, preferably liquids, such as beer. AT283245B (Patent)

Doevenspeck H (1971) Method for treating beer. US3625843A (Patent)

Doevenspeck H (1985) Elektroimpulsverfahren zur behandlung von stoffen und vorrichtung zur durchfuehrung des verfahrens (Electric pulse method for the treatment of materials and apparatus for performing the method). DE3413583A1 (Patent)

Doevenspeck H (1990) Electric impulse method and device for treating substances. DE3481977D1 (Patent)

Dolinskaya IN (1993) Improving the process of sugar extraction by electrical treatment of sugar beet slices. PhD Thesis (Candidate of technical sciences), Kiev Technological Institute of Food Industry, Ukraine

Dolinskaya IN, Dan'kevich GN, Gulyj IS et al (1992) The influence of electrical and thermal factors on efficiency of processes to extract soluble substances from plant raw materials. Elektron Obrab Mater (Surf Eng Appl Electrochem) N1:66–69

Dunn JE, Pearlman JS (1987) Methods and apparatus for extending the shelf life of fluid food product. US4695472A (Patent)

Dunn JE, Clark RW, Asmus JF, et al (1989) Methods for preservation of foodstuffs. US4871559A (Patent)

Dunne CP (2011) Future prospects for nonthermal processing technologies-linking products with technologies. In: Zhang HQ, Barbosa-Cánovas GV, Balasubramaniam VM, Dunne CP, Farkas DF, Yuan TC (eds) Nonthermal processing technologies for food. Wiley-Blackwell, Oxford, pp 571–592

Dutrochet MH (1837) Memoires pour servir a l'Histoire anatomigue et physiologique des Vegetaux et des Animaux (). J. -B. Bailliere et fils, Libraire De l'Académie Royale de Médecine, Paris, France

Fan-Jung AF, Flaumenbaum BL, Izotov AK (1956) Technology of fruits and vegetables preservation: textbook. Pishchepromizdat, Moscow

Fetterman JC (1928) The electrical conductivity method of processing milk. Agric Eng 9:107–108

Flaumenbaum BL (1941) Extraction of juice from raw plan material. PhD Thesis (Candidate of technical sciences), Affiliated branch of All-Union Scientific Research Institute for the Canning Industry (VNIIKP) , Odessa, Ukraine

Flaumenbaum B (1949) Electrical treatment of fruits and vegetables before extraction of juice. Proc Odessa Technol Inst Cann Ind (Trudy OTIKP, Odess Tehnol Instituta Konservn Promyshlennosti) 3:15–20

Flaumenbaum BL (1953) Commercial application of the method of electrical pre-processing fruits before pressing. Proc Odessa Technol Inst Food Refrig Ind (Trudy OTIPHP Odess Tehnol Instituta Pishhevoj i Holodil'noj Promyshlennosti) 5(2):37–50

Flaumenbaum BL (1960) The theoretical basis of the sterilization of canned food: textbook. Publishing house, Kiev University, Kiev

Flaumenbaum BL (1968) Anwendung der Elektroplasmolyse bei der Herstellung von Fruchtsaften. Fluss Obst 35:19–22

Flaumenbaum BL (1969) Problems of intensification of technological processes of preservation of food products. Thesis for degree of doctor science habilitat (Technical sciences), Odessa Technological Institute of the Food Industry (OTIPP)

Flaumenbaum BLI, Al-Saadi C (1966) Elektroplazmoliz Pri polucemi Tsitrusovih Sokov. Konservn i ovoshhesushil'naja promyshlennost 7:7

Flaumenbaum BL, Kazandzhii MY (1966) Electrostability of different fruits and berries. Izv vuzov Pischevaya Technol 5:76–78

Flaumenbaum BL, Shin IN (1990) Method of determining the degree of plasmolysis of fruits steeped in syrups. SU1589139A1 (Inventor's certificate)

Flaumenbaum BL, Yablochnik LM (1950) Electroplasmolizator for the treatment of vegetables and fruits. SU83502 A1 (Inventor's certificate)

Flaumenbaum BL, Rusakov VA, Kalmykova IS (1991) A method of obtaining of red wines. SU1687599 A1 (Inventor's certificate)

Fricke H, Morse S (1925) The electric resistance and capacity of blood for frequencies between 800 and 41/2 million cycles. J Gen Physiol 9:153

Fronek D (2012) Professor Jaroslav Dedek-great scientist and analyst in the field of sugar making. List Cukrov Repar 128:34

Fruengel F (1960) Method and device for electrically sterilizing and cleaning milking machines or the like. US2931947A (Patent)

Fursov SP, Bordiyan VV (1974) Certain features of electroplasmolysis of plant tissue at high frequencies. Elektron Obrab Mater (Surf Eng Appl Electrochem) N6:70–73

Gatin AI, Zelenyuk AL, Rosenberg DA, Flaumenbaum BL (1975) Electroplasmolyzer for treatment of plant materials. SU459211 A1 (Inventor's certificate)

Gatin AI, Vaishtok YG, Zelenyuk AL, Rosenberg DA (1978) Electroplasmolyzer for treatment of plant materials. SU627813 A1 (Inventor's certificate)

Gelpi AJ Jr, Devereux ED (1930) Effect of the electropure process and of the holding method of treating milk upon bacterial endospores. J Dairy Sci 13:368–371

Getchell BE (1935) Electric pasteurization of milk. Agric Eng 16:408–410

Geulen M, Teichgräber P, Knorr D, et al (1992) Elektroimpulsbehandlung von Karottenmaische. In: Symposium: Aktueller Stand und Trends in der Lebensmitteltechnologie. Stuttgart, Germany

Golden KE (1933a) Electrical sterilizing apparatus. US1934703A (Patent)
Golden KE (1933b) Electrical sterilizing apparatus. US1934704A (Patent)
Goucher JL (1909) Method of treating milk. US922134A (Patent)
Grahl T (1994) Abtöten von Mikroorganismen mit Hilfe elektrischer Hochspannungsimpulse (Killing of microorganisms with the help of electrical high-voltage pulses). PhD Thesis, Technical University Hamburg-Harburg, Germany
Grahl T, Sitzmann W, Märkl H (1992) Keimabtötung mit Hilfe elektrischer Hochspannungsimpulse in pumpfähigen Medien. In: DECHEMA-Jahrestagung der Biotechnologen. VCH Verlagsgesellschaft MBH, Weinheim , Germany
Griffith EL (1942) Apparatus for electric conductance cooking. US2299088A (Patent)
Grishko AA, Kozin VM, Chebanu VG (1991) Electroplasmolyzer for processing plant raw material. US5031521A (Patent)
Grudinovker GL (1954a) Some features of the electromechanical treatment of fruits and vegetables. PhD Thesis (Candidate of technical sciences), Moscow Technological Institute of Food Industry
Grudinovker GL (1954b) Electroplasmolyzer for processing beets and other vegetables. SU 96911 A1 (Inventor's certificate)
Hamilton WA, Sale AJH (1967) Effects of high electric fields on microorganisms: II. Mechanism of action of the lethal effect. Biochim Biophys Acta (BBA) Gen Subj 148:789–800
Heller R (1897) Beitrag zur Kenntniss der Wirkung elektrischer Ströme auf Mikroorganismen. Plant Syst Evol 47:326–331
Höber R (1910) Eine Methode, die elektrische Leitfähigkeit im Innern von Zellen zu messen (A method to measure the electrical conductivity inside cells). Pflüger's Arch für die gesamte Physiol des Menschen und der Tiere 133:237–253
Höber R (1930) The first Reynold A. Spaeth memorial lecture: the present conception of the structure of the plasma membrane. Biol Bull 58:1–17
Höber R (1933) Permeability. Annu Rev Biochem 2:1–14
Hülsheger H, Niemann E-G (1980) Lethal effects of high-voltage pulses on E. coli K12. Radiat Environ Biophys 18:281–288
Hülsheger H, Potel J, Niemann E-G (1981) Killing of bacteria with electric pulses of high field strength. Radiat Environ Biophys 20:53–65
Hülsheger H, Potel J, Niemann E-G (1983) Electric field effects on bacteria and yeast cells. Radiat Environ Biophys 22:149–162
Ilyeva ES (1996) Intensification of the process of obtaining fruitful juices by diffusion method. PhD Thesis (Candidate of technical sciences), OGAPT, Odessa, Ukraine
Jones CS, Crooker WW, Artos F (1886) Applying electricity for destroying living organisms in the bodies of slaughtered animals, US337334A (Patent)
Kalmykova IS (1992) Influence of a method of preliminary processing of grapes by electroplasmolysis on extractivity of grape juice. In: Abstracts of reports of the scientific conference devoted to the 90th anniversary of OTIPP, Odessa Technological Institute of the Food Industry, Odessa, p 52
Kalmykova IS (1993) The use of electroplasmolysis to intensify the extraction of phenolic substances from grapes in the technology of red table wines and natural juice. PhD Thesis (Candidate of technical sciences), Odessa Technological Institute of the Food Industry (OTIPP), Ukraine
Karpovich NS, Bazhal IG, Gulyi IS et al (1981) The behavior of the structural elements of a plant cell in an electric field. Sakharnaya Promyshlennost (Sugar Ind) 10:32–35
Kartashev AK, Koval ET (1956) The results of research and application of plasmolysis for the extraction of sugar from sugar beet. In: Shakin AN (ed) Proceedings of the All-Union Central Research Institute of the Sugar Industry (AUCRISI). Issue 4. Technological processes of sugar beet production, AUCRISI, Moscow, pp 44–67
Kartashov AK, Koval ET (1955) The reaction of beetroot tissue to various effects. Sakharnaya Promyshlennost (Sugar Ind) N2:12–15

Katroha IM, Matvienko AB, Vorona AG, Kupchik MP (1984) Intensification of sugar extraction from sugar beet slices in an electric field. Sugar Ind No 7:28–31
Katroha IM, Ponomarenko V V, Bezlyuda NI (1993) Method for the producing of diffusion juice. SU1789563 A1 (Inventor's certificate)
Katrokha IM, Zaets VA, Kupchik MP (1987) Effect of an electric field on the pH of a juice-cossette mixture. Sov Surf Eng Appl Electrochem 3:121–123
Kazandzhii MY (1966) Investigation of parameters of electroplasmolysis process for fruits and berries in the production of fruit juices. PhD Thesis (Candidate of technical sciences), Odessa Technological Institute of Food and Refrigeration Industry , Odessa, Ukraine (in Russian)
Kazandzhii M, Flaumenbaum BL (1977) Calculation of the parameters of electroplasmolysis. Proc High Educ institutions Food Technol (Izvestiia Vyss uchebnykh Zaved Pishchevaia tekhnologiia) 1: 162-163
Kinosita K Jr, Tsong TY (1977a) Voltage-induced pore formation and hemolysis of human erythrocytes. Biochim Biophys Acta (BBA) Biomembr 471:227–242
Kinosita K Jr, Tsong TY (1977b) Formation and resealing of pores of controlled sizes in human erythrocyte membrane. Nature 268:438
Kinosita K Jr, Tsong TY (1979) Voltage-induced conductance in human erythrocyte membranes. Biochim Biophys Acta (BBA) Biomembr 554:479–497
Kinosita K, Tsong TT (1977) Hemolysis of human erythrocytes by transient electric field. Proc Natl Acad Sci 74:1923–1927
Kleiber M (1924) Über die elektrische Konservierung von saftigem Futter (Elektrosilierung). PhD Thesis, ETH Zurich
Knorr D, Geulen M, Grahl T, Sitzmann W (1994) Food application of high electric field pulses. Trends Food Sci Technol 5:71–75
Kogan FI (1955) Electroplasmolyzer for plant materials. SU104718 A1 (Inventor's certificate)
Kogan FI (1958) Electroplasmolyzer. SU112554 (Inventor's certificate)
Kogan FY (1968) Electrophysical methods in canning technologies of foodstuff. Tekhnica (in Russian), Kiev
Kolpakov EK, Botoshan NI, Berzoj SE, Zakharchuk AV (1991) Intensification of a technological process for obtaining fish meal through application of electroplasmolysis. Elektron Obrab Mater (Surf Eng Appl Electrochem) 5:64–65
Kramer PJ, Currier HB (1950) Water relations of plant cells and tissues. Annu Rev Plant Physiol 1:265–284
Krause GA (1936) Sterilization of liquids by means of oligodynamy. US2046467A (Patent)
Kuhn EW (1909) Process of manufacturing fermented beverages. US936328A (Patent)
Kupchik MP (1986) Obtaining and purification of diffusion juice in an electric field. Sakharnaya Promyshlennost (Sugar Ind) 3:16–19
Kupchik MP (1991) Development of sugar beet technology using the low frequency electric fields. Thesis for degree of doctor science habilitat (technical sciences). Moscow Institute of Food Technology, Russia
Kupchik MP, Fischuk NU, Mihaylik TA et al (1982) Extraction of juice from raw plant material in an electric field. Elektron Obrab Mater N4(106):81–83
Kupchik MP, Fishchuk NU, Mikhailik VA et al (1984) Extraction of juice from raw plants in electrical field. Elektron Obrab Mater (Surf Eng Appl Electrochem) N4(106):81–83
Kupchik MP, Matvienko AB, Mank VV (1987) Changes in the ultrastructure of beet cells during diffusion under the influence of temperature and electric field. Sakharnaya Promyshlennost (Sugar Ind) 5:25–27
Küster E (1939) Vital-staining of plant cells. Bot Rev 5:351
Lasarenko BR, Fursow SP, Stscheglow JA, Golostschapow JW (1975) Vorrichtung zur durchfuehrung des plasmolyseverfahrens zerkleinerter pflanzlicher rohstoffe (Device for carrying out the plasmolysis process of vegetable raw materials), DE2402947A1 (Patent)
Lazarenko BR, Mamakov AA, Shheglov JA, Reshet'ko EV (1967) Method of electroplasmolysis of cells of raw plant material. SU194016 (Inventor's certificate)

Lazarenko BR, Mamakov AA, Scheglov JA, Reshet'ko E V (1968) Method of electroplasmolysis of grinded raw plant material. SU212147 (Inventor's certificate)
Lazarenko BR, Fursov SP, Scheglov YA, Goloshchapov YV (1976) Electroplasmolyzer for treatment of grinded plant materials. SU429601 A1 (Inventor's certificate)
Lazarenko BR, Fursov, Shheglov SP, et al (1977) Electroplasmolysis. Cartea Moldoveneascu, Chisinau
Lazarenko BR, Fursov SP, Scheglov YA, Goloshchapov YV (1979) Electroplasmolyzer for grinded plant materials. SU407457 A1 (Inventor's certificate)
Lepeschkin WW (1930) My opinion about protoplasm. Protoplasma 9:269–297
Lewis FC (1914) An electro-chemical apparatus for the disinfection and cleansing of cultures and slides for use in bacteriological and pathological laboratories. Epidemiol Infect 14:48–51
Lewis MJ, Neil JH (eds) (2000) Continuous thermal processing of foods: pasteurization and UHT sterilization, Food engineering series. Blackwell Publishing, Malden
Lillie RS (1909) The general biological significance of changes in the permeability of the surface layer or plasma-membrane of living cells. Biol Bull 17:188–208
Lincoln JC (1912) Process of preserving. US1044201A (Patent)
Louder EA (1933) Process for sterilization of liquids. US1900509A (Patent)
Martin HM (1905) Studies on the effect of some concentrated solutions on the osmotic activity of plants. Bull Torrey Bot Club 32:415–429
Martynenko II, Kotov BI, Litvinenko P V, Bezhenar GS (1980) Device for the electrical treatment of plant materials before drying. SU745427 A1 (Inventor's certificate)
Matvienko AB (1996a) Intensification of the extraction process of soluble substances by electrical treatment of aqueous media and vegetable raw materials. Thesis for degree of doctor science habilitat (Technical sciences). Kiev Technological Institute of Food Industry, Kiev, Ukraine (in Russian)
Matvienko AB (1996b) Intensification extraction process dissoluble substances by means of electrical treatment water medium and vegetable raw material. Thesis for degree of doctor science habilitat (Technical sciences), Ukrainian state university o: food technologies, Kyiv, Ukraine
Miklavcic D (2017) Handbook of electroporation. Springer, Cham
Mil'kov MY, Zagorul'ko AY (1949) Method of obtaining and purification of sugar beet raw juice and electroplasmolizator for implementing method. SU89009 (Inventor's certificate)
Mil'kov MY, Zagorul'ko AY (1950) Method of electroplasmolysis of sugarbeet and similar slices and electroplasmolyzer for implementing the method. SU92191 A1 (Inventor's certificate)
Mittelmann E (1951) Milk pasteurization method and apparatus. US2550584A (Patent)
Miyahara K (1986) Method of and apparatus for producing processed foodstuffs by passing an electric current. US4612199A (Patent)
Miyahara K (1988) Production of electrically processed food. JPS61132138A (Patent)
Moses BD (1938) Electric pasteurization of milk. Agric Eng 19:525–526
Nägeli C (1853) Systematische Übersicht der Erscheinungen im Pflanzenreich (Systematic overview of the phenomena in the plant kingdom). Friedrich Wagner'sche Buchhandling, Freiburg in Breisgau
Nasonov DN, Aleksandrov VY (1940) The reaction of living matter to external influences. Publishing House of the Academy of Sciences of the USSR, Moscow – Leningrad
Neumann E, Rosenheck K (1972) Permeability changes induced by electric impulses in vesicular membranes. J Membr Biol 10:279–290
Oneill HG (1895) Cooking and heating apparatus. US535072 (Patent)
Papchenko AY (1979) Investigation of the process of electropulse processing of plant raw materials in order to intensify its juice yield. PhD Thesis (Candidate of technical sciences), Odessa, Ukraine
Papchenko AY (1981) Method for producing of juice from fruits. SU799711 A1 (Inventor's certificate)
Papchenko AY, Kuharuk AS (1981) Electroplasmolyzer for plant materials. SU888921 A1 (Inventor's certificate)

Papchenko AY, Kuharuk AS, Koval NP, Shcheglov YA (1980) Electroplasmolyzer for plant materials. SU786966 A1 (Inventor's certificate)
Papchenko AY, Bologa MK, Berzoi SE, et al (1990) Procede de traitement de matieres premieres vegetales et dispositif pour sa realisation (Method for treating vegetable raw materials and device for carrying it out), FR2621456A1 (Patent)
Papchenko AY, Botoshan NI, Berzoj SE, Bologa MK (1992) Vegetable media electroplasmolysis performances. Elektron Obrab Mater (Surf Eng Appl Electrochem) N5:64–69
Parsons CH, Richardson WD (1930) Electrothermal treatment of cheese. US1774610A (Patent)
Pavlov IS (1959) The study of the electrical properties of food in connection with electrical processing. Thesis for degree of doctor science habilitat (Technical sciences), Kiev Institute of Food Industry, Ukraine
Pfeffer W (1877) Osmotische untersuchungen: studien zur zellmechanik (osmotic examinations: studies on cell mechanics). Verlag von Wilhelm Engelmann, Leipzig
Philippson M (1921) Les lois de la résistance électrique des tissus vivants (The laws of electrical resistance of living tissues). Bull l'Acad R Belg 7:387–403
Prichko VY, Kudryavtseva MI (1951) The use of electroplasmolyzer in the processing of grapes. Wine-Mak Vitic Mold 3:40–41
Pringsheim N (1854) Untersuchungen über den Bau und die Bildung der Pflanzenzelle (Studies on the construction and formation of the plant cells). Verlag Von August Hirshwald, Berlin
RAF (1893) Electrical cooking. Science (80-) 554:146–148
Raso J, Frey W, Ferrari G, Pataro G, Knorr D, Teissie J, Miklavčičh D (2016) Recommendations guidelines on the key information to be reported in studies of application of PEF technology in food and biotechnological processes. Innov Food Sci Emerg Technol 37:312–321
Reitler W (1990) Conductive heating of foods. PhD Thesis, Technical University of Munich
Reld C (1946) Manufacture and treatment of animal and vegetable materials. US2400951A (Patent)
Reshetko EV (1970) An electric method of intensifying the process of press extraction of juice from plant materials. PhD Thesis (Candidate of technical sciences), Krasnodar, Russia
Reshetko EV (1978) Electroplasmolyzer for treatment of plant materials. SU592406 A1 (Inventor's certificate)
Reznik D (1997) Electroheating of food products using low frequency current, GB2282052A (Patent)
Riemann F, Zimmermann U, Pilwat G (1975) Release and uptake of haemoglobin and ions in red blood cells induced by dielectric breakdown. Biochim Biophys Acta (BBA) Biomembr 394:449–462
Roberts IL (1900) Receptacle for sterilized perishable substances. US645569A (Patent)
Roeske H (1893) Method of and apparatus for purifying water. US501732 (Patent)
Rogov IA (1988) Electrophysical methods of foods product processing. Agropromizdat, Moscow. (in Russian)
Rogov IA, Gorbatov AV (1974) Physical methods of treatment of foods. Pishhevaja promyshlennost, Moscow (in Russian)
Rozhdestvensky IM (1955) Electroplasmolyzer for plant materials. SU100094 A1 (Inventor's certificate)
Rudd HB (1920) Apparatus for electrically treating liquids. US1360447A (Patent)
Sale A, Hamilton W (1967) Effect of high electric fields on microorganisms. I. Killing of bacteria and yeast. Biochim Biophys Acta 148:781–788
Sale AJH, Hamilton WA (1968) Effect of high electric fields on microorganisms. III. Lysis of erythrocytes and protoplasts. Biochem Biophys Acta 8(163):37–43
Sater LE (1935) Passing an alternating electric current through food and fruit juices. Res Bull (Iowa Agric Home Econ Exp Station) N181:275–312
Schade AL (1951) Prevention of enzymatic discoloration of potato. US2569075A (Patent)
Scheglov YA (1971) Study of the electrical processing of plant tissue at low potential gradients to increase the efficiency of extracting juice from raw materials . PhD Thesis (Candidate of technical sciences), Krasnodar, Russia

Scheglov YL, Chebanu VG (1974) Method of electroplasmolysis of plant raw materials. SU446266 A1 (Inventor's certificate)
Scheglov YA, Gasyuk GN, Lazarenko BR, Fursov SP (1969) Electroplasmolyzer for extracting juice from pulp. SU244880 A1 (Inventor's certificate)
Scheglov JA, Koval NP, Furer LA, et al (1986) Electroplasmolyzer for processing vegetable stock. US4608920A (Patent)
Schroeder GG (1891) Liquid purifier. US447585A (Patent)
Schweizer T (1924) A method for preserving feedstuffs juicy by the action of the electric current. DE415090C (Patent)
Schwerin B (1901) Process of extracting sugar. US Patent 687386
Schwerin B (1903) Electro-endosmotic process of extracting sugar. US Patent 723928
Schwerin G (1929) "Elektro-Osmose Lktiengesellsohaft" Graf Schwerin Gesellschaft . The method of cleaning sugar juice electric current. SU7981A1 (Inventor's certificate)
Shcheglov YA, Zelenskaya MI, Reshetko EV, Bogdan KN (1967) Electroplasmolysis of plant tissue. Elektron Obrab Mater (Surf Eng Appl Electrochem) 2:87–97
Shelmerdine A (1915) Apparatus for the sterilization of milk. US1147558A (Patent)
Shengeliia AS (1974) A study of diffusion method of obtaining of the fruit juices. PhD Thesis (Candidate of technical sciences), Odessa Technological Institute of the Food Industry (OTIPP), Odessa, Ukraine (in Russian)
Shulika VA (1988) Impact of electrical treatment of beet slices on the process of sucrose extraction. PhD Thesis (Candidate of technical sciences), Kiev Technological Institute of Food Industry, Kiev, Ukraine (in Russian)
Shulika VA, Gulyi IS, Kupchik MP, Vorona LG (1987) Electrical treatment of sugar beet slices using a combined method. Food Ind 33:61–65
Sichevoi PS (1955) Device for electroplasmolysis of sugarbeet slices. SU100624 A1 (Inventor's certificate)
Simpson DP, Stirling R (1984) Apparatus for heating electrically conductive flowable media. US4434357A (Patent)
Sitzmann W (1995) High Electric Field Pulses Treatment. In: Convegno Internazional, Il Pesche, Tecnolgia, Sicurezza e nutrizione dei Prodotti Ittici. Milano, Italy
Sitzmann W, Heinz V (1996) Die schonende Herstellung naturbelassener Säfte mit Hilfe eines physikalischen Niedertemperaturverfahrens. Food Technol Mag 7:28–32
Sitzmann W, Münch EW (1987) Verarbeitung tierischer Rohstoffe. Lipid/Fett (Eur J Lipid Sci Technol) 89:368–374
Sitzmann W, Münch EW (1988a) ELCRACK© und ELSTERIL© – Zwei neue Verfahren zur gezielten Beeinflussung von Phasengrenzflächen. In: GV– Fachausschusssitzung "Lebensmittelverfahrenstechnik." Brussels, Belgium
Sitzmann W, Münch EW (1988b) Elektrische Hochspannungsimpulse zur Abtötung von Mikroorganismen. In: Internationale DLG – Konferenz, „Sterile Prozesstechniken". Frankfurt/Main, Germany
Sitzmann W, Münch EW (1988c) Das ELCRACK® Verfahren: Ein neues Verfahren zur Verarbeitung tierischer Rohstoffe. Die Fleischmehlindustrie 40:22–28
Sitzmann W, Vorobiev E, Lebovka N (2016a) Pulsed electric fields for food industry: historical overview. In: Miklavcic D (ed) Handbook of electroporation. Springer International Publishing, Cham, pp 1–20
Sitzmann W, Vorobiev E, Lebovka N (2016b) Applications of electricity and specifically pulsed electric fields in food processing: historical backgrounds. Innov Food Sci Emerg Technol 37C:302–311
Skimbov AA, Berzoj SE (1992) The electrical processing of the raw material. Elektron Obrab Mater (Surf Eng Appl Electrochem) N4:65–67
Skimbov AA, Berzoj SE (1994) Electric treatment of fish stock in making fodder meal. Elektron Obrab Mater (Surf Eng Appl Electrochem) N1:64–66
Skimbov AA, Berzoy SE (1990) Device for electric treatment of food raw materials. SU1588363 A1 (Inventor's certificate)

Smith FS (1938) Method for treating materials and electrical treating apparatus, especially sterilization. US2132708A (Patent)
Smith KA, Grinnell AL (1951) Method and apparatus for preserving foodstuffs. US2576862A (Patent)
Spilker W, Gottstein A (1891) Ueber die Vernichtung von Mikroorganismen durch die Induktionselektricitat (On the destruction of microorganisms by induction electricite). Bakteri Parasitenk 9:77–88
Stadelmann E (1956) Plasmolyse und deplasmolyse. In: Allgemeine Physiologie der Pflanzenzelle/General physiology of the plant cell. Springer-Verlag, Berlin, Göttingen and Heidelberg, pp 71–115
Stämpfli R (1958) Reversible electrical breakdown of the excitable membrane of a Ranvier node. An Acad Bras Ciências (Ann Braz Acad Sci) 30:57–63
Stone GE (1909) Influence of electricity on micro-organisms. Bot Gaz 48:359–379
Teoreli T (1949) Permeability. Annu Rev Physiol 11:545–564
Thiele H, Wolf K (1899) Über die Einwirkung des elektrischen Stromes auf Bakterien. Zentralblatt Bakterien Parasitenkd 25:650–655
Thornton WM (1912) The electrical conductivity of bacteria, and the rate of sterilisation of bacteria by electric currents. Proc R Soc London Ser B, Contain Pap a Biol Character 85:331–344
Toepfl S, Heinz V, Knorr D (2007) History of pulsed electric field treatment. In: Lelieveld HLM, Notermans S, de Haan SWH (eds) Food preservation by pulsed electric fields from research to application. Woodhead, Cambridge, pp 9–42
Tswett M (1896) Etudes de physiologie cellulaire: contributions à la connaissance des mouvements du protoplasme, des membranes plasmiques et des chloroplastes (Cell physiology studies: contributions to the knowledge of the movements of the protoplasm, plasma membranes and chloroplasts). PhD Thesis. University of Geneva
Vologdin VV (1959) Conference on electric food processing (October 7–13, 1958, Kiev Technological Institute of the Food Industry of the Ukrainian SSR). Izv Vyss Uchebnykh Zaved Elektron 2(1):120–121. https://doi.org/10.20535/S0021347019590l0204
Vorobiev E, Lebovka N (2020) Processing of foods and biomass feedstocks by pulsed electric energy. Springer Nature, Cham
Wagner L, Marr J (1895) Electrolytic conduit for beer or other liquids. US535267A (Patent)
Weaver JC, Chizmadzhev YA (1996) Theory of electroporation: a review. Bioelectrochem Bioenerg 41:135–160
Weis A (1926) Beiträge zur Kenntnis der Plasmahaut. In: Planta. Springer, Berlin, pp 145–186
Willard RS (1944) Process of preserving fresh food products. US2365924A (Patent)
Wright MB (1924) Electric fluid sterilizer. US1479725A (Patent)
Wright MB (1937) Apparatus for pasteurizing liquids. US2081243A (Patent)
Yatsko MA, Zhuravleva IA (1970) Electro-hydraulic setup of a membrane type for processing of grape pulp. Elektron Obrab Mater (Surf Eng Appl Electrochem) N4:53–57
Yatsko MA, Zhuravleva NA, Flaumenbaum BL (1971) Determination of the effect of the electrohydraulic effect on juicy plant materials. Elektron Obrab Mater (Surf Eng Appl Electrochem) N5:76–80
Zagorul'ko AY (1957a) Method of electroplasmolysis of sugar beets during its cutting and a device for implementing the method. SU109588 A1 (Inventor's certificate)
Zagorul'ko AY (1957b) Impact of thermal plasmolysis and selective electroplasmolysis on the structure of the plasma cell membrane and permeability of beet tissues. Sakharnaya Promyshlennost (Sugar Ind) N11:67–71
Zagorul'ko AJ (1958a) Technological parameters of beet desugaring process by the selective electroplosmolysis. In: New physical methods of foods processing (in Russian). Izdatelstvo GosINTI, Moscow, pp 21–27
Zagorul'ko AY (1958b) Obtaining the diffusion juice with the help of electroplasmolysis. PhD Thesis (Candidate of technical sciences), Central Research Institute of Sugar Industry (TsNII saharnoy promyishlennosti), Kiev, Ukraine

Zagorul'ko AY (1974) Research and development of control methods and ways to reduce the losses of sugar in sugar beet production. Thesis for degree of doctor science habilitat (Technical sciences), Kiev Technological Institute of Food Industry, Ukraine

Zagorul'ko AY, Myl'kov MN (1953) Production of juice at low temperature using electroplasmolysis. Sakharnaya Promyshlennost (Sugar Ind) N10:15–18

Zangger H (1908) Über Membranen und Membranfunktionen. Ergeb Physiol 7:99–160

Zimmermann U, Pilwat G, Riemann F (1974) Dielectric breakdown of cell membranes. Biophys J 14:881–899

Zimmermann U, Pilwat G, Beckers F, Riemann F (1976a) Effects of external electrical fields on cell membranes. Bioelectrochem Bioenerg 3:58–83

Zimmermann U, Pilwat G, Holzapfel C, Rosenheck K (1976b) Electrical hemolysis of human and bovine red blood cells. J Membr Biol 30:135–152

Zimmermann U, Riemann F, Pilwat G (1976c) Enzyme loading of electrically homogeneous human red blood cell ghosts prepared by dielectric breakdown. Biochim Biophys Acta (BBA) Biomembr 436:460–474

Zimmermann U, Vienken J, Scheurich P (1980) Electric field induced fusion of biological cells. Eur Biophys Technische Universitat Berlin J 6:86

Generation and Application of High Intensity Pulsed Electric Fields

M. J. Löffler

Introduction

The generation of pulsed electric fields at high-power levels is one of the tasks of pulsed power engineers. In this chapter, we define pulsed power technologies to be those electrical devices that are used to treat organic and inorganic materials with electric fields >10 kV/cm and/or magnetic fields >10 T. As a rule, they work at power levels >0.1 GW, which are power levels that are not available from small- or medium-sized conventional energy sources (Bödecker 1985). Typical durations of single or repetitive power pulses are in the range between nanoseconds and milliseconds. Most pulsed power devices use capacitors to store electrical energy that is delivered to their respective electric loads via a high-power switch.

The energy stored in capacitors is used to generate electric or magnetic fields. Electric fields are used to accelerate charged particles, leading to thermal, chemical, mechanical, electromagnetic wave, or breakdown effects. Electromagnetic fields transfer energy as electromagnetic waves. X-ray, microwaves, and laser beam generation are typical examples. Magnetic fields facilitate the generation of extremely high pressures ranging from 0.1 GPa up to many GPa. These effects are applied to modify molecules; to remode, compress, weld, segment, fragment, or destroy materials; and to modify the surface of organic and inorganic parts and particles (Weise and Löffler 2001).

In general, those devices that are used to electrically pulse treat liquid foods or other aqueous substances have an electrode gap filled with a more or less conductive liquid. From an electrical point of view, these devices can be compared to water resistors driven either in the "normal operation mode" (e.g., electroporation) or in

M. J. Löffler (✉)
High Voltage and Pulsed Power Technology. EnergyInstitute, Westphalian University Gelsenkirchen, Neidenburger Str. 43, Gelsenkirchen, DE-45897, Germany, Germany
e-mail: markus.loeffler@w-hs.de

J. Raso et al. (eds.), *Pulsed Electric Fields Technology for the Food Industry*, Food Engineering Series, https://doi.org/10.1007/978-3-030-70586-2_2

the "failure mode" (e.g., shockwave treatment). In the first case, a current is driven across the entire cross section of the liquid, whereas in the second case an electrical breakdown occurs followed by high intensity current flowing through the small cross section of an arc. The specific power requirements of such "resistors" together with information about the required mass flux and the dimensions of the treatment zone define the type and the specific parameters of the power supply.

The objective of this chapter is to provide basic information:

1. About the power and energy requirements of PEF systems
2. About high-power sources
3. About low-power sources

Electric Load Requirements

From the point of view of an electrical engineer, organic materials in water is a bad insulator surrounded by a highly permittive medium, i.e., water. The effect of stressing organic materials under water with an electric field is shown in Fig. 1 (Loeffler et al. 2001). The rubber membrane was positioned between two electrodes under water. A voltage was applied between the electrodes. Due to the high permittivity of the water, the electric field was concentrated in the rubber membrane itself thus causing electrical breakdown at several locations on the foil. Increasing the field and the treatment time increased the number of holes in the foil.

The electrical specifications of devices used to inactivate microorganisms by perforating their cell membranes require some fundamental information about their electric load: strength of the electrical field, specific power consumption, mass flow rates, average power consumption, voltage waveforms, and treatment durations. Due to the different treatment profiles needed for different applications, a "characteristic application" with "characteristic values" shall be defined by averaging and evaluating values and information gained from a large number of experiments published in the open literature.

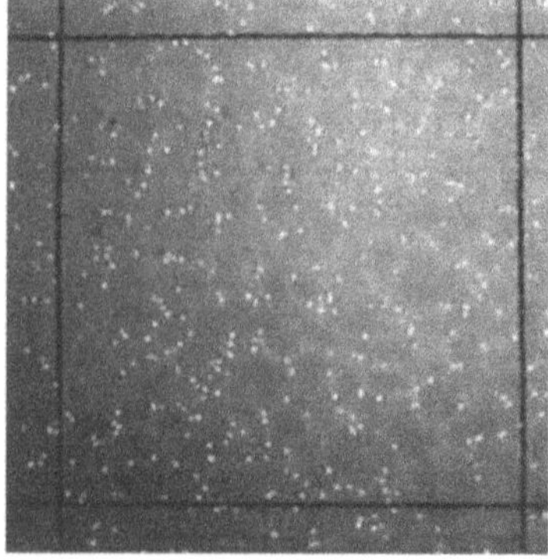

Fig. 1 Effect of pulsed electric fields on rubber membrane under water

Specific Power Consumption

The specific power consumption of liquids treated with electrical pulses is defined by:

- Their specific electrical conductivity
- The electric field strength required to kill the cells contained in the liquid

In the following, the focus will be on "electroporation devices" rather than shock wave devices. Figure 2 shows the conductivities of some selected liquid foods at different temperatures (Reitler 1990). For comparison Fig. 3 shows in a logarithmic scale the range of specific electrical conductivities σ of aqueous liquids together with the conductivities of insulating, semiconducting, resistive, and conducting materials. Depending on their ingredients, aqueous liquids have conductivities comparable to semiconducting materials. They vary between 10^{-2} and 10 S/m.

To treat aqueous liquids successfully, electrical field strengths E, which is defined to be the applied voltage u divided by the distance d between the electrodes, should be in the range between 10^0 and 10^2 kV/cm (Heinz et al. 2001; Gaudreau et al. 2004). Figure 4 gives an overview of the results from several experiments in which different microorganisms in different fluids were treated (Barbosa-Cánovas et al. 2000). This figure shows a reduction in viability by a factor of about $10^{3.5}$ when the electrical field strengths were about 30 kV/cm. Note that this data does not represent optimal values, because in most of these experiments, exponential voltage waveforms were used (for more information about the influence of voltage waveforms on the inactivation of biological material, see below).

The power densities p required to apply field strengths E across materials with conductivities σ is given by

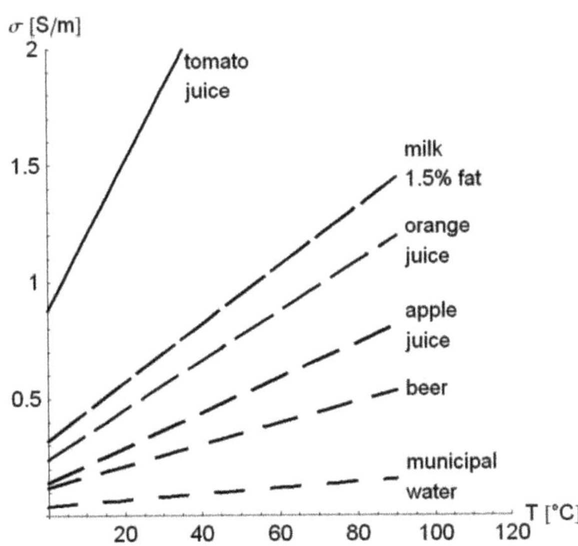

Fig. 2 Conductivities of selected liquids

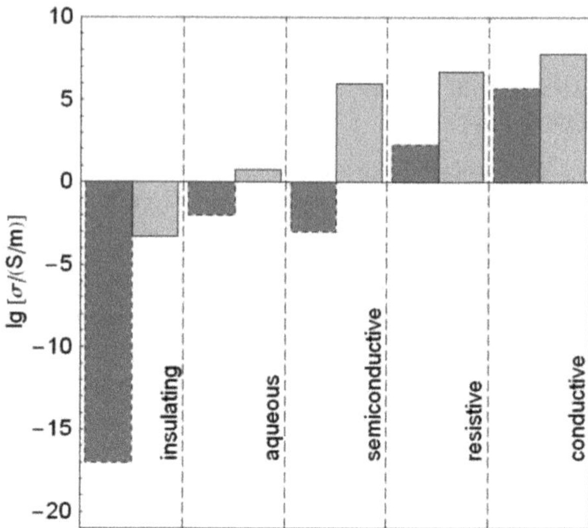

Fig. 3 Conductivities of different materials

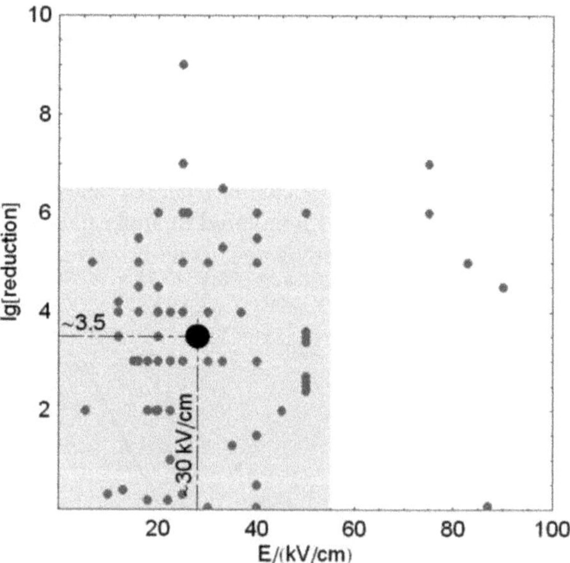

Fig. 4 Overview of the reduction effect of different electrical field strengths on different materials

$$p = \sigma \cdot E^2. \tag{1}$$

Figure 5 combines the values of the specific conductivities given in Fig. 3 with the electric field strengths given in Fig. 4. Applying formula (1) to the mean values of the conductivity and of the field strength ($\sigma_{av} \approx 0.3$ S/m, mean value taken from the logarithmic values; $E_{av} \approx 30$ kV/cm, mean value taken from the linear values) yields in an average power density of $p \approx 3$ GW/l. To treat one liter of a typical

Fig. 5 Range of specific conductivities and electric field strengths

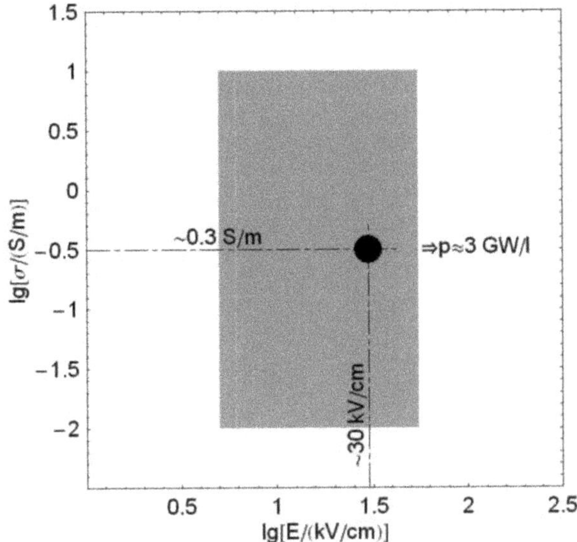

liquid requires for short times (1–1000 μs) powers at the GW level. Large power plants deliver electrical power in the GW range. By this it is obvious that power converters are essential to convert the power available from the public network into the power required at the treatment facility. In principle, comparable considerations of "shockwave devices" yield the same results.

The generation of electrical power at such levels requires capacitors and/or coils.

Average Power

Another requirement of the load is its specific energy consumption, accumulated over all pulses, together with the rate of flow of the substances to be treated. In a logarithmic scale, Fig. 6 gives information about the specific energies applied in several experiments. The data taken from (Heinz et al. 2002, 2003; Min et al. 2003a, b; Ngadi et al. 2003; Ulmer et al. 2002; Toepfl et al. 2004; Gaudreau et al. 2005) are marked with vertical lines. Note that the data are not optimum values, because energetic optimization was not the aim of these experiments. For standardization purposes, the units of the original values, given in kJ/l, were transformed into MJ/t, assuming that the density of the respective liquids was ~1 kg/l [10]. As an average value taken from the logarithmic scale, $w \approx 50$ MJ/t may be a typical specific energy. Common industrial flow rates for different applications are marked with horizontal lines in a logarithmic scale (Heinz, personal communication). As an average value taken from the logarithmic scale, $\dot{m} \approx 20$ t/h may be a typical mass flow rate. The average power $\overline{P}$ required to treat mass flows $\dot{m}$ with specific energies w can be calculated with

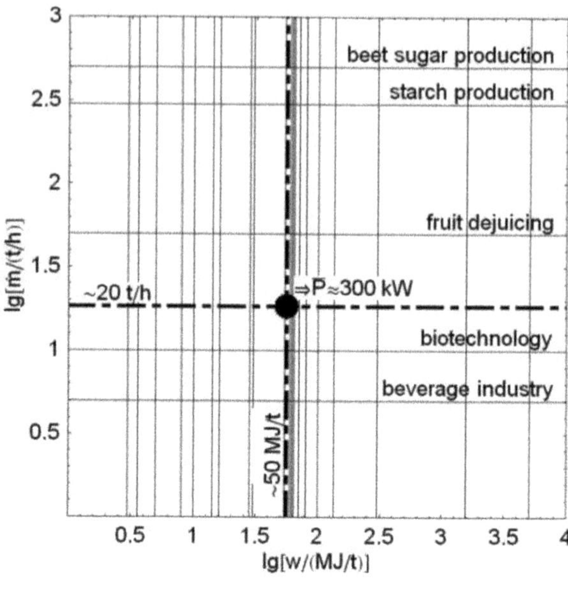

Fig. 6 Range of specific energies and of rates of flow

$$\overline{P} = \dot{m} \cdot w. \tag{2}$$

Applying Eq. (2) to the average values yields a typical value of the average power, which in this case is $\overline{P} \oplus 300\,\text{kW}$.

Voltage Waveforms

Typical voltage waveforms used to generate high intensity pulsed fields are either monopolar or bipolar (see Fig. 7). Monopolar waveforms have constant, rectangular, exponential, or mixed wave shapes, whereas bipolar waveforms are sinusoidal, triangular, trapezoidal, continuous rectangular, discontinuous rectangular, or discontinuous exponential (Lazarenko et al. 1977; Papchenko et al. 1988a, b; McLellan et al. 1991; Zhang et al. 1995; Barbosa-Cánovas et al. 2000; Bazhal 2001, Kotnik et al. 2001a, b; Jemai and Vorobiev 2002; Kotnik et al. 2003; San Martín et al. 2003).

The effects of these waveforms on several substances mainly were tested in small-scale experiments. Inactivation effects were only measured if the voltage amplitudes reached a value greater than the critical values given by the specific electric field strengths specified above. For large-scale applications as defined by the typical values given above, continuously acting waveforms without pause between single monopolar or bipolar pulses will not be practicable due to the extreme power requirements. Compared to monopolar pulses, bipolar pulses have a slight effect on the permeabilization of the cell membranes at moderate voltages, whereas none (Kotnik et al. 2001b) to some (Qin et al. 1994) significant effects were

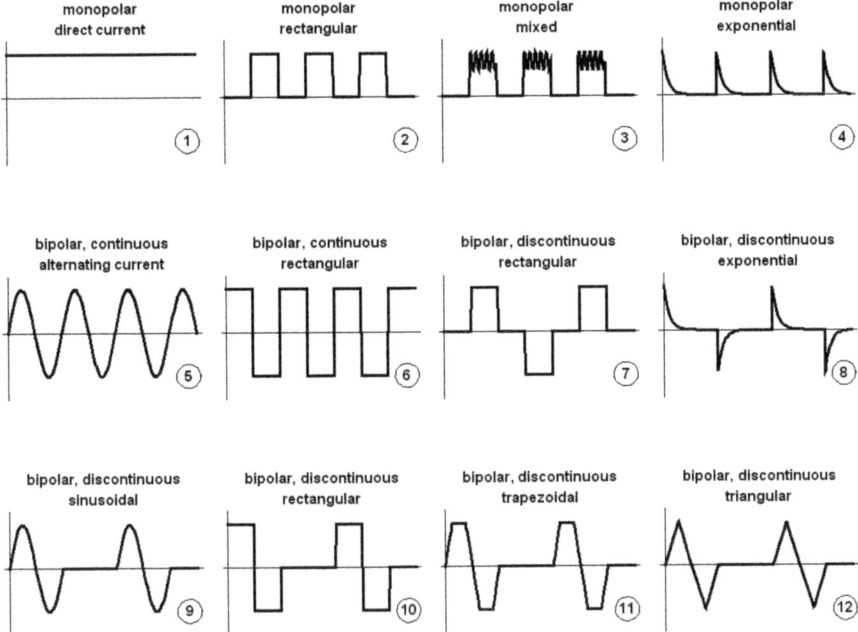

Fig. 7 Typical voltage waveforms

registered relative to the number of cells killed at voltages higher than the critical voltage. However, applying bipolar pulses, like the voltage waveform in Fig. 6, remarkably reduces the erosion of the electrodes (aluminum, stainless steel) in the treatment chamber (Kotnik et al. 2001a; Johnstone and Bodger 1997). The voltage steepness d*u*/d*t* at the beginning and at the end of rectangular pulses has no detectable influence on the efficiency of electropermeabilization, if this steepness is essentially smaller than the pulse length itself (Kotnik et al. 2003). For same pulse lengths, trapezoidal, triangular, and mixed voltage waveforms yield the same effectiveness as rectangular voltage waveforms, but only at higher peak voltages (Kotnik et al. 2003). Exponential voltage waveforms are not too energy efficient, since they have a long tail at electric field strengths lower than the required critical values. During this time, they only heat up the aqueous substance without any further effect on the biological material to be inactivated (San Martín et al. 2003).

From these findings it can be stated that voltage waveforms (2), (7), and (10) seem to be best suited for industrial applications. However, voltage waveforms (4) and (9) should also be considered due to their widespread usage, ease of generation, and relative low cost.

Pulse Lengths and Repetition Rates

Independent of the voltage waveforms, typical pulse lengths vary between 1 and 1000 μs at pulse rates between 1 and 300 and at pulse repetition rates between 1 and 2000 pps (Gaudreau et al. 2004; Barbosa-Cánovas et al. 2000; Kotnik et al. 2001a).

Conclusion

To conclude, it can be stated that industrial power modulators for the treatment of liquid foods and other biological liquids or aqueous materials should have the following typical values:

- Specific electrical data for medium treatment:
  - Specific conductivity of the medium:
    - Range: 10^{-2}–10 S/m
    - Typical value: 0.3 S/m
  - Electric field strength in the medium:
    - Range: 1–100 kV/cm
    - Typical value: 30 kV/cm
  - Power density:
    - Range: 2 MW/l–30 GW/l
    - Typical value: 3 GW/l
  - Specific energy consumed in the medium:
    - Range: 0.5–5000 MJ/t
    - Typical value: 50 MJ/t
- Specific data for facilities
  - Mass flow:
    - Range: 5–300 t/h
    - Typical value: 20 t/h
  - Average power:
    - Range: >20 kW
    - Typical value: 300 kW
  - Treatment time:
    - Range of single pulse durations: 1–1000 μs
    - Range of pulse rates: 1–3000

    - Typical value of treatment time (rectangular pulses): 10 μs
  - Pulse repetition rate:
    - Range: 1–2000 pps
    - Typical value: 1 pps (remark: this value yields for laboratory devices; for industrial applications, higher repetition rates in the order of ~100 pps are required)

For each specific application, these values have to be confirmed by experiments.

Pulsed Power Systems

Figure 8 shows the general block diagram of a pulsed power system that generates pulsed electric fields. The resistive electrical load – the PEF treatment chamber – is powered by a high-power/high-voltage source. The high-power source itself is powered by a low-power source connected to a 50 Hz (60 Hz) public or local network.

The PEF treatment chamber consists of one or more electrode gap(s) filled with the substance to be treated. The electrodes must be shaped to ensure a more or less homogeneous electrical field. Contamination of the substance to be treated by the electrode material should be as small as possible.

The high-power source has to deliver the voltage required by the load at the right amplitude (1–100 kV), in the right form and at the right time (μs-ms). The main parts of the high-power source are one or more capacitors, as primary energy storage, on-switches and/or off-switches, and inductors as secondary energy stores.

The low-power source converts the AC voltage of the network to DC voltage or DC current. The DC current charges the capacitor(s) of the high-power source in the right amount of time to the voltage required to drive the high-power source.

The public or local network has to deliver the peak power requested from the low-power source.

The following subchapters will describe in more detail the following:

- PEF treatment chambers
- High-power sources
- Low-power sources

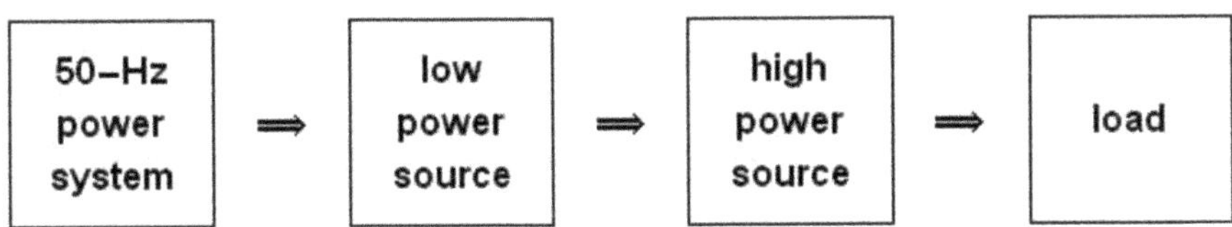

Fig. 8 Basic setup of pulsed power systems for the generation of pulsed electrical fields

Although not all the technical possibilities relative to the construction of electric pulsers can be covered in this overview, the reader will be acquainted with their technical possibilities and limitations.

One of the critical components of pulsed power generators is the switch. Therefore, we will begin with a discussion of switches.

General Remarks About Switches in Pulsed Power Generators

As a rule in in working with pulsed power devices, the load has to be electrically separated from the high-power source while it is being charged by the low-power source. After completion, the low-power source should be separated from the high-power source by switches and/or by properly chosen ohmic resistances. In general this task can be solved by using commercially existing components.

If, under large-scale industrial conditions, the load consumes energy at power levels in the multi-MW to GW range, the final switch between the high-power source and the load has to transfer this energy at this power level. If, as an example, the load consumes energy at a peak power of 1 GW and at a voltage of 100 kV, the on-/off-switch has to handle before/after switching a voltage of 100 kV and after/before switching a current of 10 kA. This task does not change, if, instead of a single switch, multiple switches are used in parallel and/or in series. This step has to be done, if switches for such voltages and currents are not available. In addition to the other parameters that need to be taken into consideration (di/dt, cost integral, recovery time, etc.), the replacement of a single switch by multiple switches can lead to other demanding tasks like the synchronization of the switches.

PEF Treatment Chambers

From the electrical point of view, the PEF treatment chamber represents the electrical load consisting of two or more electrodes filled with the liquid substance to be treated. The chamber has to be constructed in such a way that the electrical field acting on the liquid is more or less homogeneous across the entire active region. This goal can be reached in principle with planar, coaxial, and axial electrode geometries. Figure 9 shows sketches of the cross sections and of the top views of these basic geometries.

Planar electrode configurations consist of two parallel electrodes fixed by insulators. The insulators and the electrodes form a channel for the streaming liquid. Coaxial electrode configurations consist of two coaxial electrodes. The liquid streams between these electrodes that are fixed by insulators not shown in the figure. Axial electrode configurations consist of several electrode rings on alternating potentials separated by insulating rings.

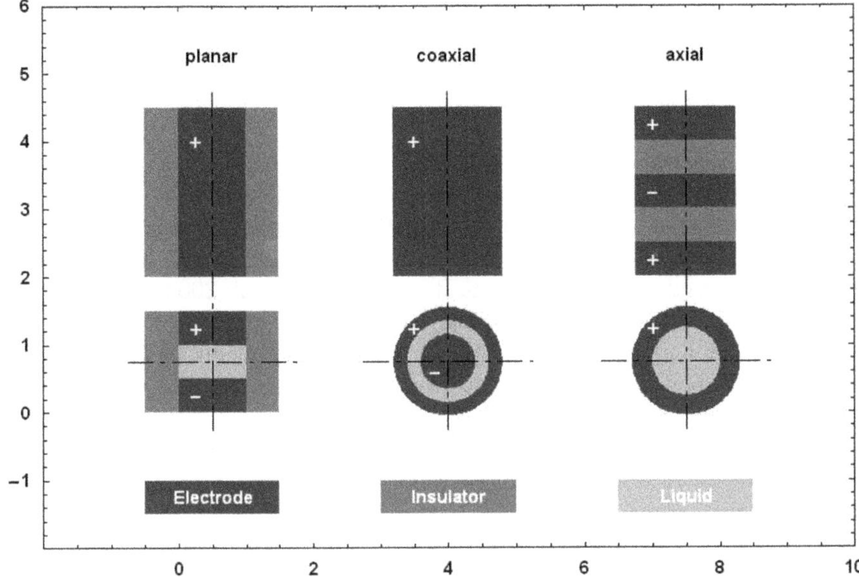

Fig. 9 Typical electrode geometries

Planar electrode configurations ensure homogeneous fields provided their separation distance is not significantly larger than their width. In the case of coaxial electrodes, the relation between the outer radius and the inner radius of the active zone should be smaller than 1.23 to ensure that the electric field does not decay by more than 10%. To ensure homogeneous fields in the main treatment zone of the axial electrode configurations, the direct proportionality of the electrode distance to the diameter of the channel should be maintained (Gaudreau et al. 2004).

In addition to geometrical effects, the homogeneity of the electrical field also strongly depends on the permittivity and on the conductivity of the liquid. The separation distance of the electrodes is restricted by the voltage available from the high-power supply. In commercial applications, voltages on the order of 100 kV are difficult to handle and, in many cases, offer a hard restriction. For exponentially damped voltages, field requirements of ~30 kV/cm lead to electrode gap separations of ~3 cm. Another restriction is the amount of power available from the high-power source. Assuming a maximum power of $\hat{P} \approx 1$ GW (e.g., this would relate to currents of ~10 kA at voltages of ~100 kV) and assuming a typical power density of $p \approx 3$ GW/l consumed by the liquid, the volume of the active treatment zone would be restricted to

$$V = \frac{\hat{P}}{p} \approx 300\,ml.$$

For electrode separation distances of ~3 cm, the active area of the electrodes would be restricted to ~100 cm^2.

An overview of some PEF treatment chambers investigated so far is given in [6]. Actually, their volume is between 0.1 and 200 ml with electrode gaps between 0.1 and 1 cm.

Electrode materials also play an essential role. If monopolar voltage waveforms are applied, electrode corrosion can become critical and the substance to be treated can be contaminated. In commercially available electroporation devices with small probes, aluminum, stainless steel, carbon, gold plated electrodes, and even silver electrodes are used (Barbosa-Cánovas et al. 2000; Puc et al. 2004). Stainless steel electrodes suffer with problems such as electrolysis, formation of deposits, electrode corrosion, and transfer of particles into the treated media (Barbosa-Cánovas et al. 2000). Carbon electrodes are better for the inactivation of cells, since they have seemingly lower erosion rates than stainless steel electrodes (Barbosa-Cánovas et al. 2000). Erosion can be minimized by applying bipolar voltage waveforms (Kotnik et al. 2001a; Johnstone and Bodger 1997).

High-Power Sources

Several methods exist for generating pulsed electric fields. Depending on the specified type of voltage waveform, as well as on the available elements (capacitors, inductors, transformers, switches), the possibilities vary between simple circuits and very sophisticated networks. In the following, several important possible pulsed power circuits will be discussed. The explanations are restricted to the essentials. Practical devices, based on the circuits presented herein, are more complex due to technical safety and failure handling necessities. There are also other possible schemes that are driven by the physical and technical needs of the components, as well as that of the continuously changing ohmic properties of the load.

Basically, these circuits can be distinguished as being:

- Basic pulsed power circuits
- Circuits with voltage multipliers
- Pulse forming circuits
- Networks with pulse forming switches

Basic Pulsed Power Circuits

Capacitive Circuits

Capacitive circuits are the most simple and popular way to generate high-power pulses. Figure 10 shows the basic scheme of a capacitive circuit.

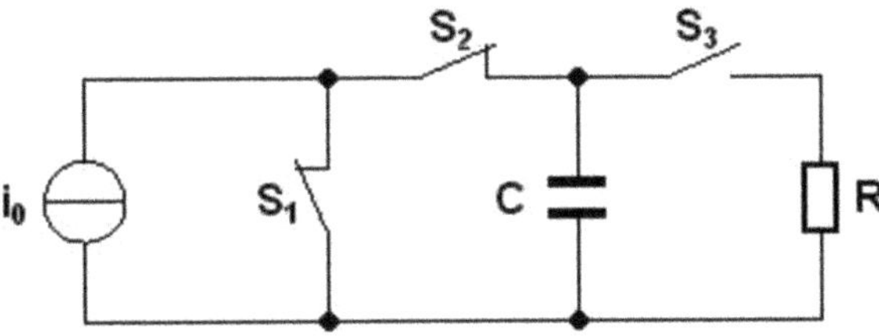

Fig. 10 Capacitive circuit

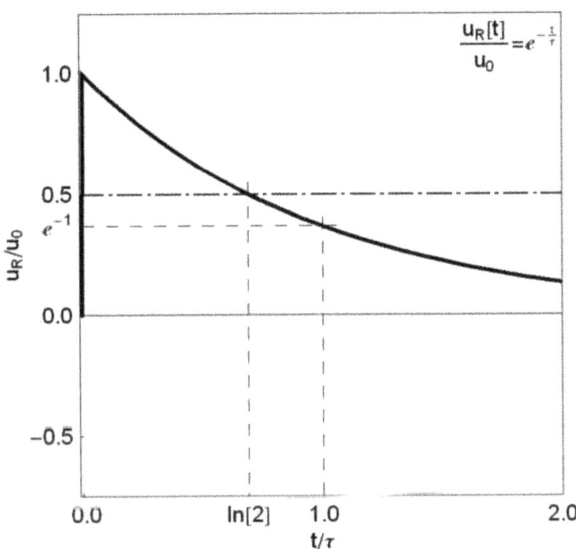

Fig. 11 Standardized load voltage in a capacitive and inductive circuits

When switch S_1 opens, capacitor C is charged by the low-power current source i_0 (e.g., a pulse width modulator, see below; charging with constant voltage via a series resistor would reduce the energy efficiency to 50%). During time t, the capacitor voltage u_C increases according to $u_C = i_0 \cdot t/C$. The power demand also increases in accordance with the formula $P_C = i_0^2 \cdot t / C$. The average power can be calculated by using $\overline{P} = i_0 \cdot u_0 / 2$, whereas the peak power of the current source is two times the average power. If voltage u_0 equals the initial voltage $u_{R,0}$ at the load, the capacitor stores energy according to this relation $W_C = C \cdot u_0^2 / 2$. Switch S_2 is opened and switch S_1 is closed. Switch S_3 is activated. The capacitor exponentially discharges via the load R. The path of the standardized voltage u_R/u_0 across the load is shown in Fig. 11, where it plotted as a function of the normalized time t/τ. The time constant τ of the discharge circuit is defined by $\tau = R \cdot C$. The voltage decays to 50% of its initial value at $t/\tau =$ ln [2] ≈ 0.693. At this time, the capacitor has delivered 75% of its energy to the load. The capacitive power and energy at this moment have decreased to 25% of its initial value.

Inductive Circuits

Figure 12 shows the basic scheme of an inductive circuit. After closing switch S_1, the coil represented by its inductance L and resistance R_L is charged by the low-power voltage source u_0 (e.g., bridge converter; charging with a constant current source would reduce the energy efficiency to 50%) via switch S_3. The coil current i_L increases according to the formula $i_L = u_0 \cdot (1 - e^{-\delta \cdot t})/R_L$, where the coil's self-damping constant is $\delta = R_L/L$. When the working current $i_0 = u_{R,0}/R$, with initial voltage $u_{R,0}$ at the load, is reached, the inductance stores an energy $W_L = L \cdot i_0^2 / 2$. Switch S_1 is opened and switch S_2 is closed (the inductance is "crowbared"). Switch S_3 is opened, commutating the current into the load. The inductance discharges exponentially via the load R. The path of the voltage across the load is the same as that shown in Fig. 11. Here the time constant τ of the circuit is defined by $\tau = L/(R + R_L)$. Again the voltage reaches 50% of its initial value at $t/\tau = \ln[2] \approx 0.693$. As in the case of capacitive circuits, at this time the inductance has delivered 75% of its energy to the load and its own resistance and the power stored in the inductor has decreased to 25% of its initial value.

The inductor can also be charged by a charged capacitor instead of by the low voltage source u_0. In general, fast repetitive pulsed inductive circuits are not as effective as capacitive circuits. The reason is that nonsuperconducting conducting coils of reasonable shape have a power consuming resistance, which decreases the overall energetic efficiency of the process.

Ringing Circuits

In addition to capacitive circuits, series and parallel ringing circuits are another simple and popular way to generate high-power pulses. Figure 13 shows the basic diagram of a series circuit. A capacitor C is charged via a low-power current source i_0 and the switches S_1 and S_2 (compare to the capacitive circuit).

If the capacitor reaches the voltage u_0, switch S_2 opens, switch S_1 closes, and switch S_3 is activated. The capacitor discharges via the coil (L, R_L) and the load R. The history of the normalized load voltage is shown in Fig. 14 as function of the normalized time t/τ, where $\tau = (R + R_L) \cdot C$ and $a = (R + R_L)/L$. Note that the formula given in the figure also holds for negative arguments under the root sign. The highest voltage can be achieved for $a \to \infty$ when the inductance L is small. In this case,

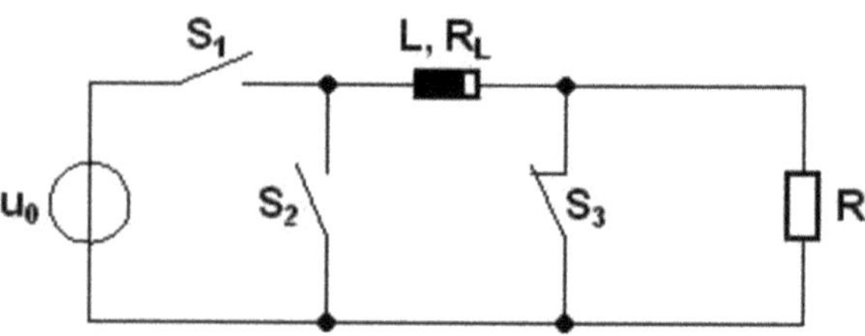

Fig. 12 Inductive circuit

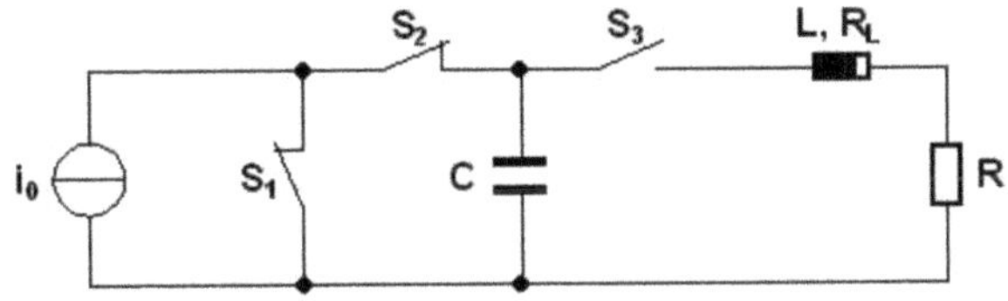

Fig. 13 Serial ringing circuit

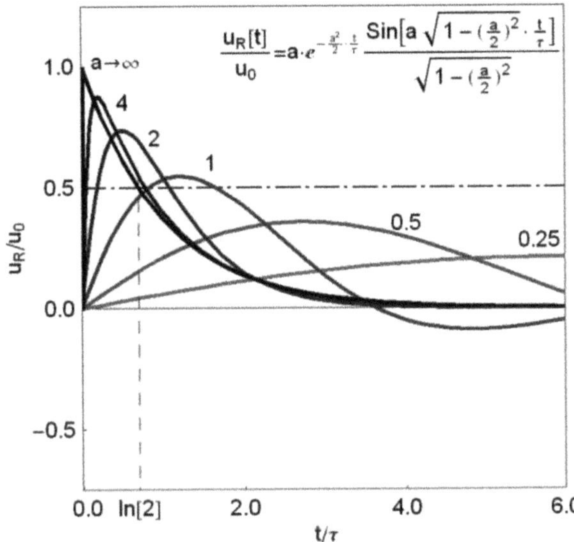

Fig. 14 Standardized load voltage in a serial ringing circuit

the ringing circuit behaves like a capacitive circuit. The load peak voltage decreases significantly when $a < 2$.

Figure 15 shows the basic diagram of a parallel circuit. After opening switch S_1, capacitor C is charged by the low-power current source i_0. After reaching voltage u_0, switch S_2 opens, switch S_1 closes, and switch S_3 closes. The capacitor discharges via the inductance L, its resistivity R_L, and the load R. The history of the normalized load voltage for $R_L << R$ is shown in Fig. 16, as a function of the normalized time $\tau = R \cdot C$ and as function of $a = 4 \cdot R^2 \cdot C/L$. The highest voltage can be achieved for $a \rightarrow 0$ when the inductance L is very large. In this case, the ringing circuit behaves like a capacitive circuit. In practice, this behavior is maintained up to values $a < 0.1$.

Ringing circuits are less effective than capacitive circuits, either due to voltage reduction in the inductor (series) or due to the reduction in time the voltage is maintained beyond the $u_0/2$ level (parallel).

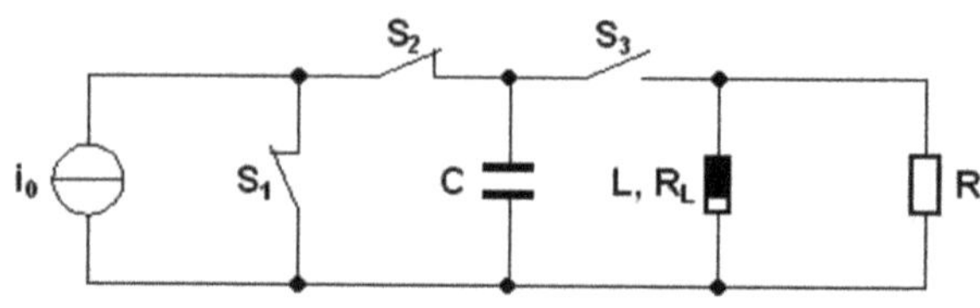

Fig. 15 Parallel ringing circuit

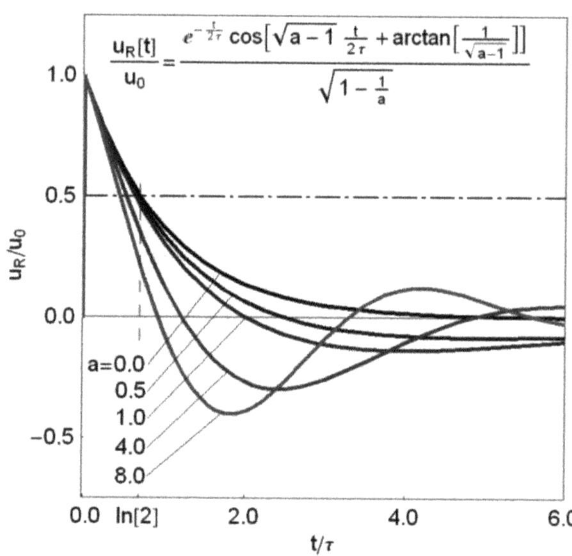

Fig. 16 Standardized load voltage in a parallel ringing circuit

Circuits with Transformers or Other Voltage Multipliers

In the circuits discussed above, the on- and off-switches have to maintain a full load voltage (capacitive and inductive circuits), that is, the voltage across the capacitors (capacitive and ringing circuits). At voltages above 100 kV and for repetition rates of ~1 Hz or more, the availability of high-lifetime switches, as well as operation of the facilities become more and more critical. In this case, transformers or voltage multiplying circuits should be taken into consideration. In the following, coils will only be represented by their inductances, assuming that their ohmic power losses are neglectible compared to the energy input into the PEF load.

Circuits with Pulse Transformers

One way to multiply voltages is to use transformers that are switched between the high-power source and the load. As an example, Fig. 17 shows the situation for capacitive circuits.

Capacitance C can be charged and discharged, as in capacitive circuits without transformer. If the windings of the transformer, with inductances L_1 and L_2, are

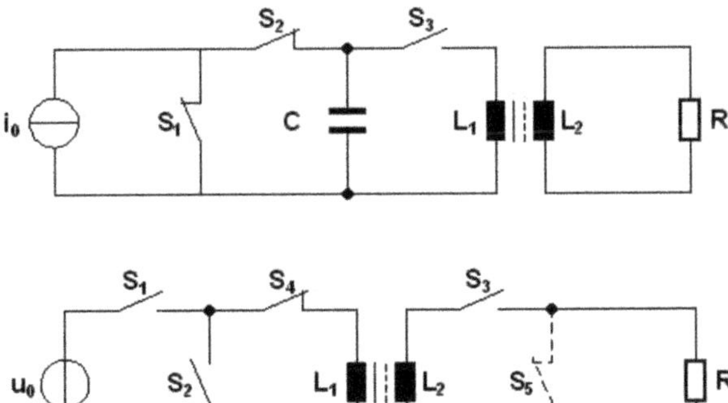

Fig. 17 Capacitive circuit with pulse transformer

Fig. 18 Circuit with storage transformer

coupled perfectly (where the mutual inductance between the windings is $M \approx \sqrt{L_1 \cdot L_2}$), the transformer amplifies the voltage by a factor of $m \oplus \sqrt{L_2 / L_1}$. Because the discharge frequencies usually exceed 1 kHz, such transformers must be either air core transformers or ferrite transformers. Looking backwards from the load to the capacitor, the circuit behaves like a parallel ringing circuit (Fig. 15), where inductance L is replaced with $m^2 \cdot L_1$ and the capacitance C is replaced with C/m^2 at the same voltage level as the load. This circuit shows a nearly capacitive behavior, if $a = 4 \cdot R^2 \cdot C/(m^4 \cdot L_1) < 0.1$ and if $L_2 = m^2 \cdot L_1 > 40 \cdot R^2 \cdot C/m^4$. Compared to the capacitive circuit, the capacitance C and the on-switch S_3 only have to achieve a voltage u_2/m.

However, in real circuits with pulse transformers, the energy is stored in the capacitor. The power consumed in the on-switch is slightly higher than in the capacitive circuits due to the unavoidable stray inductances and ohmic resistances in the pulse transformer. Furthermore, the additional pulse transformer increases the cost. Capacitive circuits with pulse transformers are only cost effective, if the on-switches with the required hold-off voltages are not available or, due to other technical problems, cannot be connected in series. They also are financially practical if the current source is not able to deliver the charging power at the required voltage level.

Circuits with Storage Transformers

Figure 18 shows a circuit with a storage transformer and no capacitor. After closing switch S_1, the low-power voltage source u_0 drives a current i_1 across the inductance L_1 of the transformer. Switch S_3 is open, so that no current is induced in the secondary coil L_2. The transformer behaves like an inductor L_1, so that the circuit behaves like an inductive circuit. After charging L_1 to the desired current level, S_1 is reopened and switches S_2 and S_3 are closed. Assuming perfect magnetic coupling between the transformer windings ($M = \sqrt{L_1 \cdot L_2}$) and after the powerless opening of switch S_4, the magnetic flux and the energy of primary inductance L_1 transfers to the secondary inductance L_2. The current across L_2 can be calculated by using $i_2 = i_1 \cdot \sqrt{L_1 / L_2}$.

Opening switch S_5 commutates i_2 into the load, resulting in an initial load voltage $u_R = R \cdot i_2$. The circuit again behaves like an inductive circuit with the time constant $\tau = L_2/R$ and with an exponentially damped current i_2 (Figs. 11 and 12). When switch S_5 is introduced into the circuit, the relationship between the primary and the secondary inductance should be chosen so that $L_2 << L_1$ which means that the switches on the primary side of the transformer can act at lower voltages, lower currents, and lower power during the charging of L_1 and during switching at the end of the charge cycle.

In this scheme, the circuit has no advantage over inductive circuits, because switch S_5 has to overtake the full voltage at the load. In principle, the circuit also works without this switch. Without switch S_5, switch S_4 would have to overtake the voltage $u_{S4} = \sqrt{L_1 / L_2} \cdot u_2$. In this case, the primary inductance, in order to act as a voltage multiplier, has to be essentially smaller than the secondary inductance: $L_1 << L_2$. Although this relieves the high voltage requirement for switch S_4, it has to be considered that the overall switching power of this switch, at least, keeps the same value as that of the switching power of switch S_5 in the preceding situation.

As in the case of capacitive circuits with pulse transformers, the transformer has stray inductances and ohmic resistance. Both increase the power requirements of the switches and of the voltage source. With respect to high voltage applications, PEF circuits with storage transformers are less effective than inductive circuits.

Voltage Multiplier I (MARX Generator)

The basic operating principles of the MARX generator were first published by E. Marx in 1923 (Marx 1923). The original application of this voltage multiplying circuit was to test isolators and other electrical high voltage devices. Figure 19 shows the original scheme. In two branches n capacitors C are connected in parallel via resistors W. Switches F connect the high potential and the low potential terminals of successive capacitors (in this case spark gap switches). The ends of the capacitor branches connect to an isolator J (respectively, the load). The capacitors are charged by a current source via the resistors to the same voltage. Igniting switches F connects the capacitors in series. By doing this, voltage multiplication occurs at the load. The resistors W discouple the capacitors if their resistance is essentially higher than the (transient) resistance of the load. The overall capacitance discharging via the load is given by C/n. The resistors can also be replaced with opening switches if necessary. In this case these switches have to open shortly before the switches F are ignited. After switching, the discharge behavior is equivalent to the discharge behavior of the capacitive circuit.

In MARX-generators, the on-switches have to maintain only $1/n$ of the voltage at the load. They have to conduct the load current. The low current source charging the capacitors only has to deliver $1/n$ of the voltage than that required for the basic capacitive circuit at n times higher charging current. During charging, the overall capacitance of the capacitors switched in parallel has to be n^2 times higher than in the capacitive circuit.

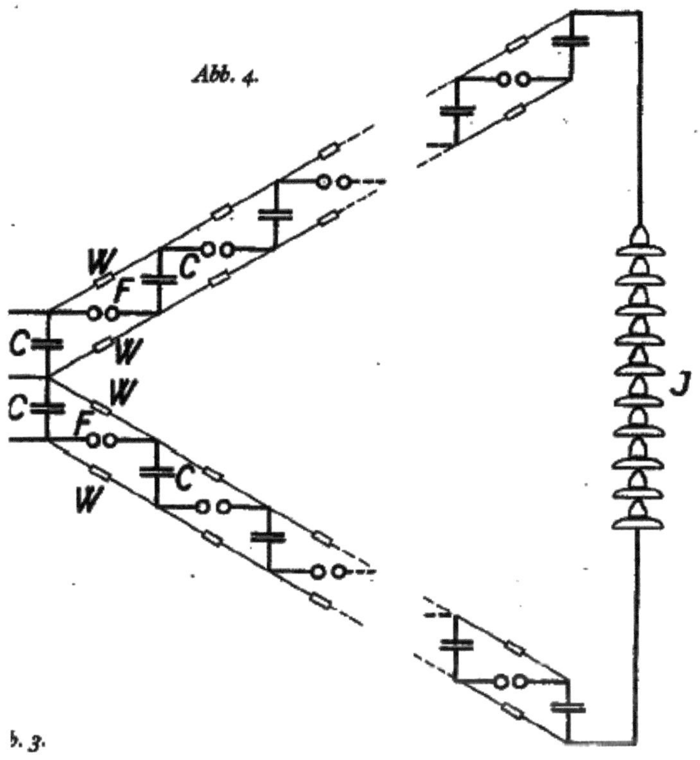

Fig. 19 MARX generator (from Marx 1923)

As opposed to capacitive circuits, the capacitance and the main on-switch are divided into several smaller switches and larger capacitances connected in series when system discharges. The overall volume, as well as the overall mass, of the capacitors will be slightly larger than that for basic capacitive circuits when the same energy is to be stored in the capacitors due to added buswork. The overall switching power remains the same. As in the case with pulse transformers, MARX-generators should only be chosen, if, with regard to the voltage requirements at the load, no adequate switches and/or capacitors and/or current sources are available.

Voltage Multiplier II (GREINACHER Cascade)

The GREINACHER cascade was invented by H. Greinacher in 1917 (Greinacher 1919). Figure 20 shows a picture from the original source, whereas Fig. 21 shows its circuit diagram as it is used today. An AC-voltage source with a voltage amplitude $\hat{u}_{\sim}$ drives, via a charging resistor R_{ch}, a column consisting of capacitors C and diodes D. After charging the capacitors via the diodes D, the $n = 3$ capacitors C on the right side are each charged to $<2 \cdot \hat{u}_{\sim}$ and are discharged in series via a switch S to

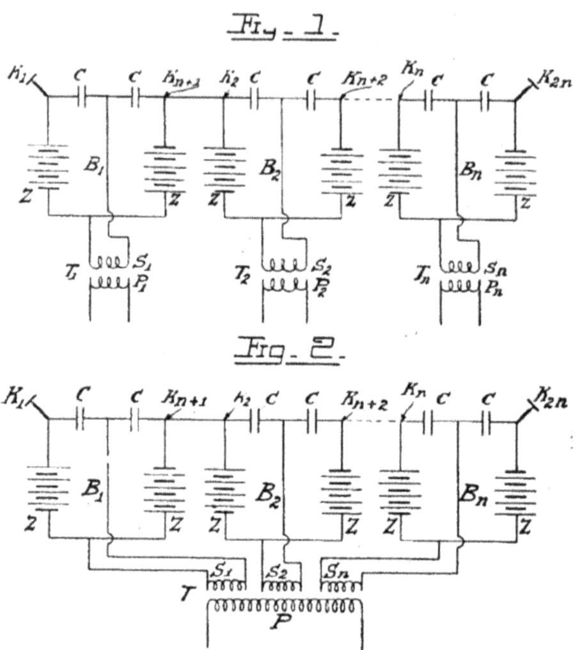

Fig. 20 Original GREINACHER cascade (from Greinacher 1919)

the load R. The resulting initial voltage on the load is $<6{\cdot}\hat{u}_{\sim}$, with a resulting capacitance of $C/3$. n modules, consisting of n capacitors and n diodes, provide an initial voltage $\hat{u}_{load} < 2{\cdot}n{\cdot}\hat{u}_{\sim}$ across the load at an overall capacitance C/n. The diodes in the circuit are driven at low power, because they only have to overtake the charging current and double the voltage amplitude at the voltage source. The main switch S has to handle the entire energy transfer at load powers. The efficiency of Greinacher cascades decreases with increasing number of stages due to the unavoidable ohmic losses in the diodes that hinders charging each capacitor to $2{\cdot}\hat{u}_{\sim}$. From this point of view, MARX generators provide higher efficiencies.

Current Multiplier (XRAM Generator)

The XRAM Generator was published by E. Marx and W. Koch in 1966 (Marx and Koch 1970). Figure 22 shows the original scheme. n coils with inductance L_1 are connected in series via opening switches (2). At the ends of each inductor. Closing switches (3) (in this case spark gaps) are connected. The inductors are charged in series by the voltage source (1) to a current $i_{1,0}$. After reaching the desired current level, switches (2a) and (2) open and switches (3) close. By doing this, the inductors are switched in parallel, generating an initial current $i_{2,0} = n{\cdot}i_{1,0}$ across the load (4). The current generates an initial load voltage $u_{R,0} = n{\cdot}R{\cdot}i_{1,0}$. After switching, the discharge behavior is equivalent to that of the basic inductive circuit.

Fig. 21 GREINACHER cascade

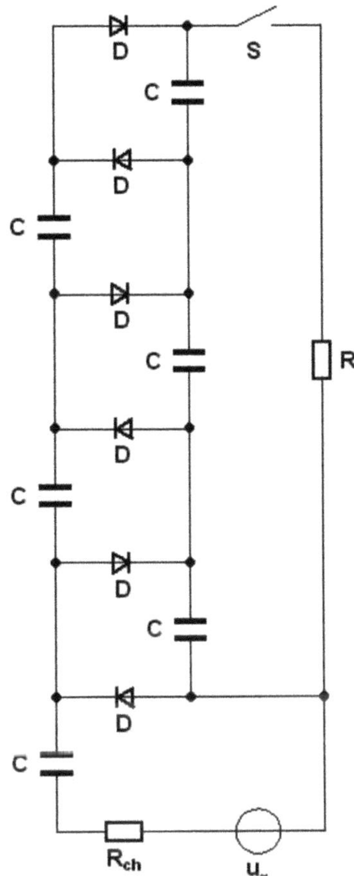

In XRAM generators, the opening switches have to maintain only $1/n$ of the current across the load. However, they do have to withstand the load voltage. The closing switches have to withstand ≤50% of the charge voltage at $1/n$ of the current level at the load. The low-power voltage source, charging the inductances, has to deliver $1/n$ of the current required in the basic inductive circuit, but at n times higher charging voltage. During charging, the overall inductance of the single coils switched in series has to be n^2 times greater than that of the inductive circuit. The overall volume, as well as the overall mass, of the coils will be the same as in the basic inductive circuits due to the same amount of energy being stored in the coils for the same charging times. The necessity of repetitive simultaneous switching of the off-switches causes problems that make the application of this type of generator very difficult. Thus, XRAM generators are not applicable to PEF applications.

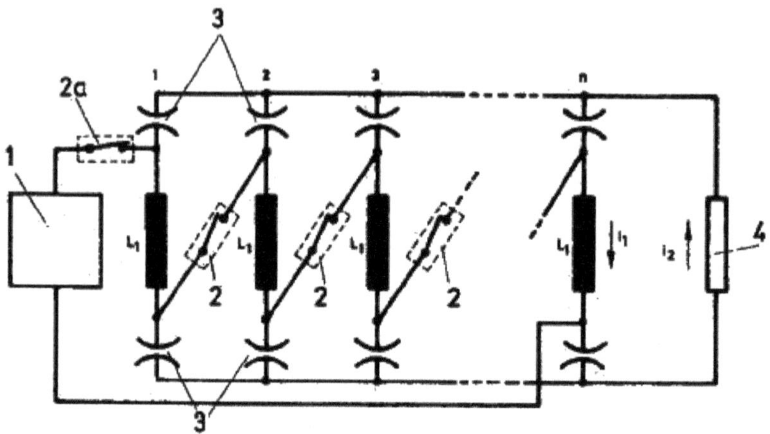

Fig. 22 XRAM generator (from Marx and Koch 1970)

Pulse Forming Networks

Pulse forming networks allow the generation of monopolar or bipolar rectangular voltage waveforms preferred for PEF treatment applications.

The properties of pulse forming networks (PFNs) are based on the fundamental principles of transmission lines. So, prior to an explanation of the electrical behavior of PFNs, the principal electrical behavior of transmission lines will be briefly explained.

Pulse Forming Lines

An ideal lossless transmission line (or pulse forming line) may be represented by a coaxial cable of length l, inductance $L = L' \cdot l$, and capacitance $C = C' \cdot l$. Charging the capacitance of this transmission line to a voltage u_0, or, respectively, to an energy $W_C = C \cdot u_0^2 / 2$, and discharging the line via a switch S into a load with resistance $R = \sqrt{L / C} = \sqrt{L' / C'}$ generates a single rectangular voltage pulse with amplitude $u_0/2$ and a duration of $\tau = 2 \cdot \sqrt{L \cdot C} = 2 \cdot l \cdot \sqrt{\varepsilon_r} / c$. ε_r is the relative permittivity of the line's insulator and c is the velocity of light. Circuit diagrams of this circuit are presented in the two pictures at the top of Fig. 23.

When the switch is triggered and if the line to the load is short-circuited at its other end, a bipolar rectangular voltage waveform with amplitude $\pm u_0/2$ is generated, where each pulse has a pulse length of $\tau/2$ (see Fig. 23, pictures in the center).

As a third possibility, the transmission line is split in two parts and a resistance $2 \cdot R$ reconnects the lines as seen in the bottom diagrams in Fig. 23. Short-circuiting the left end of the line generates a rectangular voltage waveform of duration $\tau/2$ at a voltage amplitude u_0.

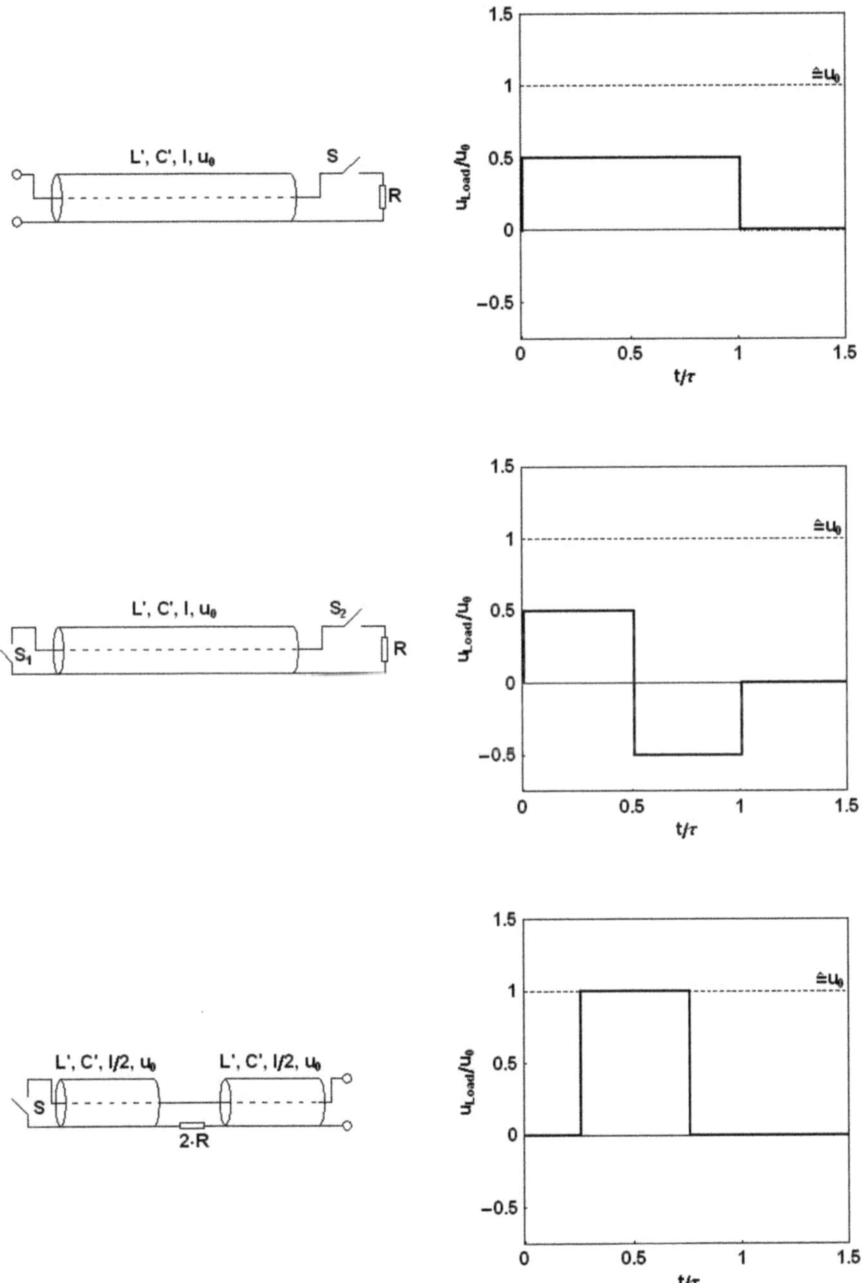

Fig. 23 Pulse forming lines and voltage waveforms

These examples show the flexibility of transmission lines with respect to the generation of rectangular voltage waveforms. In all cases, the energy originally stored in the line's capacitance is completely transferred to the load. However, the following example will show the disadvantage of this solution.

With the following reasonable values

- Load voltage $u_{R,0} = 100$ kV.
- Load resistance $R = 10\ \Omega$.
- Pulse duration $\tau \approx 15$ µs.
- Pulse energy $W = 5$ kJ.

the specification of the pulse forming line would be as described below:

For a relative permittivity of $\varepsilon_r \approx 3$, the length of the line has to be $l \approx 1.2$ km. To store 5 kJ at a voltage of 200 kV (only 50% of this voltage acts on the load), C' is calculated to be $C' = 250$ µF/km. Conventional 20 kV cables, like N(A)HKBA cables, have capacitances up to about 2 µF/km (source: Felten and Guilleaume Energietechnik AG, Taschenbuch, 1995). The cables can withstand a peak voltage of about 30 kV. So the ends of 7 cables have to be switched in series, decreasing C' to about 0.3 µF/km. To reach 250 µF/km, about 80 cable groups of 7 would have to be switched in parallel. Overall, such a device would result in the handling of 560 cables with a length of 1.2 km each. At a mass of about 4 t/km, this results in an overall mass of about 2700 t. The disadvantage of this solution is obvious. Furthermore, the ohmic resistance of the cables could become intolerably high for the high-frequency applications.

Due to this disadvantage of a cable solution, other solutions have to be considered without losing the advantage of rectangular pulse shape generation.

Pulse Forming Networks

Pulse forming networks allow one to simulate the behavior of transmission lines by using capacitors and inductors with free choice of their values. Using cables, the relation between C and L is fixed due to their fixed geometry. Pulse forming networks can be derived from transmission lines as follows:

The transmission line can be split into several smaller pieces connected in a chain. Each of these pieces has its own capacitance and inductance (as well as conductor conductivity, which usually can be neglected). In principle, the diagram of a transmission line based on this design now looks like that shown in Fig. 24. This diagram shows a chain of several identical four-terminal networks consisting of the same L-C-combinations. This chain is called a pulse forming network with a finite number of L-C combinations. A transmission line, in principle, would consist of an infinite number of these combinations with infinitesimal small inductances and capacitances.

Appendix A gives in a brief introduction to the differential equation system used to calculate the electrical behavior of pulse forming networks, without the restriction of identical inductances and capacitances. Furthermore conductivities parallel

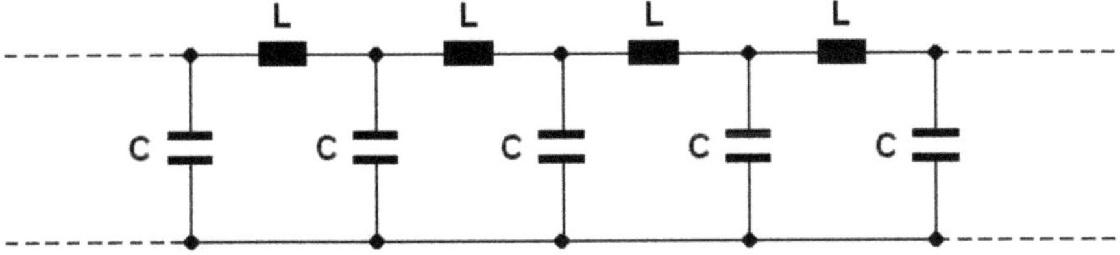

Fig. 24 Pulse forming network

to the capacitances, resistances in series with the inducivities, and mutual inductances between the inductances, as well as resistances in the capacitor lines, are all considered.

Appendix B gives a simplified version of the differential equation system, if, as shown in Fig. 24, only identical inductances and capacitances without any ohmic losses are considered. The capacitances are charged to the same initial voltage u_0. Further considerations will be restricted to this type of a pulse forming network.

To adapt a pulse forming network to a given electric load *R*, the formulas defining the behavior of a pulse forming line can be overtaken, due to the similarity of both types of pulse forming devices. Assuming that there are *n* capacitors and inductors that $C_{PFN} = n \cdot C$ is the overall capacitance, and $L_{PFN} = n \cdot L$ is the overall inductance of the network, the working equations are:

$$R = \sqrt{L_{PFN} / C_{PFN}} = \sqrt{L / C}.$$

$$\tau = 2 \cdot \sqrt{L_{PFN} \cdot C_{PFN}} = 2 \cdot n \cdot \sqrt{L \cdot C}.$$

$$W_C = C_{PFN} \cdot u_0^2 / 2.$$

$C = 2 \dfrac{W_C}{n \cdot u_0^2}$. From these equations, *L*, *C*, and τ can be calculated as

$$L = 2 \cdot R^2 \cdot \frac{W_C}{n \cdot u_0^2}.$$

$$\tau = 4 \cdot \frac{R \cdot W_C}{u_0^2}.$$

The pulse duration τ is fixed by the values of *R*, u_0, and W_C.

Figure 25 shows some basic pulse forming networks, together with their voltage waveforms for different numbers, *n*, of network elements. The networks differ slightly with respect to the switches and to the transition from the network to the load. The networks in sub-figures (a) and (b) simulate the behavior of the pulse forming line shown in Fig. 23, top. Networks in sub-figures (c) and (d) simulate the pulse forming line shown in Fig. 22, center. And the network in sub-figure (e) relates to the pulse forming line in Fig. 23, bottom. The calculation of the load voltage was performed by varying the number network elements between 1 and 10 (a–d), respectively, between 2 and 14 (e).

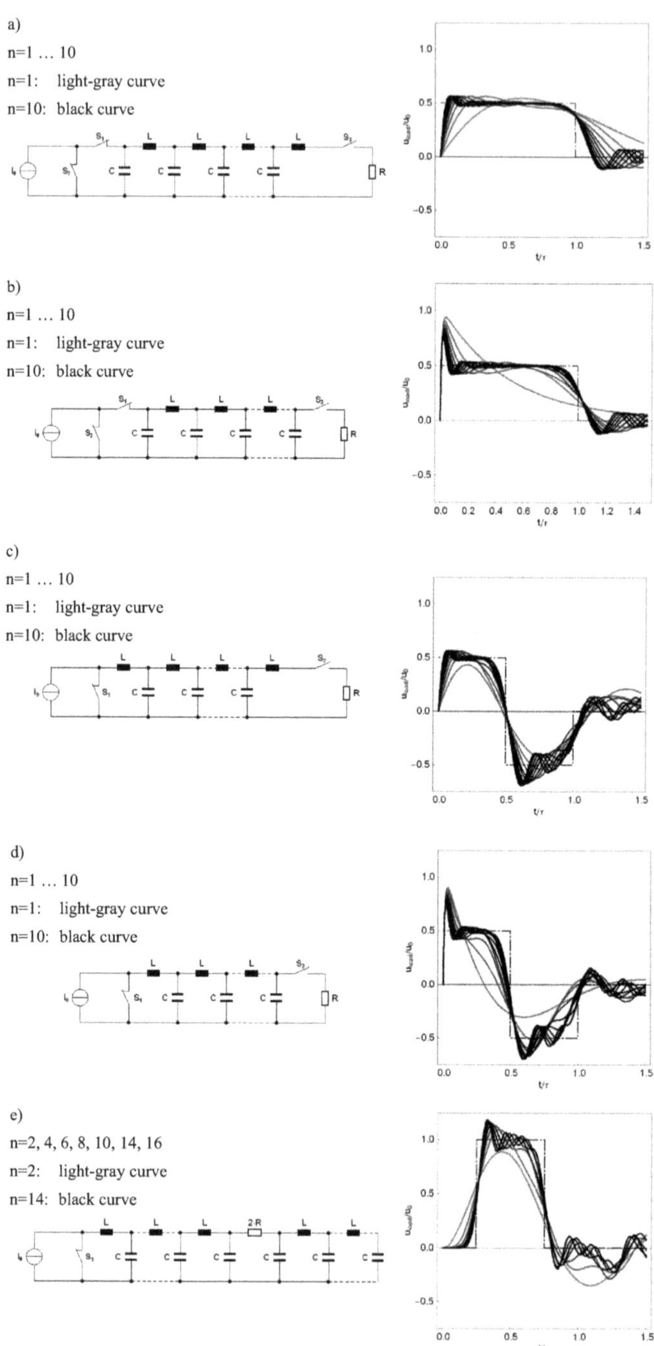

Fig. 25 Basic pulse forming networks and related load voltage waveforms

The load voltage waveforms shown in Fig. 25 are normalized with the initial capacitance voltage u_0. The time is normalized with the time constant τ. The dashed lines represent the voltage behavior of an ideal pulse forming line, that is, the behavior of a pulse forming network (PFN) with $n \rightarrow \infty$. The voltages in sub-figure (a) are calculated by using the formulas given in Appendix B, whereas the voltages in sub-figures (b)– (d), with to some minor modifications, are calculated by using the formulas given in Appendix A. In all cases, the load resistance was chosen to be $R = \sqrt{L/C}$.

In detail the figures show:

Figure 25 a):

Initially, switch S_2 is opened and the current source i_0 charges the n capacitors C via the inductors L. After charging is completed, switch S_1 is opened and switch S_2 is closed. Closing switch S_3 initiates discharge via the load R.

For $n = 1$, the network represents a well-damped ringing circuit with the respective load voltage (light-gray curve). Increasing n leads more and more to an approximation of the voltage waveform of a pulse forming line (dashed curve; compare to Fig. 23, top). The voltage keeps at about 50% of the initial voltage of the capacitors.

Figure 25 b):

The PFN differs from the PFN in Fig. 25 a) by the lack of an inductance between the last capacitor and the load. Charging and initialization of the discharge is the same as with the previous network.

For $n = 1$, the network represents a capacitive circuit with a respective load voltage (light-gray curve). The fact that the voltage does not yield the capacitance's initial voltage is due to a small inductance between the capacitance and the load, which is not shown in the diagram, but is included the calculation. Increasing n leads more and more to an approximation of the voltage waveform of a pulse forming line (dashed curve). Beside the unavoidable voltage peak at the beginning, the voltage remains at about 50% of the initial voltage of the capacitors nearly throughout the entire pulse duration.

Figure 25 c):

Initially, S_1 is open and the current source i_0 charges the n capacitors C via the inductors L. After charging is completed, switches S_1 and S_2 are closed simultaneously to initiate discharge via the load R.

For $n = 1$, the load voltage is ringing (light-gray curve). Increasing n leads more and more to an approximation of the bipolar rectangular voltage waveform of a pulse forming line short-circuited at one end (dashed curve; compare to Fig. 23, center). The voltage amplitude keeps at about 50% of the initial voltage of the capacitances. The ringing period is τ.

Figure 25 d):

The PFN differs from the PFN in Fig. 25 c) due to the lack of an inductance between the last capacitance and the load. Charging and initialization of the discharge is the same as that with the previous network.

For $n = 1$, the load voltage rings (light-gray curve). Increasing n leads more and more to an approximation of the bipolar rectangular voltage waveform of a pulse forming line short-circuited at one end (dashed curve). The voltage amplitude remains at about 50% of the initial voltage of the capacitances, despite some initial discharge. The ringing period is τ.

Figure 25 e):

This PFN differs from the previous PFNs in that the resistance is doubled $2{\cdot}R$ in the center of the PFN. Charging and initialization of the discharge is the same as with the previous networks.

For $n = 2$, the load voltage is rings slightly (light-gray curve). Increasing n leads more and more to an approximation of the monopolar rectangular voltage waveform of a split pulse forming line short-circuited at one end (dashed curve; compare to Fig. 23, bottom). The voltage amplitude remains at about 100% of the initial voltage of the capacitances. The ringing period is $\tau/2$. For increasing n, the start time of the load voltage is approximately $\tau/4$.

To increase the voltage steepness at the beginning and at the end, smaller capacitances are used at the ends of the PFNs.

Networks with Pulse Forming Switches

Another possible method for generating more or less rectangular voltage waveforms is to use capacitive or inductive circuits.

In capacitive circuits (Fig. 10), the capacitance has to be chosen in such a way that the time constant of the circuit is essentially larger than the desired pulse duration. After initializing discharge of the capacitor by closing switch S_3, this switch can, in principle, be reopened again at a specified time to interrupt the discharge of the capacitors. After this interrupt, the capacitor can be recharged to its initial voltage level and the process repeated.

As an advanced example, Fig. 26 shows the diagram of a capacitive double-circuit capable of generating the bipolar rectangular voltage waveforms as shown in Fig. 7 (Gaudreau et al. 2005). The load is connected to two independent capacitive circuits. In the upper half of the circuit, a capacitor $C_{storage}$ is connected via a solid stage switch, a serial resistance R_{series}, and a serial inductance L_{series} to the load, represented by its resistance and its capacitance. After charging $C_{storage}$ with a solid state power supply (~constant current source), the switch is activated and the capacitor discharges via R_{series}, L_{series}, and the load. After opening the switch at the specified time, the current from the capacitor is interrupted. Until this time, the load voltage is kept nearly constant. To release the switch, during opening, from dumping the energy stored in L_{series}, the diode in parallel to the inductance overtakes the inductor's current. To generate a bipolar voltage waveform, the lower half of the network can be activated in the same manner after switching off the upper half.**Fig. 26** Caption

Also, inductive circuits (Fig. 12) can allow one generate rectangular voltage waveforms. The inductance has to be chosen so that the time constant of the circuit is essentially larger than the desired pulse duration. After initializing the discharge of the inductance by opening switch S_3 (see Fig. 12), the switch can be re-closed, again at a specified time, interrupting the discharge of the inductor. After interruption, the inductance can be recharged to its initial current level and the process can be repeated.

Concluding Remarks

PEF treatment is possible with capacitive and inductive circuits, as well as with pulse forming networks. Voltage multiplying networks like MARX generators or circuits with pulse transformers have to be used if no switches are available to provide the required voltage amplitudes. The exponentially damped voltage characteristic of capacitive or inductive circuits is not changed by these measures in principle. The energy efficiency is slightly lower. Ringing circuits have no advantage if the highest possible voltages are required.

Rectangular voltage waveforms can be generated by either applying pulse forming networks or capacitive or inductive circuits with "oversized" time constants and with switches that are able to interrupt the current across the load at a specified time.

In all networks or circuits, the overall power demand of the switches remains independent of the number and the arrangement of the switches as well as the type of circuit. The same is true for the energy content of the capacitors or inductors.

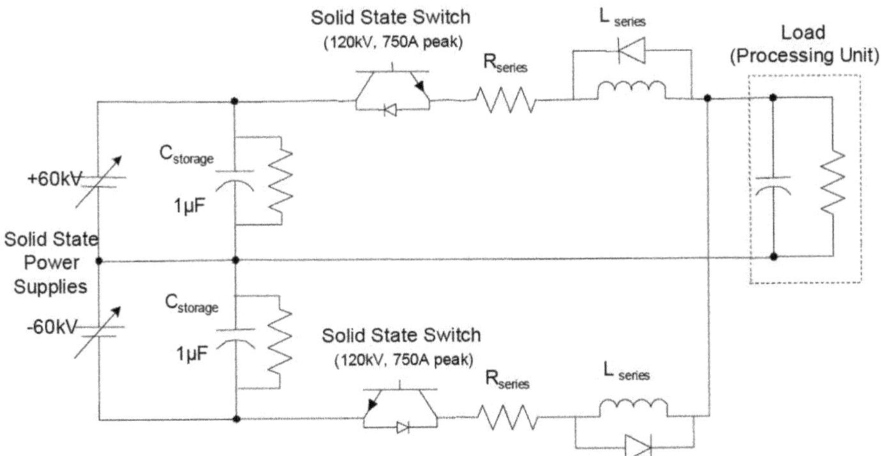

Fig. 26 Caption

Components of High-Power Sources

The main components of high-power sources are storage capacitors and on- and off-switches. Due to their relatively high ohmic power consumption, inductors in comparison to capacitors play a minor role.

Depending on the application, different capacitor types with different specific prices are on the market. One of the main requirements of capacitors is a long lifetime at the specified voltage and energy content. Based on a common design, the construction of different types of capacitors depends on the discharge current.

Many variants of fast switches are available: spark gaps, electron tubes (thyratrons), and semiconducting switches (diodes, thyristors). Depending on the technical specifications, these switches have specific merits and limitations. Mechanical switches are too slow for high repetition rate PEF generation.

High-Power Capacitors

In principle high-power capacitors consist of a couple of thin sandwich-like metallic-dielectric-metallic-dielectric strips of ~100 μm thickness. The dielectrics are made, for example, from Kraft paper, polypropylene, mixtures of Kraft paper with polypropylene, or PVDF. The electrodes are made from aluminum, either as a foil or sprayed onto the dielectric. Each sandwich strip is rolled up. The resulting windings are flattened. Each flat winding represents an elementary capacitor. They can be switched in series or in parallel to construct the final capacitor. Further information about the construction of capacitors is given in MacDougall (1996).

Together with some basic data about commercially available capacitors, Fig. 27 lists some different types of capacitors of different sizes in metallic or in plastic casings.

Figure 28, left, top, gives an overview of the range of voltage and capacitance of capacitors provided by a large capacitor supplier (Ennis et al. 2003). At voltages of about 1 kV, the highest capacitances are nearly 100 mF, whereas at voltages in the

- capacitance: up to ~0.1 F
- voltage: up to ~MV
- energy: <2.5 kJ/l
- lifetime: up to ~10^{14} c/d cycles
- specific cost: >0.05 US-$/J

Fig. 27 Capacitors
Left: different types of capacitors (source: Maxwell Laboratories Inc.)
Right: general data (taken from Fig. 26–2)

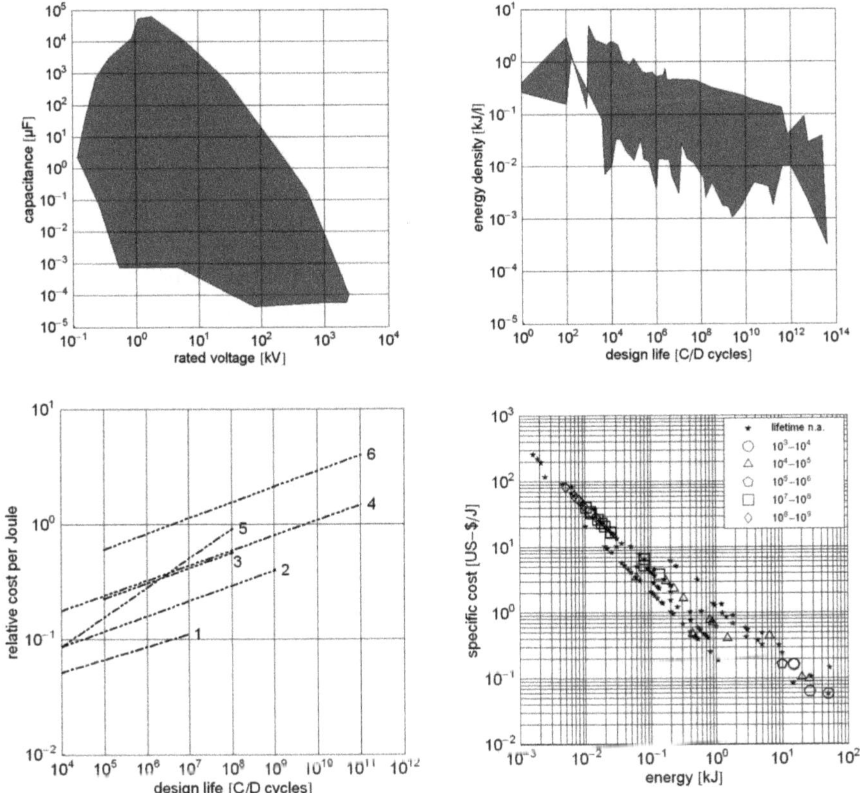

Fig. 28 Typical data of capacitors
Left, top: voltage and capacitance (MacDougall 1996)
Right, top: range of energy density and rated voltage (MacDougall 1996)
Left, bottom: relative cost per Joule versus life of different high energy capacitor technologies
Right, bottom: specific cost versus energy (Maxwell Laboratories Inc./AMS Electronic GmbH, 1997)

MV range, the capacitances have values in the 100-pF range. The figure shows that, in general, the realized capacitance of single capacitors has smaller values at higher voltages.

Figure 28, right, top, gives an overview of the range of energy densities and design life of capacitors (Ennis et al. 2003). For a design life of about 1000 charge/discharge cycles, energy densities up to about 2 kJ/l were achieved. To achieve very high lifetimes, that is, on the order of 10^{13} charge/discharge cycles, the energy density has to be reduced to about 0.1 kJ/l. Increasing the lifetime of capacitors is possible by decreasing the energy stored per unit volume. The lifetime of capacitors also depends on other factors like voltage, dielectric stress, voltage reversal, stressed area of the dielectric, and ringing frequency (Smith et al. 2004): "A traditional power scaling guideline has been used by capacitor manufacturers to predict actual component lifetimes based on known sensitivities to these important parameters. One of its most general forms is given by

$$L = L_0 \cdot \left(\frac{V}{V_0}\right)^{-4} \cdot \left(\frac{E[V]}{E_0}\right)^{-3.5} \cdot \left(\frac{Q}{Q_0}\right)^{-1.6} \cdot \left(\frac{A}{A_0}\right)^{-1/\beta} \cdot \left(\frac{f}{f_0}\right)^{-0.5} .$$

In this expression, L represents the expected shot life after scaling; V is the capacitor operating terminal voltage; E is the dielectric electric field stress; A is the stressed dielectric area of the operating unit; β is the Weibull slope or shape factor; f is the ringing frequency of the operating unit; $Q = (\pi/2)/\ln[1/R]$ is the operating circuit quality factor; and R is the actual operating reversal (%). The symbols L_0, V_0, E_0, A_0, f_0, R_0, and Q_0 are the baseline rated design parameters." The exponents in the scaling law vary depending on the materials used. Increasing/decreasing each of these parameters compared to the baseline-rated parameters decreases/increases the expected lifetime of capacitors remarkably.

Figure 28 287, left, bottom, shows the relation between the relative cost per Joule and the design life. The curves relate to different types of materials and technologies:

1. Self-healing metallized electrodes, Kraft paper with film
2. Self-healing metallized electrodes, metallized film dielectrics, high resistance electrode
3. Combined metallized electrodes with foil electrodes
4. Self-healing metallized electrodes, metallized film dielectrics, segmented electrode
5. Extended foil electrodes/tabbed foil; "single shot," up to 100 kV, up to 1 MA, more than 20% voltage reversal; paper dielectric/mixed dielectric/all-film

6. Combined metallized electrodes with foil electrodes.

"Hybrid electrode," types 3 and 6, are especially useful in long-life industrial applications, such as water and food sterilization. In the case of type 5, a cost increase by a factor of ~6 can be expected if the design life increases by a factor of ~10^6.

Figure 28, right, bottom, shows the specific cost of capacitors relative to the stored energy for different capacitors in general and for different lifetimes (price base: year 1997). As a general rule, the specific cost decreases with increasing energy. At energies of ~100 kJ/capacitor, the decrease in the cost reaches its minimum at ~0.05 US-$/J.

Switches

The following specific high-power switches will now be taken into consideration below:

1. Trigatrons (3-electrode spark gaps)
2. Ignitrons
3. Thyratrons and pseudospark switches
4. Thyristors and IGCTs

In the following, some typical electrical values of the respective switches will be given.

Figure 30, left, top, shows the typical operating voltage and peak current ranges and, on the right, top, peak powers and energy transfer rates that can be handled by trigatrons in single-shot operation. Trigatrons can work at voltages up to 500 kV, at peak currents up to 3 MA, at peak powers up to~300 GVA, and at energy transfer rates up to ~2 MJ/discharge. Here, as in the following sections, the peak power is defined to be the product of the hold-off voltage and the peak current.

The peak currents decrease to 10–30% as the repetition rate increases (up to 100 Hz). Nonetheless, these values are impressive and imply that trigatrons can be used for nearly any on-switch tasks in pulsed power technology. The drawback is the relatively short trigatron lifetime of ~10^6 discharge cycles at currents of 10 kA or more (Fig. 30, left, bottom). At repetition rates of ~10 Hz, the lifetime is on the order of only ~30 h, thus hindering the industrial applications for trigatrons. The prices of trigatrons are shown in Fig. 30, right, bottom. In addition, the price of trigger generators has to be taken into account with prices of about 3–4 thousand US-$ per single switch (source: Richardson Electronics Inc., 2005).

Trigatrons

Trigatrons or spark gaps are the work horse in laboratories. In principle, they consist of two electrodes with an additional trigger electrode between the main electrodes or implemented into one of the electrodes. The gap between the electrodes is filled with air, synthetic air, sulfur hexafluoride, SF_6-Argon, or oxygen-argon at different pressures ranging down to vacuum pressures. Due to their simple construction, trigatrons are cheap if no demands are imposed on their high lifetimes.

Figure 29 shows a selection of trigatrons of different size for different purposes together with some fundamental data.

Figure 30, left, top, shows the typical operating voltage and peak current ranges and, on the right, top, peak powers and energy transfer rates that can be handled by trigatrons in single-shot operation. Trigatrons can work at voltages up to 500 kV, at peak currents up to 3 MA, at peak powers up to~300 GVA, and at energy transfer rates up to ~2 MJ/discharge. Here, as in the following sections, the peak power is defined to be the product of the hold-off voltage and the peak current. The peak currents decrease to 10–30% as the repetition rate increases (up to 100 Hz). Nonetheless, these values are impressive and imply that trigatrons can be used for nearly any on-switch tasks in pulsed power technology. The drawback is the relatively short trigatron lifetime of ~10^6 discharge cycles at currents of 10 kA or more (Fig. 30, left, bottom). At repetition rates of ~10 Hz, the lifetime is on the order of only ~30 h,

- peak current: up to ~MA
- voltage: up to ~500 kV
- energy: <3 MJ/pulse
- peak power: <0.4 TVA
- lifetime: up to $\sim 10^9$ discharges
- specific cost: >400 US-$/GVA

Fig. 29 Trigatrons
Left: different types of trigatrons (source: Maxwell Laboratories **Inc.)**
Right: general data (taken from Fig. 30)

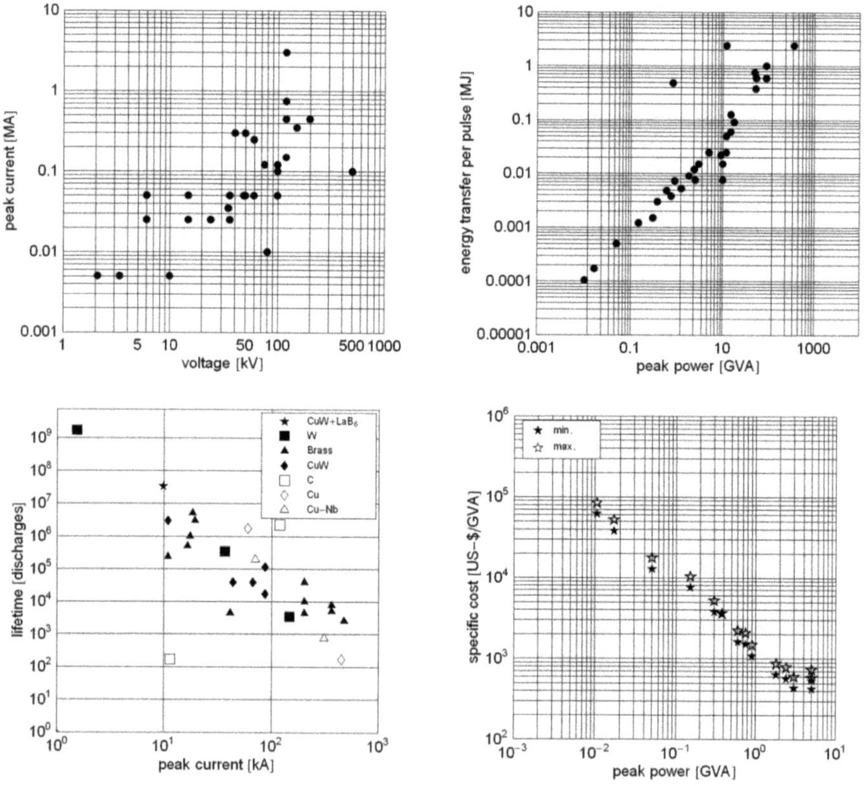

Fig. 30 Typical data of trigatrons
Left, top: range of trigatron voltage and peak currentRight, top: range of trigatron energy transfer and peak power
Left, bottom: peak current and lifetime expectancies (from (Donaldson 1990), modified)
Right, bottom: peak power and specific cost (source: Richardson Electrics Ltd., 2005)

thus hindering the industrial applications for trigatrons. The prices of trigatrons are shown in Fig. 30, right, bottom. In addition, the price of trigger generators has to be taken into account with prices of about 3–4 thousand US-$ per single switch (source: Richardson Electronics Inc., 2005).

Ignitrons

Ignitrons consist of an anode (usually graphite) and a mercury pool cathode. A semiconducting trigger electrode which dips into the mercury pool cathode and initiates the discharge. An electron emitting source is formed at the point where the trigger contacts the pool. This initiates an arcing between the anode and the cathode. As long as there is a significant current flow, the ignitron will remain conducting. Continuously operating ignitrons must be cooled to a well-defined temperatures. Ignitrons must be placed in a vertical position so that the mercury pool is in the right position in the tube.

Figure 31 shows an ignitron with a surrounding cooling coil together with some basic data.

Figure 32, left, top, shows typical trigger voltage and peak current data for a ignitron. Ignitrons can withstand voltages up to 50 kV and peak currents up to 700 kA. They can switch peak powers up to ~10 GVA (Fig. 32, right, top). The average power ignitrons as shown in Fig. 32, left, bottom, can operate at powers greater than the peak power. The average power is on the order of 0.01, that is, 0.1% of the peak power. The specific power cost of ignitrons decreases with increasing switching power. They vary between 400 and 2000 US-$/GVA peak power.

Although, in general, ignitrons don't cause trouble, they have an increasing negative image due to the mercury inside the ignitrons. For the most applications, they can be replaced by other switch devices, so their use should be avoided when possible.

Thyratrons and Pseudospark Switches

Thyratrons are switches that are filled with hydrogen, deuterium, mercury vapor, xenon, or neon at typical pressures of 1.5 to 3 kPa. They consist of a heated cathode, a cold anode, and one or more gridlike gate electrodes between the cathode and

- peak current: ≤700 kA
- voltage: ≤50 kV
- average power: <1 MVA
- peak power: <10 GVA
- specific cost: >400 US-$/GVA

Fig. 31 Ignitrons
Left: ignitron
Right: general data (taken from Fig. 32)

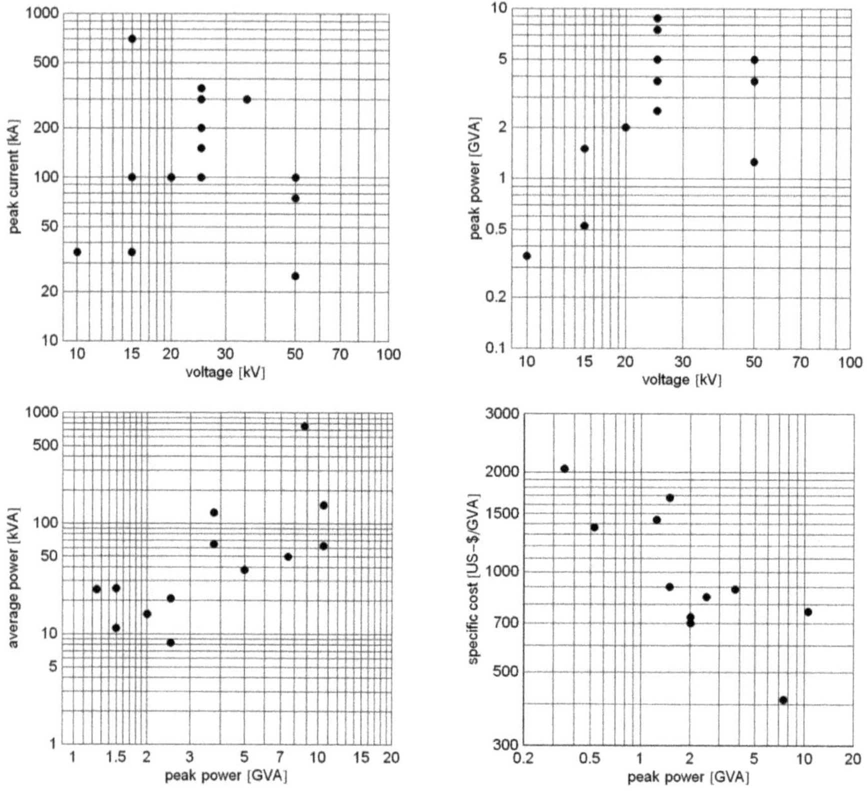

Fig. 32 Typical data of ignitrons
Left, top: voltage and peak current. Right, top: voltage and peak power
Left, bottom: peak power and average power
Right, bottom: peak power and specific cost (source: Richardson Electronis Ltd., 2005)

anode. In many designs (hydrogen thyratrons are a common exception), the gate electrode must be biased highly negative in the off-state and then biased positive to achieve switching. Once turned on, the thyratron will remain conducting as long as there is a significant current flowing through it. When the anode voltage or current falls to zero, the device switches off. Triode, tetrode, and pentode variations of the thyratron have been manufactured in the past, though most are of the triode design.

Pseudospark switches are gas-discharge devices with a hollow cathode and, in some types, also a hollow anode. In the cathode cavity, a trigger unit is used to initiate breakdown in the main gap. The characteristics of the pseudospark switches are similar to those of classical pulse hydrogen thyratrons.

Figure 33 shows a view of a thyratron together with its cross-sectional view and some basic data.

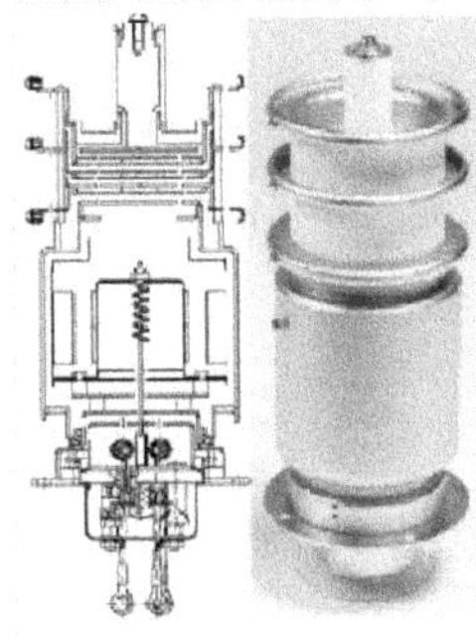

- peak current: ≤200 kA
- voltage: ≤250 kV
- average power: <2 MVA
- peak power: <3 GVA

Fig. 33 Thyratrons
Left: thyratron
Right: general data (taken from Fig. 34)

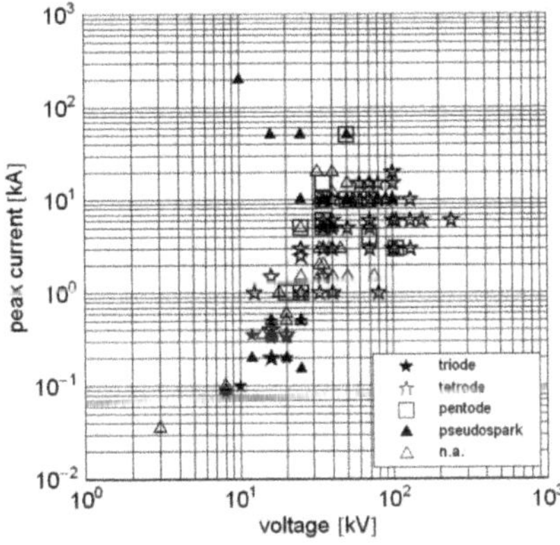

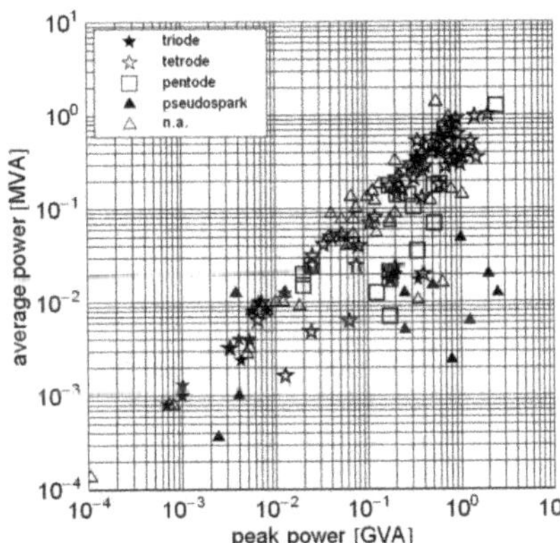

Fig. 34 Typical data of thyratrons
Left: range of thyratron voltage and peak currentRight: range of thyratron peak power and average power

Figure 34, left, shows typical ranges of thyratron voltage and peak current data. Thyratrons and pseudospark switches can handle voltages up to ~200 kV and peak currents up to ~200 kA. These switches allow energy transfers with peak powers up to ~2 GVA and average powers up to ~1 MVA (Fig. 34, right).

The current increase rate d*i*/d*t* is on the order of ~100 kA/μs (Richardson et al. 2004).

In comparison to the mounting position of other switches like trigatrons, ignitrons, or thyristors, thyratrons usually have to be mounted as shown in Fig. 35. Without an additional switch, the current used to charge a capacitor must also flow through the load. Pseudospark switches can be mounted like the other types of switches.

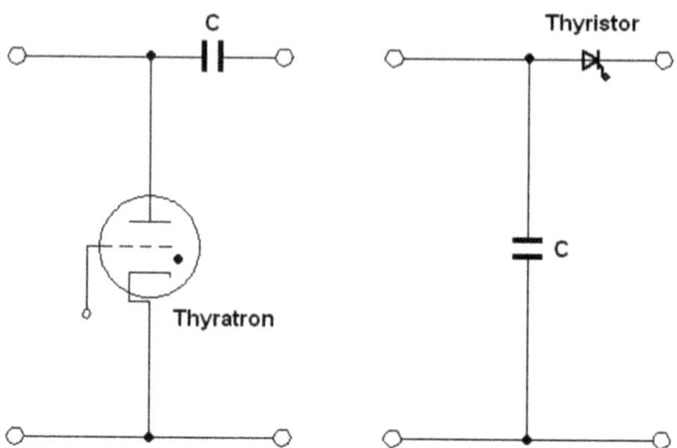

Fig. 35 Mounting position of thyratrons compared to thyristors or other switches

The average lifetime of thyratrons is on the order of 1½ for 3 years of continuous operation (Richardson et al. 2004; Welleman et al. 2004). A thyratron system that operates at a voltage of 9 kV, a pulse current of 2.5 kA, a pulse length of 10 μs, a d*i*/d*t* of 5 kA/μs, and a pulse repetition rate of 400 Hz usually costs 4000 US-$, whereas the tube itself costs 2300 US-$ (Welleman et al. 2004).

Thyristors, Diodes, and IGCT Switches

In principle, thyristors (thyratron + transistor) are four-layer semiconducting devices, with each layer consisting of an alternately N- or P-endowed material, for example, N-P-N-P. The outer layers represent the cathode (N) and the anode (P). The control terminal, or "gate," is attached to one of the middle layers. After ignition, thyristors remain conducting as long as the current does not reverse. Some thyristors are optically triggered.

Figure 36 shows some types of thyristors with either a single semiconducting wafer in a ceramic housing, multiple wafers in a ceramic housing, or combined in a stack. Their key parameters are also given.

Figure 37 shows some typical characteristics of single thyristor disks, multilayer thyristors, and stacked thyristors. Single-thyristor disks can be stressed by voltages of 4.5 kV at peak currents up to more than 100 kA (Fig. 37, left, top). Multilayer thyristors can be stressed by voltages of ~15 kV at peak currents up to 50 kA, whereas thyristor stacks are available for voltages up to 70 kV and for peak currents up to more than 100 kA. Note that for continuous operation, the currents and the power are a factor of 10 to 20 less than the peak currents and the peak powers, respectively.

The action integral

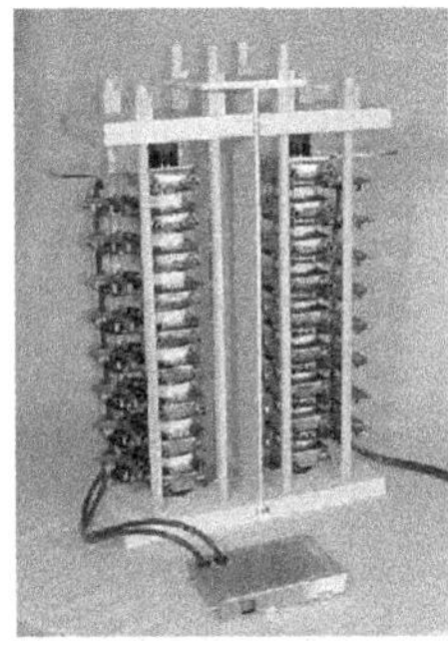

- peak current: ~100 kA
- voltage:
  - single: 4.5 kV
  - multi: ≤13.5 kV
  - stack: ≤70 kV
- rated power: <2 MVA (single)
- peak power: <5 GVA (stack)
- specific cost: ≥0.7 €/kVA

Fig. 36 Thyristors
Left: thyristor (single, multi, stack; source: ABB Semiconductors)
Right: general data (taken from Fig. 37)

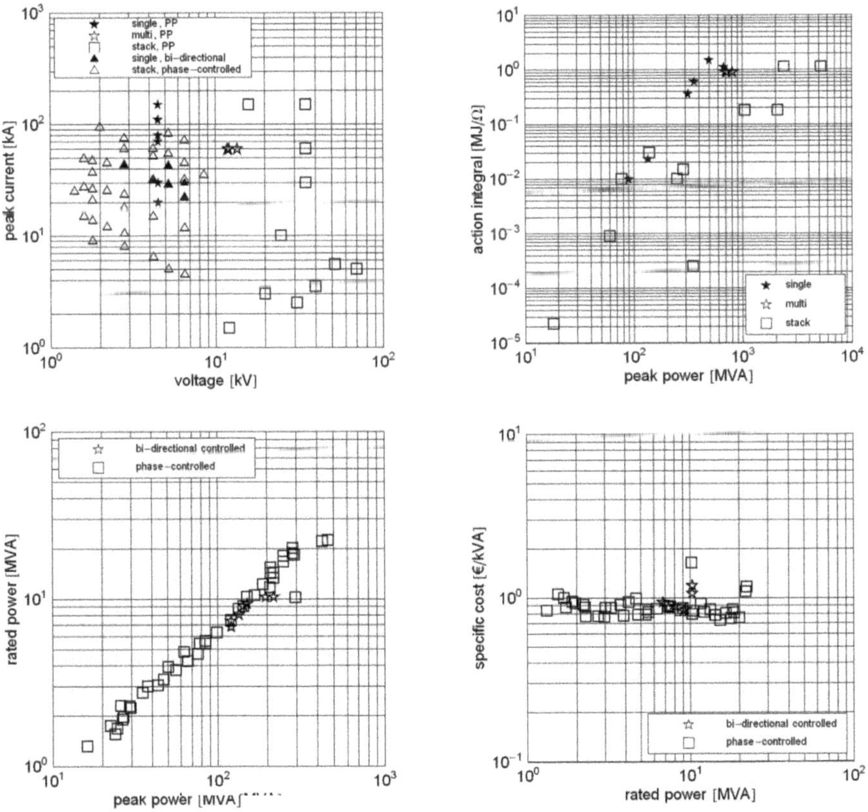

Fig. 37 Typical thyristor data (source: ABB Semiconductors)
Left, top: voltage and peak currentRight, top: power and action integral
Left, bottom: peak power and rated power
Right, bottom: specific cost and rated power (rough estimate)

$$A = \int_0^\infty i^2 dt,$$

which is proportional to the energy deposit in the thyristor and the load during one discharge at peak current, can reach values up to ~1 MJ/Ω at power levels up to 5 GVA (Fig. 37, right, top) for a single shot. Current increase rates d*i*/d*t* are limited to ~20 kA/μs. Repetition rates can be up to 100 Hz at rated power. Typical values of the rated power (or average power) normalized by the peak power are shown in Fig. 37, left, bottom. As shown in Fig. 37, right, bottom, the specific cost of thyristors (single) are in the range of 0.7–1.5 €/kVA rated power. For continuous operation, the lifetime of thyristors is on the order of 10 years and more (Welleman et al. 2004).

At a voltage of 9 kV, a pulse current of 2.5 kA, a pulse length of 10 μs, a di/dt of 5 kA/μs, and a pulse repetition rate of 400 Hz, the typical cost of a thyristor stack is 8700 US-$ (Welleman et al. 2004).

Thyristors switch off if the current reverses, but are not able to switch off the current before zeroing. Integrated gate-controlled thyristors (IGCT) have the capability to switch off currents actively. As the IGCT is the improved version of GTOs (gate turn-off thyristor), the GTO technology is not used anymore for pulsed applications (Welleman et al. 2004). Although their turn-on behavior and d*i*/d*t* is reduced compared to the discharge switches mentioned earlier, they offer the possibility to drive circuits with pulse forming switches. Single GTO switches are designed to switch on and to switch off currents up to 4 kA. Blocking voltage per device is 4.5 kV or 6 kV. They are usually used in medium- and high-power drivers and frequency converters. The switches can be stacked so that, with a reverse blocking IGCT device consisting of seven 4.5-kV IGCTs, forward/reverse blocking voltages of 31 kV at nominal 1600-μs pulse currents of 2.5 kA can be switched at repetition rates of 15 Hz (Welleman et al. 2004). Frequencies up to 400 Hz are possible.

As a perspective optically triggered emitter turn-off thyristors (ETO) can operate at currents up to 4 kA and at voltages of 6 kV at frequencies of more than 1 kHz with less power consumption. The maximum current increase rate d*i*/d*t* is 1 kA/μs.

Diodes are passive switches, which switch on automatically at negative voltages. Figure 38, left, shows the peak current and voltage of several types of diodes with a single wafer. Peak currents up to 90 kA can be switched for blocking voltages up to 6 kV. The rated power over the peak power is shown in Fig. 38, right. The rated power (average power) has values up to 25 MVA at peak power levels up to 400 MVA. As a rule of thumb, the specific power cost of diodes relative to their average power can be estimated as ¾ of the cost of comparable thyristors.

Concluding Remarks

For PEF applications, thyratrons and semiconducting switches offer the most promising possibilities. Trigatrons suffer from relatively short lifetimes, whereas ignitrons are more and more out of favor to the mercury inside the tube. The choice between

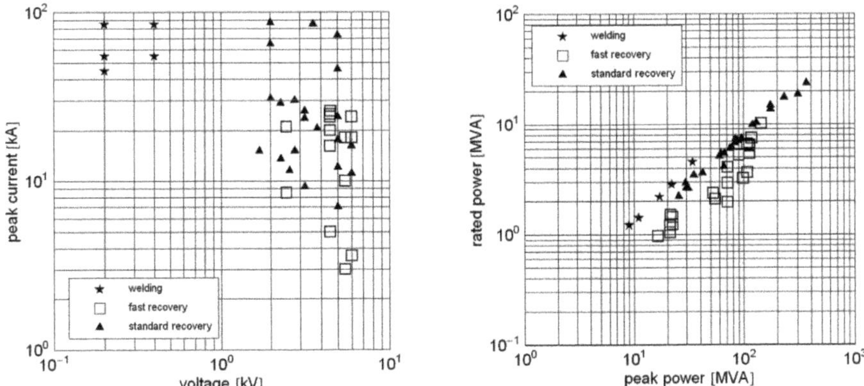

Fig. 38 Typical data of diodes (source: ABB Semiconductors)
Left: voltage and peak current. Right: peak power and operational power

thyristors and thyratrons depends on the specific requirements of the load. At voltages of ~100 kV, there are clear advantages for using thyratrons, especially with respect to the investment cost. At voltages lower than 30 kV, thyristor switches may yield lower operation cost than thyristors due to their comparably higher lifetime.

Low-Power Source

Basic Considerations

The low-power source (AC/DC-converters) converts an alternating voltage into a constant current, constant voltage, or constant power to charge the capacitors or inductors of the high-power source. The term "low power" means low power compared to the power of the high-power sources. Several types of such converters exist. Their technical details are discussed, for example, in [48]. Figure 39 shows typical values of the power and output frequencies of commercially available converters (Brosch et al. 2000). Power levels up to ~100 MVA have been realized in large facilities.

In high repetition rate circuits, like PEF circuits, the converter is the most expansive part. The cost depends on the maximum current and the maximum voltage, which defines the rated power of the device that the converter can handle.. For cost effectiveness, the power utilization factor should be as high as possible to achieve high energetic efficiency.

It is well known that the charging of capacitors with constant voltage via a resistor yields an energetic efficiency of less than 50% and a power utilization less than 25%. On the other hand, charging with constant currents yields energetic efficiencies of more than 90% (theoretically: 100%) at comparably high-power utilization

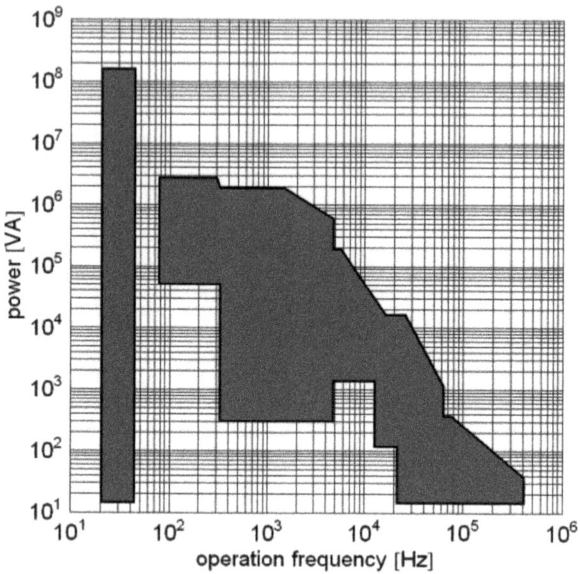

Fig. 39 Converters
Range of frequency and power (from (Brosch et al. 2000, Mitsubishi), modified in the low-frequency region with values from ABB Semiconductors)

factors (theoretically: 100%). Charging with constant power is in between these two types. From this, it clearly turns out that capacitors should be charged by a constant current source.

For constant charging current i_0, the power in the capacitor is

$$P_C = u_C \cdot i_0.$$

The capacitor voltage increases according to the following formula

$$u_C = \frac{i_0}{C} \cdot t$$

so that the power can be calculated with

$$P_C = \frac{i_0^2}{C} \cdot t$$

The power increases linearly with time. If the charging voltage u_0 is attained, the peak power at the capacitor is

$$\hat{P}_C = u_0 \cdot i_0.$$

This power, plus some additional power to account for ohmic losses, has to be installed in the constant-current power supply. The minimum cost of this device can be calculated by applying the rule of thumb that define the specific cost to be 0.5 €/W.

The average power is half the peak power. So, as an example, a device with an average power of 200 kW (compare Fig. 6, right, bottom) would need an installed power of 400 kW. The cost would be in minimum 200 k€.

Typical Devices

To close this chapter, some typical systems that have already been constructed and that are in use will be briefly described.

Figure 40 shows a power supply for PEF applications that allows one to charge capacitors up to 18 kV at an average power of 35 kJ/s and at repetition rates up to 100 Hz. On the left side of the figure, the power supply is shown as a customer usually sees it. The cost of a device is about 30 k€ according to the specific cost per unit power estimate given above. Doubling the power would increase the cost by ~50% (Neuman, personnal communication).

Figure 41 shows a solid-state pulser delivering monopolar exponentially damped pulse voltages of 20 kV. The 50 Hz pulser has an average power demand of 20 kW. Its purpose is the treatment of potatoes with PEF for the starch industry (see Fig. 41, top). The potatoes are positioned by a steady flow of water between two electrodes. After the PEF treatment, a conveyor screw transports the potatoes to the next stage, where the softened potatoes are treated mechanically. A low-voltage power and the electronic controls are located on the left side of the pulser (see Fig. 41 right, bottom). A 20 kJ/s average power supply is positioned below a capacitor and a solid state switch part on the right side. The cost of the complete system was about 150 k€.

Figure 42 shows a solid-state PEF pulser with an average power of 75 kJ/s (Gaudreau et al. 2004, 2005; Jin and Zhang 2002). Its high-power circuit is shown in Fig. 26. The pulser generates bipolar or monopolar square waveform voltages

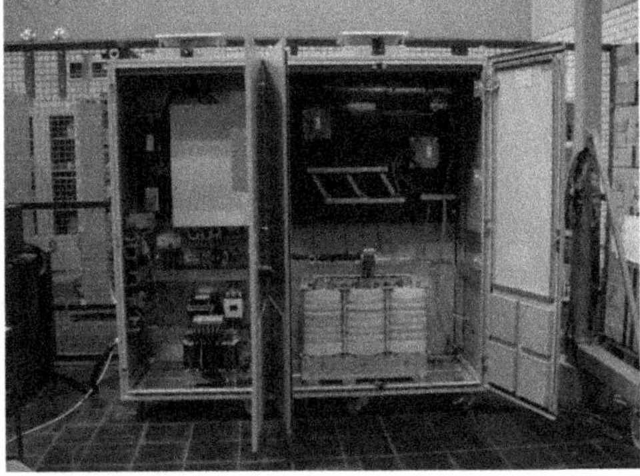

Fig. 40 35 kJ/s/18 kV power supply. (Source: BARMAG, Propuls)

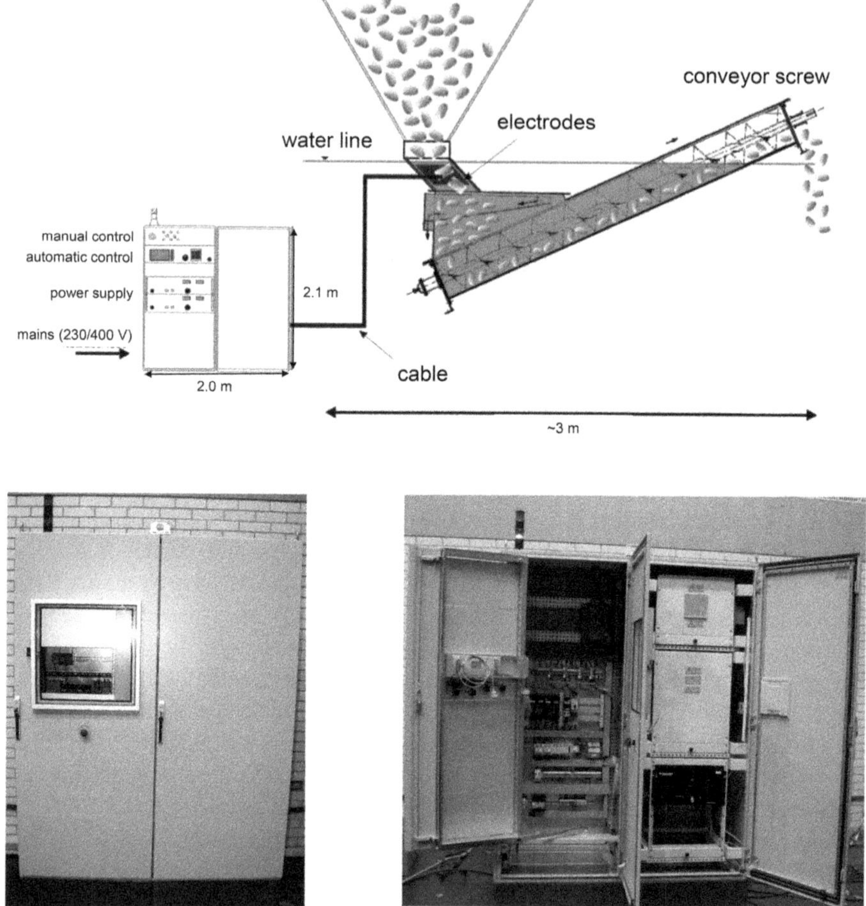

Fig. 41 20 kJ/s/20 kV monopolar solid-state PEF pulser. (Source: AMS Electronics/Propuls)

with 60 kV amplitude at currents up to 750 A. The pulse repetition rate is between 500 and 2000 pps. The pulse width varies between 2 and 8 μs. Each pulse delivers 10 to 100 J of energy. The system is nominally sized to process orange juice at a rate of 2000 l/h. For highly resistive foods, like apple juice, the unit can process up to 5000 l/h. The total cost of the system, including pulse generator, aseptic drink processing unit, PEF treatment chamber unit, and aseptic packaging machine, was about 700 k€.

Acknowledgment The author of this chapter would like to express his thanks to Dr. Larry Altgilbers who significantly improved the readability of the text.

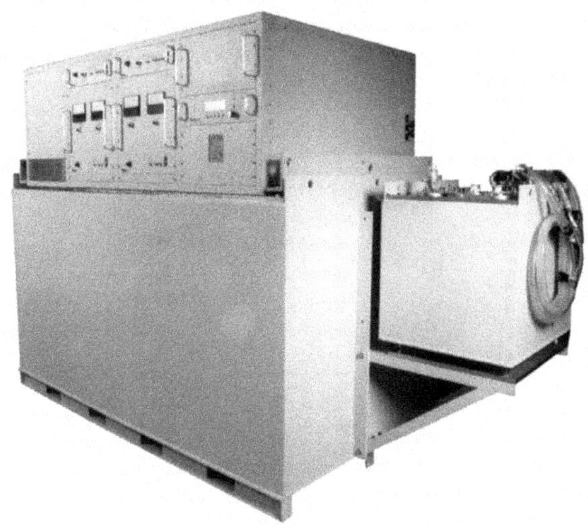

Fig. 42 75 kJ/s/60 kV bipolar solid-state PEF pulser (Gaudreau et al. 2004, 2005). (Source: Diversified Technologies)

Appendices

Appendix A: Differential Equation System of Inhomogeneous Pulse Forming Networks

Figure A-1 shows the most generalized form of a pulse forming network replacing pulse forming lines. Different admittances $C_k||G_k$ connected in parallel are connected with different resistances W_k and with switches S in serial. Different impedances L_k-R_k are connected in series. The capacitances C_k are charged to different charging voltages $u_{k,0}$. Switches S simultaneously switch on and initiate a current via the load R. The n^2-n mutual inductances $M_{k,l}$ between the single inductances are represented by the collective letter M.

To define the linear differential equation system of the circuit, the following matrices and vectors are introduced:

Matrix of capacitances:

$$C = \begin{pmatrix} C_1 & 0 & 0 & \cdots & 0 & 0 \\ 0 & C_2 & 0 & \cdots & 0 & 0 \\ 0 & 0 & C_3 & \cdots & 0 & 0 \\ \vdots & \vdots & \vdots & \ddots & \vdots & \vdots \\ 0 & 0 & 0 & \cdots & C_{n-1} & 0 \\ 0 & 0 & 0 & \cdots & 0 & C_n \end{pmatrix}$$

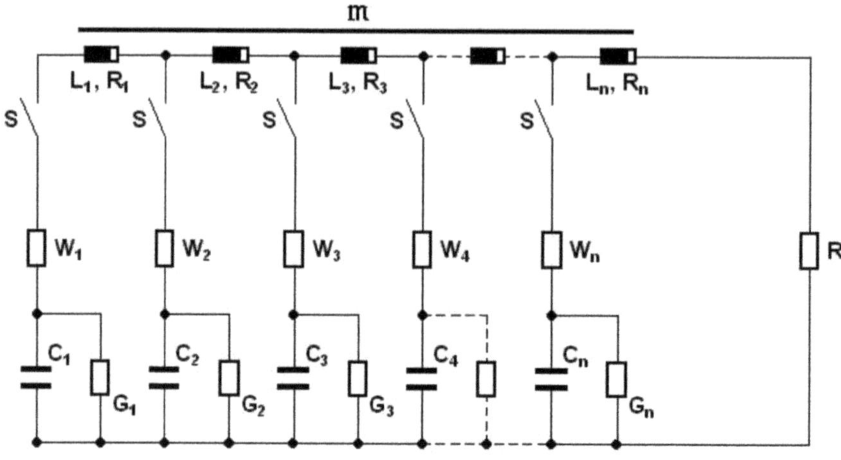

Fig. A-1 Inhomogeneous pulse forming network

1-Matrix:

$$\mathcal{E} = \begin{pmatrix} 1 & 0 & 0 & \cdots & 0 & 0 & 0 \\ -1 & 1 & 0 & \cdots & 0 & 0 & 0 \\ 0 & -1 & 1 & \cdots & 0 & 0 & 0 \\ \vdots & \vdots & \vdots & \ddots & \vdots & \vdots & \vdots \\ 0 & 0 & 0 & \cdots & -1 & 1 & 0 \\ 0 & 0 & 0 & \cdots & 0 & -1 & 1 \end{pmatrix}$$

Matrix of conductivities:

$$\mathcal{G} = \begin{pmatrix} G_1 & 0 & 0 & \cdots & 0 & 0 \\ 0 & G_2 & 0 & \cdots & 0 & 0 \\ 0 & 0 & G_3 & \cdots & 0 & 0 \\ \vdots & \vdots & \vdots & \ddots & \vdots & \vdots \\ 0 & 0 & 0 & \cdots & G_{n-1} & 0 \\ 0 & 0 & 0 & \cdots & 0 & G_n \end{pmatrix}$$

Matrix of inductances and mutual inductances:

$$\mathcal{L} = \begin{pmatrix} L_1 & M_{1,2} & M_{1,3} & \cdots & M_{1,n-1} & M_{1,n} \\ M_{1,2} & L_2 & M_{2,3} & \cdots & M_{2,n-1} & M_{2,n} \\ M_{1,3} & M_{2,3} & L_3 & \cdots & M_{3,n-1} & M_{3,n} \\ \vdots & \vdots & \vdots & \ddots & \vdots & \vdots \\ M_{1,n-1} & M_{2,n-1} & M_{2,n-1} & \cdots & M_{n-1,n-1} & M_{n-1,n} \\ M_{1,n} & M_{2,n} & M_{3,n} & \cdots & M_{n,n-1} & L_n \end{pmatrix}$$

Matrix of zeroes (n × n-matrix):

$$\mathcal{O} = \begin{pmatrix} 0 & 0 & 0 & \cdots & 0 & 0 \\ 0 & 0 & 0 & \cdots & 0 & 0 \\ 0 & 0 & 0 & \cdots & 0 & 0 \\ \vdots & \vdots & \vdots & \ddots & \vdots & \vdots \\ 0 & 0 & 0 & \cdots & 0 & 0 \\ 0 & 0 & 0 & \cdots & 0 & 0 \end{pmatrix}$$

Matrix of resistances:

$$\mathcal{W} = \begin{pmatrix} -(W_1 + R_1 + W_2) & W_2 & 0 & \cdots & 0 & 0 \\ W_2 & -(W_2 + R_2 + W_3) & W_3 & \cdots & 0 & 0 \\ 0 & W_3 & -(W_3 + R_3 + W_4) & \cdots & 0 & 0 \\ \vdots & \vdots & \vdots & \ddots & \vdots & \vdots \\ 0 & 0 & 0 & \cdots & -(W_{n-1} + R_{n-1} + W_n) & W_n \\ 0 & 0 & 0 & \cdots & W_n & -(R + R_n) \end{pmatrix}$$

Vector of capacitance voltages and initial capacitance voltages:

$$u[t] = \begin{pmatrix} u_1[t] \\ u_2[t] \\ u_3[t] \\ \vdots \\ u_{n-1}[t] \\ u_n[t] \end{pmatrix}, \quad u_0 = \begin{pmatrix} u_{1,0} \\ u_{2,0} \\ u_{3,0} \\ \vdots \\ u_{n-1,0} \\ u_{n,0} \end{pmatrix}$$

Vector of inductance currents and initial inductance currents:

$$i[t] = \begin{pmatrix} i_1[t] \\ i_2[t] \\ i_3[t] \\ \vdots \\ i_{n-1}[t] \\ i_n[t] \end{pmatrix}, \quad i_0 = \begin{pmatrix} 0 \\ 0 \\ 0 \\ \vdots \\ 0 \\ 0 \end{pmatrix}$$

Applying these matrices and vectors the differential equation system is defined with:

$$\frac{d}{dt} v[t] = \mathcal{A}^{-1} \cdot \mathcal{B} \cdot v[t], \quad v[0] = v_0$$

with the block matrices

$$\mathcal{A} = \begin{pmatrix} \mathcal{C} & \mathcal{O} \\ \mathcal{O} & \mathcal{L} \end{pmatrix}$$

and

$$\mathcal{B} = \begin{pmatrix} -\mathcal{G} & -\mathcal{E} \\ \mathcal{E}^T & \mathcal{W} \end{pmatrix}$$

and with the block vectors

$$v[t] = \begin{pmatrix} u[t] \\ i[t] \end{pmatrix}$$

and

$$v_0 = \begin{pmatrix} u_0 \\ i_0 \end{pmatrix}$$

Appendix B: Differential Equation System Describing Homogeneous Pulse Forming Networks

Figure B-1 shows the most popular form of a pulse forming network. Here all capacitances and inductances are identical: $C_k = C$, $L_k = L$. Resistances R_k and W_k, conductivities G_k, and mutual inductances $M_{k,l}$ are neglected. The initial capacity voltages are now identical: $u_{k,0} = u_0$. Hence the switches S now can be combined in a single switch at the load R.

To define the linear differential equation system of the circuit, the matrices and vectors of the inhomogeneous pulse forming network can be simplified as follows:

Matrix of capacitances:

$$C = \begin{pmatrix} C_1 & 0 & 0 & \cdots & 0 & 0 \\ 0 & C_2 & 0 & \cdots & 0 & 0 \\ 0 & 0 & C_3 & \cdots & 0 & 0 \\ \vdots & \vdots & \vdots & \ddots & \vdots & \vdots \\ 0 & 0 & 0 & \cdots & C_{n-1} & 0 \\ 0 & 0 & 0 & \cdots & 0 & C_n \end{pmatrix}$$

1-Matrix:

$$\varepsilon = \begin{pmatrix} 1 & 0 & 0 & \cdots & 0 & 0 & 0 \\ -1 & 1 & 0 & \cdots & 0 & 0 & 0 \\ 0 & -1 & 1 & \cdots & 0 & 0 & 0 \\ \vdots & \vdots & \vdots & \ddots & \vdots & \vdots & \vdots \\ 0 & 0 & 0 & \cdots & -1 & 1 & 0 \\ 0 & 0 & 0 & \cdots & 0 & -1 & 1 \end{pmatrix}$$

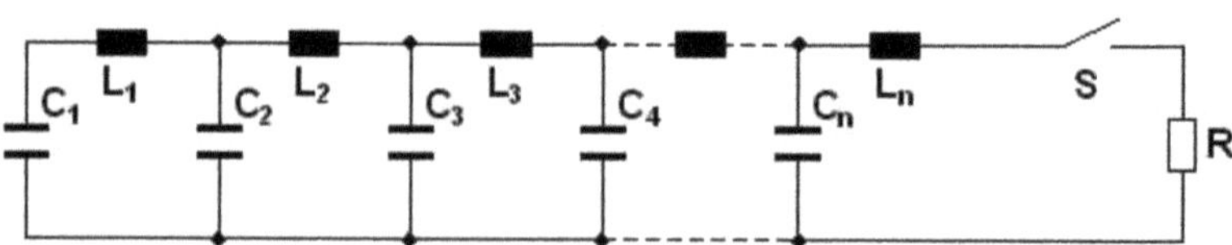

Fig. B-1 Homogeneous pulse forming network without losses

Matrix of inductances:

$$\mathcal{L} = \begin{pmatrix} L_1 & 0 & 0 & \cdots & 0 & 0 \\ 0 & L_2 & 0 & \cdots & 0 & 0 \\ 0 & 0 & L_3 & \cdots & 0 & 0 \\ \vdots & \vdots & \vdots & \ddots & \vdots & \vdots \\ 0 & 0 & 0 & \cdots & 0 & 0 \\ 0 & 0 & 0 & \cdots & 0 & L_n \end{pmatrix}$$

Matrix of zeroes (n × n-matrix):

$$\mathcal{O} = \begin{pmatrix} 0 & 0 & 0 & \cdots & 0 & 0 \\ 0 & 0 & 0 & \cdots & 0 & 0 \\ 0 & 0 & 0 & \cdots & 0 & 0 \\ \vdots & \vdots & \vdots & \ddots & \vdots & \vdots \\ 0 & 0 & 0 & \cdots & 0 & 0 \\ 0 & 0 & 0 & \cdots & 0 & 0 \end{pmatrix}$$

Matrix of resistances:

$$\mathcal{W} = \begin{pmatrix} 0 & 0 & 0 & \cdots & 0 & 0 \\ 0 & 0 & 0 & \cdots & 0 & 0 \\ 0 & 0 & 0 & \cdots & 0 & 0 \\ \vdots & \vdots & \vdots & \ddots & \vdots & \vdots \\ 0 & 0 & 0 & \cdots & 0 & 0 \\ 0 & 0 & 0 & \cdots & 0 & -R \end{pmatrix}$$

Vector of capacitance voltages and initial capacitance voltages:

$$u[t] = \begin{pmatrix} u_1[t] \\ u_2[t] \\ u_3[t] \\ \vdots \\ u_{n-1}[t] \\ u_n[t] \end{pmatrix}, \quad u_0 = \begin{pmatrix} u_0 \\ u_0 \\ u_0 \\ \vdots \\ u_0 \\ u_0 \end{pmatrix}$$

Vector of inductance currents and initial inductance currents:

$$i[t] = \begin{pmatrix} i_1[t] \\ i_2[t] \\ i_3[t] \\ \vdots \\ i_{n-1}[t] \\ i_n[t] \end{pmatrix}, \quad i_0 = \begin{pmatrix} 0 \\ 0 \\ 0 \\ \vdots \\ 0 \\ 0 \end{pmatrix}$$

Applying these matrices and vectors the differential equation system is defined with:

$$\frac{d}{dt} v[t] = \mathcal{A}^{-1} \cdot \mathcal{B} \cdot v[t], \quad v[0] = v_0$$

with the block matrices

$$\mathcal{A} = \begin{pmatrix} \mathcal{C} & \mathcal{O} \\ \mathcal{O} & \mathcal{L} \end{pmatrix}$$

and

$$\mathcal{B} = \begin{pmatrix} \mathcal{O} & -\mathcal{E} \\ \mathcal{E}^T & \mathcal{W} \end{pmatrix}$$

and with the block vectors

$$v[t] = \begin{pmatrix} u[t] \\ i[t] \end{pmatrix}$$

and

$$v_0 = \begin{pmatrix} u_0 \\ i_0 \end{pmatrix}$$

References

Barbosa-Cánovas GV, Pierson MD, Zhang QH, Schaffner DW (2000) Pulsed electric fields. J Food Sci 65:65–80
Bazhal MI (2001) Etude du mécanisme d'électroperméabilisation des tissus végétux. Application à l'extraction du jus des pommes. Thèse de Doctorat, Université de Technologie de Compiègne
Bödecker V (1985) Zum Technologiepotential der Plasmaphysik (About the technological potential of plasma physics). VDI-Verlag, Düsseldorf
Brosch PF, Landrath J, Wehberg J (2000) Leistungselektronik. Braunschweig, Wiesbaden
Donaldson AL (1990) Lifetime considerations. In: Schaefer G, Kristiansen M, Guenther A (eds) Gas discharge closing switches. Plenum Press, New York
Ennis JB, Song BM, Bushnell AH, Cooper RA, Jichetti J, MacDougall F, McDowell R, Andermatt B, Bates J (2003) Custom design of components and power supplies for pulsed power systems. General atomics energy products, engineering bulletin. Presented at IEEE Industrial Electronics Conference (IECON 2003)
Gaudreau MPJ, Hawkey T, Petry J, Kempkes M (2005) Pulsed power systems for food and wastewater processing. Diversified Technologies, Inc, Bedford. www.divtecs.com
Gaudreau MPJ, Hawkey T, Petry J, Kempkes M (2004) Design considerations for pulsed electric field processing. In: Proceedings 2nd European pulsed power symposium, Hamburg, pp 58–61
Greinacher H (1919) Hochspannungsanlage zur Erzeugung von hochgespanntem Gleichstrom aus Wechselstrom unter Verwendung von elektrolytischen Ventilzellen und Kondensatoren. Swiss Patent No. 80289

Heinz V (2005) Information provided by Dr. V. Heinz, Technical University Berlin, Department of Food Biotechnology and Process Engineering

Heinz V, Angersbach A, Knorr D (2001) Membrane permeabilization in biological materials in response to pulsed electric fields – a highly efficient process for biotechnology. In: Proceedings of international conference on pulsed power applications, Gelsenkirchen, pp B.03/1–B.03/7

Heinz V, Alvarez I, Angersbach A, Knorr D (2002) Preservation of liquid foods by high intensity pulsed electric fields – basic concepts for process design. Trends Food Sci Technol 12:103–111

Heinz V, Toepfl S, Knorr D (2003) Impact of temperature on lethality and energy efficiency of apple juice pasteurization by pulsed electric fields treatment. Innov Food Sci Emerg Technol 4(2):167–175

Jemai AB, Vorobiev W (2002) Effect of moderate electric field pulse (MFEP) on the diffusion coefficient of soluble substances from apple slices. Int J Food Sci Technol 37:73–86

Jin ZT, Zhang QH (2002)Commercial scale integrated PEF processing system and its operation cost. 2002 IFT Technical Paper Presentation 91E-21

Johnstone PT, Bodger PS (1997) High voltage disinfection of liquids. IPENZ Trans 24:30–35

Kotnik T, Miklavčič D, Mir LM (2001a) Cell membrane electropermeabilization by symmetrical bipolar rectangular pulses; Part II. Reduced electrolytic contamination. Bioelectrochemistry 54:91–95

Kotnik T, Mir LM, Flisar K, Puc M, Miklavčič D (2001b) Cell membrane electropermeabilization by symmetrical bipolar rectangular pulses; Part I. Increased efficiency of permeabilization. Bioelectrochemistry 54:83–90

Kotnik T, Pucihar G, Reberšek M, Miklavčič D, Mir LM (2003) Role of pulse shape in cell membrane electropermeabilization. Biochim Biophys Acta 1614:193–200

Lazarenko BR, Fursov SP, Scheglov YA, Bordiyan VV, Chebanu VG (1977) Electroplasmolysis. Karta Moldovenaske, Kishinev

Loeffler M, Schmidt W, Schuhmann R, Röttcring A, Neumann J, Dreesen C (2001) Treatment of sewage sludge with pulsed electric fields. In: Proceedings of international conference on pulsed power applications, Gelsenkirchen

MacDougall FW (1996) Capacitors for the pulsed power industry. Technical Paper Aerovox Div. PPC, 167 John Vertente Boulevard, New Bedford, MA 02745-1221

Marx E (1923) Verfahren zur Schlagprüfung von Isolatoren und anderen elektrischen Vorrichtungen. German Patent No. 455933, 12

Marx E, Koch W (1970) Induktiver Stoßstromgenerator. Offenlegungsschrift 1488941. Deutsches Patentamt. 19

McLellan MR, Kime RL, Lind LR (1991) Electroplasmolysis and other treatments to improve apple juice yield. J Sci Food Agr 57:77–90

Min S, Jin ZT, Min SK, Yeom H, Zhang QH (2003a) Commercial-scale pulsed electric field processing of orange juice. J Food Sci 68:1265–1271

Min S, Min SK, Zhang QH (2003b) Inactivation kinetics of tomato juice lipoxygenase by pulsed electric fields. J Food Sci 68:1995–2001

Neumann J (2005) Private communication. Propuls GmbH Bottrop

Ngadi M, Bazhal M, Raghavan GSV (2003) Engineering aspects of pulsed electroplasmolysis of vegetable tissues. Agric Eng Int 5:1–10

Papchenko AY, Bologa MK, Berzoi SE (1988a) Apparatus for processing vegetable raw material. US patent no. 4787303

Papchenko AY, Bologa MK, Berzoi SE, Paukov JN, Rudkovskaya GV, Chebanu VG (1988b) Electroplasmolyzer for processing vegetable stock. US patent no. 4723483

Puc M, Čorovič S, Flisar K, Petkovšek M, Nastran J, Miklavčič D (2004) Techniques of signal generation required for electropermeabilization. Survey of electropermeabilization devices. Bioelectrochemistry 64:113–124

Qin BL, Zhang Q, Barbosa-Canovas GV, Swanson B, Pedrow PD (1994) Inactivation of microorganisms by pulsed electric field of different voltage waveforms. IEEE Trans Dielectr Electr Insul 1:1047–1057

Reitler W (1990) Conductive heating of foods, PhD thesis, TU Munich, Munich
Richardson B, Pirrie C, Clarke K, Dunlop A (2004) Thyratron or solid state? Switching technology options for pulse power applications. In: Proceedings 2nd European pulsed power symposium, Hamburg, pp 8–13
San Martín MF, Barbosa-Cánovas GV, Swanson BG (2003) Innovations in food processing. CEP (www.cepmaganzine.org), March, pp 54–60
Smith DL, Savage ME, Ziska GR, Starnbird RL (2004) ZR MARX capacitor lifetime test results. Proceedings of the 2004 power modulator conference in San Francisco, CA, (May 2004), paper number P-HPG&HCL 1-1
Toepfl S, Heinz V, Knorr D (2004) Optimization of pulsed electric field treatment for liquid food pasteurization. In: Proceedings of 2nd European pulsed power symposium, Hamburg, pp 68–72
Ulmer HM, Heinz V, Gänzle MG, Knorr D, Vogel RF (2002) Effects of pulsed electric fields on inactivation and metabolic activity of *Lactobacillus plantarum* in model beer. J Appl Microbiol 92:326–335
Weise THGG, Löffler MJ (2001) Overview on pulsed power applications. In: Proceedings of the international conference on pulsed power applications, Gelsenkirchen, pp A.01/1–A.01/8
Welleman A, Fleischmann W, Waldmeyer J (2004) Solid state switches for repetitive pulse applications. In: Proceedings 2nd European pulsed power symposium, Hamburg, pp 176–180
Zhang Q, Barbosa-Cánovas GV, Swanson BG (1995) Engineering aspects of pulsed electric field pasteurization. J Food Eng 25:261–281

The Phenomenon of Electroporation

Samo Mahnič-Kalamiza and Damijan Miklavčič

Abbreviations

CHO	Chinese hamster ovarian
DAMP	Damage-associated molecular pattern
EIT	Electric impedance tomography
HPEF	High pulsed electric field
MD	Molecular dynamics
MEF	Moderate electric field
MREIT	Magnetic resonance electric impedance tomography
PEF	Pulsed electric fields
PFA	Pulsed field ablation
ROS	Reactive oxygen species
TMV	Transmembrane voltage

The Phenomenon and Its Applications

Exposure of biological cells and tissues to short electric pulses of sufficient amplitude to increase the permeability of the membrane was "discovered" at the end of 1950s (Stampfli 1958). Though still a subject of some controversy and debate as to whom the laurel for this discovery should go (Sitzmann et al. 2016), it is by now a well-known and extensively investigated phenomenon at the heart of a broadly applicable set of techniques. The phenomenon facilitates techniques of increasing importance in biomedicine (Yarmush et al. 2014; Bradley and Haines 2020; Geboers et al. 2020), in food science and technology (Mahnič-Kalamiza et al. 2014), as well as in biotechnology and environmental science (Kotnik et al. 2015). In different application fields, the process or technique is referred to either as electropermeabilization, electroporation, electropulsation, PEF (pulsed electric field) treatment, or PFA (pulsed field ablation). This terminology, although it may seem often arbitrary,

S. Mahnič-Kalamiza · D. Miklavčič (✉)
Faculty of Electrical Engineering, University of Ljubljana, Ljubljana, Slovenia
e-mail: damijan.miklavcic@fe.uni-lj.si

J. Raso et al. (eds.), *Pulsed Electric Fields Technology for the Food Industry*, Food Engineering Series, https://doi.org/10.1007/978-3-030-70586-2_3

reflects the emphasis on a particular property or mechanism of action of the treatment within its respective application field. Electroporation emphasizes that the phenomenon is believed to be responsible for electrically induced formation of aqueous pores in the lipid bilayer under the influence of the induced transmembrane voltage (Neumann et al. 1982; Weaver 1993; Weaver and Chizmadzhev 1996). Electropermeabilization somewhat more reservedly places the emphasis on the electrically induced increase in the membrane permeability for molecules devoid of physiological mechanisms of transmembrane transport that is observed following application of electric pulses, without alluring to any mechanisms themselves that might or might not be involved (Teissié 2014). The link and the limit between the two are being established (Mir 2020). Electropulsation, pulsed electric field treatment, and pulsed field ablation are perhaps the most generic or neutral of the terms and explicitly stand only for the process of exposing cells to electric pulses itself, with only implied consequences of this treatment leading to membranes' structural alteration and increased conductivity and/or permeability.

The underlying phenomenon is termed either electroporation or electropermeabilization. The terms are often used interchangeably and can be considered synonyms, although the former term refers only to the contribution of aqueous pores formed in the lipid bilayer to the increased permeability of the membrane (see Fig. 1), while the latter term is more general. Electropermeabilization ascribes the increased permeability of the membrane to a broader range of biophysical and biochemical mechanisms potentially involving structures other than the bare lipid bilayer (Kotnik et al. 2019). Formation of (transient) hydrophilic pores in the lipid bilayer – termed electroporation – is now a widely recognized mechanism of membrane permeabilization governed by thermodynamics. The mechanism has been

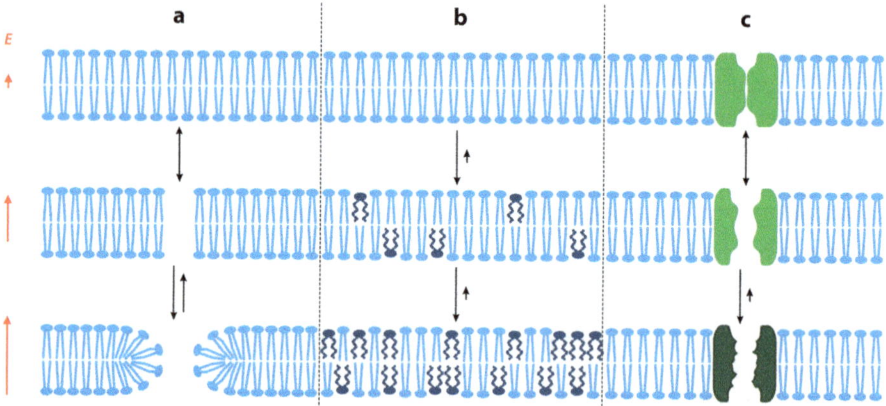

Fig. 1 The phenomenon of electroporation on a molecular level as a simplified conceptual scheme. (**a**) Aqueous pores are formed under the influence of the electric field in the lipid bilayer, shown in two stages; (**b**) electrically induced chemical changes to lipids constituting the membrane – peroxidation deforms their tails and increases the bilayer's permeability to water, ions, and small molecules; (**c**) electrically induced modulation of membrane proteins' function, shown for a voltage-gated channel. (Reproduced with permission from Kotnik et al. 2019)

corroborated by molecular dynamics (MD) simulations; however, there is increasing evidence that exposure to electric pulses also causes chemical changes to the lipids as well as alteration of the membrane proteins' function, both contributing factors to the membrane's increased permeability (Breton and Mir 2018; Kotnik et al. 2019).

Throughout the past several decades, the phenomenon of electroporation has been greatly elucidated through extensive research, in the process of which many fascinating features were revealed about its nature and mechanisms. Among other discoveries, it was found that electroporation can be used to nonselectively increase cell uptake of drugs and genetic material (Mir et al. 1988; Escoffre et al. 2012; Miklavčič et al. 2014) and extract molecules from cells (Knorr et al. 1994; Mahnič-Kalamiza et al. 2014), for membrane protein insertion (El Ouagari et al. 1993; Raffy et al. 2004), to induce cell-to-cell fusion and cell-vesicle fusion (with viable hybrids as a result) (Zimmermann 1982; Saito et al. 2014), to fuse individual cells with tissue (Heller and Grasso 1990), to initiate targeted necrotic or apoptotic cell death (Miller et al. 2005; Pakhomova et al. 2013), to induce intracellular effects (e.g., release of intracellular calcium) or indeed introduction of calcium into cells for purposes of triggering cell death (Vernier et al. 2003; Frandsen et al. 2020), and to modify the viscoelastic properties and texture of plant tissues (Mahnič-Kalamiza and Vorobiev 2015; Botero-Uribe et al. 2017).

The most appealing feature of electroporation is that it seems to be universal to the lipid bilayer and consequently is ubiquitous throughout the spectrum of various life forms, i.e., the phenomenon can be observed in all cell types (eukaryotes, bacteria, and archaea (Polak et al. 2014)), in all cell arrangements (be it the cells are in suspension, grow adhered to surfaces, clustered, or in tissue). Apart from cells, electroporation can also be observed in any other membrane bilayer systems, e.g., in planar lipid bilayers, lipid vesicles, and polymeric vesicles (Benz et al. 1979; Teissie and Tsong 1981; Aranda-Espinoza et al. 2001). This has led to the development of a considerable number of applications in diverse fields as mentioned in the first paragraph of the section; some of these applications have already reached patients and consumers. At present, the most developed and promising biomedical applications include cardiac muscle ablation for the treatment of arrhythmias (Stewart et al. 2019; Maor et al. 2019; Reddy et al. 2019; Sugrue et al. 2019), electrochemotherapy (Campana et al. 2019a, b), gene electrotransfer (Broderick and Humeau 2015; Lambricht et al. 2016; Rosazza et al. 2016b), and tumor tissue ablation by means of irreversible electroporation (Scheffer et al. 2014; Geboers et al. 2020). In food processing existing industrial applications range from pasteurization of liquid foods (Min et al. 2007; Jin 2017) to modifying structural and textural properties of raw materials (e.g., potato tubers (Fincan and Dejmek 2003; Chalermchat and Dejmek 2005)) and enhancing mass transport processes, thus facilitating extraction of valuable compounds from raw materials or by-products (biological waste materials). Electroporation for enhancing mass transport is important in, e.g., recovering sugars from sugar beet root (Mhemdi et al. 2016; Vorobiev and Lebovka 2017), improving the maceration process in winemaking (Sack et al. 2010), and enhancement of drying (Thamkaew and Gómez Galindo 2020), with additional applications ranging into the field of biorefinery where electroporation can be and has been used for valorization of various by-products (e.g., extraction of polyphenols from grape pomace (Saldaña et al. 2017), polyphenol and protein extraction from rapeseed stems and leaves (Yu et al. 2015), etc.). There is also growing interest in using electroporation for cooking (Blahovec et al. 2017).

In many of the applications listed above, electroporation is performed on cells comprising tissues. Understanding the mechanisms by which electric pulses act upon cells in such a complex environment as the biological tissue requires a multiscale approach, where we combine the insights gained from molecular dynamics models of simple lipid systems with those gained from in vitro and in vivo experiments (Rems and Miklavčič 2016). The in vitro data come primarily from mammalian cell cultures, for which the literature is most abundant and since studying electroporation at the level of single cells allows for the most efficient and straightforward approach to understanding the basic aspects of the phenomenon. Results obtained at that level can then be translated to increasingly complex levels up to the level of biological tissue in vivo. In continuation of this chapter, we present the current understanding of the electroporation phenomenon at scales that are orders of magnitude apart. Assembling and reconciling findings obtained at such disparate levels of examination remains a daunting challenge, as is briefly discussed in the concluding section on current challenges facing basic electroporation research.

Applying an electric field to the cell can have, in general, three possible outcomes. The outcome depends on the local field strength, the duration of cell's exposure to the electric field, and the membrane recovery rate. Provided that the field strength and exposure time are completely insufficient, electroporation is not achieved, and cell's permeability and consequently viability are left unaffected. If, however, the field strength exceeds the so-called reversible threshold and exposure to the sufficiently high field strength is for an adequate amount of time, *reversible electroporation* occurs (Rols and Teissié 1990; Čorović et al. 2012). In reversible electroporation, the membrane is permeabilized and remains in a state of increased permeability for a period. The phenomenon is reversible as the term would imply, and the membrane eventually returns to its original state by a process of membrane resealing. Membrane resealing entails pore closure and restoration of the cell's normal (called *resting*) transmembrane potential (see sect. 2 in continuation). This process, as is the process of pore formation and evolution, is still a subject of experimental research and modelling endeavors (molecular- and cell-scale) to determine its exact underlying mechanisms and kinetics (Saulis et al. 1991; Demiryurek et al. 2015; Kotnik et al. 2019). It should be noted that reversibility of the phenomenon of pore formation and re-establishing of the transmembrane potential are only possible if the environmental conditions are conducive to cell survival and function.

In case field strength and/or energy is too high, *irreversible electroporation* occurs (Rubinsky et al. 2007). Irreversible electroporation results in loss of cell homeostasis, effectively killing the cell. See Fig. 2 for a simplified diagram of some of the applications of reversible and irreversible electroporation. Note that though the figure gives multiple application examples for reversible electroporation, and only cell death as resulting from irreversible electroporation, the latter as a modality

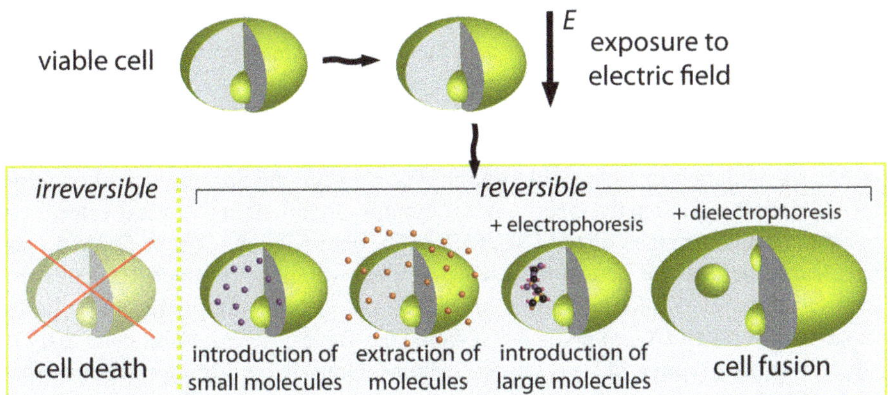

Fig. 2 Some of the possible outcomes/target processes of exposing a viable biological cell to an electric field of appropriate strength. (Redrawn based on a figure first published in Mahnič-Kalamiza et al. 2014)

of tissue disintegration and ablation does have its own broad set of uses. The reader should be aware that it proves difficult to achieve exclusively reversible electroporation without any irreversible damage and vice versa, especially when delivering electroporation to tissues. This is mainly due to finite electrode geometry and material inhomogeneities (Hjouj et al. 2012; Blumrosen et al. 2014; Zmuc et al. 2019; Polajžer et al. 2020) resulting in a heterogeneous field strength distribution and coverage. If reversible electroporation pulses are delivered to target cells/tissues in combination with other fields of appropriate form and strength, other phenomena can be facilitated, such as electrophoresis and dielectrophoresis, used for introduction of large molecules such as DNA in gene delivery applications and for cell fusion, respectively (Hu et al. 2013; Dean 2013; Rems et al. 2013). As previously evidenced by the review of electroporation applications, both reversible and irreversible electroporation have found their applications in such disparate fields such as biomedicine, food technology, and biotechnology (Toepfl et al. 2014; Yarmush et al. 2014; Kotnik et al. 2015; Geboers et al. 2020).

In line with the subject matter this book is devoted to, i.e., electroporation applications in the food industry, we dedicate the remainder of this section to developing the list of applications in this field a bit further, to give the reader a broader overview of what can be achieved when applying pulsed electric fields to food matrices.

In food processing, electroporation has been applied to either enhance or facilitate extraction of juices and other valuable compounds from tissues and microorganisms (e.g., microalgae) (Fincan et al. 2004; Donsi et al. 2010; Vorobiev and Lebovka 2010; Vanthoor-Koopmans et al. 2013), tissue dehydration (Ade-Omowaye et al. 2001; Lebovka et al. 2007; Shynkaryk et al. 2008; Thamkaew and Gómez Galindo 2020), and nonthermal preservation (of mainly liquid foods) by microbial inactivation (Wesierska and Trziszka 2007; Mosqueda-Melgar et al. 2008; Bermudez-Aguirre et al. 2012; Marsellés-Fontanet et al. 2012). In food engineering, reversible electroporation may help in preventing biological tissues from sustaining damage due to ice crystal formation during freezing (Phoon et al. 2008; Shayanfar et al. 2013; Dymek 2015) and serve as a method of cell metabolism stimulation (Ye et al. 2004; Dymek et al. 2012; Straessner et al. 2013).

Regardless whether the objective is to introduce into or extract out of tissue either water or solutes, electroporation entails applying a series of trains of electric pulses of moderate or high field strength. The reader should note that though no formal definitions exist, the electric field strengths are often, in literature, referred to per their defined ranges of field strength or by abbreviations. These fields do not necessarily reflect the actual local electric field strength to which the cells are exposed. Moderate electric field, abbr. MEF, is most often considered to be in the range of several 100 V/cm up to 1–2 kV/cm (see Table 1 in (Puertolas et al. 2012)), which is a broad range of field strength values. Note that arbitrary categorizations by field strength or any other pulse parameter are best avoided, since the augmenting effect of electroporation to mass transport has already been observed in case of low-intensity electric fields on the order of less than 100 V/cm (Asavasanti et al. 2010). It is therefore desirable to respect the general recommendation guidelines whenever reporting on research studies of electroporation technology in food and biotechnological processes (Raso et al. 2016) so as to avoid any ambiguity. Higher-intensity electric pulses (abbr. HPEF (high pulsed electric field)), i.e., on the order of about 5–50 kV/cm, are associated with irreversible damage to plant and animal cells and can possibly result in irreversible electroporation of bacterial and yeast cells and can therefore be used as a method of preserving material by inactivating and destroying these microorganisms of spoilage.

The objective of microbial inactivation by electroporation is mainly to pasteurize biological liquids: foods or sludge (Álvarez et al. 2006; Mosqueda-Melgar et al. 2008; Guerrero-Beltran et al. 2010; Bermudez-Aguirre et al. 2012). During the past two or three decades, electroporation as means of food preservation has found numerous applications, e.g., a number of microorganism species have effectively been inactivated in different liquid foods (e.g., cider, milk, beer, soups, miscellaneous fruit, and vegetable juices) and semisolid/solid food products. Moreover, synergistic effects between electroporation and other treatments, e.g., nisin, acid, low-temperature/low-energy heating, or high pressure, have been demonstrated (Kotnik et al. 2015). The combined approach of electroporation in tandem with high pressure and ultraviolet light is important as it seems to hold promise in the inactivation of bacterial spores, for which other, traditional mechanisms normally fail to achieve inactivation on their own. In fact, as many harmful microorganisms prove hard to destroy by electroporation alone, electroporation is almost exclusively applied as combined treatment with traditional or other novel techniques of food preservation, such as thermal, enzymatic, ultraviolet, chemical, or pressure-based (Shin et al. 2010; Sobrino-Lopez and Martin-Belloso 2010). Inactivation of bacterial spores is generally considered as one of the hurdles in pasteurization and sterilization processes due to the spores' capability of tolerating high field strengths, temperatures, and pressures (Pol et al. 2001; Siemer et al. 2014a, b). Note the higher field requirements and specific energy consumption in microbial inactivation applications as compared to other objectives of electroporation application in Fig. 3.

Table 1 The phenomenon of electroporation, its time evolution, and principal characteristics of its stages. *DAMPs (damage-associated molecular patterns) molecules are biomolecules associated with an inflammatory response, released from damaged and/or dying cells

Stage	Characterized by	Illustrated by	Timescale
1- Initiation of the state of elevated permeability (cell is exposed to an electric field)	• Charged carriers in the electrolyte are ions/charged molecules • Charging of the membrane • Induced membrane potential difference ≈ 500 mV over membrane thickness ≈ 5 nm • Establishing huge electric field within the membrane, V/d = 500 mV/5 nm = 100 MV/m (!); note that electrical breakdown in air occurs at E > 3 MV/m	Figure 5a	ns-ms
2- Formation and expansion of the transmembrane transport area	• Pores are formed in the lipid domain of the membrane • Phosphatidylserine externalization • ROS* generation and access to lipids in the membrane • Lipid oxidative damage • Electro-conformational change of membrane proteins	Figure 5b	ns-ms
3- Stabilization (with partial recovery); the cell remains in a state of increased membrane permeability	• Leakage of ions/ionic and osmotic imbalance • Loss of cell homeostasis • DAMP* molecule release (ATP, HMGB1, nucleic acids, etc.) • Membrane protein structure and function alteration • Cytoskeleton disassembly	Figure 5c	ms-min
4- Cell membrane repair/ resealing and cell recovery	• Exocytosis/patching/membrane repair • Reassembling cytoskeleton • Recovering protein function • Exocytosis/patching/membrane repair • Reassembling cytoskeleton • Recovering protein function • Cell response to stress(stress-related gene expression) • Re-establishing homeostasis	Figure 5dandFig. 5e	min-hrs
5- Cessation of the cells' altered physiological processes; complete cell recovery, or loss of homeostasis and cell death	• Cell response to stress(stress related gene expression) • Loss of cell homeostasis • Cell death	Figure 5f	min-hrs

To understand why significantly higher field strengths and energies are generally required for microbial inactivation, in comparison to, for instance, extraction applications, we must understand the mechanism of action of electroporation, i.e., the mechanism by which inactivation of microorganisms is achieved by pulsed electric

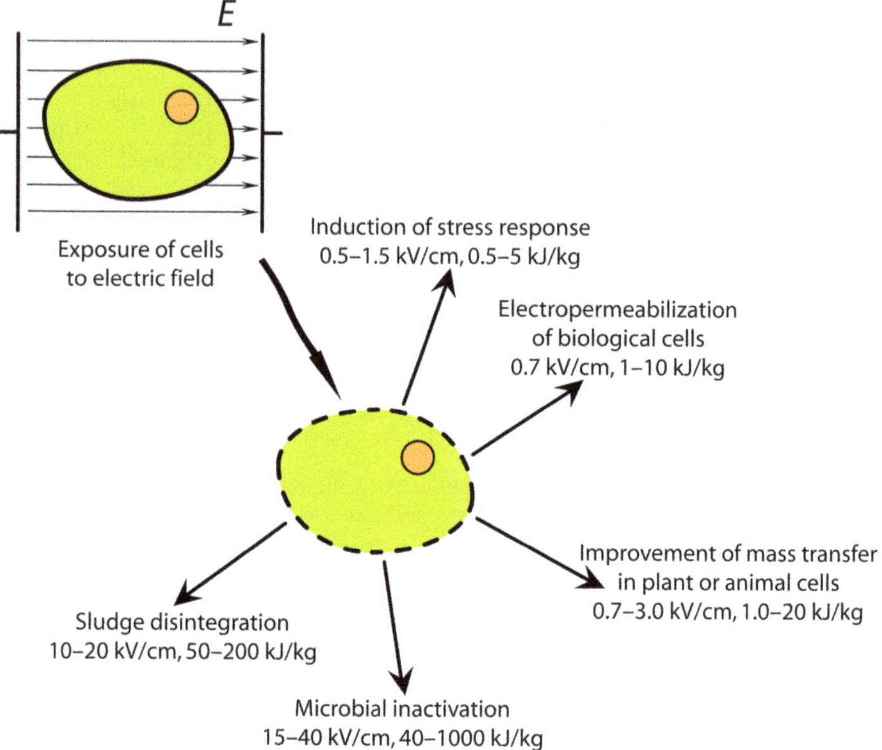

Fig. 3 Target mechanisms of action and applications of electroporation by field strength and delivered energy; a schematic representation of exposing a biological cell to an external electric field with various corresponding applications. (Redrawn based on a figure first published in Toepfl et al. 2006)

fields. If we follow the paradigm presented by Saulis (2010), the process of exposing a living and viable cell, capable of reproduction, to an external electric field can be divided into stages. The first stage involves an increase in transmembrane potential in time, as the plasma membrane is charged by the externally superimposed electric field. The final amplitude of this transmembrane potential that is reached is determined by material and geometrical properties, namely, cell size and shape, membrane, and medium ionic composition that is affecting the conductivity (intra- and extracellular), and by the pulse parameters (e.g., shape, duration, amplitude). When the critical transmembrane potential is reached, the pore initiation stage begins, and small metastable hydrophilic pores appear in the plasma membrane. Pores evolve in time in both size and number. Following the cessation of pulse application is the posttreatment stage, which can last from milliseconds to hours. During this stage leakage of intracellular compounds to the cell exterior, influx of substances from the extracellular space into the cell, and finally either pore resealing and membrane repair or cell death will occur. The final result of the treatment is then either a viable or a dead cell; the outcome is determined by the relevant conditions and processes taking place during the mentioned stages of electroporation. These conditions vary considerably for different applications (see, e.g., Figure 3 for the field strength and delivered energy aspect).

As an example, consider a pathogenic bacterial species representing a typical prokaryotic microorganism several micrometers in length. For this prokaryotic cell, the induced transmembrane voltage will be an order of magnitude lower given the same external electric field strength, compared to a small eukaryotic cell with a diameter of tens of micrometers. The presence and structure of any cell wall must likewise be considered (Aronsson et al. 2005; Golberg et al. 2012). For a more thorough treatment of microbial inactivation and considerations relevant to electroporation, refer to chapters 5, "Microbial inactivation by PEF," and 6, "Liquid food pasteurization by PEF" in the book.

In addition to the enhanced mass transport or irreversible cell damage effects of electroporation and applications facilitated by these effects, electroporation can be used at lower intensities (below 100 V/cm) for inducing a stress response in plant and yeast cells. At these low field strengths, electroporation may be achieved without affecting cell viability by carefully controlling the electric pulse parameters, resulting in predominantly reversible electroporation of cells. However, even though the electroporated cells survive the electric field treatment, they are still stressed due to the opening of pores in the membrane, triggered repair mechanisms, and various biochemical processes, as they struggle to recover their normal functionality (re-establishing homeostasis). The exact details of recovering from reversible electroporation at the level of the cell and its membranes on the molecular level, as well as physiological responses to electroporation-induced stress, remain largely unknown (Gómez Galindo 2016).

One of the challenges when electroporating plant tissue with intent of reversibly electroporating them and thereby inducing cell stress is posed by the cell's heterogeneous structures. Since cells vary in shape, size, and cell wall structure, these heterogeneities influence the effect of electroporation protocols on cells of a particular size, shape, or within a particular spatial region. If cells are successfully reversibly permeabilized, physiological responses to electroporation-induced stress will include, among others, the production of ROS (reactive oxygen species), release of stored energy, activation of regulatory genes in charge of the cell's stress response, as well as the production of secondary metabolites. Applying reversible electroporation has also been shown, as a few examples, to influence barley seed germination, increase the strength of the cell wall in potatoes, and, by consequence, alter the potato's textural properties. Reversible electroporation can also have an impact on protoplasts (i.e., plant cells with cell walls having been removed) and consequentially the regeneration of new plants. Reversible permeabilization of plant cells and tissue is not often reported upon; however, foundations have been laid for a fascinating area of academic research and industrial innovation (Gómez Galindo 2016).

Electroporation at the Membrane and Molecular Level

Researchers have endeavored to provide a theoretical description of the events underlying the phenomenon of electroporation since its discovery. A number of competing theories have emerged over the years (Pavlin et al. 2008; Yarmush et al. 2014). The set of possible explanations includes assuming a certain type of lipid deformation (Crowley 1973), lipid phase transition (Sugár 1979), breakdown of interfaces adjoining domains of different lipid compositions (Cruzeiro-Hansson and Mouritsen 1988), and/or membrane protein denaturation (Tsong 1991). These descriptions, however, by themselves, fail to provide an adequate explanation for observed phenomena associated with electroporation, and the state-of-the-art consensus is that electroporation can be best described as formation of aqueous pores in the lipid bilayer (Freeman et al. 1994; Kotnik et al. 2012; Casciola and Tarek 2016). This description is supportive of the continued use of the term that has become prevalent – *electroporation*. Another, broader term for the phenomenon is *electropermeabilization*. The latter describes the consequence, i.e., observed increased membrane permeability, rather than the underlying mechanism; as such, it is applicable to other, alternative explanations for the phenomenon in a broader sense.

The theory of aqueous pore formation is based on consideration of thermodynamic relations (Neumann et al. 1982); according to the theory, the formation of aqueous pores is initiated by water molecules penetrating the membrane's lipid bilayer. This promotes reorientation of the adjacent lipids and their polar head groups, which, being hydrophilic, begin pointing toward these penetrating water molecules (see Fig. 4). Unstable, short-lived pores (of nanosecond lifetime) can form even in the absence of an external electric field, but these rare events are of a stochastic nature. They are also forming after or when an externally imposed electric field is present, with a higher probability of occurrence. Contrary to these sporadic spontaneous pores, exposure of the membrane to an electroporating field strength reduces the energy barrier experienced by water molecules preventing

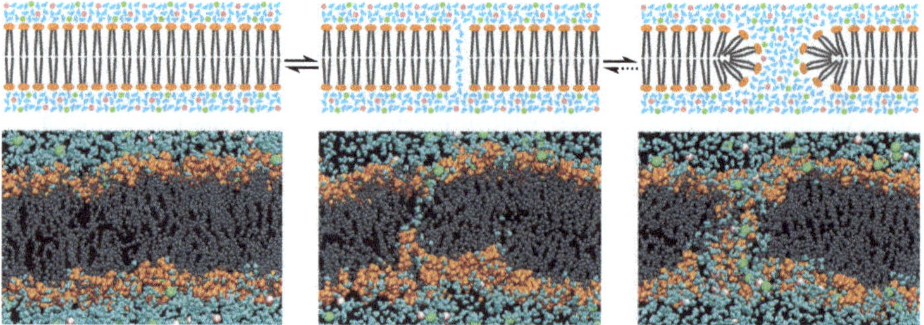

Fig. 4 An idealized molecular-level scheme (top row) and an atomic-level MD simulation (bottom row) of electroporation. The electric field is perpendicular to the bilayer plane. (Reproduced with permission from Yarmush et al. 2014)

them to penetrate the bilayer. Within nanoseconds, penetration results from the transfer of the external field to the membrane. This is followed, within a microsecond, by the transmembrane field being amplified by polarization (positive and negative charge carriers accumulating either side of the membrane), resulting in the buildup of an induced transmembrane voltage (Kotnik and Miklavčič 2006). The intrusion of water into the bilayer increases the statistical probability of a pore forming, resulting in a bulk increase in the number of pores formed in the bilayer per unit of time and unit area. Field-induced pores are more stable than the spontaneous, sporadic, and short-lived pores mentioned above. For transmembrane voltages of several hundred mV, pores become sufficiently numerous and long-lasting (up to milliseconds (Saulis and Saule 2012)) to produce a detectable increase in membrane permeability to those molecules that would otherwise not be able to cross the membrane.

These aqueous pores with radii of at most a few nm are too small for optical microscopy. Moreover, sample preparation for electron microscopy of soft matter is too detrimental to the structures in question (semi-stable nanopores in the bilayer), so much that pores cannot be distinguished from imaging artifacts. Still, there is evidence in favor of the theory of aqueous pore formation, and it comes in the form of molecular dynamics (MD) modelling and simulation. MD simulations by and large agree with the hypothesized sequence of events on the molecular-scale and also demonstrate a clear correlation between the rate of pore formation and the strength of the electric field to which the membrane is exposed (Delemotte and Tarek 2012; Ho et al. 2013; Casciola et al. 2014; Casciola and Tarek 2016).

MD simulations enable two possible disparate approaches to modelling electroporation conditions (Tarek et al. 2010; Rems and Miklavčič 2016). The first option is to model for an electric field strength $\boldsymbol{E}$, acting on charged atoms with force $\boldsymbol{F}_e = q_i\boldsymbol{E}$ (q_i is the charge of the i-th atom). Note that this electric field is not the same one as reported in experimental studies, which the bulk of material is exposed to. This imposed electric field causes the water dipoles to reorient in the electric field, an effect most pronounced at the water-lipid interface. This increases the electric field inside the bilayer, thus increasing the voltage across it (Böckmann et al. 2008; Vernier et al. 2013). The second approach in order to increase the transmembrane voltage is by imposing a charge imbalance. This can be achieved via placement of an excess number of monovalent cations on one side of the bilayer and a corresponding excess number of monovalent anions on the other side of the bilayer (Delemotte and Tarek 2012). The first approach is ordinarily simulated in the absence of ions, modelling only the dielectric response. This scenario represents ps or ns pulses, pulses too short for ion redistribution to occur, and thus there is no charging of the membrane (Vernier 2020). The alternative, second approach is thought to be more relevant for situations employing longer pulses and does allow for full charging of the membrane.

Regardless of which of the two methods is employed, evolution of pore formation and closure is similar to both (Delemotte and Tarek 2012). When the imposed electric field (or the charge imbalance) is sufficient, a conical structure composed of water molecules (a so-called water finger) penetrates into the bilayer's hydrophobic core. Eventually, a water-spanning column is formed across the bilayer as water from one side connects with water from the other side (Levine and Vernier 2010) (see Fig. 4, center column). Because this configuration exposes the hydrophobic lipid tails to water, it is known as a hydrophobic pore (Abidor et al. 1979). If the bilayer were composed of specific lipids, e.g., lipids with large head groups or negatively charged lipids, the head group reorientation energy barrier would be very high. This would cause the hydrophobic pore to expand, allowing passing of ions (Dehez et al. 2014; Polak et al. 2014). In the alternative case of a typical zwitterionic phospholipid comprising the bilayer, the lipid head group migration is in direction along the water column and establishes a so-called hydrophilic pore (Ziegler and Vernier 2008; Levine and Vernier 2010). This pore then stabilizes, increases in size, and is capable of conducting ions (Ho et al. 2013; Dehez et al. 2014). Upon removal or cessation of the external energy source (either superimposed electric field or a charge imbalance), the pore evolution then follows the formation sequence of events in reverse order and eventually closes (Gurtovenko and Vattulainen 2005; Ziegler and Vernier 2008; Levine and Vernier 2010).

Electroporation at the Level of a Single Cell and Cell Suspension

The majority of cells continue to constantly maintain an electric potential difference between their inner and outer side of their plasma membrane. This electric potential difference, called the resting transmembrane voltage (TMV), is generated and regulated by a system of ion pumps and channels in the membrane. For a typical eukaryotic cell, the resting TMV ranges from about −40 to about −70 mV. The minus sign indicates that inner potential (the potential within the cell at the membrane's inner surface) is lower than the one on the outer side of the membrane. This can be considered the natural state of a biological membrane, and both lipid and protein components of the membrane are well adapted (by evolution) to function under voltages in this range (Kotnik et al. 2019).

Upon exposure of a cell to an external electric field, an additional component is superimposed to the resting TMV and is termed the induced TMV, as previously introduced in the section on molecular dynamics (see sect. 2). Induced TMV is only sustained for the duration of the externally imposed electric field and is proportional to its strength (Ehrenberg et al. 1987; Pucihar et al. 2006). For a single spherical cell with a nonconductive plasma membrane, the induced TMV is calculated according to the steady-state Schwan equation (Pauly and Schwan 1959): induced TMV $= 1.5 \times E \times R \times \cos\theta$, where E is the electric field, to which the cell is exposed, R is the cell radius, and θ is the angle measured from the center of the cell with respect to the direction of the field (Kotnik et al. 2010). Thus, exposure to a sufficiently strong field can induce transmembrane voltages that are far exceeding their resting values, which causes both structural changes to the membrane and changes to the constituent molecules. These changes generally would not occur under normal physiological conditions and lead to among others, membrane electropermeabilization. As previously explained, electropermeabilization is the rapid onset of a substantial increase in membrane permeability, indicated to the observer by transmembrane transport of molecules for which an intact membrane is practically impermeable (Kennedy et al. 2008; Kotnik et al. 2010). In the preceding section, we attributed this effect of increased permeability to the formation of aqueous conduits or pores in the lipid bilayer.

A number of studies, both experimental and theoretical, imply that the molecular flow across the permeabilized membrane or, in other words, the heightened degree of permeability of the membrane seems to be limited to the regions of the membrane that were exposed to sufficiently high TMV (Hibino et al. 1993; Towhidi et al. 2008). This is repeatedly shown experimentally for single cells as well as clusters of cells, by monitoring both the TMV and the transmembrane transport on the same cells upon their exposure to electric pulses (Kotnik et al. 2010).

The elevated state of membrane permeability facilitates the inflow of membrane-impermeant molecules into the cell and the outflow of biomolecules from the cell. The kinetics of this transmembrane transport that is enabled by electroporation have been studied extensively, showing that membrane electrical conductivity and permeability increase considerably and measurably within less than a microsecond after the onset of the electric pulse, provided that the TMV locally (for the cell under observation) exceeds a certain critical value. This critical value is not a universal constant but rather a variable. Nevertheless, to summarize some general observations, the kinetics of transmembrane transport can roughly be divided into five stages: (1) the initiation of the state of elevated permeability, (2) the formation and expansion of the transmembrane transport area, (3) stabilization (with partial recovery), (4) the resealing/repair of the membrane, and finally (5) gradual cessation of the cells' altered physiological processes, followed by any reactions to various stressors, possibly resulting in complete cell recovery but possibly also resulting in cell death. Table 1 and Fig. 5 are in aid of illustrating these stages.

Electroporation of the cell membrane, from a theoretical perspective, is not strictly a threshold event. By a threshold event, we would mean that these processes would occur only in an electric field exceeding a certain value. On the contrary, we observe at most that the rates of these processes increase nonlinearly with the increase in the field amplitude which the cells are exposed to. Empirically, however, there is a critical value of the electric field to which a cell must be exposed for electroporation-mediated transport to become detectable. This critical value (the reversible electroporation field strength) depends on the type of cell, the type of molecule being transported/observed, the duration of the exposure, and on the particular set of environmental conditions, e.g., temperature. There is also another, higher critical value of the field (the irreversible electroporation field strength) that must not be exceeded if membrane restabilization, recovery, and resealing are to be expected. As a consequence of this thresholding nature of the phenomenon as observed in experiments, experimentalists often treat electroporation as a quasi-threshold phenomenon. These thresholds are however so ambiguous and dependent

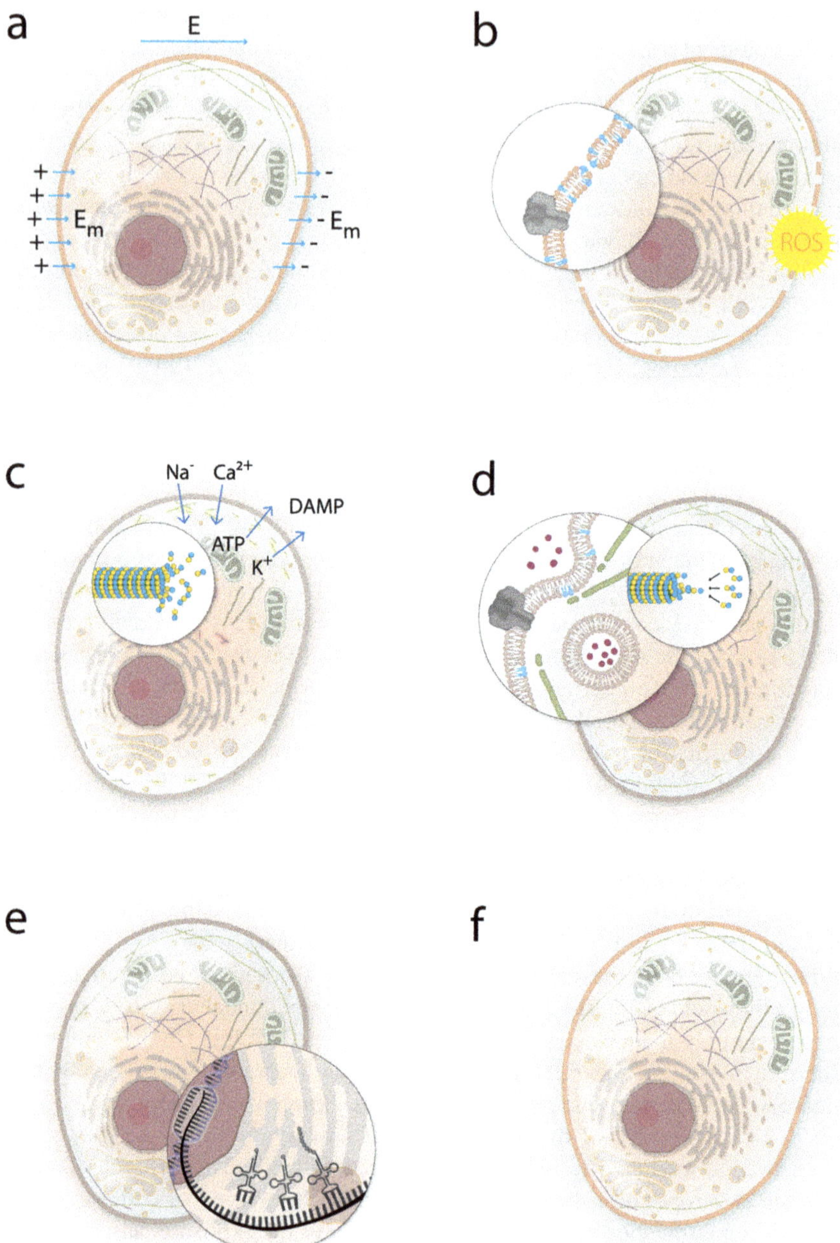

Fig. 5 Different stages of single-cell electroporation: (**a**) initiation of the state of elevated permeability (the cell is exposed to an electric field); (**b**) formation and expansion of the transmembrane transport area, transmembrane transport; (**c**) stabilization (with partial recovery) – cell remains in a state of increased membrane permeability; (**d**) and (**e**) cell membrane repair/resealing and recovery; (**f**) recovery of cell functions or loss of cell homeostasis and subsequent cell death. *ROS stands for reactive oxygen species

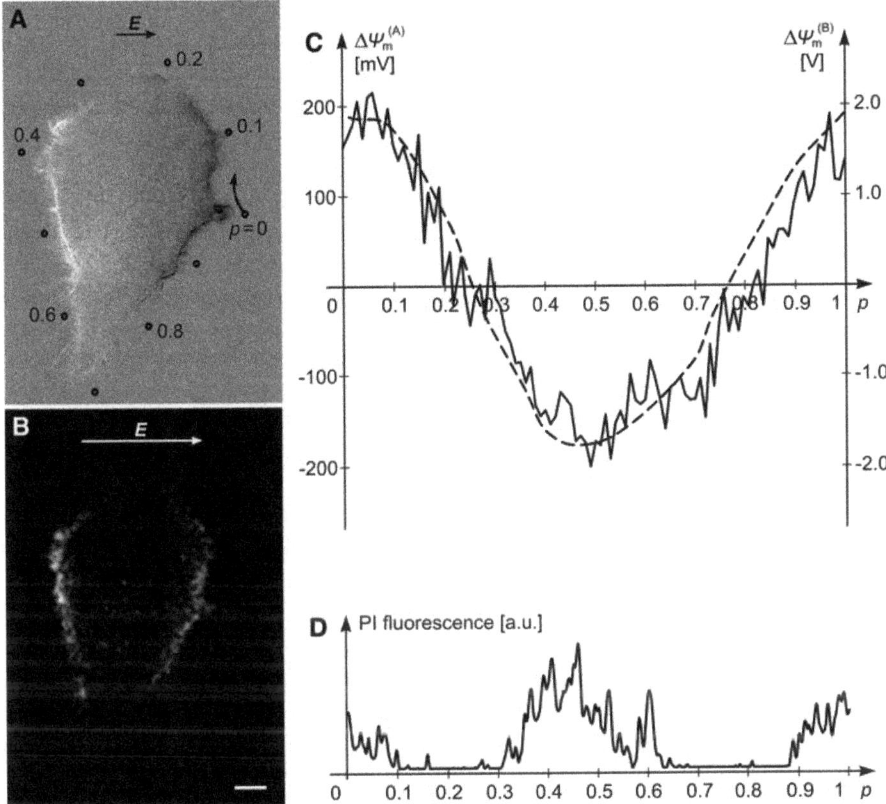

Fig. 6 Steady-state induced transmembrane voltage and electroporation of an irregular CHO (Chinese hamster ovarian) cell; (**a**) changes in fluorescence of di-8-ANEPPS (a fluorescent dye) caused by a non-porating 50 ms, 100 V/cm pulse; (**b**) transport of PI into the same cell caused by a porating 200 μs, 1000 V/cm pulse, visualized 200 ms after exposure; (**c**) steady-state induced transmembrane voltage along the path shown in (**a**) as measured (solid) and as predicted in a numerical model (dashed), the left ordinate axis scale corresponds to the 100 V/cm pulse amplitude used in (**a**), and the right ordinate axis scale to the 1000 V/cm used in (**b**); (**d**) propidium iodide fluorescence measured along the path shown in (**a**). (Reproduced with permission from Kotnik et al. 2010)

on so many factors that only the order of magnitude for their values can confidently be determined in general for a particular cell type and set of experimental conditions. In example, for a eukaryotic cell, detection occurs for electric fields resulting in TMV in hundreds of mV and irreversible damage for electric fields about three to five times higher than the minimum for detection (Teissié and Rols 1993; Towhidi et al. 2008; Kotnik et al. 2010) (Fig. 6).

As just mentioned, the critical electric field strength and the corresponding TMV for detectable electroporation depend on the cell type (Čemažar et al. 1998), transported molecule (Rols and Teissié 1998; Puc et al. 2003), and duration of exposure

(Hibino et al. 1993; Rols and Teissié 1998; Puc et al. 2003). These critical strengths are also influenced by cell size and local membrane curvature (Kakorin et al. 2003; Towhidi et al. 2008; Henslee et al. 2011), temperature (Kandušer et al. 2008; Polak et al. 2014), and osmotic pressure (Golzio et al. 1998). Our ability to accurately measure them is also subject to the sensitivity of the detection technique (Kotnik et al. 2000; Pakhomov et al. 2015; Wegner et al. 2015).

Direct microscopic observations reveal that electroporation-facilitated transport is strongly dependent on the size and charge of the molecules traversing the membrane through the electropores. Small molecules can enter the cell both during and after the pulse through the membrane regions with sufficiently high (positive or negative) TMV (Saulis and Saule 2012). For charged species, the entry is mostly electrophoretic (during the pulse as electrophoresis is field-driven) and proceeds, for the given net charge, from the side with the opposite polarity of the TMV. The transport after the pulse is mainly diffusive and takes place from both sides of the membrane (Gabriel and Teissié 1998; Gabriel and Teissie 1999; Pucihar et al. 2008). Experiments also suggest a non-negligible contribution of post-pulse TMV recovery in the transport of small charged species (Sözer et al. 2018a). Larger molecules and/or molecules exhibiting multiple charges enter only while the pulse is on and that only from the side with the opposite polarity of the TMV (e.g., negatively charged oligonucleotides enter from the side with positive TMV) (Paganin-Gioanni et al. 2011). For still larger molecules, such as plasmid DNA, electroporation initializes the transport, with longer (approx. ms duration) pulses required for sufficient electrophoretic drag on DNA to form a so-called DNA-membrane complex (Wolf et al. 1994; Rols and Teissié 1998; Mir et al. 1999). The subsequent DNA uptake is a slow process (as compared to the timescales of pore formation) and involves endocytosis for DNA uptake into the cytosol and its intracellular trafficking to the nucleus (Golzio et al. 2002; Rosazza et al. 2016a). A much more detailed account of the molecule-dependent specifics of electroporation-facilitated transport is provided by Rems and Miklavčič (2016).

Electroporation at the Level of Cell Suspensions and Tissues

This section focuses on cell electroporation in a multicellular environment. In particular, we consider cell suspensions of increasing densities and tissues, mainly from the perspective of the distribution of the electric field and the induced TMV (transmembrane voltage). The TMV is, in a multicellular environment, affected by the proximity of neighboring cells. If a cell is forcibly charged (e.g., when exposed to an electric pulse), redistribution of ions around the membrane will alter the electric field in the cell's immediate vicinity. This means that, provided the cells are in close proximity, the cells will *feel* the electric field perturbation caused by their neighboring cells, and that field will be superimposed to the already present externally applied electric field (Rems and Miklavčič 2016). The first time this has been described was in the study by Susil et al. (1998).

First, let us consider a cell suspension of ever-increasing density. When in a dilute suspension, cells can be considered sufficiently distant from one another not to, on average, sense the alteration of the electric field resulting from their neighboring cells. With increasing density of the suspension however, the induced TMV on a cell is more and more impacted by the field generated by other cells in its proximity. If spherical cells are in close contact, the maximum induced TMV will equal the product of the electric field strength and the cell's radius (TMV = E × R). This corresponds to a 1.5 times reduction with respect to the TMV on isolated cells (Susil et al. 1998; Pavlin et al. 2002). This means that cells in dense suspensions have to be electroporated at higher electric field strength than those in dilute suspensions, and that is due to the reduction in the induced TMV. This has been corroborated by both numerical calculations (Essone Mezeme et al. 2012b) and experiments (Canatella and Prausnitz 2001; Pucihar et al. 2007). Additionally, in experiments, the number of molecules that entered densely packed cells was determined to be lower, possibly due to a limited number of dye molecules available for transmembrane transport in the extracellular space. This is further potentiated by the phenomenon of cell swelling, following the cell exposure to electric pulses in a low-conductivity medium (Pavlin et al. 2005; Pucihar et al. 2007).

To elaborate a bit further on the phenomenon of cell swelling, note that electroporation of cells is accompanied by cytosolic solutes leaking into the extracellular medium and the (transient) inability of the cell to regulate homeostasis. In dense suspensions (when the volume fraction of cells is on par with the volume of the extracellular space), this leakage results in an increase in the suspension conductivity. In a very low-conductivity medium, the cell's inability to regulate cross-membrane transport results in a considerable amount of water to be taken up by the cell via the permeable membrane, thus disrupting the osmotic balance of the cell. This disruption in osmotic balance and water uptake cause the cell to swell up, while the release of intracellular ions to the cell's exterior medium results in medium conductivity changes. These changes and cell swelling were extensively studied by Pavlin et al. (Pavlin et al. 2005, 2007; Pavlin and Miklavcic 2008) during electroporation in dense suspensions of B16-F1 cells. They found a two-part increase in suspension conductivity: a large increase during the pulse cause by an increase in the conductance of the cell membranes and the second, gradual increase between the pulses caused by efflux of ions from the cells (Rems and Miklavčič 2016).

The contribution to bulk conductivity due to the increase in conductance of the individual cell membranes diminishes rapidly after the pulse during fast membrane recovery (pore closure). But since the cells are electroporated by the electric field imposed from multiple directions, this affects the conductivity of the cell suspension in an anisotropic fashion. During the pulse, the conductivity of the suspension is higher in the direction parallel to the electric field than in the direction perpendicular to it. Similar increase in conductivity anisotropic in nature can be observed in tissues (Essone Mezeme et al. 2012a).

Cells in suspensions can be brought into contact by manipulation (e.g., using electrophoresis, dielectrophoresis). The quality of thusly achieved contact is lacking as compared to the level of contact when cells are growing in monolayers or clusters. Neighboring cells in cell cultures form spontaneous contacts by connecting via different membrane structures. These contacts are formed rapidly (within 20 min) of cell plating (Ušaj et al. 2013). Kotnik et al. (Kotnik et al. 2010) studied the effect of cell connections on the induced TMV by means of an in vitro model of CHO cell clusters. Cells in such clusters are interconnected by gap junctions. These junctions form conductive pathways between the cytoplasm of one and another interconnected cell. By consequence, if the connected cells are exposed to a non-electroporative pulse, the cluster will act as a single large cell with a single cytoplasm. On the other hand, an electroporative pulse will result in a higher induced TMV across the membranes, impacting the dynamics of the opening and closing of gap junctions. As gap junctions become affected by electroporation, cells in the cluster start behaving as individual cells. This allows for molecular transport at membrane areas where the cells form connections with each other.

Although clustered cells behave as though they are individual entities with regard to electroporation, their shape and orientation in a plated cluster can vary. Cells are spread over the surface to which they are adhered and irregularly shaped, thereby increasing their size in comparison to cells in suspension. Electroporation can, in surface-adhered clusters of cells, be detected at lower electric field strength than in suspensions (Rols and Teissie 1989). But as we gradually increase the electric field strength, we notice that larger cells whose orientation is longitudinally aligned with the electric field (i.e., their long axis is parallel to the electric field vector) get electroporated at lower field strengths (Valič et al. 2003). Note that this holds for longer pulses (e.g., tens or hundreds of μs), but not for short (Dermol-Černe et al. 2020). Still, the amount of transmembrane molecular transport is lower for monolayers as compared to suspensions (Pucihar et al. 2008). Partially the reason is the effective reduction in the electric field by the neighboring cells (field shading) and partially the hindered diffusion of molecules (for both cell-to-cell and cell-to-surface contacts).

Overall, we can conclude that cells in an assembly will respond to electric pulses similarly as will single cells in dilute suspensions of nearly perfect spherical shape, provided we account for the local electric field distribution. With the latter, it is important to determine the actual field strength an individual cell is exposed to. The knowledge of basic characteristics of membrane electroporation as deduced from experiments on single cells is transferable, in a large degree, to cells in assemblies. However, the experimental studies also highlighted the importance of cell assembly heterogeneity (shape, size, orientation, and sensitivity to electric pulses), cell clustering (hindering inter-cell diffusion), and possible implications of other structures (e.g., the extracellular matrix, which is absent in suspended cells). The theoretical analysis of reversible and irreversible electroporation as correlated with the electric field distribution in tissue is therefore indispensable if we are to understand how different factors affect cell electroporation in assemblies and how to transfer (or "upscale") the knowledge obtained in vitro on cell suspensions to structures such as adhered cell monolayers or even tissues (Miklavcic et al. 1998; Miklavčič et al. 2000; Kranjc et al. 2016).

When theoretically modelling tissue in relation to studies of the phenomenon of electroporation, we most often work under the assumption that the tissue is a homogeneous structure. This homogeneity is meant in the sense of some hypothetical bulk electrical properties, i.e., conductivity and permittivity, which we can measure. By such analysis we of course completely neglect the cellular structure of tissue. Such simplifications can be justified by their benefits, chief among them being the computational cost which arises from modelling cells individually in a large volume. The second benefit stems from incomplete knowledge on how structures in tissue manifest themselves as bulk tissue properties.

In the bulk tissue model, the heterogeneity of the tissue structure is reflected in the functional dependence of these electrical properties on the frequency (Čorović et al. 2010) and on the applied field (Gabriel et al. 1996; Sel et al. 2005). The frequency dependency is tissue-specific, since tissues differ in their microscopic structure (shape, size, orientation, and cell density). Also, some tissues exhibit anisotropy due to cells' orientation in one preferential direction. Skeletal muscle tissue is a prime example of an anisotropic structure, where the long muscle fibers conduct the electric current more readily in the direction parallel than perpendicular to the fibers (Miklavčič et al. 2006).

If bulk properties of a tissue are known, it is possible to calculate the macroscopic electric field distribution inside with relative ease for an arbitrary electrode configuration. As tissues are also subject to a threshold phenomenon for electroporation with respect to the electric field strength, it is possible to do a comparison between numerically determined field strength distribution and the experimentally determined reversibly and irreversibly porated tissue area/volume, in order to determine the reversible and irreversible electroporation threshold, respectively (Miklavčič et al. 2000; Šel et al. 2007).

The daunting challenge here is, once having determined the "bulk" tissue electroporation thresholds of reversible and irreversible electroporation, how to then relate these bulk thresholds back with the micro-level, local electric field as experienced by individual cells. One simple possibility is to model a simplified, "average" representative cell in tissue and then treat the problem as though the cells are in a dense suspension. Thus it has been shown (Miklavčič et al. 2000) that the behavior of cells in tissue is not unlike the behavior of cells in dense suspensions, at least when considering the electroporation threshold.

The simplified treatment just described does however not account for local tissue conductivity changes as resulting from electroporation. If a square-wave pulse is applied by means of needle electrodes inserted into tissue, the electric field distribution will be inhomogeneous. Do note however that this heterogeneous current density is still present, although less pronounced, in other electrode geometries, including a seemingly homogenous field-producing configuration of plate electrodes. In those regions exposed to above reversible electroporation threshold field strengths, tissue conductivity will increase due to increase in cell membrane conductivity. These localized changes in conductivity consequently change the electric field distribution both during the pulse and between individual pulses (provided several are applied in a train with inter-pulse pause short enough not to permit membrane resealing). The electric field strength thus becomes higher in those regions which have not yet experienced noticeable electroporation and remain less conductive. This results in a gradual propagation (though self-limited) of the electroporated tissue area. These conductivity changes have profound and significant implications for determining thresholds for tissue electroporation based on comparison of experimental results and numerical modelling. The importance of accounting for conductivity changes and the local field strength distribution in tissue has been highlighted in numerous studies (Miklavčič et al. 2000; Sel et al. 2005; Neal 2nd et al. 2012; Neal et al. 2015; Corovic et al. 2013; Bhonsle et al. 2015). One of the difficulties in estimating tissue conductivity changes is that they form an inhomogeneous area within the target tissue and cannot be directly resolved (spatially) by measuring the electric current between the electrodes. Important progress in this direction has nevertheless been achieved recently by a novel technique permitting imaging of local tissue conductivity during pulse application using electric impedance tomography (EIT) (Davalos et al. 2002, 2004; Granot et al. 2009). This technique was then further advanced and developed into a complementary technique of magnetic resonance electric impedance tomography (MREIT) (Kranjc et al. 2015, 2016), permitting imaging of electric current density distributions in tissue during electroporation.

If considering electroporation in connection with food processing applications, by far the most common target tissue for the treatment will be plant tissue. As shown by the schematic representation of plant tissue in Fig. 7 (left-hand side), the most distinguishing feature of plant tissue compared to animal tissues is the structure of the extracellular matrix, i.e., the cell wall. The cell wall is there not only to support the plasma membrane and give the cell its shape but also as a warrant of structural integrity, enabling the cell to withstand pressure differences across the membrane that can be enormous (the cell would otherwise burst in an instance due to turgor). It acts as a selective filter, allowing water and ions to freely diffuse, while posing a limiting factor to transmembrane transport of large molecules (over approx. 20 kDa in size). However, smaller molecules (up to 10 kDa) can pass between cells of some

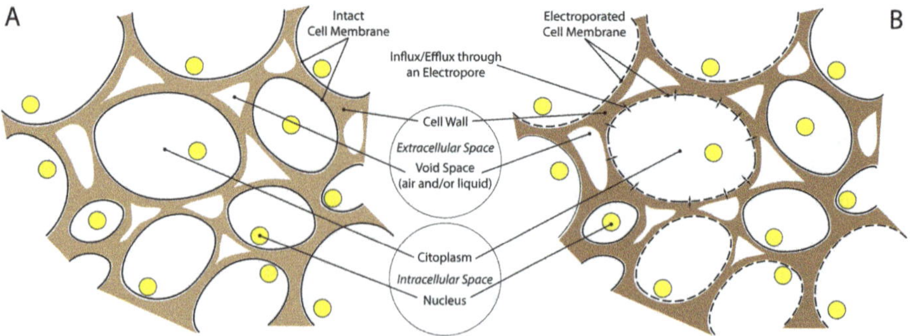

Fig. 7 A considerably simplified schematic representation of tissue comprising cells with intact membrane (left) and electroporated membrane (right). The cell membrane delineates the intracellular and the extracellular space

higher plants through structures known as plasmodesmata. Plasmodesmata are specialized cell-to-cell junctions extending through the cell wall (Lodish et al. 2008). Note that these junctions are not shown in Fig. 7.

If plant tissue is intact, its plasma membrane represents the greatest resistance to cell-to-cell transport and to (transmembrane) transport between the intracellular and the extracellular space. Moreover, there is the additional hindrance and filtering behavior of the cell wall. Both of these limiting structures must be considered when studying mass transport in plant tissues (Buttersack and Basler 1991). This must be contrasted to the problem of animal tissues (Dermol-Černe et al. 2018; Dermol-Černe and Miklavčič 2018), where, although an extracellular matrix is present and to a degree is a hinderance to transport, cells are not embedded in an extracellular structure akin to a cell wall. The cell wall in this sense presents an additional level of complexity (Dymek et al. 2015) that must be considered when studying electroporation-mediated mass transport to, from, and within plant tissues (Janositz and Knorr 2010; Janositz et al. 2011).

Figure 7 (right-hand side) is a simplified representation of electroporated plant tissue. Plasma membrane's permeability is illustrated as increased by the treatment and has lost the ability to selectively control the influx/efflux of water and of solutes that are able to pass through the membrane (subject to their hydrodynamic size). According to current understanding (Galindo et al. 2008; Ganeva et al. 2014; Stirke et al. 2014), electroporation may also affect the cell wall by either decreasing or increasing its permeability. Recent studies' results are contradictory, highlighting the need for further studies in the fundamentals of electroporation-based phenomena, something that has historically not been receiving due attention in the field of food processing applications of electroporation (this trend has recently seen a turn for the better). Nevertheless, according to recent literature, we can summarize that the limited electroporation effect on the cell wall is dependent on cell configuration (tissues, monolayers, cells in suspension), the organism (yeast or plant species), and the treatment protocol. Observed increase in permeability of electroporation-treated (plant) tissue remains attributable in the largest extent to the increase in permeability of the cell membrane.

As mentioned in preceding sections of this chapter, electroporation can irreversibly affect a cell membrane, causing it to ultimately break down. Plant cells are in no lesser extent subject to this phenomenon than animal cells and tissues. However, in case of plant tissue, the cytosol, parts of the disintegrated membrane, and the intracellular material will most likely remain entrapped in the extracellular matrix following irreversible electroporation. Given the enormous turgor pressures in some of the constituent cells in plant tissues, it is questionable whether we can speak of reversible electroporation, given that once the osmotic balance of a cell is disrupted, the cell is no longer able to control the influx or efflux of water.

Conclusions

Even though the mechanisms of electroporation have been investigated for at least four decades (and in food processing, applying electric fields to food matrices has an even longer tradition), there are still open questions remaining to be answered. One of the main reasons why understanding electroporation is posing a challenge is the wide range of spatial scales (from nm-thick membranes through µm-sized cells and up to cm-scale tissue segments) and temporal scales involved (from ns to hrs, or even days, as evident from Table 1). Therefore, investigation of electroporation necessitates a multi-scale modelling approach, ranging from MD simulations to large-scale continuum models of cells and tissues, coupled with systematic experiments at every stage of the modelling way. In recent years, such an approach has indeed resulted in significant progress (Kotnik et al. 2019).

The importance of quantifying transport, both the number of transported solutes into the cell and its temporal dynamics, is increasingly gaining recognition (Pakhomov et al. 2015; Sözer et al. 2017, 2018b). While there is a general consensus that the TMV induced by an electric field promotes formation of pores in the lipid bilayer, the contribution of other mechanisms to cell membrane permeabilization remains to be elucidated. Recently, the long-standing assumption that the pores formed due to the pulse application are also the primary conduits of transport for seconds and up to minutes after the pulse is increasingly being challenged (Smith et al. 2004; Son et al. 2014), since MD simulations failed to indicate that pores retain any stability once the induced TMV vanishes or drops to a very low level. One possible explanation could be that these "primary" pores evolve into more complex, stable pores that involve both lipids and other molecules. However, the molecular organization of these secondary, complex pores, remains elusive (Weaver and Vernier 2017). Another hypothesis is that lipid oxidation caused by exposure to the imposed electric field results in spontaneous formation of pores in areas of oxidized membrane lesions. Both possible explanations just mentioned imply that we may have to distinguish between at least two different types of pores. This has previously been proposed (Neumann et al. 1998; Pavlin and Miklavcic 2008), but a description of the underlying pore structure is lacking. Another possibility still would be that the non-transient (i.e., longer-lived) permeability following the pulse does not involve pores at all but is instead facilitated by leaky peroxidized membrane lesions and modified membrane proteins (Teissié et al. 2005; Mir 2020).

Answering these important and at least partially open questions is prerequisite to optimizing existing and developing new electroporation-based technologies and treatments, food processing applications included.

We conclude with a brief mention of the problems and challenges related to electroporation as applied to food processing (Lebovka and Vorobiev 2016). The nature of electroporation when applied to food tissues can be complicated, as it includes different scales, ranging all the way from single cells to suspensions and tissues with elaborate structure and profoundly expressed heterogeneity. What effects electroporation will have on plant material depends not only on the size of the constituent cells but also on their orientation and spatial distribution. Also important are electrophysical parameters of cells, the pH of media, and the presence of osmotic agents. This means that optimization of electroporation treatment and electroporation-assisted processing of plant tissues entails, among other, precise adaptation of electroporation protocols, selection of the optimal electric field strength, optimal pulse duration, pulse repetition time, and the most suitable temperature. Mathematical modelling can provide valuable information when optimizing electroporation protocols in terms of both optimal power consumption and preservation of the food material quality. Additionally, numerical simulations can be useful when describing electroporation effects in the presence of complex transport processes and changes in temperature, electrical conductivity, electric field strength, and spatial distribution of the electrolytes in tissue during electroporation. In recent decades, various models and simulations were performed, aiming at optimizing electroporation for treatment of liquid foods, geometrical optimization of the treatment chambers, and prediction of the electric field distribution, flow velocity, and temperature inside the treatment chamber. Figure 8 illustrates electric field and temperature distributions in a parallel-plate (a–d) and a colinear (e–f) treatment chamber at the end of a single pulse of 100 μs with the fluid flow field taken into account. Notice the inhomogeneity that is particularly present in the colinear treatment chamber and is less pronounced in the parallel-plate flow chamber, though some heterogeneity is present there as well, in particular at the electrode edges (see the zoomed-in pair of subfigures, Fig. 8b, d). Since the available computing power available to researchers for modelling and simulation has greatly increased over the last couple of decades, it is now possible to construct and analyze using very detailed mechanistic models of equipment and processes, historically considered too demanding for computation. New, multi-physics approaches (i.e., modelling of highly interdependent, coupled quantities such as flow, temperature, electric field/current and conductivity, etc.) can now be employed for studying spatial distributions in systems of interest also in the temporal domain (dynamics!) in an efficient manner, as opposed to computing only the steady state or relying on experiment-based empirical models.

In preparing such a broad overview chapter on the fundamental principles underpinning electroporation as a phenomenon, one is faced with difficulties in summarizing existing findings, as experimental detail is lacking in many reports, making comparison of results from different studies difficult, if not impossible. To address this, it is extremely important that further studies published in future articles follow the recently published recommendation papers (Campana et al. 2016; Raso et al. 2016; Cemazar et al. 2018). This holds in particular for the problem of evaluating the local electric field strength, more often than it should be simply estimated as the voltage-to-distance ratio, in spite of diverse electrode geometries for many of which such estimation is an oversimplification.

Acknowledgments The authors would like to acknowledge the financial support through research programs and projects granted by the Slovenian Research Agency (ARRS), namely, the research program P2-0249 and the postdoctoral project Z7-1886.

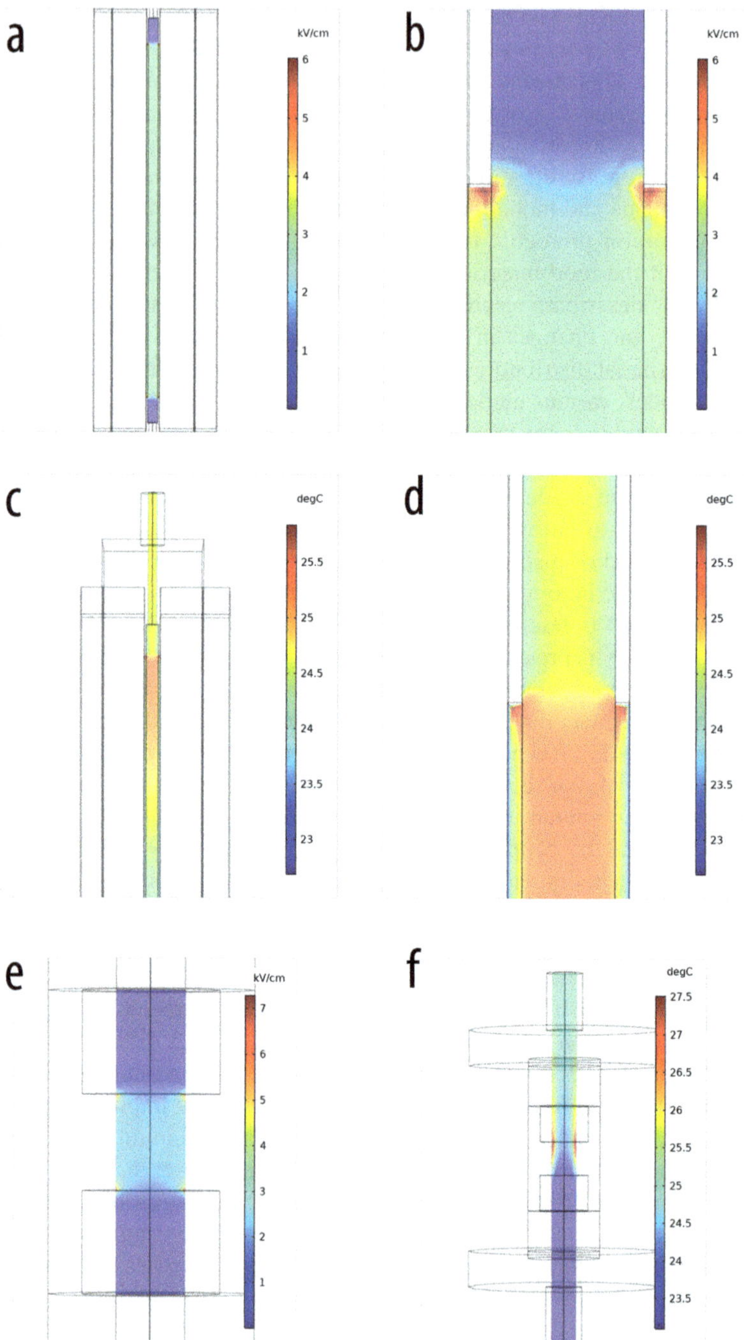

Fig. 8 Electric field strength and temperature distribution in a parallel-plate and colinear flow-through treatment chamber, in aid of illustrating one of the remaining challenges in electroporation research in the domain of food processing. The electrical, thermal, and flow relations in continuous flow treatment chambers are complex, and necessitate advanced multi-physics computational models for accurate description of conditions impacting treatment efficiency

References

Abidor IG, Arakelyan VB, Chernomordik LV et al (1979) Electric breakdown of bilayer lipid membranes: I. The main experimental facts and their qualitative discussion. J Electroanalytical Chem Interfacial Electrochem 104:37–52. https://doi.org/10.1016/S0022-0728(79)81006-2

Ade-Omowaye BIO, Angersbach A, Taiwo KA, Knorr D (2001) Use of pulsed electric field pre-treatment to improve dehydration characteristics of plant based foods. Trends Food Sci Technol 12:285–295. https://doi.org/10.1016/S0924-2244(01)00095-4

Álvarez I, Condón S, Raso J (2006) Microbial inactivation by pulsed electric fields. In: Raso J, Heinz V (eds) Pulsed electric fields technology for the food industry, Springer, Boston, pp 97–129

Aranda-Espinoza H, Bermudez H, Bates FS, Discher DE (2001) Electromechanical limits of polymersomes. Phys Rev Lett 87:208301. https://doi.org/10.1103/PhysRevLett.87.208301

Aronsson K, Rönner U, Borch E (2005) Inactivation of Escherichia coli, Listeria innocua and Saccharomyces cerevisiae in relation to membrane permeabilization and subsequent leakage of intracellular compounds due to pulsed electric field processing. Int J Food Microbiol 99:19–32. https://doi.org/10.1016/j.ijfoodmicro.2004.07.012

Asavasanti S, Ersus S, Ristenpart W et al (2010) Critical electric field strengths of onion tissues treated by pulsed electric fields. J Food Sci 75:E433–E443. https://doi.org/10.1111/j.1750-3841.2010.01768.x

Benz R, Beckers F, Zimmermann U (1979) Reversible electrical breakdown of lipid bilayer membranes: a charge-pulse relaxation study. J Membrain Biol 48:181–204. https://doi.org/10.1007/BF01872858

Bermudez-Aguirre D, Dunne CP, Barbosa-Canovas GV (2012) Effect of processing parameters on inactivation of Bacillus cereus spores in milk using pulsed electric fields. Int Dairy J 24:13–21. https://doi.org/10.1016/j.idairyj.2011.11.003

Bhonsle SP, Arena CB, Sweeney DC, Davalos RV (2015) Mitigation of impedance changes due to electroporation therapy using bursts of high-frequency bipolar pulses. Biomed Eng Online 14:S3. https://doi.org/10.1186/1475-925X-14-S3-S3

Blahovec J, Vorobiev E, Lebovka N (2017) Pulsed electric fields pretreatments for the cooking of foods. Food Eng Rev 9:71–81. https://doi.org/10.1007/s12393-017-9160-z

Blumrosen G, Abazari A, Golberg A, et al (2014) Efficient procedure and methods to determine critical electroporation parameters. In: 2014 IEEE 27th international symposium on computer-based medical systems. pp 314–318

Böckmann RA, de Groot BL, Kakorin S et al (2008) Kinetics, statistics, and energetics of lipid membrane electroporation studied by molecular dynamics simulations. Biophys J 95:1837–1850. https://doi.org/10.1529/biophysj.108.129437

Botero-Uribe M, Fitzgerald M, Gilbert RG, Midgley J (2017) Effect of pulsed electrical fields on the structural properties that affect french fry texture during processing. Trends Food Sci Technol 67:1–11. https://doi.org/10.1016/j.tifs.2017.05.016

Bradley CJ, Haines DE (2020) Pulsed field ablation for pulmonary vein isolation in the treatment of atrial fibrillation. J Cardiovasc Electrophysiol 31:2136–2147. https://doi.org/10.1111/jce.14414

Breton M, Mir LM (2018) Investigation of the chemical mechanisms involved in the electropulsation of membranes at the molecular level. Bioelectrochemistry 119:76–83. https://doi.org/10.1016/j.bioelechem.2017.09.005

Broderick KE, Humeau LM (2015) Electroporation-enhanced delivery of nucleic acid vaccines. Expert Rev Vaccines 14:195–204. https://doi.org/10.1586/14760584.2015.990890

Buttersack C, Basler W (1991) Hydraulic conductivity of cell-walls in sugar-beet tissue. Plant Sci 76:229–237. https://doi.org/10.1016/0168-9452(91)90145-X

Campana LG, Clover AJP, Valpione S et al (2016) Recommendations for improving the quality of reporting clinical electrochemotherapy studies based on qualitative systematic review. Radiol Oncol 50:1–13. https://doi.org/10.1515/raon-2016-0006

Campana LG, Edhemovic I, Soden D et al (2019a) Electrochemotherapy - emerging applications technical advances, new indications, combined approaches, and multi-institutional collaboration. Eur J Surg Oncol 45:92–102. https://doi.org/10.1016/j.ejso.2018.11.023

Campana LG, Miklavčič D, Bertino G et al (2019b) Electrochemotherapy of superficial tumors – current status: basic principles, operating procedures, shared indications, and emerging applications. Semin Oncol 46:173–191. https://doi.org/10.1053/j.seminoncol.2019.04.002

Canatella PJ, Prausnitz MR (2001) Prediction and optimization of gene transfection and drug delivery by electroporation. Gene Ther 8:1464–1469. https://doi.org/10.1038/sj.gt.3301547

Casciola M, Tarek M (2016) A molecular insight into the electro-transfer of small molecules through electropores driven by electric fields. Biochim Biophys Acta 1858:2278–2289. https://doi.org/10.1016/j.bbamem.2016.03.022

Casciola M, Bonhenry D, Liberti M et al (2014) A molecular dynamic study of cholesterol rich lipid membranes: comparison of electroporation protocols. Bioelectrochemistry 100:11–17. https://doi.org/10.1016/j.bioelechem.2014.03.009

Čemažar M, Jarm T, Miklavčič D et al (1998) Effect of electric-field intensity on Electropermeabilization and Electrosensitivity of various tumor-cell lines in vitro. Electro Magnetobiol 17:263–272. https://doi.org/10.3109/15368379809022571

Cemazar M, Sersa G, Frey W et al (2018) Recommendations and requirements for reporting on applications of electric pulse delivery for electroporation of biological samples. Bioelectrochemistry 122:69–76. https://doi.org/10.1016/j.bioelechem.2018.03.005

Chalermchat Y, Dejmek P (2005) Effect of pulsed electric field pretreatment on solid–liquid expression from potato tissue. J Food Eng 71:164–169. https://doi.org/10.1016/j.jfoodeng.2004.10.030

Čorović S, Županič A, Kranjc S et al (2010) The influence of skeletal muscle anisotropy on electroporation: in vivo study and numerical modeling. Med Biol Eng Comput 48:637–648. https://doi.org/10.1007/s11517-010-0614-1

Čorović S, Mir LM, Miklavčič D (2012) In vivo muscle electroporation threshold determination: realistic numerical models and in vivo experiments. J Membr Biol 245:509–520. https://doi.org/10.1007/s00232-012-9432-8

Corovic S, Lackovic I, Sustaric P et al (2013) Modeling of electric field distribution in tissues during electroporation. Biomed Eng Online 12:16. https://doi.org/10.1186/1475-925X-12-16

Crowley JM (1973) Electrical breakdown of bimolecular lipid membranes as an electromechanical instability. Biophys J 13:711–724

Cruzeiro-Hansson L, Mouritsen OG (1988) Passive ion permeability of lipid membranes modelled via lipid-domain interfacial area. Biochim Biophys Acta Biomembr 944:63–72. https://doi.org/10.1016/0005-2736(88)90316-1

Davalos RV, Rubinsky B, Otten DM (2002) A feasibility study for electrical impedance tomography as a means to monitor tissue electroporation for molecular medicine. IEEE Trans Biomed Eng 49:400–403. https://doi.org/10.1109/10.991168

Davalos RV, Otten DM, Mir LM, Rubinsky B (2004) Electrical impedance tomography for imaging tissue electroporation. IEEE Trans Biomed Eng 51:761–767. https://doi.org/10.1109/TBME.2004.824148

Dean DA (2013) Cell-specific targeting strategies for electroporation-mediated gene delivery in cells and animals. J Membr Biol 246:737–744. https://doi.org/10.1007/s00232-013-9534-y

Dehez F, Delemotte L, Kramar P et al (2014) Evidence of conducting hydrophobic nanopores across membranes in response to an electric field. J Phys Chem C 118:6752–6757. https://doi.org/10.1021/jp4114865

Delemotte L, Tarek M (2012) Molecular dynamics simulations of lipid membrane electroporation. J Membr Biol 245:531–543. https://doi.org/10.1007/s00232-012-9434-6

Demiryurek Y, Nickaeen M, Zheng M et al (2015) Transport, resealing, and re-poration dynamics of two-pulse electroporation-mediated molecular delivery. Biochim Biophys Acta Biomembr 1848:1706–1714. https://doi.org/10.1016/j.bbamem.2015.04.007

Dermol-Černe J, Miklavčič D (2018) From cell to tissue properties—modeling skin electroporation with pore and local transport region formation. IEEE Trans Biomed Eng 65:458–468. https://doi.org/10.1109/TBME.2017.2773126
Dermol-Černe J, Vidmar J, Ščančar J et al (2018) Connecting the in vitro and in vivo experiments in electrochemotherapy - a feasibility study modeling cisplatin transport in mouse melanoma using the dual-porosity model. J Control Release 286:33–45. https://doi.org/10.1016/j.jconrel.2018.07.021
Dermol-Černe J, Batista Napotnik T, Reberšek M, Miklavčič D (2020) Short microsecond pulses achieve homogeneous electroporation of elongated biological cells irrespective of their orientation in electric field. Sci Rep 10:9149. https://doi.org/10.1038/s41598-020-65830-3
Donsi F, Ferrari G, Pataro G (2010) Applications of pulsed electric field treatments for the enhancement of mass transfer from vegetable tissue. Food Eng Rev 2:109–130. https://doi.org/10.1007/s12393-010-9015-3
Dymek K (2015) Impregnation of leaf tissues and its consequences on metabolism and freezing; study on vacuum impregnation and pulsed electric field treatment. Dissertation, Lund University
Dymek K, Dejmek P, Panarese V et al (2012) Effect of pulsed electric field on the germination of barley seeds. LWT-Food Sci Technol 47:161–166. https://doi.org/10.1016/j.lwt.2011.12.019
Dymek K, Rems L, Zorec B et al (2015) Modeling electroporation of the non-treated and vacuum impregnated heterogeneous tissue of spinach leaves. Innovative Food Sci Emerg Technol 29:55–64. https://doi.org/10.1016/j.ifset.2014.08.006
Ehrenberg B, Farkas DL, Fluhler EN, Lojewska Z (1987) Membrane potential induced by external electric field pulses can be followed with a potentiometric dye. Biophys J 51:833–837
El Ouagari K, Gabriel B, Benoist H, Teissié J (1993) Electric field-mediated glycophorin insertion in cell membrane is a localized event. Biochim Biophys Acta Biomembr 1151:105–109. https://doi.org/10.1016/0005-2736(93)90077-D
Escoffre J-M, Nikolova B, Mallet L et al (2012) New insights in the gene Electrotransfer process: evidence for the involvement of the plasmid DNA topology. Curr Gene Ther 12:417–422
Essone Mezeme M, Kranjc M, Bajd F et al (2012a) Assessing how electroporation affects the effective conductivity tensor of biological tissues. Appl Phys Lett 101:213702–213704. https://doi.org/10.1063/1.4767450
Essone Mezeme M, Pucihar G, Pavlin M et al (2012b) A numerical analysis of multicellular environment for modeling tissue electroporation. Appl Phys Lett 100:143701–143704. https://doi.org/10.1063/1.3700727
Fincan M, Dejmek P (2003) Effect of osmotic pretreatment and pulsed electric field on the viscoelastic properties of potato tissue. J Food Eng 59:169–175. https://doi.org/10.1016/S0260-8774(02)00454-5
Fincan M, DeVito F, Dejmek P (2004) Pulsed electric field treatment for solid-liquid extraction of red beetroot pigment. J Food Eng 64:381–388. https://doi.org/10.1016/j.jfoodeng.2003.11.006
Frandsen SK, Vissing M, Gehl J (2020) A comprehensive review of calcium electroporation—a novel cancer treatment modality. Cancers (Basel) 12. https://doi.org/10.3390/cancers12020290
Freeman SA, Wang MA, Weaver JC (1994) Theory of electroporation of planar bilayer membranes: predictions of the aqueous area, change in capacitance, and pore-pore separation. Biophys J 67:42–56
Gabriel B, Teissié J (1998) Mammalian cell electropermeabilization as revealed by millisecond imaging of fluorescence changes of ethidium bromide in interaction with the membrane 1 Presented at the 14th BES symposium in Vingstedcentret (Denmark), 1998.1. Bioelectrochem Bioenerg 47:113–118. https://doi.org/10.1016/S0302-4598(98)00174-3
Gabriel B, Teissie J (1999) Time courses of mammalian cell electropermeabilization observed by millisecond imaging of membrane property changes during the pulse. Biophys J 76:2158–2165
Gabriel S, Lau RW, Gabriel C (1996) The dielectric properties of biological tissues: II. Measurements in the frequency range 10 Hz to 20 GHz. Phys Med Biol 41:2251–2269. https://doi.org/10.1088/0031-9155/41/11/002

Galindo FG, Vernier PT, Dejmek P et al (2008) Pulsed electric field reduces the permeability of potato cell wall. Bioelectromagnetics 29:296–301. https://doi.org/10.1002/bem.20394
Ganeva V, Galutzov B, Teissie J (2014) Evidence that pulsed electric field treatment enhances the cell wall porosity of yeast cells. Appl Biochem Biotechnol 172:1540–1552. https://doi.org/10.1007/s12010-013-0628-x
Geboers B, Scheffer HJ, Graybill PM et al (2020) High-voltage electrical pulses in oncology: irreversible electroporation, electrochemotherapy, gene electrotransfer, electrofusion, and electroimmunotherapy. Radiology 295:254–272. https://doi.org/10.1148/radiol.2020192190
Golberg A, Rae CS, Rubinsky B (2012) Listeria monocytogenes cell wall constituents exert a charge effect on electroporation threshold. Biochim Biophys Acta-Biomembr 1818:689–694. https://doi.org/10.1016/j.bbamem.2011.11.003
Golzio M, Mora M-P, Raynaund C et al (1998) Control by osmotic pressure of voltage - induced permeabilization and gene transfer in mammalian cells. Biophys J 74:3015–3022
Golzio M, Teissie J, Rols MP (2002) Direct visualization at the single-cell level of electrically mediated gene delivery. Proc Natl Acad Sci U S A 99:1292–1297. https://doi.org/10.1073/pnas.022646499
Gómez Galindo F (2016) Responses of plant cells and tissues to pulsed electric field treatments. In: Miklavcic D (ed) Handbook of electroporation. Springer, Cham, pp 1–15
Granot Y, Ivorra A, Maor E, Rubinsky B (2009) In vivo imaging of irreversible electroporation by means of electrical impedance tomography. Phys Med Biol 54:4927–4943. https://doi.org/10.1088/0031-9155/54/16/006
Guerrero-Beltran JA, Sepulveda DR, Gongora-Nieto MM et al (2010) Milk thermization by pulsed electric fields (PEF) and electrically induced heat. J Food Eng 100:56–60. https://doi.org/10.1016/j.jfoodeng.2010.03.027
Gurtovenko AA, Vattulainen I (2005) Pore formation coupled to ion transport through lipid membranes as induced by transmembrane ionic charge imbalance: atomistic molecular dynamics study. J Am Chem Soc 127:17570–17571. https://doi.org/10.1021/ja053129n
Heller R, Grasso RJ (1990) Transfer of human membrane surface components by incorporating human cells into intact animal tissue by cell-tissue electrofusion in vivo. Biochim Biophys Acta Biomembr 1024:185–188. https://doi.org/10.1016/0005-2736(90)90223-B
Henslee BE, Morss A, Hu X et al (2011) Electroporation dependence on cell size: optical tweezers study. Anal Chem 83:3998–4003. https://doi.org/10.1021/ac1019649
Hibino M, Itoh H, Kinosita K (1993) Time courses of cell electroporation as revealed by submicrosecond imaging of transmembrane potential. Biophys J 64:1789–1800
Hjouj M, Last D, Guez D et al (2012) MRI study on reversible and irreversible electroporation induced blood brain barrier disruption. PLoS One 7:e42817. https://doi.org/10.1371/journal.pone.0042817
Ho M-C, Casciola M, Levine ZA, Vernier PT (2013) Molecular dynamics simulations of ion conductance in field-stabilized nanoscale lipid Electropores. J Phys Chem B 117:11633–11640. https://doi.org/10.1021/jp401722g
Hu N, Yang J, Joo SW et al (2013) Cell electrofusion in microfluidic devices: a review. Sens Actuator B-Chem 178:63–85. https://doi.org/10.1016/j.snb.2012.12.034
Janositz A, Knorr D (2010) Microscopic visualization of pulsed electric field induced changes on plant cellular level. Innov Food Sci Emerg Technol 11:592–597. https://doi.org/10.1016/j.ifset.2010.07.004
Janositz A, Noack A-K, Knorr D (2011) Pulsed electric fields and their impact on the diffusion characteristics of potato slices. LWT-Food Sci Technol 44:1939–1945. https://doi.org/10.1016/j.lwt.2011.04.006
Jin TZ (2017) Pulsed electric fields for pasteurization: defining processing conditions. In: Miklavčič D (ed) Handbook of electroporation. Springer, Cham, pp 2271–2295
Kakorin S, Liese T, Neumann E (2003) Membrane curvature and high-field electroporation of lipid bilayer vesicles. J Phys Chem B 107:10243–10251. https://doi.org/10.1021/jp022296w

Kandušer M, Šentjurc M, Miklavčič D (2008) The temperature effect during pulse application on cell membrane fluidity and permeabilization. Bioelectrochemistry 74:52–57. https://doi.org/10.1016/j.bioelechem.2008.04.012
Kennedy SM, Ji Z, Hedstrom JC et al (2008) Quantification of electroporative uptake kinetics and electric field heterogeneity effects in cells. Biophys J 94:5018–5027. https://doi.org/10.1529/biophysj.106.103218
Knorr D, Geulen M, Grahl T, Sitzmann W (1994) Food application of high-electric-field pulses. Trends Food Sci Technol 5:71–75. https://doi.org/10.1016/0924-2244(94)90240-2
Kotnik T, Miklavčič D (2006) Theoretical evaluation of voltage inducement on internal membranes of biological cells exposed to electric fields. Biophys J 90:480–491. https://doi.org/10.1529/biophysj.105.070771
Kotnik T, Macek-Lebar A, Miklavcic D, Mir LM (2000) Evaluation of cell membrane electropermeabilization by means of a nonpermeant cytotoxic agent. BioTechniques 28:921–926. https://doi.org/10.2144/00285st05
Kotnik T, Pucihar G, Miklavcic D (2010) Induced transmembrane voltage and its correlation with electroporation-mediated molecular transport. J Membr Biol 236:3–13. https://doi.org/10.1007/s00232-010-9279-9
Kotnik T, Kramar P, Pucihar G et al (2012) Cell membrane electroporation – Part 1: the phenomenon. IEEE Electr Insul Mag 28:14–23. https://doi.org/10.1109/MEI.2012.6268438
Kotnik T, Frey W, Sack M et al (2015) Electroporation-based applications in biotechnology. Trends Biotechnol 33:480–488. https://doi.org/10.1016/j.tibtech.2015.06.002
Kotnik T, Rems L, Tarek M, Miklavčič D (2019) Membrane electroporation and electropermeabilization: mechanisms and models. Annu Rev Biophys 48:63–91. https://doi.org/10.1146/annurev-biophys-052118-115451
Kranjc M, Markelc B, Bajd F et al (2015) In situ monitoring of electric field distribution in mouse tumor during electroporation. Radiology 274:115–123. https://doi.org/10.1148/radiol.14140311
Kranjc M, Bajd F, Serša I et al (2016) Electric field distribution in relation to cell membrane electroporation in potato tuber tissue studied by magnetic resonance techniques. Innov Food Sci Emerg Technol 37(Part C):384–390. https://doi.org/10.1016/j.ifset.2016.03.002
Lambricht L, Lopes A, Kos S et al (2016) Clinical potential of electroporation for gene therapy and DNA vaccine delivery. Expert Opin Drug Deliv 13:295–310. https://doi.org/10.1517/17425247.2016.1121990
Lebovka N, Vorobiev E (2016) Mathematical models of pulsed electric field treatment of plant tissues and simulation of related phenomena. In: Miklavcic D (ed) Handbook of electroporation. Springer International Publishing, Cham, pp 1–25
Lebovka NI, Shynkaryk NV, Vorobiev E (2007) Pulsed electric field enhanced drying of potato tissue. J Food Eng 78:606–613. https://doi.org/10.1016/j.jfoodeng.2005.10.032
Levine ZA, Vernier PT (2010) Life cycle of an Electropore: field-dependent and field-independent steps in pore creation and annihilation. J Membrane Biol 236:27–36. https://doi.org/10.1007/s00232-010-9277-y
Lodish HF, Berk A, Kaiser CA et al (2008) Molecular cell biology. W.H. Freeman, New York
Mahnič-Kalamiza S, Vorobiev E (2015) Numerical dual-porosity model of solid-liquid expression from Electroporated bio-solids. In: Proceedings of the FILTECH 2015 conference. Cologne, Germany, p 14
Mahnič-Kalamiza S, Vorobiev E, Miklavčič D (2014) Electroporation in food processing and biorefinery. J Membr Biol 247:1279–1304. https://doi.org/10.1007/s00232-014-9737-x
Maor E, Sugrue A, Witt C et al (2019) Pulsed electric fields for cardiac ablation and beyond: a state-of-the-art review. Heart Rhythm 16:1112–1120. https://doi.org/10.1016/j.hrthm.2019.01.012
Marsellés-Fontanet AR, Elez-Martínez P, Martín-Belloso O (2012) Juice preservation by pulsed electric fields. Stewart Postharvest Rev 8:1–4. https://doi.org/10.2212/spr.2012.2.3

Mhemdi H, Bals O, Vorobiev E (2016) Combined pressing-diffusion technology for sugar beets pretreated by pulsed electric field. J Food Eng 168:166–172. https://doi.org/10.1016/j.jfoodeng.2015.07.034

Miklavcic D, Beravs K, Semrov D et al (1998) The importance of electric field distribution for effective in vivo electroporation of tissues. Biophys J 74:2152–2158

Miklavčič D, Šemrov D, Mekid H, Mir LM (2000) A validated model of in vivo electric field distribution in tissues for electrochemotherapy and for DNA electrotransfer for gene therapy. Biochim Biophys Acta Gen Subj 1523:73–83. https://doi.org/10.1016/S0304-4165(00)00101-X

Miklavčič D, Pavšelj N, Hart FX (2006) Electric properties of tissues. In: Wiley encyclopedia of biomedical engineering. Wiley, New York, pp 3578–3589

Miklavčič D, Mali B, Kos B et al (2014) Electrochemotherapy: from the drawing board into medical practice. Biomed Eng Online 13:29. https://doi.org/10.1186/1475-925X-13-29

Miller L, Leor J, Rubinsky B (2005) Cancer cells ablation with irreversible electroporation: technology in cancer research & treatment. https://doi.org/10.1177/153303460500400615

Min S, Evrendilek GA, Zhang HQ (2007) Pulsed electric fields: processing system, microbial and enzyme inhibition, and shelf life extension of foods. IEEE Trans Plasma Sci 35:59–73. https://doi.org/10.1109/TPS.2006.889290

Mir LM (2020) Electroporation and electropermeabilisation - pieces of puzzle put together. In: Kramar P, Miklavčič D (eds) Electroporation-based technologies and treatments, 1st edn. Založba FE, Ljubljana, pp 178–181

Mir LM, Banoun H, Paoletti C (1988) Introduction of definite amounts of nonpermeant molecules into living cells after electropermeabilization: direct access to the cytosol. Exp Cell Res 175:15–25. https://doi.org/10.1016/0014-4827(88)90251-0

Mir LM, Bureau MF, Gehl J et al (1999) High-efficiency gene transfer into skeletal muscle mediated by electric pulses. Proc Natl Acad Sci U S A 96:4262–4267

Mosqueda-Melgar J, Raybaudi-Massilia RM, Martin-Belloso O (2008) Non-thermal pasteurization of fruit juices by combining high-intensity pulsed electric fields with natural antimicrobials. Innov Food Sci Emerg Technol 9:328–340. https://doi.org/10.1016/j.ifset.2007.09.003

Neal RE 2nd, Garcia PA, Robertson JL, Davalos RV (2012) Experimental characterization and numerical modeling of tissue electrical conductivity during pulsed electric fields for irreversible electroporation treatment planning. IEEE Trans Biomed Eng 59:1076–1085. https://doi.org/10.1109/TBME.2012.2182994

Neal RE, Garcia PA, Kavnoudias H et al (2015) In vivo irreversible electroporation kidney ablation: experimentally correlated numerical models. IEEE Trans Biomed Eng 62:561–569. https://doi.org/10.1109/TBME.2014.2360374

Neumann E, Schaefer-Ridder M, Wang Y, Hofschneider PH (1982) Gene transfer into mouse lyoma cells by electroporation in high electric fields. EMBO J 1:841–845

Neumann E, Toensing K, Kakorin S et al (1998) Mechanism of electroporative dye uptake by mouse B cells. Biophys J 74:98–108. https://doi.org/10.1016/S0006-3495(98)77771-9

Paganin-Gioanni A, Bellard E, Escoffre JM et al (2011) Direct visualization at the single-cell level of siRNA electrotransfer into cancer cells. Proc Natl Acad Sci U S A 108:10443–10447. https://doi.org/10.1073/pnas.1103519108

Pakhomov AG, Gianulis E, Vernier PT et al (2015) Multiple nanosecond electric pulses increase the number but not the size of long-lived nanopores in the cell membrane. Biochim Biophys Acta Biomembr 1848:958–966. https://doi.org/10.1016/j.bbamem.2014.12.026

Pakhomova ON, Gregory BW, Semenov I, Pakhomov AG (2013) Two modes of cell death caused by exposure to nanosecond pulsed electric field. PLoS One 8:e70278. https://doi.org/10.1371/journal.pone.0070278

Pauly H, Schwan HP (1959) Über die Impedanz einer Suspension von kugelförmigen Teilchen mit einer Schale: Ein Modell für das dielektrische Verhalten von Zellsuspensionen und von Proteinlösungen. Zeitschrift für Naturforschung B 14:125–131. https://doi.org/10.1515/znb-1959-0213

Pavlin M, Miklavcic D (2008) Theoretical and experimental analysis of conductivity, ion diffusion and molecular transport during cell electroporation - relation between short-lived and long-lived pores. Bioelectrochemistry 74:38–46. https://doi.org/10.1016/j.bioelechem.2008.04.016
Pavlin M, Pavselj N, Miklavcic D (2002) Dependence of induced transmembrane potential on cell density, arrangement, and cell position inside a cell system. IEEE Trans Biomed Eng 49:605–612. https://doi.org/10.1109/TBME.2002.1001975
Pavlin M, Kanduser M, Rebersek M et al (2005) Effect of cell electroporation on the conductivity of a cell suspension. Biophys J 88:4378–4390. https://doi.org/10.1529/biophysj.104.048975
Pavlin M, Leben V, Miklavcic D (2007) Electroporation in dense cell suspension - theoretical and experimental analysis of ion diffusion and cell permeabilization. Biochim Biophys Acta-Gen Subj 1770:12–23. https://doi.org/10.1016/j.bbagen.2006.05.014
Pavlin M, Kotnik T, Mikilavčič D et al (2008) Electroporation of planar lipid bilayers and membranes. In: Leitmanova LA (ed) Advances in planar lipid bilayers and liposomes. Elsevier, Amsterdam, pp 165–226
Phoon PY, Galindo FG, Vicente A, Deimek P (2008) Pulsed electric field in combination with vacuum impregnation with trehalose improves the freezing tolerance of spinach leaves. J Food Eng 88:144–148. https://doi.org/10.1016/j.jfoodeng.2007.12.016
Pol IE, van Arendonk WGC, Mastwijk HC et al (2001) Sensitivities of germinating spores and Carvacrol-adapted vegetative cells and spores of Bacillus cereus to Nisin and pulsed-electric-field treatment. Appl Environ Microbiol 67:1693–1699. https://doi.org/10.1128/AEM.67.4.1693-1699.2001
Polajžer T, Jarm T, Miklavčič D (2020) Analysis of damage-associated molecular pattern molecules due to electroporation of cells in vitro. Radiol Oncol 54:317–328. https://doi.org/10.2478/raon-2020-0047
Polak A, Tarck M, Tomšič M et al (2014) Electroporation of archaeal lipid membranes using MD simulations. Bioelectrochemistry 100:18–26. https://doi.org/10.1016/j.bioelechem.2013.12.006
Puc M, Kotnik T, Mir LM, Miklavcic D (2003) Quantitative model of small molecules uptake after in vitro cell electropermeabilization. Bioelectrochemistry 60:1–10. https://doi.org/10.1016/S1567-5394(03)00021-5
Pucihar G, Kotnik T, Valič B, Miklavčič D (2006) Numerical determination of transmembrane voltage induced on irregularly shaped cells. Ann Biomed Eng 34:642–652. https://doi.org/10.1007/s10439-005-9076-2
Pucihar G, Kotnik T, Teissie J, Miklavcic D (2007) Electropermeabilization of dense cell suspensions. Eur Biophys J Biophys Lett 36:173–185. https://doi.org/10.1007/s00249-006-0115-1
Pucihar G, Kotnik T, Miklavcic D, Teissie J (2008) Kinetics of transmembrane transport of small molecules into electropermeabilized cells. Biophys J 95:2837–2848. https://doi.org/10.1529/biophysj.108.135541
Puertolas E, Luengo E, Alvarez I, Raso J (2012) Improving mass transfer to soften tissues by pulsed electric fields: fundamentals and applications. In: Doyle MP, Klaenhammer TR (eds) Annual review of food science and technology, vol 3. Annual Reviews, Palo Alto, pp 263–282
Raffy S, Lazdunski C, Teissie J (2004) Electroinsertion and activation of the C-terminal domain of colicin A, a voltage gated bacterial toxin, into mammalian cell membranes. Mol Membr Biol 21:237–246. https://doi.org/10.1080/09687680410001711632
Raso J, Frey W, Ferrari G et al (2016) Recommendations guidelines on the key information to be reported in studies of application of PEF technology in food and biotechnological processes. Innovative Food Sci Emerg Technol 37:312–321. https://doi.org/10.1016/j.ifset.2016.08.003
Reddy VY, Neuzil P, Koruth JS et al (2019) Pulsed field ablation for pulmonary vein isolation in atrial fibrillation. J Am Coll Cardiol 74:315–326. https://doi.org/10.1016/j.jacc.2019.04.021
Rems L, Miklavčič D (2016) Tutorial: electroporation of cells in complex materials and tissue. J Appl Phys 119:201101. https://doi.org/10.1063/1.4949264
Rems L, Ušaj M, Kandušer M et al (2013) Cell electrofusion using nanosecond electric pulses. Sci Rep 3. https://doi.org/10.1038/srep03382

Rols M-P, Teissie J (1989) Ionic-strength modulation of electrically induced permeabilization and associated fusion of mammalian cells. Eur J Biochem 179:109–115. https://doi.org/10.1111/j.1432-1033.1989.tb14527.x
Rols MP, Teissié J (1990) Electropermeabilization of mammalian cells. Quantitative analysis of the phenomenon. Biophys J 58:1089–1098. https://doi.org/10.1016/S0006-3495(90)82451-6
Rols MP, Teissié J (1998) Electropermeabilization of mammalian cells to macromolecules: control by pulse duration. Biophys J 75:1415–1423
Rosazza C, Deschout H, Buntz A et al (2016a) Endocytosis and endosomal trafficking of DNA after gene Electrotransfer in vitro. Mol Ther Nucleic Acids 5:e286. https://doi.org/10.1038/mtna.2015.59
Rosazza C, Meglic SH, Zumbusch A et al (2016b) Gene Electrotransfer: a mechanistic perspective. Curr Gene Ther 16:98–129
Rubinsky B, Onik G, Mikus P (2007) Irreversible electroporation: a new ablation modality - clinical implications. Technol Cancer Res Treat 6:37–48
Sack M, Sigler J, Frenzel S et al (2010) Research on industrial-scale electroporation devices fostering the extraction of substances from biological tissue. Food Eng Rev 2:147–156. https://doi.org/10.1007/s12393-010-9017-1
Saito AC, Ogura T, Fujiwara K et al (2014) Introducing micrometer-sized artificial objects into live cells: a method for cell–Giant Unilamellar vesicle electrofusion. PLoS One 9:e106853. https://doi.org/10.1371/journal.pone.0106853
Saldaña G, Cebrián G, Abenoza M et al (2017) Assessing the efficacy of PEF treatments for improving polyphenol extraction during red wine vinifications. Innovative Food Sci Emerg Technol 39:179–187. https://doi.org/10.1016/j.ifset.2016.12.008
Saulis G (2010) Electroporation of cell membranes: the fundamental effects of pulsed electric fields in food processing. Food Eng Rev 2:52–73. https://doi.org/10.1007/s12393-010-9023-3
Saulis G, Saule R (2012) Size of the pores created by an electric pulse: microsecond vs millisecond pulses. Biochim Biophys Acta-Biomembr 1818:3032–3039. https://doi.org/10.1016/j.bbamem.2012.06.018
Saulis G, Venslauskas MS, Naktinis J (1991) Kinetics of pore resealing in cell membranes after electroporation. Bioelectrochem Bioenerg 26:1–13
Scheffer HJ, Nielsen K, de Jong MC et al (2014) Irreversible electroporation for nonthermal tumor ablation in the clinical setting: a systematic review of safety and efficacy. J Vasc Interv Radiol 25:997–1011; quiz 1011. https://doi.org/10.1016/j.jvir.2014.01.028
Sel D, Cukjati D, Batiuskaite D et al (2005) Sequential finite element model of tissue Electropermeabilization. IEEE Trans Biomed Eng 52:816–827. https://doi.org/10.1109/TBME.2005.845212
Šel D, Lebar AM, Miklavčič D (2007) Feasibility of employing model-based optimization of pulse amplitude and electrode distance for effective tumor electropermeabilization. IEEE Trans Biomed Eng 54:773–781. https://doi.org/10.1109/TBME.2006.889196
Shayanfar S, Chauhan OP, Toepfl S, Heinz V (2013) The interaction of pulsed electric fields and texturizing - antifreezing agents in quality retention of defrosted potato strips. Int J Food Sci Technol 48:1289–1295. https://doi.org/10.1111/ijfs.12089
Shin JK, Lee SJ, Cho HY et al (2010) Germination and subsequent inactivation of Bacillus subtilis spores by pulsed electric field treatment. J Food Process Preserv 34:43–54. https://doi.org/10.1111/j.1745-4549.2008.00321.x
Shynkaryk MV, Lebovka NI, Vorobiev E (2008) Pulsed electric fields and temperature effects on drying and rehydration of red beetroots. Dry Technol 26:695–704. https://doi.org/10.1080/07373930802046260
Siemer C, Toepfl S, Heinz V (2014a) Inactivation of Bacillus subtilis spores by pulsed electric fields (PEF) in combination with thermal energy – I. Influence of process- and product parameters. Food Control 39:163–171. https://doi.org/10.1016/j.foodcont.2013.10.025
Siemer C, Toepfl S, Heinz V (2014b) Inactivation of Bacillus subtilis spores by pulsed electric fields (PEF) in combination with thermal energy II. Modeling thermal inactivation of B. subtilis

spores during PEF processing in combination with thermal energy. Food Control 39:244–250. https://doi.org/10.1016/j.foodcont.2013.09.067
Sitzmann W, Vorobiev E, Lebovka N (2016) Pulsed electric fields for food industry: historical overview. In: Miklavcic D (ed) Handbook of electroporation. Springer, Cham, pp 1–20
Smith KC, Neu JC, Krassowska W (2004) Model of creation and evolution of stable electropores for DNA delivery. Biophys J 86:2813–2826
Sobrino-Lopez A, Martin-Belloso O (2010) Review: potential of high-intensity pulsed electric field technology for milk processing. Food Eng Rev 2:17–27. https://doi.org/10.1007/s12393-009-9011-7
Son RS, Smith KC, Gowrishankar TR et al (2014) Basic features of a cell electroporation model: illustrative behavior for two very different pulses. J Membr Biol 247:1209–1228. https://doi.org/10.1007/s00232-014-9699-z
Sözer EB, Levine ZA, Vernier PT (2017) Quantitative limits on small molecule transport via the Electropermeome — measuring and modeling single nanosecond perturbations. Sci Rep 7:57. https://doi.org/10.1038/s41598-017-00092-0
Sözer EB, Pocetti CF, Vernier PT (2018a) Transport of charged small molecules after electropermeabilization — drift and diffusion. BMC Biophys 11:4. https://doi.org/10.1186/s13628-018-0044-2
Sözer EB, Pocetti CF, Vernier PT (2018b) Asymmetric patterns of small molecule transport after nanosecond and microsecond electropermeabilization. J Membr Biol 251:197–210. https://doi.org/10.1007/s00232-017-9962-1
Stampfli R (1958) Reversible electric breakdown of the excitable membrane of a Ranvier node. Ann Acad Bras Cien 30:57–63
Stewart MT, Haines DE, Verma A et al (2019) Intracardiac pulsed field ablation: proof of feasibility in a chronic porcine model. Heart Rhythm 16:754–764. https://doi.org/10.1016/j.hrthm.2018.10.030
Stirke A, Zimkus A, Ramanaviciene A et al (2014) Electric field-induced effects on yeast cell wall permeabilization. Bioelectromagnetics 35:136–144. https://doi.org/10.1002/bem.21824
Straessner R, Eing C, Goettel M et al (2013) Monitoring of pulsed electric field-induced abiotic stress on microalgae by chlorophyll fluorescence diagnostic. IEEE Trans Plasma Sci 41:2951–2958. https://doi.org/10.1109/TPS.2013.2281082
Sugár IP (1979) A theory of the electric field-induced phase transition of phospholipid bilayers. Biochim Biophys Acta Biomembr 556:72–85. https://doi.org/10.1016/0005-2736(79)90420-6
Sugrue A, Vaidya V, Witt C et al (2019) Irreversible electroporation for catheter-based cardiac ablation: a systematic review of the preclinical experience. J Interv Card Electrophysiol 55:251–265. https://doi.org/10.1007/s10840-019-00574-3
Susil R, Šemrov D, Miklavčič D (1998) Electric field-induced transmembrane potential depends on cell density and organizatio. Electro Magnetobiol 17:391–399. https://doi.org/10.3109/15368379809030739
Tarek M, Delemotte L, Delemotte L (2010) Electroporation of lipid membranes: insights from molecular dynamics simulations. In: Advanced electroporation techniques in biology and medicine. https://www.taylorfrancis.com/. Accessed 10 Sep 2020
Teissié J (2014) Electropermeabilization of the cell membrane. Methods Mol Biol 1121:25–46. https://doi.org/10.1007/978-1-4614-9632-8_2
Teissié J, Rols M-P (1993) An experimental evaluation of the critical potential difference inducing cell membrane electropermeabilization. Biophys J 65:409–413
Teissie J, Tsong TY (1981) Electric field induced transient pores in phospholipid bilayer vesicles. Biochemistry 20:1548–1554. https://doi.org/10.1021/bi00509a022
Teissié J, Golzio M, Rols M-P (2005) Mechanisms of cell membrane electropermeabilization: a minireview of our present (lack of ?) knowledge. Biochim Biophys Acta 1724:270–280. https://doi.org/10.1016/j.bbagen.2005.05.006

Thamkaew G, Gómez Galindo F (2020) Influence of pulsed and moderate electric field protocols on the reversible permeabilization and drying of Thai basil leaves. Innovative Food Sci Emerg Technol 64:102430. https://doi.org/10.1016/j.ifset.2020.102430

Toepfl S, Heinz V, Knorr D (2006) Applications of pulsed electric fields technology for the food industry. In: Raso J, Heinz V (eds) Pulsed electric fields technology for the food industry, Springer, Boston, pp 197–221

Toepfl S, Siemer C, Saldaña-Navarro G, Heinz V (2014) Chapter 6: Overview of pulsed electric fields processing for food. In: Sun D-W (ed) Emerging technologies for food processing, 2nd edn. Academic Press, San Diego, pp 93–114

Towhidi L, Kotnik T, Pucihar G et al (2008) Variability of the minimal transmembrane voltage resulting in detectable membrane electroporation. Electromagn Biol Med 27:372–385. https://doi.org/10.1080/15368370802394644

Tsong TY (1991) Electroporation of cell membranes. Biophys J 60:297–306

Ušaj M, Flisar K, Miklavcic D, Kanduser M (2013) Electrofusion of B16-F1 and CHO cells: the comparison of the pulse first and contact first protocols. Bioelectrochemistry 89:34–41. https://doi.org/10.1016/j.bioelechem.2012.09.001

Valič B, Golzio M, Pavlin M et al (2003) Effect of electric field induced transmembrane potential on spheroidal cells: theory and experiment. Eur Biophys J 32:519–528. https://doi.org/10.1007/s00249-003-0296-9

Vanthoor-Koopmans M, Wijffels RH, Barbosa MJ, Eppink MHM (2013) Biorefinery of microalgae for food and fuel. Bioresour Technol 135:142–149. https://doi.org/10.1016/j.biortech.2012.10.135

Vernier PT (2020) Nanoscale and multiscale membrane electrical stress and permeabilization. In: Kramar P, Miklavčič D (eds) Electroporation-based technologies and treatments, 1st edn. Založba FE, Ljubljana, pp 105–119

Vernier PT, Sun Y, Marcu L et al (2003) Calcium bursts induced by nanosecond electric pulses. Biochem Biophys Res Commun 310:286–295. https://doi.org/10.1016/j.bbrc.2003.08.140

Vernier PT, Levine ZA, Gundersen MA (2013) Water bridges in Electropermeabilized phospholipid bilayers. Proc IEEE 101:494–504. https://doi.org/10.1109/JPROC.2012.2222011

Vorobiev E, Lebovka N (2010) Enhanced extraction from solid foods and biosuspensions by pulsed electrical energy. Food Eng Rev 2:95–108. https://doi.org/10.1007/s12393-010-9021-5

Vorobiev E, Lebovka N (2017) Application of pulsed electric fields for root and tuber crops biorefinery. In: Miklavčič D (ed) Handbook of electroporation. Springer, Cham, pp 2899–2922

Weaver J (1993) Electroporation: a general phenomenon for manipulating cells and tissues. J Cell Biochem 51:426–435

Weaver JC, Chizmadzhev YA (1996) Theory of electroporation: a review. Bioelectrochem Bioenerg 41:135–160. https://doi.org/10.1016/S0302-4598(96)05062-3

Weaver JC, Vernier PT (2017) Pore lifetimes in cell electroporation: complex dark pores? arXiv:170807478. [physics]

Wegner LH, Frey W, Silve A (2015) Electroporation of DC-3F cells is a dual process. Biophys J 108:1660–1671. https://doi.org/10.1016/j.bpj.2015.01.038

Wesierska E, Trziszka T (2007) Evaluation of the use of pulsed electrical field as a factor with antimicrobial activity. J Food Eng 78:1320–1325. https://doi.org/10.1016/j.foodeng.2006.01.002

Wolf H, Rols MP, Boldt E et al (1994) Control by pulse parameters of electric field-mediated gene transfer in mammalian cells. Biophys J 66:524–531

Yarmush ML, Golberg A, Serša G et al (2014) Electroporation-based technologies for medicine: principles, applications, and challenges. Annu Rev Biomed Eng 16:295–320. https://doi.org/10.1146/annurev-bioeng-071813-104622

Ye H, Huang L-L, Chen S-D, Zhong J-J (2004) Pulsed electric field stimulates plant secondary metabolism in suspension cultures of Taxus chinensis. Biotechnol Bioeng 88:788–795. https://doi.org/10.1002/bit.20266

Yu X, Bals O, Grimi N, Vorobiev E (2015) A new way for the oil plant biomass valorization: polyphenols and proteins extraction from rapeseed stems and leaves assisted by pulsed electric fields. Ind Crop Prod 74:309–318. https://doi.org/10.1016/j.indcrop.2015.03.045

Ziegler MJ, Vernier PT (2008) Interface water dynamics and porating electric fields for phospholipid bilayers. J Phys Chem B 112:13588–13596. https://doi.org/10.1021/jp8027726

Zimmermann U (1982) Electric field-mediated fusion and related electrical phenomena. Biochim Biophys Acta Rev Biomembr 694:227–277. https://doi.org/10.1016/0304-4157(82)90007-7

Zmuc J, Gasljevic G, Sersa G et al (2019) Large liver blood vessels and bile ducts are not damaged by electrochemotherapy with bleomycin in pigs. Sci Rep 9:3649. https://doi.org/10.1038/s41598-019-40395-y

Electrochemical Reactions in Pulsed Electric Fields Treatment

Gianpiero Pataro and Giovanna Ferrari

Nomenclature

A	*Electrode area* $[m^{-2}]$
C_{dl}	Electronic double-layer capacitance [F]
E	Electric field strength $[V\ m^{-1}]$
E°_{red}	Standard reduction potential [V]
F	Faraday constant $[C\ mol^{-1}]$
f	Pulse repetition frequency [Hz]
F_v	Volumetric flow rate $[m^3\ s^{-1}]$
i	Current intensity [A]
i_f	Faradaic current [A]
i_c	Non-Faradaic charging current [A]
$\mathbf{j}$	Current density vector
j	Current density $[A\ m^{-2}]$
i_f	FARADAIC current [A]
i_c	Non-Faradaic charging current [A]
$\mathbf{j}$	Current density vector
j	Current density $[A\ m^{-2}]$
m_{rs}	Amount of species s produced by the reaction r [g]
M_s	Molecular weight of the species s $[g\ mol^{-1}]$
n_p	Number of pulses [−]
n_{rs}	Number of electrons transferred for each product s formed by the reaction r [−]
Q_p	Total charge transferred per each pulse [C]
Q_r	Cumulative electric charge transferred by the reaction r [C]
Q_{tot}	Total transferred charge [C]
R_s	Resistance of the electrolytic solution [Ω]

G. Pataro (✉)
Department of Industrial Engineering, University of Salerno, Fisciano, SA, Italy
e-mail: gpataro@unisa.it

G. Ferrari
Department of Industrial Engineering, University of Salerno, Fisciano, SA, Italy

ProdAl Scarl – University of Salerno, Fisciano, SA, Italy

J. Raso et al. (eds.), *Pulsed Electric Fields Technology for the Food Industry*,
Food Engineering Series, https://doi.org/10.1007/978-3-030-70586-2_4

t_r Residence time [s]
U Electric potential [V]
U_{th} Reaction potential [V]
v Treatment volume [m^3]
W_p Specific energy per pulse [kJ $kg^{-1}pulse^{-1}$]
W_T Total specific energy [kJ/kg]
x_{dl} Thickeners of the double layer [m]
Z_f Faradaic impedance [Ω]

Greek Letters

σ Electrical conductivity [S m^{-1}]
τ Pulse width [s]
$\Delta\phi_{dl}$ Potential drop across the electronic double layer [V]

Operators

Δ Difference

Introduction

Pulsed electric fields (PEF) is an innovative nonthermal technology which has attracted considerable interest in the last decades in food industry due to its potential to provide the food processor with the opportunity to produce new and value-added food products and ingredients with reduced energy consumption and enhanced quality attributes preferred by consumers.

In PEF processing, a food product is physically and electrically contacted with the metal electrodes of a treatment chamber, operated either in batch-wise or in continuous, and exposed to repetitive (Hz-kHz) very short (μs-ms) electric field (E) pulses of low (E < 2 kV/cm), moderate (E = 1–5 kV/cm), or high intensity (E = 10–40 kV/cm) and relatively low energy input (W_T = 0.1–150 kJ/kg) supplied by a pulse generator. The pulse shapes commonly used in PEF treatments are either exponential or square-wave pulses, monopolar, or bipolar (Raso et al. 2016).

The application of electric pulses may cause either temporary or permanent formation of pores in the membranes of microbial, plant, or animal cells, referred to as electroporation or electropermeabilization, which disturbs and damages membrane's functionality (Kotnik et al. 2012), as well as induces targeted structural modification in the food matrix (Soliva-Fortuny et al. 2009).

This make PEF technology suitable for a wide range of applications in food industry, such as those involving the inactivation of microbial cells in liquid foods (e.g., pasteurization) (Saldaña et al. 2014), or mass transport phenomena of juice, water, and bioactive compounds from plant sources (e.g., extraction, expression, drying, osmotic dehydration, freeze-drying, freezing, freeze-thawing) (Bobinaitė et al. 2015; Donsì et al. 2010; Jalte et al. 2009; Parniakov et al. 2015, 2016a, b; Pataro et al. 2020; Wiktor et al. 2013), or requiring structural modifications to facilitate some unit operation of food processing (e.g., peeling, cutting) (Arnal et al. 2018; Soliva-Fortuny et al. 2009)..

With regard to these applications, PEF processing is particularly interesting due to its relatively low energy consumption and processing temperature, easy scale-up, and implementation into existing processing line due to its continuous flow operability with very short processing time. This has led, today, to the first industrial applications especially in the potato processing industry, where the PEF is operated at low field strength and energy input ($E < 2$ kV/cm, $W_T = 0.1$–2 kJ/kg) to facilitate cutting operations providing the highest potential in terms of technical feasibility as well as economical relevance (Fauster et al., 2018). On the other hand, only very few commercial-scale operations have been achieved for applications requiring high field strength and energy input ($E > 10$ kV/cm, $W_T = 50$–150 kJ/kg), as reported for pasteurization of liquid food such as fruit juices and smoothies (Kempkes 2017). In this case, several technological issues mainly related to the long-term reliability of PEF generator and electrodes as well as high investment, service and maintenance costs, consumers' acceptance, regulatory aspects, and toxicity risks still remain and have to be addressed prior to the full exploitation of PEF technology.

Most of these limitations are related to the unavoidable occurrence of electrophoresis and electrochemical reactions accompanying the flow of electric current through the PEF treatment chamber when typical conditions for PEF processing are applied. These reactions, especially those leading to corrosion and fouling of the electrodes, electrolysis of water, migration of electrode material, and chemical changes that may occur into the food product undergoing a PEF treatment, are undesired and must be minimized since these may seriously affect food safety (other than microbial) and quality, as well as process efficiency, equipment reliability, and cost aspects.

For these reasons electrochemical reactions have found increasing attention within the last years since their occurrence obviously raises concerns about PEF commercialization.

Historical Background

Electrically driven electrochemical reactions were first reported at the end of the nineteenth century, when it was found that the bactericidal effect of direct or low-frequency alternating current resulted from thermal and electrochemical effects (Prochownick and Spaeth 1890; Krüger 1893; Thiele and Wolf 1899).

In the 1920s a process called "Electropure" was introduced in Europe and USA (Beattie and Lewis 1925; Fetterman 1928), as the first attempts to use electricity for milk pasteurization. The process was performed by pumping the milk through a carbon electrode treatment chamber connected to a (not pulsed) 220 V alternating current source. Apart from thermal effects based on the mechanism of ohmic heating, lethal effects of electrochemical reactions such as the hydrolysis of chlorine were found when the applied voltage used to treat food was 3000–4000 V (Pareilleux and Sicard 1970).

A further attempt to use electricity to kill microorganisms was performed in the 1950s in a process called electrohydraulic treatment, which consisted in a rapid discharge of high-voltage electricity (electric arcs) across two electrodes submerged in the liquid medium containing the microorganisms (Gilliland and Speck 1967). Electrochemical reactions (electrolysis of water), shock waves, and ultraviolet light forming free, highly reactive radicals were reported to be responsible for the bactericidal effect. Additionally, the contamination of treated food due to chemical products from electrolysis, electrode erosion, and the disintegration of particulates within the food (by the shock waves) inhibited an industrial application of this process except for wastewater (Jeyamkondan et al. 1999).

The possibility of electrochemical reactions (electrolysis) occurring during PEF treatment has been first mentioned in 1960s by Sale and Hamilton (1967), who carried out a pioneering work on the use of PEF for microbial inactivation and showed the insignificance of electrolysis on the lethal effect of direct current (DC) pulses.

However, only since the 1980s electrochemical reaction in PEF treatment systems found increasing attention, when Dunn and Pearlman (1987) introduced an ion permeable polymer membrane functioning as a physical barrier between the electrodes and the liquid food in order to avoid electrode material from being released in the liquid food upon the application of PEF. However, this arrangement showed several drawbacks such as the possibility of using only monopolar pulses due to the selective nature of membranes, additional cost of membrane material and high sensitivity to puncture by any electrical discharges, frequently required maintenance, and complicated arrangement for industrial implementation. Due to these complexities, it was understood that the reduction of electrode reactions, rather than a complete prevention, might be preferred as long as the metal concentration remains within the limits dictated by food regulations.

In this line, Bushnell et al. (1995a, b) patented a method to avoid or reduce electrochemical reactions and the fouling of electrodes in a PEF processing system. Based only on a theoretical analysis, the authors suggested that the rate of electrophoresis and electrochemical reactions is proportional to the net electric charge delivered to the PEF chamber, and therefore, a claim has been made that a "zero-net-charge delivery" substantially prevents fouling of electrode(s) and reduces electrochemical reactions in general. Authors explained that, by removing all the residual charge at the electrode interface during the time elapsing between two consecutive pulses (zero-net-charge concept), no cumulative buildup of charges occurs and, hence, no electrochemical reactions will take place for any pulse of a width shorter than a certain threshold.

Few years later, a broad discussion of fundamental electrochemistry related to PEF has been given by Morren et al. (2003), who developed electrical equivalent circuit models to describe the behavior of the electrode/electrolyte interface under various conditions. Moreover, the authors performed the first attempt to study the release of electrode material into liquid product undergoing a PEF treatment. They exposed stainless steel electrodes in contact with an aqueous solution of NaCl to sinusoidal waveforms of only 15 and 150 V peak-to-peak voltages, resulting in a maximum electric field of 0.15 kV/cm and a maximum current of 480 mA, which were far below the typical PEF treatment conditions. This notwithstanding, the results have a certain relevance showing that at each voltage level the amount of the different elements (iron, chromium, nickel, and manganese) released from stainless steel electrodes is proportional to the width of applied pulses after a critical width.

Since then several researchers have addressed the problem of electrochemical reactions in PEF treatment systems highlighting that it is a very complex phenomena, which is largely unavoidable in the long-term trials, especially when PEF is carried out under severe treatment conditions (Pataro et al. 2014a). Most of the attention has been focused on the phenomenon of metal release and electrode corrosion and its dependence on the main operating parameters (Gad and Jayaram 2011, 2012a, 2014; Góngora-Nieto et al. 2002; Kim and Zhang 2011; Kotnik et al. 2001; Master et al. 2007; Morren et al. 2003; Pataro et al. 2014a, 2015a, b; Roodenburg et al. 2005a, b, 2010; Saulis et al. 2007). To this regard, numerical simulation has been recently applied as a valuable tool to improve the understanding of fundamentals of the electrochemical phenomena occurring at the electrode-liquid interface of a PEF chamber, clarifying the effects of the main electrical parameters, chamber design characteristics, and treatment medium composition on electrode corrosion or release of electrode's materials (Pataro et al. 2015a, b, 2017a, b).

Very little attention has been, instead, paid to possibility that reaction products (e.g., metal ions) and food constituents can participate in secondary chemical reactions also after pulse treatment has been completed, thus negatively affecting chemical and sensorial properties of food treated products and arising possible toxicity problems (Evrendilek et al. 2004; Rodaitė-Riševičienė et al. 2014; Sun et al. 2011; Zhao et al. 2012).

Therefore, additional work is needed to obtain more detailed insights on the influence on these phenomena on the technical feasibility of PEF technology as well as on safety and quality aspects of food products.

Electrochemical Behavior of a PEF Treatment Chamber

Ionic Double Layer and Electrode Reactions

In a PEF system the two metal electrodes of the treatment chamber are electrically connected by a high voltage cable to a pulse generator and placed in direct contact with a liquid food (or any electrolyte solution). From an electrochemical point of view, this system acts as an electrolytic cell where an external source of electrical energy is responsible for driving an electrical current through the cell. The current flowing in the conductive solid electrodes consists of free electrons, while in the conductive electrolyte solution, the current is carried by ions rather than by free electrons (Roodenburg et al. 2005a).

To understand the electrochemical behavior of a PEF chamber, it is necessary to identify the characteristics of the electrode/electrolyte interface.

Initially the PEF chamber can be considered as made with equal electrodes in contact with an electrolyte with homogeneous concentration. Before starting the pulse treatment (i.e., without externally applied voltage), immediately at each electrode-electrolyte interface, in order to maintain the condition of electroneutrality, a so-called double layer will develop. This layer, which consists of charged particles and/or of orientated dipoles (Fig. 1), behaves as an electrical capacitor (Morren et al. 2003). The formation of the double layer causes a potential drop ($\Delta\phi_{dl}$) across the electrode-electrolyte interface, which tends asymptotically to zero at the imaginary boundary of the double layer. When no external voltage is applied across the electrodes, this potential drop is of the order of magnitude of mV and only low-level reactions occur at each electrode-electrolyte interface. In these conditions the two competing reactions reach an equilibrium, whereby the currents are equal with opposite sign leading to a zero net current flow (Morren et al. 2003).

During pulse treatment, due to the application of a potential difference across the electrodes, an electrical current pass through the cell occurs, and charges buildup across the double layer takes place (Bockris et al. 2002; Gileadi 1993a). In this case,

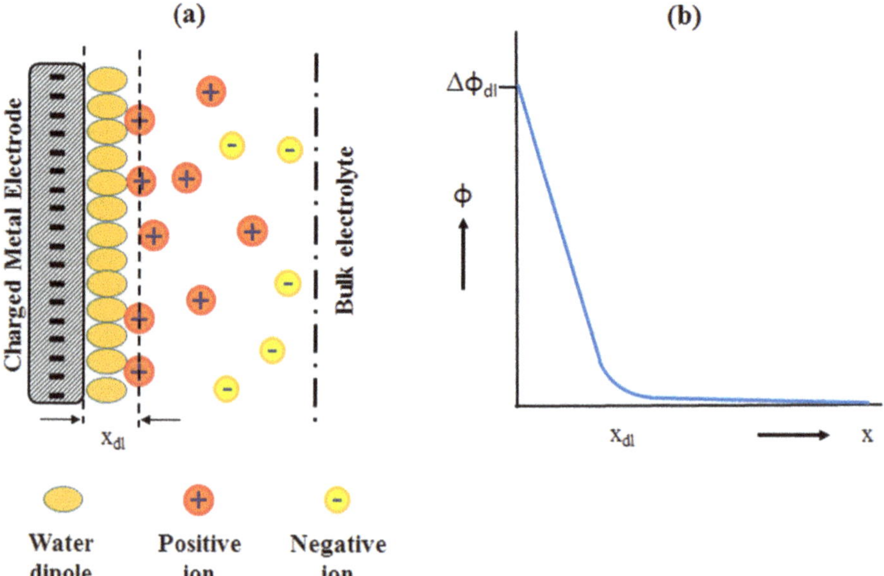

Fig. 1 (**a**) Schematic diagram of an ionic double layer at the electrode-electrolyte interface (x_{dl} is the thickness of the double layer); (**b**) variation of the electrostatic potential ϕ with distance x from the electrode surface

the charge density and the thickness (x_{dl}) of the ionic layers at the interface increase leading to an increase of the potential drop $\Delta\phi_{dl}$.

As long as the potential across the double layer remains below the typical threshold voltage (U_{th}) of the reaction potential of reacting species (~1–2 V), no electrochemical reactions occur, except some low-level reactions due to the exchange current. The threshold voltage, or standard reduction potentials (E°_{red}), may be different for each of the electrode and depends, among others, on temperature, electrode material, pH, and the chemical content of the fluid (Pataro et al. 2017a, b).

If a high enough voltage is applied across the electrodes for a long enough time to let the voltage across the double-layer capacitor exceed the threshold value, in order to preserve the charge conservation principle, two independent electrochemical (Faradaic) half-cell reactions will occur at the electrode interfaces (Eqs. 1 and 2): oxidation half-cell reactions (i.e., loss of electrons) will take place at the electrode surface behaving as anode (i.e., high voltage electrode), and reduction reactions (i.e. gain of electrons) at the electrode surface behaving as cathode (i.e., grounded electrode) (Morren et al. 2003). As a result, the overall reactions (Eq. 3) produce periodic concentration changes of redox species at the electrode-electrolyte interfaces.

Anodic half-reaction:

$$R_{ed,2} \rightarrow O_{x,2} + n \cdot e^{-} \tag{1}$$

Cathodic half-reaction:

$$O_{x,1} + n \cdot e^{-} \rightarrow R_{ed,1} \tag{2}$$

Overall redox reaction:

$$O_{x,1} + R_{ed,2} \rightarrow R_{ed,1} + O_{x,2} \tag{3}$$

where $O_{x,1}$ and $R_{ed,1}$ are, respectively, the oxidized and reduced form of the chemical species 1; $O_{x,2}$ and $R_{ed,2}$ are, respectively, the oxidized and reduced form of the species 2; and n is the number of electrons (e^{-}) exchanged during the reaction.

If during PEF treatment monopolar pulses are applied, the electrode connected to high voltage will behave as anode, while the grounded electrode will behaves as cathode. If instead bipolar pulses are applied, the cathode and anode interchange places according to the pulse repetition frequency and, therefore, oxidative and reductive half-reactions will occur alternatively at the same electrode site (Roodenburg et al. 2005a; Pataro et al. 2017a, b).

The electrode reactions involve mass transport of electro-active species to the electrode, electron transfer across the electrode interface, and the mass transport of the reaction products back to the solution. Mass transport and charge transfer are two consecutive processes whose rate depends, besides the applied potential, on electrode material, electrolyte, pulse repetition frequency, pulse polarity, temperature, and pH, among others. The slower the two processes is, therefore, the limiting step determines the overall rate (Gileadi 1993b; Morren et al. 2003; Pataro et al. 2015a).

During PEF treatment, and especially when high treatment intensities are applied, a variety of electrochemical reactions may simultaneously occur at the electrode-food interfaces involving chemical species of either electrode material or food product (Pataro et al. 2020). Since predicting all possible electrochemical reactions during the PEF processing of any real liquid food is a very complex task, for simplicity, only few examples of potential anodic and cathodic half-cell reactions related to electrolysis of water and corrosion will be presented in the following equations (Gad and Jayaram 2014).

Cathodic half-reactions:

$$Cl_{2(gas)} + 2e^{-} \Rightarrow 2Cl^{-}_{(aq)} \tag{4}$$

$$O_{2(g)} + 4H^{+}_{(aq)} + 4e^{-} \Rightarrow 2H_2O_{(liq)} \tag{5}$$

$$O_{2(g)} + 2H^{+}_{(aq)} + 2e^{-} \Rightarrow 2H_2O_{2(liq)} \tag{6}$$

$$2H^{+}_{(aq)} + 2e^{-} \Rightarrow H_{2(gas)} \tag{7}$$

$$2H_2O_{(liq)} + 2e^{-} \Rightarrow H_{2(gas)} + 2OH^{-}_{(aq)} \tag{8}$$

Anodic half-reaction:

$$2H_2O_{(liq)} \Rightarrow O_{2(g)} + 4H^{+}_{(aq)} + 4e^{-} \tag{9}$$

$$2H_2O_{(liq)} \Rightarrow 2H_2O_{2(liq)} + 2H^{+}_{(aq)} + 2e^{-} \tag{10}$$

$$v_1 M_{(s)} + v_2 H_2O_{(liq)} \Rightarrow M_{v1}O_{v2} + 2v_2 H^{+}_{(aq)} + 2v_2 e^{-} \tag{11}$$

$$M_{(s)} \Rightarrow M^{n+}_{(aq)} + ne^{-} \left(M = \text{Fe,Ni,Cr,Mn,Ti,Pt,}\ldots\right) \tag{12}$$

Which of the half-reactions occurs depends on the relative ease of each of the competing reactions.

After pulsing (i.e., when PEF system is switched off), due to either concentration differences around the electrodes or to surface material changes, the behavior of cell may shift from electrolytic to galvanic. Thus, if the treatment chamber is not dried up, some of the products of those electrochemical reactions may spontaneously initiate a number of secondary chemical reactions, which may involve food components, even with no external voltage applied (Pataro et al. 2020).

Electrical Equivalent Circuit

The relationship between an external electrical perturbation at the terminals of a PEF chamber and the response of the double layer at each electrode interface can be described by an equivalent electrical circuit. In Fig. 2, simplified equivalent electrical circuits for the PEF treatment chamber with electrode-electrolyte interfaces are

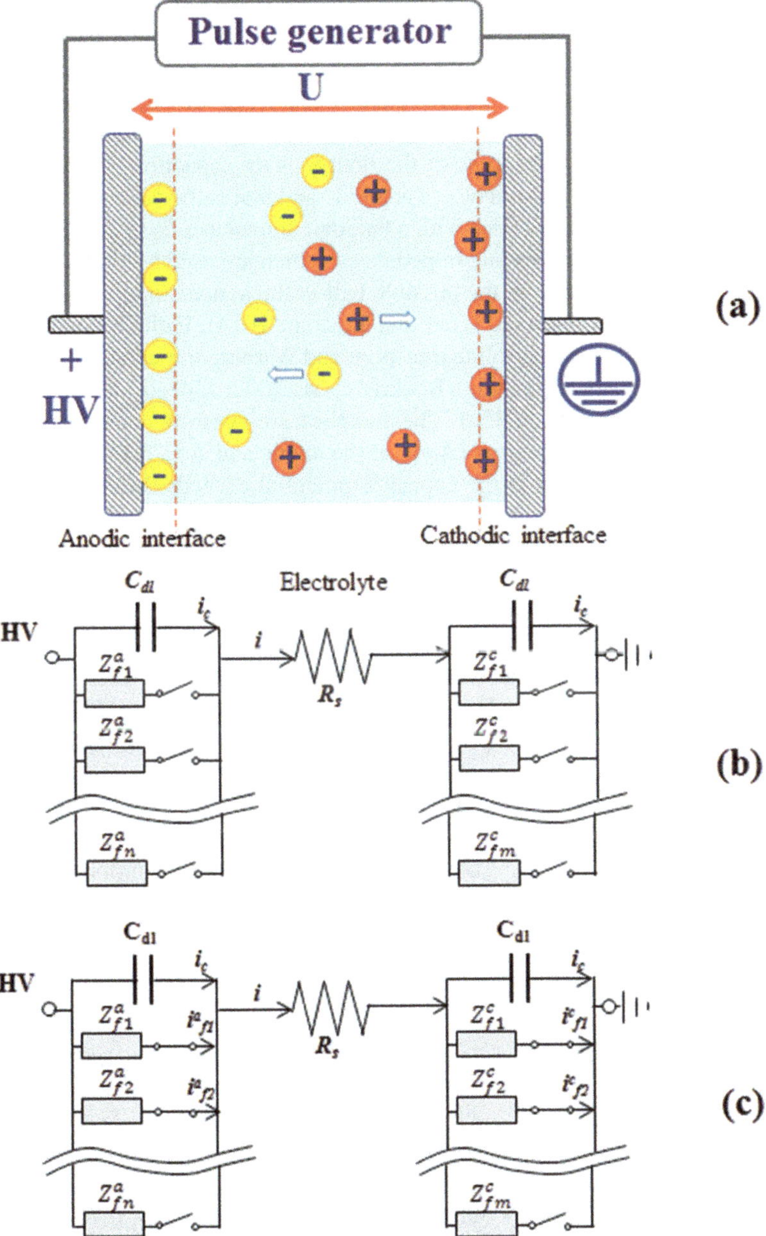

Fig. 2 Electrical equivalent circuit for the PEF treatment chamber with electrode-electrolyte interfaces. (**a**) schematic of a parallel plate PEF chamber with a pulse generator; (**b**) equivalent circuit with no Faradaic reactions; (**c**) equivalent circuit with Faradaic reactions. C_{dl} is the double-layer capacitance; R_s is the bulk resistance of the electrolyte solution; Z^a_{fk} is the anodic Faradaic impedance in the branch k (k = 1…n); Z^c_{fp} is the cathodic faradaic impedance in the branch p (p = 1…m); i^a_{fk} and i^c_{fp} represent the anodic and cathodic Faradaic current passing in the branch k and p, respectively

shown for different conditions. An external dc source of voltage (*U*) is applied to a conductive electrolyte through the metal electrodes of the PEF chamber. In the equivalent circuits, R_s represents the bulk resistance of the solution, while the ohmic resistance of both electrodes is negligible. Each electrode-electrolyte interface is analogous to a parallel combination of the double-layer capacitance C_{dl}, which is involved in the charging process of the double-layer capacitor by a non-Faradaic current, and a number of branches (*n* at anode and *p* at cathode), each consisting of a normally open switch in series with a Faradaic impedance Z_f (Z^a_{fk} at the anode and Z^c_{fp} at the cathode). The Faradaic impedances, which are used to model the Faradaic process associated to each of the possible half-cell reactions that might occur at the anodic and cathodic interface (Gad and Jayaram 2014), include resistors for the Faradaic processes and electrolyte transport, and Warburg impedance for diffusional limitations whenever applicable (Bockris et al. 2002; Gileadi 1993b, c; Morren et al. 2003; Pataro et al. 2015a). The switches are controlled through the corresponding double-layer potential ($\Delta\phi_{dl,a}$ at the anode and $\Delta\phi_{dl,c}$ at the cathode), i.e., each switch will close when the capacitor potential exceeds the threshold voltage ($U_{th,i}$) characteristic of a given electrode reaction *i* (Gad and Jayaram 2014).

These models do not take into account some peculiar factors such as the heterogeneity of electrode surface and the eventual formation of reaction intermediates absorbed at the electrode surface. Moreover, the circuit is nonlinear being C_{dl} and Z_f dependent on potential (Gileadi 1993b).

Initially, when an external voltage (*U*) is applied, and the threshold voltage at which the half-cell reactions ($U_{th,i}$) typically start is not reached yet, the model takes the form of Fig. 14.2b. In this phase, the current (*i*) starts flowing through the electrolyte resistance, and only a capacitive, non-Faradaic charging current (i_c) is flowing in C_{dl}. This current is called charging current (i_c) since it is responsible for charging the double-layer capacitor up to the reaction potential (U_{th}) of the reacting species (Morren et al. 2003). The time required for this charging process is called threshold time (t_{th}, in μs) and is given by the following equation (Morren et al. 2003):

$$t_{th} = \frac{C'_{dl} U'_{dl}}{j} \tag{13}$$

where *j* is the current density (in A cm^{-2}) through the chamber, while the quoted parameters represent the equivalent values of the specific capacitance (capacitance per unit of area, C_{dl}, in μF cm^{-2}) and reaction potential (U_{th}, in V) of the two electrode-electrolyte interfaces in series (i.e., $C'_{dl} = (1/2) \cdot C_{dl}$ and $U'_{th} = 2 \cdot U_{th}$). From Eq. 13, it can be seen that, for a given electrode reaction, the threshold time depends mainly on the current density, which in turn depends on the applied voltage, fluid conductivity, and chamber geometry, as well as on the value of the double-layer capacitance.

As soon as the potential of one capacitor ($\Delta\phi_{dl,a}$ *or* $\Delta\phi_{dl,c}$) exceeds the reaction potentials ($U_{th,i}$) of one or some of the possible anodic or cathodic half-cell reactions, the switches in the corresponding branches close, and the currents ($i_{f,i}$) start flowing through the impedance element ($Z_{f,i}$), and the circuit appearing in Fig. 2c holds. The current flowing through either Z^a_{fk} or Z^c_{fp} elements is a Faradaic current since it is involved in electrochemical half-cell reactions and represents a direct measure of the rate of the reactions taking place at the electrode-electrolyte interfaces. Dividing of the total current (i) passing through the treatment chamber among the half-cell reactions depends on their reaction potential and relative rates. Faraday's law governs the relationship between the amount m_{rs} (in g) of a species s produced by the reaction r, and the cumulative electrical charge Q_r (in C) transferred by the reaction over a period of time, i.e.,

$$m_{rs} = \frac{Q_r}{F}\frac{M_s}{n_{rs}} \tag{14}$$

where $F = 96{,}485$ C mol^{-1} is the Faraday constant, M_s is the molar mass of the substance s (in g mol^{-1}), and n_{rs} is the number of electrons transferred for each product s formed by reaction r. The ratio M_s/n_{rs} is the equivalent weight of the species s.

From the preceding discussion, the occurrence of electrode reactions might be avoided or minimized by maintaining the capacitor potential below the reaction potential of the reacting species through controlling the shape (amplitude/width) and protocol of the applied pulses, a proper design of circuit topology and treatment chamber, and the chemical-physical properties of the food product undergoing PEF process.

Factors Affecting Electrode Reactions

The extent of the electrochemical reactions occurring at the electrode-electrolyte interface of a PEF chamber is mainly related to the total amount of electrical charge passing through the electrode-electrolyte interface during the pulse treatment, which is in agreement with the Faraday's law.

The total charge transferred per each pulse Q_p (in C) depends primarily on current i (in A) and pulse duration τ (in s), where the current depends indirectly on food conductivity (σ), applied voltage (U), and chamber geometry. The transferred charge can be calculated by integrating the current density vector $\boldsymbol{j}$ at the interface area (Eq. 15). For conductive media the current density vector $\boldsymbol{j}$ equals $\sigma\cdot\boldsymbol{E}$ (Roodenburg et al. 2005a).

$$Q_p = \int_0^{\tau} |i|\,\mathrm{d}t = \int_0^{\tau}\iint_{A_i} |\boldsymbol{j}|\,\mathrm{d}\boldsymbol{A}\,\mathrm{d}t = \int_0^{\tau}\iint_{A_i} |\sigma\cdot\boldsymbol{E}|\,\mathrm{d}\boldsymbol{A}\,\mathrm{d}t \tag{15}$$

where $\boldsymbol{j}$ is the local time dependent current density (in A m^{-2}), $\boldsymbol{A}$ is the oriented surface of the interface (in m^2), σ is the electrical conductivity of the treated medium (in S m^{-1}), and $\boldsymbol{E}$ is the electric field vector (in V m^{-1}).

The total transferred charge (Q_{tot}, in C) during a PEF treatment can be determined by multiplying the charge delivered per pulse (Q_p, in C/pulse) by the total number of pulses (n_p) (Eq. 16). The number of pulses (n_p) is generally set for a batch treatment chamber, while for a continuous flow treatment chamber, it can be evaluated by multiplying the pulse repetition frequency (f, in Hz) and the residence time (t_r, in s) of the product in the chamber, which in turn can be estimated as the ratio of the treatment volume (v, L) and the volumetric flow rate (F_v, L s^{-1}).

$$Q_{tot} = n_p \cdot Q_p = f \cdot t_r \cdot Q_p = f \cdot \frac{v}{F_v} \cdot Q_p \tag{16}$$

It should be also noted that the total charge passing through the PEF chamber is strictly related to the total specific energy input W_T, which, along the field strength E, is one the main process parameters that define the PEF treatment intensity (Eq. 17):

$$W_T = n \cdot W_p = n \cdot \frac{Q_p \cdot U}{v} = \frac{f \cdot Q_p \cdot U}{F_v} = \frac{Q_{tot} \cdot U}{v} \tag{17}$$

where W_p is the specific energy input per pulse (in J L^{-1} pulse^{-1}).

However, besides the total amount of charge transferred, the way of delivering the electrical charge also affects the rate of charging and discharging of the double-layer capacitors, thus affecting the rate and the amount of electrochemical reactions.

From Eqs. (13, 15, and 16), it can be inferred that the extent and the rate of the electrochemical reactions occurring at the electrode-electrolyte interface of a PEF chamber depend on many factors, which are related to each other and that can be classified into three groups: processing parameters, design parameters, and treatment medium characteristics. An adequate knowledge of the influence of these factors and their interactions on the extent of the electrochemical reactions occurring at the electrode-electrolyte interface is necessary to establish optimal PEF treatment conditions, circuit topology of the pulse generator, geometry and material of electrodes, and food characteristics allowing to minimize any undesired side effect related to these reactions.

Processing Parameters

As it can be inferred from the Eqs. (13 and 15, 16, 17), for a given chamber geometry and treatment medium characteristics, the main PEF processing parameters that can affect the buildup of charge on the double-layer capacitors and, consequently, the extent and rate of electrode reactions are electric field strength, energy input, pulse width, and pulse polarity.

In general, at a fixed pulse width, increasing the field strength (i.e., applied voltage) and energy input (i.e., number of pulses or treatment frequency) will increase the total transferred charge through the electrode-electrolyte interface (Eqs. 15, 16 and 17), as well as the charging rate (Eq. 13) of the double-layer by increasing the current density. However, it was shown that variation in the applied voltage, within the typical PEF conditions for microbial inactivation, did not have a remarkable effect on the extent of electrochemical reactions (Gad and Jayaram 2014). On the other hand, instead, the pulse repetition frequency, which is the main processing parameter that need to be adjusted when operating in a continuous flow mode to deliver the desired amount of energy through the treatment chamber, plays a crucial role on the charging rate of the double-layer capacitors, especially when monopolar pulses are applied (Pataro et al. 2014a, 2015b). In particular, as the pulse frequency increases upon increasing the energy input (Eq. 17), the time interval between two consecutive monopolar pulses decreases resulting in only a partial discharge of the double-layer capacitor at each electrode-electrolyte interface, before the next pulse will be applied. As a consequence, charges will be accumulated on the double layer at a faster rate as the frequency increases, and when the threshold voltage is exceeded, electrochemical reactions will occur causing the corrosion of the electrodes or other undesired side effects (Pataro et al. 2014a, 2015b). This explanation agrees with the concept of charging the double-layer capacitors as well as with the need to avoid a cumulative buildup of charge discussed in Morren et al. (2003).

In addition to field strength, energy input and frequency, electrochemical reactions are also dependent on pulse width and pulse polarity since these parameters may affect the charging rate of the double-layer capacitance.

It has been shown that, when the applied voltage is kept the same, shorter pulses lead to a lower amount of electrochemical reaction (Gad and Jayaram 2014; Morren et al. 2003). To explain the effect of pulse width, Gad (2014) suggested that, from the perspective of each half-cell reaction, the width of each pulse can be divided into two parts. In the first part, the double layer is charged as long as its potential remains below the potential reaction of the reacting species. For a given electrode material (i.e., for a given double-layer capacitance), the duration of the first part depends on the current (i.e., applied voltage) and remains unchanged (Eq. 13). The second part starts when the potential reaction is reached and thus the half-cell reaction occurs. With shorter pulses, the second part of the pulse (when electrode reactions occur) is shortened considerably. This is in agreement with Eq. (13) that gives a trade-off between current density and pulse duration for a certain treatment chamber geometry. A higher field strength in the same treatment chamber (i.e., a higher applied voltage) results in a higher current density. Thus for higher field strength, the pulse duration should be decreased. Therefore, the optimum pulse should feature relatively short pulse width with an applied voltage satisfying the minimum field strength required for microbial inactivation (Gad and Jayaram 2014).

As compared to monopolar pulses, bipolar or oscillatory pulses featuring a ratio between positive and negative pulses closer to unity may prevent electrochemical reactions by the same concept of zero-net-charge delivery (Bushnell et al. 1995b; Roodenburg et al. 2005a), where both double-layer capacitors are partially but continuously charged and discharged without reaching the threshold potential (U_{th}) corresponding to the onset of anodic and cathodic Faradaic reactions (Amatore et al. 1998). Under such circumstances, the current flowing through the cell would be purely capacitive leading to only a formal displacement of charges from one double layer to the double layer of the other electrode (Gad 2014).

Design Parameters

The extent of electrophoretic phenomena and electrochemical reactions at the electrode-electrolyte interface of a PEF chamber is also affected by circuit topology of the pulse generator, treatment chamber geometry, and electrode materials.

Besides the generation of a given pulse shape, a proper design of the circuit topology can enable to force double-layer discharging based on the concept of "zero-net-charge delivery," which substantially reduces the extent of electrophoretic and electrochemical effects (Bushnell et al. 1995a, b). In this line, Gad and Jayaram (2014) designed a pulse generator characterized by a bidirectional nature of the current waveform, which was used to ensure zero-net-charge delivery with monopolar pulses. Based on this principle, the charging current of the generator's capacitor was allowed to pass through the PEF processing chamber in one direction, and then, the pulse current generated during the capacitor discharging phase flowed in the opposite direction. Thus, the current flowing through the chamber during the capacitor-discharging phase removes any charge remaining on the double layer from the capacitor-charge phase. Thus, the accumulation of charge on the double layer from one pulse to another can be avoided.

From Eq. (13), it can be inferred that, for a given electrode reaction and external applied voltage, the threshold time depends mainly on the current (density) through the electrical resistance of the PEF chamber and the value of the double-layer capacitance of the electrode material.

Thus, PEF treatment chamber geometries featuring high intrinsic electrical resistance, which, therefore, operate at a relatively low current, may limit the occurrence of electrochemical reactions (Gongora Nieto et al. 2002). Different types of electrode arrangement have been proposed both for static and continuous chamber, with the most commonly used having been parallel plate and co-field configuration. Parallel plate electrodes configuration are the simplest in design and produce the most uniform distribution of the electric field. They are typically characterized by a large electrode surface and low intrinsic electrical resistance (from few ohms to tens of ohms) (Gongora-Nieto et al. 2002; Raso et al. 2016). Therefore, this configuration generally operates at relatively high current density, which makes shorter the charging time of the double-layer capacitance. On the other hand, a co-field chamber configuration cannot develop a constant electric field across the gap (as with parallel plate), thus yielding nonuniform treatment. However, such configuration has advantageous fluid dynamics, highly desiderate for food processing and convenient for cleaning in place, as well as a high intrinsic electrical resistance (of the order of hundreds of ohms) due to the low effective area in the cross section of the tubular electrodes (Gongora-Nieto et al. 2002). Thus, this configuration typically operates at lower current density than the parallel plate configuration, which makes it suitable for limiting the occurrence of electrode reactions (Eq. 13).

Similarly, the use of high capacitance electrode materials, such as platinum (C_{dl} = 48 μF/cm^2), titanium (C_{dl} = 50 μF/cm^2), glassy carbon (C_{dl} = 260 μF/cm^2), or dimensionally stable anode (DSA) (C_{dl} = 2000 μF/cm^2), instead of the most commonly used stainless steel electrodes (C_{dl} = 35 μF/cm^2), may limit the occurrence of electrode reactions. In fact, according to Eq. (13), the use of electrode material with high C_{dl} values may slow down the charging process of the double-layer capacitor, avoiding its full charging before the next pulses is applied (Amatore et al. 1998; Góngora-Nieto et al. 2002; Pataro and Ferrari 2020). Moreover, on the basis of the same principle of avoiding full charging of the double-layer capacitors, it has been found that the extent of the electrochemical reactions may be limited by using electrode materials featuring higher resistance to electrochemical reactions such as titanium and platinized titanium (Amatore et al. 1998; Gad et al. 2014; Kim and Zhang 2011).

Treatment Medium Characteristics

The composition as well as chemical physical properties of the treatment medium can significantly affect the type and the extent of the electrophoretic phenomena and electrochemical reactions occurring at electrode-electrolyte interface.

In rich-protein foods such as milk, a film of material can collect or agglomerate on at least one electrode via electrophoretic process leading to the so-called fouling of the electrodes.

On the other hand, many food products contain halides such as chlorides, which are aggressive anions which may damage the thin oxide layer on the passivated metal electrode surface protecting the metals against corrosion. Thus the presence of these active species may contribute to the enhancement of the oxidation half-cell reactions at the anode, including those causing the release of metals from the electrodes (Pataro et al. 2014a, 2015b).

Also electrical conductivity and pH of the treatment medium may affect the extent of electrode reactions. Acidic food products, such as most fruit juices, may face higher problems with electrode material migration than nonacidic products, especially when PEF is carried out under severe processing conditions using electrochemically active electrode material (e.g., stainless steel) (Amatore et al. 1998; Gad and Jayaram 2011, 2012a; Pataro et al. 2012, 2014a, b; Roodenburg et al. 2005b). This can be explained by knowing that lower pH means higher concentration of hydrogen ions at the cathode. As a result, the cathodic half-reaction (Eq. 18) could couple with the anodic half-reaction (Eq. 19) resulting in accelerated corrosion and hydrogen generation (Eq. 20) compared to those at the neutral pH value (Amatore et al., 1998).

Cathodic half-reaction:

$$2H^{+}_{(aq)} + 2e^{-} \leftrightarrow H_{2(g)} \tag{18}$$

Anodic half-reaction:

$$M_{(s)} \leftrightarrow M^{2+}_{(aq)} + 2e^{-} \tag{19}$$

Overall reaction:

$$M_{(s)} + 2H^{2+}_{(aq)} \leftrightarrow M^{2+}_{(aq)} + H_{2(g)} \tag{20}$$

where M = Fe, Cr, Ni or any other metallic components of the electrode material.

Thus, for treating of acidic products, including beverages such as orange juice, the use of more inert materials may be an interesting way for markedly reducing the extent of electrode corrosion and electrochemical phenomena that occur at the electrode-solution interface (Gad and Jayaram 2012b). The effect of the type of acid, however, requires further investigation.

On the other hand, highly conductive foods such as tomato juice, milk, and orange juice may experience a higher rate of electrode material migration (Amatore et al. 1998; Gad and Jayaram 2012b; Pataro et al. 2012), due to the high current flow passing through the medium with high conductivity, which lead to a higher charging rate of the double-layer capacitors (Eq. 13). As a result, according to Eq. (13), beverages such as orange juice, which has a typical conductivity of 0.3 S/m, can be treated with a longer pulse than tomato juice, which has a typical conductivity of 2 S/m.

Side Effects of Electrochemical Phenomena and Challenges

During PEF treatment electrophoresis-based phenomena (electrode fouling) as well as various electrochemical reactions can potentially occur, including those involving chemical changes in food products, electrode corrosion, and electrolysis of water. In addition, some of the products of those electrochemical reactions may initiate a number of secondary chemical reactions involving food components, even after pulse treatment has been completed. These reactions are unavoidable, especially for those applications requiring high treatment intensity (e.g., microbial inactivation) and lead to undesired effects, which may hamper commercialization of the PEF process through technological, safety, quality, and costs aspects (Pataro and Ferrari, 2020), as depicted in Table 1.

Electrode corrosion is one of the main consequences of the electrochemical reactions occurring at the electrode-electrolyte interface of a PEF treatment chamber. The dissolution of the anode material due to oxidation of the metal of the electrode

Table 1 Side effects of electrochemical phenomena occurring in the operation of a PEF treatment chamber (Pataro and Ferrari, 2020)

Electrophoresis, electrochemical phenomena, secondary reactions	Process and technological issues	Effects on Food safety and quality
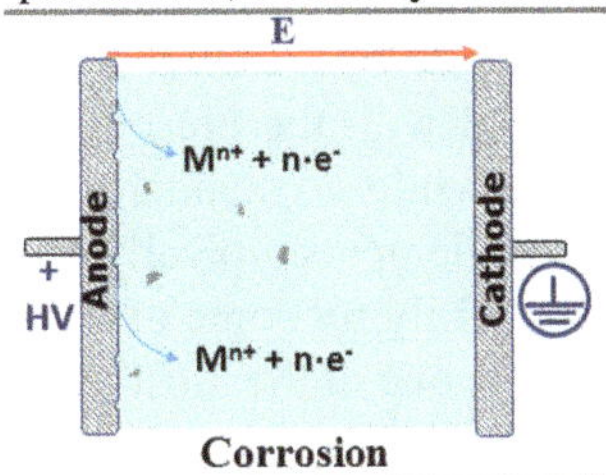 E; Anode; Cathode; + HV; $M^{n+} + n{\cdot}e^-$; $M^{n+} + n{\cdot}e^-$ **Corrosion**	Distortion or local enhancements of the electric field Electric arcs formation Short electrode lifetime Reduced reliability and life time of pulse generator	Health safety regulation Alteration in color, flavor, and taste (metallic mouth feeling) Toxic compounds Formation of high reactive species (free radicals) Food disintegration (by-products)
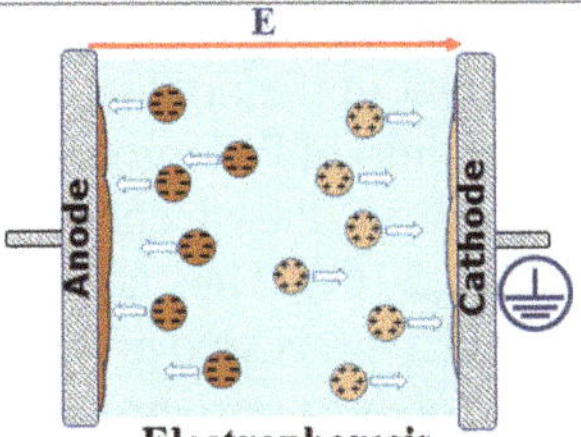 E; Anode; Cathode **Electrophoresis**	Electrode fouling Electric arcs formation Short electrode lifetime Reduced reliability and life time of pulse generators Interruption of the flow of the product	Contamination of the processed food Formation of high reactive species (free radicals)
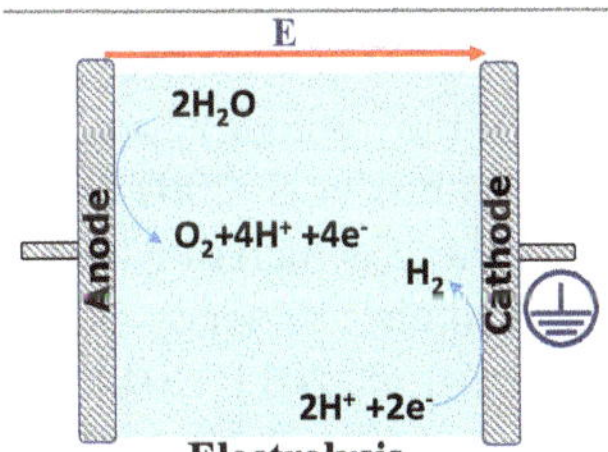 E; Anode; Cathode; $2H_2O$; $O_2+4H^+ +4e^-$; H_2; $2H^+ +2e^-$ **Electrolysis**	Gas bubble generation Electrode corrosion Electric arcs formation Reduced reliability of and lifetime of pulse generators	Oxidative degradation of food components Formation of toxic compounds (H_2O_2) Under-processing of food
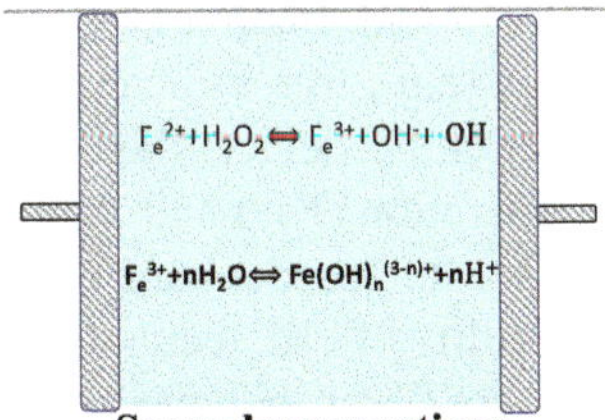 $F_e^{2+} + H_2O_2 \Leftrightarrow F_e^{3+} + OH^- + OH$; $F_e^{3+} + nH_2O \Leftrightarrow Fe(OH)_n^{(3-n)+} + nH^+$ **Secondary reactions**	Contamination of the system (rust formation)	Local changes of pH Generation of reactive oxygen species (ROS) Oxidative degradation of food constituents Production of active chlorine species (ACS) Toxicity problems

M^{n+} migrated metal ions, piece of anode material (pitting corrosion), negative-charged particles, positive charged particles

(Eq. 12) along with detachment of small particles of electrode material (pitting corrosion) is responsible for this phenomenon.

The electrode material released into the PEF processed foods are basically contaminants and may have some toxic potential, as well as affect food quality, equipment reliability, treatment efficiency, and electrode lifetime (Evrendilek et al. 2004; Morren et al. 2003; Pataro et al. 2014a, b, 2015b; Roodenburg et al. 2005a,b; Saulis et al. 2007).

For the safety aspect of PEF processing, the type and amount of electrode material released in the processed products must be within the health safety regulations before introducing it as processed food to the market (Gad and Jayaram 2012a; Roodenburg et al. 2005a). By reviewing the regulations for food industries dictated by governments in Europe and USA (CODEX STAN 193–1995; CODEX STAN 247–2005; USDA 2018), they have been mainly concerned with the presence of heavy/toxic metals, namely, cadmium, lead, and mercury, while no standards are given for the maximum concentration of dissolved metallic elements typically constituting the materials used as electrodes in a PEF chamber. Only for some of them, such as the main metallic elements of stainless steel, which is one of the most popular electrode materials of PEF chamber, guidelines could be taken from drinking water regulations (Directive No 98/83/EG; U.S. EPA 2009), as shown in Table 2.

Apart from safety aspects, the release of electrode material in the product may cause alteration in color, flavor, and taste (metallic mouth feeling) of PEF processed foods, which may stand against consumers' acceptance to this technology (Evrendilek et al. 2004; Gad and Jayaram 2012b; Sun et al. 2011). Thus, sensory analyses should be performed to evaluate the acceptance of PEF processed products against either conventionally or unconventionally processed foods or fresh products.

Finally, corrosion can cause serious damages to the electrodes, whose surface roughness can markedly increase as a consequence of the metal release from the anode and, in part, as partial deposition of metal oxides on the cathode (Kim and Zhang 2011; Pataro et al. 2014a; Roodenburg et al. 2005a; Saulis et al. 2007). This might create distortion or local enhancements of the electric field, which may markedly impair PEF treatment uniformity as well as reduce treatment efficiency during the operation (Donsì et al. 2007; Jaeger et al. 2009; Pataro et al. 2011, 2014a, b; Saulis et al. 2007). Moreover, the increase in surface roughness may trigger arcs formation (dielectrical breakdown of food) within the treatment chamber, which may likely promote the formation of high reactive species (e.g., free radicals), as well as reduce the reliability of pulse generator (e.g., switching devices) and the lifetime of electrodes (pitting corrosion) to few hours of operation (Kim and Zhang 2011; Gad et al. 2014; Saulis et al. 2007; Pataro et al. 2014a; Roodenburg et al. 2005b; Toepfl et al. 2007), thus reducing the technical feasibility of PEF technology.

Fouling of the electrodes is another undesirable side effect that is believed to be due to electrophoresis accompanying the flow of electric current during PEF processing. Specifically, under the influence of the external electric field, charged particles (e.g., relatively large protein molecules or other fouling agent) suspended in the liquid food product are moved toward the oppositely charged electrode where they are collected, or agglomerated, to form a film of food particles (Bushnell et al. 1995a, b). The formation of such a film on the electrode(s) during extended processing periods can cause local electric field distortion, arcing, contamination of the system, and, in some cases, the interruption of the flow of the product (Bushnell et al. 1995a, b).

Electrolysis of water, which is largely contained in food subjected to PEF treatment, may also occur generating H_2 and O_2 gases (Eqs. 7, 8, and 9) at the electrode-solution interfaces. In addition, especially when there is a significant amount of

Table 2 Maximum contamination levels of metallic elements of stainless steel dictated by European legislation (Directive No 98/83/EG) and US regulations for drinking water

Concentration level of metallic elements (μg/L)				
Fe	Cr	Ni	Mn	Regulation
200	50	20	50	*Directive No 98/83/EG*
300[a]	100[b]	NA	50[a]	*U.S. EPA 2009*

[a]National Secondary Drinking Water Regulations are non-enforceable guidelines regarding contaminants that may cause cosmetic effects (such as skin or tooth discoloration) or aesthetic effects (such as taste, odor, or color) in drinking water
[b]National Primary Drinking Water Regulations (NPDWRs) are enforceable standards that are established to protect the public against consumption of drinking water contaminants that present a risk to human health

chloride ions in the treated medium, cathodic half-reactions for electrolysis (Eqs. 7 and 8) may be also coupled with the anodic half-reaction (Eq. 4), resulting in $H_{2(g)}$ and $Cl_{2(g)}$ generation. The liberation of oxygen by electrolysis may further promote electrode corrosion (Tzedakis et al. 1999) as well as induce degradation of food components (e.g., phenolic compounds, lipids, and vitamins like ascorbic acid) and the formation of toxic compounds, such as H_2O_2 (Eq. 6) (Morren et al. 2003). Moreover, the presence of gas bubbles formed by electrolysis increases the chance of the occurrence of dielectric breakdown and arcing, thus causing technological and food safety problems and reducing the treatment efficiency (Góngora-Nieto et al. 2003; Toepfl et al. 2007).

Finally, products generate from electrochemical reactions during PEF treatments may initiate a number of secondary chemical reactions by reacting each other or involving food constituents, which may lead to changes in chemical composition, organoleptic attributes, and physical properties of processed food products, as well as the formation of toxic compounds even after the pulse treatment has been completed (Morren et al. 2003; Pataro and Ferrari 2020).

For example, secondary reactions triggered by corrosion and electrolysis products may induce local changes of pH in the vicinity of electrode surfaces, which may affect activity and stability of enzymes and microbial resistance to PEF treatments (Meneses et al. 2011; Rodaitė-Riševičienė et al. 2014). Moreover, the migrated metal ions along with pH changes can affect degradation of natural antioxidant compounds such as ascorbic acid (Assiry et al. 2006), as well as the stability of pigments such as anthocyanins (Sun et al. 2011), thus potentially altering color and nutritional and functional properties of food products. Corrosion and electrolysis products may also generate reactive oxygen species (ROS) such as hydroxyl (•OH), hydroperoxyl (•OOH), superoxide anion (O_2•−) radicals, H_2O_2, and singlet oxygen ($1O_2$), which may cause oxidative degradation of lipids, vitamins, and amino acids/proteins, thus potentially leading to production of undesirable flavor, toxic, and color compounds, which make foods less acceptable or unacceptable to consumers (Zhao et al. 2012).

In product containing chloride compounds, as many food products do, secondary reactions triggered by electrolysis products (e.g., Cl_2, Eq. 4) can result in the production of active chlorine species (ACS) with high redox potentials, such as hypochlorous acid (HOCl) and hypochlorite anions (OCl). The formation of these compounds can cause toxicity problems, even though they may also act as bactericides, thus improving the preservation efficiency of PEF treatment (Zhao et al. 2012).

Based on the above discussion, there is a challenge to assess whether the side effects are within the acceptable limits or to provide feasible strategies to eliminate, or at least reduce, the extent of electrochemical phenomena that cause these undesired effects. Specifically, although dissolution of electrode materials and other electrochemical reactions are largely unavoidable in the long term, their limitation is possible by either avoiding full charging of the double-layer capacitors, or using chemically inert electrode materials.

To this regard, for the PEF treatment of a given food product, strategies such as reduction of the buildup area (as in the case of co-field chambers) keep the applied voltage, pulse length, and frequency as the minimum value required for generating sufficient electric field and electrical energy for the specific application, as well as using bipolar pulses and electrode material having a high specific double-layer capacitance can help to significantly reduce the amount of electrochemical reactions (Amatore et al. 1998; Gad and Jayaram 2012a, 2014; Góngora-Nieto et al. 2002; Kotnik et al. 2001; Pataro et al. 2014a, 2015b; Roodenburg et al. 2005a, b). Similarly, proper design of pulse generator equipped with switching devices with a very low leakage current or by inserting a pulse transformer into the system may also reduce the amount of electrochemical reactions (Mastiwijk 2007). On the other hand, replacing commonly used stainless steel electrodes by more inert material such as titanium electrodes may be one solution to overcome this problem (Kim and Zhang 2011; Gad and Jayaram 2014; Toepfl et al. 2007). Also homogeneity of the electric field within the treatment zone, reduction and generation of enhanced field area, polishing of electrode surface after the mechanical machining (i.e. lathe), as well as degassing the food product before PEF treatment are other important design criteria and strategies for avoiding or limiting undesired electrochemical effects at the electrode-media interface (Pataro and Ferrari 2020).

Conclusions

Today, the first successful industrial applications of PEF processing in food industry have been achieved, especially as a pre-processing step to facilitate cutting operation in potato processing industry, where low or moderate PEF treatment intensities are required. In the food sector of fruit juice preservation, instead, where high treatment intensity along with and highly homogeneous treatment are required, several disadvantages and limitations still remain, such as those related with the electrophoretic and electrochemical phenomena occurring at the electrode-food interface of a PEF treatment chamber, which are unavoidable in long-term trials or when large amount of electrical charge is transferred through the electrodes.

The occurrence of electrochemical phenomena, especially those leading to corrosion and fouling of the electrodes, electrolysis of water, migration of electrode material components, and chemical changes, may seriously affect PEF commercialization through safety, quality, and technological and cost aspects.

The present challenge is to reduce the amount of electrochemical reactions by modifying the pulse generator systems, improving treatment design, and selecting more inert electrode materials and minimum treatment intensity, while taking into account chemical-physical properties of food under treatment, in order to improve the technical feasibility of PEF technology as well as to maintain chemical food safety.

References

Amatore C, Berthou M, Hebert S (1998) Fundamental principles of electrochemical ohmic heating of solutions. J Electroanal Chem 457:191–203

Arnal ÁJ, Royo P, Pataro G, Ferrari G, Ferreira VJ, López-Sabirón AM, Ferreira GA (2018) Implementation of PEF treatment at real-scale tomatoes processing considering LCA methodology as an innovation strategy in the Agri-food sector. Sustainability 10:979

Assiry AM, Sastry SK, Samaranayake CP (2006) Influence of temperature, electrical conductivity, power and pH on ascorbic acid degradation kinetics during ohmic heating using stainless steel electrodes. Bioelectrochemistry 68:7–13

Beattie JM, Lewis FC (1925) The electric current (apart from the heat generated) a bacteriological agent in the sterilization of milk and other fluids. J Hyg 24:123–137

Bobinaitė R, Pataro G, Lamanauskas N, Šatkauskas S, Viškelis P, Ferrari G (2015) Application of pulsed electric field in the production of juice and extraction of bioactive compounds from blueberry fruits and their by-products. J Food Sci Technol. https://doi.org/10.1007/s13197-014-1668-0

Bockris JOM, Reddy AKN, Gamboa-Aldeco M (2002) The electrified interface. In: Bockris JOM, Reddy AKN, Gamboa-Aldeco M (eds) Modern electrochemistry 2A. Fundamentals of electrodics. Kluwer Academic Publishers, New York, pp 771–1015

Bushnell AH, Clark RW, Dunn JE, Lloyd SW (1995a) Prevention of electrode fouling in high electric field systems for killing microorganisms in food products. United States Patent, US 5,393-541

Bushnell AH, Clark RW, Dunn JE Lloyd SW (1995b) Prevention of electrochemical and electrophoretic effects in high-strength-electric-field pumpable-food-product treatment systems. United States Patent, US 5,447-733

CODEX STAN 193-1995. http://www.fao.org/fao-who-codexalimentarius/codex-texts/tr/

CODEX STAN 247-2005. www.fao.org/input/download/standards/10154/CXS_247e.pdf

Donsì G, Ferrari G, Pataro G (2007) Inactivation kinetics of Saccharomyces cerevisiae by pulsed electric fields in a batch treatment chamber: the effect of electric field unevenness and initial cell concentration. J Food Eng 78:784–792

Donsì F, Ferrari G, Pataro G (2010) Applications of pulsed electric field treatments for the enhancement of mass transfer from vegetable tissue. Food Eng Rev 2:109–130

Dunn JE, & Pearlman JS (1987) Methods and apparatus for extending the shelf life of fluid food products. United States Patent, vol. US 4,695,472, 09/22

European legislation on the quality of water intended for human consumption for drinking water. (Directive No 98/83/EG). http://ec.europa.eu/environment/water/water-drink/legislation_en.html

Evrendilek GA, Dantzer S, Li WR, Zhang QH (2004) Pulsed electric field processing of beer: microbial, sensory, and quality analyses. J Food Sci 69:228–232

Fetterman JC (1928) The electrical conductivity method of processing milk. Agric Eng 9(4):107–108

Fauster T, Schlossnikl D, Rath F, Ostermeier R, Teufel F, Toepfl S, & Jaeger H (2018) Impact of pulsed electric field (PEF) pretreatment on process performance of industrial French fries production. Journal of Food Engineering, 235, 16–22

Gad A (2014) A study of electrode material performance during food processing by pulsed electric fields. PhD thesis. University of Waterloo, Waterloo, ON, Canada

Gad A, Jayaram SH (2011) Electrode material migration during Pulsed Electric Field (PEF) treatment. In: Conference ESA annual meeting on electrostatics 2011, Cleveland, Ohio, USA, pp 1–9. http://www.electrostatics.org/esa2011proceedings.html

Gad A, Jayaram SH (2012a) Effect of electric pulse parameters on releasing metallic particles from stainless steel electrodes during pulsed electric field processing of milk. In: Conference electrostatics joint conference 2012, Cambridge, ON (Canada), pp 1–7. http://www.electrostatics.org/esa2012proceedings.html)

Gad A, Jayaram SH (2012b) Effect of food composition and pH on electrode material migration during PEF application. In: conference BFE 2012, International conference bio & food electrotechnologies, Salerno, Italy, pp 49–52

Gad A, Jayaram SH (2014) Effect of electric pulse parameters on releasing metallic particles from stainless steel electrodes during PEF processing of milk. IEEE Trans Ind Appl 50(2):1402–1409. https://doi.org/10.1109/TIA.2013.2278424

Gad A, Jayaram SH, Pritzker M (2014) Performance of electrode materials during food processing by pulsed electric fields. IEEE Trans Plasma Sci 42(10):3161–3166. https://doi.org/10.1109/TPS.2014.2312711

Gileadi E (1993a) The ionic double-layer capacitance Cdl. In: Gileadi E (ed) Electrode kinetics for chemists, chemical engineers, and material scientists. VCH Publishers, New York, pp 185–224

Gileadi E (1993b) Electrode kinetics: some basic concepts. In: Gileadi E (ed) Electrode kinetics for chemists, chemical engineers, and material scientists. VCH Publishers, New York, pp 51–57

Gileadi E (1993c) Experimental techniques. In: Gileadi E (ed) Electrode kinetics for chemists, chemical engineers, and material scientists. VCH Publishers, New York, pp 349–454

Gilliland SE, Speck ML (1967) Mechanism of the bactericidal action produced by electrohydraulic shock. Appl Microbiol 15:1038–1044

Góngora-Nieto MM, Sepúlveda DR, Pedrow P, Barbosa-Cánovas GV, Swanson BG (2002) Food processing by pulsed electric fields: treatment delivery, inactivation level, and regulatory aspects. Lebensmittel-Wissenschaft Und-Technologie 35:375–388

Góngora-Nieto MM, Pedrow PD, Swanson BG, Barbosa-Cánovas GV (2003) Impact of air bubbles in a dielectric liquid when subjected to high electric field strengths. Innov Food Sci Emerg Technol 4:57–67

Jaeger H, Meneneses N, Knorr D (2009) Impact of PEF treatment inhomogeneity such as electric field distribution flow characteristics and temperature effects on the inactivation of *E. coli* and milk alkaline phosphatase. Innov Food Sci Emerg Technol 10:470–480

Jalte M, Lanoiselle JL, Lebovka N, Vorobiev E (2009) Freezing of potato tissue pretreated by pulsed electric fields. Food Sci Technol 42(2):576–580

Jeyamkondan S, Jayas DS, Holley RA (1999) Pulsed electric field processing of foods: a review. J Food Prot 62(9):1088–1096

Kempkes M (2017) Industrial pulsed electric field systems. In: Miklavcic D (ed) Handbook of electroporation. Springer, Cham, pp 1–21. https://doi.org/10.1007/978-3-319-26779-1_211-1

Kotnik T, Miklavčič D, Mir LM (2001) Cell membrane Electropermeabilization by symmetrical bipolar rectangular pulses. Part II. Reduced electrolytic contamination. Bioelectrochemistry 54:91–95

Kotnik T, Kramar P, Pucihar G, Miklavčič D, Tarek M (2012) Cell membrane electroporation — Part 1: The phenomenon. IEEE Electrical Insulation Magazine, 28(5), 14–23

Krüger S (1893) Über den Einfluss des constanten elektrischen Stromes auf Wachsthum und Virulenz der Bakterien. Zeitschrift klinische Medizin 22:191–207
Kim M, Zhang HQ (2011) Improving Electrode Durability of PEF Chamber by Selecting Suitable Material. In Nonthermal Processing Technologies for Food, H. Q. Zhang, G.V. Barbosa-Cánovas, V.M. Balasubramaniam, C.P. Dunne, D.F. Farkas, and J.T.C. Yuan (Ed.s), pp (201-212), Blackwell Publishing Ltd. and Institute of Food Technologists, Chichester, United Kingdom
Master AM, Schuten HJ, Mastwijk HC (2007) In Lelieveld HLM, Notermans S, de Haan SWH (eds) Food preservation by pulsed electric fields. CRC Press, New York, pp 201–211
Mastwijk HC (2007) Pulsed power systems for PEF in food industry. In: Raso J, Heinz V (eds) Pulsed electric fields technology for the food industry. Springer, New York, pp 223–237
Meneses N, Jaeger H, Knorr D (2011) pH-changes during pulsed electric field treatments – Numerical simulation and in situ impact on polyphenoloxidase inactivation. Innov Food Sci Emerg Technol 12:499–504
Morren J, Roodenburg B, de Haan SWH (2003) Electrochemical reactions and electrode corrosion in pulsed electric field (PEF) treatment chambers. Innov Food Sci Emerg Technol 4:285–295
Pareilleux A, Sicard N (1970) Lethal effects of electric current on *Escherichia coli*. Appl Microbiol 19(3):421–424
Parniakov O, Lebovka N, Bals O, Vorobiev E (2015) Effect of electric field and osmotic pre-treatments on quality of apples after freezing-thawing. Innov Food Sci Emerg Technol 29:23–30
Parniakov O, Bals O, Lebovka N, Vorobiev E (2016a) Effects of pulsed electric fields assisted osmotic dehydration on freezing-thawing and texture of apple tissue. J Food Eng 183:32–38
Parniakov O, Bals O, Lebovka N, Vorobiev E (2016b) Pulsed electric field assisted vacuum freeze-drying of apple tissue. Innov Food Sci Emerg Technol 35:52–57
Pataro G, Ferrari G (2020) Limitations of pulsed electric field utilization in food industry. In: Barba F, Parniakov O, Wiktor A (eds) Pulsed electric field to obtain healthier and sustainable food for tomorrow. Academic Press, London, pp 283–310
Pataro G, Senatore B, Donsi G, Ferrari G (2011) Effect of electric and flow parameters on PEF treatment efficiency. J Food Eng 105:79–88
Pataro G, Donsì G, Ferrari G (2012) Metal release from stainless steel electrodes of a PEF treatment chamber. In: International conference Bio & Food Electrotechnologies (BFE2012), Salerno, Italy, pp 29–33. ISBN 978-88-903261-8-9
Pataro G, Falcone M, Donsì G, Ferrari G (2014a) Metal release from stainless steel electrodes of a PEF treatment chamber: effects of electrical parameters and food composition. Innov Food Sci Emerg Technol 21:58–65
Pataro G, Barca GMJ, Pereira RN, Vicente AA, Teixeira JA, Ferrari G (2014b) Quantification of metal release from stainless steel electrodes during conventional and pulsed Ohmic heating. Innov Food Sci Emerg Technol 21:66–73
Pataro G, Barca GMJ, Donsì G, Ferrari G (2015a) On the modeling of electrochemical phenomena at the electrode solution interface in a PEF treatment chamber: methodological approach to describe the phenomenon of metal release. J Food Eng 165:34–44
Pataro G, Barca GMJ, Donsì G, Ferrari G (2015b) On the modelling of the electrochemical phenomena at the electrode solution interface of a PEF treatment chamber: effect of electrical parameters and chemical composition of model liquid food. J Food Eng 165:34–44
Pataro G, Donsì G, Ferrari G (2017a) Modeling of electrochemical reactions during pulsed electric field treatment. In: Miklavcic D (ed) Handbook of electroporation. Springer, Cham, pp 1–30. https://doi.org/10.1007/978-3-319-26779-1_5-1
Pataro G, Donsì G, Ferrari G (2017b) Modeling of electrochemical reactions during pulsed electric field treatment. In: Miklavcic D (ed) Handbook of electroporation. Springer, Cham, pp 1–30. https://doi.org/10.1007/978-3-319-26779-1_5-1
Pataro G, Carullo D, Falcone M, Ferrari G (2020) Recovery of lycopene from industrially derived tomato processing byproducts by pulsed electric fields-assisted extraction. Innov Food Sci Emerg Technol 63:102369
Prochownick L, Spaeth F (1890) Über die keimtötende Wirkung des galvanischen Stroms. Dtsch Med Wochenschr 26:564–565

Raso J, Frey W, Ferrari G, Pataro G, Knorr D, Teissie J, Miklavčič D (2016) Recommendations guidelines on the key information to be reported in studies of application of PEF technology in food and biotechnological processes. Innov Food Sci Emerg Technol 37:312–321

Rodaitė-Riševičienė R, Saule R, Snitka V, Saulis G (2014) Release of iron ions from the stainless steel anode occurring during high-voltage pulses and its consequences for cell electroporation technology. IEEE Trans Plasma Sci 42(1):249

Roodenburg B, de Haan SWH, van Boxtel LBJ, Hatt V, Wouters PC, Coronel P, & Ferreira JA (2010) Conductive plastic film electrodes for Pulsed Electric Field (PEF) treatment—A proof of principle. Innovative Food Science and Emerging Technologies, 11, 274–282

Roodenburg B, Morren J, Berg HE, de Haan SWH (2005a) Metal release in a stainless steel Pulsed Electric Field (PEF) system. Part I. Effect of different pulse shapes; theory and experimental method. Innov Food Sci Emerg Technol 6:327–336

Roodenburg B, Morren J, Berg HE, de Haan SWH (2005b) Metal release in a stainless steel pulsed electric field (PEF) system. Part II. The treatment of orange juice; related to legislation and treatment chamber lifetime. Innov Food Sci Emerg Technol 6:337–345

Sale AJH, & Hamilton WA (1967) Effects of high electric fields on microorganisms I. Killing of bacteria and yeast. Biochimica Biophysica Acta, 148, 781–788

Saldaña G, Álvarez I, Condón S, Raso J (2014) Microbiological aspects related to the feasibility of PEF technology for food pasteurization. Crit Rev Food Sci Nutr 54(11):1415–1426

Saulis G, Rodaitė-Riševičienė R, Snitka V (2007) Increase of the roughness of the stainless-steel anode surface due to the exposure to high-voltage electric pulses as revealed by atomic force microscopy. Bioelectrochemistry 70:519–523

Soliva-Fortuny R, Balasa A, Knorr D, Martín-Belloso O (2009) Effects of pulsed electric fields on bioactive compounds in foods: a review. Trends Food Sci Technol 20:544–556

Sun J, Bai W, Zhang Y, Liao X, Hu X (2011) Effects of electrode materials on the degradation, spectral characteristics, visual colour, and antioxidant capacity of cyanidin-3-glucoside and cyanidin-3-sophoroside during pulsed electric field (PEF) treatment. Food Chem 128(3):742–747

Thiele H, Wolf K (1899) Über die Einwirkung des elektrischen Stroms auf Bakterien. Centralblatt Bakterien und Parasitenkunde 25:650–655

Toepfl S, Heinz V, Knorr D (2007) High intensity pulsed electric fields applied for food preservation. Chem Eng Process 46(6):537–546

Tzedakis T, Basseguy R, Comtat M (1999) Voltammetric and coulometric techniques to estimate the electrochemical reaction rate during ohmic sterilization. J Appl Electrochem 29(7):821–828

U.S. Environmental Protection Agency (EPA), National Primary Drinking Water Regulations. https://www.epa.gov/ground-water-and-drinking-water/nationalprimary-drinking-water-regulations

U.S. Environmental Protection Agency (EPA), National Drinking Water Regulations, 2009. Secondary drinking water standards: guidance for nuisance chemicals. https://www.epa.gov/sdwa/secondary-drinking-water-standards-guidance-nuisance-chemicals

U.S. Food and Drug Administration (USDA) (2018) Guidance for industry: action levels for poisonous or deleterious substances in human food and animal feed. https://www.fda.gov/regulatory-information/search-fda-guidance-documents/guidance-industry-action-levels-poisonous-or-deleterious-substances-human-food-and-animal-feed

Wiktor A, Iwaniuk M, Sledz M, Nowacka M, Chudoba T, Witrowa-Rajchert D (2013) Drying kinetics of apple tissue treated by pulsed electric field. Dry Technol 31(1):112–119

Zhao W, Yang R, Liang Q, Zhang W, Hua X, Tang Y (2012) Electrochemical reaction and oxidation of lecithin under Pulsed Electric Fields (PEF) processing. J Agric Food Chem 2012(60):12204–12209

Part II
Effects of Interest of Pulsed Electric Fields in the Food Industry

Microbial Inactivation by Pulsed Electric Fields

Carlota Delso, Juan Manuel Martínez, Guillermo Cebrián, Santiago Condón, Javier Raso, and Ignacio Álvarez

Abbreviations

CFU	colony-forming units
RTV	resting transmembrane voltage
ITV	induced transmembrane voltage
E	electric field strength
Ec	critical electric field strength
a_w	water activity
R1	the short radius of a cell
R2	the long radius of a cell
θ	the angle measured from the vector created between the center of the cell and the evaluated point with respect to the electric field direction
D value	decimal reduction time
MNIC	Maximum noninhibitory concentration

Introduction

The safety of foodstuffs is an indispensable requirement that is taken for granted by consumers; nevertheless, sensory and nutritional quality is affected by the presence and/or growth of microorganisms in food. For these reasons, microorganisms are one of the main concerns for the food industry. Food can be contaminated by microorganisms along the entire food chain. Microbial growth and metabolism can bring about quality changes such as pH modification, off-odors, gas, or slime formation leading to food spoilage (Huis in't Veld 1996). While microbial food spoilage is a huge economic problem, foodborne illnesses caused by pathogenic microorganisms likewise represent an enormous public health concern with severe direct and indirect economic consequences (Baird-Parker 2000). In spite of immense efforts

C. Delso · J. M. Martínez · G. Cebrián · S. Condón · J. Raso · I. Álvarez (✉)
Tecnología de los Alimentos, Facultad de Veterinaria, Instituto Agroalimentario de Aragón-IA2, Universidad de Zaragoza, Zaragoza, Spain
e-mail: ialvalan@unizar.es

J. Raso et al. (eds.), *Pulsed Electric Fields Technology for the Food Industry*, Food Engineering Series, https://doi.org/10.1007/978-3-030-70586-2_5

undertaken by the food industry and by food safety authorities, the number of reported outbreaks caused by pathogenic microorganisms shows that foodborne illness cases continue to represent a major public health problem worldwide. WHO estimated a total of 600 million foodborne illnesses and 420,000 deaths for the year 2010: the most frequent agents were norovirus, *Campylobacter* spp., and nontyphoidal *Salmonella enterica* (NTC), along with others such as *Salmonella typhi*, *Taenia solium*, hepatitis A virus, and mycotoxins, especially aflatoxins (WHO 2015).

The most effective means of ensuring microbial safety is by preventing the contamination of raw material and foods at the primary production stage, as well as throughout the entire food production chain. However, the large number of possible microbial contamination sources makes microbial control by prevention enormously difficult. Once microorganisms have contaminated food, preservation technologies are essential in order to guarantee food safety: either by preventing microbial growth or by achieving microbial inactivation.

Traditional preservation technologies based on the inhibition of microbial growth include temperature control, water activity reduction, acidification, the addition of preservatives, and packaging in modified atmosphere. However, as the infection dose of some pathogenic microorganisms is very low (Blackburn and McClure 2009), microbial growth inhibition does not suffice to ensure food safety, and thus, in many cases, it is necessary to apply inactivation technologies.

Due to its simplicity and effectiveness, heating is the most frequently used method of microbial destruction in the food industry. Heat-pasteurized foods are free of vegetative pathogens, and thermal sterilization supposes the destruction of microorganisms that would be able to grow in food under normal non-refrigerated conditions, including bacterial spores capable of causing significant dangers to public health. However, heat treatments also have repercussions on the nutritional and sensory characteristics of food such as protein denaturation, non-enzymatic browning, and loss of vitamins and volatile flavor compounds (Ling et al. 2015). As a result, heat processing is not suitable for the obtention of foodstuffs with the characteristics currently demanded on the market: consumers increasingly tend to favor foods that are minimally processed, fresh-like, and natural, without preservatives (Augustin et al. 2016). Therefore, the interest in developing nonthermal food preservation technologies as an alternative to heat processing has greatly increased over the last decades. Such nonthermal food processing technologies should be able to improve shelf-life (by inactivation of pathogen and spoilage microorganisms) while preserving nutritional and organoleptic properties; at the same time, they should be economically feasible for industrial implementation (Barba et al. 2017).

As a gentle method of food preservation, pulsed electric field (PEF) is one of the most promising nonthermal technologies. PEF technology consists in applying high-voltage electric fields (5–50 kV/cm) in short pulses (μs-ms) to a matrix located between two electrodes. PEF treatments have been shown to cause microbial inactivation of vegetative cells of bacteria, yeast, and molds; however, bacterial spores are resistant to this technology at near-ambient temperatures (Pol et al. 2001; Siemer et al. 2014; Wouters et al. 2001). Thus, the main application of PEF treatments, as a preservation technology, is as an alternative to thermal pasteurization.

In order to implement PEF technology as a pasteurization process, it is necessary to prove that it ensures a level of microbial safety equivalent to traditional pasteurization processing (Saldaña et al. 2014). To achieve this objective, a series of studies need to be conducted with the purpose of understanding the mechanisms involved in microbial PEF inactivation, identifying the critical factors that affect microbial inactivation, and describing the microbial inactivation kinetics. Furthermore, it is necessary to carry out those studies on microorganisms of concern for public health, as well as to identify the pathogens that are most resistant to PEF (NACMCF 2006).

The present chapter reviews the current state of knowledge regarding microbial inactivation by PEF. First, the mechanisms of microbial inactivation by PEF and the major factors affecting PEF lethality are discussed. Then, the kinetics of microbial inactivation using this technology are described. Finally, the chapter concludes by summarizing those aspects that require further investigation for the development of PEF processes capable of supplying safe food products of high organoleptic and nutritional quality.

Mechanisms of Microbial Inactivation by PEF

The Electroporation Phenomenon

The electroporation phenomenon is generally accepted as the main mechanism of microbial inactivation by PEF. For decades, this phenomenon of permeabilization has been repeatedly demonstrated in cells, in lipid vesicles, and in planar lipid bilayers. It is accepted that the underlying physical mechanism should occur in the lipid bilayers, because they are the only molecular component common to all membrane types (Kotnik et al. 2012). Electroporation consists in the formation of stable pores in the lipid bilayer which can be transitory or permanent, depending on treatment conditions and cell characteristics. Pore formation in microbial cytoplasmic membranes generates a loss in their selective permeability, which produces an imbalance in cell homeostasis and can lead to cell death.

Although different theories have been advanced to explain the electroporation phenomenon, the theory of aqueous pore formation (Weaver and Chizmadzhev 1996) is currently the one most widely accepted (Kotnik et al. 2019). The spontaneous formation of transient aqueous pores in the lipid bilayer occurs under normal conditions in a cell, thereby allowing water molecules, ions, and other hydrophilic molecules to cross over. According to this theory, when cells are subjected to an external electric field, the energy required to form such aqueous pores is reduced (Kotnik et al. 2012), thereby facilitating the formation of larger and more stable pores. However, and in spite of sophisticated microscopy techniques already available, it has not been possible to directly observe the pores formed in the lipid bilayers or distinguish them from artifacts derived from sample preparation (Kotnik et al. 2012; Spugnini et al. 2007). This theory has only been clearly demonstrated by molecular dynamics simulations (MDS), which show the spontaneous formation of aqueous pores in lipid membranes, along with a significant increment in the rate of pore formation and pore stabilization when a sufficiently strong electric field is applied (Delemotte and Tarek 2012; Leontiadou et al. 2004; Tarek 2005; Tieleman et al. 2003).

The Microbial Cell in an Electric Field

The composition and structure of microbial membranes vary among microbial groups, species, and strains. Gram-positive bacteria present a thick peptidoglycan layer surrounding the phospholipid membrane, which provides them with rigidity and physical resistance. Conversely, Gram-negative bacteria possess a thinner peptidoglycan layer surrounded by an outer membrane made of lipopolysaccharides, which gives them special resistance against the entry of different chemical compounds. In yeasts, the cytoplasmic membrane is surrounded by a rigid cell wall composed by polysaccharides, which provides mechanical strength. Although membrane characteristics vary among microbial groups, the cytoplasmic membrane is a structure common to all microorganisms, thus making it the target component for electroporation. Apart from separating the cytoplasm from the surrounding media and maintaining osmotic equilibrium between the cell and the exterior, the cytoplasmic membrane is also involved in a number of metabolic processes (Rogers et al. 1980).

The cytoplasmic membrane mainly consists of a lipid bilayer made up of phospholipids, with the nonpolar (hydrophobic) portion oriented toward the inside and the polar (hydrophilic) portion oriented toward the outside. Thanks to this makeup, the cytoplasmic membrane is a good dielectric structure with low intrinsic electrical conductivity. Conversely, both extracellular and intracellular media are good electrical conductors. Therefore, the structure formed by the extracellular media, the cytoplasmic membrane, and the intracellular media is a conductor-dielectric-conductor structure and would behave as a capacitor (Ivorra 2010; Teissie et al. 2005). This means that charges are accumulated on both sides of the cytoplasmic membrane as in an electrical capacitor (Fig. 1).

Under normal conditions, cells maintain an irregular distribution of positive and negative ions on both sides of the membrane. This distribution generates an electric potential difference between the inside and the outside of the membrane called resting transmembrane voltage (RTV). When cells are subjected to an external electric field, ions are redistributed and accumulate on both sides of the membrane. This phenomenon supposes an increase in the electric potential difference of the membrane, which is designated as induced transmembrane voltage (*ITV*) (Tsong 1989). When *ITV* achieves a determined threshold, the phenomenon of electroporation of the cytoplasmic membrane occurs (Zimmermann et al. 1974).

Before electroporation takes place, *ITV* is proportional to the applied electric field strength and the size and shape of the cell. For a single spherical cell with a

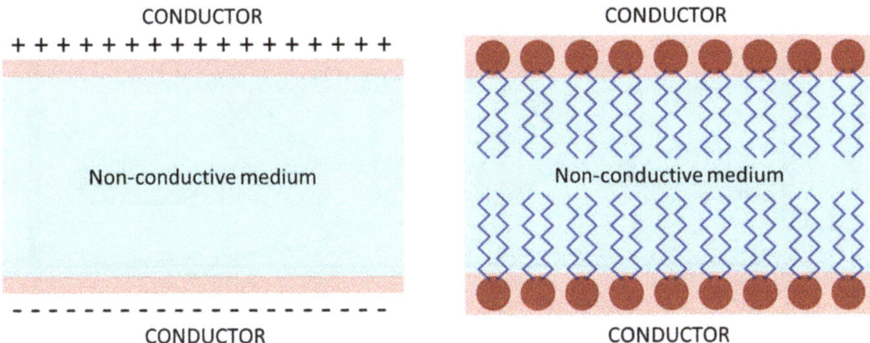

Fig. 1 The lipid bilayer as a capacitor

nonconductive cytoplasmic membrane, *ITV* can be calculated by the Laplace equation (Schwan 1957):

$$ITV = \frac{3}{2}|E|r(cos\theta) \quad \text{(Eq. 1)}$$

where *ITV* is the induced transmembrane voltage, *E* is the electric field strength applied, *r* is the cell radius, and θ is the angle measured from the vector created between the center of the cell and the evaluated point with respect to the electric field direction.

Therefore, according to this equation, the electric field strength required to induce electroporation is greater for smaller cells. In fact, data reported show that microbial cells (1–10 μm), in order to be electroporated, require higher electric field strengths (>10 kV/cm) than eukaryotic plant cells (40–200 μm; < 5 kV/cm) (Donsì et al. 2010).

Furthermore, the equation implies that electroporation does not occur in a uniform way across the entire cytoplasmic membrane. Those zones of the membrane that are perpendicular to the electric field ($\theta \approx 0°$) will be electroporated more easily because they are subjected to higher *ITV*. This fact has been demonstrated by monitoring the change in *ITV* during PEF treatment of a single cell using a potentiometric dye (Kotnik et al. 2010).

Based on *Eq. 1*, further complex equations have been developed in order to describe the *ITV* induced in nonspherical cells. Figure 2 shows the theoretical *ITV* achieved at two different points of the cytoplasmic membrane of *Escherichia coli* depending on its orientation with respect to the electric field (oblate and prolate, parallel or perpendicular, respectively). The figure clearly shows that the highest transmembrane potential would be reached at the point corresponding to the largest radius, and *ITV* would be at its highest when that point of the membrane is perpendicular to the electric field direction ($\theta = 0°$).

Once the transmembrane potential threshold is achieved and metastable hydrophilic pores are created, the latter start to evolve (Saulis 2010). The number of pores

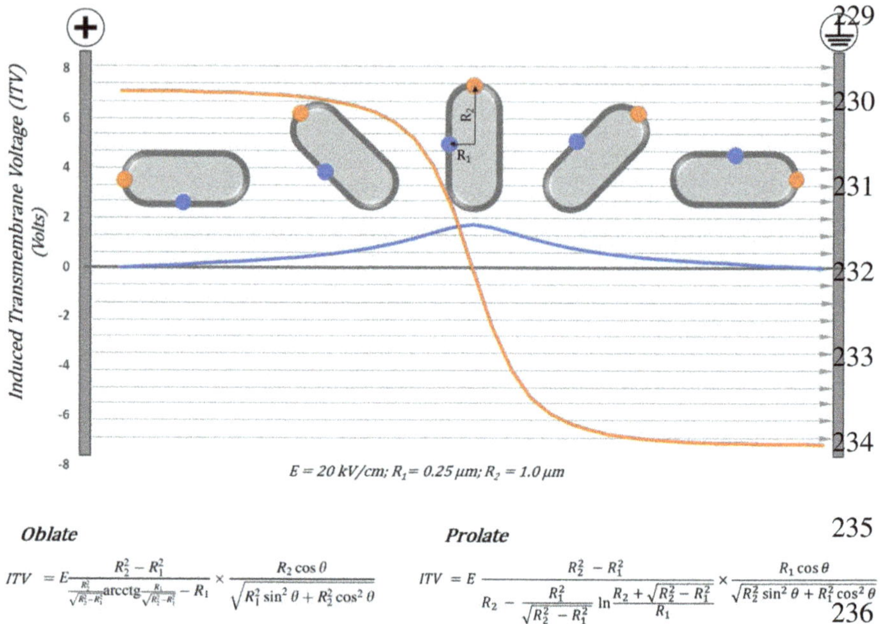

$$ITV = E\frac{R_2^2 - R_1^2}{\frac{R_1^2}{\sqrt{R_2^2-R_1^2}}\operatorname{arcctg}\frac{R_1}{\sqrt{R_2^2-R_1^2}} - R_1} \times \frac{R_2 \cos\theta}{\sqrt{R_1^2 \sin^2\theta + R_2^2 \cos^2\theta}}$$

$$ITV = E\frac{R_2^2 - R_1^2}{R_2 - \frac{R_1^2}{\sqrt{R_2^2 - R_1^2}}\ln\frac{R_2 + \sqrt{R_2^2 - R_1^2}}{R_1}} \times \frac{R_1 \cos\theta}{\sqrt{R_2^2 \sin^2\theta + R_1^2 \cos^2\theta}}$$

Fig. 2 Theoretical *ITV* (volts) reached at two points of the cytoplasmic membrane of a microbial cell ($R_1 = 0.25$ μm; $R_2 = 1.0$ μm) subjected to 20 kV/cm of field strength depending on its orientation within the electric field (from oblate to prolate and again oblate). Data calculation based on equations taken from Kotnik et al. (2010), where E is the electric field strength, R_1 is the short radius, R_2 is the long radius, and θ is the angle measured from the center of the cell to the evaluated point with respect to the direction of the electric field

and/or their size varies during this stage. At the point when the external electric field ceases to be applied, pores that were previously formed can be transitory or permanent, causing reversible or irreversible electroporation, respectively (Tsong 1989). Pore resealing, and, therefore, the cell's survival, will depend on various factors, including the ambient (recovery) conditions, which will be explained below.

Potential Cell Outcomes After PEF

After a PEF treatment, several scenarios can occur which are summarized in Fig. 3. On the one hand, cells can remain intact or non-electroporated, and therefore survive the PEF treatment, or they can be permeabilized. After a PEF treatment, as previously indicated, pores can remain permanently open (irreversible electroporation) or reseal (reversible electroporation). This behavior has been widely demonstrated in microbial cells using different techniques including propidium iodide (PI) staining (Aronsson et al. 2005; Cebrián et al. 2015; García et al. 2007; Zhao et al. 2011).

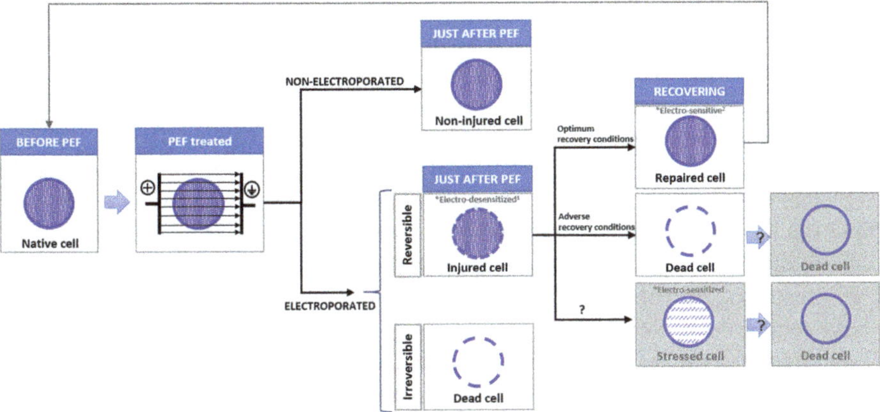

Fig. 3 Different microbial cell outcomes immediately after the application of PEF treatments and after a recovery period. *(?, hypothesis to explain some non-understood outcomes; *, New concepts proposed)*

It can be assumed that when a microbial cell is electroporated, it is damaged to a certain degree. Cell survival chances will differ depending on the severity of damages (number of pores, size, etc.) as well as on recovery conditions (Wang et al. 2018). Thus, if ambient conditions are adverse, pores might not be resealed and cells would end up dying. On the other hand, results have shown that sublethally injured cells might be able to repair the pores generated by a PEF treatment and survive if recovery conditions are favorable. At this point it should be noted that the occurrence of sublethal damage after exposing microbial cells to PEF has been the object of controversy, since contradictory results have been reported (Aronsson et al. 2004; Simpson et al. 1999; Wuytack et al. 2003). The explanation for these discrepancies might be that the occurrence of sublethal damage would highly depend on the medium pH and on the microbial strain, as demonstrated by García et al. (2003, 2005a).

The pore resealing process after PEF has been scarcely investigated in microbial cells. In eukaryotic as well as in prokaryotic cells, it is generally reported that this process depends to a great extent on the recovery conditions (medium composition, medium pH, temperature, conductivity) (Saulis 2010). As a consequence, the required period for pores to reseal has been observed to vary from seconds to hours (Kotnik et al. 2019). For example, *Saccharomyces cerevisiae* repaired sublethal damages after 1 and 4 h when the incubation was in Sabouraud Broth or peptone water, respectively, while cell recovery in a neutral buffer was not observed even after 4 h of incubation (Somolinos et al. 2008b). Moreover, it has been observed that cytoplasmic membranes can recover their impermeability to propidium iodide (PI) but remain permeable to small ions such us Cl^- (Saulis 2010). Cebrián et al. (2015) observed in *Staphylococcus aureus* that this phenomenon was highly dependent on the characteristics of the recovery media. Under optimal conditions, *S. aureus* completely recovered its impermeability to PI and NaCl simultaneously. However, under adverse conditions, cells once more became impermeable to PI, but not to NaCl, and ended up dying. This would support the theory that the resealing of pores consists in an initial rapid reduction in pore size (seconds/minutes) without the occurrence of important metabolic processes, followed by a slow one (minutes/hours) that continues until the repair process is complete (Saulis 2010).

A detailed study by García et al. (2006) reported that the repair mechanisms for sublethal membrane damage in *Escherichia coli* were associated with energy and lipid synthesis requirements in 95% of the injured population. However, the remaining 5% of the damaged population repaired their membranes immediately after the PEF treatment, probably just by means of a physical process without any biosynthetic requirement. Regarding Gram-positive bacteria, Somolinos et al. (2010a, b) found that *Listeria monocytogenes* only required energy production for membrane repair; however 10% of the population repaired their damage quickly without any biosynthetic requirement. These small percentages of cells that repaired their damage rapidly could also be responsible for the detection of a proportion of non-electroporated dead cells after PEF (Zhao et al. 2011).

On the other hand, several theories proposing other potential cell outcomes after PEF have been recently advanced. It has been suggested that in eukaryotic cells an *electro-desensitization* phenomenon occurs. It would lead to the appearance of cells which, once electroporated, are not able to become repolarized again because they cease to act as a capacitor (Silve et al. 2014). This would imply that further treatments would not result in additional electroporation: consequently, the injuries suffered by the cells would be repairable, and they would be able to survive under adequate recovery conditions. Paradoxically, once these cells had recovered their membrane integrity, they could act again as a native cell, thereby becoming electrosensitive once more. This phenomenon might also occur in prokaryotic cells, which would explain the tailing kinetics typically observed in microbial survival curves obtained by PEF. The opposite phenomenon, *electrosensitization*, has also been described in mammalian cells: cells subjected to PEF become sensitive to subsequent PEF treatments due to underlying physiological mechanisms (cell swelling, influx of Ca^{2+}, oxidative damage, cytoskeleton rearrangement, ATP leakage, and depletion) (Muratori et al. 2017).

In summary, the exposure of microbial cells to PEF can lead to thoroughly different outcomes depending on various factors mainly associated with the type of microorganism, the environment, or the PEF process parameters. Taking into account the different scenarios in which cells can end up injured or stressed, a number of PEF-based combined treatments have been proposed, as will be discussed below. These include, for example, the addition of antimicrobial compounds or storage under acid conditions.

Techniques for the Detection of Electroporation in Microbial Cells

The alteration of the integrity of the microbial cytoplasmic membrane has been demonstrated by means of a series of techniques and procedures. These techniques can be divided into (1) measurement of the uptake of membrane non-permeant exogenous compounds, (2) measurement of the leakage of intracellular compounds, (3) microscopy visualization techniques, (4) selective medium plating technique, and (5) measurement of osmotic response (García-Gonzalo and Pagán 2017). The main advantages and disadvantages of each technique are summarized in Table 1.

Propidium iodide (PI), a fluorescent dye, has been the molecule most commonly used to assess membrane permeabilization by measuring its uptake. PI is a strongly hydrophilic, small molecule (660 Da) that binds with DNA but is unable to enter intact cells. Microbial cells can be brought in contact with PI after a PEF treatment to monitor irreversible electroporation or before a PEF treatment to monitor irreversible and reversible electroporation. The number of permeabilized cells can be monitored by epifluorescence microscopy, spectrofluorometry, or flow cytometry. Other fluorescent dyes used for this purpose are ethidium bromide (EtBr), cyanine monomers, and SYTOX®/SYTO®. These fluorescent compounds have also been used to label certain dextrans of different molecular weight in order to evaluate pore size depending on the permeability of the cytoplasmic membrane to those dextrans in microalgae and other eukaryotic cells (Bodénès et al. 2019).

The leakage of intracellular compounds was one of the first techniques to evidence electroporation phenomena because of its simplicity. Proof was achieved, for example, by absorbance measurements of nucleic acids and proteins at 260 and 280 nm, respectively, or by detecting the presence of adenosine trisphosphate (ATP) via luciferin-luciferase assay. Although sensitivity can be increased by using fluorescent probes that bind to leaked compounds, this is an indirect and semiquantitative technique that is difficult to correlate with the number of electroporated cells. The release of different ions has also been used to quantify membrane electroporation, K+ being the most suitable one for this purpose. Cebrián et al. (2015) reported to have precisely determined the percentage of *S. aureus* electroporated cells (up to 95–99%) by K+ release measurement.

Electron and atomic microscopy techniques should allow researchers to detect pores from 1 nm in diameter. However, the resulting images of electroporated microbial cells are contradictory. It is thought that the observed pores are a consequence of artifacts associated with sample preparation. These techniques are useful in order to observe changes in the extra- and intracellular morphology of affected cells, such as increase in surface roughness, craters, elongation, disruption of organelles, etc., rather than to observe the presence of pores in the membrane (García-Gonzalo and Pagán 2017). No correlation has been found between observed changes in morphology and the level of microbial electroporation or inactivation due to PEF.

Another widely used method to detect electroporation consists in recovering survivors after PEF by plating onto the same culture media with a compound that

Table 1 Advantages and limitations of the different techniques used to detect and quantify electroporation on microbial cells. Based on García-Gonzalo and Pagán (2017)

Technique		Advantages	Limitations	References[a]
Uptake of extracellular compounds	Fluorescent dyes	Rapid outcomes Direct evidence of electroporation	Potentially hazardous Specific equipment required to evaluate fluorescence	Zhao et al. (2011) Cebrián et al. (2015)
	Dextrans of different sizes labelled with fluorochromes	Estimating pore size Direct evidence of electroporation	Specific equipment required to evaluate fluorescence Not useful in bacteria	Bartoletti et al. (1989) Bodénès et al. (2019)
Intracellular compound leakage		Rapid outcomes, simplicity	Semiquantitative technique Difficult correlation between the amount of leaked components and the number of electroporated cells	Aronsson et al. (2005) Martínez et al. (2016)
Electron and atomic force microscopy		Direct evaluation of morphological changes	May be affected by methodological artefacts during sample preparation Uncertainty of its capability to visualize pores. Low precision and poor detection limits Specific equipment is required	Marx et al. (2011) Picart et al. (2002)
Selective medium recovery		Simple protocol, no specific equipment required Higher sensitivity, even in ranges up to 90% of population	Long time required to obtain results (>48 h) Previous MNIC studies are required[b]	Kethireddy et al. (2016) Novickij et al. (2018)
Osmotic response		Rapid outcomes Simplicity Inexpensive	Low sensitivity Indirect measurement of electroporation Semiquantitative Not useful in gram-positive bacteria	Korber et al. (1996)

[a]*Examples of references where the technique has been used*
[b]*MNIC (Maximum noninhibitory concentration)*

prevents the growth of sublethally damaged cells (selective medium) as well as without the compound (nonselective medium) (Mackey 2000). If it is assumed that reversibly electroporated cells are sublethally damaged, their number may be calculated by the difference between the number of survivors grown in nonselective and selective media, respectively. The selective media most commonly used to detect damage to the cytoplasmic membrane is agar with sodium chloride added (Mackey 2000). Bile salts are another commonly used selective agent, but they are only suitable for the detection of damages in the outer membrane of Gram-negative bacteria (Arroyo et al. 2010a). Further antimicrobials unable to access the cytoplasm or the cytoplasmic membrane of this group of bacteria under normal conditions are kanamycin, nisin, and citral, and they have been investigated with the same purpose (Arroyo et al. 2010b; Martens et al. 2020; Novickij et al. 2018). This technique requires previous studies to detect the maximum noninhibitory concentration (MNIC) of the selective compound which permits the growth of untreated microbial cells since it varies among microorganisms.

The measurement of the optical density (OD_{600}) of PEF-treated suspensions is an indirect technique designed to determine the loss of osmotic response when cells are suspended in a hypertonic medium, i.e., a strong salt solution. This semiquantitative technique thus evaluates membrane integrity alterations through the loss of the cells' ability to plasmolyze. In a hypertonic medium, native cells respond by diffusing water from the cytoplasm to the external medium, thereby causing a strong condensation of cytoplasmic material that can be detected as an increment in OD_{600}. This response will not be detected in electroporated cells. However, this technique is only of limited use for Gram-positive bacteria and for the quantification of the number of electroporated cells.

Although all of these techniques have been suggested for the evaluation of the electroporation phenomenon, most of them do not provide a direct measurement thereof. In microbial studies, levels of inactivation from 1.0 to 5.0 Log_{10} cycles (90 to 99.999% of the population) are commonly achieved, while most of these techniques are only useful for detecting electroporation in a range lying between 0 and 95–99% of the population, except for those based on plate counting in selective and nonselective media.

Other complementary techniques, such as Fourier transform infrared spectroscopy, the measurement of changes in membrane potential, and those based on molecular tools, can reveal interesting information regarding the mechanisms of the electroporation phenomenon in microbial cells (Chueca et al. 2015; Gelaw et al. 2013; Kotnik et al. 2010). More detailed information on all these techniques can be found in some studies (i.e., Batista-Napotnik and Miklavčič 2018; Cebrián 2016; García-Gonzalo and Pagán 2017).

Factors Determining Microbial Inactivation by PEF

Microbial inactivation by PEF is affected by several factors that can be grouped as (1) processing parameters, (2) microbial characteristics, and (3) treatment medium characteristics (Wouters et al. 2001). It is essential to understand how each of these factors affects microbial inactivation in order to obtain useful data, prevents experimental artifacts, and compares published results obtained by different research groups.

Due to the substantial variety of PEF equipment operating in different laboratories as well as the variety of protocols used to generate data, it is rather challenging to attempt to compare results with the purpose of achieving definitive conclusions. As a standardization of experimental procedures used in different laboratories is quite difficult to achieve, Raso et al. (2016) listed the critical treatment conditions that should be reported in studies and that would be particularly beneficial for comparison of results.

Processing Parameters

PEF treatments are defined by the electric field strength (kV/cm), the number of pulses along with their shape and width (μs), the treatment time (s), the frequency (Hz) of pulse delivery, and the total specific energy (kJ/kg). Furthermore, treatment temperature can affect the ultimate lethal effect on the microbial population.

Electric Field Strength

Electric field strength is one of the factors that most strongly determine microbial resistance and is one of the processing parameters that fundamentally characterize PEF treatments. It has been widely reported that in order to inactivate microbial cells, a threshold value called "critical electric field strength" (*Ec*) should be reached. It has been defined as the minimum external electric field strength required in order to observe microbial inactivation. It should be noted that the *Ec* may vary depending on other treatment parameters, such as treatment time or treatment temperature; therefore, this parameter should be clearly and accurately indicated. For example, the *Ec* for *Listeria monocytogenes* can vary from 5 to 22 kV/cm when treatment time increases from 40 to 2000 μs (Fig. 4). Other authors have defined *Ec* as the minimal field strength required to achieve the transmembrane potential threshold that causes electroporation (Bazhal et al. 2003; Ho and Mittal 1996; Vernhes et al. 2002). This approach could be more useful for the characterization of microbial resistance to PEF, because it is possible to theoretically estimate this *Ec* depending on microbial size, as previously described (Fig. 2), and not depending on the moment when inactivation is achieved. Additionally, the final death of an electroporated cell depends on several factors, and the *Ec* for a level of inactivation could thus vary, depending, for example, on recovery conditions. It thus seems to be more adequate to establish the *Ec* threshold on the basis of the electroporation outcome, which is more consistent. On the other hand, it has also been extensively demonstrated that, above the *Ec*, inactivation levels increase with the electric field strength, at least up to a certain point (Raso and Álvarez 2016; Toepfl et al. 2014).

A critical point when studying microbial inactivation, but also other effects/applications of PEF, is the distribution of the electric field strength inside the treatment chamber. Thus, the geometry and configuration of treatment chambers have

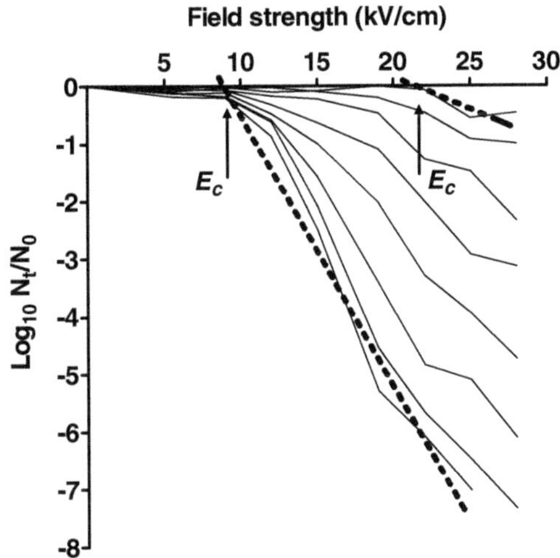

Fig. 4 Influence of the electric field strength on the PEF inactivation of *E. coli* at constant treatment times (thin lines). The thick lines represent the regression line of the straight part of the curve at a certain treatment time (40 and 2000 μs). Arrows indicate the *Ec* after 40 and 2000 μs. Treatment conditions: McIlvaine buffer pH 7.0, 2 mS/cm; 2 μs square wave pulses; frequency, 1 Hz

been pointed out as essential for accurate control of the applied electric field strength (Huang and Wang 2009). Batch chambers with parallel electrode configuration are considered the most suitable for the generation of almost uniform electric fields. Consequently, such chambers should be used in basic studies on microbial inactivation in order to obtain accurate data, particularly regarding those focused on mechanisms. Although parallel-electrode chambers should also be used for continuous processing, aspects such as fluid dynamics or electric current requirements might prevent their use. In any case, this all makes clear that PEF treatment chamber design should be based on mathematical simulations in order to clearly establish the distribution of field strength (Gerlach et al. 2008; Wölken et al. 2017).

Treatment Time

Treatment time is defined as the effective total time during which microorganisms are subjected to the electric field. It is one of the most critical parameters: along with electric field strength, it is one of the two main factors that determine the lethality of PEF treatments. Treatment time can be prolonged either by increasing the pulse length or by increasing the total number of pulses. In general, inactivation increases with treatment time; however, after a certain number of pulses, a "saturation" phenomenon appears, called the tailing effect. This behavior pronounced influence on the effectiveness of PEF treatments and will be discussed below in further detail.

Pulse Shape

As indicated, treatment time is based on the repetitive application of pulses of determined shape and width. The pulses' various characteristics can affect the microbial response to the treatment. Either exponential or square wave pulses can be applied for inactivation purposes (de Haan 2007). It has been observed that both pulse shapes have an equivalent lethal effect at the same specific energy input (Ou et al. 2017). However, square wave pulses are preferred because it is easier to calculate the total treatment time by knowing the number of pulses and the pulse width. Furthermore, exponential decay pulses partially involve the application of low voltages (below the *Ec*) which give rise to an ohmic heating effect, which can complicate the interpretation of results.

Both types of pulses can be delivered in monopolar or bipolar mode, the latter consisting in the application of alternatively positive and negative pulses. Differences in inactivation levels have been reported when comparing monopolar and bipolar pulses. Qin et al. (1994) observed higher inactivation levels with bipolar pulses, as well as a lesser degree of electrolysis. By contrast, Beveridge et al. (2005) obtained superior inactivation levels with bipolar pulses when they were wider than 3 μs and with monopolar pulses narrower than 2 μs of width. This aspect still requires more investigation in order to clarify its effect on microbial inactivation. In addition, it should be remarked that recent studies have reported the existence of a "cancellation phenomenon" in mammalian cells when applying bipolar pulses (Ibey et al. 2014; Pakhomov et al. 2014; Polajžer et al. 2020). The cancellation phenomenon consists in the annulment of a pulse effect by the next one, due to their opposite effects on ion redistribution in cells. Because of this, unipolar pulses resulted in being more efficient in cell killing. This phenomenon, only observed with pulses of a duration of nanoseconds (60–300 ns) or microseconds (1–10 μs), has not been described in the case of microbial cells.

Pulse Width

The influence of pulse width on the microbial lethality of PEF is still unclear. Some studies reported higher inactivation rates when wider pulses were applied (Aronsson et al. 2001; Wouters et al. 1999). However, they were obtained in conditions without temperature control. Longer pulses involve the delivery of a higher amount of energy in the treatment zone per unit of time and hence an increase in temperature, which may sensitize microorganisms to PEF. Other authors did not observe differences in inactivation levels when varying pulse width without modifying treatment time or specific energy under isothermal conditions (Álvarez et al. 2003a; Fox et al. 2008; Frey et al. 2013; Raso et al. 2000). To prevent misinterpretation of results when studying the influence of pulse width, Raso et al. (2016) thus recommend the use of lower frequencies to avoid the Joule heating effect. Furthermore, it should also be pointed out that long pulses (longer than 20 μs) have been reported to generate a greater number of electrochemical reactions, resulting in the release of chemical species that are potentially toxic for microorganisms, as well as in fouling and corrosion of the electrode surfaces (Ho and Mittal 2000; Pataro et al. 2017).

On the other hand, Vernhes et al. (2002) reported that for pulses of the same specific energy, shorter pulses applied at higher field strengths were more effective in microbial inactivation than wider pulses at lower field strengths. These data indicate that, as pointed out by Toepfl et al. (2014), pulse characteristics cannot be studied independently from other parameters.

Frequency

The frequency of pulse application has also been studied in terms of microbial inactivation response, in ranges from 0.5 to 500 Hz. The majority of studies reported that microbial inactivation is independent of the frequency applied for an equivalent treatment (Hülsheger et al. 1981; Mosqueda-Melgar et al. 2007; Raso et al. 2000). However, similarly to what has been described above for pulse width, it is essential to take into account that the frequency-dependent increase in temperature in the treatment medium may enhance the PEF inactivation effect, thereby leading to unreliable or noncomparable results.

Specific Energy

Microbial inactivation by PEF increases with the total specific energy (energy per mass unit) of the treatment (Fig. 5a). This parameter depends on applied voltage, total treatment time, resistance of the treatment chamber, and amount of treated product. The resistance of the treatment chamber depends on its geometry and on the degree of conductivity of the treatment medium. Due to its broad significance, specific energy is also an essential parameter to be included when characterizing a PEF treatment, together with field strength and treatment time. This is critical, because although PEF microbial inactivation increases along with energy, greater

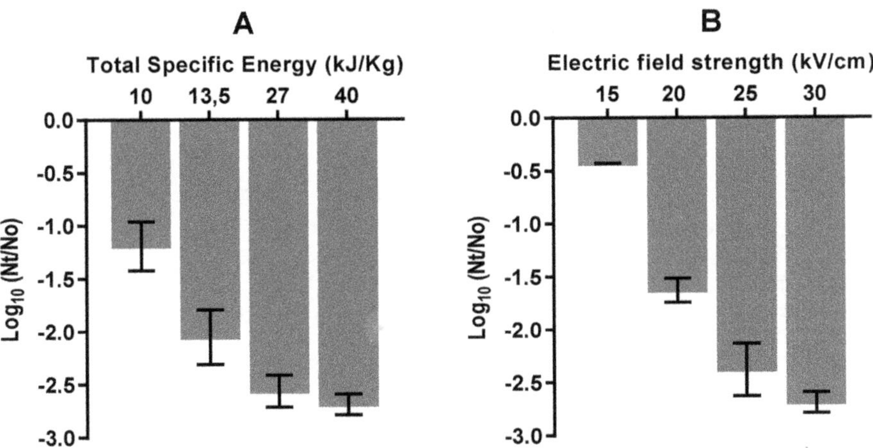

Fig. 5 Influence of total specific energy at 30 kV/cm (**a**) and the influence of electric field strength at the same total specific energy of 40 kJ/kg (**b**) in the inactivation of *Salmonella typhimurium* by PEF

lethal effects are achieved at the same specific energy by increasing the field strength (Fig. 5b) (Álvarez et al. 2003a; Huang et al. 2014; Puértolas et al. 2009; Zhong et al. 2005). Furthermore, the information provided by the specific energy factor is essential when comparing the lethal effect of PEF with that of other technologies, both thermal and nonthermal. In continuous processing with PEF, specific energy is also commonly used to estimate the temperature increment due to the Joule effect and thereby the outlet temperature of the treated product.

Temperature

PEF technology emerged as a nonthermal technique for food preservation as a replacement for traditional thermal treatments in order to avoid heat effects on food properties. However, the lethal effect of PEF has been shown to increase notably with temperature, even when combined with nonlethal temperatures (< 50 °C) (Katiyo et al. 2017; Saldaña et al. 2010a; Sharma et al. 2014; Timmermans et al. 2019) (Fig. 6). For example, the same PEF treatment (30 kV/cm; 150 μs; pH 3.5) increased the inactivation of *Salmonella typhimurium* by 4.0 Log_{10} cycles when applied at 38 °C instead of at 4 °C (Saldaña et al. 2010a). Molecular dynamic simulations (MDS) indicate that the electroporation threshold decreases as temperature is increased (Song et al. 2011), a hypothesis already advanced by Coster and Zimmermann (1975). This synergetic behavior has been attributed to a change in the lipid bilayer from a gel-like consistency to a more liquid crystalline state as a consequence of temperature rise, thereby facilitating the formation of pores (Stanley and Parkin 1991). Additionally, Cebrián et al. (2016a) pointed out that other causes might also be involved in this effect, such as changes in other membrane components or structures or a higher ion/component diffusion rate. Therefore, control of this parameter or of those aspects that affect it is essential in order to obtain accurate data concerning microbial resistance to PEF, especially when inactivation mechanisms are being studied.

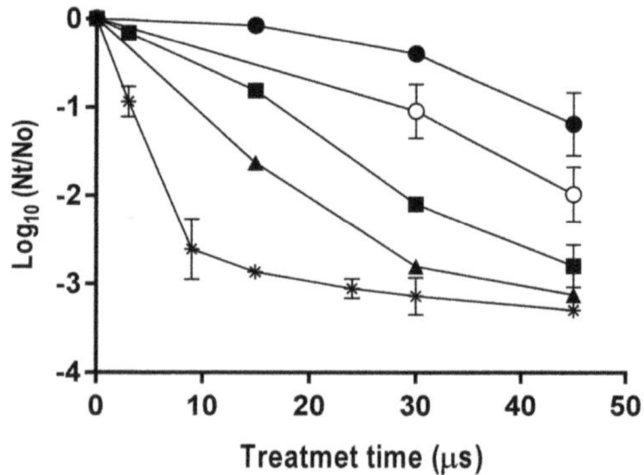

Fig. 6 Inactivation of *Saccharomyces cerevisiae* by PEF treatments at 20 kV/cm under pH 4.0 at different treatment temperatures: 10 °C (●), 25 °C(○), 30 °C (■), 40 °C (▲), and 50 °C (⁎)

Microbial Characteristics

Microbial resistance to PEF is highly reliant on intrinsic microbial characteristics, especially on the type of microorganism, cell size and shape, and growth conditions. Furthermore, the microbial resistance observed against PEF treatments does not always correlate with the resistance to other inactivation technologies (Cebrián et al. 2016b).

Type of Microorganism

For decades it has been known that although PEF is rather successful in achieving inactivation of vegetative cells of bacteria, molds, and yeast (Hamilton and Sale 1967), bacterial spores are resistant to this technology. The complex spore envelopes, and mainly the cortex, probably prevent the electroporation phenomenon from taking place (Pol et al. 2001). However, recent studies have reported that bacterial spores of *Bacillus subtilis* and *Geobacillus stearothermophilus* can be inactivated by the combination of moderate preheating and PEF treatments with elevated final temperatures (Cregenzán-Alberti et al. 2017; Reineke et al. 2015; Siemer et al. 2014). Regarding other nonbacterial spores, certain ascospores and conidiospores of yeast and molds have been proven sensitive to PEF; however, ascospores of *Neosartorya fischeri* were resistant (Raso et al. 1998).

Resistance to PEF is not only greatly dependent on the type of microorganism (Gram-positive bacteria are generally more resistant than Gram-negative), but there is also a great degree of variability among the most sensitive strains of the same microbial species. Saldaña et al. (2010) studied five strains of *Listeria monocytogenes*, *Staphylococcus aureus*, *Escherichia coli*, and *Salmonella typhimurium*, and they found differences in inactivation levels from less than 1.0 to 4.0 Log_{10} cycles. On the other hand, Cebrián et al. (2007) observed that different pigmented and nonpigmented enterotoxigenic strains of *S. aureus* showed similar PEF resistance. In general, interspecific differences in microbial PEF resistance are greater than intraspecific ones (Cebrián et al. 2016b).

Cell Size and Shape

From a theoretical point of view, and taking the mechanism of electroporation into account (Eq. 1 and Fig. 2), cell size and shape should be one of the most significant intrinsic parameters affecting microbial resistance to PEF. Yeast and molds are generally considered more sensitive than bacteria due to their size. Regarding cell shape, the orientation within the electric field is determinant for the susceptibility of nonspherical cells, which are generally more resistant (Fig. 2). Thus, it can be speculated whether the stochastic spatial orientation of microorganisms inside an electric field could also have an impact on a treatment's efficiency, as is the case with diversity in size (Heinz et al. 2001). Nevertheless, these general assumptions are not always corroborated; other intrinsic microbial characteristics may influence microbial resistance to PEF, for example, the structure and composition of the cellular envelopes (Kandušer et al. 2006; Liu et al. 2017; Wang et al. 2019).

Culture Conditions

Microbial culture conditions such as growth phase and growth temperature have been proven to generate modifications in microbial resistance to PEF. Most of the investigations point out that cells in exponential phase are more sensitive than in stationary phase (Alvarez et al. 2002; Arroyo et al. 2010a; Somolinos et al. 2008a; Wang et al. 2019). During the exponential phase, most of the cells are in the division stage, which may increase the lipid bilayer's susceptibility to be electroporated due to the cells' greater size (Jacob et al. 1981). In a growing culture, cells are observed to elongate and then form a partition that eventually separates the cell into two daughter cells. Conversely, the fact that cells are shortened when they have reached the stationary phase (Deforet et al. 2015) could also explain their higher resistance to PEF. Furthermore, it is known that certain bacterial species express a series of general stress sigma factors when they achieve stationary growth phase (Abee and Wouters 1999). Some authors have found a relation between the expression of the sigma factors *rpoS* (Somolinos et al. 2008a) and σ^B (Cebrián et al. 2009) and higher PEF resistance in stationary phase.

Regarding growth temperature, the literature shows different behaviors. An increase in microbial sensitivity as growth temperature decreases has been reported (Alvarez et al. 2002; Liu et al. 2017; Russell 2002; Wang et al. 2019; Yun et al. 2016). In this sense it has also been reported that the proportion of unsaturated fatty acids in *Escherichia coli* and *Salmonella* Typhimurium increases at low growth temperatures (Liu et al. 2017; Wang et al. 2019). This variation in the proportion of fatty acids might affect the fluidity of the membrane, which would become more susceptible to electroporation. However, when cells were treated during exponential phase, Wang et al. (2019) did not observe any influence of growth temperature on their resistance to PEF. Other data indicated that growth temperature did not alter microbial resistance to PEF (Álvarez et al. 2003b; Arroyo et al. 2010a). Similarly, Cebrián et al. (2016a) found no differences in the resistance of *S. aureus* at stationary phase when increasing the growth temperature from 10 to 42 °C, except in cells grown at 30 °C. Finally, even the opposite effect has been reported: *E. coli* displayed higher PEF resistance at the lowest growth temperatures (Cebrián et al. 2008). These discrepancies thus seem to indicate that other factors and/or mechanisms might also be involved in the effect of growth temperature on microbial resistance to PEF.

Other culture settings, such as the presence of ethanol or acid conditions, have likewise been shown to affect microbial resistance to PEF. *Acetobacter* sp. was more vulnerable to subsequent PEF treatments with a higher concentration of ethanol in the growth medium (Niu et al. 2019). Similarly, *E. faecium* grown under acid conditions showed a higher sensitivity to PEF when treated at neutral pH than nonacid-adapted cells (Fernández et al. 2018). However, the same authors observed that the acid-adapted cells were more resistant to PEF when treated at pH 4.0 than control cells. Adaptive responses when cells are transferred from the growth media to a different media for treatment might also be occurring. Both protective and sensitizing effects against PEF treatments have been detected in Gram-negative as well as in Gram-positive bacteria after having been pre-exposed to different acid, heat, or alkaline shocks (Arroyo et al. 2012; Cebrián et al. 2012; Somolinos et al. 2008a). However, different outcomes observed according to strain, treatment parameter, and type of shock treatment make it difficult to draw general conclusions on the influence of these shocks on microbial resistance to PEF.

Characteristics of the Treatment Medium

Food products are complex matrixes whose physical and chemical characteristics strongly determine microbial growth and PEF sensitivity. PEF technology has been applied to a wide range of different products, mainly liquids (e.g., buffers, milk, fruit juices, beer, liquid egg, etc.) but also solid products (e.g., agar, gellan-gum gel, beef burgers, etc.). Thus, in theory, by knowing how each of the main factors such as electrical conductivity, pH, water activity, treatment medium composition, etc., tends to exert an influence on microbial resistance to PEF, it would be possible to predict and adequately design PEF treatments for application to all these different food products. However, it should be borne in mind that the food components might be interacting with one another, thus causing effects in a different way, or to another extent, than what was theoretically predicted in terms of microbial inactivation by PEF. Therefore, the optimized PEF process parameters previously calculated in buffers or food models should still be validated in the target food matrix.

Electrical Conductivity

Due to the PEF technique's nature, the electrical conductivity of the medium is also a key parameter. The electrical conductivity of the treatment medium determines its resistivity and thus the current intensity required to generate the electric field. Certain authors did not observe an influence of conductivity (in a range from 0.05 up to 0.45 S/m) on microbial inactivation when equivalent treatments of specific energy were applied (Álvarez et al. 2003a; Gachovska et al. 2013; Timmermans et al. 2019). Other studies, on the contrary, reported greater inactivation rates at the lowest conductivity (Jayaram et al. 1993; Sensoy et al. 1997; Vega-Mercado et al. 1996; Wouters et al. 1999). Although certain authors speculate that conductivity could influence the membrane's permeability, this behavior might be attributed to the setup conditions and equipment used. What is more, those studies evaluated various dependent parameters, and each particular factor might bear an influence on microbial inactivation in a different way, even in an opposite direction (Saulis 2010). Thus, it has to be reminded that at constant input voltage, the higher the conductivity, the lower the chamber resistance, which implies an increase in the temperature of the medium, thereby potentially affecting microbial resistance to PEF. It is therefore difficult to discern whether conductivity actually exerts an influence on the electroporation phenomenon or whether the observed effects are a result of the influence of conductivity on other process parameters.

pH

The influence of pH on microbial resistance to PEF has been widely evaluated: it has been shown to be especially dependent on the type of microorganism, as indicated above. Generally, Gram-positive bacteria are more resistant in neutral pH media than in acid conditions, whereas Gram-negative behave in the opposite way (García et al. 2005a, 2007; Saldaña et al. 2010, b, c). Saldaña et al. (2010) confirmed this effect comparing five strains of two Gram-positive bacteria (*Listeria monocytogenes* and *Staphylococcus aureus*) and two Gram-negative (*Escherichia coli* and *Salmonella typhimurium*) at pH 4.0 and 7.0. The differences in PEF resistance according to medium pH were very marked for *L. monocytogenes* and *E. coli* but less pronounced for *S. aureus* and *S. typhimurium*. In general terms, the bacterial foodborne pathogen most resistant to PEF treatments in neutral pH would be *Listeria monocytogenes*. By contrast, at pH 4.0 the most resistant bacteria would be *E. coli*, *Cronobacter sakazakii*, and *Salmonella* (Cebrián et al. 2016b). Regarding yeast, Somolinos et al. (2007) did not find a statistically significant effect of pH on the PEF resistance of two yeast species (*Saccharomyces cerevisiae* and *Dekkera bruxellensis*), although they observed that *S. cerevisiae* was slightly more sensitive at pH 5.0 and 6.0. Similar results were found by Aronsson and Rönner (2001).

The higher sensitivity of Gram-positive bacterial cells at acidic pH has been attributed to a change in their capability to maintain a transmembrane pH gradient due to membrane electroporation (Vega-Mercado et al. 1996). Loss of membrane continuity would be expected to impair pH homeostasis, thereby modifying intracellular pH and affecting the critical components of the cell (DNA, enzymes, etc.). This mechanism would explain the behavior of Gram-positive bacteria, but not that of Gram-negative ones which are more sensitive at neutral pH. García et al. (2007) attributed this difference in behavior to the capacity of each bacterial group to repair sublethal damages in the cytoplasmic membrane after PEF depending on the medium pH. They only observed the existence of sublethal injury in the populations of two Gram-negative bacterial species and two Gram-positive ones after PEF when treated in acidic and neutral pH media, respectively (García et al. 2005a). The influence of pH on microbial resistance to PEF therefore seems to be related to the recovery capability of sublethal injured cells. The underlying causes of this phenomenon nevertheless still need to be elucidated. On the other hand, Somolinos et al. (2010) noticed that the apparently greater resistance of *E. coli* at pH 4.0 was actually related to the presence of higher concentrations of citric acid in the McIlvaine buffers at acid pH; similar effects were obtained with other organic acids. More research is required to completely explain the mechanisms involved in the influence of pH on microbial resistance, particularly due to their practical implications for the applicability of this technology.

Water Activity (a_w)

This parameter is one of those least investigated in terms of its influence on microbial resistance to PEF. In general, a reduction in a_w increased the PEF resistance of several bacteria and yeast (Alvarez et al. 2002, 2003b; Aronsson and Rönner 2001; Arroyo et al. 2010a). It has been speculated that microorganisms in an environment with low a_w suffer a release of inlet water to the exterior, resulting in a cell size reduction that augments their resistance to PEF. Furthermore, cell shrinkage probably causes a thickening of the cell membrane, as well as a decrease in membrane permeability and fluidity (Neidhardt et al. 1990).

However, the effect of a_w on resistance is highly dependent on the solute. For example, Fig. 7 shows the PEF survival curves of *Salmonella* Senftenberg and *L. monocytogenes* obtained at the same a_w (0.93) adjusted with 50% (w/v) of sucrose and 10% (w/v) of glycerol. Glycerol makes both microorganisms more sensitive to PEF treatments, while sucrose behaves the opposite way. This could be related to these solutes' different properties and membrane permeability among others.

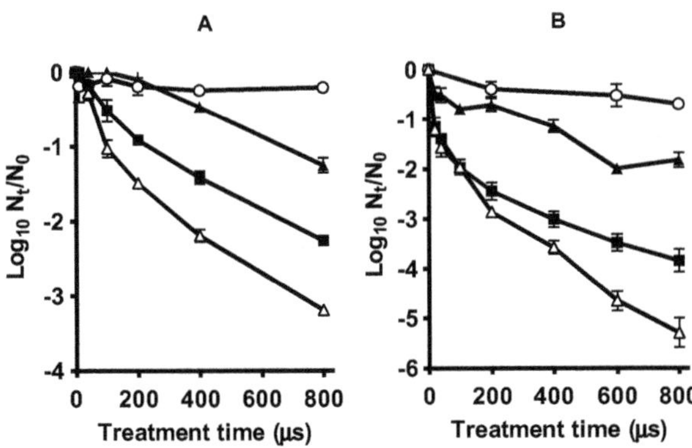

Fig. 7 Influence of the a_w of the treatment medium on the PEF inactivation of *Salmonella* serovar Senftenberg (**a**) and *L. monocytogenes* (**b**). Treatment conditions, McIlvaine buffer pH 7.0 and > 0,99 (■), a_w = 0,93 and 50% sucrose (O), a_w = 0,93 and 10% glycerol (Δ), a_w = 0,80 and 50% glycerol (▲); field strength, 22 kV/cm, 2,2 kJ/kg/pulse (*Salmonella*), 28 kV/cm, 3,6 kJ/kg/pulse (*L. monocytogenes*); 2 µsec square wave pulses; frequency, 1 Hz; conductivity, 2 mS/cm

Composition of the Treatment Medium

Several studies have been carried out to evaluate the influence of treatment media components on the efficacy of PEF. However, the wide variety of possible conditions and combinations makes it difficult to reach general conclusions.

Protein concentration has been reported to exert a protective effect during inactivation (Jaeger et al. 2009). However, Schottroff et al. (2019) found no effect of protein concentration when the treatment was carried out at neutral pH. Similarly, Mañas et al. (2001) observed non-protective behavior of emulsified lipids or soluble proteins in the microbial inactivation of *E. coli* by PEF at neutral pH.

Kinetics of Microbial Inactivation by PEF

Microbial death kinetics describes the evolution of the number of microorganisms that are subjected to a lethal agent for a long time. Survival curves corresponding to microbial inactivation by PEF are generally represented as the Log_{10} of the number of survivors to a constant electric field strength against time. It has been traditionally assumed that first-order kinetics governs this relation, which means that the number of survivors decreases exponentially with treatment time and a straight line describes the survival curves. This approach is very useful for the characterization of microbial resistance to PEF using the D values (decimal reduction time) and for the prediction of inactivation rates under different conditions (electric field strength, pH, temperature) by developing secondary models.

Some early studies have found such a linear relation in survival curves by PEF (Heinz et al. 1999; Reina et al. 1998; Sensoy et al. 1997) but only at inactivation levels of less than 4.0 Log_{10} cycles and when treatments are applied at low electric field strengths or for short times. If the treatment time is prolonged, this linear correlation tends to deviate, resulting in curves with a concave upward tail, describing rapid inactivation in the first instants of the treatment until reaching a point when the inactivation rate decreases.

Survival curves with a tailing effect are the ones most commonly reported in studies of microbial resistance to PEF treatments (Martínez et al. 2016; Monfort et al. 2010a, b; Puértolas et al. 2009; Qin et al. 2015; Walter et al. 2016). In order to describe this particular inactivation kinetics obtained by PEF treatments, several alternative models to the traditional first-order kinetics have been proposed. Table 2 summarizes different models (primary models) that have been suggested to describe PEF survival curves, from purely empirical equations to the most widely used model based on the Weibull distribution. The Weibull model has been reported as one of the best fits for the death kinetics of microorganisms subjected to PEF; several studies have used it to characterize microbial death under such circumstances. On the basis of their parameters, secondary models have been developed, mainly using quadratic response equations to describe the effect of different factors such as the electric field, treatment medium pH, and temperature on the parameters of the

Table 2 Mathematical equations used to describe the kinetics of PEF microbial survival curves

Model	Mathematical equation	Parameters	Significance
First-order kinetics	$\log_{10}S(t) = -kt$	S, survival fraction; k, death rate; t; treatment time	The velocity of inactivation is proportionally dependent on microbial concentration along treatment time
Log-logistic model based on Cole et al. (1993)	$\log_{10} S(t) = \frac{\alpha + (\omega - \alpha)}{1 + e^{(4\sigma(\tau - \log_{10} t))/(\omega - \alpha)}}$	S, survival fraction; α, upper asymptote (log cfu/mL); ω, lower asymptote log (cfu/mL); σ, maximum slope of the inactivation curve; τ, log time when the slope is maximum	Implies a distribution of resistances within the bacterial population
Based on Hülsheger's model (Hülsheger et al. 1981)	$S = \left(\frac{t}{t_c}\right)^{-(E-E_c)/k}$	S, survival fraction; E, electric field strength; Ec, critical electric field strength; k, kinetic constant; t, treatment time; tc, critical treatment time	Assumes a linear relation between the *log S* and the E for a given t, and between the *log S* and the *Log t* for a given E. Ec and tc allow to determine which treatments start to become lethal. Parameters are characteristic for each microorganism
Based on Fermi's model (Peleg 1995)	$S(E,n) = \frac{100}{1 + e^{(E - E_c(n))/a_c(n)}}$	S, survival fraction; E, electric field strength; Ec, electric field strength necessary to inactivate 50%; n, number of pulses; a_c, slope of the sigmoid curve in Ec	Describes the relationship between the percentage of survivors (S) and the electric field strength (E) for a given number of pulses
Based on Pruitt and Kamau (Pruitt and Kamau 1993)	$S(t) = pe^{-k_1 t} + (1-p)e^{-k_2 t}$	S, survival fraction; p, fraction of survivors in population 1 (PEF sensitive); $1-p$, fraction of survivors in population 2 (PEF resistant); k_1, specific death rate of subpopulation 1; k_2, specific death rate of subpopulation 2; t, treatment time	Assumes the existence of two different microbial subpopulations, one sensitive (p) and the other one resistant ($1-p$)
Based on the Weibull distribution model	$\log_{10}S(t) = -bt^n$	S, survival fraction; b, scale parameter; n, shape parameter; t, treatment time	Shape parameter represents the form of the curve (<1, concave upward; >1, concave downward; =1, linear). Assumes a distribution of resistances within the microbial population, and the survival curve is just the cumulative form of lethal agents
	$\log_{10} S(t) = -\left(\frac{t}{\delta}\right)^p$	S, survival fraction; δ, *scale* parameter; p, shape parameter; t, treatment time	
	$\log_{10} S(t) = -\left(\frac{1}{2.303}\right)\left(\frac{t}{\alpha}\right)^\beta$	S, survival fraction; α, scale parameter; β, shape parameter; t, treatment time	

primary model (Huang et al. 2013, 2014; Saldaña et al. 2010a, b, c; Sampedro et al. 2011; Walter et al. 2016).

Several theories have been proposed to explain deviations from first-order kinetics. While the mechanistic theory assumes that a microbial population has identical resistance to a lethal agent, the vitalistic theory suggests the existence of a distribution of microbial sensitivity to inactivation processes (Stewart et al. 2002). According to the latter, survival curves would describe the cumulative form of a temporal distribution of lethal effects: the tailing would correspond to the death of the more resistant ones. Likewise, a mix of two fractions or subpopulations with different degrees of PEF resistance could also cause concave upward survival curves. The first portion would mainly describe the death of the least resistant microorganisms, whereas the second one would describe the death of those that are most resistant.

Although the application of the Weibull model is based on the assumption that different degrees of PEF resistances exist within the same microbial population (Mafart et al. 2002), it should be empirically tested, since the presence of microorganisms with different PEF sensitivity within a microbial suspension has not been demonstrated. Given the Weibull model's lack of mechanistic significance, conclusions drawn from its application should not be extrapolated; the same applies to all the models featured in Table 2. In any case, these models are helpful in order to quantify and compare the magnitude of factors affecting microbial inactivation already described in this chapter.

Methodological Artifacts Affecting Microbial Inactivation Kinetics by PEF

As pointed out above, deviations from linearity are frequent in microbial survival curves to PEF; there is still no agreement on the causes or mechanisms responsible for this particular behavior. As described above, many factors affect the efficiency of PEF treatments in a variety of different directions. Moreover, the non-standardization of microbial growth protocols, of PEF treatment processing, or of recovery conditions might lead to methodological artifacts that could also exert an effect on inactivation kinetics. Table 3 summarizes the main sources of artifacts that can occur during PEF treatment processing for microbial inactivation purposes and which may affect inactivation kinetics.

Strain and Culture Conditions

Particular care must be applied regarding inoculum preparation and culture conditions of microorganisms to obtain rigorous data, especially in view of the wide distribution of degrees of PEF resistance within microbial groups and even among strains of the same microbial species. It is known that the electric field threshold

Table 3 Experimental artifacts that can influence the kinetics of PEF microbial inactivation

Artifact source	Causes
Strain and culture conditions	Lack of standardization regarding the inoculum preparation
	Mixtures of cells at different stages of growth
	Mixtures of strains with different degrees of resistance
PEF treatment	Inaccurate measurement of treatment parameters
	Nonuniform distribution of the electric field strength in the treatment chamber
	Variations in the electric field strength during the treatment
	Variation of sample temperature during the treatment
Treatment medium	Inaccurate measurement of treatment medium properties
	Generation of electrochemical reaction products
Recovery conditions	Use of selective media
	Short incubation times
	Holding of the microorganisms in the treatment medium before enumeration

required to achieve electroporation is directly dependent on cell size or on cell orientation (of nonspherical forms) within the electric field. Through computational models, Lebovka and Vorobiev (2004, 2007) validated that the variability in cell size and/or morphology within a population can justify the deviation from first-order kinetics. Therefore, microbial kinetics experiments must be performed with a single strain at a homogeneous stage of growth in order to avoid variance in the size and morphology of cells. For validation experiments, it could be more effective to use cocktails of representative strains instead of single strains.

On the other hand, heterogeneity in the microbial population due to methodological errors can also change the shape of survival curves and lead to erroneous conclusions regarding inactivation kinetics and the concept of resistance distribution. Accordingly, and in spite of the numerous studies published on microbial inactivation by PEF, further work is still necessary to elucidate if the actual underlying cause of nonlinearity in survival curves is due to a distribution of innate or acquired microbial resistance. Such research should focus on the role of microbial mechanisms triggered by exposure to PEF and on the evolution of the microbial population after treatment.

PEF Treatment Parameters

Control of PEF treatment parameters is critical to enable comparison of results, since there are huge dependencies among the different process parameters. For example, an underestimation of temperature rise during the application of the PEF treatment may imply an increase in electrical conductivity and thus a reduction of the chamber's resistance (Ω), which means that a higher intensity (A) is required to deliver the same voltage (V) between the electrodes, as previously pointed out.

Temperature, moreover, is one of the main parameters affecting microbial inactivation. This implies that if temperature is not meticulously controlled, the proportion of inactivation can be affected, thereby leading researchers to obtain false tendencies in the kinetics, either by making it concave or lineal. The same would apply to any other relevant treatment parameters.

Therefore, to obtain accurate data parameters such as actual electric field strength, treatment time, temperature, pulse width, and intensity, they should be monitored continuously throughout the treatment. It is more suitable to use square wave pulses, since they permit to calculate the actual electric field and treatment time more accurately than exponential decay pulses.

Nonuniform distribution of the electric field in the treatment chamber is probably the most important operational artifact that can occur in the estimation of the microbial resistance to PEF. This artifact has been reported by different authors as the cause of the tailing effect. Mañas et al. (Mañas et al. 2001) observed the presence of dead spaces in the head of chambers where the electric field was ineffective, and they observed an enhancement of inactivation rates when the content of the chamber was homogenized between pulses by mixing. Donsì et al. (2007) treated *S. cerevisiae* cells immobilized in agar; after visualizing colony growth, they concluded that the electric field was lower in the peripheral zones of the chamber. However, even after applying the mixing protocol, inactivation kinetics did not become lineal; the authors attributed the tail to the formation of cell clusters, which made cells more resistant. By contrast, Raso et al. (2000) found a good uniformity of the electric field using a batch chamber of parallel electrodes. No differences in the inactivation of *Salmonella* Senftenberg were obtained by applying mixing protocols, and observation of treated agar discs showed homogeneously distributed colonies. Moreover, the uniformity of the electric field in a batch treatment chamber of parallel electrodes was also confirmed by numerical simulation (Saldaña et al. 2010a). Consequently, basic studies of microbial inactivation kinetics require previous investigation of the electric field uniformity, which is generally better in static chambers of parallel electrodes rather than in continuous ones (Toepfl et al. 2014).

The deviation from first-order kinetics is not so evident when PEF treatments are applied in continuous flow conditions (Picart et al. 2002; Walkling-Ribeiro et al. 2009). This could be related to the fact that methodological artifacts such as dead spaces in the chamber or nonuniformity of the electric field are minimized by turbulent flow (Meneses et al. 2011). Thereby, if such artifacts were the cause of the nonlinearity of survival curves, they would be drastically reduced in continuous treatment chambers, thanks to the turbulent flow that homogenizes samples. Moreover, in continuous flow treatments, the residence time in the treatment chambers is extremely short, and higher frequencies are required as the treatment time is extended. These facts suppose a dissipation of the thermal energy throughout the product, leading to an increase in temperature due to ohmic heating (Lindgren et al. 2002; Saldaña et al. 2014). Hence, without temperature control, the synergistic killing effect of heat combined with PEF would make the survival curves linear. Lastly, numerical simulation has shown that the temperature rise in continuous flow treatments varies widely in different parts of the treatment chamber, especially in collinear ones (Buckow et al. 2010, 2011). It is thus difficult to discern the actual individual effect of the electric field independently of the influence of temperature, which extremely affects microbial resistance to PEF. Furthermore, an underestimation of temperature increment would imply changes in the electrical conductivity of the medium and thus a change in the actual intensity delivered as well as in the specific energy applied. Therefore, in order to obtain adequate data regarding microbial death kinetics, it is preferable to use static chambers rather than continuous flow treatments, in which it is complex to keep all parameters under control.

Treatment Medium

An accurate characterization of the treatment medium properties is essential in order to understand and predict their influence on processing parameters and avoid methodological artifacts. As discussed, microbial inactivation by PEF has been found to depend on the pH and a_w of the treatment medium. The medium's electrical conductivity has likewise been shown to be one of the most important factors, due to the particular nature of the PEF technique and its direct influence on the other electrical parameters.

Electrochemical reactions have been proven to occur during the application of PEF, resulting in partial electrolysis of the solution, electrode corrosion, and introduction of particles of electrode material into the liquid (Pataro et al. 2017). By introducing undesired toxic constituents, such phenomena can enhance the microbial killing effect and provoke an overestimation of inactivation. Furthermore, the degradation of electrode surfaces may imply a nonhomogeneous distribution of the electric field strength along the treatment chamber. These effects would be intensified with longer treatment times and thus might deform the survival curves. Such electric reactions should therefore be minimized by reducing pulse width, using suitable electrode materials, and periodically carrying out adequate maintenance of the electrodes.

Recovery Conditions

After the application of PEF to cells, different outcomes can be observed, and their evolution is highly dependent on ambient conditions (Fig. 2). For example, sublethally damaged cells can survive or die depending on the incubation time after PEF or on the temperature and pH of the medium (Raso et al. 2016). The quantification of survivors should therefore be adequately carried out with standardized procedures. Holding microorganisms in the treatment medium before enumeration, applying an incubation time that is too short, or using selective media may lead researchers to overestimate the treatment's effectiveness and hence the safety margin that has been achieved.

Although anaerobic conditions or low oxygen concentrations may affect microbial recovery, Cebrián et al. (2007) did not find differences in the recovery of

Staphylococcus aureus after PEF when anaerobic conditions were applied. On the other hand, Marcén et al. (2019) observed a greater number of *E. coli* survivors when PEF-treated cells recovered under anaerobic conditions in a medium with low redox potential, thus suggesting that oxidative damage might affect survival capacity. This effect needs to be investigated more thoroughly.

In conclusion, there are many sources of potential methodological artifacts which, in turn, can cause survival curves to deviate from first-order kinetics. Once the methodological artifacts are under control, such curve behavior can be explained by the hypothesis of a resistance distribution within a population. More basic in-depth studies should thus be carried out to elucidate the actual microbial inactivation kinetics with PEF under highly controlled conditions, as well as the underlying mechanisms. In all such studies, it would likewise be important to adequately describe the PEF treatment conditions in order to avoid misunderstandings (Raso et al. 2016).

Combined Treatments

Even though the efficacy of PEF to inactivate vegetative cells has been widely demonstrated, in most cases it is necessary to apply intense PEF treatments (high voltages and long times) to obtain substantial microbial inactivation capable of guaranteeing food safety and stability (Saldaña et al. 2014). From a practical point of view, such high-intensity treatments have many drawbacks associated with equipment requirements, their impact on food properties, and the total cost of the process. Thus the industrial scale-up potential is quite limited for microbial inactivation purposes. What is more, bacterial spores and enzymes are resistant to PEF; this could represent an additional limitation of the technology. Therefore, in order to increase the lethal effect of PEF treatments or to prevent subsequent damage by deteriorative agents, certain combinations of PEF with other preservation strategies have been investigated (Garner 2019). Combined treatments consist in the application of various preservation techniques in a successive or simultaneous form. Thanks to the additive or even synergistic lethal effect achieved by such combinations, PEF treatments can be applied with mild intensity, thereby causing considerably less impact on food properties. This approach, known as Hurdle technology, has already been successfully applied by combining PEF with traditional food preservation techniques.

One of the most successful strategies can be found in the application of PEF in combination with mild temperatures during pulse delivery. The inactivation levels obtained with a determined PEF treatment can be significantly enhanced by means of moderate temperatures (< 50 °C) that should not affect food properties (Katiyo et al. 2017; Saldaña et al. 2014; Sharma et al. 2014; Timmermans et al. 2019).

On the other hand, the acidification of foods is one of the approaches most widely applied with the purpose of reducing the undesirable effects of thermal processing, since low pH reduces microbial thermal resistance and inhibits the germination of surviving spores. Acidic food products are thus especially interesting candidates for PEF-based combined processes. Such strategies would also prevent sublethally damaged or stressed cells from recovering their viability, as might otherwise occur after a PEF treatment (Fig. 3). Although the effect of treatment medium pH on microbial resistance has been shown to be highly variable, a further incubation of PEF-resistant microorganisms under acidic conditions can lead to a progressive decrease in their numbers. For example, *E. coli* O157:H7 in apple juice was inactivated less than 90% by a PEF treatment (19 kV/cm, 400 μs); however, after just 6 hours of incubation in the juice, the inactivation level increased up to 99.99% of the population (García et al. 2005b). Therefore, although microbial sensitivity to PEF does not always increase by lowering the treatment medium pH, the key in this combination is the inhibitory effect of low pH on the germination of bacterial spores and on the growth of nonacid-tolerant microorganisms, as well as the inactivating action of low pH and/or low temperatures on microorganisms sublethally injured by PEF.

The combination of PEF with antimicrobial compounds in the treatment medium has also been proven to be an effective strategy capable of increasing the lethal effect of PEF. Some of the investigated antimicrobials with positive results are nisin (Saldaña et al. 2012), lysozyme (Smith et al. 2002), EDTA (Monfort et al. 2011), organic acids (Mosqueda-Melgar et al. 2008), ozone (Unal et al. 2001), and essential oils (Ait-Ouazzou et al. 2013; Espina et al. 2014). The effect of antimicrobials has been proven as a simultaneous and as a successive treatment. However, their effect on the viability of microorganisms after PEF treatments has not been investigated.

Studies in this area indicate that the combination of PEF with other preservation factors seems to be a very promising approach for the nonthermal preservation of foods with a low impact on food quality, while ensuring food safety and stability. However, recent studies have reported that microbial cells can occasionally develop adaptive responses to certain stresses (e.g., heat or alkaline shocks, acid growth conditions, or the presence of ethanol), thereby increasing their subsequent resistance to PEF (Cebrián et al. 2012; Fernández et al. 2018; Niu et al. 2019). In order to optimize a combined treatment with PEF, it is thus necessary to evaluate the possible emergence of cross-resistances.

Conclusions

To ensure continued consumer confidence in the food industry, it is essential that high levels of food safety and quality be maintained. There is an increasing demand for foods that contain lower levels of preservatives and have been subjected to less severe processing; however, such requirements are not always compatible with an improvement in microbial safety and stability. In order to reduce the foodborne microbiological hazard and to extend the shelf-life of minimally processed foods, alternative nonthermal technologies for microbial inactivation have been widely investigated.

Pulsed electric fields have widely proven to be effective in killing vegetative bacteria, yeast, and molds at mild temperatures and are thus becoming a promising technology that can be applied to obtain safe, fresh-like food products. Prior to an implementation on an industrial scale, it is necessary to establish adequate process parameters capable of successfully inactivating the target pathogenic microorganisms.

The application of an external electric field causes the formation of stable aqueous pores in the cytoplasmic membrane, with the corresponding loss of osmotic equilibrium, leading to microbial cell death. Several further varieties of cell outcome can be also observed after the application of a PEF treatment, including sublethally damaged cells that are capable of surviving under optimal conditions. Although other potential cell phenomena have been theorized, further research is still necessary to better understand the variety of different cell responses after having been exposed to PEF. The knowledge thus gained would allow researchers to design improved PEF experiments and to adequately investigate the inactivation kinetics.

A wide range of techniques are currently available for the detection and measurement of microbial electroporation. Although the one most widely used is the measurement of the uptake of fluorescence dyes, other techniques present advantages in certain particular scenarios. Those based on plate count display a much higher sensitivity at high levels of electroporation, which is especially useful in microbial inactivation studies. Different techniques should therefore be chosen, depending on the objectives and characteristics of the study.

Electric field strength, treatment time, and total specific energy stand out as the main parameters that characterize PEF treatments and determine the degree of final microbial death, which increases with treatment temperature. On the other hand, the actual influence of pulse shape and width as well as of frequency on inactivation is still to be fully elucidated. In any case, all these process parameters should be considered, monitored, and correctly reported in research on microbial inactivation by PEF. The influence of microbial type, of cell size and shape, as well as of culture conditions are all immensely relevant. Microbial cells tend to be more sensitive to PEF as their size increases, during exponential growth phase. Gram-positive bacteria are generally the most resistant ones, followed by Gram-negative bacteria, yeast, and molds; however, media composition can drastically change this assumption, since medium pH is one of the most relevant factors.

The kinetics of microbial PEF inactivation usually deviates from linearity, resulting in the appearance of tails in the survival curves. Several mathematical models have been proposed to describe this inactivation kinetics, whereby Weibull is the one the most used. However, these models' lack of biological significance limits their extrapolability. Although the cause of this tailing has been attributed to different phenomena (resistance distribution within a microbial population on the one hand or methodological artifacts on the other), there is no consensus on this subject, and further work is still necessary. Understanding the actual underlying mechanism of this deviation would allow for an optimization of PEF process parameters for purposes of microbial inactivation.

In summary, further research is required in order to clarify the underlying mechanism of PEF microbial inactivation. Future studies will also help to understand how different parameters influence PEF lethality and to develop mathematical models with biological significance that would allow for the prediction of inactivation of foodborne pathogens as well as spoilage microorganisms of concern under different treatment conditions. All of this would contribute toward the definitive implementation of PEF as an alternative to thermal treatments for food pasteurization processes in continuous flow conditions at industrial scale for liquid – and perhaps in the near future also for solid – food products.

References

Abee T, Wouters JA (1999) Microbial stress response in minimal processing. Int J Food Microbiol 50(1–2):65–91

Ait-Ouazzou A, Espina L, García-Gonzalo D, Pagán R (2013) Synergistic combination of physical treatments and Carvacrol for *Escherichia Coli* O157:H7 inactivation in apple, mango, Orange, and tomato juices. Food Control 32(1):159–167

Alvarez I, Pagan R, Raso J, Condon S (2002) Environmental factors influencing the inactivation of *Listeria Monocytogenes* by pulsed electric fields. Lett Appl Microbiol 35(6):489–493

Álvarez I, Pagán R, Condón S, Raso J (2003a) The influence of process parameters for the inactivation of *Listeria Monocytogenes* by pulsed electric fields. Int J Food Microbiol 87(1–2):87–95

Álvarez I, Raso J, Sala FJ, Condón S (2003b) Inactivation of *Yersinia Enterocolitica* by pulsed electric fields. Food Microbiol 20(6):691–700

Aronsson K, Rönner U (2001) Influence of PH, water activity and temperature on the inactivation of *Escherichia Coli* and *Saccharomyces Cerevisiae* by pulsed electric fields. Innovative Food Sci Emerg Technol 2(2):105–112

Aronsson K, Lindgren M, Johansson BR, Rönner U (2001) Inactivation of microorganisms using pulsed electric fields: the influence of process parameters on *Escherichia Coli, Listeria Innocua, Leuconostoc Mesenteroides* and *Saccharomyces Cerevisiae*. Innovative Food Sci Emerg Technol 2(1):41–54

Aronsson K, Borch E, Stenlöf B, Rönner U (2004) Growth of pulsed electric field exposed *Escherichia Coli* in relation to inactivation and environmental factors. Int J Food Microbiol 93(1):1–10

Aronsson K, Rönner U, Borch E (2005) Inactivation of *Escherichia Coli, Listeria Innocua* and *Saccharomyces Cerevisiae* in relation to membrane Permeabilization and subsequent leakage of intracellular compounds due to pulsed electric field processing. Int J Food Microbiol 99(1):19–32

Arroyo C, Cebrián G, Pagán R, Condón S (2010a) Resistance of *Enterobacter Sakazakii* to pulsed electric fields. Innovative Food Sci Emerg Technol 11(2):314–321

Arroyo C, Somolinos M, Cebrián G, Condón S, Pagán R (2010b) Pulsed electric fields cause sublethal injuries in the outer membrane of *Enterobacter Sakazakii* facilitating the antimicrobial activity of Citral. Lett Appl Microbiol 51(5):525–531

Arroyo C, Cebrián G, Condón S, Pagán R (2012) Development of resistance in *Cronobacter Sakazakii* ATCC 29544 to thermal and nonthermal processes after exposure to stressing environmental conditions. J Appl Microbiol 112(3):561–570

Augustin MA, Riley M, Stockmann R, Bennett L, Kahl A, Lockett T, Osmond M, Sanguansri P, Stonehouse W, Zajac I, Cobiac L (2016) Role of food processing in food and nutrition security. Trends Food Sci Technol 56:115–125

Baird-Parker TC (2000) The production of microbiologically safe and stable foods. In: Microbiological safety and quality of food. Aspen Publishers, Gaithersburg, pp 3–18

Barba FJ, Koubaa M, do Prado-Silva L, Orlien V, de Souza Sant'Ana A (2017) Mild processing applied to the inactivation of the Main foodborne bacterial pathogens: a review. Trends Food Sci Technol 66:20–35

Bartoletti DC, Harrison GI, Weaver JC (1989) The number of molecules taken up by Electroporated cells: quantitative determination. FEBS Lett 256(1–2):4–10

Batista-Napotnik T, Miklavčič D (2018) In vitro electroporation detection methods – an overview. Bioelectrochemistry 120:166–182

Bazhal M, Lebovka N, Vorobiev E (2003) Optimisation of pulsed electric field strength for Electroplasmolysis of vegetable tissues. Biosyst Eng 86(3):339–345

Beveridge JR, MacGregor SJ, Anderson JG, Fouracre RA (2005) The influence of pulse duration on the inactivation of Bacteria using monopolar and bipolar profile pulsed electric fields. IEEE Trans Plasma Sci 33(4):1287–1293

Blackburn CDW, McClure PJ (2009) Foodborne pathogens hazards, risk analysis and control, 2nd edn. Woodhead Publishing Limited

Bodénès P, Bensalem S, Français O, Pareau D, Le Pioufle B, Lopes F (2019) Inducing reversible or irreversible pores in *Chlamydomonas Reinhardtii* with electroporation: impact of treatment parameters. Algal Res 37:124–132

Buckow R, Schroeder S, Berres P, Baumann P, Knoerzer K (2010) Simulation and evaluation of pilot-scale pulsed electric field (PEF) processing. J Food Eng 101(1):67–77

Buckow R, Baumann P, Schroeder S, Knoerzer K (2011) Effect of dimensions and geometry of co-field and co-linear pulsed electric field treatment chambers on electric field strength and energy utilisation. J Food Eng 105(3):545–556

Cebrián G (2016) Techniques to detect electroporation in microbial cells. In 3rd PEF school of pulsed electric field processing of food, Dublin (Ireland). Book of proceedings. 3rd PEF School of Pulsed Electric Field Processing of Food, Dublin (Ireland). Book of proceedings

Cebrián G, Sagarzazu N, Pagán R, Condón S, Mañas P (2007) Heat and pulsed electric field resistance of pigmented and non-pigmented Enterotoxigenic strains of *Staphylococcus aureus* in exponential and stationary phase of growth. Int J Food Microbiol 118(3):304–311

Cebrián G, Sagarzazu N, Pagán R, Condón S, Mañas P (2008) Resistance of *Escherichia coli* grown at different temperatures to various environmental stresses. J Appl Microbiol 105(1):271–278

Cebrián G, Sagarzazu N, Aertsen A, Pagán R, Condón S, Mañas P (2009) Role of the alternative sigma factor σB on *Staphylococcus aureus* resistance to stresses of relevance to food preservation. J Appl Microbiol 107(1):187–196

Cebrián G, Raso J, Condón S, Mañas P (2012) Acquisition of Pulsed Electric Fields Resistance in *Staphylococcus aureus* after exposure to heat and alkaline shocks. Food Control 25(1):407–414

Cebrián G, Mañas P, Condón S (2015) Relationship between membrane Permeabilization and sensitization of *S. aureus* to sodium chloride upon exposure to pulsed electric fields. Innovative Food Sci Emerg Technol 32:91–100

Cebrián G, Condón S, Mañas P (2016a) Influence of growth and treatment temperature on *Staphylococcus aureus* resistance to pulsed electric fields: relationship with membrane fluidity. Innovative Food Sci Emerg Technol 37:161–169

Cebrián G, Mañas P, Condón S (2016b) Comparative resistance of bacterial foodborne pathogens to non-thermal Technologies for Food Preservation. Front Microbiol 7

Chueca B, Pagán R, García-Gonzalo D (2015) Transcriptomic analysis of *Escherichia coli* MG1655 cells exposed to pulsed electric fields. Innovative Food Sci Emerg Technol 29:78–86

Cole MB, Davies KW, Munro G, Holyoak CD, Kilsby DC (1993) A Vitalistic model to describe the thermal inactivation Of Listeria Monocytogenes. J Ind Microbiol 12(3–5):232–239

Coster HGL, Zimmermann U (1975) The mechanism of electrical breakdown in the membranes of Valonia Utricularis. J Membr Biol 22(1):73–90

Cregenzán-Alberti O, Arroyo C, Dorozko A, Whyte P, Lyng JG (2017) Thermal characterization of *Bacillus subtilis* endospores and a comparative study of their resistance to high temperature pulsed electric fields (HTPEF) and thermal-only treatments. Food Control 73:1490–1498

de Haan SWH (2007) Circuitry and pulse shapes in pulsed electric field treatment of food. In: Food preservation by pulsed electric fields. Elsevier, pp 43–69

Deforet M, van Ditmarsch D, Xavier JB (2015) Cell-size homeostasis and the incremental rule in a bacterial pathogen. Biophys J 109(3):521–528

Delemotte L, Tarek M (2012) Molecular dynamics simulations of lipid membrane electroporation. J Membr Biol 245(9):531–543

Donsì G, Ferrari G, Pataro G (2007) Inactivation kinetics of *Saccharomyces cerevisiae* by pulsed electric fields in a batch treatment chamber: the effect of electric field unevenness and initial cell concentration. J Food Eng 78(3):784–792

Donsì F, Ferrari G, Pataro G (2010) Applications of pulsed electric field treatments for the enhancement of mass transfer from vegetable tissue. Food Eng Rev 2(2):109–130

Espina L, Monfort S, Álvarez I, García-Gonzalo D, Pagán R (2014) Combination of pulsed electric fields, mild heat and essential oils as an alternative to the Ultrapasteurization of liquid whole egg. Int J Food Microbiol 189:119–125

Fernández A, Cebrián G, Álvarez-Ordóñez A, Prieto M, Bernardo A, López M (2018) Influence of acid and low-temperature adaptation on pulsed electric fields resistance of *Enterococcus faecium* in Media of Different PH. Innovative Food Sci Emerg Technol 45:382–389

Fox MB, Esveld DC, Mastwijk H, Boom RM (2008) Inactivation of L. Plantarum in a PEF microreactor. The effect of pulse width and temperature on the inactivation. Innov Food Sci Emerg Technol

Frey W, Gusbeth C, Schwartz T (2013) Inactivation of *Pseudomonas putida* by pulsed electric field treatment: a study on the correlation of treatment parameters and inactivation efficiency in the short-pulse range. J Membr Biol 246(10):769–781

Gachovska TK, Subbiah J, Thippareddi H, Marx D, Williams F (2013) Inactivation of *E. coli* affected by medium conductivity in pulsed electric field. In: 2013 19th IEEE pulsed power conference (PPC). IEEE, pp 1–7

García D, Gómez N, Condón S, Raso J, Pagán R (2003) Pulsed electric fields cause sublethal injury in *Escherichia coli*. Lett Appl Microbiol

García D, Gómez N, Mañas P, Condón S, Raso J, Pagán R (2005a) Occurrence of sublethal injury after pulsed electric fields depending on the micro-organism, the treatment medium pH and the intensity of the treatment investigated. J Appl Microbiol 99(1):94–104

García D, Hassani M, Mañas P, Condón S, Pagán R (2005b) Inactivation of *Escherichia coli* O157:H7 during the storage under refrigeration of apple juice treated by pulsed electric fields. J Food Saf 25(1):30–42

García D, Manas P, Gomez N, Raso J, Pagan R (2006) Biosynthetic requirements for the repair of sublethal membrane damage in *Escherichia coli* cells after pulsed electric fields. J Appl Microbiol 100(3):428–435

García D, Gómez N, Mañas P, Raso J, Pagán R (2007) Pulsed electric fields cause bacterial envelopes Permeabilization depending on the treatment intensity, the treatment medium pH and the microorganism investigated. Int J Food Microbiol 113(2):219–227

García-Gonzalo D, Pagán R (2017) Detection of electroporation in microbial cells: techniques and procedures. In: Handbook of electroporation. Springer, Cham, pp 1359–1373

Garner AL (2019) Pulsed electric field inactivation of microorganisms: from fundamental biophysics to synergistic treatments. Appl Microbiol Biotechnol 103(19):7917–7929

Gelaw TK, Espina L, Pagán R, García-Gonzalo D, De Lamo-Castellví S (2013) Prediction of injured and dead inactivated *Escherichia coli* O157:H7 cells after heat and pulsed electric field treatment with attenuated total reflectance infrared microspectroscopy combined with multivariate analysis technique. Food Bioprocess Technol 7:2084–2092

Gerlach D, Alleborn N, Baars A, Delgado A, Moritz J, Knorr D (2008) Numerical simulations of pulsed electric fields for food preservation: a review. Innovative Food Sci Emerg Technol 9(4):408–417

Hamilton W, Sale A (1967) Effects of high electric fields on microorganisms II. Mechanism of action of the lethal effect. Biochim Biophys Acta Gen Subj 148(3):789–800

Heinz V, Phillips ST, Zenker M, Knorr D (1999) Inactivation of *Bacillus subtilis* by high intensity pulsed electric fields under close to isothermal conditions. Food Biotechnol 13(2):155–168

Heinz V, Alvarez I, Angersbach A, Knorr D (2001) Preservation of liquid foods by high intensity pulsed electric fields—basic concepts for process design. Trends Food Sci Technol 12(3–4):103–111

Ho SY, Mittal GS (1996) Electroporation of cell membranes: a review. Crit Rev Biotechnol 16(4):349–362

Ho SY, Mittal GS (2000) High voltage pulsed electrical field for liquid food pasteurization. Food Rev Intl 16(4):395–434

Huang K, Wang J (2009) Designs of pulsed electric fields treatment chambers for liquid foods pasteurization process: a review. J Food Eng 95(2):227–239

Huang K, Yu L, Liu D, Gai L, Wang J (2013) Modeling of yeast inactivation of PEF-treated Chinese Rice wine: effects of electric field intensity, treatment time and initial temperature. Food Res Int 54(1):456–467

Huang K, Yu L, Wang W, Gai L, Wang J (2014) Comparing the pulsed electric field resistance of the microorganisms in grape juice: application of the Weibull model. Food Control 35(1):241–251

Huis in't Veld JHJ (1996) Microbial and biochemical spoilage of foods: an overview. Int J Food Microbiol 33(1):1–18

Hülsheger H, Potel J, Niemann EG (1981) Killing of Bacteria with electric pulses of high field strength. Radiat Environ Biophys 20(1):53–65

Ibey BL, Ullery JC, Pakhomova ON, Roth CC, Semenov I, Beier HT, Tarango M, Xiao S, Schoenbach KH, Pakhomov AG (2014) Bipolar nanosecond electric pulses are less efficient at Electropermeabilization and killing cells than monopolar pulses. Biochem Biophys Res Commun 443(2):568–573

Ivorra A (2010) Tissue electroporation as a bioelectric phenomenon: basic concepts. In: Irreversible electroporation. Springer, Berlin, pp 23–61

Jacob HE, Förster W, Berg H (1981) Microbiological implications of electric field effects II. Inactivation of yeast cells and repair of their cell envelope. Z Allg Mikrobiol 21(3):225–233

Jaeger H, Schulz A, Karapetkov N, Knorr D (2009) Protective effect of Milk constituents and sublethal injuries limiting process effectiveness during PEF inactivation of *Lb. rhamnosus*. Int J Food Microbiol 134(1–2):154–161

Jayaram S, Castle GSP, Margaritis A (1993) The effects of high field DC pulse and liquid medium conductivity on survivability of *Lactobacillus brevis*. Appl Microbiol Biotechnol 40(1)

Kandušer M, Šentjurc M, Miklavčič D (2006) Cell membrane fluidity related to electroporation and resealing. Eur Biophys J 35(3):196–204

Katiyo W, Yang R, Zhao W (2017) Effects of combined pulsed electric fields and mild temperature pasteurization on microbial inactivation and physicochemical properties of cloudy red apple juice (Malus Pumila Niedzwetzkyana (Dieck)). J Food Saf 37(4):e12369

Kethireddy V, Oey I, Jowett T, Bremer P (2016) Critical analysis of the maximum non inhibitory concentration (MNIC) method in quantifying sub-lethal injury in *Saccharomyces cerevisiae* cells exposed to either thermal or pulsed electric field treatments. Int J Food Microbiol 233:73–80

Korber DR, Choi A, Wolfaardt GM, Caldwell DE (1996) Bacterial plasmolysis as a physical Indicator of viability. Appl Environ Microbiol 62(11):3939–3947

Kotnik T, Pucihar G, Miklavčič D (2010) Induced transmembrane voltage and its correlation with electroporation-mediated molecular transport. J Membr Biol 236(1):3–13

Kotnik T, Kramar P, Pucihar G, Miklavcic D, Tarek M (2012) Cell membrane electroporation- part 1: the phenomenon. IEEE Electr Insul Mag 28(5):14–23

Kotnik T, Rems L, Tarek M, Miklavčič D (2019) Membrane electroporation and Electropermeabilization: mechanisms and models. Annu Rev Biophys 48(1):63–91

Lebovka NI, Vorobiev E (2004) On the origin of the deviation from the first-order kinetics in inactivation of microbial cells by pulsed electric fields. Int J Food Microbiol 91(1):83–89

Lebovka N, Vorobiev E (2007) The kinetics of inactivation of spheroidal microbial cells by pulsed electric fields. ArXiv Preprint

Leontiadou H, Mark AE, Marrink SJ (2004) Molecular dynamics simulations of hydrophilic pores in lipid bilayers. Biophys J 86(4):2156–2164

Lindgren M, Aronsson K, Galt S, Ohlsson T (2002) Simulation of the temperature increase in pulsed electric field (PEF) continuous flow treatment chambers. Innovative Food Sci Emerg Technol 3(3):233–245

Ling B, Tang J, Kong F, Mitcham EJ, Wang S (2015) Kinetics of food quality changes during thermal processing: a review. Food Bioprocess Technol 8(2):343–358

Liu Z-W, Zeng X-A, Ngadi M, Han Z (2017) Effect of cell membrane fatty acid composition of *Escherichia coli* on the resistance to pulsed electric field (PEF) treatment. LWT Food Sci Technol 76:18–25

Mackey BM (2000) Injured Bacteria. In: Lund B, Baird-Parke T (eds) Microbiological safety and quality of food, vol 1, Berlin, pp 315–341

Mafart P, Couvert O, Gaillard S, Leguerinel I (2002) On calculating sterility in thermal preservation methods: application of the Weibull frequency distribution model. Int J Food Microbiol 72(1–2):107–113

Mañas P, Barsotti L, Cheftel JC (2001) Microbial inactivation by pulsed electric fields in a batch treatment chamber: effects of some electrical parameters and food constituents. Innovative Food Sci Emerg Technol 2(4):239–249

Marcén M, Cebrián G, Ruiz-Artiga V, Condón S, Mañas P (2019) Cellular events involved in *E. coli* cells inactivation by several agents for food preservation: a comparative study. Food Microbiol 84:103246

Martens SL, Klein S, Barnes RA, TrejoSanchez P, Roth CC, Ibey BL (2020) 600-ns pulsed electric fields affect inactivation and antibiotic susceptibilities of Escherichia Coli and Lactobacillus acidophilus. AMB Express 10(1):55

Martínez JM, Cebrián G, Álvarez I, Raso J (2016) Release of Mannoproteins during *Saccharomyces cerevisiae* autolysis induced by pulsed electric field. Front Microbiol 7

Marx G, Moody A, Bermúdez-Aguirre D (2011) A comparative study on the structure of *Saccharomyces cerevisiae* under nonthermal technologies: high hydrostatic pressure, pulsed electric fields and Thermo-sonication. Int J Food Microbiol 151(3):327–337

Meneses N, Jaeger H, Moritz J, Knorr D (2011) Impact of insulator shape, flow rate and electrical parameters on inactivation of *E. coli* using a continuous co-linear PEF system. Innovative Food Sci Emerg Technol 12(1):6–12

Monfort S, Gayán E, Raso J, Condón S, Álvarez I (2010a) Evaluation of pulsed electric fields Technology for Liquid Whole egg Pasteurization. Food Microbiol 27(7):845–852

Monfort S, Gayán E, Saldaña G, Puértolas E, Condón S, Raso J, Álvarez I (2010b) Inactivation of *Salmonella* typhimurium and *Staphylococcus aureus* by pulsed electric fields in liquid whole egg. Innovative Food Sci Emerg Technol 11(2):306–313

Monfort S, Gayán E, Condón S, Raso J, Álvarez I (2011) Design of a Combined Process for the inactivation of *Salmonella enteritidis* in liquid whole egg at 55°C. Int J Food Microbiol 145(2–3):476–482

Mosqueda-Melgar J, Raybaudi-Massilia RM, Martín-Belloso O (2007) Influence of treatment time and pulse frequency on *Salmonella* Enteritidis, *Escherichia coli* and *Listeria monocytogenes* populations inoculated in melon and watermelon juices treated by pulsed electric fields. Int J Food Microbiol 117(2):192–200

Mosqueda-Melgar J, Raybaudi-Massilia RM, Martín-Belloso O (2008) Inactivation of *Salmonella enterica* ser. Enteritidis in tomato juice by combining of high-intensity pulsed electric fields with natural antimicrobials. J Food Sci 73(2):M47–M53

Muratori C, Casciola M, Pakhomova O (2017) Electric pulse repetition rate: sensitization and desensitization. In: Handbook of electroporation. Springer, Cham, pp 353–367
NACMCF (2006) Requisite scientific parameters for establishing the equivalence of alternative methods of pasteurization†. J Food Prot 69(5):1190–1216
Neidhardt FC, Ingraham JL, Schaechter M (1990) The effects of temperature, pressure, and pH. In: Physiology of the bacterial cell. A molecular approach. Sinauer Associates, Sunderland, pp 226–246
Niu D, Wang L-H, Zeng X-A, Wen Q-H, Brennan CS, Tang Z-S, Wang M-S (2019) Effect of ethanol adaption on the inactivation of *Acetobacter sp.* by pulsed electric fields. Innovative Food Sci Emerg Technol 52:25–33
Novickij V, Zinkevičienė A, Stanevičienė R, Gruškienė R, Servienė E, Vepštaitė-Monstavičė I, Krivorotova T, Lastauskienė E, Sereikaitė J, Girkontaitė I, Novickij J (2018) Inactivation of *Escherichia coli* using nanosecond electric fields and Nisin nanoparticles: a kinetics study. Front Microbiol 9:3006
Ou Q-X, Nikolic-Jaric M, Gänzle M (2017) Mechanisms of inactivation of *Candida humilis* and *Saccharomyces cerevisiae* by pulsed electric fields. Bioelectrochemistry 115:47–55
Pakhomov AG, Semenov I, Xiao S, Pakhomova ON, Gregory B, Schoenbach KH, Ullery JC, Beier HT, Rajulapati SR, Ibey BL (2014) Cancellation of cellular responses to Nanoelectroporation by reversing the stimulus polarity. Cell Mol Life Sci 71(22):4431–4441
Pataro G, Donsì G, Ferrari G (2017) Modeling of electrochemical reactions during pulsed electric field treatment. In: Handbook of electroporation. Springer, Cham, pp 1059–1088
Peleg M (1995) A model of microbial survival after exposure to pulsed electric fields. J Sci Food Agric 67(1):93–99
Picart L, Dumay E, Cheftel JC (2002) Inactivation of *Listeria innocua* in dairy fluids by pulsed electric fields: influence of electric parameters and food composition. Innovative Food Sci Emerg Technol 3(4):357–369
Pol IE, van Arendonk WGC, Mastwijk HC, Krommer J, Smid EJ, Moezelaar R (2001) Sensitivities of germinating spores and Carvacrol-adapted vegetative cells and spores of *Bacillus cereus* to Nisin and pulsed-electric-field treatment. Appl Environ Microbiol 67(4):1693–1699
Polajžer T, Dermol–Černe J, Reberšek M, O'Connor R, Miklavčič D (2020) Cancellation effect is present in high-frequency reversible and irreversible electroporation. Bioelectrochemistry 132:107442
Pruitt KM, Kamau DN (1993) Mathematical models of bacterial growth, inhibition and death under combined stress conditions. J Ind Microbiol 12(3–5):221–231
Puértolas E, López N, Condón S, Raso J, Álvarez I (2009) Pulsed electric fields inactivation of wine spoilage yeast and Bacteria. Int J Food Microbiol 130(1):49–55
Qin B-L, Qinghua Z, Barbosa-Canovas GV, Swanson BG, Pedrow PD (1994) Inactivation of microorganisms by pulsed electric fields of different voltage waveforms. IEEE Trans Dielectr Electr Insul 1(6):1047–1057
Qin S, Timoshkin IV, Maclean M, Wilson MP, Given MJ, Wang T, Anderson JG, Macgregor SJ (2015) Pulsed electric field treatment of *Saccharomyces cerevisiae* using different waveforms. IEEE Trans Dielectr Electr Insul 22(4):1841–1848
Raso J, Álvarez I (2016) Pulsed electric field processing: cold pasteurization. In: Reference module in food Science. Elsevier, Amsterdam, pp 1–8
Raso J, Calderón María L, Góngora M, Barbosa-Cánovas G, Swanson BG (1998) Inactivation of Mold Ascospores and Conidiospores suspended in fruit juices by pulsed electric fields. LWT Food Sci Technol 31(7–8):668–672
Raso J, Alvarez I, Condón S, Sala Trepat FJ (2000) Predicting inactivation of Salmonella Senftenberg by pulsed electric fields. Innovative Food Sci Emerg Technol 1(1):21–29
Raso J, Frey W, Ferrari G, Pataro G, Knorr D, Teissie J, Miklavčič D (2016) Recommendations guidelines on the key information to be reported in studies of application of PEF Technology in Food and Biotechnological Processes. Innovative Food Sci Emerg Technol 37:312–321

Reina LD, Jin ZT, Zhang QH, Yousef AE (1998) Inactivation of *Listeria monocytogenes* in Milk by pulsed electric field. J Food Prot 61(9):1203–1206

Reineke K, Schottroff F, Meneses N, Knorr D (2015) Sterilization of liquid foods by pulsed electric fields-an innovative ultra-high temperature process. Front Microbiol 6

Rogers HJ, Perkins HR, Ward JB (1980) In: Science S, Media B (eds) Microbial cell walls and membranes. Springer, Dordrecht

Russell NJ (2002) Bacterial membranes: the effects of chill storage and food processing. An overview. Int J Food Microbiol 79(1–2):27–34

Saldaña G, Puértolas E, López N, García D, Álvarez I, Raso J (2010) Comparing the PEF resistance and occurrence of sublethal injury on different strains of *Escherichia coli*, *Salmonella* typhimurium, *Listeria monocytogenes* and *Staphylococcus aureus* in Media of PH 4 and 7. Innovative Food Sci Emerg Technol 10(2):160–165

Saldaña G, Puértolas E, Álvarez I, Meneses N, Knorr D, Raso J (2010a) Evaluation of a static treatment chamber to investigate kinetics of microbial inactivation by pulsed electric fields at different temperatures at quasi-isothermal conditions. J Food Eng 100(2):349–356

Saldaña G, Puértolas E, Condón S, Álvarez I, Raso J (2010b) Modeling inactivation kinetics and occurrence of sublethal injury of a pulsed electric field-resistant strain of *Escherichia coli* and *Salmonella* typhimurium in media of different PH. Innovative Food Sci Emerg Technol 11(2):290–298

Saldaña G, Puértolas E, Condón S, Álvarez I, Raso J (2010c) Inactivation kinetics of pulsed electric field-resistant strains of *Listeria monocytogenes* and *Staphylococcus aureus* in media of different PH. Food Microbiol 27(4):550–558

Saldaña G, Monfort S, Condón S, Raso J, Álvarez I (2012) Effect of temperature, pH and presence of Nisin on inactivation of *Salmonella* typhimurium and Escherichia Coli O157:H7 by pulsed electric fields. Food Res Int 45(2):1080–1086

Saldaña G, Álvarez I, Condón S, Raso J (2014) Microbiological aspects related to the feasibility of PEF technology for food pasteurization. Crit Rev Food Sci Nutr 54(11):1415–1426

Sampedro F, Rodrigo D, Martínez A (2011) Modelling the effect of PH and pectin concentration on the PEF inactivation of *Salmonella enterica* serovar typhimurium by using the Monte Carlo simulation. Food Control 22(3–4):420–425

Saulis G (2010) Electroporation of cell membranes: the fundamental effects of pulsed electric fields in food processing. Food Eng Rev 2(2):52–73

Schottroff F, Gratz M, Krottenthaler A, Johnson NB, Bédard MF, Jaeger H (2019) Pulsed electric field preservation of liquid whey protein formulations – influence of process parameters, pH, and protein content on the inactivation of Listeria Innocua and the retention of bioactive ingredients. J Food Eng 243:142–152

Schwan HP (1957) Electrical properties of tissue and cell suspensions. In: Advances in biological and medical physics. Academic Press, New York, pp 147–209

Sensoy I, Zhang QH, Sastry SK (1997) Inactivation kinetics of *Salmonella dublin* by pulsed electric field. J Food Process Eng 20(5):367–381

Sharma P, Bremer P, Oey I, Everett DW (2014) Bacterial inactivation in whole Milk using pulsed electric field processing. Int Dairy J 35(1):49–56

Siemer C, Toepfl S, Heinz V (2014) Inactivation of *Bacillus subtilis* spores by pulsed electric fields (PEF) in combination with thermal energy – I. influence of process- and product parameters. Food Control 39:163–171

Silve A, Guimerà Brunet A, Al-Sakere B, Ivorra A, Mir LM (2014) Comparison of the effects of the repetition rate between microsecond and nanosecond pulses: Electropermeabilization-induced electro-desensitization? Biochim Biophys Acta Gen Subj 1840(7):2139–2151

Simpson RK, Whittington R, Earnshaw RG, Russell NJ (1999) Pulsed high electric field causes 'all or nothing' membrane damage in *Listeria monocytogenes* and *Salmonella* typhimurium, but membrane H+–ATPase is not a primary target. Int J Food Microbiol 48(1):1–10

Smith K, Mittal GS, Griffiths MW (2002) Pasteurization of Milk using pulsed electrical field and antimicrobials. J Food Sci 67(6):2304–2308

Somolinos M, García D, Condón S, Mañas P, Pagán R (2007) Relationship between sublethal injury and inactivation of yeast cells by the combination of Sorbic acid and pulsed electric fields. Appl Environ Microbiol 73(12):3814–3821

Somolinos M, García D, Mañas P, Condón S, Pagán R (2008a) Effect of environmental factors and cell physiological state on pulsed electric fields resistance and repair capacity of various strains of *Escherichia coli*. Int J Food Microbiol 124(3):260–267

Somolinos M, Mañas P, Condón S, Pagán R, García D (2008b) Recovery of *Saccharomyces cerevisiae* sublethally injured cells after pulsed electric fields. Int J Food Microbiol 125(3):352–356

Somolinos M, Espina L, Pagán R, Garcia D (2010a) SigB absence decreased *Listeria monocytogenes* EGD-e heat resistance but not its pulsed electric fields resistance. Int J Food Microbiol 141(1–2):32–38

Somolinos M, García D, Mañas P, Condón S, Pagán R (2010b) Organic acids make *Escherichia coli* more resistant to pulsed electric fields at acid pH. Int J Food Microbiol 136(3):381–384

Song J, Joshi RP, Schoenbach KH (2011) Synergistic effects of local temperature enhancements on cellular responses in the context of high-intensity, ultrashort electric pulses. Med Biol Eng Comput 49(6):713–718

Spugnini EP, Arancia G, Porrello A, Colone M, Formisano G, Stringaro A, Citro G, Molinari A (2007) Ultrastructural modifications of cell membranes induced by 'electroporation' on melanoma xenografts. Microsc Res Tech 70(12):1041–1050

Stanley DW, Parkin KL (1991) Biological membrane deterioration and associated quality losses in food tissues. Crit Rev Food Sci Nutr 30(5):487–553

Stewart CM, Tompkin RB, Cole MB (2002) Food safety: new concepts for the new millennium. Innovative Food Sci Emerg Technol 3(2):105–112

Tarek M (2005) Membrane electroporation: a molecular dynamics simulation. Biophys J 88(6):4045–4053

Teissie J, Golzio M, Rols MP (2005) Mechanisms of cell membrane Electropermeabilization: a Minireview of our present (lack of ?) knowledge. Biochim Biophys Acta Gen Subj 1724(3):270–280

Tieleman DP, Leontiadou H, Mark AE, Marrink S-J (2003) Simulation of pore formation in lipid bilayers by mechanical stress and electric fields. J Am Chem Soc 125(21):6382–6383

Timmermans RAH, Mastwijk HC, Berendsen LBJM, Nederhoff AL, Matser AM, Van Boekel MAJS, Nierop Groot MN (2019) Moderate intensity pulsed electric fields (PEF) as alternative mild preservation technology for fruit juice. Int J Food Microbiol 298:63–73

Toepfl S, Siemer C, Saldaña-Navarro G, Heinz V (2014) Overview of pulsed electric fields processing for food. In: Emerging Technologies for Food Processing. Elsevier, pp 93–114

Tsong TY (1989) Electroporation of cell membranes. In: Electroporation and electrofusion in cell biology. Springer, Boston, pp 149–163

Unal R, Kim J-G, Yousef AE (2001) Inactivation of *Escherichia coli* O157:H7, *Listeria monocytogenes*, and *Lactobacillus leichmannii* by combinations of ozone and pulsed electric field. J Food Prot 64(6):777–782

Vega-Mercado H, Pothakamury UR, Chang F-J, Barbosa-Cánovas GV, Swanson BG (1996) Inactivation of *Escherichia coli* by combining pH, ionic strength and pulsed electric fields hurdles. Food Res Int 29(2):117–121

Vernhes M, Benichou A, Pernin P, Cabanes P, Teissié J (2002) Elimination of free-living amoebae in fresh water with pulsed electric fields. Water Res 36(14):3429–3438

Walkling-Ribeiro M, Noci F, Riener J, Cronin DA, Lyng JG, Morgan DJ (2009) The impact of Thermosonication and pulsed electric fields on *Staphylococcus aureus* inactivation and selected quality parameters in Orange juice. Food Bioprocess Technol 2(4):422–430

Walter L, Knight G, Ng SY, Buckow R (2016) Kinetic models for pulsed electric field and thermal inactivation of *Escherichia coli* and *Pseudomonas fluorescens* in whole Milk. Int Dairy J 57:7–14

Wang M-S, Wang L-H, Bekhit AE-DA, Yang J, Hou Z-P, Wang Y-Z, Dai Q-Z, Zeng X-A (2018) A review of sublethal effects of pulsed electric field on cells in food processing. J Food Eng 223:32–41

Wang R-Y, Yun O, Zeng X-A, Guo C-J (2019) Membrane fatty acids composition and fluidity modification in *Salmonella* typhimurium by culture temperature and resistance under pulsed electric fields. Int J Food Sci Technol 54(6):2236–2245

Weaver JC, Chizmadzhev YA (1996) Theory of electroporation: a review. Bioelectrochem Bioenerg 41(2):135–160

WHO. 2015. "World Health Organization global estimates and regional comparisons of the burden of foodborne disease in 2010" edited by L. von Seidlein. PLoS Med 12(12):e1001923

Wölken T, Sailer J, Maldonado-Parra FD, Horneber T, Rauh C (2017) Application of numerical simulation techniques for modeling pulsed electric field processing. In: Handbook of Electroporation. Springer, Cham, pp 1237–1267

Wouters PC, Dutreux N, Smelt JPPM, Lelieveld HLM (1999) Effects of pulsed electric fields on inactivation kinetics of *Listeria innocua*. Appl Environ Microbiol 65(12):5364–5371

Wouters PC, Alvarez I, Raso J (2001) Critical factors determining inactivation kinetics by pulsed electric field food processing. Trends Food Sci Technol 12(3–4):112–121

Wuytack EY, Thi Phuong l D, Aertsen A, Reyns KMF, Marquenie D, De Ketelaere B, Masschalck B, Van Opstal I, Diels AMJ, Michiels CW (2003) Comparison of sublethal injury induced in *Salmonella enterica* serovar typhimurium by heat and by different nonthermal treatments. J Food Prot 66(1):31–37

Yun O, Liu Z-W, Zeng X-A, Han Z (2016) *Salmonella* typhimurium resistance on pulsed electric fields associated with membrane fluidity and gene regulation. Innovative Food Sci Emerg Technol 36:252–259

Zhao W, Yang R, Zhang HQ, Zhang W, Hua X, Tang Y (2011) Quantitative and real time detection of pulsed electric field induced damage on *Escherichia coli* cells and sublethally injured microbial cells using flow cytometry in combination with fluorescent techniques. Food Control 22(3–4):566–573

Zhong K, Chen F, Wang Z, Wu J, Liao X, Xiaosong H (2005) Inactivation and kinetic model for the *Escherichia coli* treated by a co-axial pulsed electric field. Eur Food Res Technol 221(6):752–758

Zimmermann U, Pilwat G, Riemann F (1974) Dielectric breakdown of cell membranes. Biophys J 14(11):881–899

Cell Membrane Permeabilization by Pulsed Electric Fields for Efficient Extraction of Intercellular Components from Foods

E. Vorobiev and N. I. Lebovka

Abbreviations

AA	antioxidant activity
AC	antioxidant capacity
AFM	atomic force microscopy
CI	color intensity
DM	dry matter
EE	ethanol extraction
FM	fresh material
GAE	gallic acid equivalent
HVED	high voltage electrical discharges
ICUMSA	International Commission for Uniform Methods of Sugar Analysis
MW	microwave
OD	osmotic dehydration
OH	ohmic heating
PEF	pulsed electric fields
SAE	supplementary aqueous extraction
SEM	scanning electron microscopy
SG	solid gains
TAC	total anthocyanin content
TEM	transmission electron microscopy
TPC	total phenolic compounds/content
TPI	total polyphenol index

E. Vorobiev (✉)
Sorbonne Universités, Université de Technologie de Compiègne, Laboratoire de Transformations Intégrées de la Matière Renouvelable, EA 4297, Centre de Recherches de Royallieu, Compiègne Cedex, France
e-mail: eugene.vorobiev@utc.fr

N. I. Lebovka
Institute of Biocolloidal Chemistry named after F.D. Ovcharenko, NAS of Ukraine, Kyiv, Ukraine

J. Raso et al. (eds.), *Pulsed Electric Fields Technology for the Food Industry*, Food Engineering Series, https://doi.org/10.1007/978-3-030-70586-2_6

US ultrasound
WE aqueous extraction
WL water losses

Symbols

°Brix Brix value
A area of cell
A_P area of a single pore
a modulus of elasticity
b consolidation coefficient
C solute concentration in the solvent
C_m specific capacitance of membrane
D diffusivity
d Diameter of a sample
d_m thickness of membrane
E electric field strength
Fi Fick number
f_e electroporation factor
f_i relative fraction
f_p density of pores in a membrane
G compressibility modulus
h height or thickness of a sample
k electrical conductivity contrast, $k = \sigma_d/\sigma_u$
k_e hydraulic permeability of the extracellular space
k_i hydraulic permeability of the intracellular space
k_p hydraulic permeability of the pore
k_BT thermal energy
m mass
n number of pulses
N_p number of pores per cell
P pressure
P_{max} fracture pressure
R radius of a cell
r_c critical radius of a pore in a membrane
r_p radius of a pore in a membrane
S firmness (stiffness) coefficient
S_c specific surface of the cell
T temperature
T pressing time
t_c time of a cell charging
$t_{c,m}$ time of a membrane charging
t_p pulse duration
t_{PEF} total time of PEF treatment, $t_{\mathrm{PEF}} = nt_{\mathrm{p}}$

U	voltage
u_m	transmembrane potential (voltage)
W	specific energy consumption
W_a	activation energy
W_c	maximum energy of pore formation in membrane
Y	diffusion (juice) yield
Y_c	consolidation ratio
Z_e	fraction of electroporated cell
Z	disintegration index
Z_c	electrical conductivity disintegration index
Z_d	diffusivity disintegration index
α	transmembrane flow coefficient
α_r	aspect ratio
Δt	distance between pulses (
ε	relative deformation
γ	surface tension
η	viscosity
θ	angle
ρ	density
σ	electrical conductivity
τ	characteristic time
τ_{D}	consolidation time
τ_{d}	diffusion time
τ_{r}	retardation time
ω	line tension

Introduction

During the last few decades, there has been a growing interest in application of pulsed electric fields (PEF) for processing of food and agricultural materials. The recent efforts have been aimed at applying PEF for extraction of bioactive molecules, osmotic dehydration, expressing of juices, drying, freezing, frying, cooking of foods, fermentation, inactivation of microorganisms, etc. (Vorobiev and Lebovka 2020).

The early interest on effects of electric field on plant growth, seed germination, and living organisms arose in the middle of the eighteenth century, starting from the works of Abbot Nollet (Heilbron 1979). The bactericidal and sterilization effects of electricity in food products were revealed at the end of the nineteenth century (Cohn and Mendelsohn 1879). The "electropure process" for heating the milk was accepted as a safe milk pasteurization technology in Europe and the USA in the 1930s (Moses 1938). In the middle of the twentieth century, the DC and AC electrical treatments were applied for assistance of sugar beet processing (Zagorul'ko 1958), juice extraction from fruits and vegetables (grapes, apples, carrots) (Flaumenbaum 1949), and processing of vegetable raw materials, meat, and fish (Kogan 1968; Matov and Reshetko 1968; Lazarenko et al. 1977; Rogov 1988). The advantages of DC and AC techniques were the short time of treatment and their relatively easy implementation to assist different processes, like diffusion, pressing, drying, etc. However, the applications of DC and AC electrical treatments were limited by significant elevation of temperature (due to the ohmic heating), thermal degradation of food and biomaterials, significant electrode erosion, and high energy costs. Thus DC and AC methods have found only limited usages.

In this line, the PEF treatment presents a good nonthermal alternative for processing of food and agricultural products without significant elevation temperature and food quality deterioration and with moderate power consumption. Commonly, the PEF causes selective damage of cell membranes in cells without significant effects on cell walls. In plant tissues, an important damage of cell membranes can be observed at the electric field strengths of $E = 100–1000$ V/cm, using pulses with duration between 1 and 10^3μs and total PEF treatment time within $t_{PEF} = 10^{-4}–10^{-1}$ s (Vorobiev and Lebovka 2008). For microbial killing larger fields ($E = 20–50$ kV/cm) and smaller treatment times ($t_{PEF} = 10^{-5}–10^{-4}$ s) are required (Barbosa-Cánovas et al. 1998). In pioneering works, the application of PEF treatment (with electric field strength up to $E = 20$ kV/cm and exponential pulses with duration of $t_p = 20$μs) was shown to be effective for the selective damage of membranes of sugar beet tissue without significant heating of surrounding media (Zagorul'ko 1958). This phenomenon was called electroplasmolysis. The PEF treatment was also used for inactivation of *Salmonella* in egg powder suspensions and processing of fish products by Doevenspeck in the early 1960s (Doevenspeck 1961) (for more details see the reviews (Sitzmann et al. 2016a, b)). More active research efforts on PEF treatments of food and agricultural materials started about two decades ago (Gulyi et al. 1994; Knorr et al. 1994a, b; Barbosa-Cánovas et al. 1998), and they are basically grounded on the electroporation concept (Pakhomov et al. 2010; Akiyama and Heller 2017; Miklavčič 2017).

This chapter presents the short review on PEF applications for the extraction of intercellular components from biological tissues. The concept of electroporation, main mechanisms of PEF actions, protocols of PEF treatments, techniques to detect PEF-induced changes, recent applications of PEF to assist solid/liquid expression, and solute extraction from different food and biomaterials are presented.

Electroporation, Its Detection, and PEF Protocols

Electroporation of Membranes

The academic interest on mechanism of PEF actions on biological cell has started in the middle of the twentieth century. The experimentally observed efficiency of high-frequency electric currents for inactivation of bacteria and viruses (Nyrop

1946) was explained by the effects of local overheating (Ingram and Page 1953). The effects of PEF-induced damage of cell membrane (reversible and irreversible) were experimentally observed (Stämpfli and Willi 1957; Stämpfli 1958), and these effects were explained by losses of the membrane's function as semipermeable barriers between the bacterial cell and its environment (Hamilton and Sale 1967; Sale and Hamilton 1967, 1968). The further works performed in the period between 1960 and 1990 revealed detailed mechanisms of PEF action on breakdown of cell membranes (Neumann and Rosenheck 1972; Zimmermann et al. 1974; Riemann et al. 1975; Zimmermann et al. 1976a, b, c, 1980) and formation of pores in membranes (Kinosita Jr. and Tsong 1977a, b; Kinosita and Tsong 1977). The knowledge on the sublethal and lethal effects of PEF in function of pulse intensity, duration, frequency, temperature, and other parameters has been collected in a series of papers (Hülsheger and Niemann 1980; Hülsheger et al. 1981, 1983).

Based on these early works, the electroporation concept was theoretically grounded (Weaver and Chizmadzhev 1996; Miklavčič 2017). Electric field selectively modifies structure of cell membranes. The lipid membranes have significantly smaller electrical conductivity (σ_m) as compared to internal cytoplasmic (σ_i) and external extracellular media (σ_e). Therefore, electric field is predominantly concentrated on the biological membrane, which leads to its alteration and damage.

The transmembrane potential required for the breakdown of biological membrane was experimentally estimated to be in the order of $u_m \approx 1.0$ V (Weaver and Chizmadzhev 1996). Electric field strength applied for the breakdown of membrane with thickness of $d_m \approx 5$ nm can be estimated as $E = u_m/d_m \approx 2 \times 10^{10}$ V/cm, which is a typical value required for the breakdown of insulating materials (Rumble 2019).

For the description of cell membrane breakdown, different physical models have been proposed. These models are based on the considerations of electrically induced thermal, osmotic, mechanical, hydrodynamic, and viscous-elastic instabilities (Chen et al. 2006). Historically, the first speculation on selective breaking of cell membranes was based on electroosmotic effect during the PEF application (Zagorul'ko 1958). Later on it was speculated that the fast molecular exchange induced by electroosmosis could cause chemical imbalances and provoke cell lysis (Dimitrov and Sowers 1990). The ohmic heating could cause selective heating of the membrane as the most resistive part of the cell. However, simulations show that heat diffusion effects effectively prevent the Joule overheating of cell membrane owing to its small thickness ($d_m \approx 5$ nm) (Lebovka et al. 2000).

Electroporation model, based on the formation of electropores inside cell membrane, is actually popular to explain PEF effects (Fig. 1). In this model cell membrane is treated as a condenser (a homogeneous slab of lipids) with a specific capacitance, which can be calculated as $C_m = \varepsilon_m \varepsilon_0 / d_m$ ($\approx 3.5 \times 10^{-3}$ F/m^2), where ε_m (≈ 2) is the dielectric constant of a lipid membrane and $\varepsilon_0 = 8.85 \times 10^{-12}$ F/m is the permittivity o free space. The external electric field stimulates creation and growth of pores in a membrane. The energy of aqueous pore formation represents parabolic function of r for the balance between line tension ($2\pi r \omega$) and surface tension ($\pi r^2 \gamma$) contributions (Weaver and Chizmadzhev 1996):

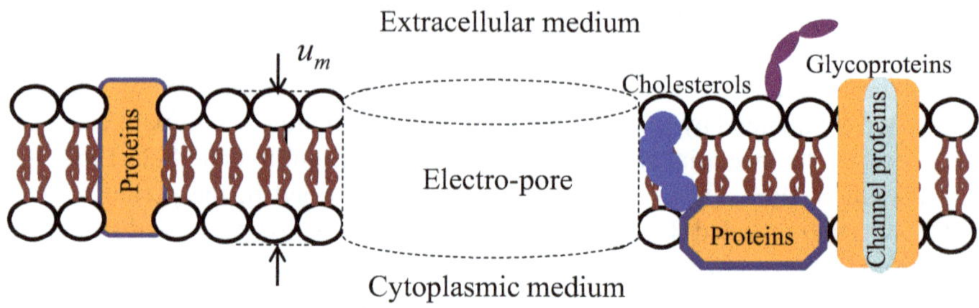

Fig. 1 Pore induced by external electric field in cell membrane. The real membranes contain a variety of molecules like proteins, glycoproteins, glycolipids, sphingolipids, cholesterol, etc. (Tien and Ottova 2003)

$$W = 2\pi r\omega - \pi r^2\gamma \tag{1}$$

Thus the formation of pores is supported by the surface tension forces and is suppressed by the line tension forces. Inside the pore the lipid part of membrane with low dielectric constant ε_m ($\approx$2) is replaced by water or ionic solution with high dielectric constant ε_w ($\approx$80).

The induced transmembrane voltage (u_m) increases the effective surface tension and promotes the opening of pores

$$\gamma^* = \gamma\left(1 + \left(u_m / u_o\right)^2\right), \tag{2}$$

where is the voltage parameter ($u_o \approx 0.17$ V for the typical values $\gamma \approx 2 \times 10^{-3}$ N/m and $\omega \approx 2 \times 10^{-11}$ N for lipid membranes (Winterhalter and Helfrich 1987).

The critical radius of the pore, r_c, and maximum energy of pore formation $W(r_c)$ can be estimated from the condition $\partial W/\partial r = 0$. It gives for the critical radius of the pore $r_c = \omega/\gamma^*$ and for the activation energy $W_c(r_c) = \pi\omega^2/\gamma^* = \pi r_c\omega$. The equilibrium density of pores induced in cell membrane (or pore surface fraction) can be estimated as

$$f_p \approx \exp\left(-W_c / k_B T\right). \tag{3}$$

In absence of the external electric field ($u_m = 0$), the estimations give the values $r_c \approx 10$ nm, where $W_c(r_c) \approx 109kT$ ($k_BT \approx 4.11 \times 10^{-21}$ J at $T = 298$ K) is the thermal energy and $f_p \approx 4.6 \times 10^{-48}$. Therefore formation of pores is a highly improbable event.

However, in presence of the external electric field, the value of r_c decreases and W_c increases with increasing of u_m. For example, for the transmembrane potential of $u_m = 1$ V we get

$$r_c = \omega / \gamma* = \left(\omega / \gamma\right) / \left(1 + \left(u_m / u_o\right)^2\right) \approx 0.28\,\text{nm}, \tag{4a}$$

$$W_c\left(r_c\right) = \pi r_c \omega \approx 4.28kT, \tag{4b}$$

and $f_p \approx 0.014$; therefore the pore formation can be initiated by thermal fluctuations.

The transient aqueous pore model gives the following relation for estimation of the lifetime τ of a single membrane (Weaver and Chizmadzhev 1996).

$$\tau\left(u_m\right) = \tau_\infty \exp\left(W_c / kT\right) = \tau_\infty \exp\left(\frac{\pi\omega^2 / \left(kT\gamma\right)}{1 + \left(u_m / u_o\right)^2}\right), \tag{5}$$

where experimental estimation gives $\tau_\infty \approx 3.7 \times 10^{-7}$ s (at $T = 298$ K) for the lipid membrane (Lebedeva 1987).

The electropore model presented in Fig. 1 corresponds to the formation of simple pore with hydrophilic internal coating. The electroporation effects for pores with complicated structures were also discussed (Neu and Krassowska 1999). In general, within the temperature range of 20–55 °C, the complex phase behavior and melting transitions were observed for the lipid bilayers (Heimburg 1998; Holl 2008). It can affect the changes in transmembrane potential with temperature elevation (Zimmermann 1986). In general, electroporation evolves changes in the size and number of pores induced by PEF, interactions between pores, aggregation of pores, and transport processes of biomolecules (dyes, vitamins, disaccharides, anticancer drugs, peptides, and genes) through the membrane (Rols 2006). Moreover, electroporation behavior can be influenced by complex composition of real membranes containing a variety of molecules like proteins, glycoproteins, glycolipids, sphingolipids, cholesterol, etc. (Tien and Ottova 2003) (Fig. 1).

In dependence of PEF treatment intensity, sublethal, intermediate, and overlethal electroporation effects can occur. At moderate PEF treatment, the sublethal injury with recovery or resealing effects can be observed (Saulis 1997). Typically, the duration of resealing is of order of 1–100μs, and it significantly depends on the temperature and composition of lipid membrane (Sugar et al. 1987). However, in some cases very prolonged resealing duration (seconds, minutes, and even hours) was also observed (Rols and Teissié 1990; Pakhomov et al. 2009) that can be explained by the lipid peroxidation (oxidative degradation of lipids) mechanism (Rems et al. 2019).

During the last decades, the formation of electropores was intensively studied by computer simulations (Rems and Miklavčič 2016). Particularly, at high electric fields, formation of the bilayer patches containing hydroperoxide lipid derivatives was demonstrated (Rems et al. 2019). Detailed reviews on this subject can be found in the literature (Apollonio et al. 2012; Delemotte and Tarek 2012; Casciola and Tarek 2016).

Electroporation of Single Cell and Ensembles of Cells and Tissues

Spherical Cell

The electroporation of single cell is complicated by the distribution of transmembrane potential u_m over cell surface. For the idealized spherical cell (Fig. 2a), the distribution of transmembrane potential u_m can be evaluated using Schwan's equation (Grosse and Schwan 1992):

$$u_m = 1.5 f_e ER \cos\theta \left[1 - \exp\left(-t / t_c\right)\right], \tag{6}$$

where f_e is the electroporation factor, θ is the angle between the external electric field E and the radius-vector on the membrane surface, R is the radius of the cell, and t_c is the time of a cell charging.

The highest drops of transmembrane potential occur at the cell poles, and it decreases up to a zero at $\theta = \pm\pi/2$. The factor [1-exp($-t/t_c$)] accounts for the membrane charging, and $f_e(\leq 1)$ is an electroporation factor (Kotnik et al. 1997, 1998):

$$f_e = \frac{\sigma_e\left[3d_m R^2 \sigma_i + \left(3d_m^2 R - d_m^3\right)\left(\sigma_m - \sigma_i\right)\right]}{R^3\left(\sigma_m + 2\sigma_e\right)\left(\sigma_m + 0.5\sigma_i\right) - \left(R - d_m\right)^3\left(\sigma_e - \sigma_m\right)\left(\sigma_i - \sigma_m\right)}, \tag{7}$$

where d_m is the thickness of the membrane and σ_m, σ_e, σ_i are the electrical conductivities of the membrane, extracellular, and intracellular media, respectively.

The time of a cell charging can be evaluated using the equation

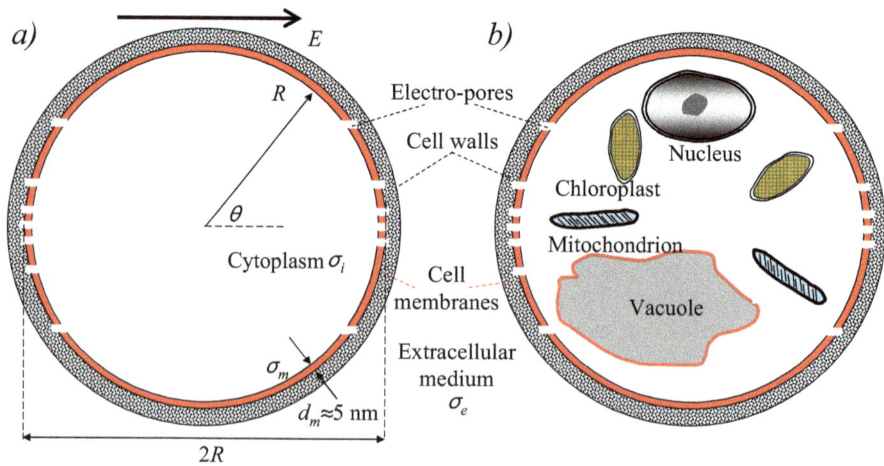

Fig. 2 A biological cell in the external electric field. Schematic presentations of the idealized spherical cell covered by cell membrane (**a**) and a structure of the eukaryotic plant cell (**b**)

$$t_c = \frac{RC_m}{2\sigma_e\sigma_i/(2\sigma_e+\sigma_i)+\sigma_m R-d_m)} = \frac{t_{c,m}}{\frac{2\sigma_i d_m}{\sigma_m R(2+\sigma_i/\sigma_e)}+1}, \tag{8}$$

where $t_{c,m} = d_m C_m/\sigma_m$ is the time of a membrane charging.

In the limit of nonconducting membrane ($\sigma_m = 0$), we have $f_e = 1$. In general case, the value of f_e depends significantly on the conductivity ratio σ_e/σ_i and radius of the cell, R. For biological cells the typical values of parameters are $\sigma_m \approx 5 \times 10^{-7}$ S/m, $\sigma_i \approx 2 \times 10^{-1}$ S/m, $\varepsilon_m \approx 2$, $d_m \approx 5$ nm (Kotnik et al. 1997), $R \leq 10\mu$m for yeasts and microbial cells, $R = 10$–30μm for animal cells, and $R = 10$–100μm for plant cells. Using these values we can estimate $t_{c,m} \approx 35.4\mu$s, $C_m \approx 3.5 \times 10^{-3}$ F/m^2. In vivo conditions (i.e., at $\sigma_e/\sigma_i \approx 1$), $f_e \approx 1$ and $t_c << t_{c,m}$. For external medium with low electrical conductivity (i.e., at $\sigma_e/\sigma_i. << 1$), the value of can be f_e is significantly smaller than 1 and $t_c \approx t_{c,m}$.

The real biological cells have more complicated internal structure (e.g., see Fig. 2b). The effect of electroporation is dependent on the size of membrane envelope. Therefore, with moderate electric fields (<1–5 kV/cm), only electroporation of the external cell membrane is expected. The pronounced electroporation of the membranes of intracellular formations can be achieved at more strong electric fields (≈10–100 kV/cm) applied with nanosecond durations (Tekle et al. 2005; Buchmann and Mathys 2019).

Cells with Different Shapes and Sizes

The real cells show a vast variety of shapes and sizes. For the elongated cell, u_m has its maximum or minimum value when the longest axis of the cell is parallel or perpendicular to the electric field, respectively.

For the prolate spheroid oriented with its long axis along the external field, the stationary value of u_m can be evaluated using the following equation (Kotnik and Miklavčič 2000):

$$u_m = ER\frac{e^3}{e-(1-e^2)\ln(a_r(1+e))}\frac{\cos\theta}{\sqrt{a_r^2\sin^2\theta+\cos^2\theta}}, e=\sqrt{1-a_r^{-2}}, \tag{9a}$$

where α_r (≥1) is an aspect ratio of spheroid (major/minor axis ratio).

In the limiting case of $\alpha_r = 1$ (i.e., for the sphere), this equation reduces to the $u_m = 1.5ER$ that is equivalent to Eq. (6). In the limiting case of large aspect ratio ($a \geq 10$), it reduces to

$$u_m(\theta=0)/ER = 1-(1-\ln 2+\ln a_r)/a_r^2. \tag{9b}$$

For the arbitrary-oriented spheroidal cells, the values of u_m can be found in the literature (Kotnik and Pucihar 2010). The more complicated electroporation problems account for the irregular geometry of cells, local heterogeneities in the membrane structure (Gowrishankar and Weaver 2003), nonuniform membrane conductance (Zudans et al. 2007), and changes in the membrane permeabilization and electrical conductivity during the pulsing (Mercadal et al. 2016).

Ensembles of Cells

In the ensembles of cells (e.g., bio-suspension), electroporation phenomena are complicated by different cell size distributions and different orientations of cells. For example, for the ensemble of spherical cells of different size, simulation predicts for the fraction (Z) of electroporated (damaged) cells, the Weibull's kinetics $Z_e = 1\text{-exp-}[(t/\tau_c)^n]$, where τ_c is the empirical parameter and n is the shape parameter dependent upon the width of the cell size distribution (Lebovka and Vorobiev 2004).

For the elongated cells, the nonexponential kinetics can also reflect the disorder in cell orientations (Lebovka and Vorobiev 2007). The simulations predict upward concavity, $n < 1$, for the prolate spheroids and near-exponential kinetics for the oblate spheroids. The most pronounced deviations from the exponential electroporation kinetics were observed for the disordered suspensions of prolate spheroids having large aspect ratios (a_r) and treated by the small electric field strength (E). For the partially oriented suspensions, efficiency of electroporation increases with increasing of order parameter and field strength.

In the concentrated ensembles of cells (dense suspensions), the effects of deformation of the external electric field by the adjacent cells are also important. These effects can result in significant violation of Schwan's equation (Eq. (6) (Susil et al. 1998; Pavlin et al. 2002; Qin et al. 2005; Ramos et al. 2006). The corrections were evaluated for different cell packings (Susil et al. 1998; Pavlin et al. 2002; Qin et al. 2005; Ramos et al. 2006; Henslee et al. 2014), and the theoretical estimates were used for explanation of the experimental results on electroporation in dense suspensions of Chinese hamster ovary cells (Pucihar et al. 2007).

Tissues

In the plant tissues, electroporation phenomena are even more complicated due to the different distributions of local electric fields, different solute concentrations, electrical conductivities, etc. (Pucihar et al. 2007). The tissue electroporation can be monitored by changes of some material property, P (such as the electrical conductivity, textural characteristics, diffusivity, acoustic response, etc.) in the course of PEF treatment (Fig. 3).

It is useful to introduce the disintegration index (Z) defined as

$$Z = (P - P_i)/(P_d - P_i), \tag{10}$$

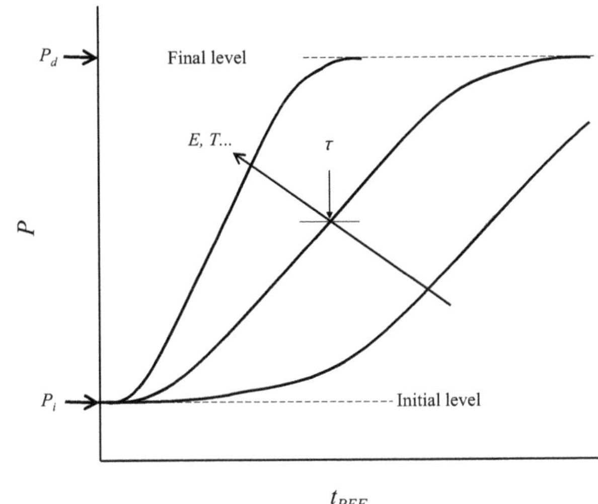

Fig. 3 Schematical time dependence of some material property (P), versus time of PEF treatment (t_{PEF}). Here, P_i P_d, are the initial (before the PEF treatment) and final (after the long time of PEF treatment) levels of P, and τ is the characteristic damage time. The arrow shows direction of increase of the electric field strength E, temperature, T, or other processing parameters

where the initial level, P_i, corresponds to the intact tissue and the final level, P_d, corresponds to the long time of PEF treatment and maximum tissue damage. This equation gives $Z = 0$ for the intact and $Z - 1$ for the completely electroporated material. The characteristic damage time, τ, defined as the PEF duration needed for attaining a half of the material damage ($Z/2$) is also useful for the characterization of electroporation phenomena (Bazhal et al. 2003).

Figure 4a shows typical dependencies of the characteristic damage time τ versus electric field strength E experimentally estimated for the potato tissue preheated to two different temperatures (Lebovka et al. 2002; Bazhal et al. 2003). The data were obtained by analyzing the changes in low-frequency electrical conductivity during PEF treatment. The value of τ decreased with increasing of both the temperature T and the electric field strength E. The empirical equation similar in shape to Eq. (5) has been proposed to fit the experimental data (Lebovka et al. 2005a)

$$\tau = \tau_\infty \exp \frac{W / kT}{1 + \left(E / E_o\right)^2}. \tag{11}$$

Here, τ_∞, W and E_0 are adjustable parameters.

The observed noticeable decrease of the characteristic electrical damage time after preheating of cell tissues can be explained by the structural transitions (melting) inside the cell membranes and the more pronounced electroporation efficiency at high temperatures (Zimmermann 1986). For the studied case of potato tissue, the direct thermal damage (thermal plasmolysis) was insignificant at temperatures below 50 °C, and the observed effects can only reflect electroporation effects. Note that even at small electric fields (E < 100 V/cm), the noticeable electroporation of

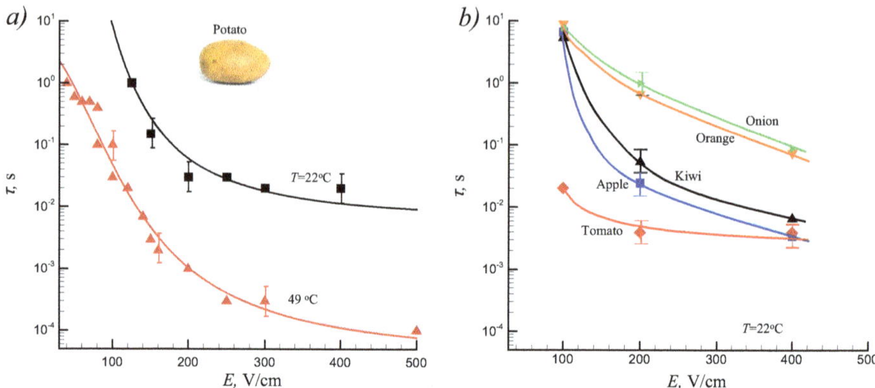

Fig. 4 Characteristic damage time (τ) versus the electric field strength (E) for the potato tissue at two different temperatures (**a**) (Lebovka et al. 2005a) and for the different tissues at room temperature (**b**) (Grimi 2009). The values of τ were estimated from the measurements of tissue electrical conductivity (**a**) or tissue acoustic properties (**b**)

cell tissues is still possible at the room temperature after prolonged treatment duration, $t_{PEF} \approx 10$–1000 s.

The $\tau(E)$ dependencies for different plant tissues may be rather different (Fig. 4b), and they can be greatly influenced by the presence of spatial variations in the density, porosity, electrical conductivity, sizes, and shapes of cells. For example, the extracellular porosity of apple tissue is high (14–25%), but the extracellular porosity of carrot (4%), potato (2%), and many other tissues is considerably smaller (Karathanos et al. 1996). Usually, the outer parenchyma porosity is higher than that of the inner parenchyma (Mavroudis et al. 1998).

PEF-treated cells of the plant tissue can be intact (with vital metabolic activity), partially disintegrated (with incomplete metabolic activity), or completely disintegrated (with lost metabolic activity) (Jaeger et al. 2009). In the vicinity of electroporated cells, the local electrical conductivity is high, whereas it is low near the intact cells. This inhomogeneous distribution of electrical conductivity has an impact on the distributions of local transmembrane potentials. The computer simulations (Lebovka et al. 2001) based on the distribution of electrical conductivities of electroporated (σ_d) and intact (σ_u) cells ($\sigma_u << \sigma_d$) revealed formation of the conductive paths of electroporated cells between electrodes and correlations between extent of electroporation and changes in electrical conductivity. The data of simulation were in qualitative correspondence with electroporation experiments for the PEF-treated apples (Lebovka et al. 2001). Experimentally observed effects of the pulse frequency were explained accounting for the presence of resealing and mass transfer phenomena.

Simulations have been also used to explain the correlation between electroporation efficiency and electrical conductivity contrast ($k = \sigma_d/\sigma_u$) experimentally observed for different fruit and vegetable plants (apple, potato, carrot, courgette, orange, and banana) (Ben Ammar et al. 2011a). The simulated data were in good correspondence with existing experimental data (Lebovka et al. 2002). For example, the estimated values of the optimal electric field strength were $E_o \approx 360$ V × cm^{-1} ($\sigma_d/\sigma_u \approx 14.3$) for the potato and $E_o \approx 976$ V × cm^{-1} ($\sigma_d/\sigma_u \approx 5.6$) for the banana (Ben Ammar et al. 2011a).

Techniques to Detect Electroporation

Disintegration index Z defined by Eq. (10) (sometimes called as electroporation degree or electropermeabilization index) is a very useful characteristic to estimate the degree of electroporation. However, determination of Z requires direct analysis of the microscopic structure of cells or examination of modifications in material properties induced by PEF, e.g., changes of electrical and mass transport characteristics and textural, acoustic, and other properties.

The estimation of extent of electroporation is very important for the optimization of different processing operations assisted by PEF, e.g., pressing and extraction. Unfortunately, it is not a simple or definite task. In principle, disintegration induced by PEF reflects selective damage of cell membranes without direct impact on the cell walls. However, electroporation can also affect the cell walls owing to changes in their moisture content and other secondary effects. Moreover, the electroporation is statistical process, and reversible or irreversible pore formation can develop depending on the PEF protocol. The supplementary effects related to pore resealing, mass transfer processes, and inhomogeneities in tissue structure can also be important.

Nowadays, practically all proposed techniques to detect electroporation are indirect. Moreover, they can be destructive (invasive), affecting the structure of electroporated material. In general, each technique requires a careful adaptation for the studied type of material (Lebovka and Vorobiev 2014, 2016).

Light and Other Microscopies

Light microscopy is a simple, attractive, demonstrable, and decisive technique. It allows application of modern computerization tools and can be used for the visualization of food material structure and microstructure and quantitative evaluation of changes in size, shape, and other characteristics of biological cell (Chanona-Pérez et al. 2008; Russ and Neal 2016). The effects of disintegration and spatial distributions of different components inside the PEF-treated material may be easily revealed. The light microscopy was used to reveal PEF-induced changes and counting the number of electroporated cells in onion epidermis (Fincan and Dejmek 2002) and in apple parenchyma (Chalermchat et al. 2010). Light microscopy can be used for the visualization of modifications induced by diffusion, osmotic impregnation, drying, and freezing (Ben Ammar 2011; Loginova 2011). Disintegration index Z can be evaluated from image analysis as the ratio of damaged and total numbers of cells. The disadvantages of this method may be related to problems of tissue preparation, staining strategies, and individual adaptation of pH for selected type of materials.

TEM, SEM, and AFM techniques are also very illustrative and powerful tools for checking the morphological changes in membranes and cell walls (Condello et al. 2013). These methods were tested for the quantification of PEF-induced changes in membrane structure (Chang and Reese 1990), formation of electropores (Chen et al. 2006), and other effects of electroporation (Lee et al. 2012). Particularly, SEM and TEM have been widely applied to study the PEF effects in materials of biological origin such as food materials (Fazaeli et al. 2012). AFM technique has been used for characterization of changes in the membrane and cell walls provoked by electroporation (Suchodolskis et al. 2011; Garcia-Gonzalo and Pagán 2016; Pillet et al. 2016; Napotnik and Miklavčič 2018). In several works magnetic resonance imaging (MRI) techniques have been applied to study the electroporation in vegetable tissues (Hjouj and Rubinsky 2010; Rondeau et al. 2012; Kranjc and Miklavčič 2017; Dellarosa et al. 2018; Suchanek and Olejniczak 2018). However, all these techniques are destructive and require applications of individual methods for specimen preparation (staining, fixation, dehydration, ultrathin sectioning, etc.) or high-vacuum conditions.

Electrical Impedance Techniques

The electrical impedance techniques are the most popular to detect electroporation. These techniques are straightforward, simple, and non-expensive and can be easily adapted for the continuous monitoring of electroporation during the PEE treatment. In these techniques the electrical conductivity of PEF-treated materials is used as the material property (see Eq. (10)). In a low-frequency (≈1–10 kHz) impedance technique, the electrical conductivity disintegration index Z_c is defined as (Lebovka et al. 2002)

$$Z_c = (\sigma - \sigma_i)/(\sigma_d - \sigma_i). \tag{12a}$$

Electrical conductivity (σ) increases from the value of σ_i (intact or untreated tissue) to the value of σ_d (completely disintegrated tissue) during the PEF treatment. The value of σ_d can be estimated using a long treatment time ($t_{PEF} \approx 0.1$–1 s) at high electric field strength ($E > 1000$ V/cm). However, this method can evolve electrochemical reactions and generation of gas bubbles during the prolonged PEF treatment, which can affect the value of σ_d. Such effects were clearly demonstrated in experiments with sugar beet tissue (Lebovka et al. 2007b). Another method to produce a completely disintegrated tissue is based on the application of freeze-thawing. However, this procedure can violate the structure of cell walls and affect the proper estimation of σ_d.

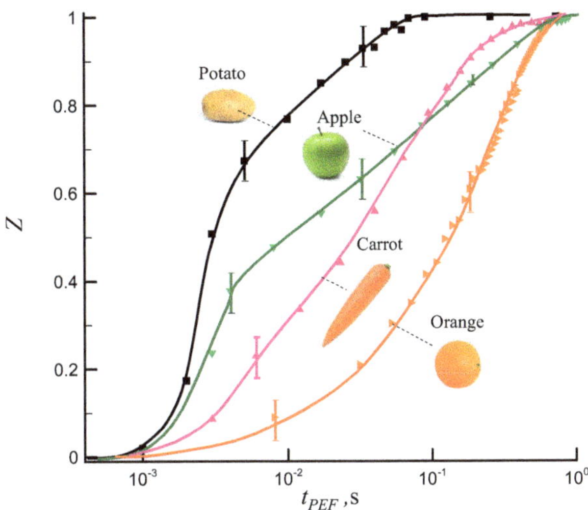

Fig. 5 Electrical conductivity disintegration index, Z_c, versus the time of PEF treatment, t_{PEF}, for different food materials obtained using a low-frequency impedance technique. PEF treatment was done at $E = 400$ V/cm using bipolar pulses with duration of $t_p = 1000\mu s$ and pulse repetition time of $\Delta t = 10$ ms. (Adapted from the data presented in (Ben Ammar 2011))

Figure 5 shows temporal behaviors of the electrical conductivity disintegration index Z_c for different food materials (potato, apple, carrot, and orange) obtained on the base of low-frequency impedance technique (Ben Ammar 2011).

In low-high frequency technique, the value of Z_c is defined as (Angersbach et al. 2002)

$$Z_c = (\alpha\sigma - \sigma_i)/(\sigma_d - \sigma_i), \tag{12b}$$

using the values of σ and σ_i (for the partially electroporated and intact materials, respectively) measured at low frequency (≈1–10 kHz); the value of $\sigma_d = \sigma_i^\infty$ is measured for the intact material at high frequency (50 MHz). In this equation the measured value of σ for PEF-treated material is corrected using the correction factor $\alpha = \sigma_i^\infty/\sigma^\infty$.

However, electrical impedance techniques are destructive, and they require preliminary cutting of samples to assure good contact with electrodes. Moreover, the electrochemical reactions and generation of gas bubbles inside the treated tissues and near electrodes may affect the measured values of electrical conductivity.

Diffusivity

These techniques are based on the mass transfer measurements using solid-liquid extraction or drying experiments. They are straightforward, relatively simple, and non-expensive. The diffusivity disintegration index, Z_d, can be evaluated by Eq. (10) using solute or moisture diffusivity (D) as a material property (P) (Barba et al. 2015). However, these techniques are destructive, and estimated values of D may depend on the used experimental procedure.

Textural Tests

Different textural disintegration indexes, Z_t, can be evaluated by Eq. (10) using the cutting force, stress-deformation, or relaxation tests (Lu and Abbott 2004) for the characterization of material property (P). These techniques are straightforward, relatively simple, and non-expensive, and they were widely applied for the quantification of PEF-induced changes in different fruit and vegetable tissues (Chalermchat and Dejmek 2005; Grimi et al. 2009a; Liu et al. 2021). However, all these techniques are destructive, the cutting or puncture tests require the penetration of a probe (nail or blade) inside the sample, and in stress deformation or relaxation tests, the sample is compressed.

Acoustic Test

The test is straightforward, relatively simple, and less destructive and can be used for the detection of electroporation in PEF-treated whole roots or fruits, e.g., sugar beet, potato, tomato, and apple (Grimi et al. 2010a, 2011). The acoustic disintegration index is defined as (Grimi et al. 2010a).

$$Z_a = (S - S_i)/(S_d - S_i), \qquad (12c)$$

where S is the firmness (stiffness coefficient) $S = f^2_m m^{2/3} \rho^{1/3}$ (here f_m is the frequency corresponding to the maximum acoustic amplitude, A, in the acoustic spectrum, m and ρ are the sample mass and density, respectively).

PEF Protocols, Treatment Chambers, and Optimization of the Treatment

In food processing, the capacitor discharge, square wave, and analog PEF generators can be used (Reberšek and Miklavčič 2011). PEF generators include power source, capacitors, coils of inductance, resistors, and quick switches of different types (Pourzaki and Mirzaee 2008; Toepfl et al. 2014). The electric energy is stored in a bank of capacitors.

PEF Protocols

Figure 6 demonstrates the most popular exponential decay and square wave shapes for bipolar pulses. The bipolar pulses offer minimum energy consumption with reduced dissolution of the electrodes and reduced electrolysis (Qin et al. 1994; Wouters and Smelt 1997). Note that square wave generators are more expensive and more complex than the exponential decay generators. However, the square wave

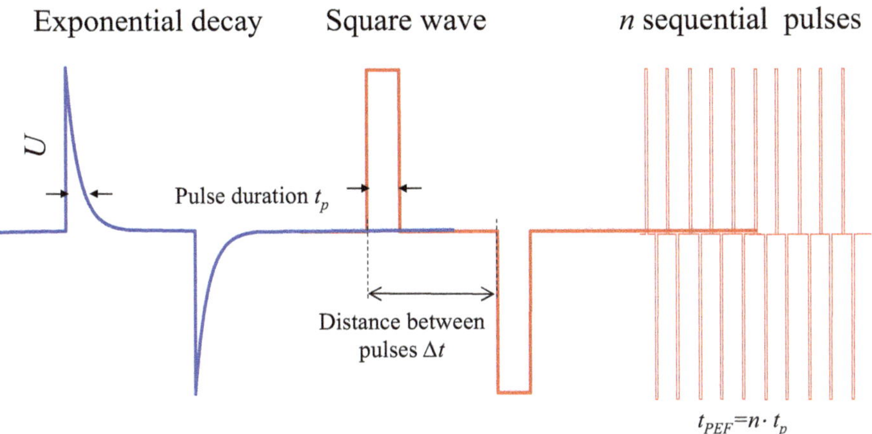

Fig. 6 The exponential decay and square wave shapes for bipolar pulses and the simple protocol of PEF treatment including n sequential pulses with total duration of $t_{PEF} = nt_p$

generators have better disintegrating and energy efficiencies as compared with the exponential decay ones.

The pulse protocol is defined by the applied amplitude, U (peak voltage); pulse duration, t_p; time interval between pulses, Δt (or pulse repetition rate $f = 1/\Delta t$); and number of pulses, n (Canatella et al. 2001). The total time of PEF treatment is defined as $t_{PEF} = nt_p$. In more complicated protocols, the trains with long pause between pulses are applied to avoid significant ohmic heating during the PEF treatment.

Treatment Chambers

Different designs of treatment chambers were used in laboratory and large-scale applications. Nano- and microfluidic systems have been used in electro-transfection experiments for studying electroporation phenomena (Chang et al. 2016), for cell handling and manipulation (Čemažar et al. 2013, 2017), and for in situ observation of electroporation under the optical microscope (Bodénès 2017; Bodénès et al. 2019). Different designs of microscale fluidic chips have been used for analyzing cellular properties or intracellular content of bacteria and yeast (Fox et al. 2006) and studying extraction of cell compounds (Rockenbach et al. 2019). In small-scale laboratory experiments, the cuvettes with parallel configuration of electrodes and 1–4 mm gaps are rather popular (Bio-Rad Laboratories 2019; BTX 2019; Eppendorf 2019).

In food processing and bio-recovery applications, different chambers were proposed for batch and continuous PEF treatment. The batch chambers typically use plate-to-plate electrode geometry. In continuous flow operations, the electrode configurations with parallel plates and coaxial and collinear cylinders were tested

(Reberšek et al. 2014; Toepfl et al. 2014; Raso et al. 2016). Note that for parallel plate or coaxial electrode configurations, the electric field within the interelectrode space can be rather homogeneous. However, the collinear chambers with more inhomogeneous field can be easier inserted in processing lines, and they are more preferable for continuous treatment of solid/liquid suspensions with large particles.

Optimization of the PEF Treatment

The numerical studies have shown that PEF-induced damage in fruit and vegetable tissues is mainly controlled by the electric field strength E and the treatment time t_{PEF} ($t_{PEF} = nt_p$) (Barba et al. 2015). In general, the electric fields with higher value of E and longer duration of the treatment t_{PEF} lead to the better electroporation efficiency and higher values of disintegration index Z.

However, the optimum PEF treatment permitting minimized energy consumption can be achieved using the proper combination of the values of E and t_{PEF}. Typically, the values of specific energy consumption, required for the effective PEF damage of fruit and vegetable materials, are situated in the interval of 1–16 kJ/kg, and they are noticeably smaller than the values of specific energy consumption required for the mechanical ($W = 20$–40 kJ/kg), enzymatic ($W = 60$–100 kJ/kg), and thermal ($W > 100$ kJ/kg) disintegration (Toepfl 2006).

For the protocol of PEF treatment including n sequential pulses, the specific energy, W (kJ/kg), can be evaluated using the summing of pulse energy inputs

$$W_i = \int_0 \sigma(t) E^2(t) dt / \rho \quad (13a)$$

over the n sequential pulses

$$W = \sum_{n}^{i=1} W_i. \quad (13b)$$

Here, ρ is the density of the treated material and $\sigma(t)$ is the electrical conductivity of the treated material, which is a growing function of PEF treatment time (Raso et al. 2016).

Therefore, the total specific energy is mainly determined by the behavior of the electrical conductivity ($\sigma(t)$) or the electrical conductivity disintegration index ($Z_c(t)$) of the material during PEF treatment. The total specific energy is approximately proportional to the product of the characteristic damage time ($\tau(E)$) and the square of electric field strength (E^2) (Lebovka et al. 2002). The treatment time decreases significantly with E increase, approaching the τ_∞ value in the limit of high fields ($E \approx 1000$ V/cm) (Fig. 7a), and the product $\tau(E)E^2$ corresponding to the energy consumption goes through a minimum.

This minimum of $\tau(E)E^2$ function determines the optimum value of the electric field strength $E \approx E_0$ (Bazhal et al. 2003). For apple, carrot, and potato, the experimental curves $\tau(E)E^2$ passed through the minimum at $E = 200$–400 V/cm (Lebovka

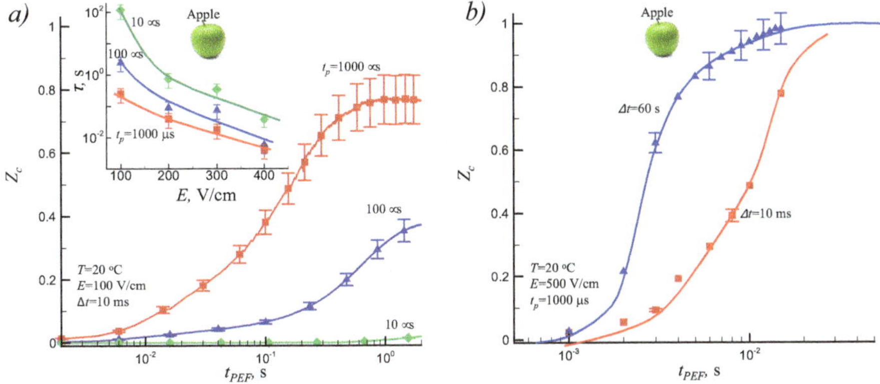

Fig. 7 Electrical conductivity disintegration index (Z_c) for apple tissue (determined by low-frequency impedance technique) versus the time of PEF treatment (t_{PEF}): (**a**) at different pulse durations (t_p = 10, 100 and 1000μs), E = 100 V/cm and Δt = 10 ms. Inset shows characteristic damage time, τ, versus E at different values of t_p (Adapted from the data presented in (De Vito et al. 2008)); (**b**) at two pulse repetition times Δt = 10 ms and 60 s, E = 500 V/cm, and t_p = 1000μs. (Adapted from the data presented in (Lebovka et al. 2001))

et al. 2002). Typically, for many vegetables and fruits, the optimum values of E are situated in the interval between 200 and 1000 V/cm. The PEF treatment at excess values of E above E_o can result in elevated energy consumption without additional electroporation effect. In a general case, the power consumption can be rather complex function of the size and shape of cells, electrical properties of material, and the treatment protocol (Ben Ammar et al. 2010).

The pulse duration (t_p) and the distance between pulses (Δt) may also affect the efficiency of electroporation in plant tissues. For example, the influence of pulse duration (10–1000μs), on the electroporation behavior of grapes, apples, and potatoes, was experimentally demonstrated (De Vito et al. 2008; Vorobiev and Lebovka 2010; Ben Ammar 2011) (Fig. 7a). Longer pulses were more effective, particularly at room temperature and moderate electric fields (E = 100–300 V/cm). The application of PEF protocols with longer distance between pulses (Δt) at fixed values of E and t_{PEF} also resulted in more accelerated kinetics of disintegration of the apple tissue (Lebovka et al. 2001) (Fig. 7b). Similarly, a cell permeabilization of an onion tissue was significantly higher above the critical distance between pulses, Δt = 1 s (Asavasanti et al. 2011b). It was suggested that at low frequencies, convection and cytoplasmic streaming play a significant role in distributing more conductive fluid throughout the tissue (Asavasanti et al. 2012).

During the PEF treatment, the ohmic heating can be also significant (Blahovec et al. 2015). The superposition of the PEF treatment and moderate ohmic heating give synergetic effect for enhancement of electroporation (Lebovka et al. 2005b; Lebovka and Vorobiev 2011). Such synergetic effect can be revealed by analyzing of the behavior of characteristic damage time (τ) at different temperatures (Lebovka et al. 2005a)). The synergy PEF and thermal treatments at moderate temperatures

($T \approx 50$ °C) give a unique opportunity to reach high tissue disintegration at moderate electric fields without noticeable product quality losses.

Solid/Liquid Expression and Solvent Extraction of Valuable Compounds from Food Plants Enhanced by PEF

Solid/Liquid Expression: Principles and Models

Solid/liquid expression (often named as pressing or pressure extraction) is commonly used for liquid expelling from food and biomass materials. Solid/liquid expression can be produced by movable elements, for example, piston, screw, rolls, etc. (Couper et al. 2012). In hydraulic presses, used to process fruits, oilseeds, and nuts on a small scale, the pressure is applied via an adjustable piston, and the resultant juice is passed through a fine wire screen. In belt presses, the liquid is removed by squeezing the press-cake between compressing rollers (Cheremisinoff 2017). This technique is used for the dewatering of biological sludge, juice production from fruits, etc. In the screw presses, a slowly rotating screw transports and compresses solid-liquid mixtures (Virkutyte 2017; Qingwen et al. 2018). These presses are widely used on the large scale for the oil recovery from oilseeds, dewatering of sugar beet pulp, production of corn fibers and starches, fruit and vegetable juicing, green tea leaf dewatering, winemaking, etc. (Mushtaq 2018). The pressing allows producing fresh-like juices and extracts without undesirable effects of thermal degradation with special interest in applications for temperature or solvent sensible materials (Ashurst 2016).

Solid/liquid expression can be realized using constant pressure or constant rate regimes or their combinations. Constant pressure solid/liquid expression can be performed on industrial pressing equipment like filter press and tube and hydraulic piston presses. In this regime the pressure is suddenly increased up to some given value, which may cause a sudden compression of the material. For biological materials, gentler regime of a constant rate expression with gradually increasing pressure may be preferable. Such regime may be implemented using hydraulic pistons or screw presses. Combined regimes with a constant rate and with a constant pressure for the respectively initial and final pressing periods can be also applied.

Commonly, the juice yield obtained by pressing of coarse (natural) food particles is insufficient. Efficiency of the solid/liquid expression depends greatly on the tissue integrity. It can be significantly improved using preliminary operations like fine grinding of raw material, thermal treatment (heating or freeze-thawing), or enzyme maceration (Monteith and Parker 2016; Sheikh and Kazi 2016; To et al. 2016; Cheremisinoff 2017). However, these operations lead to the passage of cell components like pectins, cellulose, and other impurities into the extracts. As a result, cloudy juices are often produced (Taylor 2016). For further processing of such juices, costly multistage clarification and stabilization procedures are needed

(Albagnac et al. 2002). For example, in sugar production, the fine grinding of sugar beet into the mash was not industrially implemented because of the difficulties for purification of extracted juices (Van der Poel et al. 1998). To overcome the difficulties for juice clarification and purification, coarse fragmentation of raw material (e.g., by slicing) is widely used.

Conventional solid/liquid expression from cellular tissue is accompanied by air expulsion; collapse and different transformations of cellular structure; breakage and separation of cell walls; liquid flowing between intracellular, extracellular, and extraparticle spaces; and many other effects (Das et al. 2018). In rigid materials, the fluid flow is determined by two main characteristics: porosity and permeability. The porosity is a measure of available pore space, and permeability reflects ability of the fluid to flow through the medium under applied pressure.

In compressible porous materials, fluid flow is influenced by the compressibility that is the measure of the mechanical strength (Jönsson and Jönsson 1992). In such materials, the mechanisms of liquid-solid expression are determined by the filtration and consolidation phenomena (Lanoisellé et al. 1996). Moreover, for compressible cellular materials, the cell rupture during liquid expulsing can significantly modify the tissue compressibility. In this case, the simultaneous cell cracking and filtration/consolidation phenomena are manifested. Solid/liquid expression can be also complicated by secondary (creep) consolidation (Lanoisellé et al. 1996). Initial compaction of sliced or grinded particles is needed to start solid/liquid expression. For the description of such phenomena, different mechanistic models have been developed (Lanoisellé et al. 1996; Schwartzberg 1997).

For a constant pressure regime, the model with a spring (a_1) and one viscoelastic element (a_2, η_2) (Fig. 8a) predicts three retardation processes related to the liquid flow resistance of the Terzaghi element (time of τ_1), liquid flow resistance of the

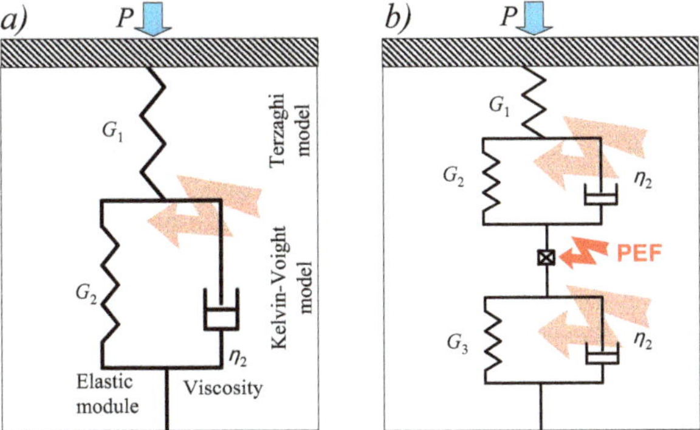

Fig. 8 Viscoelastic rheological model of cellular tissue with one (**a**) and two (**b**) Kelvin-Voigt elements. Here, Hook's spring a_1 is supplemented by viscoelastic elements with the modulus of elasticity (a_2, a_3) and viscosity (η). P is the applied pressure. PEF treatment can affect the rheological properties of cellular tissue, and several viscoelastic elements can be used for description of the model

Voigt element (time of τ_2), and creep deformation of the particulate bed (time of τ_3), and it is assumed that $\tau_3 >> \tau_1, \tau_2$ (Shirato et al. 1986).

In rheological models the tissue can be represented by Terzaghi's spring with the compressibility modulus G_1 supplemented by one (a) or several (b) Kelvin-Voigt (G-η) elements. The preliminary treatment of tissue (e.g., by application of slicing, fine grinding, heating, or freeze-thawing) causes initial cell damage. The solid/liquid expression starts after compaction of sliced or grinded particles. The initial expression and primary filtration-consolidation behavior are dominated by the compression of the spring a_1 (Shirato et al. 1986). The deformation of Kelvin-Voigt elements corresponds to the secondary (creep) consolidation (Lanoisellé et al. 1996).

In general case, biological tissues can include two (Fig. 8b) and even more Kelvin-Voigt elements representing different consolidation periods. Preliminary fragmentation and application of PEF treatment can affect the viscoelastic properties and number of Kelvin-Voigt elements. PEF treatment during the solid/liquid expression can also affect the characteristics of rheological model. For example, PEF application before pressing may influence the behavior of primary consolidation period. PEF application during the pressing may prolong this primary consolidation period or even induce additional consolidation periods (Lanoisellé et al. 1996).

For characterization of compression the consolidation ratio, Y_c, defined as

$$Y_c = \frac{h - h_o}{h_\infty - h_0} = \frac{\varepsilon}{\varepsilon_\infty}, \tag{14}$$

is used. Here h, h_0, and h_∞ are the actual, initial, and final (obtained after the infinitively long time of pressing) thicknesses of the sample, and $\varepsilon = 1\text{-}h/h_o$ is the relative deformation.

Important characteristics of the compression behavior are the infinitive relative deformation (ε_∞) and the consolidation coefficient, (b) expressed in m^2/s, which is the analog of diffusion coefficient and represents filtration diffusivity during the press-cake consolidation under applied pressure.

The compression characteristics may significantly depend on the pressure and applied treatment (Grimi et al. 2010b). Figure 9 shows examples of the limiting (infinitive) relative deformation (ε_∞) (a) and consolidation coefficient (b) (b) versus the applied pressure (P) for the untreated, PEF-treated (400 V/cm, t_{PEF} = 0.1, $Z_c \approx 0.9$), and freeze-thawed sugar beet tissues (disks with diameter of 25 mm and height of 10 mm) (Grimi et al. 2010b). The pressure behaviors ε_∞ *(P)* were rather similar for PEF-treated and freeze-thawed tissues, and the value of $\varepsilon_\infty \approx 0.95$ seems to be maximally attainable for the sugar beet tissue. However, for the untreated samples, the values of ε_∞ are significantly smaller (e.g., $\varepsilon_\infty = 0.4$ even at high applied pressure of 40 bar). The estimated fracture pressure of the untreated sample was of order of $P_{max} \approx 84$ bar (Grimi et al. 2010b).

The consolidation coefficients (b) increase with P for both PEF-treated and freeze-thawed tissues (Fig. 9b). The higher values of b for the freeze-thawed tissue as compared with PEF-treated tissue were observed. This reflects weaker and softer structure of the freeze-thawed tissue.

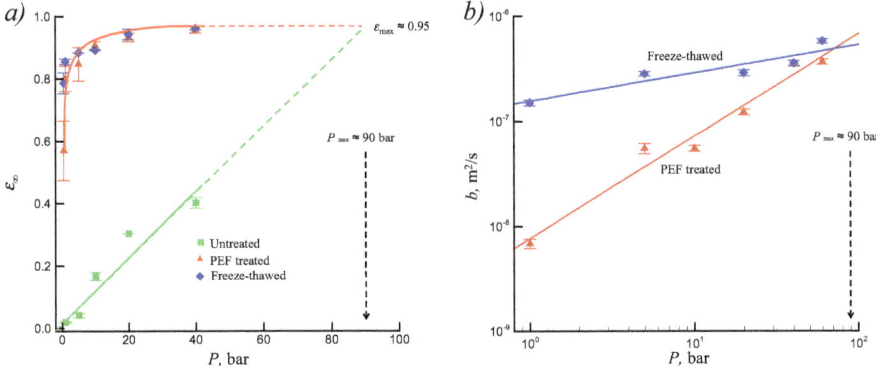

Fig. 9 Infinitive relative deformation (ε_∞) ε_∞ (***a***) and consolidation coefficient (***b***) (***b***) determined at different applied pressures P for the untreated, PEF-treated (400 V/cm, $t_{PEF} = 0.1$, $Z_c \approx 0.9$), and freeze-thawed sugar beet tissues. (Adopted from (Grimi et al. 2010b))

In the generalized model, the primary and creep consolidation periods of the extraparticle volume and the consolidations of both extracellular and intracellular volumes were accounted (Lanoisellé et al. 1996). This model represents the consolidation ratio (Y_c) as the sum of different retardation processes:

$$Y_c(t) = f_1 Y_b(t) + \sum_{4}^{i=2} f_i\left(1 - \exp(-t/t)\right), \tag{15}$$

where

$$Y = 1 - \left(8/\pi^2\right) \sum_{\infty}^{n=0} \frac{\exp\left(-(2n+1)^2 t/\tau\right)}{2_n+1)^2}, \tag{16}$$

is the primary consolidation term with a consolidation time of $\tau = \tau_b = 4\, h^2{}_0/(\pi^2 b)$ (here b is the consolidation coefficient), $f_i = (1/G_i)/\Sigma\,(1/G_i)$ (i = 1–4) are the corresponding fractions of the retardation processes related to Hookean springs (G_i), and τ_i are the corresponding retardation times. The compressibility modulus (G_1) is attributed to primary consolidation of the tissue, and other terms in Eq. (15) are attributed to the secondary (creep) consolidation of the tissue (i = 2) and to the primary (i = 3) and secondary (i = 4) consolidation of the extraparticle channels.

At large values of Y (when $Y > 0.415$ or t is large), the primary consolidation (Eq.16) can be well fitted (with less than 1% error) by the first term in the series (Schwartzberg 1997):

$$Y(t) = 1 - \frac{8}{\pi^2}\exp(-t/\tau). \tag{16a}$$

However, the large numbers of terms are needed when Y is small (or t is small). In this case (when $Y < 0.62$), the primary consolidation can be well fitted (with less than 1% error) by the equation (Olson 1986):

$$Y(t) = \left(\frac{16}{\pi^3} t / \tau \right)^{0.5}. \tag{16b}$$

For full diapason, $0 < Y < 1$, the primary consolidation can be well fitted (with less than 0.3% error) by the following equation (Shirato et al. 1980):

$$Y_p(t) \frac{\left(\frac{16}{\pi^3} t / \tau \right)^{0.5}}{\left[1 + \left(\frac{16}{\pi^3} t / \tau \right)^{\nu} \right]^{0.5/\nu}}, \tag{16c}$$

where $\nu \approx 2.8$–2.85 is an empirical parameter.

Figure 10 illustrates $Y(t/\tau)$ dependencies obtained using the exact theory (Eq. (16)) and different approximations (Eqs. 16a–16c).

For more precise analysis of the deviations of solid/liquid expression from the classical filtration-consolidation behavior, the dual-porosity model was developed

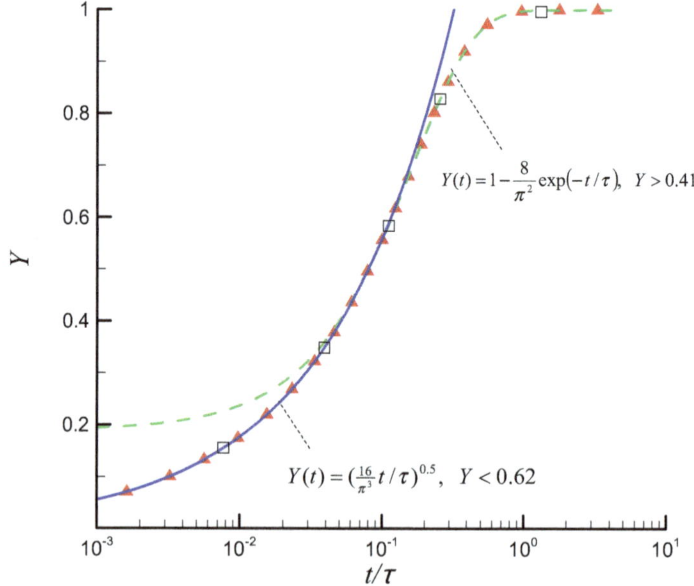

Fig. 10 Primary consolidation ratio (Y) versus the reduced time (t/τ) obtained using the exact theory (solid triangles, Eq. (16)) and different approximations (dashed line, Eq. (16a), solid line, Eq. (16b), and open squares, Eq. (16c), $\nu = 2.8$)

(Mahnič-Kalamiza and Vorobiev 2014; Mahnič-Kalamiza et al. 2015; Mahnič-Kalamiza 2016). In this model the biological tissue was represented by the intracellular (symplast) and the extracellular (apoplast) spaces (Fig. 11). The model postulated that liquid flows from the intracellular to the extracellular space under the pressure gradient. The cells were supposed to be spherical with radius of R, the specific surface of the cell was calculated as $S_c = 3/R$, and the cell area was calculated as $A = 4\pi R^2$. The cells are covered by semipermeable cell membranes with permeability, which is a function of the size and the quantity of pores appeared during the PEF treatment. The pore surface fraction was evaluated as $f_p = N_p \cdot A_p / A$, where N_p is the number of pores per cell and A_p is the average single pore area.

The transmembrane flow coefficient (α) was calculated as $\alpha = k_p S^2 f_p$, where k_p is the permeability of the pore of radius (r_p) induced in cell membrane by PEF. The value of k_p can be estimated based on the Hagen-Poiseuille equation as $k_p = r_p^2/8$. The compressibility of the intracellular space was assumed to be considerably higher than that of the extracellular space. Extracellular space with hydraulic permeability (k_e) is formed by the cell wall, extracellular liquid, and air (Fig. 11). Liquid flow in this space was supposed to obey the Darcy filtration law. The permeability of the intracellular space (e.g., $k_i \approx 10^{-24}$ for sugar beet tissue) was neglected

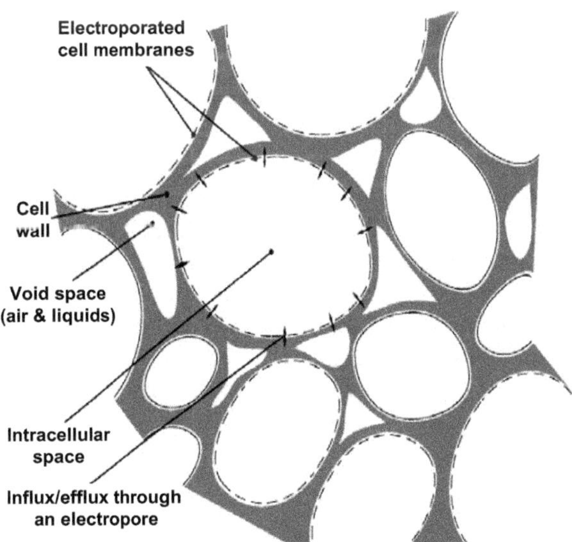

Fig. 11 A schematic representation of dual-porosity model for description of solid/liquid expression from electroporated cellular tissue. The tissue includes intracellular and extracellular media with semipermeable boundaries (membranes) between them. The membrane has its own hydraulic permeability, which is a function of electroporation parameters. The extracellular space with own hydraulic permeability has intricate structure formed by the cell wall, extracellular liquid, and air. (Mahnič-Kalamiza and Vorobiev 2014)

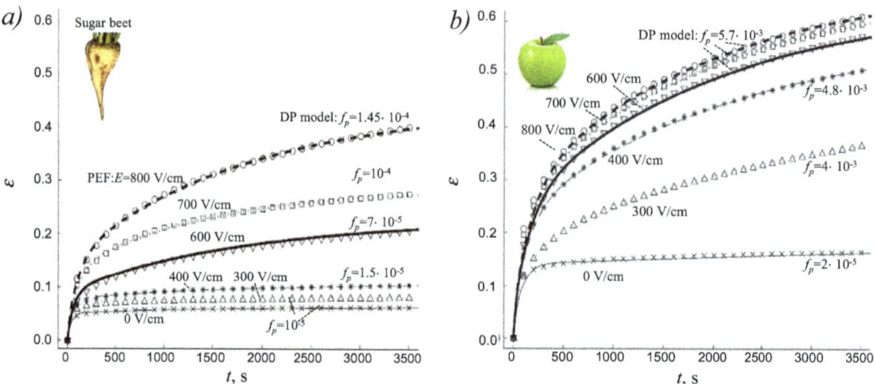

Fig. 12 Relative deformation, ε, versus the pressing time t for sugar beet (**a**) and apple (**b**) disks. Symbols correspond to the experimental data at different values of electric field strength (E), and solid lines correspond to the simulation data obtained using the dual-porosity (DP) model at different values of the pore surface fraction (f_p). (Adapted from (Mahnič-Kalamiza et al. 2015))

in comparison to the permeability of the extracellular space ($k_e \approx 10^{-17}$) (Mahnič-Kalamiza and Vorobiev 2014).

The theory based on the dual-porosity model has been tested for checking of the PEF treatment effects on solid/liquid expression (Mahnič-Kalamiza and Vorobiev 2014; Mahnič-Kalamiza et al. 2015; Mahnič-Kalamiza 2016). The disk-like samples of sugar beet and apple (with diameter of 25 mm and height of 5 mm) were PEF treated by bipolar rectangular pulses (E = 300–800 V/cm, t_p = 100μs, t_{PEF} = 1.6 ms) and then pressed at the constant pressures of P = 5.82 bar for sugar beet and P = 2.91 bar for apple.

Figure 12 presents examples of the relative deformation (ε) versus the pressing time (t) for sugar beet (a) and apple (b) disks. Symbols correspond to the experimental data at different values of electric field strength (E), and solid lines correspond to the simulation data obtained using the dual-porosity (DP) model at different values of the pore surface fraction (f_p) (Mahnič-Kalamiza et al. 2015). At the same PEF treatment conditions (the same E and t_{PEF}), the estimated values of f_p for sugar beet (Fig. 12a) were an order of magnitude smaller than those for apple (Fig. 12b). It can be explained by the smaller size of cells for the sugar beet.

Solvent Extraction of Soluble Substances: Principles and Models

Transfer of solute from biological solids to solvent is a traditional unit operation in different food applications (extraction of sucrose from sugar beets, of phytochemicals from plants, of lipids from oilseeds, of valuable biomolecules from algae, and many others) (Aguilera 2003). For example, the conventional hot water extraction at 70–74 °C is widely used in industrial production of sucrose from sugar beet. Such processing permits denaturation of cell membranes (thermal plasmolysis) and accelerates the sucrose extraction from sugar beet slices. Unfortunately, during thermal plasmolysis the cell walls are also subjected to the thermal destruction, and different undesirable substances (like pectin and other colloids) can also be extracted into the juice (Van der Poel et al. 1998). The purification of juice needs numerous operations and is a very costly process.

In recent decades, the solvent extraction from biological solids assisted by different unconventional techniques (high pressure, ultrasound, microwaves, and pulsed electric fields) has been tested (Aguilera 2003; Puri 2017). Application of PEF treatment is of special interest. The early studies demonstrated that tissue treatment even at low gradient electric fields (DC or AC) in the range of 100–200 V/cm could significantly enhance diffusion of soluble substances (Flaumenbaum 1949; Zagorul'ko 1958). However, such treatments can result in undesirable electrolysis and loss of quality of the products (Bazhal and Gulyi 1983), as well as in significant energy consumption and temperature elevation (ohmic heating effects) (Jemai 1997). The PEF treatment is a good alternative to the DC or AC applications. It is a nonthermal cost-effective technique that has shown high potential for extraction purposes. In recent two decades, many studies have demonstrated the PEF effects on solute extraction from plant tissues (Vorobiev and Lebovka 2013; Chan et al. 2014; Vorobiev and Lebovka 2016a, b, 2019).

Solvent extraction goal is the diffusion of biomolecules from a sample into the external solvent. It is a multiphase and unsteady state mass transfer operation (Aguilera 2003). In the food industry, the commonly used solvents are water, ethanol (or ethanol-water mixtures), and other solvents with preference to the so-called green solvents (Nelson 2003).

The most popular theory for the description of the diffusion processes is based on the analytical solution of Fick's second law (Crank 1979). The kinetics of extraction depends upon the shape of the sample, temperature, type of the solvent, and agitation of the solution. For the infinite slab (plane sheet), the theory gives the following equation for the time evolution of the relative solute concentration in solution (diffusion yield):

$$Y = C(t)/C_{\infty} = 1 - \left(8/\pi^2\right) \sum_{\infty}^{n=0} \frac{\exp\left(-(2n+1)^2 t/\tau\right)}{(2n+1)^2}, \tag{17}$$

where C is the solute concentration in the solvent, C_{∞} is its equilibrium value in the limit of long (infinitive) extraction time, $\tau = \tau_d = h^2/(\pi^2 D)$ is the diffusion time, h is the thickness of a slab, and D is the effective diffusion coefficient (solute diffusivity). It is assumed that at the initial moment of time ($t = 0$), the solute is homogeneously distributed inside the slab, and there is instantaneous equilibrium between the surface of the slab and surrounding solvent. A dimensionless time $\mathrm{Fi} = tD/h^2$ is called Fick number (Gekas 1992).

The time dependence of the diffusion yield (Y), with time of τ_d (Eq. 17), is equivalent to the time dependence of the primary consolidation ratio (Y), with time of τ_b (Eq. 16). Therefore, the consolidation coefficient b can be considered as the filtration-consolidation analog of the solute diffusivity (D).

The approximate relations represented by Eqs. (16a)–(16c) can be used for estimation of the diffusion time (τ_d) and solute diffusivity (D). Fick's solution is frequently used for the estimation of solute diffusivity in different biological tissues (Gekas 1992; Zogzas and Maroulis 1996; Varzakas et al. 2005; Azarpazhooh and Ramaswamy 2009). The values of solute diffusivities in foods have been reviewed extensively in the literature (Schwartzberg and Chao 1982; Aguilera 2003; Varzakas et al. 2005; Saravacos and Krokida 2014).

Single exponential law with coefficient $f = 8/\pi^2$ (and relaxation time τ_d) can be used at the large values of Y (>0.415) for the good fitting of experimental data (see Fig. 10, Eq. 16a). For more precise fitting in whole diapason of $0 < Y < 1$, the relation represented by Eq. (16c) with fitting parameter ν can be used. Note that for the fitting of the experimental curves of $Y(t)$ other empirical relations were also tested (Varzakas et al. 2005; Chan et al. 2014). We can refer the extended exponential relation (also known as the Weibull or Kohlrausch law (Elton 2018))

$$Y = 1 - fexp\left[\left(-t/\tau\right)^{\beta}\right], \tag{18}$$

or multi-exponential relation

$$Y = \sum_{n}^{i=1} f_i(1 - \exp\left(-t/\tau_i\right). \tag{19}$$

Here, f, β are the fitting parameters, and f_i corresponds to the fraction of the exponential process with relaxation time of τ_i.

Two-exponential law can be used for the fitting of extraction curves, which includes the first step of a fast extraction (washing stage) and the second step related to a slow extraction (solute diffusion) (Chan et al. 2014). Figure 13 presents such situation for the diffusion yield (Y) obtained using two-exponential equation (Eq. 19) with fast (washing) contribution ($f_1 = 0.5$, $\tau_1 = 0.1\tau$) and slow bulk diffusion contribution ($f_2 = 0.5$, $\tau_2 = \tau$).

The theory based on the dual-porosity model has been also tested for checking of the PEF treatment effects on diffusion extraction of soluble substances (Mahnič-Kalamiza and Vorobiev 2014; Mahnič-Kalamiza et al. 2015; Mahnič-Kalamiza 2016).

Some Examples of PEF-Assisted Processes for Different Foods

In recent decades PEF-assisted solid/liquid expression and solvent extraction have been tested for different food and biomass materials, like apple, citrus, tomato, sugar crops, potato and carrot crops, grapes and residues of wine industry, other

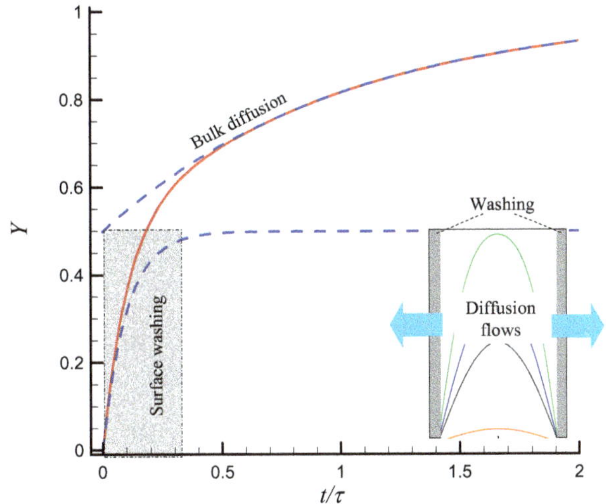

Fig. 13 Diffusion yield ($Y{=}C/C_{\infty}$) versus the normalized time of extraction (t/τ) obtained using two-exponential equation (Eq. (19)) with fast (washing) contribution ($f_1 = 0.5$, $\tau_1 = 0.1\tau$) and slow bulk diffusion contribution ($f_2 = 0.5$, $\tau_2 = \tau$)

fruits and vegetables, microalgae, mushrooms, leaves, etc. Additional recovery of valuable cell compounds, e.g., carbohydrates, polyphenols, proteins, etc., was demonstrated.

Potatoes and Apples

Potato and apple were frequently used in early studies as model tissues for testing different electroporation effects. For instance, juice expression and solute extraction can be significantly improved when these tissues are electroporated by PEF.

As example, experimental studies revealed complex character of solid/liquid expression from PEF-treated potato slices (Lebovka et al. 2003). Figure 14 presents the juice yield (relative mass losses) $Y = \Delta m/m$ (open squares) versus pressing time t for the untreated potato slices at constant pressure of $P = 5$ bar (symbols). The value of Y is proportional to the consolidation ratio (Y_c) defined in Eq. (14). The dashed line represents the least square fit of the consolidation data using the multi-exponential relation

$$Y = \sum_{m}^{i=1} f_i \left[1 - \exp\left(-t/\tau_i\right)\right], \tag{20}$$

where $i = 1$–3, τ_i, and f_i are the expression (consolidation) periods and corresponding relative fractions.

The time of the initial expression period was $\tau_1 = 10 \pm 2$ s. During this initial period, about 20% of the juice was expressed ($p_1 = 0.20 \pm 0.05$). The initial period may be attributed to the expression of external juice and air. The value of τ_1 for this initial expression period may depend upon the type of tissue and the degree of tissue fragmentation (Schwartzberg 1997). The parameters for the second expression

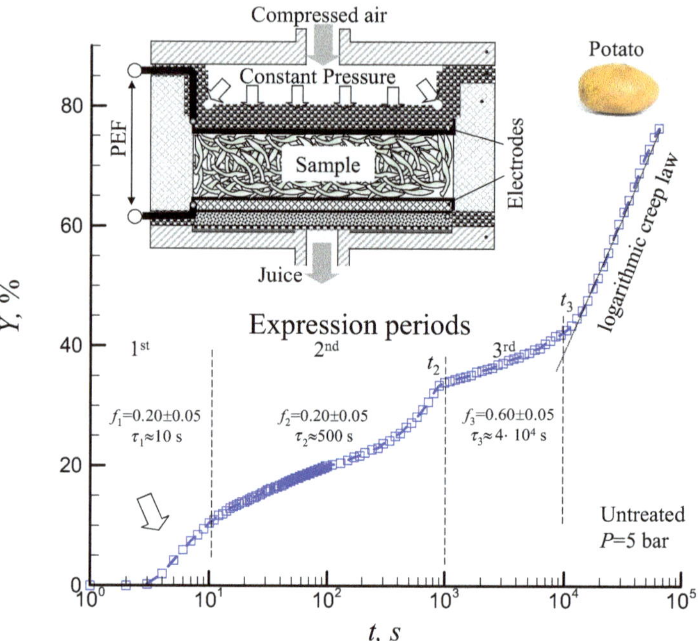

Fig. 14 Juice yield (Y) versus time (t) for the constant pressure expression ($P = 5$ bar) from potato slices. Here, τ_i-($i = 1$–3) are the different expression (consolidation) periods, and f_i are the corresponding relative fractions. Open squares correspond to the experimental data, and the dashed line was obtained using the multi-exponential relation (Eq. (20)). Inset shows experimental device. (Adapted from (Lebovka et al. 2003))

period were $\tau_2 \approx 500 \pm 150$ s and $f_2 = 0.20 \pm 0.05$ (Fig. 14). The third expression period was more prolonged with time of $\tau_3 \approx (4 \pm 1) \times 10^4$ s, and biggest quantity of liquid was expressed during this period ($f_3 = 0.60 \pm 0.05$). After certain transition time, kinetics of the longest expression period may be approximated by Buisman's logarithmic creep law, $Y \propto \ln t$ (Buisman 1936).

Potato was used for checking the effects of PEF protocols on the juice expression behavior. Figure 15a shows typical juice expression curves obtained from the untreated and PEF-treated potato slices at constant pressure ($P = 5$ bar). In these experiments, the electric field strength was fixed at $E = 200$ V/cm, pulse duration was $t_p = 100\mu s$, pulse repetition time was $\Delta t = 10$ ms, number of pulses (n) was varied, and total time of the PEF treatment $t_{PEF} = nt_p$ was lasted from 5 ms to 3 s. PEF was applied at $t = 20$ s just after the first consolidation stage when about 20% of juice was expressed (Lebovka et al. 2003). After the PEF application, additional quantity of juice was released from the electroporated cells, and the consolidation process was developed more intensively. Increase of the PEF treatment duration (t_{PEF}) leads to a more rapid expression kinetics.

Difference between the behaviors of second consolidation period (lasted from ≈20 s to ≈10^3 s) and third consolidation period (lasted from ≈10^3 s to ≈10^4 s) was

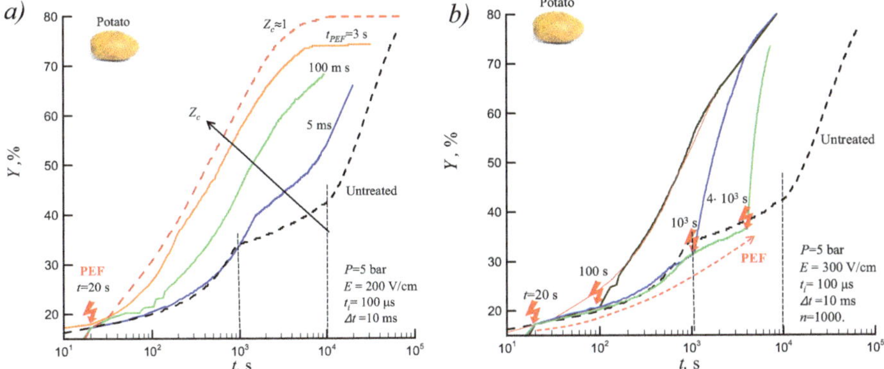

Fig. 15 Juice yield, Y, from PEF-treated potato slices versus time, t, for the constant pressure expression (P = 5 bar). PEF treatment was applied at t = 20 s (a) or at the different times t = 20–4000 s (b). PEF treatment parameters were t_p = 100µs, Δt = 10 ms, E = 200 V/cm, (**a**), and E = 300 V/cm (**b**), T = 25 °C. (Adapted from (Lebovka et al. 2003))

noticeable for the untreated tissue, and it completely disappeared after the prolonged PEF treatment (t_{PEF} > 100 ms) (Fig. 15a). Figure 15b shows the curves of juice expression $Y(t)$ before and after the PEF treatment of potato slices. PEF treatment was applied at the different moments of time directly during expression. The treatment protocol was E = 300 V/cm, t_p = 100µs, Δt = 10 ms, and the fixed number of pulses was n = 1000 (t_{PEF} = 0.1 s) (Lebovka et al. 2003). The delay in PEF application resulted in the retardation of juice release from the electroporated cells and lead to the longer duration of pressing. From an industrial point of view, it is important to reduce the pressing time as much as possible. So, it is desirable to apply the intermediate PEF treatment at the initial period of compression, predominantly before or during the second consolidation period.

Potato was also used for testing the PEF effects on the characteristics of tissue at constant velocity regime of compression. Figure 16 shows the deformation behavior $\varepsilon(t) = 1\text{-}h(t)/h_0$ (a) and the final deformation ε_f (at t_f = 1.3 × 10⁴ s) (b) of the untreated and PEF-treated potato disks (diameter d = 25 mm and initial thickness h_0 = 10 mm) when compression is started at the constant velocity (V = 0.1 mm/s) and continued at the constant pressure (P = 10, 20 or 40 bar) (Grimi et al. 2009a). During the constant velocity stage, at $t < t_t$, the deformation increases linearly with time, and pressure also increases up to attain the fixed value of P = 10, 20, or 40 bar (at $t = t_t$) (Fig. 16a).

The difference in the compression behavior for the untreated and PEF-treated potato samples appeared just at $t > t_t$, when pressing is continued at the constant pressure. For the untreated samples, two different effects were observed: (a) the pressure-induced rupture (cracking) of intact cells with release of juice from ruptured cells and (b) the juice expression outside the tissue by filtration-consolidation. At the high level of electroporated cells (conductivity disintegration index Z_c = 0.95), the temporal variation of tissue deformation reflects mainly filtration-consolidation

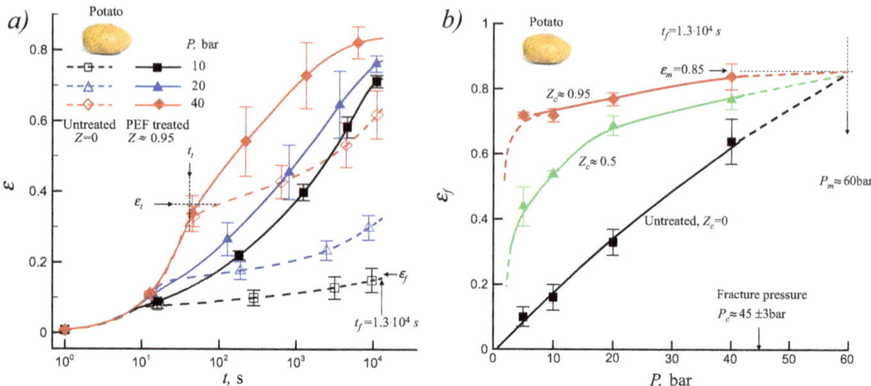

Fig. 16 Deformation (ε) versus the pressing time (t) (**a**) and the final deformation, ε_f, (at $t_f = 1.3 \times 10^4$ s) versus the pressure (P) (**b**) for untreated and PEF-treated potato disks. The compression was started at the constant velocity ($V = 0.1$ mm/s) and continued at the constant pressure (P = 10, 20 and 40 bar). The conductivity disintegration index (Z_c) of PEF-treated specimens was fixed ($Z_c = 0.95$) (*a*) or varied ($Z_c = 0.5$, $Z_c = 0.95$) (*b*). (Adapted from (Grimi et al. 2009a))

behavior (Lebovka et al. 2003). Juice expression from the electroporated cells was much faster than from the mechanically cracked cells for the same values of pressure. For example, 3.5 h and just 70 s of compression at $P = 10$ bar were required to attain the same deformation ($\varepsilon \approx 0.15$) of the untreated and PEF-treated ($Z_c = 0.95$) potato disks, respectively. Final deformation (ε_f) of the untreated and PEF-treated potato disks was estimated after the long compression time ($t_f = 1.3 \cdot 10^4$ s $\approx$ 3.5 h). The values of ε_f obtained at the different pressures are presented in Fig. 16b. It can be seen that extrapolation of the $\varepsilon_f(P)$ curves obtained for the different degrees of cell disintegration (Z_c = 0, 0.5, 0.95) gives the same hypothetical pressure value of $P_m \approx 60$ bar, which exceeds the experimentally measured fracture pressure for the potato samples (Grimi et al. 2009a).

PEF-assisted juice expression from apple (Golden Delicious) slices was studied on a laboratory filter-press cell at the constant pressure (Bazhal and Vorobiev 2000). The slices were obtained using a 6 mm grater. For the untreated slices, juice yield was increased from 28% to 61% when pressure, P, was increased from 1 to 30 bar. PEF treatment (electric field strength of E = 200–1000 V/cm, pulse duration of t_p = 20–100μs, pulse repetition time of Δt = 10 ms, number of pulses of n = 3–1000) permitted enhance importantly the juice yield, which attained about 80% at E = 1000 V/cm t_p = 100μs, n = 1000, and P = 3 bar.

Correlations between the PEF energy consumption and apple juice yield were established. The maximum juice yield was attained at optimum energy consumption of 3 kJ/kg. The juice obtained without PEF treatment was turbid and required filtration. The juice obtained for PEF-assisted pressing was significantly clearer and did not require additional filtration.

Note that PEF treatment may significantly affect the structure of cell membranes, whereas it seems to retain the cell wall architecture in plant tissues (Fincan and

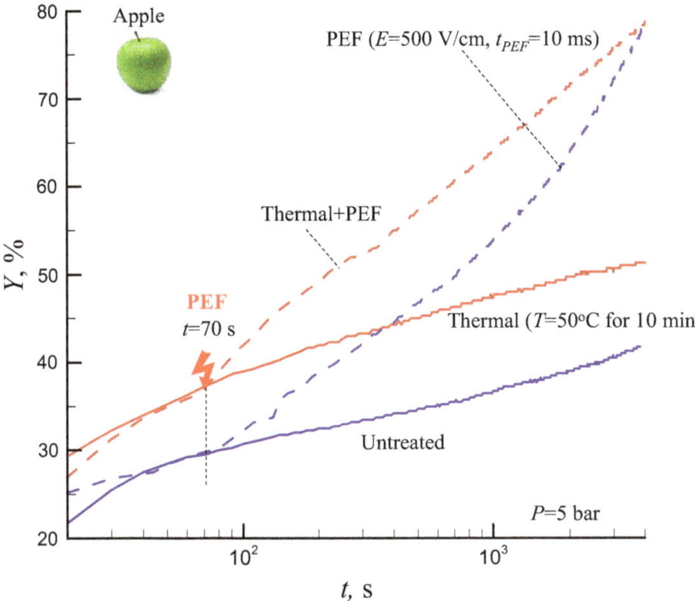

Fig. 17 Juice yield (Y) versus time (t) for the constant pressure expression (P = 5 bar) of apple slices. Thermal pretreatment was done at T = 50 °C for 10 min. PEF treatment (E = 500 V/cm, t_p = 10μs, Δt = 10 ms, t_{PEF} = 0.01 s) was applied during the pressing at t = 70 s. (Adapted from (Lebovka et al. 2004b))

Dejmek 2003). However, the rigidity of cell walls may hinder the consolidation behavior during solid/liquid expression. The studies have revealed significant effects of mild heating (e.g., at 50 °C) on the textural softening of different food tissues (apple, carrot, potato) (Lebovka et al. 2004a) and juice expression behavior (Lebovka et al. 2004b).

Figure 17 presents the juice yield curves obtained at the constant pressure expression (P = 5 bar) for fine-cut apple slices (1.5 mm in width, 40–50 mm in length). The data for untreated, preheated (thermal), PEF-treated (E = 500 V/cm, t_p = 10μs, Δt = 10 ms, t_{PEF} = 0.01 s), and preheated and PEF-treated samples can be compared. Slices were preheated at $T \approx$ 50 °C for 10 min, and then they were treated or not by PEF at the ambient temperature. For the preheated sample, the juice yield was significantly higher than for the untreated one. PEF treatment accelerated the juice yield (Y) for both untreated and preheated samples. However, the most significant effect of PEF was observed for the thermal + PEF sample.

Ohmic heating (OH) and PEF treatment were applied to enhance the juice yield from potato (1.5 mm in width, 1 mm in thickness, and 30–40 mm in length) and apple (6 mm in width, 1.5 mm in thickness, and 30–40 mm in length) slices (Fig. 18a) (Praporscic et al. 2005). Before OH slices were pre-compacted to ensure the good contact with electrodes. Ohmic heating was realized at E = 30 or E = 50 V/cm for 45 s. The interrupted heating-cooling mode of OH treatment was used when

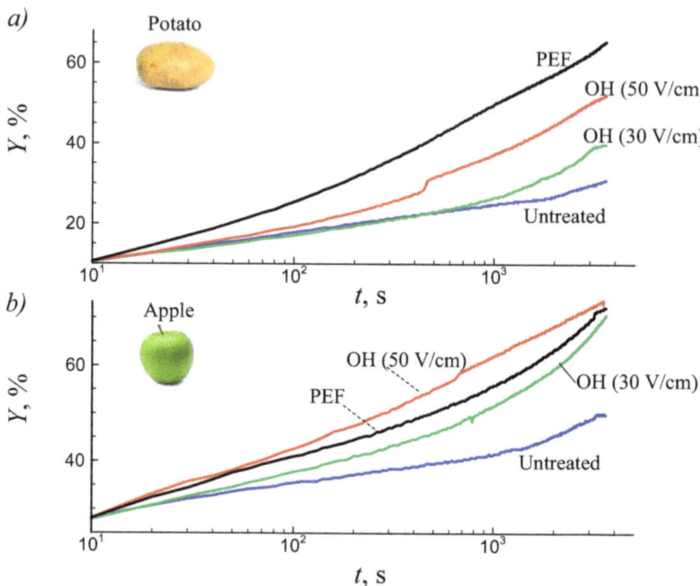

Fig. 18 Juice yield Y expressed from potato (**a**) and apple (**b**) slices versus pressing time, t, at the constant pressure regime ($P = 5$ bar) for the untreated, OH-treated (at $E = 30$ V/cm or $E = 50$ V/cm for 45 s), and PEF-treated ($E = 850$ V/cm and $t_{PEF} = 10$ ms, $t_p = 100$μs, $\Delta t = 10$ ms) slices. OH and PEF treatments were done before pressing. (Adapted from (Praporscic et al. 2005))

the temperature of slices reached 50 °C. Therefore, the final temperature of slices never exceeded 50 °C. Before PEF treatment, slices were also pre-compacted to ensure the good contact with electrodes. PEF treatment was done at $E = 850$ V/cm and $t_{PEF} = 10$ ms ($t_p = 100$μs, $\Delta t = 10$ ms) and ambient temperature. For the potato samples, the highest juice yield was observed after the PEF treatment (Fig. 18a).

However, for apple samples, the highest juice yield was observed after the OH at $E = 50$ V/cm (Fig. 18b). This result can be explained by synergetic effect of thermal and electric field treatments that was more significant for the apple than for the potato (Lebovka et al. 2004a, b).

Aqueous extraction of solutes from apple tissue was also significantly enhanced by PEF (Jemai and Vorobiev 2002, 2003). Different pretreatments were compared to enhance extraction kinetics from apple disks (Golden Delicious): thermal denaturation (plasmolysis) (75 °C, 2 min) followed by diffusion at different temperatures (20–75 °C); (2) PEF treatments ($E = 100$–650 V/cm, $t_p = 50$–200μs, $n = 1000$) followed by diffusion at 20 °C; and (3) PEF treatments ($E = 500$ V/cm, $t_p = 100$μs, $n = 1000$) followed by diffusion (60 min) at $T = 20$–75 °C. Detectable enhancement of diffusion kinetics was observed beginning from the field strengths of $E = 100$–150 V/cm. Further increase of E and t_{PEF} resulted in more rapid diffusion kinetics. The temperature dependence of solute diffusivity D followed the Arrhenius law with activation energy $W_a \approx 28$ kJ/mole (for samples without thermal pretreatment) and $W_a \sim 13$ kJ/mole (for thermally denaturized samples). For the PEF-treated samples, only one regime with intermediate activation energy, $W_a \approx 20$ kJ/mole, was observed.

PEF treatment ($E = 100$–1100 V/cm and $t_{PEF} = 10$–100 ms) was used for accelerating the osmotic dehydration (OD) of apples (*Golden* variety) (Amami et al. 2005). PEF-treated apple disks (2.8 cm in diameter and 0.85 cm in thickness) were placed in stirred hypertonic solution of sucrose. The best results were obtained at $E = 900$ V/cm, $t_{PEF} = 75$ ms, and energy input of $W = 13.5$ kJ/kg. PEF treatment noticeably increased water losses (WL) (on 50%) and solids (sugar) gain (SG) (on 6%) compared to the control samples. Therefore, PEF treatment allowed obtaining the high dehydration effect with minimal sugar uptake by the apple. With increase of sucrose concentration from 44.5 to 55 and 65 w/w, the WL and SG values became higher, especially for the PEF-treated apples (Amami et al. 2006). PEF-assisted OD of apple (*Idared* variety) has been studied (Wiktor et al. 2014). For PEF-treated samples, the increase in WL up to 36–46% was observed.

Sugar Crops

Sugar crops include sugar beet, sugarcane, chicory, sugar maple, sugar palm, etc. The most intensive studies have been devoted to the PEF-assisted solid/liquid expression and water extraction processes applied to sugar beets. Conventional sugar beet processing employs the hot water extraction followed by very complex multistage juice purification. This is a time- and energy-consuming processing that can be significantly improved by application of PEF treatment.

Sugar Beet

In the early works, the effect of PEF protocols ($E = 215$–427 V/cm, $t_p = 100 \mu s$, $\Delta t = 10$ ms, $n = 250$–500) on juice expression from sugar beet cossettes (slices) was studied at the constant pressure regime ($P = 5$ bar) (Bouzrara and Vorobiev 2000, 2001). An intermediate PEF treatment applied during the pressing enhanced markedly the juice yield, and a correlation between the PEF energy input and juice yield was demonstrated. In addition to the parameters of PEF protocol, other factors, such as pressure and size of sliced particles, have been reported to have significant effects. For the untreated sugar beet slices, pressure increase from 0.5 bar to 10 bar enhanced the juice extraction yield from 1.6% to 34%. PEF treatment ($E = 500$ V/cm, $t_p = 100 \mu s$, $\Delta t = 10$ ms, $n = 1000$) followed by pressing at $P = 10$ bar enhanced the juice yield up to 62.3%. The combined process with an intermediate PEF application was most effective and enhanced the juice yield up to 78.5% (at $P = 5$ bar) and up to 82.4% (at $P = 10$ bar). Effect of the size of PEF-treated sliced particles (width of 1.5–7 mm, length of 5 mm, and thickness of 1.5 mm) on the kinetics of solid/liquid expression was also evaluated. As expected, the quantity of juice expressed (at $P = 5$ bar) from small untreated particles was higher (up to 40% for slices with width of 1.5 mm) than the quantity of juice expressed from large untreated particles (15% for slices with width of 7 mm). Nevertheless, the total juice yield after the PEF application (E = 500 V/cm, t_p = 100μs, Δt = 10 ms, n = 1000) was nearly the same for particles with sizes in the range of 1.5–6 mm (76–78%), and it was slightly smaller for the largest particles of 7 mm (73%).

The attractive feature of pressing with an intermediate PEF treatment was a better purity and a paler color of sugar beet juice (Bouzrara and Vorobiev 2001). Moreover, the juice obtained with assistance of PEF had higher sugar concentration and did not contain any pectin substances, which is advantageous for the subsequent filtration and purification processing.

Experiments on PEF-assisted solid/liquid expression from sugar beet cossettes (grated with 6 mm grater) at the constant rate regimes were also performed (Praporscic et al. 2004). During the pressing of sliced particles at the same constant velocity V (V = 0.65 or 5.5 mm/min), the loading pressure increased continuously with the time up to the maximum value of 25 bar, and then pressing was continued at this pressure of P_m = 25 bar. The PEF treatment (E = 0–1000 V/cm, t_p = 100μs, Δt = 10 ms, t_{PEF} = 0–0.1 s) was applied before pressing (P = 0) or during pressing at the different moments of time (corresponding to the pressures of P = 1.5–25 bar). It was demonstrated that PEF application to the non-pressurized sugar beet cossettes leads to a higher energy consumption. On the other side, PEF treatment of pressurized cossettes resulted in a delayed kinetics of juice expression. The optimal PEF treatment results minimizing energy consumption were obtained after the pressurization of cossettes at P = 1.5–5 bar.

Figure 19 shows the kinetics of juice yield $Y(t)$ expressed from sugar beet cossettes during the constant rate solid/liquid expression (V = 0.65 mm/min, P_m = 25 bar) for different modes of PEF treatment (Praporscic et al. 2004). In these experiments

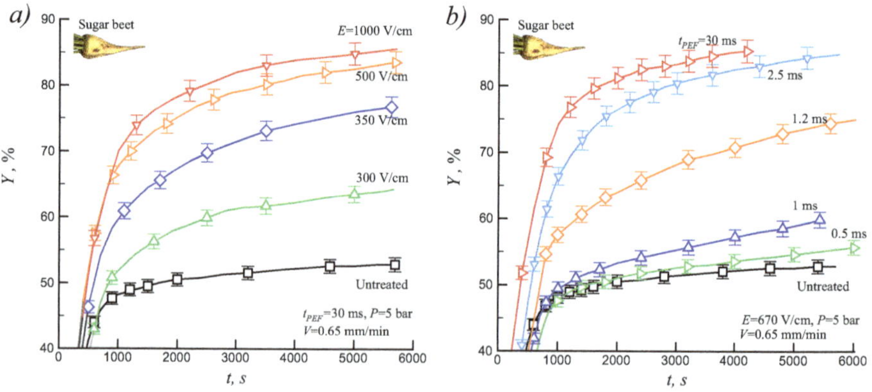

Fig. 19 Juice yield (Y) expressed from sugar beet cossettes versus expression time (t) at the constant rate regime (V = 0.65 mm/min, P_m = 25 bar). The PEF treatment (t_{PEF} = 30 ms and different values of electric field strength, (E)) (**a**) or (E = 670 V/cm and different values of t_{PEF}) (**b**) were applied when pressure attained 5 bar. (Adapted from (Praporscic et al. 2004))

the PEF treatment was applied when pressure attained the value of 5 bar. However, the PEF treatment parameters were different: $t_{PEF} = 30$ ms and different values of E (Fig. 19a) or $E = 670$ V/cm and different values of t_{PEF} (Fig. 19b). At the fixed value of $t_{PEF} = 30$ ms, the noticeable PEF effect on juice yield was only observed for E above 250–300 V/cm (Fig. 19a). However, above 500 V/cm increase in E gave only insignificant supplementary effect. At the fixed value of $E = 670$ V/cm, the PEF effect was significant for t_{PEF} duration between 1 and 2.5 ms (Fig. 19b).

Enhanced sucrose extraction from PEF-treated sugar beet cossettes has been also revealed (Jemai and Vorobiev 2003). A laboratory-scale solid-liquid extractor was used in these experiments. Soluble matter concentration of the extract and electrical conductivity of the solid-liquid mixture were simultaneously measured during extraction. Significant enhancement of soluble matter release was observed for the PEF treatments above $E = 150$ V/cm. Later on, PEF-assisted sucrose extraction from sugar beet disks (30 mm in diameter and 8.5 mm in thickness) treated at $E = 235$–1180 V/cm, $t_p = 100$μs, $\Delta t = 10$ ms, and $n = 100$–1000 were studied in details (El Belghiti and Vorobiev 2004). It was demonstrated that Fick's diffusion law was insufficient to explain the overall mass transfer kinetics from the PEF-treated sugar beet tissue. The extraction kinetics was modeled using a two-exponential model presented by Eq. (19). This model involves two simultaneous processes related to rapid convection of solutes from the tissue surface to the solution (washing process) and with slower transport of solutes from the interior to the exterior of electroporated tissue (diffusion process). During the rapid washing stage of extraction, up to 20–25% of surface sugar was recovered under the optimal PEF treatment conditions. The kinetics of the prolonged diffusion stage of extraction can be accelerated by intensive stirring.

Figure 20 shows kinetics of aqueous sucrose extraction from sugar beet cossettes at different electric field strengths (E). Extraction was performed in well-stirred water at 25 °C for 2 h. Just $Y \approx 48\%$ was attained in the absence of PEF in saturated state after 2 h of extraction. This may indicate extraction of sucrose from mechanically broken cells obtained during the slicing of the beet. The yield of solute was significantly increased with PEF treatment attaining $Y \approx 93\%$ for the PEF-treated cossettes at 670–800 V/cm. The influence of PEF treatment time (t_{PEF}) and elevated extraction temperatures (30, 40, and 50 °C) were also evaluated. At the fixed electric field strength of $E = 670$ V/cm, the effective electroporation of tissue was obtained using $t_{PEF} = 0.025$ s.

The effect of temperature ($T = 20$–80 °C)) on the aqueous extraction of sucrose from sugar beet cossettes electroporated by PEF ($E = 400$ V/cm, $t_{PEF} = 0.1$ s) was studied (Lebovka et al. 2007a, b). The extraction yield was characterized by changes in °Brix measured during extraction:

$$Y_B = \left(^\circ\text{Brix} - ^\circ\text{Brix}_i\right) / \left(^\circ\text{Brix}_f - ^\circ\text{Brix}_i\right), \tag{21}$$

where °Brix, $^\circ\text{Brix}_i$, and $^\circ\text{Brix}_f$ are, respectively, the actual, the initial, and the final values of °Brix of the extract.

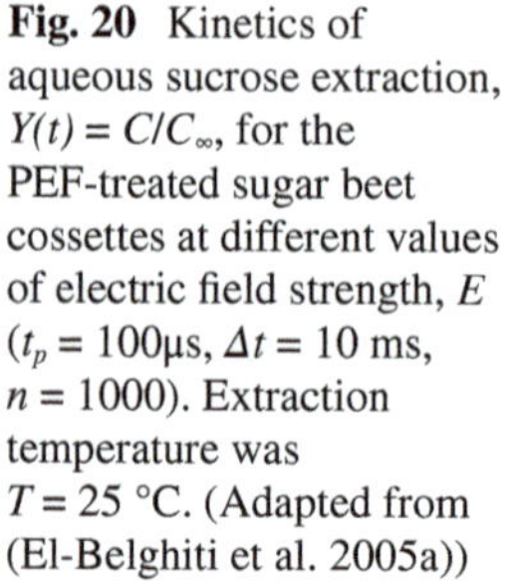

Fig. 20 Kinetics of aqueous sucrose extraction, $Y(t) = C/C_\infty$, for the PEF-treated sugar beet cossettes at different values of electric field strength, E ($t_p = 100 \mu s$, $\Delta t = 10$ ms, $n = 1000$). Extraction temperature was $T = 25$ °C. (Adapted from (El-Belghiti et al. 2005a))

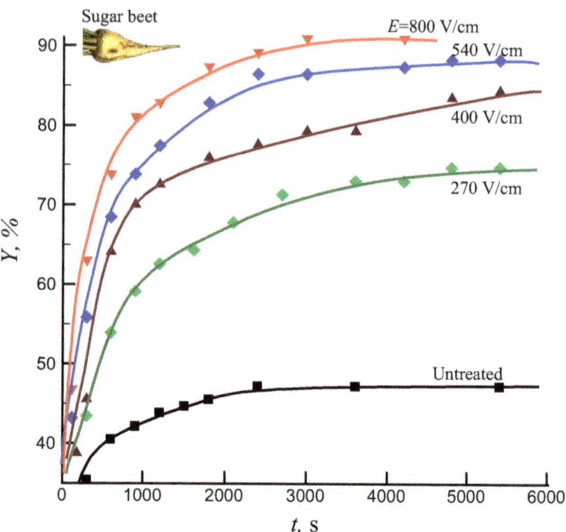

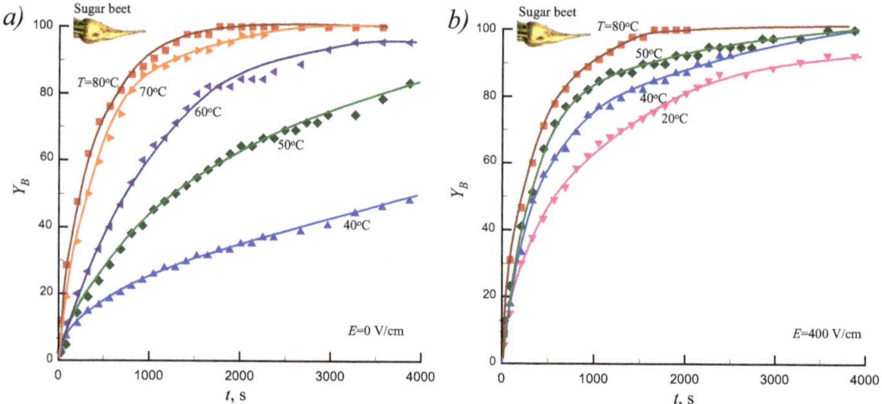

Fig. 21 Solute extraction yield, Y_B (normalized soluble solid content according to Eq. 21) at different temperatures versus time, t, for the untreated (**a**) and electroporated at $E = 400$ V/cm (**b**) sugar beet cossettes. (Lebovka et al. 2007a, b)

Figure 21 shows the kinetics of soluble solid diffusion from the untreated (*a*) and electroporated (*b*) cossettes at different temperatures (Lebovka et al. 2007a, b). For the untreated cossettes, diffusion duration of 50–60 min was needed at 70–80 °C to attain the maximal value of °Brix, $Y_B = 1$ (Fig. 21a). At lower temperatures of 40–50 °C, very important time would be needed to extract such quantity of soluble solids. However, for the electroporated cossettes, the maximal value of °Brix was attained more rapidly even at lower extraction temperature of 60 °C. Moreover, effective diffusion can be achieved at lower temperatures of 40–50 °C (Fig. 21b), and even cold diffusion at temperatures of 20–30 °C seemed to be attainable.

Aqueous extraction of sugar from the PEF-treated sugar beet cossettes was studied using the laboratory counter-current extractor with 14 extraction sections (Loginova et al. 2011a, c). Perforated plastic baskets filled with 500 g of cossettes were used for the manual transportation. The most concentrated solution (diffusion juice) was produced in sect. 1 in the contact with fresh cossettes, and the exhausted cossettes (pulp) were obtained in the last section, 14. The total extraction time was 70 min. The draft (juice to cossettes ratio) was varied from 120% to 90%. Exhausted pulp was pressed at 5 bar. The PEF intensity was fixed at 600 V/cm. It was demonstrated that even the cold and warm diffusion at 30–50 °C permitted effective exhausting of electroporated cossettes (Loginova et al. 2011a, c). Cold and warm diffusion led to the lower quantity of colloidal and pectin compounds released from sugar beet tissue to the extract. At such conditions the juice turbidity decreased by 10%, and it was less colored by 27%. The purification of diffusion juice obtained from the electroporated cossettes was evaluated (Loginov et al. 2011, 2012). Filtration properties of the juice of first carbonatation obtained from electroporated cossettes by cold and warm diffusion (at 30 and 50 °C) were considerably better than those of the juice obtained by hot diffusion (at 70 °C). Moreover, the thin juice of the second carbonatation obtained from electroporated cossettes had significantly paler coloration than corresponding thin juice obtained by hot diffusion from the untreated cossettes.

Later on, the characteristics of juices expressed at $P = 5$ bar from untreated, preheated (70 °C), and electroporated ($E = 600$ V/cm, $t_{PEF} = 7$ ms) sugar beet cossettes were compared (Mhemdi et al. 2014). Juice yield expressed from the untreated cossettes was rather low ($Y < 25\%$). Juice yields obtained from the preheated and PEF-treated cossettes were significantly higher (70–75% for the 30 min of pressing). However, the quality of juice expressed from the preheated cossettes was rather poor (purity of 92.5% and color of 7800 ICUMSA units) as compared to the quality of juice expressed from PEF-treated cossettes (purity of 93.5% and color of 5600 ICUMSA units). The worse quality of the juice obtained from preheated cossettes (70 °C) can be explained by the dissolution of hydrosoluble cell pectins and the enhanced enzymatic reactions at this elevated temperature. The better quality of juice obtained by cold pressing of electroporated sugar beet cossettes permits simplification of juice purification with decreased use of lime during carbonatation. Purification of juice obtained by PEF-assisted cold pressing has been studied (Mhemdi et al. 2015). The raw juice was heated to 45 °C and pre-limed with addition of CaO, and then the main liming was heated to 85 °C. The juices of first carbonatation obtained from electroporated cossettes demonstrated better filtration properties than those obtained from thermally treated cossettes (Mhemdi et al. 2015). PEF-assisted procedure permitted to decrease significantly the quantity of lime used for the juice purification and to reduce the quantity of wastes generated by the filtration station of sugar beet factory. Thin juices of second carbonatation obtained from the electroporated cossettes had significantly better quality than those obtained from the preheated to 70 °C cossettes. The purity of juice obtained from PEF-treated cossettes increased from 93.5% (raw juice) to 95.5% (thin juice) (Mhemdi et al. 2015).

The high quality of raw juice obtained by cold pressing of electroporated sugar beet cossettes allowed effective application of membrane purification (Mhemdi et al. 2014). In the proposed method, the juice was clarified by preliminary centrifugation and then filtrated at the laboratory Amicon filter under the pressure of 2 bar using the polyether sulfone (PES) membranes with pore sizes of 10–100 кDa. Ultrafiltration of juices obtained from electroporated sugar beet cossettes using dynamic filter with rotating disk (1000 tr/min) was also investigated (Zhu et al. 2015). The juice purity increased up to 96.4% using PES membrane with pore size of 10 kDa. The juice purity was even higher when lime was added for the cossette pressing (Almohammed et al. 2016a, b).

Combined pressing-diffusion experiments with sugar beet cossettes treated by PEF were conducted in a small pilot scale (Mhemdi et al. 2016). PEF-treated cossettes ($E = 600$ V/cm, $t_{PEF} = 10$ ms) were pressed at 5 bar at ambient temperature to obtain 50% of juice from the mass of cossettes. Then pressed pulp was loaded to the laboratory counter-current extractor with 12 extraction sections (Loginova et al. 2011a, c). Different quantities of water were used for the extraction, and the draft (water to cossettes ratio) was varied in the experiments from 80% to 100%. The extraction temperature was fixed at 30 °C or at 70 °C.

Diffusion juice (extract) was mixed with pressing juice to obtain the mixed juice. Cossettes exhausted in the extractor (pulp) were subjected to pressing in the laboratory batch press at 5 bar. Results were compared with those obtained after the conventional aqueous extraction (70 °C, draft of 120%) from the untreated sugar beet cossettes. Mixed juice obtained by pressing-diffusion technology from the electroporated sugar beet cossettes was 30% less colored than the juice obtained by conventional diffusion at 70 °C from the untreated cossettes (approximately 7000 units vs. 10,000 units ICUMSA) (Mhemdi et al. 2016). The purity of mixed juices (92.5–92.8) was higher than the purity of juice obtained by conventional diffusion (91.8%). This was explained by the very high purity of the juice obtained by cold expression from the electroporated cossettes. Decreasing diffusion temperature from 70 °C to 30 °C increased even more the quality of the mixed juice. Such juice had lowest coloration ($\approx$ 5500 units ICUMSA) and highest purity ($\approx$ 93.3%).

Cold pressing of PEF-treated sugar beet tails was studied to produce a fermentable juice (Almohammed et al. 2017a, b). Optimal PEF treatment protocol ($E = 450$ V/cm, $t_{PEF} = 10$ ms, and $W = 6.84$ kJ/kg) allowed increasing the yield of soluble solutes from 16.8% to 79.85%, and the dryness of press-cake was increased from 15% to 24%. Juice expressed from the PEF treated tails was more concentrated (10 vs. 5.2°Brix) and contained more sucrose (8.9 vs. 4.5 °S) than the juice expressed from the untreated tails.

Sugarcane

The effects of PEF treatment ($E = 400$–1000 V/cm, $t_{PEF} = 4$ ms) on sugar extraction from sugarcane stems have been studied (Almohammed et al. 2016a, b). The stems of sugar cane (diameter of 28 mm, thickness of 4 mm) were PEF treated and then immersed in water for 1 h to extract sugar at different temperatures (20–80 °C). For the untreated samples, solutes were almost completely extracted at 80 °C ($Y_B \approx 0.98$ in Eq. (21)), while just 30% of solutes ($Y_B = 0.3$) were extracted at 20 °C. PEF treatment significantly enhanced extraction yield at lower temperatures. For instance, 60% of solutes were extracted from the electroporated stems at 20 °C (Almohammed et al. 2016a, b).

Samples of sugarcane were PEF treated ($E = 2$ kV/cm and $W = 2.4$ kJ/kg) and then pressed with intermediate water soakings at the temperatures of 45–90 °C (Eshtiaghi and Yoswathana 2012). The yield of sugar extracted from the electroporated samples at the temperature of 45 °C was nearly the same as that extracted from the untreated samples at higher temperatures of 80–90 °C. Moreover, PEF treatment allowed reduction of the extraction time by 20%.

Chicory

The PEF-assisted ($E = 400$ V/cm, $t_{PEF} = 0.1$ s) extraction of solutes from chicory root slices (2 mm × 10 mm × 20 mm) at different temperatures ($T = 20$–80 °C) has been studied (Loginova et al. 2010). Kinetics of solute diffusion from chicory slices was similar to the kinetics of solute diffusion from sugar beet slices. For the untreated chicory slices, the maximum extraction yield ($Y_B \approx 1$) was obtained at $T = 70$–80 °C for ≈1 h. The same extraction yield was obtained for the electroporated samples at lower temperature $T = 40$ °C. Pulsed ohmic heating combining both electrical and thermal treatments has been tested to assist inulin extraction from chicory (Zhu et al. 2012). Inulin juice extracted from the electroporated chicory slices had better quality characteristics. Dynamic ultrafiltration with rotating disk was studied for the purification of inulin extracted from PEF-treated and untreated chicory slices (Zhu et al. 2013). A high rotational speed (2000 rpm) and a membrane with large pore size (0.45μm) were selected. Significant differences in the appearance and color were observed for feed juice, permeate, and retentate. Aqueous extraction of inulin from the PEF-treated ($E = 600$ V/cm, $t_{PEF} = 10$–50 ms) chicory slices (2 mm × 4 mm × 67 mm) was studied using a laboratory counter-current extractor (Zhu et al. 2012). Extraction temperature was varied from 30 to 80 °C. Inulin content in the juice extracted at 50–60 °C from the PEF-treated chicory slices was comparable with that extracted at 70–80 °C from the untreated slices. The purity of juice extracted from the electroporated chicory slices was higher compared to the purity of juice extracted from the untreated slices.

Carrots

The effect of PEF treatment ($E = 180$–360 V/cm, $t_p = 100$μs, $\Delta t = 10$ ms, $n = 250$–500, $t_{PEF} = nt_p = 5$ s) on the solid/liquid expression of carrot slices was studied. Slices were obtained with 6 mm grater and pressed at the constant pressure of $P = 5$ bar (Bouzrara and Vorobiev 2001). Without PEF treatment, the juice yield was 25.6%.

With PEF treatment it was increased up to 38.3% (at E = 180 V/cm) and up to 72.4% (at E = 360 V/cm). Moreover, PEF treatments permitted obtaining a lighter and more concentrated soluble matter juice.

Effects of PEF treatment on the solute extraction kinetics from carrot slices have been investigated by El-Belghiti and Vorobiev (2005). Coarse slices (thickness of 1.5 mm, obtained with 6 mm grater) and fine slices (thickness of 0.5 mm, obtained with 2 mm grater) were treated by PEF (E = 200–650 V/cm, t_p = 100μs, Δt = 1 ms, n = 200–1000).

Figure 22a shows kinetics of aqueous solute extraction from coarse and fine carrot slices treated by PEF applied with different energies (W). Extraction occurred in well-stirred water at 25 °C for 8 h. In absence of the PEF treatment, only around 45% of solutes were released from the coarse slices after 8 h of extraction. With increase of the energy provided by PEF treatment, the quantity of extracted solutes increased accordingly, until a threshold value of $W \approx 9$ kJ/kg was reached. No further increase of the solute yield was observed above this threshold. Figure 22b compares extraction kinetics for different types of carrot samples (disk-like with diameter of 30 mm and thickness of 8.5 mm, coarse and fine slices). The energy provided by PEF treatment was the same ($W = 9$ kJ/kg). In these conditions, similar extraction kinetics were observed for both coarse and fine slices, and ≈2 h were needed to attain nearly stabilized solute yield ($Y \approx 93\%$). However, the extraction was considerably slower for the disk-like samples.

Effects of centrifugal force on the extraction from PEF-treated carrot tissue were also studied (El-Belghiti et al. 2005b). Figure 23 presents kinetics of aqueous solute extraction from the untreated and PEF-treated (E = 670 V/cm, t_p = 100μs, Δt = 10 ms, n = 300 pulses) carrot slices at different centrifugal accelerations, (G) (El-Belghiti et al. 2005b). In the absence of PEF treatment, a yield of around 58% was obtained

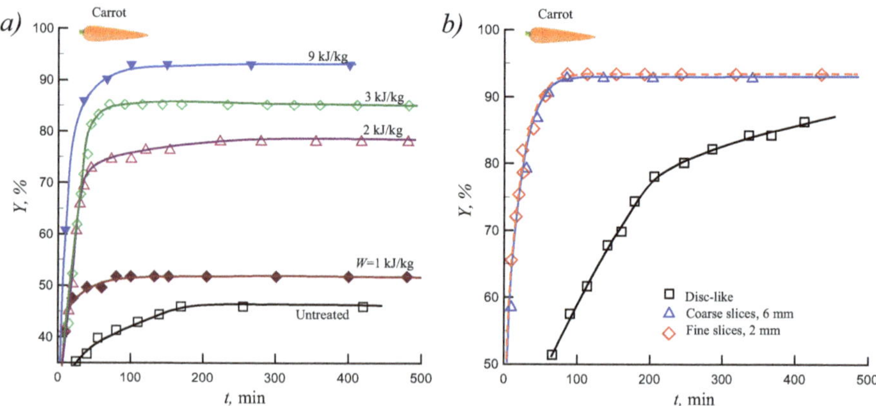

Fig. 22 Kinetics of aqueous solute extraction ($Y(t) = C/C_\infty$) from PEF-treated carrot slices. (a) Extraction from the coarse carrot slices treated by PEF (E = 300–800 V/cm) with different energies, (b) extraction from the coarse and fine carrot slices treated by PEF (E = 800 V/cm) with fixed energy (9 kJ/kg). Extraction temperature was T = 25 °C. (Adapted from (El-Belghiti and Vorobiev 2005))

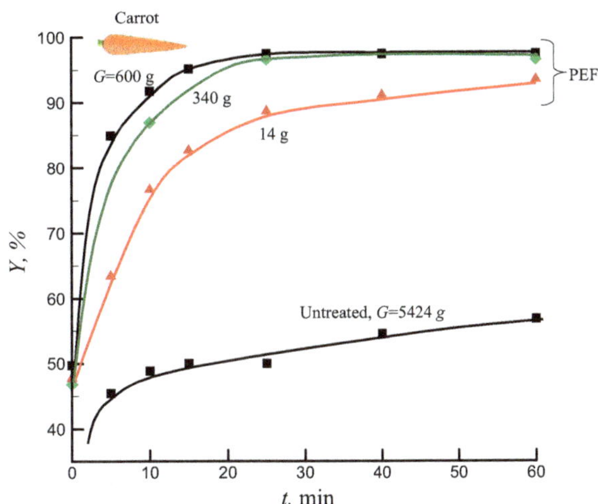

Fig. 23 Kinetics of aqueous solute extraction ($Y_{(t)} = C/C_{\infty}$), from the untreated and PEF-treated ($E = 670$ V/cm, $n = 300$, $t_p = 100$µs, $t_{PEF} = 0.03$ s, $\Delta t = 10$ ms) carrot slices at $T = 20$ °C and different centrifugal accelerations (G). (Adapted from (El-Belghiti et al. 2005b))

after 60 min of centrifugal extraction under the high centrifugal acceleration of 5434 g. This yield corresponded to the extracted solutes from cells, which were broken mechanically during slicing. After the PEF treatment, solute yield was significantly higher. Extraction yield increased up to around 92% after 60 min of centrifugal extraction even at low centrifugal acceleration ($G = 14$ g). For the centrifugal accelerations above $G \approx 150$ g, no further enhancement of solute concentration was observed. At this acceleration threshold, the solute concentration of around 97% was reached for 25 min of centrifugation. For the fixed centrifugal acceleration ($G = 150$ g), the final yield of solutes was approximately the same ($Y \approx 97\%$) at various extraction temperatures in the interval between 18 and 35 °C. However, this yield was attained for 15 min at 35 °C and for 40 min at 18 °C.

Red Beets

PEF-assisted extraction of pigments from red beets has been studied in several works (López et al. 2009; Fincan 2017). Water-soluble betalains (red-violet betacyanins and yellow betaxanthins) are the main red beet pigments (Jackman and Smith 1996; Rodriguez-Amaya 2019). The important fraction of pigments (75–95% mass) is related to betanine. These pigments have application in the food, medical, and pharmaceutical industries, owing to their visual and health-promoting functional properties (Tiwari and Cullen 2012). However, the temperature factor is very important for the extraction of these heat-sensitive pigments (Leong et al. 2018). A PEF treatment of 1 kV/cm allowed releasing ≈ 90% of the total red pigment during 1 h of aqueous extraction (Fincan et al. 2004). The kinetics of PEF-assisted extraction from red beet slices was described using a bimodal Fickian diffusion model (Chalermchat et al. 2004). PEF treatment (1–9 kV/cm) was applied to extract betanine from red beet roots at different temperatures (T = 10–60 °C) and pH values

(3.0–6.5) (López et al. 2009). At optimal operational conditions (E = 7 kV/cm, t_{PEF} = 10μs, T = 30 °C and pH 3.5), ≈90% of total betanine was extracted after 300 min. Effects of PEF treatment (E = 375–1500 V/cm) and temperature (T = 30–80 °C) on the kinetics of extraction and degradation of colorants of red beet have been studied (Loginova et al. 2011b). PEF treatment was found to be effective for the acceleration of colorant extraction and reducing operational time. Increase of the temperature caused acceleration of both extraction and colorant degradation processes. The colorant degradation was rather important at $T \geq 60$ ° C. PEF treatment allowed effective "cold" extraction with a high level of colorant extraction and a small level of colorant degradation. For example, for extraction at T = 30 °C, PEF treatment allowed reaching of the high yield ($Y \approx 0.95$) at the low level of degradation ($D \approx 0.10$).

The efficiencies of PEF-assisted extraction of colorant from red beets have been compared for different protocols (Luengo et al. 2016). For PEF protocols applied at E = 0.6 kV/cm, t_{PEF} = 40 ms, W = 43.2 kJ/kg and at E = 6 kV/cm, t_{PEF} = 150μs, W = 28.8 kJ/kg, the betanine extraction yield increased by 6.7 and 7.2 times, respectively, compared to the untreated samples.

Onions

In many studies, onions have been used as a model material to investigate effects of PEF on the cell membrane permeabilization (Fincan and Dejmek 2002; Ben Ammar et al. 2011b). For PEF treatments above a threshold level of E = 350 V/cm, damage of onion cells was observed (Fincan and Dejmek 2002). The permeabilization occurred along preferential paths connecting the electrodes. Effects of PEF protocols (electric field strength, pulse width, total pulse duration, and frequency) on the onion damage efficiency was studied in several works (Asavasanti et al. 2010; Ersus et al. 2010; Asavasanti et al. 2011a, 2012). Irreversible cell rupture was observed for PEF treatment of 333 V/cm (Ersus and Barrett 2010).

At constant total PEF treatment time, the degree of tissue disintegration was increased with shorter pulse widths and a larger number of pulses (Ersus et al. 2010). With application of ten pulses of 100μs, two critical values of electric field strength E were established: a lower value of E = 67 V/cm (for cell membrane breakdown) and a higher value of E = 200 V/cm (for tonoplast membrane breakdown) (Asavasanti et al. 2010). The effect of PEF frequency on the integrity of onion tissues was studied. It was established that lower frequencies (f < 1 Hz) can cause more important disintegration effect than higher frequencies (f = 1 to 5000 Hz) at the same value of E = 333 V/cm (Asavasanti et al. 2011a). This effect was explained by the influence of PEF frequency on the cytoplasmic streaming (Asavasanti et al. 2012).

Effects of PEF treatment (E = 0.3, 0.7 and 1.2 kV/cm) on the volatile compounds produced in onion cultivars were studied (Nandakumar et al. 2018). Significant increases in the concentrations of alkanes, aldehyde (2-methyl-2-pentenal), and the sulfur-containing compounds were observed. These changes were explained by

PEF-induced electroporation and facilitation of the enzyme-substrate reactions after the PEF treatment. PEF-assisted (2.5 kV/cm) aqueous extraction of phenolic (PC) and flavonoid (FC) compounds from the onion tissue was also investigated (Liu et al. 2018). Under the optimal treatment conditions, the content of PC and FC in extracts was increased by 2.2 and 2.7 times, respectively, in comparison to the control (untreated) sample. PEF treatment (0.3–1.2 kV/cm) was applied to bunching onion bulbs (Liu et al. 2019). The carbohydrate leakage and fructan leakage from onion bulbs were observed.

Tomatoes

PEF treatment (E = 3–7 kV/cm, t_p = 3μs, t_{PEF} = 0–300μs) was used to enhance the carotenoid extraction from pulp and peels of tomato (*Canario* variety) (Luengo et al. 2014a). Carotenoid extraction from tomato pulp was not significantly increased by the PEF treatment. However, the PEF treatment (E = 5 kV/cm, t_{PEF} = 90μs) of tomato peels was more successful and permitted improvement of the carotenoid extraction yield by 39% as compared to the control extraction in a mixture of hexane:ethanol:acetone (50:25:25).

Moreover, PEF treatment permitted reduced hexane proportion in the mixture and decreased extraction time without affecting extraction yield. Extracts obtained with PEF treatment had a higher antioxidant capacity than the control ones due to their higher carotenoid concentration. The extractability of carotenoids from tomato peels (*Pachino* variety) treated by PEF (E = 0.25–0.75 kV/cm, 1 kJ/kg) and steam blanched (1 min at mild temperatures 50–70 °C) was also studied (Pataro et al. 2018). Extraction was realized in acetone. PEF treatment alone increased the carotenoid extractability by 44%, 144%, and 189%, respectively, at E = 0.25, 0.50, and 0.75 kV/cm. Steam blanching alone also increased the carotenoid extractability (60–188%). Meanwhile, a synergetic effect of combined PEF and steam blanching was observed (Pataro et al. 2018).

PEF treatment (E = 4–24 kV/cm) applied for the assistance of lycopene extraction from tomato pulp (*Savoura* variety) permitted improvement of the extraction yield by 68.8% using optimal value of E = 16 kV/cm (Gachovska et al. 2013).

Citruses

The citrus peels resulted from industrial production of different juices (orange, lemon, grapefruit, pomelo, and lime) constitute about 50% of fresh fruit weight, and they represent the reach source of polyphenols, pigments, and essential oils (Putnik et al. 2017). Recovery of these components can be significantly improved with assistance of PEF. The effects of PEF treatment at E = 1, 3, 5, and 7 kV/cm on the extraction by pressing of total polyphenols and flavonoids (naringin and hesperidin) from orange peel have been investigated (Luengo et al. 2013). For treatment at E = 7 kV/cm after pressing for 30 min at P = 5 bar, the total polyphenol extraction yield increased by 159%, and the antioxidant activity of the extract increased by 192%. PEF treatment at $E = 10$ kV/cm has been applied for the assistance of polyphenol extraction from orange peels in ethanol-water solutions (0–50% for 1 h) (Kantar et al. 2018). With 50% of ethanol in solution, the extraction yields were 12 and 22 mg GAE/g DM, respectively, for the untreated and PEF-treated samples. PEF treatment at $E = 3$–9 kV/cm and 0–300μs on the extraction of polyphenol from lemon peel residues by pressing has been also investigated (Peiró et al. 2019). The effects of PEF were independent of lemon residue size. The treatment at $E = 7$ kV/cm increased the efficiency of polyphenol extraction by 300%.

Winemaking from Grapes

The red and white grape berries are rich in bioactive molecules such as sugars, carbohydrates, fruit acids (mainly tartaric and malic), mineral elements, and vitamins (Rousserie et al. 2019). During the last decades, different applications of PEF treatment have been tested for the assistance of maceration/extraction processes in red and white winemaking and for the recovery of bioactive molecules from pomace, grape seeds, and vine shoots. Different examples of PEF applications for the improvement of winemaking in batch and continuous flow modes were described (Praporscic et al. 2007; López et al. 2008a, b; Grimi 2009; Noelia et al. 2009; Grimi et al. 2009b; Donsi et al. 2010; Puértolas et al. 2010a; Sack et al. 2010; Puértolas et al. 2010b; Donsi et al. 2011; Delsart et al. 2012; El Darra et al. 2013a; Garde-Cerdán et al. 2013; López-Alfaro et al. 2013; El Darra et al. 2013b; Delsart et al. 2014; Luengo et al. 2014b; López-Giral et al. 2015; El Darra et al. 2016a, b; Saldaña et al. 2017; Vicaş et al. 2017; Comuzzo et al. 2018).

Mushrooms

Mushrooms are rich in antioxidant, anticancer, and anti-inflammatory components including polysaccharides, antioxidants, polyphenols, proteins, lipids, and vitamins (Muszyńska et al. 2018). They also contain bioactive molecules like β-glucans and triterpenoids that can act as immune-modulators (Rathore et al. 2017). However, the conventional solid/liquid extraction of these bioactive molecules involves organic solvents and requires elevated temperatures. This may result in the noticeable degradation of valuable cell compounds.

PEF-assisted solid/liquid extraction of polysaccharides from mushroom *Inonotus obliquus* (commonly known as chaga) was investigated (Yin et al. 2008). PEF treatment was applied at $E = 30$ kV/cm at the solid/liquid ratio of 1:25. The data were compared with alkali extraction, microwave-assisted extraction, and ultrasonic-assisted extraction. The PEF-assisted method gave the highest extraction yield and better purity of extracts. The PEF-assisted solid/liquid extraction of exopolysaccharides from Tibetan spiritual mushroom was also studied (Zhang et al. 2011). The

PEF treatment at $E = 40$ kV/cm under the optimal conditions allowed increasing the extraction yield by 84.3% compared to the control sample.

Later on, the efficiency of extraction and stability of extracts obtained from mushroom (*Agaricus bisporus*) by pressure extraction (PE) and by pressure extraction combined with PEF (PE + PEF) was studied (Parniakov et al. 2014). Extraction was conducted at room temperature at a constant pressure of 5 bar. In PE + PEF experiments, the PEF treatment was done in nonstationary conditions, because the thickness of mushroom cake decreased and the applied electric field strength increased during pressing (from 800 V/cm up to 1333 V/cm). The obtained data were compared with hot water extraction (WE) at elevated temperature $T = 70$ °C for 2 h and with ethanol extraction (EE) at $T = 25$ °C for 24 h.

Application of WE or EE methods resulted in high contents of proteins, total polyphenols, and polysaccharides in extracts. However, these extracts were cloudy and unstable. The extracts, produced by PE and PE + PEF methods, were clear and their colloid stability was high. Moreover, application of PE allows selective extraction of bioactive compounds with high purity, fresh-like proteins and polysaccharides. Both PE and PE + PEF methods allowed selective separation of different components. Application PE + PEF method gave highest nucleic acid/proteins ratio and allowed production of mushroom extracts with high contents of fresh-like proteins and polysaccharides. Supplementary EE from solid residues (cakes) allowed additional increase of yield of bioactive compounds.

For example, Fig. 24 compares concentration of total polyphenols obtained by PE, PE + PEF, aqueous (WE), and ethanol extraction (EE) methods. Supplementary ethanol extraction was also evaluated.

PEF-assisted solid/liquid extraction of polysaccharides from mushroom (*Agaricus bisporus*) was studied (Xue and Farid 2015). PEF treatment was applied

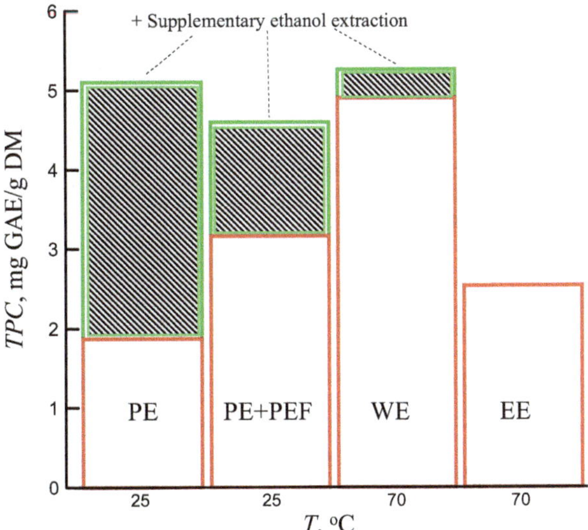

Fig. 24 Concentration of total polyphenols (TPC), in the mushroom extracts obtained by pressure extraction (PE), PEF-assisted pressure extraction (PE + PEF), aqueous extraction (WE), and ethanol extraction (EE). The effects of supplementary ethanol extraction ($T = 25$ °C, 24 h) from mushroom cakes are shown by dashed areas. (Adapted from (Parniakov et al. 2014))

to a mushroom suspension (9% w/w) in continuous mode in flowing chamber. PEF treatment ($E = 38.4$ kV/cm, $t_{PEF} = 272$µs, and a treatment temperature of $T = 85$ °C) permitted extract ≈98% polysaccharides, ≈51% total polyphenols, and ≈49% proteins, whereas the conventional extraction (at 95 °C for 1 h) resulted in recovery of only ≈56% polysaccharides, ≈25% total polyphenols, and ≈45% proteins.

PEF-assisted solid/liquid extraction of Morchella esculenta polysaccharide (MEP) from morel mushroom (*Morchella esculenta*) was studied (Liu et al. 2016). PEF treatment was applied at E = 15–25 kV/cm at the solid/liquid ratio of S/L = 1:20–1:40. For PEF-assisted procedure, the extracted polysaccharides were less degraded compared to those obtained by heat extraction.

A general review on the nonconventional methods of valuable compound recovery from mushrooms (including enzyme-assisted extraction, PEF, ultrasound, microwaves, subcritical and supercritical fluid extraction) was recently published (Roselló-Soto et al. 2016).

Conclusions

Pulsed electric fields (PEF) can be used as effective tool for the permeabilization of cell membrane by electroporation mechanism. Efficiency of electroporation depends critically on pulse protocols and properties of food materials. Under properly adapted PEF protocols, the efficient and highly selective extraction of useful molecules from food plants and residues is possible. Moreover, PEF treatment allows the "pure" extraction with preservation of quality, color, flavor, vitamins, and important nutrients of products. PEF-assisted extraction is in a full correspondence with the green extraction concept, i.e., it allows excluding the toxic organic solvents. Many promising examples of the PEF-assisted extraction of juices, sugars, proteins, polyphenolic substances, and oils from different foods were already demonstrated. However, more investigations at pilot and preindustrial scales are still required to evaluate the advantages of PEF treatment taking into account quality of products, economic efficiency, and environmental aspects.

References

Aguilera JM (2003) Solid-liquid extraction. In: Extraction optimization in food engineering. Markel Dekker Inc., New York, pp 51–70

Akiyama H, Heller R (eds) (2017) Bioelectrics. Springer, Japan

Albagnac G, Varoquaux P, Montigaud J-C (2002) Technologies de transformation des fruits. TecDoc/Lavoisier, Paris

Almohammed F, Mhemdi H, Vorobiev E (2016a) Pulsed electric field treatment of sugar beet tails as a sustainable feedstock for bioethanol production. Appl Energy 162:49–57. https://doi.org/10.1016/j.apenergy.2015.10.050

Almohammed F, Mhemdi H, Vorobiev E (2016b) Several-staged alkaline pressing-soaking of electroporated sugar beet slices for minimization of sucrose loss. Innov Food Sci Emerg Technol 36:18–25. https://doi.org/10.1016/j.ifset.2016.05.011

Almohammed F, Koubaa M, Khelfa A, Nakaya M, Mhemdi H, Vorobiev E (2017a) Pectin recovery from sugar beet pulp enhanced by high-voltage electrical discharges. Food Bioprod Process 103:95–103. https://doi.org/10.1016/j.fbp.2017.03.005

Almohammed F, Mhemdi H, Vorobiev E (2017b) Purification of juices obtained with innovative pulsed electric field and alkaline pressing of sugar beet tissue. Sep Purif Technol 173:156–164. https://doi.org/10.1016/j.seppur.2016.09.026

Amami E, Vorobiev E, Kechaou N (2005) Effect of pulsed electric field on the osmotic dehydration and mass transfer kinetics of apple tissue. Dry Technol 23:581–595. https://doi.org/10.1081/DRT-200054144

Amami E, Vorobiev E, Kechaou N (2006) Modelling of mass transfer during osmotic dehydration of apple tissue pre-treated by pulsed electric field. LWT – Food Sci Technol 39:1014–1021. https://doi.org/10.1016/j.lwt.2006.02.017

Angersbach A, Heinz V, Knorr D (2002) Evaluation of process-induced dimensional changes in the membrane structure of biological cells using impedance measurement. Biotechnol Prog 18:597–603. https://doi.org/10.1021/bp020047j

Apollonio F, Liberti M, Marracino P, Mir L (2012) Electroporation mechanism: review of molecular models based on computer simulation. In: 2012 6th European Conference on Antennas and Propagation (EUCAP). České Vysoké Učení Technické, Prague, pp 356–358

Asavasanti S, Ersus S, Ristenpart W, Stroeve P, Barrett DM (2010) Critical electric field strengths of onion tissues treated by pulsed electric fields. J Food Sci 75:E433–E443. https://doi.org/10.1111/j.1750-3841.2010.01768.x

Asavasanti S, Ristenpart W, Stroeve P, Barrett DM (2011a) Permeabilization of plant tissues by monopolar pulsed electric fields: effect of frequency. J Food Sci 76:E98–E111. https://doi.org/10.1111/j.1750-3841.2010.01940.x

Asavasanti S, Ristenpart W, Stroeve P, Barrett DM (2011b) Permeabilization of plant tissues by monopolar pulsed electric fields: effect of frequency. J Food Sci 76(1):E96–E111. https://doi.org/10.1111/j.1750-3841.2010.01940.x

Asavasanti S, Stroeve P, Barrett DM, Jernstedt JA, Ristenpart WD (2012) Enhanced electroporation in plant tissues via low frequency pulsed electric fields: Influence of cytoplasmic streaming. Biotechnol Prog 28:445–453. https://doi.org/10.1002/btpr.1507

Ashurst PR (2016) Chemistry and technology of soft drinks and fruit juices. Wiley, Chichester

Azarpazhooh E, Ramaswamy HS (2009) Evaluation of diffusion and Azuara models for mass transfer kinetics during microwave-osmotic dehydration of apples under continuous flow medium-spray conditions. Dry Technol 28:57–67. https://doi.org/10.1080/07373930903430694

Barba FJ, Parniakov O, Pereira SA, Wiktor A, Grimi N, Boussetta N, Saraiva JA, Raso J, Martin-Belloso O, Witrowa-Rajchert D, Lebovka N, Vorobiev E (2015) Current applications and new opportunities for the use of pulsed electric fields in food science and industry. Food Res Int 77:773–798. https://doi.org/10.1016/j.foodres.2015.09.015

Barbosa-Cánovas GV, Gongora-Nieto MM, Pothakamury UR, Swanson BG (1998) Preservation of foods with pulsed electric fields. Academic Press, San Diego

Bazhal IG, Gulyi IS (1983) Extraction of sugar from sugar-beet in a direct-current electric field. Food Technol Pishchevaya Tekhnologiya, Russ 5:49–51

Bazhal M, Vorobiev E (2000) Electrical treatment of apple cossettes for intensifying juice pressing. J Sci Food Agric 80:1668–1674. https://doi.org/10.1002/1097-0010(20000901)80:11<1668::AID-JSFA692>3.0.CO;2-7

Bazhal M, Lebovka N, Vorobiev E (2003) Optimisation of pulsed electric field strength for electroplasmolysis of vegetable tissues. Biosyst Eng 86:339–345. https://doi.org/10.1016/S1537-5110(03)00139-9

Ben Ammar J (2011) Etude de l'effet des champs electriques pulses sur la congelation des produits vegetaux. PhD Thesis, Compiegne: Universite de Technologie de Compiegne, France

Ben Ammar J, Lanoisellé J-L, Lebovka NI, Van Hecke E, Vorobiev E (2010) Effect of a pulsed electric field and osmotic treatment on freezing of potato tissue. Food Biophys 5:247–254. https://doi.org/10.1007/s11483-010-9167-y

Ben Ammar J, Lanoisellé J-L, Lebovka NI, Van Hecke E, Vorobiev E (2011a) Impact of a pulsed electric field on damage of plant tissues: effects of cell size and tissue electrical conductivity. J Food Sci 76:E90–E97. https://doi.org/10.1111/j.1750-3841.2010.01893.x

Ben Ammar J, Van Hecke E, Lebovka N, Vorobiev E, Lanoisellé J-L 2011b Freezing and freeze-drying of vegetables: benefits of a pulsed electric fields pre-treatment. In: CIGR Section VI International Symposium on "Towards a Sustainable Food Chain Food Process, Bioprocessing and Food Quality Management." Nantes, pp. 1–6

Bio-Rad Laboratories I (2019) Electroporation. http://www.bio-rad.com

Blahovec J, Kouřím P, Kindl M (2015) Low-temperature carrot cooking supported by pulsed electric field—DMA and DETA thermal analysis. Food Bioprocess Technol 8:2027–2035. https://doi.org/10.1007/s11947-015-1554-4

Bodénès P (2017) Etude de l'application de champs électriques pulsés sur des microalgues en vue de l'extraction de lipides neutres. PhD Thesis, L'Universite Paris-Saclay preparee à l'Ecole Normale Superieure de Cachan (Ecole Normale Superieure Paris-Saclay)

Bodénès P, Wang H-Y, Lee T-H, Chen H-Y, Wang C-Y (2019) Microfluidic techniques for enhancing biofuel and biorefinery industry based on microalgae. Biotechnol Biofuels 12:33. https://doi.org/10.1186/s13068-019-1369-z

Bouzrara H, Vorobiev E (2000) Beet juice extraction by pressing and pulsed electric fields. Int Sugar J 102:194–200

Bouzrara H, Vorobiev E (2001) Non-thermal pressing and washing of fresh sugarbeet cossettes combined with a pulsed electrical field. Zuckerindustrie 126:463–466

BTX HA (2019) Electroporation & Electrofusion Products. http://www.harvardapparatus.com

Buchmann L, Mathys A (2019) Perspective on pulsed electric field treatment in the bio-based industry. Front Bioeng Biotechnol 7:265

Buisman K (1936) Results of long duration settlement tests. In: Proceedings of the First International Conference on Soil Mechanics and Foundation Engineering, Cambridge, MA, pp 103–107

Canatella PJ, Karr JF, Petros JA, Prausnitz MR (2001) Quantitative study of electroporation-mediated molecular uptake and cell viability. Biophys J 80:755–764

Casciola M, Tarek M (2016) A molecular insight into the electro-transfer of small molecules through electropores driven by electric fields. Biochim Biophys Acta (BBA)-Biomembranes 1858:2278–2289

Čemažar J, Miklavčič D, Kotnik T (2013) Microfluidic devices for manipulation, modification and characterization of biological cells in electric fields–a review. Electron Components Mater 43:143–161

Čemažar J, Ghosh A, Davalos RV (2017) Electrical manipulation and sorting of cells. In: Microtechnology for cell manipulation and sorting. Springer, pp 57–92

Chalermchat Y, Dejmek P (2005) Effect of pulsed electric field pretreatment on solid--liquid expression from potato tissue. J Food Eng 71:164–169

Chalermchat Y, Fincan M, Dejmek P (2004) Pulsed electric field treatment for solid-liquid extraction of red beetroot pigment: mathematical modelling of mass transfer. J Food Eng 64:229–236. https://doi.org/10.1016/j.jfoodeng.2003.10.002

Chalermchat Y, Malangone L, Dejmek P (2010) Electropermeabilization of apple tissue: effect of cell size, cell size distribution and cell orientation. Biosyst Eng 105:357–366. https://doi.org/10.1016/j.biosystemseng.2009.12.006

Chan C-H, Yusoff R, Ngoh G-C (2014) Modeling and kinetics study of conventional and assisted batch solvent extraction. Chem Eng Res Des 92:1169–1186

Chang DC, Reese TS (1990) Changes in membrane structure induced by electroporation as revealed by rapid-freezing electron microscopy. Biophys J 58:1–12

Chang L, Li L, Shi J, Sheng Y, Lu W, Gallego-Perez D, Lee LJ (2016) Micro–/nanoscale electroporation. Lab Chip 16:4047–4062

Chanona-Pérez J, Quevedo R, Aparicio AJ, Chávez CG, Pérez JM, Dominguez GC, Alamilla-Beltrán L, Gutiérrez-López GF (2008) Image processing methods and fractal analysis for quantitative evaluation of size, shape, structure and microstructure in food materials. In: Gutiérrez-López GF, Welti-Chanes J, Parada-Arias E (eds) Food engineering: integrated approaches. Springer-Verlag, New York, pp 277–286

Chen C, Smye SW, Robinson MP, Evans JA (2006) Membrane electroporation theories: a review. Med Biol Eng Comput 44:5–14

Cheremisinoff NP (2017) Industrial liquid filtration equipment. In: Fibrous filter media. Woodhead Publishing, Duxford, pp 27–50

Cohn F, Mendelsohn B (1879) Ueber Einwirkung des electrischen Stromes auf die Vermehrung von Bacterien (about the effect of electric current on the growth of bacteria). Beitrage Biol der Pflauzen 3:141–162

Comuzzo P, Marconi M, Zanella G, Querze M (2018) Pulsed electric field processing of white grapes (cv. Garganega): Effects on wine composition and volatile compounds. Food Chem 264:16–23

Condello M, Caraglia M, Castellano M, Arancia G, Meschini S (2013) Structural and functional alterations of cellular components as revealed by electron microscopy. Microsc Res Tech 76:1057–1069

Couper JR, Penney WR, Fair JR, Walas SM (eds) (2012) Chemical process equipment-selection and design. Butterworth-Heinemann, Waltham

Crank J (1979) The mathematics of diffusion. Oxford University Press, London

Das MK, Mukherjee PP, Muralidhar K (2018) Modeling transport phenomena in porous media with applications. Springer Nature Switzerland AG

De Vito F, Ferrari G, Lebovka NI, Shynkaryk NV, Vorobiev E (2008) Pulse duration and efficiency of soft cellular tissue disintegration by pulsed electric fields. Food Bioprocess Technol 1:307–313

Delemotte L, Tarek M (2012) Molecular dynamics simulations of lipid membrane electroporation. J Membr Biol 245:531–543

Dellarosa N, Laghi L, Ragni L, Dalla Rosa M, Galante A, Ranieri B, Florio TM, Alecci M (2018) Pulsed electric fields processing of apple tissue: Spatial distribution of electroporation by means of magnetic resonance imaging and computer vision system. Innov Food Sci Emerg Technol 47:120–126

Delsart C, Ghidossi R, Poupot C, Cholet C, Grimi N, Vorobiev E, Milisic V, Peuchot MM (2012) Enhanced extraction of phenolic compounds from Merlot grapes by pulsed electric field treatment. Am J Enol Vitic 63:205–211

Delsart C, Cholet C, Ghidossi R, Grimi N, Gontier E, Gény L, Vorobiev E, Mietton-Peuchot M (2014) Effects of pulsed electric fields on cabernet sauvignon grape berries and on the characteristics of wines. Food Bioprocess Technol 7:424–436. https://doi.org/10.1007/s11947-012-1039-7

Dimitrov DS, Sowers AE (1990) Membrane electroporation - fast molecular exchange by electroosmosis. Biochim Biophys Acta (BBA) – Biomembranes 1022:381–392

Doevenspeck H (1961) Influencing cells and cell walls by electrostatic impulses. Fleischwirtschaft 13:968–987

Donsi F, Ferrari G, Fruilo M, Pataro G (2010) Pulsed electric field-assisted vinification of Aglianico and Piedirosso grapes. J Agric Food Chem 58:11606–11615

Donsi F, Ferrari G, Fruilo M, Pataro G (2011) Pulsed electric fields-assisted vinification. Procedia Food Sci 1:780–785

El Belghiti K, Vorobiev E (2004) Mass transfer of sugar from beets enhanced by pulsed electric field. Food Bioprod Process 82:226–230

El Darra N, Grimi N, Maroun RG, Louka N, Vorobiev E (2013a) Pulsed electric field, ultrasound, and thermal pretreatments for better phenolic extraction during red fermentation. Eur Food Res Technol 236:47–56. https://doi.org/10.1007/s00217-012-1858-9

El Darra N, Grimi N, Vorobiev E, Louka N, Maroun R (2013b) Extraction of polyphenols from red grape pomace assisted by pulsed ohmic heating. Food Bioprocess Technol 6:1281–1289. https://doi.org/10.1007/s11947-012-0869-7
El Darra N, Rajha HN, Ducasse M-A, Turk MF, Grimi N, Maroun RG, Louka N, Vorobiev E (2016a) Effect of pulsed electric field treatment during cold maceration and alcoholic fermentation on major red wine qualitative and quantitative parameters. Food Chem 213:352–360. https://doi.org/10.1016/j.foodchem.2016.06.073
El Darra N, Turk MF, Ducasse M-A, Grimi N, Maroun RG, Louka N, Vorobiev E (2016b) Changes in polyphenol profiles and color composition of freshly fermented model wine due to pulsed electric field, enzymes and thermovinification pretreatments. Food Chem 194:944–950. https://doi.org/10.1016/j.foodchem.2015.08.059
El-Belghiti K, Vorobiev E (2005) Modelling of solute aqueous extraction from carrots subjected to a pulsed electric field pre-treatment. Biosyst Eng 90:289–294
El-Belghiti K, Rabhi Z, Vorobiev E (2005a) Kinetic model of sugar diffusion from sugar beet tissue treated by pulsed electric field. J Sci Food Agric 85:213–218
El-Belghiti K, Rabhi Z, Vorobiev E (2005b) Effect of centrifugal force on the aqueous extraction of solute from sugar beet tissue pretreated by a pulsed electric field. J Food Process Eng 28:346–358
Elton DC (2018) Stretched exponential relaxation. ArXiv Prepr arXiv 1808(00881):1–9
Eppendorf VDG (2019) Eppendorf Eporator®, Operating manual. https://geneseesci.com/wp-content/uploads/2013/12/Eporator-Electroporator-User-Manual.pdf
Ersus S, Barrett DM (2010) Determination of membrane integrity in onion tissues treated by pulsed electric fields: use of microscopic images and ion leakage measurements. Innov Food Sci Emerg Technol 11:598–603. https://doi.org/10.1016/j.ifset.2010.08.001
Ersus S, Oztop MH, McCarthy MJ, Barrett DM (2010) Disintegration efficiency of pulsed electric field induced effects on onion (Allium cepa L.) tissues as a function of pulse protocol and determination of cell integrity by 1H-NMR relaxometry. J Food Sci 75:E444–E452. https://doi.org/10.1111/j.1750-3841.2010.01769.x
Eshtiaghi MN, Yoswathana N (2012) Laboratory scale extraction of sugar cane using high electric field pulses. World Acad Sci Eng Technol 65:1217–1222
Fazaeli M, Tahmasebi M, Djomeh EZ (2012) Characterization of food texture: application of microscopic technology. In: Méndez-Vilas A, Rigoglio NN, Mendes Silva MV, Guzman AM, Swindell WR (eds) Current Microscopy Contributions to Advances in Science and Technology. Formatex Research Center, Badajoz, pp 855–871
Fincan M (2017) Potential application of pulsed electric fields for improving extraction of plant pigments. In: Miklavcic D (ed) Handbook of electroporation. Springer, Cham, pp 2171–2192
Fincan M, Dejmek P (2002) In situ visualization of the effect of a pulsed electric field on plant tissue. J Food Eng 55:223–230. https://doi.org/10.1016/S0260-8774(02)00079-1
Fincan M, Dejmek P (2003) Effect of osmotic pretreatment and pulsed electric field on the visco-elastic properties of potato tissue. J Food Eng 59:169–175
Fincan M, DeVito F, Dejmek P (2004) Pulsed electric field treatment for solid-liquid extraction of red beetroot pigment. J Food Eng 64:381–388. https://doi.org/10.1016/j.jfoodeng.2003.11.006
Flaumenbaum B (1949) Electrical treatment of fruits and vegetables before extraction of juice. Proc Odessa Technol Inst Cann Ind (Trudy OTIKP, Odess Tehnol Instituta Konservn Promyshlennosti) 3:15–20 (in Russian)
Fox MB, Esveld DC, Valero A, Luttge R, Mastwijk HC, Bartels PV, van den Berg A, Boom RM (2006) Electroporation of cells in microfluidic devices: a review. Anal Bioanal Chem 385:474
Gachovska TK, Ngadi M, Chetti M, Raghavan GS V (2013) Enhancement of lycopene extraction from tomatoes using pulsed electric field. In: 2013 19th IEEE Pulsed Power Conference (PPC), pp 1–5
Garcia-Gonzalo D, Pagán R (2016) Detection of electroporation in microbial cells: techniques and procedures. In: Miklavcic D (ed) Handbook of electroporation. Springer, Cham, pp 1–15
Garde-Cerdán T, González-Arenzana L, López N, López R, Santamaría P, López-Alfaro I (2013) Effect of different pulsed electric field treatments on the volatile composition of Graciano,

Tempranillo and Grenache grape varieties. Innov Food Sci Emerg Technol 20:91–99. https://doi.org/10.1016/j.ifset.2013.08.008
Gekas V (1992) Transport phenomena of foods and biological materials. CRC Press, Taylor & Francis Group, Boca Raton
Gowrishankar TR, Weaver JC (2003) An approach to electrical modeling of single and multiple cells. Proc Natl Acad Sci 100:3203–3208
Grimi N (2009) Vers l'intensification du pressage industriel des agroressources par champs électriques pulsés: étude multi-échelles. PhD thesis, Compiegne: Universite de Technologie de Compiegne, France
Grimi N, Lebovka N, Vorobiev E, Vaxelaire J (2009a) Compressing behavior and texture evaluation for potatoes pretreated by pulsed electric field. J Texture Stud 40:208–224
Grimi N, Lebovka NI, Vorobiev E, Vaxelaire J (2009b) Effect of a pulsed electric field treatment on expression behavior and juice quality of Chardonnay grape. Food Biophys 4:191–198. https://doi.org/10.1007/s11483-009-9117-8
Grimi N, Mamouni F, Lebovka N, Vorobiev E, Vaxelaire J (2010a) Acoustic impulse response in apple tissues treated by pulsed electric field. Biosyst Eng 105:266–272. https://doi.org/10.1016/j.biosystemseng.2009.11.005
Grimi N, Vorobiev E, Lebovka N, Vaxelaire J (2010b) Solid-liquid expression from denaturated plant tissue: filtration–consolidation behaviour. J Food Eng 96:29–36
Grimi N, Mamouni F, Lebovka N, Vorobiev E, Vaxelaire J (2011) Impact of apple processing modes on extracted juice quality: pressing assisted by pulsed electric fields. J Food Eng 103:52–61
Grosse C, Schwan HP (1992) Cellular membrane potentials induced by alternating fields. Biophys J 63:1632–1642
Gulyi IS, Lebovka NI, Mank V V, Kupchik MP, Bazhal MI, Matvienko AB, Papchenko AY (1994) Scientific and practical principles of electrical treatment of food products and materials. UkrINTEI (in Russian), Kiev
Hamilton WA, Sale AJH (1967) Effects of high electric fields on microorganisms: II. Mechanism of action of the lethal effect. Biochim Biophys Acta (BBA)-General Subj 148:789–800
Heilbron JL (1979) Electricity in the 17th and 18th centuries: A study of early modern physics. Univ of California Press
Heimburg T (1998) Mechanical aspects of membrane thermodynamics. Estimation of the mechanical properties of lipid membranes close to the chain melting transition from calorimetry. Biochim Biophys Acta (BBA) – Biomembranes 1415:147–162
Henslee BE, Morss A, Hu X, Lafyatis GP, Lee LJ (2014) Cell-cell proximity effects in multi-cell electroporation. Biomicrofluidics 8:52002
Hjouj M, Rubinsky B (2010) Magnetic resonance imaging characteristics of nonthermal irreversible electroporation in vegetable tissue. J Membr Biol 236:137–146
Holl MMB (2008) Cell plasma membranes and phase transitions. In: Pollack GH, Chin W-C (eds) Phase transitions in cell biology. Springer, pp 171–181
Hülsheger H, Niemann E-G (1980) Lethal effects of high-voltage pulses on E. coli K12 Radiat Environ Biophys 18:281–288
Hülsheger H, Potel J, Niemann E-G (1981) Killing of bacteria with electric pulses of high field strength. Radiat Environ Biophys 20:53–65
Hülsheger H, Potel J, Niemann E-G (1983) Electric field effects on bacteria and yeast cells. Radiat Environ Biophys 22:149–162
Ingram M, Page LJ (1953) The survival of microbes in modulated high-frequency voltage fields. In: Proceedings of the Society for Applied Bacteriology, pp 69–87
Jackman RL, Smith JL (1996) Anthocyanins and betalains. In: Hendry GAF, Houghton JD (eds) Natural food colorants. Springer, Dordrecht, pp 244–309
Jaeger H, Schulz A, Karapetkov N, Knorr D (2009) Protective effect of milk constituents and sublethal injuries limiting process effectiveness during PEF inactivation of Lb. rhamnosus. Int J Food Microbiol 134:154–161

Jemai A (1997) Contribution à l'étude de l'effet d'un traitement électrique sur les cossettes de betterave à sucre: incidence sur le procédé d'extraction. Thèse de Doctorat, Université de Technologie de Compiègne, France

Jemai AB, Vorobiev E (2002) Effect of moderate electric field pulses on the diffusion coefficient of soluble substances from apple slices. Int J Food Sci Technol 37:73–86

Jemai AB, Vorobiev E (2003) Enhanced leaching from sugar beet cossettes by pulsed electric field. J Food Eng 59:405–412. https://doi.org/10.1016/S0260-8774(02)00499-5

Jönsson KA-S, Jönsson BTL (1992) Fluid flow in compressible porous media: I: Steady-state conditions. AICHE J 38:1340–1348. https://doi.org/10.1002/aic.690380904

Kantar S El, Boussetta N, Lebovka N, Foucart F, Rajha HN, Maroun RG, Louka N, Vorobiev E (2018) Pulsed electric field treatment of citrus fruits: improvement of juice and polyphenols extraction. Innov Food Sci Emerg Technol 46:153–161. https://doi.org/10.1016/j.ifset.2017.09.024

Karathanos VT, Kanellopoulos NK, Belessiotis VG (1996) Development of porous structure during air drying of agricultural plant products. J Food Eng 29:167–183

Kinosita K Jr, Tsong TY (1977a) Voltage-induced pore formation and hemolysis of human erythrocytes. Biochim Biophys Acta (BBA) – Biomembranes 471:227–242

Kinosita K Jr, Tsong TY (1977b) Formation and resealing of pores of controlled sizes in human erythrocyte membrane. Nature 268:438

Kinosita K, Tsong TT (1977) Hemolysis of human erythrocytes by transient electric field. Proc Natl Acad Sci 74:1923–1927

Knorr D, Geulen M, Grahl T, Sitzmann W (1994a) Energy cost of high electric field pulse treatment: reply. Trends Food Sci Technol 5:265

Knorr D, Geulen M, Grahl T, Sitzmann W (1994b) Food application of high electric field pulses. Trends Food Sci Technol 5:71–75

Kogan FY (1968) Electrophysical methods in canning technologies of foodstuff. Tekhnica (in Russian), Kiev

Kotnik T, Miklavčič D (2000) Analytical description of transmembrane voltage induced by electric fields on spheroidal cells. Biophys J 79:670–679

Kotnik T, Pucihar G (2010) Induced transmembrane voltage-theory, modeling, and experiments. In: Pakhomov AG, Miklavčič D, Markov MS (eds) Advanced electroporation techniques in biology and medicine. CRC Press, Taylor & Francis Group, Boca Raton, pp 51–70

Kotnik T, Bobanovic F, Miklavčič D (1997) Sensitivity of transmembrane voltage induced by applied electric fields: a theoretical analysis. Bioelectrochem Bioenerg 43:285–291

Kotnik T, Miklavčič D, Slivnik T (1998) Time course of transmembrane voltage induced by time-varying electric fields: a method for theoretical analysis and its application. Bioelectrochem Bioenerg 45:3–16

Kranjc M, Miklavčič D (2017) Electric field distribution and electroporation threshold. In: Miklavčič D (ed) Handbook of electroporation. Springer, Cham, pp 1043–1058

Lanoisellé J-L, Vorobyov EI, Bouvier J-M, Pair G (1996) Modeling of solid/liquid expression for cellular materials. AICHE J 42:2057–2068

Lazarenko BR, Fursov SSP, Bordijan JA, Chebanu VG (1977) Electroplasmolysis. Cartea Moldoveneascu, Chisinau

Lebedeva NE (1987) Electric breakdown of bilayer lipid membranes at short times of voltage action. Biol Membr (Biochem Moscow Ser A Membr Cell Biol) 4:994–998 (in Russian)

Lebovka NI, Vorobiev EI (2004) On the origin of the deviation from the first-order kinetics in inactivation of microbial cells by pulsed electric fields. Int J Food Microbiol 91:83–89

Lebovka N, Vorobiev E (2007) The kinetics of inactivation of spheroidal microbial cells by pulsed electric fields. ArXiv Prepr arXiv 0704(2750):1–18

Lebovka N, Vorobiev E (2011) Food and biomaterials processing assisted by electroporation. In: G PA, Miklavcic D, Markov MS (eds) Advanced electroporation techniques in biology and medicine. CRC Press, Taylor & Francis Group, Boca Raton, pp 463–490

Lebovka N, Vorobiev E (2014) Techniques and procedures to detect electroporation in food cellular tissues. In: Raso J, Álvarez I (eds) Proceedings of the COST TD1104 school on applications of pulsed electric fields for food processing. Servicio de publicaciones, Zaragoza, pp 39–50

Lebovka N, Vorobiev E (2016) Techniques to detect electroporation in food tissues. In: Miklavcic D (ed) Handbook of electroporation. Springer, Cham, pp 1–23

Lebovka NI, Melnyk RM, Kupchik MP, Bazhal MI, Serebrjakov RA (2000) Local generation of ohmic heat on cellular membranes during the electrical treatment of biological tissues. Sci Pap Naukma 18:51–56

Lebovka NI, Bazhal MI, Vorobiev E (2001) Pulsed electric field breakage of cellular tissues: visualisation of percolative properties. Innov Food Sci Emerg Technol 2:113–125

Lebovka NI, Bazhal MI, Vorobiev E (2002) Estimation of characteristic damage time of food materials in pulsed-electric fields. J Food Eng 54:337–346

Lebovka NI, Praporscic I, Vorobiev E (2003) Enhanced expression of juice from soft vegetable tissues by pulsed electric fields: consolidation stages analysis. J Food Eng 59:309–317

Lebovka N, Praporscic I, Vorobiev E (2004a) Effect of moderate thermal and pulsed electric field treatments on textural properties of carrots, potatoes and apples. Innov Food Sci Emerg Technol 5:9–16

Lebovka N, Praporscic I, Vorobiev E (2004b) Combined treatment of apples by pulsed electric fields and by heating at moderate temperature. J Food Eng 65:211–217

Lebovka NI, Praporscic I, Ghnimi S, Vorobiev E (2005a) Temperature enhanced electroporation under the pulsed electric field treatment of food tissue. J Food Eng 69:177–184

Lebovka NI, Praporscic I, Ghnimi S, Vorobiev E (2005b) Does electroporation occur during the ohmic heating of food? J Food Sci 70:E308–E311

Lebovka N, Shynkaryk N, El-Belghiti K, Benjelloun H, Vorobiev E (2007a) Plasmolysis of sugarbeet: pulsed electric fields and thermal treatment. J Food Eng 80:639–644

Lebovka NI, Shynkaryk M, Vorobiev E (2007b) Moderate electric field treatment of sugarbeet tissues. Biosyst Eng 96:47–56

Lee EW, Wong D, Prikhodko SV, Perez A, Tran C, Loh CT, Kee ST (2012) Electron microscopic demonstration and evaluation of irreversible electroporation-induced nanopores on hepatocyte membranes. J Vasc Interv Radiol 23:107–113

Leong HY, Show PL, Lim MH, Ooi CW, Ling TC (2018) Natural red pigments from plants and their health benefits: a review. Food Rev Int 34:463–482

Liu C, Sun Y, Mao Q, Guo X, Li P, Liu Y, Xu N (2016) Characteristics and antitumor activity of Morchella esculenta polysaccharide extracted by pulsed electric field. Int J Mol Sci 17:986

Liu Z-W, Zeng X-A, Ngadi M (2018) Enhanced extraction of phenolic compounds from onion by pulsed electric field (PEF). J Food Process Preserv 42:e13755

Liu T, Burritt DJ, Oey I (2019) Understanding the effect of Pulsed Electric Fields on multilayered solid plant foods: Bunching onions (Allium fistulosum) as a model system. Food Res Int 120:560–567. https://doi.org/10.1016/j.foodres.2018.11.006

Liu C, Grimi N, Bals O, Lebovka N, Vorobiev E (2021) Effects of pulsed electric fields and preliminary vacuum drying on freezing assisted processes in potato tissue. Food Bioprod Process 125:126–133. https://doi.org/10.1016/j.fbp.2020.11.002

Loginov M, Loginova K, Lebovka N, Vorobiev E (2011) Comparison of dead-end ultrafiltration behaviour and filtrate quality of sugar beet juices obtained by conventional and "cold" PEF-assisted diffusion. J Memb Sci 377:273–283

Loginova K (2011) Mise en oeuvre de champs électriques pulsés pour la conception d'un procédé de diffusion à froid à partir de betteraves à sucre et d'autres tubercules alimentaires (étude multi-échelle). Universite de Technologie de Compiegne, France, Compiegne

Loginova KV, Shynkaryk MV, Lebovka NI, Vorobiev E (2010) Acceleration of soluble matter extraction from chicory with pulsed electric fields. J Food Eng 96:374–379

Loginova K, Loginov M, Vorobiev E, Lebovka NI (2011a) Quality and filtration characteristics of sugar beet juice obtained by "cold" extraction assisted by pulsed electric field. J Food Eng 106:144–151

Loginova KV, Lebovka NI, Vorobiev E (2011b) Pulsed electric field assisted aqueous extraction of colorants from red beet. J Food Eng 106:127–133. https://doi.org/10.1016/j.jfoodeng.2011.04.019

Loginova KV, Vorobiev E, Bals O, Lebovka NI (2011c) Pilot study of countercurrent cold and mild heat extraction of sugar from sugar beets, assisted by pulsed electric fields. J Food Eng 102:340–347

Loginova K, Loginov M, Vorobiev E, Lebovka NI (2012) Better lime purification of sugar beet juice obtained by low temperature aqueous extraction assisted by pulsed electric field. LWT – Food Sci Technol 46:371–374

López N, Puértolas E, Condón S, Álvarez I, Raso J (2008a) Effects of pulsed electric fields on the extraction of phenolic compounds during the fermentation of must of Tempranillo grapes. Innov Food Sci Emerg Technol 9:477–482

López N, Puértolas E, Condón S, Álvarez I, Raso J (2008b) Application of pulsed electric fields for improving the maceration process during vinification of red wine: influence of grape variety. Eur Food Res Technol 227:1099

López N, Puértolas E, Condón S, Raso J, Alvarez I (2009) Enhancement of the extraction of betanine from red beetroot by pulsed electric fields. J Food Eng 90:60–66. https://doi.org/10.1016/j.jfoodeng.2008.06.002

López-Alfaro I, González-Arenzana L, López N, Santamaria P, Lopez R, Garde-Cerdan T (2013) Pulsed electric field treatment enhanced stilbene content in Graciano, Tempranillo and Grenache grape varieties. Food Chem 141:3759–3765. https://doi.org/10.1016/j.foodchem.2013.06.082

López-Giral N, González-Arenzana L, González-Ferrero C, López R, Santamaría P, López-Alfaro I, Garde-Cerdán T (2015) Pulsed electric field treatment to improve the phenolic compound extraction from Graciano, Tempranillo and Grenache grape varieties during two vintages. Innov Food Sci Emerg Technol 28:31–39. https://doi.org/10.1016/j.ifset.2015.01.003

Lu R, Abbott JA (2004) Force/deformation techniques for measuring texture. In: Kilcast D (ed) Texture in food: Solid foods. Woodhead Publishing, Sawston, Cambridge, pp 109–145

Luengo E, Álvarez I, Raso J (2013) Improving the pressing extraction of polyphenols of orange peel by pulsed electric fields. Innov Food Sci Emerg Technol 17:79–84

Luengo E, Álvarez I, Raso J (2014a) Improving carotenoid extraction from tomato waste by pulsed electric fields. Front Nutr 1:12. https://doi.org/10.3389/fnut.2014.00012

Luengo E, Franco E, Ballesteros F, Álvarez I, Raso J (2014b) Winery trial on application of pulsed electric fields for improving vinification of Garnacha grapes. Food Bioprocess Technol 7:1457–1464. https://doi.org/10.1007/s11947-013-1209-2

Luengo E, Martinez JM, Álvarez I, Raso J (2016) Comparison of the efficacy of pulsed electric fields treatments in the millisecond and microsecond range for the extraction of betanine from red beetroot. In: Jarm T, Kramar P (eds) International Federation for Medical and Biological Engineering (IFMBE) Proceedings. Springer, pp 375–378

Mahnič-Kalamiza S (2016) Dual-porosity model of liquid extraction by pressing from plant tissue modified by electroporation. In: Handbook of electroporation. Springer International Publishing AG, Cham, pp 1–25

Mahnič-Kalamiza S, Vorobiev E (2014) Dual-porosity model of liquid extraction by pressing from biological tissue modified by electroporation. J Food Eng 137:76–87

Mahnič-Kalamiza S, Miklavčič D, Vorobiev E (2015) Dual-porosity model of mass transport in electroporated biological tissue: simulations and experimental work for model validation. Innov Food Sci Emerg Technol 29:41–54

Matov BI, Reshetko E V (1968) Electrophysical methods in food industry. Kartja Moldavenjaske, Kishinev (in Russian)

Mavroudis NE, Gekas V, Sjöholm I (1998) Osmotic dehydration of apples. Shrinkage phenomena and the significance of initial structure on mass transfer rates. J Food Eng 38:101–123

Mercadal B, Vernier PT, Ivorra A (2016) Dependence of electroporation detection threshold on cell radius: an explanation to observations non compatible with Schwan's equation model. J Membr Biol 249:663–676

Mhemdi H, Bals O, Grimi N, Vorobiev E (2014) Alternative pressing/ultrafiltration process for sugar beet valorization: impact of pulsed electric field and cossettes preheating on the qualitative characteristics of juices. Food Bioprocess Technol 7:795–805

Mhemdi H, Almohammed F, Bals O, Grimi N, Vorobiev E (2015) Impact of pulsed electric field and preheating on the lime purification of raw sugar beet expressed juices. Food Bioprod Process 95:323–331

Mhemdi H, Bals O, Vorobiev E (2016) Combined pressing-diffusion technology for sugar beets pretreated by pulsed electric field. J Food Eng 168:166–172

Miklavčič D (ed) (2017) Handbook of electroporation. Springer, Cham

Monteith H, Parker W (2016) Emissions from wastewater treatment plants. In: Tata P, Witherspoon J, Lue-Hing C (eds) VOC emissions from biosolid's dewatering processes. CRC Press, Taylor & Francis Group, Boca Raton, pp 171–188

Moses BD (1938) Electric pasteurization of milk. Agric Eng 19:525–526

Mushtaq M (2018) Extraction of fruit juice: An overview. In: Rajauria G, Tiwari BK (eds) Fruit juices. Elsevier, pp 131–159

Muszyńska B, Grzywacz-Kisielewska A, Kała K, Gdula-Argasińska J (2018) Anti-inflammatory properties of edible mushrooms: a review. Food Chem 243:373–381

Nandakumar R, Eyres GT, Burritt DJ, Kebede B, Leus M, Oey I (2018) Impact of pulsed electric fields on the volatile compounds produced in whole onions (*Allium cepa* and *Allium fistulosum*). Foods 7:1–15. https://doi.org/10.3390/foods7110183

Napotnik TB, Miklavčič D (2018) In vitro electroporation detection methods–an overview. Bioelectrochemistry 120:166–182

Nelson WM (2003) Green solvents for chemistry: perspectives and practice. Oxford University Press

Neu JC, Krassowska W (1999) Asymptotic model of electroporation. Phys Rev E 59:3471

Neumann E, Rosenheck K (1972) Permeability changes induced by electric impulses in vesicular membranes. J Membr Biol 10:279–290

Noelia V, Puértolas E, Hernández-Orte P, Álvarez I, Raso J (2009) Effect of a pulsed electric field treatment on the anthocyanins composition and other quality parameters of Cabernet Sauvignon freshly fermented model wines obtained after different maceration times. LWT – Food Sci Technol 42:1225–1231

Nyrop JE (1946) A specific effect of high-frequency electric currents on biological objects. Nature 157:51. https://doi.org/10.1038/157051a0

Olson RE (1986) State of the art: consolidation testing. In: Yong RN, Townsend FC (eds) Consolidation of Soils: Testing and Evaluation. ASTM Special technical publication 892. American Society for Testing and Materials, Philadelphia, pp 7–70

Pakhomov AG, Bowman AM, Ibey BL, Andre FM, Pakhomova ON, Schoenbach KH (2009) Lipid nanopores can form a stable, ion channel-like conduction pathway in cell membrane. Biochem Biophys Res Commun 385:181–186

Pakhomov AG, Miklavčič D, Markov MS (2010) Advanced electroporation techniques in biology and medicine. CRC Press, Boca Raton

Parniakov O, Lebovka NI, Van Hecke E, Vorobiev E (2014) Pulsed electric field assisted pressure extraction and solvent extraction from mushroom (Agaricus bisporus). Food Bioprocess Technol 7:174–183

Pataro G, Carullo D, Siddique MAB, Falcone M, Donsì F, Ferrari G (2018) Improved extractability of carotenoids from tomato peels as side benefits of PEF treatment of tomato fruit for more energy-efficient steam-assisted peeling. J Food Eng 233:65–73. https://doi.org/10.1016/j.jfoodeng.2018.03.029

Pavlin M, Pavselj N, Miklavčič D (2002) Dependence of induced transmembrane potential on cell density, arrangement, and cell position inside a cell system. Biomed Eng IEEE Trans 49:605–612

Peiró S, Luengo E, Segovia F, Raso J, Almajano MP (2019) Improving polyphenol extraction from lemon residues by pulsed electric fields. Waste and Biomass Valorization 10:889–897

Pillet F, Formosa-Dague C, Baaziz H, Dague E, Rols M-P (2016) Cell wall as a target for bacteria inactivation by pulsed electric fields. Sci Rep 6:19778

Van der Poel PW, Schiweck H, Schwartz T (1998) Sugar technology. Beet and Cane Sugar Manufacture. Bartens KG, Berlin

Pourzaki A, Mirzaee H (2008) Pulsed electric field generators in food processing. In: 18-th National Congress on Food Technology in Mashhad (Iran), pp 1–7

Praporscic I, Muravetchi V, Vorobiev E (2004) Constant rate expressing of juice from biological tissue enhanced by pulsed electric field. Dry Technol 22:2395–2408

Praporscic I, Ghnimi S, Vorobiev E (2005) Enhancement of pressing of sugar beet cuts by combined ohmic heating and pulsed electric field treatment. J Food Process Preserv 29:378–389

Praporscic I, Lebovka N, Vorobiev E, Mietton-Peuchot M (2007) Pulsed electric field enhanced expression and juice quality of white grapes. Sep Purif Technol 52:520–526. https://doi.org/10.1016/j.seppur.2006.06.007

Pucihar G, Kotnik T, Teissié J, Miklavčič D (2007) Electropermeabilization of dense cell suspensions. Eur Biophys J 36:173–185

Puértolas E, Hernández-Orte P, Sladaña G, Álvarez I, Raso J (2010a) Improvement of winemaking process using pulsed electric fields at pilot-plant scale. Evolution of chromatic parameters and phenolic content of Cabernet Sauvignon red wines. Food Res Int 43:761–766. https://doi.org/10.1016/j.foodres.2009.11.005

Puértolas E, López N, Sladaña G, Álvarez I, Raso J (2010b) Evaluation of phenolic extraction during fermentation of red grapes treated by a continuous pulsed electric fields process at pilot-plant scale. J Food Eng 98:120–125. https://doi.org/10.1016/j.jfoodeng.2009.12.017

Puri M (2017) Food bioactives: extraction and biotechnology applications. Springer, Cham

Putnik P, Bursać Kovačević D, Režek Jambrak A, Barba FJ, Cravotto G, Binello A, Lorenzo JM, Shpigelman A (2017) Innovative "green" and novel strategies for the extraction of bioactive added value compounds from citrus wastes: a review. Molecules 22(5):680. https://doi.org/10.3390/molecules22050680

Qin B-L, Zhang Q, Barbosa-Cánovas GV, Swanson BG, Pedrow PD (1994) Inactivation of microorganisms by pulsed electric fields of different voltage waveforms. IEEE Trans Dielectr Electr Insul 1:1047–1057. https://doi.org/10.1109/94.368658

Qin Y, Lai S, Jiang Y, Yang T, Wang J (2005) Transmembrane voltage induced on a cell membrane in suspensions exposed to an alternating field: a theoretical analysis. Bioelectrochemistry 67:57–65

Qingwen Q, Xiaoqing H, Chunmei L, Dejun Z, Ya L (2018) Study on modular design and key technology of screw pressing for sludge treatment system. J Eng Manuf Technol 6(1):1–7

Ramos A, Suzuki DOH, Marques JLB (2006) Numerical study of the electrical conductivity and polarization in a suspension of spherical cells. Bioelectrochemistry 68:213–217

Raso J, Frey W, Ferrari G, Pataro G, Knorr D, Teissie J, Miklavčič D (2016) Recommendations guidelines on the key information to be reported in studies of application of PEF technology in food and biotechnological processes. Innov Food Sci Emerg Technol 37:312–321

Rathore H, Prasad S, Sharma S (2017) Mushroom nutraceuticals for improved nutrition and better human health: a review. PharmaNutrition 5:35–46

Reberšek M, Miklavčič D (2011) Advantages and disadvantages of different concepts of electroporation pulse generation. Automatika 52:12–19

Reberšek M, Miklavčič D, Bertacchini C, Sack M (2014) Cell membrane electroporation-Part 3: the equipment. IEEE Electr Insul Mag 30:8–18

Rems L, Miklavčič D (2016) Tutorial: Electroporation of cells in complex materials and tissue. J Appl Phys 119:201101

Rems L, Viano M, Kasimova MA, Miklavčič D, Tarek M (2019) The contribution of lipid peroxidation to membrane permeability in electropermeabilization: a molecular dynamics study. Bioelectrochemistry 125:46–57

Riemann F, Zimmermann U, Pilwat G (1975) Release and uptake of haemoglobin and ions in red blood cells induced by dielectric breakdown. Biochim Biophys Acta (BBA) – Biomembranes 394:449–462

Rockenbach A, Sudarsan S, Berens J, Kosubek M, Lazar J, Demling P, Hanke R, Mennicken P, Ebert BE, Blank LM, Others (2019) Microfluidic irreversible electroporation - a versatile tool to extract intracellular contents of bacteria and yeast. Meta 9:211
Rodriguez-Amaya DB (2019) Update on natural food pigments-A mini-review on carotenoids, anthocyanins, and betalains. Food Res Int 124:200–205
Rogov IA (1988) Electrophysical methods of foods product processing. Agropromizdat, Moscow (in Russian)
Rols M-P (2006) Electropermeabilization, a physical method for the delivery of therapeutic molecules into cells. Biochim Biophys Acta (BBA) – Biomembranes 1758:423–428
Rols M-P, Teissié J (1990) Electropermeabilization of mammalian cells. Quantitative analysis of the phenomenon. Biophys J 58:1089–1098. https://doi.org/10.1016/S0006-3495(90)82451-6
Rondeau C, Le Quéré J-M, Turk M, Mariette F, Vorobiev E, Baron A (2012) The de-structuration of parenchyma cells of apple induced by pulsed electric fields: a TD-NMR investigation. In: Proceeding of the International Conference Bio & Food Electrotechnologies (BFE 2012), Book of Full Papers. ProdAl Scarl, Società consortile a responsabilità limitata, c/o University of Salerno, Fisciano, Italy, p 5
Roselló-Soto E, Parniakov O, Deng Q, Patras A, Koubaa M, Grimi N, Boussetta N, Tiwari BK, Vorobiev E, Lebovka N, Others (2016) Application of non-conventional extraction methods: Toward a sustainable and green production of valuable compounds from mushrooms. Food Eng Rev 8:214–234
Rousserie P, Rabot A, Geny-Denis L (2019) From flavanols biosynthesis to wine tannins: what place for grape seeds? J Agric Food Chem 67:1325–1343
Rumble JR (ed) (2019) CRC handbook of chemistry and physics. CRC Press, Taylor & Fransis Group, Boca Raton
Russ JC, Neal FB (2016) The image processing handbook. CRC press, Boca Raton
Sack M, Sigler J, Frenzel S, Eing C, Arnold J, Michelberger T, Frey W, Attmann F, Stukenbrock L, Müller G (2010) Research on industrial-scale electroporation devices fostering the extraction of substances from biological tissue. Food Eng Rev 2:147–156
Saldaña G, Luengo E, Puértolas E, Álvarez I, Raso J (2017) Pulsed electric fields in wineries: Potential applications. In: Miklavcic D (ed) Handbook of electroporation. Springer, Cham, pp 2825–2842
Sale A, Hamilton W (1967) Effect of high electric fields on microorganisms. I. Killing of bacteria and yeast. Biochim Biophys Acta 148:781–788
Sale AJH, Hamilton WA (1968) Effects of high electric fields on micro-organisms: III. Lysis of erythrocytes and protoplasts. Biochim Biophys Acta (BBA) – Biomembranes 163:37–43
Saravacos GD, Krokida M (2014) Mass transfer properties of foods. In: Engineering properties of foods. CRC Press, Boca Raton, pp 349–402
Saulis G (1997) Pore disappearance in a cell after electroporation: theoretical simulation and comparison with experiments. Biophys J 73:1299–1309
Schwartzberg HG (1997) Expression of fluid from biological solids. Sep Purif Methods 26:1–213
Schwartzberg HG, Chao RY (1982) Solute diffusivities in leaching processes. Food Technol 36:73–86
Sheikh SM, Kazi ZS (2016) Technologies for oil extraction: a review. Int J Environ Agric Biotechnol 1(2):106–110
Shirato M, Murase T, Atsumi K (1980) Simplified computational method for constant pressure expression of filter cakes. J Chem Eng Japan 13:397–401
Shirato M, Murase T, Iwata M, Nakatsuka S (1986) The Terzaghi-Voigt combined model for constant-pressure consolidation of filter cakes and homogeneous semi-solid materials. Chem Eng Sci 41:3213–3218
Sitzmann W, Vorobiev E, Lebovka N (2016a) Applications of electricity and specifically pulsed electric fields in food processing: Historical backgrounds. Innov Food Sci Emerg Technol 37:302–311

Sitzmann W, Vorobiev E, Lebovka N (2016b) Pulsed electric fields for food industry: historical overview. In: Miklavčič D (ed) Handbook of electroporation. Springer, Cham, pp 1–20

Stämpfli R (1958) Reversible electrical breakdown of the excitable membrane of a Ranvier node. An da Acad Bras Ciências (Annals Brazilian Acad Sci) 30:57–63

Stämpfli R, Willi M (1957) Membrane potential of a Ranvier node measured after electrical destruction of its membrane. Experientia 13:297–298

Suchanek M, Olejniczak Z (2018) Low field MRI study of the potato cell membrane electroporation by pulsed electric field. J Food Eng 231:54–60

Suchodolskis A, Stirke A, Timonina A, Ramanaviciene A, Ramanavicius A (2011) Baker's yeast transformation studies by Atomic Force Microscopy. Adv Sci Lett 4:171–173

Sugar IP, Förster W, Neumann E (1987) Model of cell electrofusion: membrane electroporation, pore coalescence and percolation. Biophys Chem 26:321–335

Susil R, Semrov D, Miklavčič D (1998) Electric field-induced transmembrane potential depends on cell density and organization. Electro- and Magnetobiology 17:391–399

Taylor B (2016) Fruit and juice processing. In: Ashurst PR (ed) Chemistry and technology of soft drinks and fruit juices. John Wiley & Sons, Ltd., Chichester, pp 31–64

Tekle E, Oubrahim H, Dzekunov SM, Kolb JF, Schoenbach KH, Chock PB (2005) Selective field effects on intracellular vacuoles and vesicle membranes with nanosecond electric pulses. Biophys J 89:274–284

Tien HT, Ottova A (2003) The bilayer lipid membrane (BLM) under electrical fields. IEEE Trans Dielectr Electr Insul 10:717–727

Tiwari BK, Cullen PJ (2012) Extraction of red beet pigments. In: Neelwarne B (ed) Red Beet Biotechnology: food and Pharmaceutical Applications. Springer, New York, pp 373–391

To VHP, Nguyen TV, Vigneswaran S, Ngo HH (2016) A review on sludge dewatering indices. Water Sci Technol 74:1–16

Toepfl S (2006) Pulsed Electric Fields (PEF) for permeabilization of cell membranes in food- and bioprocessing: applications, process and equipment design and cost analysis. PhD thesis. von der Fakultät III Prozesswissenschaften der Technischen Universität Berlin, Germany

Toepfl S, Siemer C, Saldaña-Navarro G, Heinz V (2014) Chapter 6 – Overview of pulsed electric fields processing for food. In: Sun D-W (ed) Emerging technologies for food processing (second edition), 2nd edn. Academic Press, San Diego, pp 93–114

Varzakas TH, Leach GC, Israilides CJ, Arapoglou D (2005) Theoretical and experimental approaches towards the determination of solute effective diffusivities in foods. Enzym Microb Technol 37:29–41

Vicaş SI, Bandici L, Teuşdea AC, Turcin V, Popa D, Bandici GE (2017) The bioactive compounds, antioxidant capacity, and color intensity in must and wines derived from grapes processed by pulsed electric field [Compuestos bioactivos, capacidad antioxidante e intensidad de color en mosto y vinos derivados de uvas sometidas a tratamiento de campos eléctricos pulsados]. CYTA J Food 15:553–562. https://doi.org/10.1080/19476337.2017.1317667

Virkutyte J (2017) Aerobic treatment of effluents from pulp and paper industries. In: Current developments in biotechnology and bioengineering. Elsevier, pp 103–130

Vorobiev EI, Lebovka NI (eds) (2008) Electrotechnologies for extraction from food plants and biomaterials. Springer, New York

Vorobiev E, Lebovka N (2010) Enhanced extraction from solid foods and biosuspensions by pulsed electrical energy. Food Eng Rev 2:95–108

Vorobiev E, Lebovka N (2013) Enhancing extraction from solid foods and biosuspensions by electrical pulsed energy (pulsed electric field, pulsed ohmic heating and high voltage electrical discharge). In: Yanniotis S, Taoukis P, Stoforos NG, Karathanos VT (eds) Advances in food process engineering research and applications. Springer, pp 415–428

Vorobiev E, Lebovka N (2016a) Selective Extraction of Molecules from Biomaterials by Pulsed Electric Field Treatment. In: Miklavcic D (ed) Handbook of Electroporation. Springer, Cham, pp 1–16

Vorobiev E, Lebovka N (2016b) Applications of pulsed electric energy for biomass pretreatment in biorefinery. In: Mussatto SI (ed) Biomass fractionation technologies for a lignocellulosic feedstock based biorefinery. Elsevier, pp 151–168

Vorobiev E, Lebovka N (2019) Pulsed Electric Field in green processing and preservation of food products. In: Chemat F, Vorobiev E (eds) Green Food Processing techniques. Preservation, transformation, and extraction. Academic Press, London, pp 403–430

Vorobiev E, Lebovka N (2020) Processing of foods and biomass feedstocks by pulsed electric energy. Springer, Switzerland AG

Weaver JC, Chizmadzhev YA (1996) Theory of electroporation: a review. Bioelectrochem Bioenerg 41:135–160

Wiktor A, Śledź M, Nowacka M, Chudoba T, Witrowa-Rajchert D (2014) Pulsed electric field pretreatment for osmotic dehydration of apple tissue: experimental and mathematical modeling studies. Dry Technol 32:408–417. https://doi.org/10.1080/07373937.2013.834926

Winterhalter M, Helfrich W (1987) Effect of voltage on pores in membranes. Phys Rev A 36:5874

Wouters PC, Smelt JPPM (1997) Inactivation of microorganisms with pulsed electric fields: potential for food preservation. Food Biotechnol 11:193–229. https://doi.org/10.1080/08905439709549933

Xue D, Farid MM (2015) Pulsed electric field extraction of valuable compounds from white button mushroom (Agaricus bisporus). Innov Food Sci Emerg Technol 29:178–186

Yin YG, Cui YR, Wang T (2008) Study on extraction of polysaccharide from Inonotus obliquus by high intensity pulsed electric fields. Trans Chinese Soc Agric Mach 39:89–92

Zagorul'ko AY (1958) Obtaining of the diffusion juice with the help of electroplasmolysis. PhD thesis (Candidate of technical sciences), All-USSR Central Research Institute of Sugar Industry, Kiev, Ukraine (in Russian)

Zhang T-H, Wang S-J, Liu D-R, Yuan Y, Yu Y-L, Yin Y-G (2011) Optimization of exopolysaccharide extraction process from Tibetan spiritual mushroom by pulsed electric fields. Jilin Daxue Xuebao (Gongxueban)/(J Jilin Univ Eng Technol Ed) 41:882–886

Zhu Z, Bals O, Grimi N, Vorobiev E (2012) Pilot scale inulin extraction from chicory roots assisted by pulsed electric fields. Int J Food Sci Technol 47:1361–1368

Zhu Z, Luo J, Ding L, Bals O, Jaffrin MY, Vorobiev E (2013) Chicory juice clarification by membrane filtration using rotating disk module. J Food Eng 115:264–271

Zhu Z, Mhemdi H, Ding L, Bals O, Jaffrin MY, Grimi N, Vorobiev E (2015) Dead-end dynamic ultrafiltration of juice expressed from electroporated sugar beets. Food Bioprocess Technol 8:615–622

Zimmermann U (1986) Electrical breakdown, electropermeabilization and electrofusion. In: Reviews of physiology, biochemistry and pharmacology, vol 105. Springer, pp 175–256

Zimmermann U, Pilwat G, Riemann F (1974) Dielectric breakdown of cell membranes. Biophys J 14:881–899

Zimmermann U, Pilwat G, Beckers F, Riemann F (1976a) Effects of external electrical fields on cell membranes. Bioelectrochem Bioenerg 3:58–83

Zimmermann U, Pilwat G, Holzapfel C, Rosenheck K (1976b) Electrical hemolysis of human and bovine red blood cells. J Membr Biol 30:135–152

Zimmermann U, Riemann F, Pilwat G (1976c) Enzyme loading of electrically homogeneous human red blood cell ghosts prepared by dielectric breakdown. Biochim Biophys Acta (BBA) Biomembranes 436:460–474

Zimmermann U, Vienken J, Scheurich P (1980) Electric field induced fusion of biological cells. Eur Biophys J 6(86)

Zogzas NP, Maroulis ZB (1996) Effective moisture diffusivity estimation from drying data. A comparison between various methods of analysis. Dry Technol 14:1543–1573

Zudans I, Agarwal A, Orwar O, Weber SG (2007) Numerical calculations of single-cell electroporation with an electrolyte-filled capillary. Biophys J 92:3696–3705

Effect of Pulsed Electric Fields on Food Quality

Olga Martín-Belloso, Robert Soliva-Fortuny, and Mariana Morales-de la Peña

Abbreviations

E	Electric field strength
GC/MS	Gas chromatography/mass spectrometry
HIPEF	High-intensity pulsed electric fields
LIPEF	Low-intensity pulsed electric fields
MIPEF	Moderate-intensity pulsed electric fields
PEF	Pulsed electric fields
PG	Polygalacturonase
PME	Pectin methylesterase
POD	Peroxidase
PPO	Polyphenoloxidase
TBARS	Thiobarbituric acid reactive substances

Introduction

Consumers' demand for high-quality foods is continuously evolving. High nutritional health-related value, reduction of synthetic additives and minimal processing are among the most challenging demands and constitute a critical reason to invest time and resources in the search for novel processes that can be industrially implemented to obtain products gathering those characteristics. In addition, there is an imperative need for the industry to enhance processing efficiency and productivity, without sacrificing overall quality attributes such as safety, flavour and appearance of the final products. In this regard, different emerging technologies have been investigated and developed along the last decades, including pulsed electric fields (PEF).

O. Martín-Belloso (✉) · R. Soliva-Fortuny
Department of Food Technology – Agrotecnio Center, University of Lleida, Lleida, Spain
e-mail: olga.martin@udl.cat

M. Morales-de la Peña
Tecnológico de Monterrey, Escuela de Ingeniería y Ciencias, Centro de Bioingeniería, Santiago de Querétaro, Querétaro, Mexico

J. Raso et al. (eds.), *Pulsed Electric Fields Technology for the Food Industry*, Food Engineering Series, https://doi.org/10.1007/978-3-030-70586-2_7

PEF treatment involves the application of pulses of high voltages (1–45 kV/cm) to foods placed between two electrodes. The process is conducted at ambient, sub-ambient or slightly above ambient temperature for a short time (μs), so that the energy losses due to heating on the food matrix are negligible (Odriozola-Serrano et al. 2013). In this sense, it is considered a non-thermal process able to retain most of the thermolabile compounds, causing minimal impact on the physicochemical properties related to food quality and product acceptance. Results from several studies have demonstrated that PEF is an advantageous technology in front of conventional processes, which can be implemented by itself or as an assistant treatment in different food manufacturing processes (Wang et al. 2018) resulting in high-quality products that are able to satisfy current consumers' claims.

A large number of studies have demonstrated the feasibility of PEF application for different purposes in the food industry: microbial or enzyme inactivation, functionality enhancement, increasing extractability and recovery of nutritionally valuable compounds in a diverse variety of foods. PEF application is currently divided into high-intensity pulsed electric fields (HIPEF, 20–40 kV/cm), moderate-intensity pulsed electric fields (MIPEF, 1–20 kV/cm) and low-intensity pulsed electric fields (LIPEF – 0.5–1.5 kV/cm) (Oey et al. 2016; Toepfl et al. 2006). Most of the effects occurring during PEF processing, regardless of the treatment intensity, have been related to the electroporation mechanism leading to the disruption and/or permeabilization of cell membranes, which facilitates transfer of intracellular compounds. However, this mechanism is still under debate (Wang et al. 2018). Barbosa-Canovas et al. (1998) stated that electroporation causes damage to cellular structural integrity of microorganisms, leading to final inactivation. Likewise, this electroporation and membrane disintegration phenomena also result in improved extraction yield of diverse target compounds from food matrices (Puértolas et al. 2016; Vinceković et al. 2017), in an increase of oil extraction of pre-treated seeds (Zhang et al. 2017; Shortskii et al. 2017), or even in a promotion of the antioxidant content in fruits (Soliva-Fortuny et al. 2017).

The efficiency of PEF processing is related to different factors such as processing parameters (pulse shape, width, polarity, electric field strength, frequency, treatment time and temperature) and food matrix properties such as electrical conductivity, pH and proximal composition. Likewise, for microbial and enzyme inactivation purposes, the treatment effectiveness will also be related to the target microorganism and/or enzyme characteristics. In this sense, each food must be separately tested to identify the optimal set of PEF processing conditions depending on the treatment objective. Regardless of the great number of PEF studies conducted over the last century and the promising results indicating that quality attributes of PEF-treated products show a better performance compared to those of conventionally treated foods, there are still some limitations and challenges to be faced, such as the inactivation of bacterial spores and the scaling-up difficulties. However, the absence of industrial manufacturers, which in the past posed a serious drawback, is not a problem anymore. Several manufacturers are now building PEF systems for industrial use, easing the development of PEF applications at the commercial level.

This chapter discusses the potential of PEF technology for promoting food quality attributes by providing selected examples in different food matrices and by critically evaluating the main advantages and drawbacks of each application. Furthermore, a detail of the main effects on different food components that are determinant to food quality is provided.

Pulsed Electric Fields as a Way to Improve Food Quality

Food quality is well-correlated with fresh-like characteristics, flavour, colour, nutritional value, safety and health-related compound content. In this sense, the effectiveness of a food manufacturing process will rely on its impact on these parameters, immediately after processing and during shelf life. As earlier mentioned, PEF processing is considered a non-thermal process. Depending on the treatment intensity, temperature increase is possible, but in most cases far below those temperatures used in heat treatments. Additionally, total treatment time is usually short, since the process involves tiny but intense pulses with high efficiency (Gabrić et al. 2018). These characteristics represent great advantages in front of conventional processes, because of the minimal impact in heat-sensible compounds and processing optimization, saving energy and, consequently, operational costs. Additionally, the application of PEF maintains or minimally modifies the natural appearance of foods, such as flavour, colour and health-related attributes, thus allowing to obtain high-quality products that meet consumers' demands.

Several research works have been focused in the study of PEF processing as an emerging technology applied to diverse food matrices such as fruit and vegetable juices, dairy products, liquid egg, tiger nut milk, kombucha, red wine, soymilk, soy protein isolate, potato, oilseeds, mixed beverages and fruit and vegetable by-products, among others, with different purposes, as seen in Table 1.

In general, the application of PEF could be focused in two areas of food processing: (i) high-intensity pulsed electric field (HIPEF) treatments applied to preserve liquid foods and (ii) moderate-intensity pulsed electric field (MIPEF) treatments with the purpose of increasing the extraction yield of bioactive substances from plant-based foods or to improve the efficiency of other unit operations such as drying, frying, oil extraction, thawing and winemaking, among others. In both cases, although the impact on the food quality attributes is minimal, it is correlated with the treatment conditions and food characteristics, as it will be described in the following sections of the chapter.

Table 1 Application of PEF processing in foods

Process	Conditions	Matrix	Objective	Results	References
HIPEF	13.8–25.26 kV/cm, 1033–1206 μs	Orange juice	Compare thermal process vs HIPEF processing in orange juice phenolic compounds content	HPEF treated samples had more stable flavonoids and phenolic acids than those thermally treated. HIPEF orange juices had higher sensory scores	Agcam et al. (2014)
	35 kV/cm, 4 μs-bipolar pulses, 200 Hz, 1800 μs	Fruit juice-milk mixed beverage	Compare thermal process vs HIPEF processing in mixed beverage phenolic compounds content	Concentration of individual phenols increased after HIPEF processing. During the storage only hesperidin significantly increase 19–61% in HIPEF-treated beverages	Morales-de la Peña et al. (2017)
	35 kV/cm, 3 μs-monopolar pulses, 45 Hz	Cactus fruit juice	Compare the effects of HIPEF, HHP and thermal pasteurization on microbial inactivation, bioactive compounds and physicochemical characteristics of cactus fruit juice	Thermal and non-thermal processes reduced microbial population to less than 10 cfu/ml; both HIPEF and HHP resulted in a better retention of ascorbic acid than heat processing	Moussa-Ayoub et al. (2017)
	15, 25, 35 kV/cm; 500, 1250, 2000 μs	Broccoli juice	Evaluate the effects of HIPEF at different processing conditions on bioactive compounds and antioxidant capacity of broccoli juice having as a reference a thermally treated juice	HIPEF-treated broccoli juice exhibited greater concentration of carotenoids, vitamin C, total phenols and antioxidant capacity than juice treated by heat	Sánchez-Vega et al. (2015)

Process	Conditions	Matrix	Objective	Results	References
MIPEF	1, 3, 5 kV/cm and energy input of 10 kJ/kg	Blueberry fruits	Applied MIPEF as pre-treatment to increase extraction yield and antioxidant properties of the obtained juice	MIPEF pre-treated berried resulted in 28% more juice yield than untreated samples. Obtained juice had a significant higher content of total phenols, anthocyanins and antioxidant activity	Bobinaite et al. (2015)
	3 kV/cm, 56 pulses	Cactus fruit	Evaluate the impact of MIPEF pre-treatment on technological characteristics of obtained cactus fruit juice and compare results with microwave heating or use of pectinase	Processing cactus fruit mash through MIPEF increased juice yield and improved juice characteristics, such as higher flavonol content	Moussa-Ayoub et al. (2016)
	13.3 kV/cm, 1–2000 pulses	Papaya peels	Enhance the extraction of nutritionally valuable and antioxidant compounds by MIPEF processing and compare the effects with HVED and conventional aqueous extraction	HVED-assisted technique showed a higher extraction efficiency of high-added value compounds compared to PEF-assisted extraction. Application of MIPEF + supplementary aqueous extraction at 50 °C allowed an enhancement of the yields and antioxidant capacities of the extracted components	Parniakov et al. (2014)
	0.25–1 kV/cm	White grapes	Evaluate the effects of MIPEF on mechanical expression and characteristics of white grape juice	There was a noticeable improvement of grape juice quality and juice yield enhancement after the MIPEF application	Praporscic et al. (2007)

Process	Conditions	Matrix	Objective	Results	References
MIPEF/HIPEF	MIPEF: 1 kV/cm, 16 pulses HIPEF: 35 kV/cm, 1500 μs, 4 μs-bipolar pulses, 100 Hz	Tomato: fruit and juice	Produce tomato juice with high content of carotenoids from MIPEF-treated tomatoes, and apply thermal and HIPEF process to increase their shelf life	MIPEF treatment of tomatoes increased the content of carotenoids in tomato juices. HIPEF-treated juices maintained higher carotenoid content through the storage time than heated and untreated juices	Varvelldú-Queralt (2013)
	MIPEF: 1 kV/cm, 1, 20, 80 pulses HIPEF: 35 kV/cm, 1500 μs, 4 μs-bipolar pulses, 100 Hz	Tomato: fruit and juice	To enhance the lycopene in vitro bioaccessibility in tomato fruit through MIPEF and in the obtained juice treated with HIPEF	A treatment of blanching followed by HIPEF showed a significant release of trans- and cis-lycopene, achieving 15.6% total lycopene bioaccessibility	Jayathunge et al. (2017)

High-Intensity Pulsed Electric Fields (HIPEF)

The basic principle of PEF at high intensity for microbial and enzyme inactivation relies in the application of electric fields at voltages higher than 20 kV/cm during short periods of time (μs-ms) to pumpable foods passing through two electrodes. It is important to highlight that only homogeneous liquids, without suspended particles and bubble-gas free, can be treated by HIPEF, in order to avoid dielectric breakdown during processing. According to Morales-de la Peña et al. (2019), the critical parameters to be considered for HIPEF processing optimization are the electric field strength (*E*), treatment time, pulse shape, pulse width, pulse frequency, pulse polarity and temperature. Usually, the process effectiveness is mainly influenced by (a) *E* and treatment time; (b) the type of target microorganism or enzyme; and (c) treatment medium characteristics (Li and Farid 2016).

The main mechanism causing cell death and enzyme inactivation is the electroporation phenomenon, which basically causes electrical breakdown of cell membranes resulting in microbial inactivation and changes in the protein structure leading to enzyme inactivation. The effects caused by electroporation could be reversible or irreversible mainly depending on HIPEF treatment conditions (Barba et al. 2015).

Despite the high effectiveness of HIPEF in microbial inactivation reported by different authors (Niakousari et al. 2018; Li and Farid 2016; Masood et al. 2018), the existing technology does not allow food sterilization, since spore inactivation is not achieved under any condition, unless it is combined with thermal processes. However, HIPEF has been proposed as a feasible alternative for pasteurization of acidic products, such as fruit juices and fruit-based beverages, with excellent quality attributes, extended shelf life, fresh-like characteristics and, additionally, better functional properties than heat-treated products (Bermudez-Aguirre 2018). According to Wang et al. (2018), the optimum level of *E* to achieve significant reduction of pathogenic and spoilage microorganisms in fruit and vegetable juices is in the range 20–45 kV/cm. HIPEF processes applied at these conditions allow the inactivation of different pathogenic microorganisms such as *Escherichia coli*, *Listeria innocua*, *Staphylococcus aureus*, *Enterobacteriaceae* and *Pseudomonas fluorescens* (Niakousari et al. 2018; Li and Farid 2016; Masood et al. 2018) without major changes in food quality attributes.

Regarding enzyme inactivation by HIPEF, usually larger specific energy inputs are required for their inactivation compared to those necessary for microorganisms, since they have demonstrated to be more resistant. The existing evidence suggests that HIPEF affects enzyme activity by changing mainly the secondary (α-helix, β-sheets, etc.), tertiary (spatial conformation), and quaternary (number and arrangement of protein subunits) structures. In this case, different factors such as the electrochemical effects, physicochemical properties of the target enzyme, processing conditions and media characteristics such as composition, pH, viscosity, conductivity, and ionic strength, among others, are well-correlated to the impact of HIPEF in the structure enzyme changes.

HIPEF may activate/enhance, inactivate, limit or cause no changes in the activity of enzymes. The stability of enzymes towards HIPEF mainly depends on its inherent properties and treatment conditions. To date, most of the works conducted to evaluate enzyme activity affected by HIPEF are focused on the effects on inactivation in order to obtain shelf-stable products. Results indicate that HIPEF is able to significantly inactivate different food quality-related enzymes such as alkaline phosphatase (Jaeger et al. 2009), peroxidase (Morales de la Peña et al. 2010), lipoxygenase (Aguiló-Aguayo et al. 2009), lipase (Bendicho et al. 2002), protease (Bendicho et al. 2003), pectin methylesterase (Elez-Martínez et al. 2007; Espachs-Barroso et al. 2006) and polyphenol oxidase (Riener et al. 2008), among others.

In addition to achieving microbial safety and significant enzyme inactivation levels by applying HIPEF, this process can adequately preserve the quality attributes of foods, such as flavour, colour and nutrient contents, even causing an increase in the concentrations of minor constituents such as health-related compounds (Masood et al. 2018).

Moderate-Intensity Pulsed Electric Fields (MIPEF)

MIPEF treatments are applied following the same principle than HIPEF treatments, but at a lower intensity (<20 kV/cm). In this case, the treatments can also be applied to solid foods suspended in liquid media (Bobinaitė et al. 2015). Under the effects of MIPEF treatments, cell membranes may be irreversibly or reversibly permeabilized (Barba et al. 2015). According to Soliva-Fortuny et al. (2009), MIPEF has the potential to induce non-thermal cell permeability and stress reactions at cellular level in plant-based foods. These effects can allow the generation of plant secondary metabolites, the selective recovery of high-added value compounds from different matrices with high purity, the improvement of mass transfer rates and the enhancement of the efficiency of some processes such as osmotic dehydration, drying, freezing and increasing oil, juice and bioactive compound extraction yields from different foods and food by-products. In this sense, MIPEF processing has the capability of being implemented as a pre-treatment or as an assistant treatment for diverse unit operations, thus allowing the development of high-quality products.

Some of the main advantages of MIPEF-assisted processes against conventional processes are the decrease of processing times, of solvent consumption and of the intensity of the conventional extraction parameters, which results in a reduction of energy cost and environmental impact. A parallel benefit of MIPEF is the reduction of heat-sensitive compound degradation, obtaining value-added products with improved functional properties (Boussetta et al. 2012; Puértolas et al. 2013). Interesting information about the MIPEF-assisted processing in osmotic dehydration, extraction by pressing, drying, freezing, extraction of valuable compounds and acceleration of winemaking treatment can be consulted in reviews conducted by Barba et al. (2015) and Wang et al. (2018).

During the last decades, different studies have been conducted to proof the efficiency of MIPEF in a great variety of food matrices with different purposes. Namely, Praporscic et al. (2007) evaluated the effects of MIPEF at 0.25–1 kV/cm applied in white grapes. They concluded that the treatment is able to accelerate juice extraction, increase the juice yield and improve the quality characteristics of the extracted juice. Consequently, avoiding the oxidation process commonly occurred in conventional juice extraction process that serves as raw material for winemaking industry, obtaining a final product with better quality attributes. Likewise, a significant increase in antioxidant capacity of apple juice was observed when apple mash was processed at 3 kV/cm. Authors argue that this effect was due to a rise in the release of phenolics from the mash into the juice (Schilling et al. 2008). Other matrices such as rapeseeds have been treated by MIPEF (5–7 kV/cm), leading to a significant increase in the oil extraction yield and its concentration of health-related compounds (Guderjan et al. 2007).

The feasibility of MIPEF to be adapted as a pre-treatment to assist other processes has attracted the attencion of the food industry. According to Siemer et al. (2018), MIPEF has been currently implemented in potato processing at industrial levels. Today, more than 50 industrial MIPEF systems have been set up worldwide with successful results in products such as French fries, in which superior colour and texture can be attained in the final product and approximately 10% lower oil absorption during the deep-frying process. These results have caught the attention of other food manufacturers, such as fruit processors, who regard MIPEF as a potential pre-treatment to improve the quality of dried fruits or juice-based beverages.

Effects of PEF on Food Constituents Related to Food Quality Attributes

Proteins

Proteins are one of the most important food constituents. Physical, chemical and structural properties of proteins have a deep impact on their functional properties during both processing and storage and thus on the quality of the end products. Changes in protein functionality are linked to physicochemical stresses occurring during food processing such as thermal, chemical, pressure and mechanical bruising (Zhao et al. 2014). Such conditions promote the unfolding of protein structures, affecting their biological activity and stability, which has a major influence on food quality. Among the consequences of PEF-induced effects on protein modifications with an impact on quality, it is worthwhile highlighting (i) changes in the catalytic activity of certain enzymes, having a determinant impact on important food quality attributes such as colour/visual appearance, texture/rheology and flavour, and (ii) modification of the thermal and physicochemical stability of proteins, having a major impact on important properties such as water-holding capacity, gel formation, gel- or foam-forming capacity and stability, emulsification, solubility and rheological behaviour.

Proteins with enzymatic activity can be inactivated under HIPEF treatment conditions, while milder conditions often only result into an up- or downregulation of enzyme activities (Soliva-Fortuny et al. 2009). Most studies identify pulse characteristics, peak intensity, shape and width, together with temperature, as the most critical process parameters having an influence on enzyme residual activities. Other food matrix-related parameters, such as composition, enzyme and food structure and pH, are as well important. For instance, the stability of enzymes in complex food systems is more difficult to predict than in a model solution, as enzyme activity and aggregation behaviour under PEF treatment could be largely affected. Scientific literature provides several examples of enzyme aggregation with other proteins or polysaccharides (Ho et al. 1997; Wu et al. 2015, Jin et al. 2020).

Enzymes are generally less sensible to HIPEF compared to microorganisms. Some studies have related the PEF-induced changes in the activity of enzymes to the association or dissociation of functional groups, movements of charges and changes in the alignment of the α-helix (Tson and Astunian 1986). It has been shown that PEF treatments may affect the conformation of the secondary structure of enzymes, causing a loss in α-helix and an increase in β-sheet contents. Although the key enzyme responsible for the loss of a specific quality trait may vary from one food commodity to another, countless examples are available illustrating the effect of PEF on enzymes in several food matrices.

In dairy technology, alkaline phosphatase is probably the most relevant enzyme, as it is often used as an indicator of the effectiveness of a heat treatment. Other enzymes such as xanthine oxidase, proteases and lipases are intimately related to biochemical modification leading to the release of flavour compounds. HIPEF treatments have been shown to exert only a partial effect on milk enzymes (Soliva-Fortuny et al. 2006; Sharma et al. 2014); hence the remaining active enzymes could generate either desirable or undesirable flavour components that would in turn limit the end product shelf life. Sharma et al. (2018) compared a PEF treatment delivering 25.7 kV/cm for 34 μs in combination with mild preheating to 55 °C for 24 s. Alkaline phosphatase activity was comparable to that of heat-treated milk (60 °C for 30 min or 73 °C for 15 s). Xanthine oxidase and plasmin activities were only slightly affected by PEF processing, whereas lipolytic activity increased throughout storage in a similar way than for thermally pasteurized milk. This lack of enzyme inactivation suggests that PEF could only be applied for the pasteurization of raw milk in combination with heat treatments, in order to allow a reduction in the intensity of the thermal treatment. On the other hand, low bacterial counts and increased enzyme activities suggest that PEF could be a suitable alternative for cheese making, especially when enzyme activities are required for flavour development.

HIPEF treatments have been also proposed as a way to inactivate endogenous lipoxygenase (LOX) in different food products. LOX is the main responsible enzyme for the deoxygenation of polyunsaturated fatty acids, thus promoting the generation of off-flavours in plant-based foods, such as soy hydrolysates. Li et al. (2013) observed a maximum soybean LOX inactivation of 88% for treatments of 42 kV/cm for 1036 μs, whereas physicochemical quality-related parameters, e.g.

colour, pH and electrical conductivity, did not substantially change. Combination of PEF and mild thermal treatments can be an alternative to reduce treatment time while maintaining enzyme inactivation. Riener et al. (2008) reported a similar maximum inactivation rate (84.5%) when applying a PEF treatment at 40 kV/cm for 100 μs in combination with a pre-treatment at 50 °C.

Effects of HIPEF on other quality-related enzymes such as polyphenol oxidase (PPO), peroxidase (POD), pectin methylesterase (PME) and polygalacturonase (PG) have been extensively evaluated especially in fruit and vegetable beverages. Enzyme inactivation in fruit juices has been usually shown to decrease following exponential decay patterns that depend on the peak intensity of the treatment and the overall treatment time (Ho et al. 1997; Giner et al. 2000; Aguiló-Aguayo et al. 2010a, b; Andreou et al. 2016).

Several examples of the application of PEF to inactivate oxidative enzymes are available in literature. PPO and POD are especially detrimental to the quality of many processed plant-based foods, as they are involved in a series of reactions leading to browning, off-flavour generation and loss of antioxidant compounds. Therefore, POD and PPO activities often stand as a major quality indicator in processed fruit and vegetable beverages. The application of HIPEF with the aim of inactivating PPO and POD at low temperatures is feasible. Inactivation rates may greatly vary between two different food matrices. For instance, almost complete inactivation of POD was observed in PEF-treated orange juice (35 kV/cm for 1500 μs using 4-μs bipolar pulses), whereas only a 27% reduction of POD activity was found in carrot juice submitted to a treatment at 35 kV/cm for 1000 μs (Elez-Martínez et al. 2006; Quitao-Teixeira et al. 2008). Similarly, PPO is usually only partially inactivated. Sánchez-Vega et al. (2019) reported a maximal 64% inactivation in broccoli juice subjected to 35 kV/cm for long treatment times (2000 μs). The stability of the green colour of broccoli juice was significantly improved compared to that of a thermally treated juice. As in other cases, combination of PEF and mild heat seems to be a feasible way of achieving high inactivation rates for PPO and POD enzymes. Riener et al. (2008) compared the effect of PEF processing and a commercial thermal pasteurization treatment on apple juice and reported that 71% and 68% decrease for PPO and POD activities could be achieved by combining preheating to 50 °C and a short PEF treatment (40 kV/cm for 100 μs). The reported inactivation levels were by far higher than those observed in a conventional mild thermal treatment (72 °C for 26 s).

On the other hand, PEF treatments have also been shown to significantly decrease pectinolytic enzyme activities. The extent of this decrease depends on process conditions, especially the delivered specific energy and the treatment temperature. Andreou et al. (2016) proposed the application of PEF to selectively inactivate pectic enzymes. Although prolonged PEF treatment times allow obtaining nearly complete inactivation of pectic enzymes such as PME and PG, combined PEF-heat treatments stand as the most feasible strategy for attaining this objective. Proper combinations of mild thermal treatments and PEF also provide an alternative to develop treatments for specific products such as 'cold break' tomato juice, thus leading to end products with improved quality characteristics.

PEF application also stands as a good alternative to the thermal processing of protein-based foods as protein structures are less affected by PEF than by thermal treatments achieving similar rates of microbial inactivation (Perez and Pilosof 2004; Monfort et al. 2010). Although food preservation studies have demonstrated that microbial inactivation can be attained without significantly damaging protein functionality, PEF treatments have been shown to cause changes, either deliberately or not, in protein structure and functionality. The proposed mechanisms involved in the loss of protein functionality include Joule heating damage and electro-conformational changes. There is no evidence to suggest that PEF treatments affect the primary structure of proteins (Garde-Cerdán et al. 2007). Sánchez-Vega et al. (2020) observed that even under intense and prolonged exposure to HIPEF (35 kV/cm for 2000 μs), free amino acid contents in broccoli juice did not significantly decrease and were higher than those found in a thermally treated juice. However, it is likely that electric field stresses are able to ionize some chemical groups while disturbing electrostatic interactions between polypeptide chains. This could in turn have an impact on protein secondary structures due to the interaction between electric field and dipole moments (Zhao et al. 2014).

Different research works regarding the effect of PEF on protein structure have been carried out on egg products, given their high protein content and good performance as gelling and foaming agents in food formulations. Barsotti et al. (2002) reported an enhanced ionization of sulfhydryl groups and a partial but transient unfolding of protein structures in ovalbumin subjected to 20 exponential-wave pulses of 31.5 kV/cm. Authors concluded that sulfhydryl reactivity was significantly influenced by the amount of energy applied per pulse. As well, HIPEF stresses have been shown to induce sulfhydryl-disulphide interchange reactions promoting protein aggregation and subsequent decrease in protein solubility (Zhao et al. 2009). PEF treatments applied to liquid whole egg have been shown to cause less modification in the final product than commercial heat pasteurization treatments (Fig. 1). Results show that the lipoprotein matrix is less affected by PEF treatments, which has a direct effect on the preservation of microstructure, colour and rheological behaviour of the egg product (Marco-Molés et al. 2011). Furthermore, heat pasteurization treatments cause a significant reduction in protein solubility (ca. 20%), a reduction in water-holding capacity and a consistent increase in the mechanical strength of egg gels. Consistently, Monfort et al. (2010) reported that a combination of HIPEF treatments (25 kV/cm) with mild heat (52–60 °C) applied to liquid whole egg significantly increased microbial inactivation while reducing the effect on the soluble protein content compared to a conventional heat pasteurization (60–64 °C).

Available literature offers some information regarding the effects of PEF on dairy proteins. HIPEF treatments achieving near pasteurization effects (2–5 log reductions in *E. coli*) were not shown to cause dramatic changes in milk proteins, monitored through the unfolding of β-lactoglobulin molecules (Barsotti et al. 2002). However, as enzymes can be inactivated by PEF processing, changes in other protein structures and functions after intense treatments should be expected. A modification of up to 40% of the native structure of β-lactoglobulin after applying 10 pulses of 12.5 kV/cm was reported by Perez and Pilosof (2004). These changes,

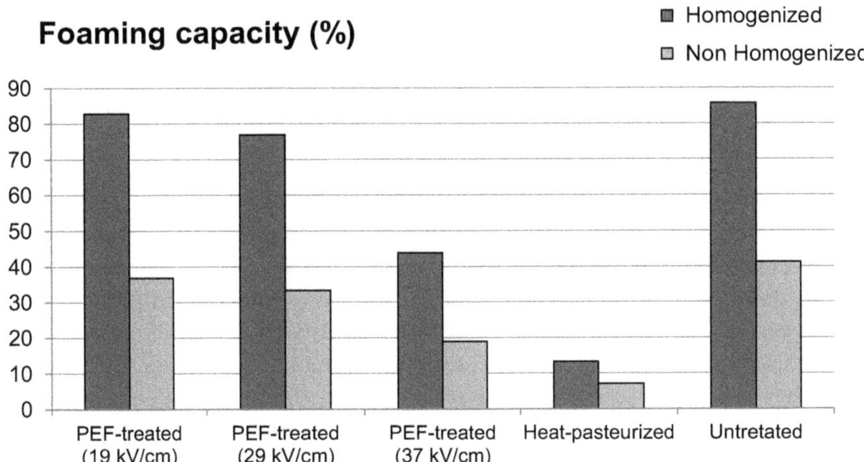

Fig. 1 Effect of HIPEF treatments (19 kV/cm, 32 kV/cm, 37 kV/cm) and heat pasteurization on the foaming capacity of liquid whole egg, subjected or not to previous homogenization. (With permissions from Marco-Molés et al. 2011)

which can be probably related to aggregate formation, are though less significant than those occurring in thermal processes. Insignificant differences between PEF-treated and heat-treated milk have been reported regarding their α-lactoglobulin, β-lactoglobulin and serum albumin contents (Odriozola-Serrano et al. 2006). No changes in the structure or in thermal stability of bovine immunoglobulins were reported by Li et al. (2005). Indeed, the coagulation properties of PEF-pasteurized milk have been compared with those of raw and heat-pasteurized milk, showing that PEF-treated milk has a better rennetability compared to thermally treated milk.

Lipids

Commercial thermal treatments ensure the safety of food products, but they may, as well, trigger or accelerate some phenomena that are deleterious to food quality from both nutritional and sensorial points of view. One of the most relevant reactions is the oxidation of lipids, which may lead to changes in colour, flavour and nutritional and bioactive content, even leading to the generation of toxic compounds. This reaction is initiated via the formation of fatty acid alkyl radicals and is strongly promoted by heat and enzyme activities, among other factors. Being a non-thermal technology, PEF may be advantageous against thermal treatments.

The effects of PEF treatments on foods with a high lipidic content have been evaluated with different aims. Arroyo et al. (2015) investigated the potential negative effects on off-flavour development of various MIPEF treatments aimed at enhancing tenderization of turkey breasts. PEF treatments of different number of pulses, voltage and frequency were applied, and authors could not report any difference regarding lipid oxidation. Suwandy et al. (2015) evaluated the effects of PEF treatments (10 kV for 20 μs) on the tenderness, cooking loss, lipid oxidation and protein profile of cold-boned beef loins. No changes in lipid composition were reported, while a decrease in the shear force and an increase in proteolysis were observed after the application of one single pulse. In a similar way, Ma et al. (2016) submitted lamb shoulder, rib and loin cooked muscles to electric field strengths of 1–1.4 kV/cm for 19,280 μs. The treatment did not immediately impact the product quality. However, after prolonged storage, an increase in volatile compounds due to lipid oxidation was observed. Faridnia et al. (2015) also found a significant increase in the lipid oxidation of frozen-thawed PEF-treated beef semitendinosus muscle. Combined freezing-thawing and PEF resulted in changes in the microstructure of meat, thus leading to improved tenderness without significant cooking loss. PEF itself did neither affect the polyunsaturated/saturated and omega 6/omega 3 fatty acid ratios nor the free fatty acids profiles. However, freezing, with or without PEF application, greatly affected the amount of volatiles directly related to lipid oxidation and accounting for off-flavour formation in the treated meat samples.

Guderjan et al. (2007) proposed MIPEF (5–10 kV/cm) as a pre-treatment to be applied before oil separation to facilitate permeabilization of cell structures of rapeseed. The treatment led to an increase not only in the oil yield but also in the content of functional food ingredients, namely, total antioxidants, tocopherols, polyphenols and phytosterols. No effect on unsaturated properties and saponification values was found, although an increase in free fatty acids and chlorophyll content was found. Zeng et al. (2010) evaluated the effects of PEF treatments (20–50 kV/cm for 40 μs) directly applied to peanut oil on its physicochemical properties. GC/MS analysis showed little changes in fatty acid composition. At the same time, treated oils exhibited less increase in acidity throughout storage, whereas carbonyl group values of PEF-treated oils were less than that of untreated reference. It was thus suggested that the PEF treatment could extend shelf life of fatty products by reducing lipid oxidation rates.

The impact of HIPEF treatments applied for the pasteurization of beverages on lipid fractions has been also assessed by different researchers. No modification of the total fat content, the peroxide index and the TBARS (thiobarbituric acid reactive substances) value, the most commonly used assay to test lipid oxidation, was reported by Cortés et al. (2005) in PEF-treated tiger nut milk. As well, Morales-de la Peña et al. (2011) did not find any major modification in the concentration of most fatty acids of a PEF-treated (35 kV/cm for 800–1400 μs) fruit-juice-soy milk beverage during storage. Only polyunsaturated fatty acid, namely, linoleic, eicosapentaenoic and docosahexaenoic, concentrations were found to decrease as it happened in the thermally treated samples. Consistently, no changes in the content of free fatty acids have been found in PEF-pasteurized whole milk (Odriozola-Serrano et al. 2006; Zhang et al. 2011). The increase in the free fatty acid content throughout storage can be likely explained by the product spoilage caused by microorganisms but not to the direct effect of PEF treatments on the lipid content.

Carbohydrates

PEF treatments do not have a major impact on low molecular weight carbohydrates as most of them are not charged. No significant chemical changes in mono- and disaccharides have been reported (Garde-Cerdán et al. 2007). Consequently, PEF stands as a promising physical technique for enhancing the extraction of sucrose from sugar beet, as it allows releasing the intracellular content without affecting the characteristics of the final product (López et al. 2009a). However, it has been recently reported that PEF may have some effect on high molecular weight polysaccharides. Han et al. (2009) found that treatments of up to 50 kV/cm led to structural changes in starch granules, which cause a decrease in some starch properties such as gelatinization temperature and enthalpy as electric field increased. Zeng et al. (2016) observed a reduction in the molecular weight of amylopectin in rice starches. Changes in molecular weight in starch and other polysaccharides are likely to be associated with a higher susceptibility to enzymatic hydrolysis induced by PEF treatments. Further studies are still needed to clarify the mechanisms of polysaccharide modification by PEF.

Minor Compounds

Many research efforts have been made to assess the potential of PEF treatments to preserve or even promote the content of minor compounds in foods. Different minor phytochemicals, e.g. vitamins, free-radical scavengers and flavour compounds, have been related with good health and a reduced risk of suffering from major chronic diseases, thus having a huge impact on product quality. Literature offers a comprehensive picture of the potential applications of PEF to preserve and promote bioactive compound in different food matrices (Elez-Martínez et al. 2017).

Increased vitamin C retention has been reported in PEF-treated juices and beverages compared to heat-pasteurized juices (Odriozola-Serrano et al. 2008; Quitão-Teixeira et al. 2009; Morales-de la Peña et al. 2010; Marsellés-Fontanet et al. 2013; Elez-Martínez et al. 2017). Vitamin C contents have been shown to depend mostly on the intensity and duration of the treatment. However, as a labile compound in the presence of oxygen at high temperatures, vitamin C can be outstandingly preserved when the product is deaerated and temperature is not increased for too long. However, higher vitamin C degradation has been shown to be triggered in other more complex food matrices, e.g. fruit juice-milk beverages, which may be a consequence of PEF-induced changes in the structure and composition of the food product (Zulueta et al. 2013). No major changes in other vitamin contents have been reported. For instance, vitamin B concentrations, namely, niacin and thiamine, have not been shown to be affected by intense PEF treatments (35 kV/cm for 1800 μs) in fruit juice-milk mixed beverages, whereas riboflavin levels were significantly higher than those found in heat-treated beverages (Salvia-Trujillo et al. 2011). Furthermore, as PEF treatments may cause changes in the food matrix, some investigations are currently being aimed at determining their impact on nutrient bioaccessibility. In the case of fruit juices, no positive or negative changes in vitamin C bioaccessibility have been reported (Rodríguez-Roque et al. 2015).

Among bioactive substances, phenolic compounds are the most widespread class of secondary metabolites in plant-based foods. These compounds are usually linked to mono- and polysaccharides or found as ester or methyl ester derivatives (Harborne et al. 1999). Phenolic acids, flavonoids and tannins are the main phenolic compounds found in the human diet. HIPEF treatments, generally applied as a non-thermal alternative to the heat pasteurization of liquid foods, have been reported to have a protective impact on the phenolic composition of fruit juices and vegetable purées. Several examples can be found in different plant-based products such as fruit juices, vegetable purees or fruit juice-milk mixed beverages (Noci et al. 2008; Quitão-Teixeira et al. 2009; Rodríguez-Roque et al. 2015). Indeed, the content of other antioxidants has been found to be well preserved or even increased in HIPEF-treated food commodities. Frandsen et al. (2014) found little differences in the glucosinolate content of broccoli juices after applying intense PEF conditions (35 kV/cm), whereas higher losses were reported after milder treatments (15 and 25 kV/cm) probably as a consequence of a lower inactivation rate for myrosinase enzymes. Consistently, several research works report an enhancement of carotenoid content in PEF-processed juices. Some examples are a 18.5% increase in the total carotenoid content observed in a papaya-mango fruit juice blend treated at 35 kV/cm with an overall energy input of 256 kJ/kg (Carbonell-Capella et al. 2016) or a 46.2% increase in the lycopene content in a PEF-pasteurized tomato juice (Odriozola-Serrano et al. 2007). These increases in carotenoids are most likely due to the physical damage infringed to cell structures, this way facilitating carotenoid release from cell membrane structures, namely, from chloroplast and chromoplast membranes, in a similar way than what occurs after low-temperature thermal processes (<100 °C).

As well, MIPEF treatments have been proposed as intensification methods for improving the nutritional and antioxidant content of different food products while enhancing extraction yields as well as the sensory attributes of different food products. The optimization of critical parameters, namely, treatment intensity, food matrix structure, mode of processing, process scale and type of product, is necessary in order to further improve the beneficial effects of PEF pre-treatments for obtaining food commodities with enhanced health-related properties. In the specific case of fruit juice extraction processes, the adjustment of milling and pressing conditions, in concomitance with mash electroporation, can be used to increase the content of total phenolics in the pressed juices (Jaeger et al. 2012). Higher β-carotene and vitamin C values were reported in PEF-treated paprika mash (1.7 kV/cm, 5 pulses) (Ade-Omowaye et al. 2001). Similarly, beverages with enhanced antioxidant content may be obtained after applying selected conditions to the fruit mash or even directly to the intact raw material (Fig. 2). Literature provides some interesting examples in apple juice (3 kV/cm and 10 kJ/kg; Schilling et al. 2008), white (0.4 kV/cm, 100 ms, 15 kJ/kg; Grimi et al. 2009) and red grape musts (1.5 kV/cm, 15–70 kJ/kg; Leong et al. 2016) and berry juices (1–5 kV/cm, 10 kJ/kg; Lamanauskas et al.

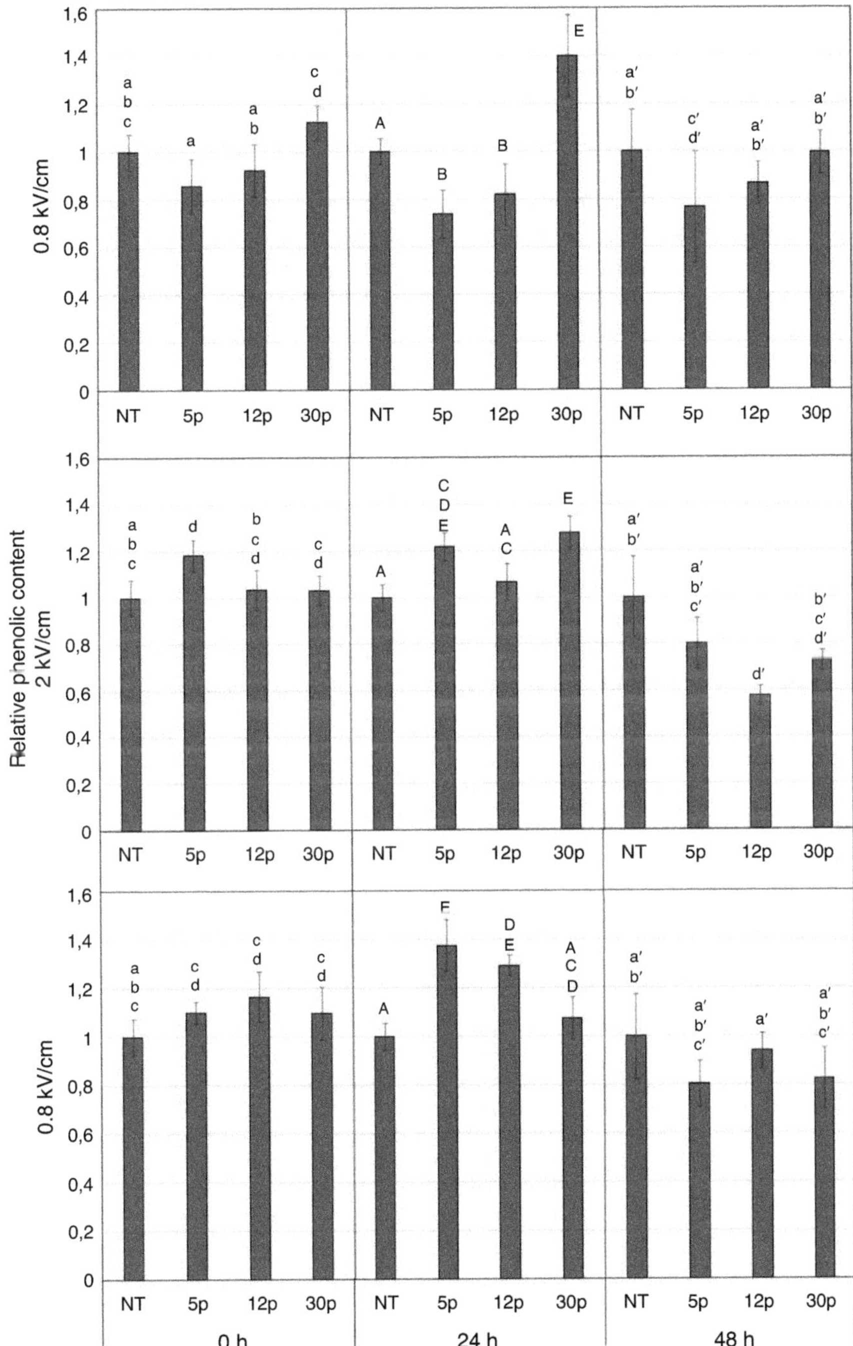

Fig. 2 Relative phenolic content in non-treated (NT) and PEF-treated carrots throughout post-treatment storage. Different letters indicate significant ($p < 0.05$) differences among groups at the same post-treatment time: 0 h, lowercase letters; 24 h, uppercase letters; 48 h, lowercase letters. (With permissions from López-Gámez et al. 2020)

2016). Furthermore, the potential use of MIPEF treatments to promote phenolic extraction from red grape skins during the maceration-fermentation step of the winemaking process has been deeply investigated (López et al. 2008, 2009b; Puértolas et al. 2010). Results demonstrate that PEF can be used to enhance the anthocyanin and total polyphenol content in red wines, hence leading to a remarkable increase of colour intensity.

Regarding oil extraction, MIPEF treatments have been shown to increase yields while leading to increase in phytosterol contents (Guderjan et al. 2005). Similarly, the recovery of tocopherols, polyphenols and phytosterols in rapeseed oil was boosted after a PEF treatment (5–7 kV/cm and 42–84 kJ/kg), as it happened in oil obtained from PEF-treated olive paste (2 kV/cm and 5.22 kJ/kg). The concentration of antioxidants in the oil was 11.5%, 9.9% and 15.0% higher than in control oils for phenolic compounds, total phytosterols and total tocopherols, respectively (Puértolas and Martínez de Marañón 2015). Table 2 provides selected examples of the application of PEF treatments to assist the extraction of oils.

Applications of PEF to assist drying processes have also revealed changes in some minor constituents in the final product. Taiwo et al. (2003) reported lower amounts of vitamin C in osmotically dehydrated apple slices, especially after treatments of relatively high electric field strength (2 kV/cm). However, as cell membrane permeabilization in vegetable tissues causes an increase in mass transfer phenomena, some compounds may be significantly affected, either in a positive or in a negative way. Compared to osmotic dehydration, combining the application of

Table 2 Selected examples of the application of PEF treatments to assist oil extraction

Type of oil	Treatment conditions	Outcome	References
Maize germ	0.6–7.3 kV/cm; 0.62 and 91.4 kJ/kg	Up to 88.4% higher yield after milder treatment intensities Phytosterol contents peaked by 32.4%	Gurderjan et al. (2005)
Rapeseed	4–7 kV/cm; 42–84 kJ/kg	Yield increase from 23% to 32% Higher concentrations of total antioxidants, tocopherols, polyphenols and phytosterols No effects on unsaturation degree and saponification values Increase in fatty acids and chlorophyll content	Guderjan et al. (2007)
Olive	1–2 kV/cm; 1.47–5.22 kJ/kg	Yield increase by 14.1% Treatments allowed reducing the malaxation temperature from 26 to 15 °C without impairing the extraction yield and quality parameters established for extra virgin oils	Abenoza et al. (2013)
Sesame	20 kV/cm; 40–240 kJ/kg	Yield increase by 4.9% No generation of lignans	Sarkis et al. (2015)
Sunflower	7 kV/cm; 6,1 kJ/kg	Yield increase by 2.3% Minor effects on main chemical characteristics (acidity and peroxide values), colour parameters and tocopherol content Increase in phenolic content by 9%	Shortskii et al. (2020)

MIPEF treatments and air drying could contribute to better preserve some heat-labile compounds such as vitamin C, as drying times are significantly reduced (Ade-Omowaye et al. 2003).

More recently, MIPEF treatments have been proposed for increasing the content of some phytochemical compounds in metabolically active vegetable tissues, through the induction of secondary metabolic pathways leading to their accumulation in plant cells. Namely, the application of electric fields of ca. 1 kV/cm is able to induce a transmembrane potential difference of 0.2–1 V that is able to induce the formation of pores. This damage can be lethal or sublethal depending on the treatment conditions as well as on food matrix characteristics. In the case of sublethal treatments, reversible damage likely induces the generation of reactive oxygen species eliciting the synthesis of secondary metabolites. Different examples of the application of MIPEF treatments to increase the antioxidant content of fresh fruit and vegetables are found in scientific literature. Ribas-Agustí et al. (2019) proposed the application of MIPEF treatments to enhance the antioxidant content in fresh whole apples. Total phenolic content was increased by 26% after mild treatments of 0.01 kJ/kg, while physicochemical quality attributes of the fresh commodity were not affected. Optimization of processing conditions allowed to dramatically increasing the accumulation of total carotenoids and lycopene without significantly compromising the fresh-like quality of tomato fruits (González-Casado et al. 2018a). In other cases, some changes in the general appearance and texture (loss of turgidity) of the fresh fruits are observed. However, the treatments may be regarded as a pre-treatment applied for increasing the antioxidant content in fruits prior to further processing operations. For instance, a 36% increase in phenolic compounds and lycopene content has been reported to increase by 20% after applying 16 pulses of 1 kV/cm to whole tomatoes (Vallverdú-Queralt et al. 2012). Similar results have been recently obtained to increase the carotenoid content in tomato fruits and carrots to be used as industrial raw materials (González-Casado et al. 2018b; López-Gámez et al. 2020). An accurate selection of treatment intensities and post-treatment storage is required in order to obtain products with a higher antioxidant content (Fig. 2).

The flavour of many food products is easily affected during thermal processing and subsequent storage. Changes in the volatile profile of PEF treated products have been shown to depend on the product characteristics and intensity of the PEF treatment. Nevertheless, the sensory acceptability of PEF treated products, such as fruit and vegetable juices, does not much differ from the freshly squeezed products. Consistently, the volatile compounds either in PEF-treated high acidic juices (orange juice, 35–40 kV/cm for 59–97 μs) or low acidic juices (watermelon, 35 kV/cm for 1737 μs) have been shown to be much better preserved compared to heat-pasteurized juices (Min et al. 2003a, b; Aguiló-Aguayo et al. 2010a). Similar results were reported by Cserhalmi et al. (2006) in a variety of citrus juices. Indeed, Min and Zhang (2003) reported a much better preservation of flavour in tomato juice. This observation was corroborated by Aguiló-Aguayo et al. (2010b) who concluded that HIPEF-treated (35 kV/cm for 1500 μs) tomato juices exhibited higher levels of compounds contributing to tomato aroma than heat-treated juices and even higher than untreated juices, which could be related to the action of residual fractions of enzyme activities related to flavour development in the juice.

Final Remarks

Scientific research provides vast evidence of the potential of PEF, applied either with a preservation aim or as a pre-treatment to improve the quality of food products. Application of HIPEF treatments to achieve microbial inactivation at temperatures that do not cause major deleterious effects on colour, flavour or nutritional values allows very sound applications of this technology in the processing of plant-based foods. As well, the use of MIPEF treatments as a pre-treatment to assist unit operations in food processing such as infusion, drying, extraction, fermentation and cooking, among many others, opens many ways for the future implementation of this technology at the industrial scale.

References

Abenoza M, Benito M, Saldaña G, Álvarez I, Raso J, Sánchez-Gimeno AC (2013) Effects of pulsed electric field on yield extraction and quality of olive oil. Food Bioprocess Technol 6:1367–1373

Ade-Omowaye BIO, Angersbach A, Taiwo KA, Knorr D (2001) The use of pulsed electric fields in producing juice from paprika (*Capsicum annuum* L.). J Food Process Preserv 25:353–365

Ade-Omowaye BIO, Taiwo KA, Eshtiaghi NM, Angersbach A, Knorr D (2003) Comparative evaluation of the effects of pulsed electric field and freezing on cell membrane permeabilisation and mass transfer during dehydration of red bell peppers. Innov Food Sci Emerg 4:177–188

Agcam E, Akyıldız A, & Evrendilek GA (2014) Comparison of phenolic compounds of orange juice processed by pulsed electric fields (PEF) and conventional thermal pasteurisation. Food Chemistry 143:354–361

Aguiló-Aguayo I, Soliva-Fortuny R, Martín-Belloso O (2009) Effects of high–intensity pulsed electric fields on lipoxygenase and hydroperoxide lyase activities in tomato juice. J Food Sci 74:C595–C601

Aguiló-Aguayo I, Soliva-Fortuny R, Martín-Belloso O (2010a) High-intensity pulsed electric fields processing parameters affecting polyphenoloxidase activity of strawberry juice. J Food Sci 75:C641–C646

Aguiló-Aguayo I, Soliva-Fortuny R, Martín-Belloso O (2010b) Impact of high-intensity pulsed electric field variables affecting peroxidase and lipoxygenase activities of watermelon juice. LWT-Food Sci Technol 43:897–902

Andreou V, Dimopoulos G, Katsaros G, Taoukis P (2016) Comparison of the application of high pressure and pulsed electric fields technologies on the selective inactivation of endogenous enzymes in tomato products. Innov Food Sci Emerg 38:349–355

Arroyo C, Eslami S, Brunton NP, Arimi JM, Noci F, Lyng JG (2015) An assessment of the impact of pulsed electric fields processing factors on oxidation, color, texture, and sensory attributes of Turkey breast meat. Poult Sci 94:1088–1095

Barba FJ, Parniakov O, Pereira SA, Wiktor A, Grimi N, Boussetta N, Saraiva JA, Raso J, Martín-Belloso O, Witrowa-Rajchert D, Lebovka N, Vorobiev E (2015) Current applications and new

opportunities for the use of pulsed electric fields in food science and industry. Food Res Int 77:773–798
Barbosa-Canovas GV, Gongora-Nieto MM, Pothakamury UR, Swanson BG (eds) (1998) Preservation of foods with pulsed electric fields. Academic, London
Barsotti L, Dumay E, Mu TH, Fernandez-Diaz MD, Cheftel JC (2002) Effects of high voltage electric pulses on protein based food constituents and structures. Trends Food Sci Technol 12:136–144
Bendicho S, Estela C, Giner J, Barbosa-Cánovas GV, Martín O (2002) Effects of high intensity pulsed electric field and thermal treatments on a lipase from Pseudomonas fluorescens. J Dairy Sci 85:19–27
Bendicho S, Barbosa-Cánovas GV, Martín O (2003) Reduction of protease activity in milk by continuous flow high–intensity pulsed electric field treatments. J Dairy Sci 86:697–703
Bermudez-Aguirre D (2018) Technological hurdles and research pathways on emerging technologies for food preservation. In: Barba FJ, Sant'Ana AS, Orlien V, Koubba M (eds) Innovative technologies for food preservation. Academic, London, pp 277–303
Bobinaitė R, Pataro G, Lamanauskas N, Šatkauskas S, Viškelis P, Ferrari G (2015) Application of pulsed electric field in the production of juice and extraction of bioactive compounds from blueberry fruits and their by-products. J Food Sci Technol 52:5898–5905
Boussetta N, Voroviev LH, Cordin-Falcimaigne A, & Lanoisellé JL (2012) Application of electrical treatments in alcoholic solvent for polyphenols extraction from grape seeds. LWT–Food Science and Technology 46:127–134
Carbonell JM, Buniowska M, Braba FJ, Grimi N, Vorobiev E, Esteve MJ, Frígola A (2016) Changes of antioxidant compounds in a fruit juice-Stevia rebaudiana blend processed by pulsed electric technologies and ultrasound. Food Bioprocess Technol 9:1159–1168
Cortés C, Esteve MJ, Frigola A, & Torregrosa F (2005) Quality characteristics of horchata (a Spanish vegetable beverage) treated with pulsed electric fields during shelf-life. Food Chemistry 91(2):319–325
Cserhalmi Z, Sass-Kiss Á, Tóth-Markus M, Lechner N (2006) Study of pulsed electric field treated citrus juices. Innov Food Sci Emerg Technol 7:49–54
Elez-Martínez P, Aguiló-Aguayo I, Martín-Belloso O (2006) Inactivation of orange juice peroxidase by high-intensity pulsed electric fields as influenced by processing parameters. J Sci Food Agric 86:71–81
Elez-Martínez P, Suárez-Recio M, Martín-Belloso O (2007) Modeling the reduction of pectin methyl esterase activity in orange juice by high intensity pulsed electric fields. J Food Eng 78:184–193
Elez-Martínez P, Odriozola-Serrano I, Oms-Oliu G, Soliva-Fortuny R, Martín-Belloso O (2017) Effects of pulsed electric fields processing strategies on health-related compounds of plant-based foods. Food Eng Rev 9:213–225
Espachs-Barroso A, Van Loey A, Hendrickx M, Martín-Belloso O (2006) Inactivation of plant pectin methylesterase by thermal or high intensity pulsed electric field treatments. Innov Food Sci Emerg Technol 7:40–48
Faridnia F, Ma QL, Bremer PJ, Burritt DJ, Hamid N, Oey I (2015) Effect of freezing as pre-treatment prior to pulsed electric field processing on quality traits of beef muscles. Innov Food Sci Emerg Technol 29:31–40
Frandsen HB, Markedal KE, Martín-Belloso O, Sánchez-Vega R, Soliva-Fortuny R, Sørensen H, Sørensen S, Sørensen JC (2014) Effects of novel processing techniques on glucosinoales and membrane associated myrosinases in broccoli. Polish J Food Nutr Sci 64:17–25
Gabrić D, Barba F, Roohinejad S, Gharibzahedi SMT, Radojčin M, Putnik P, Bursać KD (2018) Pulsed electric fields as an alternative to thermal processing for preservation of nutritive and physicochemical properties of beverages: a review. J Food Process Eng 41:1–14
Garde-Cerdán T, Arias-Gil M, Marsellés-Fontanet AR, Ancín-Azpilicueta C, Martín-Belloso O (2007) Effects of thermal and non-thermal processing treatments on fatty acids and free aminoacids of grape juice. Food Control 18:473–479

Giner J, Gimeno V, Espachs A, Elez P, Barbosa-Cánovas GV, Martín O (2000) Inhibition of tomato (*Lycopersicon esculentum* mill.) pectin methylesterase by pulsed electric fields. Innov Food Sci Emerg 1:57–67

González-Casado S, Elez-Martínez P, Martín-Belloso O, Soliva-Fortuny R (2018a) Induced accumulation of individual carotenoids and quality changes in tomato fruits treated with pulsed electric fields and stored at different post-treatments temperatures. Postharvest Biol Technol 146:117–123

González-Casado S, Elez-Martínez P, Martín-Belloso O, Soliva-Fortuny R (2018b) Application of pulsed electric fields to tomato fruit for enhancing the bioaccessibility of carotenoids in derived products. Food Funct 9:2282–2289

Grimi N, Lebovka N, Vorobiev E, Vaxelaire J (2009) Effect of a pulsed electric field treatment on expression behavior and juice quality of Chardonnay grape. Food Biophys 4:191–198

Guderjan M, Töpfl S, Angersbach A, Knorr D (2005) Impact of pulsed electric field treatment on the recovery and quality of plant oils. J Food Eng 67:281–287

Guderjan M, Elez-Martínez P, Knorr D (2007) Application of pulsed electric fields at oil yield and content of functional food ingredients at the production of rapeseed oil. Innov Food Sci Emerg 8:55–62

Han Z, Zeng XA, Zhang BS, Yu SJ (2009) Effects of pulsed electric fields (PEF) treatment on the properties of corn starch. J Food Eng 93:318–323

Harborne JB, Baxter H, Moss GP (1999) Phytochemical dictionary: a handbook of bioactive compounds from plants, 2nd edn. Taylor & Francis, London

Ho SY, Mittal GS, Cross JD (1997) Effects of high field electric pulses on the activity of selected enzymes. J Food Eng 31:69–84

Jaeger H, Meneses N, Knorr D (2009) Impact of PEF treatment inhomogeneity such as electric field distribution, flow characteristics and temperature effects on the inactivation of E. coli and milk alkaline phosphatase. Innov Food Sci Emerg 10:470–480

Jaeger H, Schulz M, Lu P, Knorr D (2012) Adjustment of milling, mash electroporation and pressing for the development of a PEF assisted juice production in industrial scale. Innov Food Sci Emerg 14:46–60

Jayathunge KGLR, Stratakos AC, Cregenzán-Albertia O, Grant IR, Lyng J, & Koidis A (2017) Enhancing the lycopene in vitro bioaccessibility of tomato juice synergistically applying thermal and non-thermal processing technologies. https://doi.org/10.1016/j.foodchem.2016.11.117

Jin W, Wang Z, Peng D, Shen W, Zhu Z, Cheng S, Li B, Huang Q (2020) Effect of pulsed electric field on assembly structure of a-amylase and pectic electrostatic complexes. Food Hydrocoll 101:105547

Lamanauskas N, Pataro G, Bobinas C, Satkauskas S, Viskelis P, Bobinaite R, Ferrari G (2016) Impact of pulsed electric field treatment on juice yield and recovery of bioactive compounds from raspberries and their by-products. Zemdirbyste-Agriculture 103:83–90

Leong SY, Burritt DJ, Oey I (2016) Evaluation of the anthocyanin release and health promoting properties of Pinot Noir grape juices after pulsed electric fields. Food Chem 196:833–841

Li X, Farid M (2016) A review on recent development in non-conventional food sterilization technologies. J Food Eng 182:33–45

Li S-Q, Bomser JA, Zhang QH (2005) Effects of pulsed electric fields and heat treatment on stability and secondary structure of bovine immunoglobulin G. J Agric Food Chem 53:663–670

Li Y-Q, Tian W-L, Mo H-Z, Zhang Y-L, Zhao X-Z (2013) Effects of pulsed electric field processing on quality characteristics and microbial inactivation of soymilk. Food Bioprocess Technol 6:1907–1916

López N, Puértolas E, Condón S, Álvarez I, Raso J (2008) Application of pulsed electric fields for improving the maceration process during vinification of red wine: influence of grape variety. Eur Food Res Technol 227:1099–1107

López N, Puértolas S, Condón S, Raso J, Álverez I (2009a) Enhancement of the solid-liquid extraction of sucrose from sugar beet (*Beta vulgaris*) by pulsed electric fields. LWT-Food Sci Technol 42:1674–1680

López N, Puértolas E, Hernández-Orte P, Álvarez I, Raso J (2009b) Effect of a pulsed electric field treatment on the anthocyanins composition and other quality parameters of Cabernet Sauvignon freshly fermented model wines obtained after different maceration times. LWT–Food Sci Technol 42:1225–1231

López-Gámez G, Elez-Martínez P, Martín-Belloso O, Soliva-Fortuny R (2020) Enhancing phenolic content in carrots by pulsed electric fields during post-treatment time: effects on cell viability and quality attributes. Innov Food Sci Emerg 59:102252

Ma QL, Hamid N, Oey I, Kantono K, Faridnia F, Yoo M, Farouk M (2016) Effect of chilled and freezing pre-treatments prior to pulsed electric field processing on volatile profile and sensory attributes of cooked lamb meats. Innov Food Sci Emerg 37:359–374

Marco-Molés R, Rojas-Graü MA, Hernando I, Pérez-Munuera I, Soliva-Fortuny R, Martín-Belloso O (2011) Physical and structural changes in liquid whole egg treated with high-intensity pulsed electric fields. J Food Sci 76:257–264

Marsellés-Fontanet AR, Puig-Pujol A, Olmos P, Mínguez-Sanz S, Martín-Belloso O (2013) A comparison of the effects of pulsed electric field and thermal treatments on grape juice. Food Bioprocess Technol 6:978–987

Masood H, Trujillo FJ, Knoerzer K, Juliano P (2018) Designing, modeling, and optimizing processes to ensure microbial safety and stability through emerging technologies. In: Barba FJ, Sant'Ana AS, Orlien V, Koubba M (eds) Innovative technologies for food preservation. Academic, London, pp 187–229

Min S, & Zhang QH (2003) Effects of commercial-scale pulsed electric field processing on flavor and color of tomato juice. Journal of Food Science 68(5):1600–1606

Min S, Jin ZT, Min SK, Yeom H, Zhang QH (2003a) Commercial-scale pulsed electric field processing of orange juice. J Food Sci 68:1265–1271

Min S, Jin ZT, Zhang QH (2003b) Commercial scale pulsed electric field processing of tomato juice. J Agric Food Chem 51:3338–3344

Monfort S, Gayán E, Saldaña G, Puértolas E, Condón S, Raso J (2010) Inactivation of *Salmonella* typhimurium and *Staphylococcus aureus* by pulsed electric fields in liquid whole egg. Innov Food Sci Emerg 11:306–313

Morales de la Peña M, Salvia-Trujillo L, Rojas-Graü MA, Martín-Belloso O (2011) Impact of high intensity pulsed electric fields or heat treatments on the fatty acid and mineral profiles of a fruit juice-soymilk beverage during storage. Food Control 22:1975–1198

Morales-de la Peña M, Salvia-Trujillo L, Rojas-Graü MA, Martín-Belloso O (2010) Impact of high intensity pulsed electric field on antioxidant properties and quality parameters of a fruit juice-soymilk beverage in chilled storage. LWT-Food Sci Technol 43:872–881

Morales-de la Peña M, Salvia-Trujillo L, Rojas-Graü MA, & Martín-Belloso O (2017) Effects of high intensity pulsed electric fields or thermal pasteurization and refrigerated storage on antioxidant compounds of fruit juice-milk beverages. Part I: Phenolic acids and flavonoids. Journal of Food Processing and Preservation 41(3): e12912

Morales-de la Peña M, Welti-Chanes J, Martín-Belloso O (2019) Novel technologies to improve food safety and quality. Curr Opin Food Sci 30:1–7

Moussa-Ayoub TE, Jäger H, Knorr D, El-Samahy SK, Kroh LW, & Rohn S (2017) Impact of pulsed electric fields, high hydrostatic pressure, and thermal pasteurization on selected characteristics of Opuntia dillenii cactus juice. LWT-Food Science and Technology 79:534–542

Moussa-Ayoub TE, Jaeger H, Youssef K, Knorr D, El-Samahy S, Kroh LW, & Rohn S (2016) Technological characteristics and selected bioactive compounds of Opuntia dillenii cactus fruit juice following the impact of pulsed electric field pre-treatment. Food chemistry 210:249–261

Niakousari M, Gahruie HH, Razmjooei M, Roohinejad S, Greiner R (2018) Effects of innovative processing technologies on microbial targets based on food categories: comparing traditional and emerging technologies for food preservation. In: Barba FJ, Sant'Ana AS, Orlien V, Koubba M (eds) Innovative technologies for food preservation. Academic, London, pp 133–185

Noci F, Riener J, Walkling-Ribeiro M, Cronin DA, Morgan DJ, Lyng JG (2008) Ultraviolet irradiation and pulsed electric fields (PEF) in a hurdle strategy for the preservation of fresh apple juice. J Food Eng 85:141–146

Odriozola-Serrano I, Bendicho-Porta S, Martín-Belloso O (2006) Comparative study on shelf-life of whole milk processed by high-intensity pulsed electric field or heat treatment. J Dairy Sci 89:905–911

Odriozola-Serrano I, Aguiló-Aguayo I, Soliva-Fortuny R, Gimeno-Añó V, Martín-Belloso O (2007) Lycopene, vitamin C, and antioxidant capacity of tomato juice as affected by high-intensity pulsed electric fields critical parameters. J Agric Food Chem 55:9036–9042

Odriozola-Serrano I, Soliva-Fortuny R, Gimeno-Añó V, Martín-Belloso O (2008) Modeling changes in health-related compounds of tomato juice treated by high-intensity pulsed electric fields. J Food Eng 89:210–216

Odriozola-Serrano I, Aguiló-Aguayo I, Soliva-Fortuny R, Martín-Belloso O (2013) Pulsed electric fields processing effects on quality and health-related constituents of plant-based foods. Trends Food Sci Technol 29:98–107

Oey I, Roohinejad S, Leong SY, Faridnia F, Lee PY, Kethireddy V (2016) Pulsed electric field processing: its technological opportunities and consumer perception. In: Jaiswal AK (ed) Food processing technologies. CRC Press, Boca Raton, pp 447–516

Parniakov O, Barba FJ, Grimi N, Lebovka N, & Vorobiev E (2014) Impact of pulsed electric fields and high voltage electrical discharges on extraction of high-added value compounds from papaya peels. Food Research International 65:337–343

Perez OE, Pilosof AMR (2004) Pulsed electric fields effects on the molecular structure and gelation of β-lactoglobulin concentrate and egg white. Food Res Int 37:102–110

Praporscic I, Lebovka N, Vorobiev E, Mietton-Peuchot M (2007) Pulsed electric field enhanced expression and juice quality of white grapes. Sep Purif Technol 52:520–526

Puértolas E, Cregenzán O, Luengo E, Álvarez I, & Raso J (2013) Pulsed-electric-field-assisted extraction of anthocyanins from purple-fleshed potato. Food Chemistry 136(3-4):1330–1336

Puértolas E, Martínez de Marañón I (2015) Olive oil pilot-production assisted by pulsed electric field: impact on extraction yield, chemical parameters and sensory properties. Food Chem 167:497–502

Puértolas E, Hernández-Orte P, Saldaña G, Álvarez I, Raso J (2010) Improvement of winemaking process using pulsed electric fields at pilot plant scale. Evolution of chromatic parameters and phenolic content of Cabernet Sauvignon red wines. Food Res Int 43:761–766

Puértolas E, Koubaa M, Barba FJ (2016) An overview of the impact of electrotechnologies for the recovery of oil and high-value compounds from vegetable oil industry: energy and economic cost implications. Food Res Int 80:19–26

Quitão-Teixeira LJ, Aguiló-Aguayo I, Ramos AM, Martín-Belloso O (2008) Keeping carrot juice color through the inactivation of oxidative enzymes by high-intensity pulsed electric field. Food Bioprocess Technol 1:364–373

Quitão-Teixeira LJ, Odriozola-Serrano I, Soliva-Fortuny R, Mota-Ramos A, Martín-Belloso O (2009) Comparative study on antioxidant properties of carrot juice stabilised by high-intensity pulsed electric field or heat treatments. J Sci Food Agric 89:2363–2642

Ribas-Agustí A, Martín-Belloso O, Soliva-Fortuny R, Elez-Martínez P (2019) Enhancing hydroxycinnamic acids and flavan-3-ol contents by pulsed electric fields without affecting quality attributes of apple. Food Res Int 121:433–440

Riener J, Noci F, Cronin DA, Morgan DJ, Lyng JG (2008) Combined effect of temperature and pulsed electric fields on apple juice peroxidase and polyphenoloxidase inactivation. Food Chem 109:402–407

Rodríguez-Roque MJ, de Ancos B, Sánchez-Moreno C, Cano MP, Elez-Martínez P, Martín-Belloso O (2015) Impact of food matrix and processing on the in vitro bioaccessibility of vitamin C, phenolic compounds, and hydrophilic antioxidant activity from fruit juice-based beverages. J Funct Foods 14:33–43

Salvia-Trujillo L, Morales-de la Peña M, Rojas-Graü A, Martín-Belloso O (2011) Changes in water soluble vitamins and antioxidant capacity of fruit juice-milk beverages as affected by high-intensity pulsed electric field (HIPEF) or heat during chilled storage. J Agric Food Chem 59:10034–10043

Sánchez-Vega R, Elez-Martínez P, & Martín-Belloso O (2015) Influence of high-intensity pulsed electric field processing parameters on antioxidant compounds of broccoli juice. Innovative Food Science & Emerging Technologies 29:70–77

Sánchez-Vega R, Rodríguez-Roque MJ, Elez-Martínez P, Martín-Belloso O (2019) Impact of critical high-intensity pulsed electric field processing parameters on oxidative enzymes and color of broccoli juice. J Food Process Preserv 44:e14362

Sánchez-Vega R, Garde-Cerdán T, Rodríguez-Roque MJ, Elez-Martínez P, Martín-Belloso O (2020) High-intensity pulsed electric fields or thermal treatment of broccoli juice: the effects of processing on minerals and free amino acids. Eur Food Res Technol 246:539–548

Sarkis JR, Boussetta N, Tessaro IC, Markzac LDF, Voroviev E (2015) Application of pulsed electric fields and high voltage electrical discharges for oil extraction from sesame seeds. J Food Eng 153:20–27

Schilling S, Toepfl S, Ludwig M, Dietrich H, Knorr D, Neidhart S, Schieber A, Carle R (2008) Comparative study of juice production by pulsed electric field treatment and enzymatic maceration of apple mash. Eur Food Res Technol 226:1389–1398

Sharma P, Oey I, Bremer P, Everett DW (2014) Reduction of bacterial counts and inactivation of enzymes in bovine whole milk using pulsed electric fields. Int Dairy J 39:146–156

Sharma P, Oey I, Bremer P, Everett DW (2018) Microbiological and enzymatic activity of bovine whole milk treated by pulsed electric fields. Int J Dairy Technol 71:10–19

Shorstkii I, Mirshekarloo MS, Koshevoi E (2017) Application of pulsed electric field for oil extraction from sunflower seeds: electrical parameter effects on oil yield. J Food Process Eng 40:e12281

Shorstkii I, Khudyakov D, Mirshekarloo MS (2020) Pulsed electric field assisted sunflower oil pilot production: impact on oil yield, extraction kinetics and chemical parameters. Innov Food Sci Emerg Technol 60:102309

Siemer C, Töpfl S, Witt J, Otermeier R (2018) Use of pulsed electric fields (PEF) in the food industry. DLG Expert Rep 5:1–12

Soliva-Fortuny R, Balasa A, Knorr D, Martín-Belloso O (2009) Effects of pulsed electric fields on bioactive compounds in foods: a review. Trends Food Sci Technol 20:544–556

Soliva-Fortuny R, Bendicho-Porta S, Martín-Belloso O (2006) Modeling high-intensity pulsed electric field inactivation of a lipase from Pseudomonas fluorescens. J Dairy Sci 89(11): 4096–4104

Soliva-Fortuny R, Vendrell-Pacheco M, Martin-Belloso O, Elez-Martinez P (2017) Effect of pulsed electric fields on the antioxidant potential of apples stored at different temperatures. Postharvest Biol Technol 132:195–201

Suwandy V, Carne A, van de Ven R, Bekhit AEA, Hopkins DL (2015) Effect of repeated pulsed electric field treatment on the quality of cold-boned beef loins and topsides. Food Bioprocess Technol 8:1218–1228

Taiwo KA, Angersbach A, Knorr D (2003) Effects of pulsed electric field on quality factors and mass transfer during osmotic dehydration of apples. J Food Process Eng 26:31–48

Toepfl S, Heinz V, Knorr D (2006) Applications of pulsed electric fields technology for the food industry. In: Raso J, Heinz V (eds) Food engineering series. Springer, New York, pp 197–221

Tson TY, Astunian RD (1986) Absorption and conversion of electric field energy by membrane bound ATPases. Bioelectric Bioenerg 15:457–476

Vallverdú-Queralt A, Oms-Oliu G, Odriozola-Serrano I, Lamuela-Raventós RM, Martín-Belloso O, Elez-Martínez P (2012) Effects of pulsed electric fields on the bioactive compound content and antioxidant capacity of tomato fruit. J Agric Food Chem 60:3126–3134

Vallverdú-Queralt A, Odriozola-Serrano I, Oms-Oliu G, Lamuela-Raventós RM, Elez-Martínez P, & Martín-Belloso O (2013) Impact of high-intensity pulsed electric fields on carotenoids profile of tomato juice made of moderate-intensity pulsed electric field-treated tomatoes. Food chemistry 141(3):3131–3138

Vinceković M, Viskić M, Jurić S, Giacometti J, Bursać Kovačević D, Putnik P, Režek Jambrak A (2017) Innovative technologies for encapsulation of Mediterranean plants extracts. Trends Food Sci Technol 69:1–12

Wang Q, Li Y, Sun DW, Zhu Z (2018) Enhancing food processing by pulsed and high voltage electric fields: principles and applications. CRC Crit Rev Food Sci 58:2285–2298

Wu L, Zhao W, Yang R, Yan W (2015) Pulsed electric field (PEF)-induced aggregation between lysozyme, ovalbumin and ovotransferrin in multi-protein system. Food Chem 175:115–120

Zeng XA, Han Z, Zi ZH (2010) Effects of pulsed electric field treatments on quality of peanut oil. Food Control 21:611–614

Zeng F, Gao QY, Han Z, Zeng XA, Yu SJ (2016) Structural properties and digestibility of pulsed electric field treated waxy rice starch. Food Chem 194:1313–1319

Zhang S, Yang R, Hua X, Zhang W, Zhang Z (2011) Influence of pulsed electric field treatments on the volatile compounds of milk in comparison with pasteurized processing. J Food Sci 76(1):C127–C132

Zhang L, Wang LJ, Jiang W, Qian JY (2017) Effect of pulsed electric field on functional and structural properties of canola protein by pretreating seeds to elevate oil yield. LWT-Food Sci Technol 84:73–81

Zhao W, Yang RJ, Tang YL, Zhang WB, Hua X (2009) Investigation of the protein-protein aggregation of egg white proteins under pulsed electric fields. J Agric Food Chem 57:3571–3577

Zhao W, Tang Y, Lu L, Chen X, Li C (2014) Pulsed electric fields processing of protein-based foods. Food Bioprocess Technol 7:114–125

Zulueta A, Barba FJ, Esteve MJ, Frígola A (2013) Changes in quality and nutritional parameters during refrigerated storage of orange juice-milk beverage treated by equivalent thermal and non-thermal processes for mild pasteurization. Food Bioprocess Technol 6(8):2018–2030

Part III
Applications of Pulsed Electric Fields in the Food Industry

Liquid Food Pasteurization by Pulsed Electric Fields

Rian Timmermans, Masja Nierop Groot, and Ariette Matser

Abbreviations

CFU	colony forming unit
E	electric field strength, electric field intensity (kV/cm)
LOX	lipoxygenase
LWE	liquid whole egg
n	number of pulses
PEF	pulsed electric field
PG	polygalucturonase
PME	pectin methylesterase
POD	peroxidase
PPO	polyphenol oxidase
PU	pasteurization unit
T_{max}	maximum temperature (°C)
w	specific energy, electric energy (kJ/kg)
τ	pulse width, pulse duration (μs)

Introduction

Classical pasteurization is a thermal process that is widely used in the food industry to inactivate vegetative cells of food pathogens or spoilage micro-organisms and enzymes. Thermal pasteurization processes are relatively mild and do not act on bacterial spores; hence the resulting products usually require chilled storage to ensure quality and safety throughout the shelf life of the product. Thermal pasteurization processes apply temperature regimes between 65 and 95 °C with holding

R. Timmermans (✉) · M. Nierop Groot · A. Matser
Wageningen Food and Biobased Research, Wageningen University & Research, Wageningen, The Netherlands
e-mail: rian.timmermans@wur.nl; masja.nieropgroot@wur.nl; ariette.matser@wur.nl

J. Raso et al. (eds.), *Pulsed Electric Fields Technology for the Food Industry*, Food Engineering Series, https://doi.org/10.1007/978-3-030-70586-2_8

times between seconds and minutes (Silva and Gibbs 2009), followed by a cooling step. Pasteurization can either be directly in package or in bulk followed by hot fill or an aseptic filling step. Product properties, target micro-organisms, and enzymes determine the treatment intensity required to ensure product quality and safety.

Today, thermal processes are widely used for pasteurization as they are efficient and robust, and inactivation effects on micro-organisms are established. Moreover, safety boundaries for treatment have been established for various product categories (Bean et al. 2012). The drawback of thermal pasteurization is that it may also affect the fresh characteristics of a product (i.e., loss of fresh flavor, texture, color) or nutritional value (i.e., loss of vitamin content). Furthermore, thermal processes may induce an off-flavor or browning of a product or lead to a change in structure due to, e.g., denaturation of proteins or gelatinization of starch.

Milder processes that are based on different mechanisms to target micro-organisms and enzymes received a lot of attention during the last decades, because these may avoid extensive heating, resulting in better food quality. Pulsed electric field (PEF) is a mild preservation technology for non-thermal pasteurization of liquid food products.

Many PEF studies have been published examining the impact of PEF on microbial inactivation, taste, enzymes, proteins, and quality (see Chapters "Microbial Inactivation by Pulsed Electric Fields" and "Effect of Pulsed Electric Fields on Food Quality"). These studies confirm that PEF effectively inactivates a diverse range of microbial cells when combined with low to moderate heat ($T < 60$ °C), which makes it an interesting alternative to conventional thermal preservation processes for liquid foods that contain heat-labile components like specific nutrients and flavors. This chapter focuses on the possibilities and constraints of using PEF for non-thermal pasteurization of liquid foods. Specific focus is on PEF conditions using high voltage pulses with a duration of microseconds (E = 10–40 kV/cm, τ = 1–100μs, w = 50–8000 kJ/kg).

Principle of PEF for Pasteurization

The principle of PEF pasteurization is based on the electroporation of microbial cell membranes in the liquid food product, by application of short, but very high voltage pulses, which is explained in more detail in Chapter "Microbial Inactivation by Pulsed Electric Fields". Depending on the applied pulse conditions and temperature, target micro-organism, and product characteristics, electroporation of microbial cells can be either reversible or irreversible (Wouters et al. 2001). Irreversible electroporation results in cell inactivation. Reversible electroporated cells are damaged but not inactivated; hence depending on the intrinsic and extrinsic product properties, cells may recover. PEF is often combined with mild heat or other preservation hurdles to enhance inactivation effects or prevent/delay outgrowth of surviving cells. The pH of the food matrix largely determines the inactivation effect of the PEF treatment with higher reductions in more acidic environment. The effect of an acidic environment is twofold: (1) an influx of acid during the electroporation process causes additional stress to the cell, and (2) at low pH outgrowth of surviving cells in food products is delayed or avoided (Guan and Liu 2020; Lund et al. 2014). This explains why PEF treatment is more effective in high-acid than in low-acid liquid products. Combinations of PEF with anti-microbials that act on the cell membrane, for example, nisin or citral, have been shown more effective by displaying synergistic effects (Montanari et al. 2019; Pol et al. 2000). However, for development of clean label concepts, these approaches may be of less interest.

A typical PEF pasteurization process consists of the following steps for a continuous flow of liquid food product that is (1) mildly preheated (typically to temperatures ranging between 20 and 55 °C), (2) exposed to an electric field in a PEF treatment chamber to obtain electroporation of the cells, (3) cooled down, and (4) aseptically packed to reduce the risk of recontamination after the process (Fig. 1). Because the PEF process is a volumetric process, no holding section is required as is the case for thermal processes, meaning that the liquid product can directly be cooled down after leaving the treatment chambers(s).

Mild preheating of the liquid enhances pore formation in the cell membrane by changing the cell membrane fluidity and thereby reduces the amount of electrical

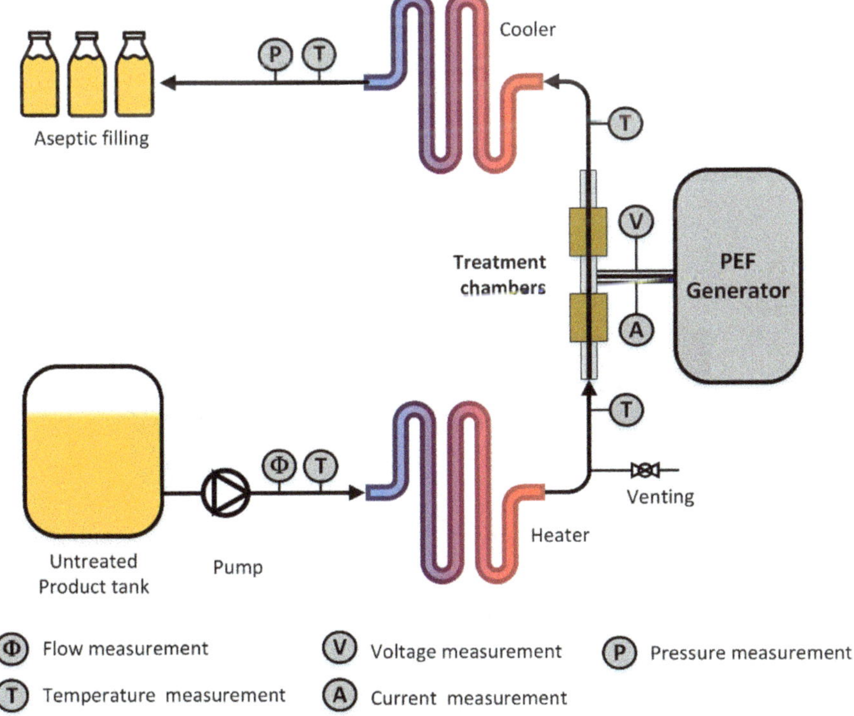

Fig. 1 Schematic representation of a continuous process of pulsed electric field (PEF) for pasteurization of a liquid food product, including measurement points for adequate process control

energy required by the PEF process (Timmermans et al. 2014; Toepfl 2011) (Fig. 1). Just before the liquid product enters the treatment chamber(s), often an inline venting step is included to reduce the number of air bubbles, as these air bubbles could lead to flashovers in the treatment chambers. Next, liquid product enters the treatment chambers that are vertically positioned to assure that the chambers are homogenously filled, which is a prerequisite for sufficient contact between liquid product and electrodes. Liquid food product is subjected to short electric pulses with a duration (τ) in the range of 1–100μs that generate electric field strengths (E), typically between E = 10–40 kV/cm. Multiple pulses are given at a selected frequency to ensure that every part of the liquid received the desired PEF treatment intensity during its passage through the treatment chamber(s).

Different configurations of treatment chambers have been used in various studies as reviewed by Huang and Wang (2009). The configuration of treatment chambers most frequently used for commercial PEF pasteurization is a layout of two or more co-linear treatment chambers (Fig. 2a). The co-linear treatment chamber consists of an electrical insulating tube through which the liquid flows. On each side of this chamber, electrodes are positioned. Electrodes consist of metal pipes that can also serve as the entrance and exit of the fluid and can be integrated in the circulating pipes used for preheating and cooling. Electrodes can be either grounded or connected to the high voltage pulse generator that produces the electric pulses. Usually,

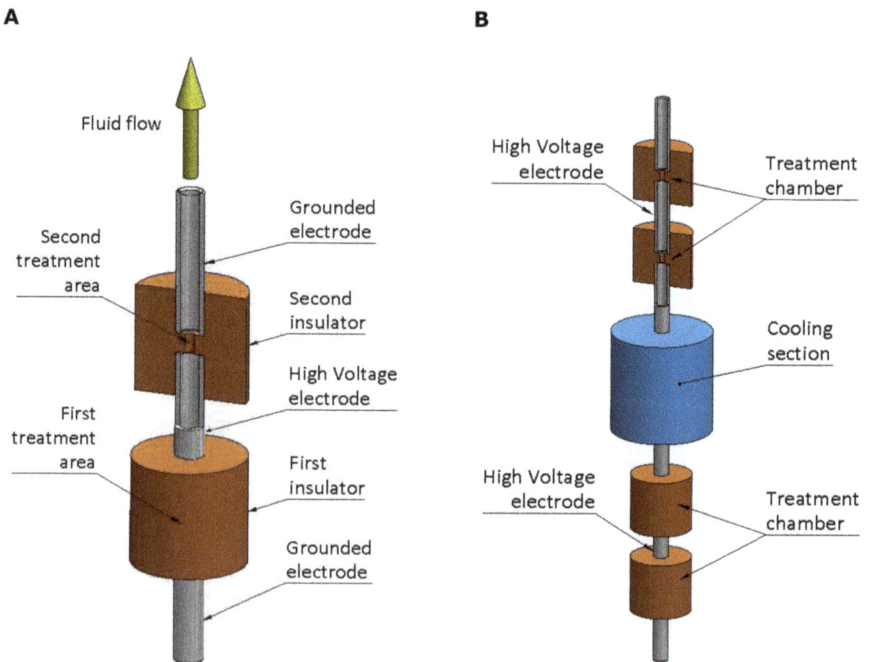

Fig. 2 Schematic representation of a PEF treatment chamber configuration with two co-linear chambers without (**a**) and with (**b**) intermediate cooling section

monopolar or bipolar pulses are used in industrial systems (Alirezalu et al. 2019). The combination of electric field strength or electric field intensity (E), pulse width (τ), and number of pulses (n) determines the specific electric energy (w) delivered. This amount of energy, the product properties and the starting temperature of the PEF treatment determine the temperature increase due to joule heating (ohmic heating). The main challenge of this co-linear configuration is to cope with the inhomogeneity in electric field and temperature distribution in the treatment chamber during PEF processing. Optimization of the treatment chamber geometry by design and generation of turbulent flow can improve treatment uniformity (Jaeger et al. 2009a; Meneses et al. 2011; Schroeder et al. 2009). To avoid overheating of the product, cooling sections between the treatment chambers can be inserted (Sharma et al. 2014a) (Fig. 2b). This will lead to more energy use for applying the treatment.

Potential Liquid Products for PEF Pasteurization

Target micro-organisms and enzymes relevant for PEF treatment are dependent on the type of liquid product to be treated. Dependent how the PEF process is designed (geometry of treatment chambers, additional cooling), this leads to a certain maximum temperature in the treatment chamber(s) (T_{max}). The thermal load of a treatment is determined by the temperature/time exposure during the entire process from preheating until cooling. The thermal load affects the residual microbial and enzyme activity, nutritional value, and visual and sensorial aspects of the treated product.

For the commercial application of PEF pasteurization, the safety of the product is of key importance. Secondly, applying PEF to a selected product must also bring added value, e.g., shelf life and product quality, with respect to a conventional thermal process. Important in the comparison of PEF toward thermal pasteurization processing is that this is on equivalent processing impact, with respect to microbial safety and spoilage micro-organisms (Matser et al. 2010), i.e., comparison of conditions to achieve a 5 or 6 $\log_{10}$ inactivation of a relevant target micro-organism. In this section, product categories will be discussed that are either applied industrially or potentially interesting to be treated with PEF, including the added value of the PEF treatment and challenges to these product groups.

Fruit Juices

In this section a discrimination between high-acid fruit juices, with $pH < 4.6$, and low-acid fruit juices with $pH > 4.6$ is made, as the pH determines the relevant target organisms and PEF process conditions required.

High-Acid Fruit Juices

PEF targets vegetative micro-organisms, but not spore-formers. For acidic fruit juices, the combination of acid conditions and refrigerated storage of these products control the germination and outgrowth of microbial spores; hence acidic fruit juices are highly suitable for PEF processing. Inactivation studies showed that treating fresh fruit juices with a low pH (pH < 4.6), such as apple juice, orange juice, and blends of juices with PEF at E ~ 20 kV/cm, τ = 2–3μs, w = 50–120 kJ/kg, and T_{max} = 60 °C, leads to 4–6 $\log_{10}$ reduction of *Escherichia coli*, *Salmonella* Panama, *Listeria monocytogenes*, *Bacillus megaterium*, and *Saccharomyces cerevisiae* (Timmermans et al. 2014; Toepfl 2011). Reduction of electric field strength from 20 to 10 kV/cm while keeping preheating (36 °C), pulse duration (τ = 2μs), and energy input (w = 75 kJ/kg) at similar levels significantly reduces the inactivation of bacterial cells to ~1 $\log_{10}$ (*E. coli*, *L. monocytogenes*, *Salmonella* Senftenberg, and *Lactobacillus plantarum*) in orange juice, while the PEF effectivity for the larger sized yeast cells of *S. cerevisiae* did not change (Timmermans et al. 2019). Further increase of the electrical energy input at E = 10 kV/cm would end up in a similar heat load as given during a thermal treatment. This implies that it is essential to evaluate the PEF process on relevant target organisms (pathogens and spoilage micro-organisms), instead of evaluation of only spoilage micro-organisms such as yeast.

The quality parameters of orange juice were evaluated after PEF processing (E = 23 kV/cm, τ = 2μs, w = 76 kV/cm, and T_{max} = 58 °C) showing no changes in sugar profile, organic acid profile, bitter compounds, vitamin C, and carotenoid profile after processing or during storage of 58 days at 4 °C. Formation of furfural and 5-hydroxymethylfurfural was not detected (Vervoort et al. 2011). The only difference observed after applying these PEF process conditions compared to untreated juice and thermal treatment (72 °C/20 s) was on the residual activity of the enzymes pectin methylesterase (PME) and peroxidase (POD) (Vervoort et al. 2011). After this PEF treatment, still residual enzyme activity was present, leading to cloud instability of orange juice (Timmermans et al. 2011). Although cloud instability has seen as a defect in pasteurized juices for decades (Braddock 1999), nowadays marketers use this as argument to the consumer that the product is pure without use of any heat treatment that causes loss of important nutrients.

An in-house trained panel of a commercial fruit juice producer did not detect changes in taste and flavor during the extended shelf life of orange juice from 7 to 27 days when stored at refrigerated conditions (Fig. 3) (WFBR, unpublished data), indicating that the quality of the PEF treated products is very close to untreated products.

To produce ambient stable, high-acid juices, process conditions should ensure complete inactivation of the enzymes naturally present in juice, such as pectin methylesterase, peroxidase, and polyphenol oxidase (PPO). Substantial inactivation (78.0–92.7%) of PME in orange juice can be obtained using PEF based on literature data reviewed by Buckow et al. (2013); however very high specific energy inputs (up to about 8 MJ/kg) were required to obtain this reduction, and still no complete

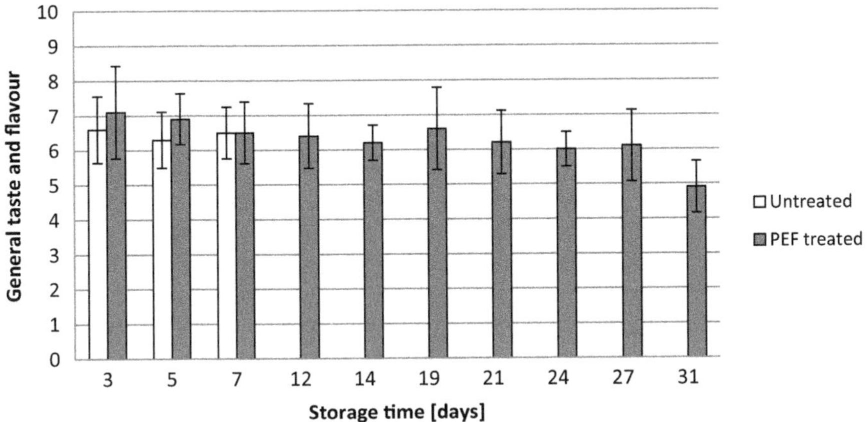

Fig. 3 Sensorial evaluation of untreated and PEF-treated orange juice by a trained panel, during shelf life (7 °C). Scores range from 0 (dislike extremely) to 10 (like extremely), and scores <5 are considered as not acceptable. Untreated juice showed microbial spoilage at $t = 12$ days; microbial counts for PEF-treated juice at $t = 31$ days were still below the detection level (<10 CFU/mL). PEF conditions and material and methods were similar as discussed in Timmermans et al. 2011 (WFBR, unpublished results)

inactivation was obtained. This energy requirement is by far higher than typically required for inactivation of vegetative micro-organisms (w = 50–120 kJ/kg). In addition to the enzyme inactivation, also conidia and ascospores of yeasts and molds, as well as endospores formed by *Alicyclobacillus* species, should be inactivated in ambient stable high-acid juices (Silva and Gibbs 2004). Studies of Raso et al. (1998) showed that conidia of *Byssochlamys fulva* could be inactivated by 3–6 $\log_{10}$ using PEF conditions of E = 30–36 kV/cm in high-acid juices, while ascospores of *Neosartorya fischeri* were not affected. A few studies describe the combination of PEF with high temperature to inactivate bacterial spores (Reineke et al. 2015; Siemer et al. 2014), but more research is required to evaluate if PEF can be used to obtain safe ambient stable high-acid fruit juices with added value compared to a thermal reference process.

During the last decade, PEF lab equipment has evolved to pilot scale equipment and industrial scale equipment (Irving 2012), and the first PEF processed fruit juices with shelf life up to 21 days are commercially available in different countries, including Germany, the Netherlands, the United Kingdom, and Austria (Buckow et al. 2013).

To conclude, a combination of moderate heat and PEF can be used for the inactivation of vegetative pathogenic and spoilage micro-organisms in high-acid fruit juices. Product quality of PEF treated juice is very close to untreated, while the shelf life could be extended when treated juice is stored at refrigerated conditions. A relatively low amount of energy (w = 50–120 kJ/kg) is required to achieve microbial inactivation, resulting in a remaining enzyme activity. This remaining enzyme activity is often accepted and used as an argument that the product is pure without heat treatment.

Low-Acid Fruit Juices

Some fruit juices including melon and watermelon juice, lychee juice, and coconut water have a pH >4.6. As pathogenic micro-organisms can grow at this pH under chilled conditions, a PEF process design that assures sufficient inactivation of pathogens is extremely important for these products. For example, application of PEF pasteurization (E = 20 kV/cm, w = 80 kJ/kg) in watermelon juice of pH 5.3 resulted in ~1 $\log_{10}$ inactivation of human pathogens *E. coli*, *L. monocytogenes*, and *S.* Panama, which is by far lower than the 4–5 $\log_{10}$ reduction reached for the same bacteria in apple or orange juice with pH 3.5 and 3.7, respectively (Timmermans et al. 2014). Blending low-acid juices with high-acid juices to reduce the pH is therefore an effective measure to improve the effectivity of a PEF treatment at E = 20 kV/cm.

Alternatively, an increase of the electric field strength to E = 35 kV/cm and increase of energy input to 7.6 MJ/kg were effective to inactivate 3–4 $\log_{10}$ *Salmonella enteritis*, *E. coli*, and *L. monocytogenes* in melon and watermelon juice (Mosqueda-Melgar et al. 2007). At these severe PEF conditions (E = 35 kV/cm, total treatment time of 2050μs, T_{max} = 40 °C), vitamin C content and antioxidant capacity reduced to 46% and 78% of initial values, respectively, and no effect on lycopene content was measured in watermelon juice (Oms-Oliu et al. 2009). No noticeable changes in color and 5-hydroxymethyl furfural (HMF) concentration were reported after applying PEF (E = 35 kV/cm, total treatment time = 1000μs, T_{max} = 40 °C) in watermelon juice, while significant changes in the thermal reference (90 °C/60 s) were observed (Aguiló-Aguayo et al. 2009). Also during shelf life study (56 days at 4 °C), the PEF-treated (E = 35 kV/cm, total treatment time 1727μs, T_{max} = 35 °C, w = 9.5 MJ/kg) watermelon juice showed brighter red color than the thermally treated (90 °C/30 or 60 s) (Aguiló-Aguayo et al. 2010a). Activity of lipoxygenase (LOX), pectin methylesterase (PME), and polygalacturonase (PG) was reduced after PEF treatment with a residual activity of 85%, 35%, and 86%, respectively, which was similar to activity of thermal treatment applied, while residual activity of peroxidase (POD) was lower after PEF (1.7%) than after thermal treatment (12%). Due to partial inactivation of enzymes, the viscosity was better retained after PEF and thermal treatment compared to untreated juice during storage (Aguiló-Aguayo et al. 2010a). Despite the decrease in overall flavor compounds observed during storage irrespective of the treatment applied, PEF-treated juices showed better flavor retention than heat-treated samples for at least 21 days of storage (Aguiló-Aguayo et al. 2010b).

To conclude, the use of PEF for pasteurization of low-acid juice is promising, although very high energy input up to some MJ/kg is required to inactivate pathogens and enzymes. Mixing low-acid juice with high-acid juice might be an approach to reduce the amount of energy. Better retention of nutritional values was obtained after PEF treatment compared to a thermal reference process. More research is required to establish conditions that guarantee the food safety of the pasteurized product and meanwhile demonstrating the added value of the PEF process.

Smoothies and Fruit Puree

Another food category that may be pasteurized with PEF is smoothies and fruit purees, although their high viscosity compared to juice is challenging for PEF processing. Smoothies are blended drinks consisting of a number of ingredients including fruit or vegetables, fruit juice, yoghurt, and/or milk (Safefood 2009). Three main types of smoothies can be distinguished all having a high viscosity in common: products containing only fruit, products containing fruit in combination with dairy ingredients, and functional smoothies that contain probiotics. The high viscosity influences the type of flow profile in the PEF treatment chamber (laminar, transitional, or turbulent). At high viscosity, a lower flow velocity will occur near the insulator wall of the treatment chamber, leading to longer cumulative treatment times of each individual particle and high (peak) temperatures, while a shorter total treatment time occurs in the center of the treatment chamber where a higher flow velocity exists (Jaeger et al. 2009a). This inhomogeneity in the treatment chamber can be improved by increasing the flow velocity, resulting in a better mixing of the fluid in the treatment chamber (Meneses et al. 2011). This explains why this effect will be observed more frequently in laminar conditions obtained at smaller-sized PEF equipment, than at turbulent flow conditions that are often reached at industrial equipment. Optimization of the geometry of the treatment chambers is another option to improve the inhomogeneity of the process to avoid over- or under-processing (Gerlach et al. 2008).

The number of scientific studies describing the use of PEF as preservation step for smoothies and fruit purees is limited. In general, for the inactivation of pathogens in smoothies by PEF, a higher energy input is required compared to high-acid juices. For example, in a study by Palgan et al. (2012), the effect of PEF on the inactivation of *Listeria innocua* in a milk-based tropical smoothie was evaluated. At $E = 18$ kV/cm and $w = 178$ kJ/L, no inactivation was observed, while increase of the intensity of the PEF treatment ($E = 32$ kV/cm, $\tau = 32$μs and $w = 900$ kJ/L) leads to 2.7 $\log_{10}$ inactivation. Similar observations in relation to the effect of the electric field strength were reported by Walkling-Ribeiro et al. (2008), who observed 3 $\log_{10}$ inactivation of *E. coli* in a tropical smoothie at $E = 24$ kV/cm (650 kJ/L) and 5 $\log_{10}$ reduction at $E = 34$ kV/cm (650 kJ/L). Geveke et al. (2015) obtained a 7.3 $\log_{10}$ inactivation of *E. coli* in strawberry puree using PEF ($E = 24.0$ kV/cm, $T_{max} = 52.5$ °C), but energy required to obtain this inactivation was not reported.

A combination of moderate heat and PEF ($E = 34$ kV/cm, T_{max} 55 °C and $w = 650$ KJ/L) in a tropical fruit smoothie (pineapple, banana, apple, coconut, orange juice, pH = 4.3) reduced the total plate count of the smoothie by ~3.4 $\log_{10}$, which was similar to the microbial reduction obtained for the thermal reference process (72 °C/15 s) (Walkling-Ribeiro et al. 2010). Microbial shelf life of PEF and thermally treated smoothie had expired at 28 and 21 days storage at 4 °C, as the total plate counts were larger than the criteria of 3.0 $\log_{10}$ CFU/mL. Sensorial evaluation indicated comparable product quality in overall acceptability for both PEF and thermal treatment, and no differences were observed in pH and conductivity during storage. PEF-treated smoothie showed a better stability of the soluble solids and viscosity, while thermal treatment showed a better stability of the color (Walkling-Ribeiro et al. 2010).

A study by Timmermans et al. (2016) evaluated the effect of variable electric field strength ranging from 13.5 to 24.0 kV/cm and a fixed amount of energy (~ 53 kJ/kg) and T_{max} = 58 °C on a fruit smoothie containing strawberry, apple, and banana (pH = 3.6). For untreated smoothie, outgrowth of yeasts led to spoilage after 8 days when stored at 4 or 7 °C, whereas in PEF-treated smoothie, yeasts were (partly) inactivated, but this provided outgrowth opportunities for spoilage molds naturally present in the smoothie/ingredients. PEF treatment led to a shift in the spoilage causing organisms from yeast to molds after 14 days (7 °C storage) or 18 days (4 °C storage). Increase of the electric field strength from 13.5 to 24.0 kV/cm resulted in more inactivation of yeasts naturally present, up to 2.8 $\log_{10}$ at most intense condition; however no complete inactivation was observed. At similar PEF conditions (E = 20 kV/cm), >7 $\log_{10}$ inactivation of *S. cerevisiae* was observed in a low viscous fruit juice with similar conductivity and pH (Timmermans et al. 2014). A possible explanation might be that the size of the yeast species naturally present in the mixed population is smaller than the cultivated *S. cerevisiae* studied in less viscous juices. Also the increased viscosity and presence of seeds might influence the homogeneity of the electric field as discussed before.

The conidiospores of *Penicillium* present in the strawberry, apple and banana smoothie of Timmermans et al. (2016) were further evaluated in a study by Nierop Groot et al. (2019) who determined the effect of PEF (E = 17 kV/cm and various T_{max}) on these isolated spores in orange juice. Effectivity of the PEF treatment showed to be species dependent. Conidia of *Penicillium expansum* are less sensitive toward mild PEF process conditions (T_{max} = 56 °C), compared to conidia formed by *Penicillium bialowiezense*, whereas at more elevated temperature (T_{max} = 63 °C), more than $\log_{10}$ inactivation was obtained for both species. This might be an explanation why this specific mold is detected in PEF-treated juices and smoothies. Further measures to assure shelf life are avoiding contamination of juices with conidia after PEF treatment and aseptic packaging to prevent outgrowth of molds.

To conclude, a combination of moderate heat and PEF can be used to extent the shelf life of low-acid, refrigerated fresh fruit smoothies and purees when compared to untreated smoothies. In general, a very high amount of energy (w = 650–900 kJ/kg) is required for inactivation of pathogens, probably due to the high viscosity. A comparison between this combined process of moderate heat and PEF with high energy input toward a thermal treatment on microbial inactivation, enzyme activity, quality, and shelf life remains to be determined.

Vegetable Juices and Soups

Vegetable juices are becoming increasingly popular due to their high nutritional content, low sugar content, and health-promoting characteristics (Shishir and Chen 2017). A large range of vegetable juice products are commercially available, often blended with fruit juices. Vegetables frequently used for single juices or blends are tomato, carrot, spinach, kale, avocado, beet root, cucumber, fennel, and celery. Most target micro-organisms in vegetables are similar to fruits, including *E. coli*, *Salmonella*, and *L. monocytogenes*; however vegetables can be contaminated with micro-organisms present in soil as well, such as the spore-forming micro-organisms *Bacillus cereus* and *Clostridium botulinum* (ICMSF 2005). The pH of vegetables juices is in general relatively high and requires additional measures to prevent out-growth during shelf life. PEF combined with moderate heat (T_{max} around 60 °C) is not effective for spore inactivation. High acidic condition (pH <4.6) prevents germination and outgrowth of *B. cereus* and *Cl. botulinum*, and acidification by blending it, for example, with fruit juice, can be used to control these spore-formers. This acidification also prevents or delays growth of some other vegetative pathogens and spoilage micro-organisms.

Enzymes often present in vegetables are PPO and POD and can lead to changes in color (browning) and flavor or off-flavor development. Other enzymes that can be present are PG and PME, which are involved in the degradation of pectin and can affect the product viscosity and texture, leading to, for example, thinning of purees or cloud loss in juices. The enzyme LOX is a primary contributor to the development of off-flavor and off-odors.

PEF has limited effect on enzyme activity at processing conditions sufficient for microbial inactivation (50–1000 kJ/kg), and quality degradation may arise from the residual activity of the enzymes, which can be controlled to a certain extent through appropriate packaging and refrigerated storage (Terefe et al. 2013). At sufficient high-specific energy input (1000–44,000 kJ/kg), PEF causes significant inactivation of enzymes at ambient and mild temperature conditions, including PME (up to 97%), PG (up to 76.5%), PPO (up to 97%), POD (up to 97%), and LOX (up to 64%) (Terefe et al. 2013). Terefe and co-workers (2013) suggest in their review that both electrochemical effects and ohmic heating contribute to the observed inactivation but that the contribution of each of them depends on the design of the PEF treatment chamber as well as treatment conditions used. It has been estimated that the specific energy input required for enzyme inactivation by PEF is four times higher compared to thermal processing, as the PEF process requires energy for both the electric pulses and the intermediate cooling sections (Espachs-Barrosso et al. 2006) (Fig. 2b).

Although the specific energy input might be very high to reduce enzyme activity, application of PEF for pasteurization of vegetable juices or soups can have some benefits compared to a thermal treatment. PEF (E = 35 kV/cm, w = 8.2 MJ/L)-treated tomato juice maintained better nutritional value than those thermally pasteurized (90 °C/30 or 60 s), with higher total and individual carotenoids (lycopene, b-carotene, and phytofluene), vitamin C, and quercetin directly after processing and during storage (56 days at 4 °C) (Odriozola-Serrano et al. 2008, 2009). Similar results were obtained when PEF treatment (E = 35 kV/cm, w = 5.5 MJ/L) and thermal treatment (90 °C/60 s) were applied to gazpacho (a cold vegetable soup), showing better retention of vitamin C and anti-oxidant capacity in the PEF-treated soup (Elez-Martínez and Martín-Belloso 2007).

In vitro studies with a static simulation of human gastric and small intestinal model have demonstrated that PEF (T_{max} < 40 °C) increases lycopene bioaccessibility up to 40% as compared to raw tomato juice (González-Casado et al. 2018; Jayathunge et al. 2017; Vallverdú-Queralt et al. 2013), while a combination of heat (90 °C/2 min) and PEF (E = 35 kV/cm, 1500μs) can improve total lycopene bioaccessibility from tomatoes up to 238% relative to raw juice (Jayathunge et al. 2017). Recent work of Zhong and co-workers (2019) evaluated the effect of PEF and thermal treatment on the bioaccessibility of lutein, lycopene, and beta-carotene following in vitro digestion of tomato juice (pH 3.9) and kale-based juice (pH 3.9) in a model for human intestinal cells (Caco-2 human intestinal cell line). PEF increased lycopene bioaccessibility by 150% but reduced β-carotene bioaccessibility by 44%, relative to raw tomato juice. Lutein uptake was increased in both PEF and thermal treatment. Kale-based juice showed no difference in bioaccessibility or cell uptake by the PEF and thermal treatment (Zhong et al. 2019).

The conductivity of tomato juice or puree is 3–5 times higher than the conductivity of other fruits or vegetable juices (Heinz et al. 2002), leading to a high increase of the electric current and temperature. Although additional cooling sections can be added in laboratory, pilot, or industrial equipment (Min et al. 2003), this intermittent cooling step introduces extra costs when the process will be commercially applied, which should be balanced with the added value brought by the application of PEF.

Another aspect that should be considered is the use of multiple ingredients with different conductivity, i.e., different small vegetable pieces or chunks in a soup, which could potentially lead to a heterogeneous distribution of the electric field, therefore leading to local hotspots and cold spots. More research is required to evaluate how variations in particle size distribution and conductivity affect microbial inactivation by PEF treatment.

To conclude, PEF can be used to extend the shelf life of vegetable juices and soups, although very high energy input up to some MJ/kg is required to control the enzyme activity. Control of the pH of the product, for example, by acidification, is important to prevent or delay the growth of vegetative pathogens and spoilage micro-organisms. Better retention of nutritional values was obtained after PEF treatment compared to a thermal reference process, which might be interesting for specific or exclusive applications as costs are four times higher than a thermal treatment.

Milk and Milk-Derived Products

Milk is one of the most consumed and nutritious liquid foods around the world (Alirezalu et al. 2019), as it is rich in essential proteins, amino acids, vitamins, and minerals. Due to this high nutrient content and its low acidity (pH 6.5–6.7), milk is an ideal growth environment for many micro-organisms, and consequently, fresh milk is a perishable and easy to spoil product (Ritota et al. 2017). According to the

European regulation, raw milk or dairy products must undergo a heat treatment to satisfy the requirements of the hygiene of foodstuff (EC 853/2004). The same regulation specifies the temperature and time combination to obtain correct pasteurization to obtain a 5 $\log_{10}$ reduction of *Coxiella burnetii* which is the most heat-resistant bacterium of public health concern in milk (Bean et al. 2012). Pasteurization must be carried out by (i) a high temperature for a short time (at least 72 °C/15 s) or (ii) a low temperature for a long time (at least 63 °C/30 min) or (iii) any other combination of temperature-time conditions to obtain an equivalent effect, such that foods show negative reaction to alkaline phosphatase immediately after the heat treatment (EC 853/2004).

As milk contains heat-labile (immune active) proteins, such as α-lactalbumin, β-lactoglobulin, and immunoglobulins, PEF could potentially be used as pasteurization process to inactivate micro-organisms and some enzymes while preserving heat-labile molecules and proteins. Many studies have been carried out to evaluate the effect of PEF on milk, and contrasting effects on inactivation of micro-organisms and enzymes in milk and derived products have been reported, mainly due to the various PEF conditions studied and heat involved. Moreover, milk components can also influence the effectivity of the PEF treatment. Presence of fat globules showed a protective effect toward PEF, as demonstrated for xanthine oxidase activity in whole milk fat (Sharma et al. 2014b). Also milk proteins influence the effectivity of PEF pasteurization, as shown for different types of milk proteins (Jaeger et al. 2009b), although this was also dependent on the concentration of milk proteins (Schottroff et al. 2020; Jaeger et al. 2009b).

Most important pathogenic and spoilage micro-organisms in milk include *Coxiella burnetii*, *E. coli*, *Bacillus cereus*, *Cronobacter sakazakii*, *Mycobacterium paratuberculosis*, *Staphylococcus aureus*, *L. monocytogenes*, *Pseudomonas* spp., and *Salmonella* spp. (Buckow et al. 2014; Alirezalu et al. 2019). Target organism to be inactivated by thermal processing is *C. burnetii* as this is the most thermal resistant one, although it is possible that a different bacterium is target for PEF processing as another inactivation mechanism is involved. Optimization of PEF processing conditions to inactivate micro-organisms is the focus of many research groups, who aim to explore the synergistic effect of electric pulses with moderate heat (see reviews by Buckow et al. 2014; Alirezalu et al. 2019). PEF technology for milk processing is a delicate balance between the electrical and thermal effect (Alirezalu et al. 2019), where more intense heat dosage might lead to more microbial inactivation (Fleischman et al. 2004; Walter et al. 2016), but also represents a process closer to a thermal treatment. Inactivation of spoilage micro-organism *Pseudomonas* and pathogens *Listeria*, *Salmonella*, and *E. coli* in milk could be obtained at a combination of moderate heat (T_{max}: 40–70 °C) and a PEF process with E = 25–35 kV/cm with total treatment time of 20–40μs, which is equivalent to a specific energy input of approximately 120 and 240 kJ/kg, respectively (Buckow et al. 2014; Sharma et al. 2014a; Walter et al. 2016). Total energy input of the entire PEF process including preheating and intermediate cooling was 314 kJ/kg, which is higher than a thermal process 63 °C/30 min (232 kJ/kg) or 73 °C/15 s (262 kJ/kg) (Sharma et al. 2018). PEF treatment under these conditions would also require at least one intermediate cooling step to avoid overheating of the product. Longer treatment times are unlikely to be commercially viable or feasible, because of the high demand for electrical and cooling energy (Buckow et al. 2014).

Raw milk contains a large number of enzymes that display different functions and impact on dairy products (Kelly and Fox 2006). Moreover different enzymes have different stability toward processing steps, and this feature is used as indicator for thermal processing, such as alkaline phosphatase or lactoperoxidase, whereas others are of interest for their either positive or negative effect on quality of dairy products (e.g., plasmin, lipoprotein lipase, microbial protease). The effect of PEF processing parameters on the activity of milk enzymes was reviewed by Buckow et al. in 2014. The sensitivity of dairy enzymes to PEF varied substantially between the studies reported in literature, and it is dependent on the enzyme studied, applied treatment parameters, and intensities (electric field strength, treatment time, temperature) and dairy matrix (Buckow et al. 2014). The large differences in reported enzyme activity could partly be ascribed by the different enzyme assays used, the lack of accurate temperature control reported in the PEF studies, and the variations of treatment efficiency by the different PEF treatment systems used (Buckow et al. 2014). In general, for commercially or economical feasible PEF treatment conditions, the effect of PEF treatment on enzymes was limited, but effects improved when more intense PEF treatments ($E > 10$ kV/cm or treatment time > 250μs) were combined with higher temperatures ($T > 40$ °C) (Terefe et al. 2013).

Several studies report on changes in the size and structure of the main milk proteins and their functional properties induced by PEF (Eberhard and Sieber 2006; Perez and Pilosof 2004), while others reported no changes in functional properties (De Luis et al. 2009; Grahl and Markl 1996). Treatment of milk or whey at relatively high PEF intensities ($E = 30$ kV/cm, $w =$ up to 2460 kJ/L) decreased the clotting time and increased its curd firmness or rennetability (Perez and Pilosof 2004; Yu et al. 2009, 2012). Contrary, at less intense PEF conditions with lower energy input, no physiochemical changes in pH, color, rennetability, and gelation parameters of milk were observed when PEF treatment ($E = 30$ kV/cm, $w = 90$ kJ/L, $T_{max} = 53$ or 63 °C) was compared to a thermal treatment (63 °C or 72 °C for 15 s) (McAuley et al. 2016). Similar observations were reported for sensory and nutrient retention, where intense PEF treatment conditions resulted in moderate changes, but mild treatments did not show any effect (Buckow et al. 2014).

For milk fat globules, contradicting effects of PEF treatment have been reported. Several studies report that PEF showed no effect (Barsotti et al. 2001; Sampedro et al. 2005), while others do report changes (Bermudez-Aguirre et al. 2011; Garcia-Amezquita et al. 2009). A reduction in the milk fat globule size and zeta-potential after applying PEF ($E = 20$ kV/cm, $w = 100$ kJ/kg) was reported by Sharma et al. (2015), although larger effects were observed for thermal treatments (63 °C/30 min or 73 °C/15 s) than for the PEF treatment. Bermudez-Aguirre et al. (2011) observed deposition of proteins and fat on the electrodes (fouling or electrodeposition) at conditions of $E = 31$–54 kV/cm, 60μs, and preheating 20–40 °C, resulting in arcing and blocking of the system at most intense conditions ($E = 53$ kV/cm, preheating 30 or 40 °C).

The effect of PEF treatment on shelf life of milk products has been reported by several studies. The shelf life of PEF-treated milk ($E = 35$ kV/cm, $T_{max} = 40$ °C) was followed when stored at 4 °C. After 5 days the milk was spoiled, which was defined as either a detection of total mesophilic count of 4.3 $\log_{10}$ or more (FDA 2007) or decrease of pH (Bermudez-Aguirre et al. 2011; Odriozola-Serrano et al. 2006). However, when moderate preheating for 24 s at 55 °C was combined with PEF ($E = 25$ kV/cm, 34μs, $w = 122$ kJ/kg), the shelf life could be extended up to 21 days at 4 °C, similar to thermally treated milk (63 °C/30 s or 73 °C/15 s) (Sharma et al. 2018). Earlier study showed that these PEF conditions resulted in >6 $\log_{10}$ inactivation of *E. coli*, *S. aureus*, and *L. innocua* in whole milk (Sharma et al. 2014a). Alkaline phosphatase inactivation decreased by 96–97% immediately after PEF or thermal treatment, whereas the activities of xanthine oxidase and plasmin were reduced to 30% and 7% immediately after PEF treatment but were similar to untreated milk at the end of the storage period (Sharma et al. 2018). The pH level decreased up to pH = 6.2 at the end of the storage period due to production of lactic acid by mesophilic bacteria or formation of fatty acids by the action of microbial lipases (Sharma et al. 2018).

Since spoilage and pathogenic micro-organisms require less intense PEF conditions for inactivation compared to the proteolytic enzymes, it seems feasible to select conditions that target micro-organisms, but leaving enzymes largely unaffected. This may lead to new dairy products or production of high-quality cheese with flavor and texture characteristics close to cheese produced from raw milk, without compromising the safety as found in cheese made of pasteurized milk. A study of Yu et al. (2012) showed that a PEF treatment ($E = 30$ kV/cm, $T_{max} = 50$ °C) that caused 5 $\log_{10}$ reduction of *Salmonella enteritidis* improved milk proteolytic profile in terms of peptide and free amino acid concentration compared to heat-pasteurized milk. Further studies evaluating effect of lipolysis and other reactions on cheese made from PEF-treated milk remain to be studied.

To conclude, a combination of moderate preheating (~55 °C) and PEF can be used to extend the shelf life of raw milk and inactivate pathogens, including *E. coli*, *S. aureus*, and *L. innocua.* Important is that at least one cooling step is included to avoid overheating of the product, as high amounts of energy are needed to inactivate pathogens and enzymes. Large differences between process conditions are reported in different studies, making it important to screen the effect of microbial inactivation, shelf life, enzymes, and nutrients under the same conditions to evaluate the added value of PEF to a specific product. For process validation it is important to select the pathogen that is most resistant toward PEF. This might be a different pathogen than used for thermal validation: for instance, *Listeria* is sensitive toward a thermal process but resistant toward PEF. Finally, implementation of PEF as pasteurization treatment for milk requires adaptation of legislation in place for pasteurization of milk.

Eggs

Eggs and egg products such as egg white or yolk are a nutritious part of our diet and useful ingredients in foods due to good functional properties for gelling, foaming, and emulsification (Monfort et al. 2010; Strixner and Kulozik 2011). As eggs can be contaminated with *Salmonella*, the European regulation EC 2073/2005 has laid down food safety criterion for *Salmonella* for egg products requiring testing for absence in five samples of 25 g of egg products. *Salmonella* is most heat resistant in the pH range between 5 and 6, and therefore a higher resistance is observed in egg yolk (pH = 6.0) and whole egg (pH = 7.6) compared to egg white (pH = 9.1). However, besides the pH influence, the dry matter content and egg constituents influence heat resistance of *Salmonella* (Cunningham 1995).

To ensure food safety, the treatments classically used to pasteurize whole egg vary from 65 to 68 °C for 2–5 min to guarantee a 5 to 6 $\log_{10}$ reduction of *Salmonella* enteritidis and *Listeria monocytogenes* (Lechevalier et al. 2017). However, when temperature rises above 60 °C, the functional properties of the egg product decrease, and their viscosity increases (Hamid-Samini et al. 1984; Lechevalier et al. 2017; Strixner and Kulozik 2011). For egg white, viscosity increases when heated >56 °C, and protein coagulates at 60 °C (Cunningham 1995); for whole egg viscosity increases at temperatures >56–66 °C, and protein coagulates at 73 °C; and egg yolk showed a viscosity increase beginning at 65 °C with a progressive increase in rigidity at 85 °C (Froning et al. 2002). PEF might be an interesting alternative to achieve non-thermal reduction of *Salmonella*, but with retention of functional protein properties.

Most of the microbial inactivation studies evaluating PEF focused on three different *Salmonella* serovars in liquid whole egg (LWE), egg white, or egg yolk as reviewed by Yogesh (2016). In general, about 1–2 $\log_{10}$ reduction was observed for *Salmonella* enteritidis, *Salmonella* Senftenberg 775 W, and *Salmonella typhimurium* in LWE at $E = 25$ kV/cm, $w = 250$ kJ/kg, and $T_{max} < 35$ °C (Hermawan et al. 2004; Jin et al. 2009; Monfort et al. 2010). The effectiveness of the PEF application could be enhanced by increasing the temperature; however the risk of protein coagulation limits the temperature increase that can be used (Martín-Belloso et al. 1997). With the use of intermediate cooling sections between treatment chambers, a 6 $\log_{10}$ reduction of *E. coli* in LWE could be obtained at $E = 26$ kV/cm at $T_{max} = 37$ °C, although energy requirements were very high (in order of 12 MJ/kg) (Martín-Belloso et al. 1997). A similar set-up using an intermediate cooling section was used to study the reduction of *Salmonella* in diluted egg white and showed 4.8 $\log_{10}$ inactivation of *Salmonella* ($E = 30$ kV/cm, total treatment time of 800μs, $T_{max} = 20$ °C) (Wu et al. 2016). Using similar PEF conditions, 4 $\log_{10}$ inactivation of *S.* enteritidis in egg white was achieved at 20 °C, while increase of the maximum temperature to 40 °C resulted in 6.3 $\log_{10}$ inactivation (Zhao et al. 2007).

Another option to increase reduction of *Salmonella* is the combination of a PEF process with thermal treatment and the use of a holding period. When after the PEF pulses a holding period at 55 °C for 3.0 or 3.5 min was applied, inactivation increased up to 3.5 $\log_{10}$ for *S. typhimurium* (Jin et al. 2009) and 4.3 $\log_{10}$ for *S.* enteritis (Hermawan et al. 2004), while a single PEF treatment at 20 °C or a heat treatment at 55 °C/3 min showed about 1 $\log_{10}$ inactivation (Jin et al. 2009). This combined application of PEF and holding period was further investigated by Monfort et al. (2012) who also added additive triethyl citrate upon treatment of LWE. PEF (E = 25 kV/cm, w = 75–100 kJ/kg) followed by heat (52 °C/3.5 min, 55 °C/2 min or 60 °C/1 min) with 2% triethyl citrate showed more than 5 $\log_{10}$ reduction of various *Salmonella* serovars tested (Monfort et al. 2012).

The combined treatment of PEF (E = 25 kV/cm, total treatment time = 250μs with intermediate cooling) and holding period (55 °C/3 min) of LWE showed no significant changes in viscosity, electric conductivity, color, pH, and soluble solids when compared to a control (Hermawan et al. 2004). Shelf life of the treated LWE could be extended to over 60 days (4 °C), whereas the control sample spoiled after a few days with total counts exceeding 7 $\log_{10}$ within 5 days (4 °C) (Hermawan et al. 2004). A similar combined process of PEF and thermal processing and the addition of additives showed a lower decrease in soluble protein (<5%) compared to the thermal treatment at 60 °C/3.5 min and 64 °C/2.5 min (Monfort et al. 2012). Minor effects of PEF (E = 30 kV/cm, 800μs, T_{max} = 20 °C) on the structure of egg white proteins were observed compared to a thermal treatment, as only insoluble aggregates were formed by PEF and both soluble and insoluble aggregates were formed after thermal treatment. Electrostatic interaction played an important role in aggregate formation between lysozyme, ovotransferrin, and ovalbumin (Wu et al. 2016). Foaming properties of egg white or LWE did not change after application of PEF up to E = 32 kV/cm (Marco-Moles et al. 2011; Zhao et al. 2007); however at E = 37 kV/cm, detrimental effects on the foaming capacity of LWE were observed (Marco-Moles et al. 2011). PEF processing up to E = 37 kV/cm showed no or minor effects on the gel hardness or water-holding capacity of LWE and egg white, whereas thermal treatment (66 °C/4.5 min) displayed a significant impact on these two attributes in LWE (Marco-Moles et al. 2011).

To conclude, reported studies on PEF processing for egg products show several advantages over traditional heat processing in inactivation of pathogens while preserving the functional characteristics of the proteins, although adequate selection of process parameters is necessary. Inactivation of *Salmonella* in egg products using PEF is challenging, as egg products have a high electric conductivity and coagulation of proteins already starts at temperatures above 56 °C resulting in blocking of the treatment chamber. PEF treatment followed by a mild thermal treatment (T_{max} = 55 °C/3 min) and optionally combined with addition of some additives showed improved inactivation of *Salmonella*, without losing protein functionality. The best results in microbial inactivation and retention of protein functionality are expected for a combination of PEF with intermediate cooling steps.

Beer

Thermal pasteurization is the final step in an industrial production process of beer, aiming to increase the beer shelf life and stability (Milani et al. 2015). Two main methods are used to pasteurize beer: tunnel pasteurization, where bottles or cans are passed through a series of water jets applying heat, and flash pasteurization, where the beer is heated rapidly in a plate heat exchanger and holding tube before packaging (Wray 2015). The combination of different preservation hurdles present in beer such as presence of ethanol, low pH (generally 3.7–4.1), low nutrient content, antimicrobial properties of hop resins, and low oxygen levels due to the presence of carbon dioxide contributes to the control of spoilage, and therefore only a mild thermal treatment is required (Wray 2015). The aim of pasteurization is to inactivate micro-organisms such as yeasts and spoilage lactobacilli. Heat-resistant spore-forming bacteria are not a target since their spores are unable to germinate and to grow in beer (Wray 2015). The pasteurization measure for beer is expressed in pasteurization unit (PU), and 1 PU is equivalent to 1 min at 60 °C with $z = 7$ °C for beer spoilage vegetative yeast (Wray 2015). Thermal treatment for pilsner and lager beer requires the lowest thermal treatment (15–25 PU), whereas higher PU values are applied for low and non-alcoholic beer (40–60 PU and 80–120 PU, respectively) or beers that contain lemonade and fruit juice (300–500 PU and 3000–5000 PU, respectively) (Wray 2015). Applying the minimal amount of heat necessary to achieve stability ensures that the effect on both flavor and energy use is minimized (Wray 2015). PEF might be an interesting alternative to thermal pasteurization, especially to the alcohol-free beers and less bitter beers, as spoilage micro-organisms and pathogens *E. coli* and *S. typhimurium* were more thermal resistant in these types of beers (Silva and Gibbs 2009).

A technological challenge for PEF application for beer is the amount of carbon dioxide present in the product. Therefore, PEF should be applied under pressure to reduce the risk of arcing (Góngora-Nieto et al. 2003). In a study by Evrendilek et al. (2004), keg beer was PEF-treated (E = 22 kV/cm, 154μs total treatment time) and showed approximately 0.5 $\log_{10}$ reduction of aerobic and 0.6 $\log_{10}$ of anaerobic plate counts. When yeast cultures were added, 4 $\log_{10}$ reduction of both *Saccharomyces uvarum* and *Rhodotorula rubra* was obtained (Evrendilek et al. 2004). However, more intense PEF conditions (E = 41 kV/cm, 175μs total treatment time) were required to inactivate bacteria of *L. plantarum*, *Pediococcus damnosus*, and *B. subtilis* (Evrendilek et al. 2004). Micro-organisms that are typically found in brewing (*S. cerevisiae*, *L. plantarum*, *B. subtilis*, and *Salmonella choleraesuis*) were subjected to a PEF treatment (E = 35–45 kV/cm, w = 530–2339 kJ/L) in beer with 0.5% or 3.5% alcohol in degassed or carbonated form (Walkling-Ribeiro et al. 2011). Up to 8 $\log_{10}$ inactivation was achieved after PEF processing, which was comparable to thermal processing at 76 °C/30 s. In beer at higher alcohol content (3.5%), more inactivation of *L. plantarum* by PEF was observed compared to low alcohol beer (0.5%). Other micro-organisms present in beer were not affected by the ethanol content. Moreover, the reduction of *S. choleraesuis* by PEF at a higher electric field was more effective in carbonated than in degassed beer. On the other hand, a change in carbon dioxide volumes per volume beer had no effect on PEF inactivation of yeast and gram-positive bacteria (Walkling-Ribeiro et al. 2011).

The study of Milani et al. (2015) evaluated effect of PEF ($E = 45$ kV/cm, 70μs, with $T_{max} = 53$ °C) on *S. cerevisiae* ascospores in nine different beers comprising ale, lager, dark, low alcohol, and no alcohol. Neither the beer production method nor the different constituents seemed to affect microbial reduction. At most intense PEF conditions, 0.9 $\log_{10}$ inactivation of ascospores in alcohol-free beer was observed, while up to 4.0 $\log_{10}$ inactivation in beer with 7% alcohol was obtained, indicating that alcohol percentage played an important factor in the efficiency of the process (Milani et al. 2015).

Studies on the effect of added value of application of PEF toward beer are scarce. Evrendilek et al. (2004) detected degradation of flavor and mouthfeel of PEF-treated beers, although this effect was ascribed to the corrosion of the stainless steel electrodes. Today, electrodes are often made of titanium, which possibly avoids the corrosion of the electrodes and might lead to better products.

Although PEF processing of carbonated drinks is challenging, the overall findings confirmed that PEF could be an alternative to thermal processing for microbiologically safe beer. A comparison of the PEF process toward a thermal process on quality and shelf life remains to be determined.

Conclusions and Outlook

An overview of the possibilities and constraints of PEF for non-thermal pasteurization of different liquid foods was provided in this chapter. Specific focus was on PEF conditions using high voltage pulses with a duration of microseconds ($E = 10$–40 kV/cm, $\tau = 1$–100μs, $w = 50$–8000 kJ/kg) and moderate heating (T_{max} 60 °C). For industrial scale applications, the electric field strength and specific energy applied are usually at the lower end of this range, ranging between $E = 10$ and 20 kV/cm (Toepfl 2012) and energy input up to 200 kJ/kg to balance between effectiveness of the treatment and costs (Kempkes 2017). These PEF conditions are industrially applied for preservation of fresh fruit juices and acidified vegetable juices and could be potentially interesting for other liquid food products.

Specific product characteristics will influence the efficiency of the PEF process to inactivate micro-organisms and enzymes, as is summarized in Table 1 and discussed in the previous sections per product category. Optimization of the PEF process layout including preheating and intermediate cooling sections is required to (i) guarantee microbial and enzyme inactivation, (ii) bring added value with respect to shelf life and product quality, and (iii) have acceptable process and energy costs. Adaptations to the product recipe, e.g., lowering the pH, could be also an option for optimization of the efficiency of PEF pasteurization.

In addition to the use of microsecond pulses at high voltages, developments are going on in the use of pulses of other duration and intensity. A PEF process with

Table 1 Product characteristics that have a positive effect (+) or reduce (−) the efficacy or applicability of a PEF process ranged for different product categories

		Product characteristic influencing efficiency and applicability of PEF					
		pH	Viscosity	Conductivity	Proteins	Fat	Dissolved gas
Product category	Fruit juices pH < 4.6	+					
	Fruit juice pH > 4.6	−					
	Smoothies and puree		−				
	Vegetable juice	−		−			
	Soup and sauce		−	−			
	Milk and milk-derived products	−	−	−	−	−	
	Eggs	−	−	−	−	−	
	Beer						−

pulses of tens of nanoseconds at very high electric field strength ($E \sim 100$ kV/cm) showed inactivation of *E. coli* in water (Guionet et al. 2014). It could be interesting to evaluate if these conditions can be used for food pasteurization at industrial scale.

Contrary to these very short and intense conditions, recent reporting of long pulses of milliseconds at relatively low electric field strength ($E = 2.7$ kV/cm) showed >5 $\log_{10}$ inactivation of *E. coli*, *L. monocytogenes*, *Salmonella*, *Lactobacillus*, and *S. cerevisiae* in both high-acid and low-acid fruit juices (Timmermans et al. 2019). Further research to these conditions that exhibit a non-thermal pulse effect responsible to inactivation needs to be done, as these conditions have the potential to be industrially applied in high volumes and might be applicable to solve some challenges summarized in Table 1 at lower energy consumption.

Acknowledgments Sincere thanks go to Remco Hamoen for his assistance to produce the figures.

References

Aguiló-Aguayo I, Soliva-Fortuny R, Martín-Belloso O (2009) Avoiding non-enzymatic browning by high-intensity pulsed electric fields in strawberry, tomato and watermelon juices. J Food Eng 92:37–43

Aguiló-Aguayo I, Soliva-Fortuny R, Martín-Belloso O (2010a) Color and viscosity of watermelon juice treated by high-intensity pulsed electric fields or heat. Innov Food Sci Emerg Technol 11:299–305

Aguiló-Aguayo I, Montero-Calderón M, Soliva-Fortuny R, Martín-Belloso O (2010b) Changes on flavor compounds throughout cold storage of watermelon juice processed by high-intensity pulsed electric fields or heat. J Food Eng 1:43–49

Alirezalu K, Munekata PES, Parniakov O, Barba FJ, Witt J, Toepfl S, Wiktor A, Lorenzo JM (2019) Pulsed electric field and milk heating for processing: a review on recent advances. J Sci Food Agric 100:16–24

Barsotti L, Dumay E, Mu TH, Diaz MDF, Cheftel JC (2001) Effects of high voltage electric pulses on protein-based food constituents and structures. Trends Food Sci Technol 12:136–144

Bean D, Bourdichon F, Bresnahan D, Davies A, Geeraerd A, Jackson T, Membré J-M, Pourkomailian B, Richardson P, Striker M, Uyttendaele M, Zwietering M (2012) Risk assessment approaches to setting thermal processes in food manufacture. ISLI Europe, Brussels

Bermúdez-Aguirre D, Fernández S, Esquivel H, Dunne PC, Barbosa-Cánovas GV (2011) Milk processed by pulsed electric fields: evaluation of microbial quality, physicochemical characteristics, and selected nutrients at different storage conditions. J Food Sci 76:S289–S299

Braddock RJ (1999) Single strength orange juice and concentrate. In: Braddock RJ (ed) Handbook of citrus by-products and processing technology. Wiley, New York, pp 53–83

Buckow R, Ng S, Toepfl S (2013) Pulsed electric field processing of orange juice: a review on microbial, enzymatic, nutritional, and sensory quality and stability. Compr Rev Food Sci Food Saf 12:455–467

Buckow R, Chandry PS, Ng SY, McAuley CM, Swanson BG (2014) Opportunities and challenges in pulsed electric field processing of dairy products. Int Dairy J 34:199–212

Cunningham FE (1995) Egg-product pasteurization. In: Stadelman WJ, Cotterill OJ (eds) Egg science and technology, 4th edn. The Haworth Press, Binghamton

De Luis R, Arias O, Puertolas E, Benede S, Sanchez L, Calvo M, Perez MD (2009) Effect of high-intensity pulse electric fields on denaturation of bovine whey proteins. Milchwissenschaft 64:422–426

Eberhard P, Sieber R (2006) Treatment of milk with pulsed electric fields – an alternative heat treatment. Mitt Lebensmittelunters Hyg 97:407–432

EC (853/2004) European Regulation laying down specific hygiene rules for on the hygiene of food stuffs

EC (2073/2005) European regulation on microbiological criteria for foodstuffs, 15 November 2005

Elez-Martínez P, Martín-Belloso O (2007) Effects of high intensity pulsed electric field processing conditions on vitamin C and antioxidant capacity of orange juice and *gazpacho*, a cold vegetable soup. Food Chem 1:201–209

Espachs-Barroso A, Van Loey A, Hendrickx M, Martín-Belloso O (2006) Inactivation of plant pectin methylesterase by thermal or high intensity pulsed electric field treatments. Innov Food Sci Emerg Technol 7:40–48

Evrendilek GA, Li S, Dantzer WR, Zhang QH (2004) Pulsed electric field processing of beer: microbial, sensory, and quality analysis. J Food Sci 69:M228–M232

FDA (2007) Grade "A" Pasteurized Milk Ordinance (PMO). Edition 2. Section 7: standards for Grade "A" milk and milk products. US Department of Health and Human Services, Public Health Services, Washington, DC

Fleischman GJ, Ravishankarb S, Balasubramaniam VM (2004) The inactivation of *Listeria monocytogenes* by pulsed electric field (PEF) treatment in a static chamber. Food Microbiol 21:91–95

Froning GW, Peters D, Muriana P, Eskridge K, Travnicek D, Sumner SS (2002) International egg pasteurization manual. United Egg Association, Alpharetta

Garcia-Amezquita LE, Primo-Mora AR, Barbosa-Cánovas GV, Sepulveda DR (2009) Effect of nonthermal technologies on the native size distribution of fat globules in bovine cheese-making milk. Innov Food Sci Emerg Technol 10:491–494

Gerlach D, Alleborn N, Baars A, Delgado A, Moritz J, Knorr D (2008) Numerical simulations of pulsed electric fields for food preservation: a review. Innov Food Sci Emerg Technol 9:408–417

Geveke DJ, Aubuchon I, Zhang HQ, Boyd G, Sites JE, Bigley ABW (2015) Validation of a pulsed electric field process to pasteurize strawberry purée. J Food Eng 166:384–389

Góngora-Nieto MM, Pedrow PD, Swanson BG, Barbosa-Cánovas GV (2003) Impact of air bubbles in a dielectric liquid when subjected to high field strengths. Innov Food Sci Emerg Technol 4:57–67

González-Casado S, Martín-Belloso O, Elez-Martínez P, Soliva-Fortuny R (2018) Application of pulsed electric fields to tomato fruit for enhancing the bioaccessibility of carotenoids in derived products. Food Funct 9:2282–2289

Grahl T, Märkl H (1996) Killing of microorganisms by pulsed electric fields. Appl Microbiol Biotechnol 45:148–157

Guan N, Liu L (2020) Microbial response to acid stress: mechanisms and application. Appl Microbiol Biotechnol 104:51–65

Guionet A, Joubert-Durigneux V, Packan D, Cheype C, Garnier J-P, David F, Zaepffel C, Leroux R-M, Teissie J, Blanckaert V (2014) Effect of nanosecond pulsed electric field on *Escherichia coli* in water: inactivation and impact on protein changes. J Appl Microbiol 117:721–728

Hamid-Samimi M, Swartzel KR, Ball HR (1984) Flow behaviour of liquid whole egg during thermal treatments. J Food Sci 49:132–136

Heinz V, Alvarez I, Angersbach A, Knorr D (2002) Preservation of liquid foods by high intensity pulsed electric fields – basic concepts for process design. Trends Food Sci Technol 12:103–111

Hermawan N, Evrendilek GA, Dantzer WR, Zhang QH, Richter ER (2004) Pulsed electric field treatment of liquid whole egg inoculated with *Salmonella* Enteritidis. J Food Saf 24:71–85

Huang K, Wang J (2009) Designs of pulsed electric fields treatment chambers for liquid foods pasteurization process: a review. J Food Eng 95:227–239

ICMSF (2005) Micro-organisms in foods 6: microbial ecology of food commodities, 2nd edn. Kluwer Academic/Plenum Publishers, New York

Irving D (2012) We zijn nu al aan het opschalen. VMT Voedingsmiddelenindustrie 16(17):11–13

Jaeger H, Meneses N, Knorr D (2009a) Impact of PEF treatment inhomogeneity such as electric field distribution, flow characteristics and temperature effects on the inactivation of *E. coli* and milk alkaline phosphatase. Innov Food Sci Emerg Technol 10:470–480

Jaeger H, Schulz A, Karapetkov N, Knorr D (2009b) Protective effect of milk constituents and sublethal injuries limiting process effectiveness during PEF inactivation of *Lb. rhamnosus*. Int J Food Microbiol 134:154–161

Jayathunge K, Stratakos AC, Cregenzán-Albertia O, Grant IR, Lyng J, Koidis A (2017) Enhancing the lycopene in vitro bioaccessibility of tomato juice synergistically applying thermal and non-thermal processing technologies. Food Chem 221:698–705

Jin T, Zhang H, Hermawan N, Dantzer W (2009) Effects of pH and temperature on inactivation of *Salmonella typhimurium* DT104 in liquid whole egg by pulsed electric fields. Int J Food Sci Technol 44:367–372

Kelly AL, Fox PF (2006) Indigenous enzymes in milk: a synopsis of future research requirements. Int Dairy J 6:707–715

Kempkes MA (2017) Industrial pulsed electric field systems. In: Miklavcic D (ed) Handbook of electroporation. Springer, Cham, p 2017

Lechevalier V, Guérin-Dubiard C, Anton M, Beaumal V, Briand ED, Gillard A, Le Gouar Y, Musikaphun N, Tanguy G, Pasco M, Dupont D, Nau F (2017) Pasteurization of liquid whole egg: optimal heat treatments in relation to its functional, nutritional and allergenic properties. J Food Eng 195:137–149

Lund P, Tramonti A, De Biase D (2014) Coping with low pH: molecular strategies in neutralophilic bacteria. FEMS Microbiol Rev 38:1091–1125

Marco-Móles R, Rojas-Graü M, Hernando I, Pérez-Munuera I, Soliva-Fortuny R, Martín-Belloso O (2011) Physical and structural changes in liquid whole egg treated with high-intensity pulsed electric fields. J Food Sci 76:C257–C264

Martín-Belloso O, Vega-Mercado H, Qin BL, Chang FJ, Barbosa-Cánovas GV, Swanson BG (1997) Inactivation of *Escherichia coli* suspended in liquid egg using pulsed electric fields. J Food Process Preserv 21:193–208

Matser AM, Mastwijk HC, Bánáti D, Vervoort L, Hendrickx M (2010) How to compare novel and conventional processing methods in new product development: a case study on orange juice. New Food 5:35–49

McAuley CM, Singh TK, Haro-Maza JF, Williams R, Buckow R (2016) Microbiological and physicochemical stability of raw, pasteurised or pulsed electric field-treated milk. Innov Food Sci Emerg Technol 38:365–373

Meneses N, Jaeger H, Moritz J, Knorr D (2011) Impact of insulator shape, flow rate and electrical parameters on inactivation of *E. coli* using a continuous co-linear PEF system. Innov Food Sci Emerg Technol 12:6–12
Milani EA, Alkhafaji S, Silva FVM (2015) Pulsed electric field continuous pasteurization of different types of beers. Food Control 50:223–229
Min S, Jin T, Zhang HQ (2003) Commercial scale pulsed electric field processing of tomato juice. J Agric Food Chem 51:3338–3344
Monfort S, Gayán E, Raso J, Condón S, Álvarez I (2010) Evaluation of pulsed electric fields technology for liquid whole egg pasteurization. Food Microbiol 27:845–852
Monfort S, Saldana G, Condón S, Raso J, Álvarez I (2012) Inactivation of *Salmonella* spp. in liquid whole egg using pulsed electric fields, heat, and additives. Food Microbiol 30:393–399
Montanari C, Tylewicz U, Tabanelli G, Berardinelli A, Rocculi P, Ragni L, Gardini F (2019) Heat-assisted pulsed electric field treatment for the inactivation of *Saccharomyces cerevisiae*: effect of presence of citral. Front Microbiol. https://doi.org/10.3389/fmicb.2019.01737
Mosqueda-Melgar J, Raybaudi-Massilia RM, Martín-Belloso O (2007) Influence of treatment time and pulse frequency on *Salmonella* enteritidis, *Escherichia coli* and *Listeria monocytogenes* populations inoculated in melon and watermelon juices treated by pulsed electric fields. Int J Food Microbiol 117:192–200
Nierop Groot M, Abee T, Van Bokhorst-van de Veen H (2019) Inactivation of conidia from three *Penicillium* spp. isolated from fruit juices by conventional and alternative mild preservation technologies and disinfection treatments. Food Microbiol 81:108–114
Odriozola-Serrano I, Bendicho-Porta S, Martın-Belloso O (2006) Comparative study on shelf life of whole milk processed by high-intensity pulsed electric field or heat treatment. J Dairy Sci 89:905–911
Odriozola-Serrano I, Soliva-Fortuny R, Martín-Belloso O (2008) Changes of health-related compounds throughout cold storage of tomato juice stabilized by thermal or high intensity pulsed electric field treatments. Innovative Food Sci Emerg Technol 9:272–279
Odriozola-Serrano I, Soliva-Fortuny R, Hernandez-Jover T, Martín-Belloso O (2009) Carotenoid and phenolic profile of tomato juices processed by high intensity pulsed electric fields compared with conventional thermal treatments. Food Chem 112:258–266
Oms-Oliu G, Odriozola-Serrano I, Soliva-Fortuny R, Martín-Belloso O (2009) Effects of high-intensity pulsed electric field processing conditions on lycopene, vitamin C and antioxidant capacity of watermelon juice. Food Chem 115:1312–1319
Palgan I, Muñoz A, Noci F, Whyte P, Morgan DJ, Cronin DA, Lyng JG (2012) Effectiveness of combined Pulsed Electric Field (PEF) and Manothermosonication (MTS) for the control of *Listeria innocua* in a smoothie type beverage. Food Control 25:621–625
Perez OE, Pilosof AMR (2004) Pulsed electric fields effect on the molecular structure and gelation of beta-lactoglobulin concentrate and egg white. Food Res Int 37:102–110
Pol IE, Mastwijk HC, Bartels PV, Smid EJ (2000) Pulsed-electric field treatment enhances the bactericidal action of nisin against *Bacillus cereus*. Appl Environ Microbiol 66:428–430
Raso J, Calderón ML, Góngora M, Barbosa-Cánovas GV, Swanson BG (1998) Inactivation of mold ascospores and conidiospores suspended in fruit juices by pulsed electric fields. Lebensm Wiss Technol 31:668–672
Reineke K, Schottroff F, Menesis N, Knorr D (2015) Sterilization of liquid foods by pulsed electric fields-an innovative ultra-high temperature process. Front Microbiol 6:400
Ritota M, Di Costanzo MG, Mattera M, Manzi P (2017) New trends for the evaluation of heat treatments of milk. J Anal Methods Chem:Article ID 1864832. https://doi.org/10.1155/2017/1864832
Safefood (2009). Consumer knowledge, attitudes and beliefs around the nutritional content of smoothies. ISBN: 978-1-905767-03-8
Sampedro F, Rodrogo M, Martinez A, Rodrigo D, Barbosa-Cánovas GV (2005) Quality and safety aspects of PEF application in milk and milk products. Crit Rev Food Sci Nutr 45:25–47

Schottroff F, Johnson K, Johnson NB, Bédard MF, Jaeger H (2020) Challenges and limitations for the decontamination of high solids protein solutions at neutral pH using pulsed electric fields. J Food Eng 268:109737

Schroeder S, Buckow R, Knoerzer K (2009) Numerical simulation of pulsed electric fields (PEF) processing for chamber design and optimisation. Seventh international conference on CDF in the minerals and process industries, CSIRO, Melbourne, Australia, 9–11 December 2009

Sharma P, Bremer P, Oey I, Everett DW (2014a) Bacterial inactivation in whole milk using pulsed electric field processing. Int Dairy J 35:49–56

Sharma P, Oey I, Bremer P, Everett DW (2014b) Reduction of bacterial counts and inactivation of enzymes in bovine whole milk using pulsed electric fields. Int Dairy J 39:146–156

Sharma P, Oey I, Everett DW (2015) Interfacial properties and transmission electron microscopy revealing damage to the milk fat globule system after pulsed electric field treatment. Food Hydrocoll 47:99–107

Sharma P, Oey I, Bremer P, Everett DW (2018) Microbiological and enzymatic activity of bovine whole milk treated by pulsed electric fields. Int J Dairy Techonol 71:10–19

Shishir MRI, Chen W (2017) Trends of spray drying: a critical review on drying of fruit and vegetable juices. Trends Food Sci Technol 65:49–67

Siemer C, Toepfl S, Heinz V (2014) Inactivation of *Bacillus subtilis* spores by pulsed electric fields (PEF) in combination with thermal energy – I. Influence of process- and product parameters. Food Control 39:163–171

Silva FVM, Gibbs P (2004) Target selection in designing pasteurization processes for shelf-stable high-acid fruit products. Crit Rev Food Sci Nutr 44:353–360

Silva FVM, Gibbs P (2009) Principles of thermal processing: pasteurization. In: Simpson R (ed) Engineering aspects of thermal food processing, contemporary food engineering series. CRC Press, Boca Raton

Strixner T, Kulozik U (2011) Egg proteins. In: Philips GO, Williams PA (eds) Handbook of food proteins, Woodhead Publishing series in food science, technology and nutrition. Woodhead Publishing, Amsterdam

Terefe NS, Buckow R, Versteeg C (2013) Quality related enzymes in plant based products: effects of novel food processing technologies. Part 2: pulsed electric field processing. Crit Rev Food Sci Nutr 55:1–15

Timmermans RAH, Mastwijk HC, Knol JJ, Quataert MCJ, Vervoort L, Van der Plancken I, Hendrickx ME, Matser AM (2011) Comparing equivalent thermal, high pressure and pulsed electric field processes for mild pasteurization of orange juice. Part I: impact on overall quality attributes. Innov Food Sci Emerg Technol 12:235–243

Timmermans RAH, Nierop Groot MN, Nederhoff AL, Van Boekel MAJS, Matser AM, Mastwijk HC (2014) Pulsed electric field processing of different fruit juices: impact of pH and temperature on inactivation of spoilage and pathogenic micro-organisms. Int J Food Microbiol 173:105–111

Timmermans RAH, Nederhoff AL, Nierop Groot MN, Van Boekel MAJS, Mastwijk HC (2016) Effect of electrical field strength applied by PEF processing and storage temperature on the outgrowth of yeasts and moulds naturally present in a fresh fruit smoothie. Int J Food Microbiol 230:21–30

Timmermans RAH, Mastwijk HC, Berendsen LBJM, Nederhoff AL, Matser AM, Van Boekel MAJS, Nierop Groot MN (2019) Moderate intensity Pulsed Electric Fields (PEF) as alternative mild preservation technology for fruit juice. Int J Food Microbiol 298:63–73

Toepfl S (2011) Pulsed electric field food treatment – scale up from lab to industrial scale. Procedia Food Sci 1:776–779

Toepfl S (2012) Pulsed electric field food processing – industrial equipment design and commercial applications. Stewart Postharvest Rev 8:1–7

Vallverdú-Queralt A, Odriozola-Serrano I, Oms-Oliu G, Lamuela-Raventós RM, Elez-Martínez P, Martín-Belloso O (2013) Impact of high-intensity pulsed electric fields on carotenoids profile of tomato juice made of moderate-intensity pulsed electric field-treated tomatoes. Food Chem 141:3131–3138

Vervoort L, Van der Plancken I, Grauwet T, Timmermans RAH, Mastwijk HC, Matser AM, Hendrickx ME, Van Loey A (2011) Comparing equivalent thermal, high pressure and pulsed electric field processes for mild pasteurization of orange juice. Part II: impact on specific chemical and biochemical quality parameters. Innov Food Sci Emerg Technol 12:466–477

Walkling-Ribeiro M, Noci F, Cronin DA, Lyng JG, Morgan DJ (2008) Inactivation of *Escherichia coli* in a tropical fruit smoothie by a combination of heat and pulsed electric fields. Food Microbiol Saf 73:M395–M399

Walkling-Ribeiro M, Noci F, Cronin DA, Lyng J, Morgan DJ (2010) Shelf life and sensory attributes of a fruit smoothie-type beverage processed with moderate heat and pulsed electric fields. LWT Food Sci Technol 43:1067–1073

Walkling-Ribeiro M, Rodríguez-González O, Jayaram SH, Griffiths MW (2011) Processing temperature, alcohol and carbonation levels and their impact on pulsed electric fields (PEF) mitigation of selected characteristic microorganisms in beer. Food Res Int 44:2524–2533

Walter L, Knight G, Ng SY, Buckow R (2016) Kinetic models for pulsed electric field and thermal inactivation of *Escherichia coli* and *Pseudomonas fluorescens* in whole milk. Int Dairy J 57:7–14

Wouters PC, Alvarez I, Raso J (2001) Critical factors determining inactivation kinetics by pulsed electric field food processing. Trends Food Sci Emerg Technol 12:112–121

Wray E (2015) Reducing microbial spoilage of beer using pasteurization. In: Hill AE (ed) Brewing microbiology. Managing microbes, ensuring quality and valorising waste, Woodhead Publishing series in food science, technology and nutrition. Woodhead Publishing, Amsterdam

Wu L, Zhao W, Yang R, Yan W, Sun Q (2016) Aggregation of egg white proteins with pulsed electric fields and thermal processes. J Sci Food Agric 96:3334–3341

Yogesh K (2016) Pulsed electric field processing of egg products: a review. J Food Sci Technol 53:934–945

Yu LJ, Ngadi M, Raghaven V (2009) Effect of temperature and pulsed electric field treatment on rennet coagulation properties of milk. J Food Eng 95:115–118

Yu LJ, Ngadi M, Raghaven V (2012) Proteolysis of cheese slurry mad from pulsed electric field-treated milk. Food Bioprocess Technol 5:47–54

Zhao W, Yang R, Tang Y, Lu R (2007) Combined effects of heat and PEF on microbial inactivation and quality of liquid egg whites. Int J Food Eng 4:article 12

Zhong S, Vendrell-Pacheco M, Heskitt B, Chitchumroonchokchai C, Failla M, Sastry SK, Francis DM, Martín-Belloso O, Elez Martinez P, Kopec RE (2019) Novel processing technologies as compared to thermal treatment on the bioaccessibility and Caco-2 cell uptake of carotenoids from tomato and kale-based juices. J Agric Food Chem 67:10185–10194

Pulsed Electric Fields in the Potato Industry

Kevin Hill, Robin Ostermeier, Stefan Töpfl, and Volker Heinz

Abbreviations

PEF	Pulsed electric field
PPO	Polyphenol oxidase
ROI	Return of investment

Usage of Pulsed Electric Field (PEF) in the Potato Industry

The first pulsed electric field (PEF) system was installed in the potato processing industry in 2010. Nowadays, over 100 machines are in use in the industry, most of them for French fries and potato chips (Elea Gmbh 2020). In the past, a preheater was applied to soften the raw potato making it processable for French fry production. This resulted in high energy and water consumptions, as well as long startup and shutdown times. With a PEF system, it is possible to eliminate this preheating step, reducing the energy and water consumption by up to 90%, and terminate startup and shutdown times, making it a standard in the French fry industry. The potatoes need about 5 to 8 seconds to traverse the PEF system, and the treatment itself only takes a few microseconds. After the treatment, it is possible to produce French fries with fewer fines during cutting and to reduce breakage as PEF leads to tissue softening due to a loss of turgor pressure induced by the electroporation of the cell membrane. The softer raw material results in a smoother cutting surface, reducing the oil absorption by approx. 10%. In addition to that, the breakage can be reduced resulting in longer French fries. The second biggest industrial field of

K. Hill (✉) · R. Ostermeier · S. Töpfl
Elea Vertriebs- und Vermarktungsgesellschaft mbh, Quakenbrück, Germany
e-mail: k.hill@elea-technology.com; r.ostermeier@elea-technology.com; s.toepfl@elea-technology.com

V. Heinz
Deutsches Institut für Lebensmitteltechnik e.V, Quakenbrück, Germany
e-mail: v.heinz@dil-ev.de

J. Raso et al. (eds.), *Pulsed Electric Fields Technology for the Food Industry*,
Food Engineering Series, https://doi.org/10.1007/978-3-030-70586-2_9

application for PEF is the potato chips production. Here the main benefits are quality based since in potato chips industry no preheating is used. Possible quality improvements include an increase in yield as a result of improved cutting and reduced starch loss and oil uptake reductions of up to 20%. The reduced oil uptake in chips is more pronounced compared to French fries due to their larger surface area and the increased frying time. This relatively long frying time (3–10 min) also allows frying time and temperature adjustments. Additionally, the open cell structure increases the water diffusion rate positively impacting the oil content, color, and crispness of the final product. This is even more pronounced in vegetable chips with high sugar content, such as carrots, beetroot, or sweet potatoes. Here the frying adjustment that is enabled by PEF can help to reduce the acrylamide content and to improve the color of the final product. Furthermore, innovative cuts with increased yields such as spiral for French fries or waffle cutting for chips are possible. The greater flexibility of the raw material also offers new possibilities for hard-to-cut raw materials like sweet potato, cassava, or taro.

Production of French Fries

French fries are cut from whole, raw potatoes or other tubers. They undergo a par-frying step in vegetable oil before being deep-frozen. The potatoes used for French fry production need to have a certain size in order to cut even potato strips from them. Usually French fries are either straight cut, characterized by a smooth surface, or crinkle cut leading to a corrugated surface (United States Department of Agriculture 1967). In addition, potatoes for French fry production are characterized by a high starch content of about 14–18% leading to improved texture, crispiness, and yield. Additionally, a reducing sugar content below 1.5% is desirable to avoid strong browning and acrylamide formation (European Snack Association 2014). Potato varieties that fit these characteristics are, for example, Agria, Marena, or Fontane (Krause et al. 2005). In Fig. 1 the industrial processing steps for French fries can be seen.

Potatoes are stored in temperature-controlled warehouses prior to being delivered to the factories. Upon arriving at the factory, they are sorted, washed, and peeled with either abrasive or steam peeling. Potatoes not suited for French fries are placed in a side stream used for potato mash or other products such as flakes. Peeling can be conducted prior to or after the PEF treatment as the treatment does not

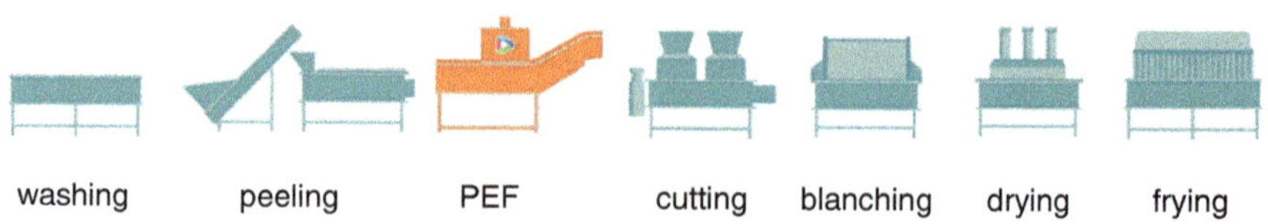

Fig. 1 Industrial French fry processing line including PEF. (Elea Gmbh 2020)

influence the peeling. However, removing the peel before PEF treatment helps to reduce the amount of dirt that is brought into the PEF system, leading to a greater reduction in water consumption. Therefore, most of today's processing lines place PEF unit after peeling. The PEF treatment is followed by cutting using a hydrojet cutter. Any remaining potato pieces with defects are removed by an optical sorter and can be used for side stream products such as puree or flakes. French fries fulfilling the quality standards are then blanched by scalding in hot water. This leads to gelatinization of starch and a partial destruction of the cell walls leading to an increased permeability (Botero-Uribe et al. 2017). Blanching is conducted in two steps. First a short blanching at high temperatures of about 80–100 °C for 2–5 minutes is applied. This high temperature inactivates the polyphenol oxidase (PPO) and peroxidase (Sanz et al. 2007). PPO can lead to browning in the potato due to oxidizing phenolic compounds to quinones (Botero-Uribe et al. 2017). For the second step, lower temperatures of approx. 65–70 °C for 15–20 min are applied to wash out reducing sugars that promote the formation of undesired color and acrylamide (Botero-Uribe et al. 2017). Agblor and Scanlon (2000) stated that blanching also helps in improving the texture of French fries. After blanching the product enters the dryer at 70–90 °C. The evaporation of water increases the dry matter content in the French fries ensuring a mealy texture of the product (van Loon et al. 2007). As water is removed during drying, less water is brought into the fryer, improving the frying performance. Furthermore, the shrinking of the pores in the potato cells reduces water transport through the product during frying. During frying, the evaporating water is replaced by oil. As less water is brought into the fryer, the frying time can be reduced, and less oil is absorbed into the product (Botero-Uribe et al. 2017). The par-frying step is performed for two main reasons. Firstly, the evaporation of water leads to an increase in dry mass, allowing a more uniform crust formation during frying. The crust is formed due to the exchange of water with the hot oil. The high energy transfer from the oil leads to gelatinization and formation of a gel. This results in a supporting structure around the cells (Botero-Uribe et al. 2017). The second reason for par-frying is to reduce the oil absorption during the second frying by shrinking the pores and reducing the water content. This leads to an improved texture of the final product (van Loon et al. 2007). The par-frying step is followed by cooling, freezing, and packaging of the product.

Challenges for the French Fry Production

The consumers or food service frying process is mainly responsible for the quality of the final product. Since this cannot be influenced by the producer, a high-quality, easy to prepare product is important. Therefore, a uniformly gelatinized par-fried product, relatively low in fat, is desirable. The high energy and water consumption necessary to soften potatoes by preheating are another challenge for the French fry producers. With an hourly consumption of 1.0–1.5 kW and 30 L water per tonne(s) of raw material for a PEF system, the needed energy and water can be reduced by

Fig. 2 Potato outlet of an industrial PEF system in a French fry processing line. (Elea Gmbh 2020)

up to 90%. In addition to that, a PEF system operates at ambient temperatures, reducing the microorganism load in the process water and the ambient heat. Due to the fact that long startup and shutdown times from the preheater are eliminated (up to 45 min) using a PEF system, it is possible to increase the total production time by 2 ½ days over the span of a year, depending on the production schedule. Figure 2 shows the potato outlet of a PEF system installed in the French fry industry.

The application of PEF can help to manage those quality concerns. The PEF-induced reduction in turgor pressure triggered by the loss of water leads to a softer product that is easier to cut and shows less breakage and, due to the smoother cutting surface, less oil absorption. The water evaporation facilitated by the open cell structure leads to the formation of a protective vapor layer, further blocking oil from entering the product. This increase in evaporation also allows a shortening of the frying time without affecting the final moisture content of the product. Additionally, the yield can be increased by up to 1.5% due to the easier-to-cut raw material and reduced losses.

Due to the homogeneous water distribution and enhanced water diffusion, the separation of the crust is less pronounced applying PEF, which is considered as a quality improvement (see Fig. 3).

Implementation in the Industrial French Fry Process

Based on the abovementioned advantages, the main benefits for using a PEF system are the increase in production yield, the quality improvement, and the elimination of the preheater which significantly reduces energy and water consumption. Today's

Fig. 3 Level of crust separation after frying on PEF-treated (left) and untreated (right) French fries. (Elea Gmbh 2020)

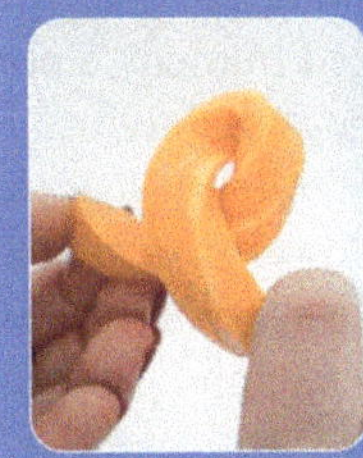

Fig. 4 Processing and quality benefits for French fry production achieved by PEF pretreatment. (Fauster et al. 2018; Heat and Control 2019; Elea Gmbh 2020)

PEF systems are able to process up to 70 t/h making them a perfect fit for the French fry industry, which is why they have already become a standard in this branch with over 80 installed machines worldwide. In Fig. 4 the main advantages of PEF, observed by researchers and in the industry, are displayed. Of course, additional costs are associated with the implementation of a PEF system into an existing processing line. The costs for a 26 t/h PEF plant with an annual production of 200,000 t raw material would amount to approx. 6 €/t raw material (depreciation of 3 years) including additional costs such as water and electricity (Elea Gmbh 2020). The return of investment (ROI) is positively influenced by the combination of processing advantages and quality improvements. The exact monetary savings are dependent on optimization of the current processing line and layout. With a state-of-the-art processing line, savings are lower compared to an older production line. An average ROI of about 1–2 years on a PEF system can be expected (Elea Gmbh 2020). Fauster et al. (2018) observed a reduction in feathering (small cuts on the French fry surface) by about 78.0% and breakage by 42.7% investigating the impact of PEF in industrial French fry production. It was also possible to reduce the starch loss by up to 16.9%. These figures show the potential benefits of PEF treatment in the French fry production.

Potato Chips Processing

Potato chips are prepared from specially bred potato varieties, fried in vegetable oil. For 1 tonne(s) of finished product, approx. 4 tonne(s) of raw material are used. The harvest season for potatoes is around June–July in Europe. Afterward, the potatoes are stored for the remaining season. The potatoes are stored under controlled conditions to keep them as high in quality as possible, preventing mold and controlling reducing sugars (European Snack Association 2014). Figure 5 shows the general production steps of potato chips processing.

Before washing, small and spoiled potato tubers are sorted out. Either before or after peeling, the potatoes pass the PEF system where they are treated within seconds. The peeling process is important because it greatly influences the quality and yield of the final product. The two main peeling methods utilized in today's potato industry are steam and abrasive peeling of which the second one is mainly used for chips processing lines as steam peeling might result in the formation of a cooking ring (Potatobuisness. 2018). Furthermore, abrasive peelers are lower in cost, do not require heat supply, and result in a good overall appearance. However, an abrasive peeler has about 25% more peeling loss compared to a steam peeler and produces a high amount of diluted waste (Fellows 2000).

After peeling, the potatoes are cut. The cutting blades have to be changed about every 60–120 minutes to ensure their sharpness and even cut slices of about

Fig. 5 Industrial potato chips processing line including PEF. (Elea Gmbh 2020)

1.5–2.0 mm. For special cuts such as wave or V-cut, different knifes are needed. After cutting the chips are fried using one of two possible frying methods. In batch frying, the chips are introduced into the fryer directly after cutting. In continuous processing lines, a washing step is added before frying to remove residues (such as starch and fines) from the chip surface and to wash out reducing sugars (Moreira et al. 1999). For lower-quality potatoes with higher sugar contents, blanching can be used to further enhance the leakage of reducing sugars, improving the color of the final product. However, blanching can lead to a less crunchy texture of the chip (De Meulenaer et al. 2016).

For batch frying temperatures around 145–165 °C are used. When the slices enter the fryer, the temperature drops by about 20 °C. This is caused by the lower temperature of the raw material and the evaporating water from the product. In most industrial batch fryers, the heater is working at full force during the whole frying process. After the initial drop, the temperature raises again, up to the initial frying temperature at which point the frying is considered to be finished. Batch-fried chips are characterized by a rougher surface and a crunchier texture as the surface starch is not washed off before frying (Riaz 2016). The other processing method opposed to batch frying is continuous frying. Here a higher initial temperature of about 175 °C is used, and the slices are washed and blanched before entering the fryer. The continuous frying curve is characterized by a linear drop of about 20 °C to a final temperature of about 155 °C. An optional deoiling step can be added at the end of the process before the chips are packaged.

Challenges for the Potato Chips Production

A variety of challenges arise with the production of potato chips at industrial level. Due to the required long shelf life, it is necessary to have an end product with a moisture content no higher than 2.5%. This results in long frying times or high temperatures as the initial moisture content of a potato is around 70%. This can cause excessive browning in the chips which is accompanied by the formation of acrylamide. It is especially important since manufactures have to control acrylamide levels due to the new EU regulation (Knott and Hill 2018). Besides selecting low-sugar raw material, the easiest way to control acrylamide levels is to fry at lower temperatures. This however brings unwanted changes to the final product such as reduced crunchiness and higher oil content. Another possible step is to introduce a blancher or hot water washing step in the frying line which helps to reduce the acrylamide precursors (reducing sugars and asparagine) but also negatively influences the texture of the final product. Furthermore, blanching is accompanied by high water and energy consumption.

PEF can help to handle the before mentioned challenges. Prior to frying, reducing sugars and asparagine can be washed out from the potato due to the electroporated cell membranes. This allows an elimination of the blanching step (Lindgren 2005; Ulrich et al. 2008; Heat and Control 2019) as up to 50% more glucose can be washed out after the application of PEF (Janositz et al. 2011; Genovese et al. 2019). Additionally, the frying temperature or time can be reduced without affecting the final product quality, as the water diffusion rate is increased due to PEF. Both the increased diffusion rate and the washing out of sugars can help to reduce the acrylamide content in the final product and to improve the color.

Another point where improvement can be achieved by the application of PEF is the fat content of the final product. As consumers are striving for healthier snack products, reduced fat products, including chips, are becoming more important. For the chip producer, this is also a further advantage because oil is more expensive than potatoes in most countries. By reducing the oil content, more potato is sold which helps to reduce production costs. This reduction in oil absorption is mainly achieved by the alteration of the raw material due to PEF: As PEF treatment results in a more flexible product, the surface area after cutting is reduced, and less fines and breakage are observed. This allows a lower oil adhesion on the chip surface. Furthermore, due to the increased water diffusion rate, a protective steam layer is formed around the chip during frying, blocking the oil from entering the product. In total, the oil content in the finished product can be reduced applying PEF by about 15%. This steam layer further prevents slices from sticking together (doubles) which would tend to have higher moisture content and would be rejected by optical sorting. The reduction in fines and starch losses during cutting also positively influences the process increasing the yield by 2%. The softening of the raw material enables new, innovative cuts such as waffle or deep rich cuts. Furthermore, hard-to-process raw material like cassava, manioc, or sweet potato can be considered for chips production.

Technical Implementation in a Chips Processing Line

Taking into consideration the abovementioned benefits, PEF is rapidly becoming a standard in the snack industry. With the fast rise in demand of PEF systems for chips producers, systems are being developed that fit the needs of the producer. As chips production lines are usually smaller compared to French fry lines, compact all-in-on PEF systems have been developed that are able to handle up to 6 t/h of raw material (see Fig. 6) with small and compact designs. Due to their short processing times of less than 10 s, PEF systems are easy to implement into existing processing lines.

The averaged energy consumption of a PEF system is quite low (1 kWh per tonne(s) of raw material), allowing relatively low operating costs. Aside from peeling, every other processing step benefits from the implementation of PEF. Due to the reduction in turgor pressure, the potato is softer and easier to cut, allowing for a longer use of the cutting knives and a smoother cutting surface. In Fig. 7 industrial scale benefits of a PEF system for chip processing are depictured. The results are gathered from manufacturers who have been using PEF for several years as well as researchers. It is however important to mention that any values on benefits are depended on raw material, the production process, and the layout of the processing line.

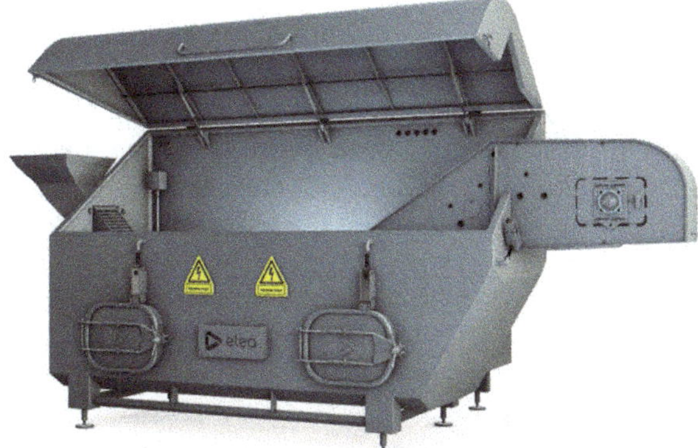

Fig. 6 PEF Advantage Belt One system with integrated generator, processing up to 6 t/h. (Elea Gmbh 2020)

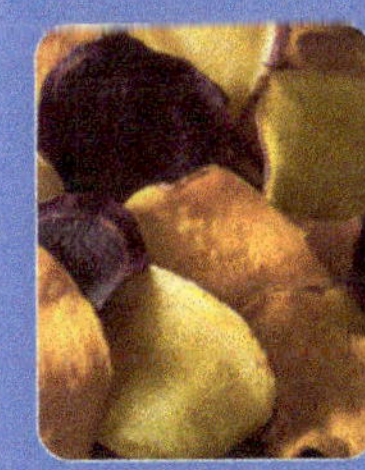

Fig. 7 Process and quality benefits for industrial chips processing achieved by the integration of PEF. (Fauster et al. 2018; Heat and Control 2019; Elea Gmbh 2020)

Naturally the implementation of PEF is accompanied by investment and running costs. Considering a processing capacity of 24,000 t raw potatoes per year, the running costs amount to about 6 €/t of raw material with a depreciation of 3 years. Additionally, the operating costs and electricity consumption amount to total cost of about 7 €/t raw material (Elea Gmbh 2020). However, installations of over 20 PEF systems in chips processing have shown a relatively fast ROI of 1–2 years due to cost savings such as 2% yield increase, reduction in doubles by 4%, and up to 15% reduction in oil content.

Conclusion

In summary, the application of PEF provides a broad spectrum of quality and processing benefits, making its implementation a standard for the potato processing industry. Here it has to be differentiated between quality benefits like reduction of acrylamide or improvement of color that have no direct influence on the ROI calculation of a PEF system and direct process and monetary benefits, like less water consumption or increased yield. The main driving factor using PEF pre-treatment for French fry processing is the possibility to replace a conventional preheater, reducing the energy consumption by almost 90%, leading to a ROI of below 2 years. For chips, the main monetary savings result out of an increase in yield and processing capacity. That is the reason why the application of PEF entered the French fry market first. But as quality is becoming a more important factor day by day, the implementation of PEF in chips processing lines has become more common nowadays. For chips processing however, it is important to consider that besides the PEF system, the following processing steps, such as cutting and frying, also have to be adjusted in order to obtain all benefits and a good, high-quality product.

References

Agblor A, Scanlon MG (2000) Processing conditions influencing the physical properties of French fried potatoes. Potato Res 43:163–177. https://doi.org/10.1007/BF02357957

Botero-Uribe M, Fitzgerald M, Gilbert RG, Midgley J (2017) Effect of pulsed electrical fields on the structural properties that affect French fry texture during processing. Trends Food Sci Technol 67:1–11. https://doi.org/10.1016/j.tifs.2017.05.016

De Meulenaer B, Medeiros R, Mestdagh F (2016) Acrylamide in potato products. In: In. Second Edition, Advances in Potato Chemistry and Technology

Elea Gmbh (2020) Pulsed electric field technology

European Snack Association (2014) Chip/Crisp Manufacturing - Preparation

Fauster T, Schlossnikl D, Rath F et al (2018) Impact of pulsed electric field (PEF) pretreatment on process performance of industrial French fries production. J Food Eng 235:16–22. https://doi.org/10.1016/J.JFOODENG.2018.04.023

Fellows PJ (2000) Food processing technology: principles and practice. CRC press, Cambridge

Genovese J, Tappi S, Luo W, Tylewicz U (2019) Important factors to consider for acrylamide mitigation in potato crisps using pulsed electric fields. Innov Food Sci Emerg Technol 55

Heat and Control (2019) Want your potato chips to be healthier?

Janositz A, Noack A-K, Knorr D (2011) Pulsed electric fields and their impact on the diffusion characteristics of potato slices. LWT - Food Sci Technol 44:1939–1945. https://doi.org/10.1016/j.lwt.2011.04.006

Knott M, Hill J (2018) European Snapshot - what are the forces shaping our industry today. Snack Mag

Krause T, Böhm H, Loges R (2005) Kartoffeln für Pommes und Chips Title

Lindgren M (2005) Potato treatment

Moreira RG, Castell-Perez E, Barrufet MA (1999) Deep fat frying: fundamentals and applications

Potatobuisness. (2018) The latest innovations in cutting, peeling, slicing

Riaz MN (2016) Snack foods: processing. Encycl Food Grains 3:414–422

Sanz T, Primo-Martín C, van Vliet T (2007) Characterization of crispness of French fries by fracture and acoustic measurements, effect of pre-frying and final frying times. Food Res Int 40:63–70. https://doi.org/10.1016/j.foodres.2006.07.013

Ulrich B, Haferkamp R, Kern M (2008) Method for the removal of acrylamide and/or Melanoidin-forming cell content from starchy plant material and plant material with reduced acrylamide and/or melanoidin content

United States Department of Agriculture (1967) United States Standards for Grades of Frozen French Fried Potatoes

van Loon WAM, Visser JE, Linssen JPH, et al (2007) Effect of pre-drying and par-frying conditions on the crispness of {French} fries. Eur Food Res Technol 225:929–935. https://doi.org/10.1007/s00217-006-0463-1

Applications of Pulsed Electric Fields in Winemaking

Marcos Andrés Maza, Ignacio Álvarez, and Javier Raso

Abbreviations

CI Color intensity
PEF Pulsed electric fields
TAC Total anthocyanin content
TPI Total polyphenol index

Introduction

Grapes are the most important fresh fruit crop worldwide. Global grape production in 2017 amounted to ca. 73.3 million metric tons; approximately 52% of the production was fermented into wine (OIV 2018). Global wine production in 2018 (excluding juice and musts) was around 279 Mhl (OIV 2018). More than 70% of that total production corresponded to red wine.

The main step in white and red winemaking is shown in Fig. 1. After the grapes are brought into the winery, the first step in winemaking involves removing leaves and stems from the bunches of grapes. In the case of red and rosé wines, the fruit is then crushed to release the juice, and the process of maceration of the resulting juice with solid part of the grapes (skins and seeds) begins. In the case of white wine, although maceration is maintained for a few hours in some cases, fermentation is nevertheless always restricted to the juice extracted from the grapes immediately upon crushing and pressing.

M. A. Maza
Departamento de Ciencias Enológicas y Agroalimentarias, Facultad de Ciencias Agrarias, Universidad Nacional de Cuyo, Mendoza, Argentina

Tecnología de los Alimentos, Facultad de Veterinaria, Instituto Agroalimentario de Aragón-IA2, Universidad de Zaragoza, Zaragoza, Spain

I. Álvarez · J. Raso (✉)
Tecnología de los Alimentos, Facultad de Veterinaria, Instituto Agroalimentario de Aragón-IA2, Universidad de Zaragoza, Zaragoza, Spain
e-mail: jraso@unizar.es

J. Raso et al. (eds.), *Pulsed Electric Fields Technology for the Food Industry*, Food Engineering Series, https://doi.org/10.1007/978-3-030-70586-2_10

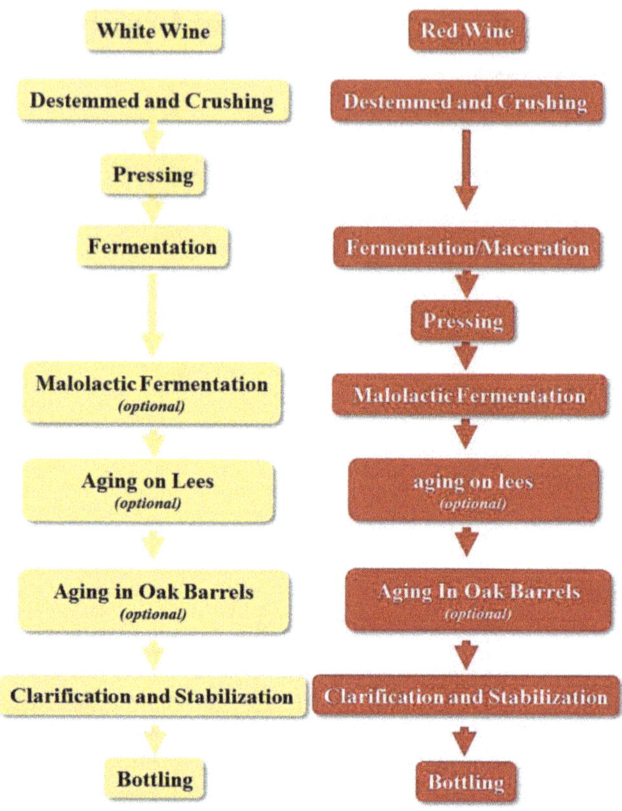

Fig. 1 Comparison of the stages for obtaining white and red wine

During red winemaking, the maceration that occurs simultaneously with alcoholic fermentation is extended from several days to weeks. The alcohol generated during fermentation enhances the extraction of polyphenols from the seeds and skins. These compounds give red wines their appearance, taste, and flavor, as well as their capability of aging.

Rosé wines are obtained from red grapes exposed to brief pre-fermentative maceration. Grape pomace is maintained in contact with the juice for a short period of time (12–24 h) at cool temperatures until sufficient color has been extracted. The juice is subsequently drawn and fermented in a manner similar to white wine.

Although wine fermentation has been traditionally conducted by yeast derived from the grapes or present in the winery, the current standard is to inoculate a commercial preparation of yeasts of which the characteristics are well known. During fermentation, yeast not only transforms the juice's sugar in ethanol, but it also engenders a wide array of aromatic compounds that contribute to the flavor attributes of wine. Alcoholic fermentation is regarded as completed once sugar levels drop to a low, undetectable level, generally below 2 $g.L^{-1}$. After having undergone alcoholic fermentation, the wine may be treated to foster a second process called malolactic fermentation. In this procedure that generally occurs in red wines, malic, a grape acid, is converted to lactic acid by lactic acid bacteria that can be indigenous to the winery or stem from commercial preparations.

After several weeks finishing the stage of fermentation, the wine is racked: it is separated from solids (yeast and bacterial cells, precipitated tannins, proteins, tartrate crystals) that settle during spontaneous or induced clarification. To enhance clarification and stability, wines are commonly chilled and filtered prior to bottling. Bottled wines are normally aged at the winery for a period of several months. Some red wines are aged in oak barrels prior to bottling for a period ranging from a few months to several years. This aging period allows for an important chemical transformation that leads to flavor modification and integration. Following in-barrel maturation, the wines are typically subjected to further in-bottle aging at the winery before release.

Another aging technique traditionally conducted in the production of sparkling wines but more recently applied in the production of white and red wines is aging on the lees. It consists in maintaining the wine in contact with the yeast after the fermentation process is over. During this process, yeast autolysis occurs, and polysaccharides of the yeast cell wall, mainly mannoproteins, are released. These compounds improve the wine's aromatic and gustatory complexity.

In summary, winemaking is a multistep procedure involving a series of microbial and biochemical processes wherein grape components are transformed into taste and aroma characteristics of wine. Wineries can make good use of PEF by applying the technology to different steps with the purpose of impacting the above-described processes involved in winemaking. A PEF-induced irreversible electroporation of the cytoplasmic membrane of grape skin cells may facilitate the release of polyphenols located in the skin cells during the process of maceration, thereby improving extraction yields or reducing maceration times. On the other hand, the irreversible electroporation of the cytoplasmic membranes of bacteria, yeast, and molds causes them to lose their selective permeability, thereby leading to microbial death. The capability of PEF to inactivate microbial cells at low temperatures, thereby avoiding the harmful effect of heat on the characteristics of wine, may make it an alternative to the use of SO_2 and filtration for controlling microbial spoilage in wineries. Finally, electroporation accelerates the release of mannoproteins from the cell wall of yeast during "aging on the lees."

The Application of PEF to Improve Red Winemaking

Phenolic compounds play a highly significant role in red winemaking. Sensory characteristics of red wine such as color, taste, and aging properties are strongly dependent on a wine's polyphenolic composition (Cheynier et al. 2006). On the other hand, in vitro and in vivo studies have shown that certain phenolic compounds of red wine possess antioxidant properties that may be beneficial to human health, including eventual protection against cardiovascular diseases and cancer (Nichenametla et al. 2006).

Two kinds of phenolic compounds are present in wine: non-flavonoids (phenolic acids and stilbenes) and flavonoids (anthocyanins, flavan-3-ols, and flavonols). The monomeric forms of anthocyanins are the main agent responsible for the red color of young red wines, and they also play a role in the development of red polymeric pigments during wine aging thanks to the association between anthocyanin pigments and other polyphenols, especially tannins and phenolic acids (Salas et al. 2003). Flavan-3-ols (tannins) are a large family of polyphenolic compounds mainly responsible for the astringency, bitterness, and structure of wine. The final group of flavonoids is flavonols: they contribute to bitterness, they display antioxidant activity, and they affect red wine color by acting as copigments of anthocyanins. Among other non-flavonoid compounds, stilbenes have also been recognized as compounds bearing antioxidant and anti-inflammatory properties (Tomera 1999).

Aroma is also an important quality attribute of wines. The aroma of young wine emerges mainly in the course of fermentation. Although grapes and must from non-floral grape varietals do not possess intense or explicit flavors, fresh fermented wine has pleasant aromas associated with its varietal origin (Delfini et al. 2001). It is estimated that more than 40 different aromatic molecules such as acids, alcohols, esters, aldehydes, terpenols, lactones, norisoprenoids, and volatile phenols are formed or released from precursors during fermentation (Loscos et al. 2007).

The juice of red grapes has a low amount of phenolic compounds and aromatic molecules. Maceration is therefore required in order to enrich the fermenting must with polyphenols and aromatic precursors located in the pomace. Whereas tannins are located in the cells of both grape skins and seeds, anthocyanins and aromatic precursors are solely located inside the grape skin cells (Amrani-Joutei and Glories 1995). Maceration is therefore one of the most critical technological operations for wineries, not only in view of its major influence on wine quality but also because it is the stage with the highest energy and manpower requirements (Togores 2011).

In traditional maceration, the solid parts of grape berries generally remain in contact with the fermenting must during the entire fermentation process (7–10 days); maceration time can even be extended for several days beyond fermentation in order to obtain a wine with adequate polyphenol content. This long period of required contact is due to the presence of cellular barriers such as the cytoplasmic membrane and the cell walls, which hamper the mass transfer of compounds of interest from inside the cells to the fermenting must.

Several drawbacks are involved in maintaining the grape pomace in contact with the must during fermentation. During maceration, approximately 20% of the fermentation tanks is occupied by the solid parts of the grapes, thereby reducing the tanks' effective volume and, as a consequence, the winerys' production capacity. This problem becomes particularly significant during the harvesting peak, because reducing maceration time with the purpose of increasing production capacity may compromise wine quality. Other negative side effects are related with the difficulty of controlling fermentation temperature when the solid parts are present in the fermentation tanks and with the workforce and energy consumption required to periodically pump the wine over the skin mass that rises to the top of the fermentation tanks.

The current techniques used in wineries to improve extraction of polyphenols and to reduce maceration time are either enzymatic treatments or procedures based on heating, such as thermovinification and flash expansion. Commercial pectolytic enzymes that act on the wall of the grape skin cells are often used by wineries to improve color and aroma. However, certain contradictory results attributed to the different nature and activities of the commercial enzyme preparations, as well as the presence of unwanted enzymes such as β-glucosidase in the preparations, have been reported (Bautista-Ortín et al. 2005). Thermal techniques, on the other hand, present a series of problems, such as the difficulty involved in stabilizing the extracted phenolic compounds (Morel-Salmi et al. 2006; Samoticha et al. 2017), a loss of varietal aromas (Geffroy et al. 2015), and a consumption of high quantities of energy.

Effect of PEF on the Extraction of Phenolic Compounds During Maceration in Red Winemaking

The application of PEF to accelerate and/or increase the extraction of phenolic compounds during the maceration stage in red winemaking has been thoroughly investigated. Some of the studies have pointed out that PEF also improves the extraction of aromatic compounds (Delsart et al. 2012; Garde-Cerdán et al. 2013). The treatment is generally applied to crushed and destemmed grapes with the objective of causing irreversible electroporation of the cytoplasmic membrane of grape skin cells, which contain the majority of the compounds required for obtaining quality red wine. Pores formed in the cell membrane improve mass transfer rates of the compound of interest without having to entirely disintegrate the cell envelopes.

The potential of PEF for improving the extraction of polyphenols in red winemaking was initially evaluated using batch treatment chambers featuring parallel electrodes with a capacity ranging from a few grams to a few kilograms. The benefits of electroporation were demonstrated for several varieties of grapes including *Tempranillo*, *Grenache*, *Graciano*, *Mazuelo*, *Merlot*, *Aglianico*, *Cabernet Sauvignon*, and *Cabernet Franc*, harvested in Spain, France, Lebanon, and Italy. The main results obtained in those experiments in terms of improvement in color intensity (CI), total anthocyanin content (TAC), and total polyphenol index (TPI) are shown in Table 1.

The improvements obtained in different oenological parameters depending on polyphenol extraction by application of PEF treatments ranged from 20% to 60% as compared with wine obtained from untreated grapes. This effect seemed to depend on the intensity of the treatment, the grape variety, and grape maturity. PEF treatments were generally more effective at higher electric field strengths, although for some varieties such as *Graciano* the most suitable treatment was at the lowest

Table 1 Improvements in color intensity (CI), anthocyanin content (TAC), and total polyphenol index (TPI) in wines obtained from grapes treated by PEF in batch

Grape variety	PEF treatment conditions	Δ CI (%)	ΔTAC (%)	Δ TPI (%)	Refs.
Tempranillo	5 kV.Cm^{-1}; 50 pulses; 1.8 kJ.Kg^{-1}	13	15	18	López et al. (2008b)
	10 kV.Cm^{-1}; 50 pulses; 6.7 kJ.Kg^{-1}	23	26	24	
Grenache	2 kV.Cm^{-1}; 50 pulses; 0.4 kJ.Kg^{-1}	26	11	18	López et al. (2008a)
	5 kV.Cm^{-1}; 50 pulses; 1.8 kJ.Kg^{-1}	21	32	14	
	10 kV.Cm^{-1}; 50 pulses; 6.7 kJ.Kg^{-1}	29	24	16	
Graciano	2 kV.Cm^{-1}; 50 pulses; 0.4 kJ.Kg^{-1}	19	18	14	
	5 kV.Cm^{-1}; 50 pulses; 1.8 kJ.Kg^{-1}	12	14	13	
	10 kV.Cm^{-1}; 50 pulses; 6.7 kJ.Kg^{-1}	6	7	12	
Mazuelo	2 kV.Cm^{-1}; 50 pulses; 0.4 kJ.Kg^{-1}	29	16	14	
	5 kV.Cm^{-1}; 50 pulses; 1.8 kJ.Kg^{-1}	51	35	31	
	10 kV.Cm^{-1}; 50 pulses; 6.7 kJ.Kg^{-1}	62	38	36	
Cabernet Sauvignon	5 kV.Cm^{-1}; 50 (exp. pulses); 2.1 kJ.Kg^{-1}	48	43	45	López et al. (2009)
Aglicano	1.0 kV.Cm^{-1}, 10^4 pulses; 50 kJ.Kg^{-1}	6	9	13	Donsì et al. (2010)
	1.5 kV.Cm^{-1}, 10^3 pulses, 10 kJ.Kg^{-1}	12	54	32	
	1.5 kV.Cm^{-1}, 2.5 x 10^3 pulses, 25 kJ.Kg^{-1}	19	76	38	
Merlot	0.5 kV.Cm^{-1}, 100 ms	ns	25	18	Delsart et al. (2012)
Cabernet Franc	0.8 kV.Cm^{-1}; 100 ms, 42 kJ.Kg^{-1}	20	49	51	El Darra et al. (2013)
	5 kV.Cm^{-1}; 1 ms, 53 kJ.Kg^{-1}	23	60	62	

ns no significant difference

intensity applied (in the range of 2–10 kV.cm^{-1}) and in *Cabernet Sauvignon* the intensity was intermediate (5 kV.cm^{-1}) (López et al. 2008a). These results indicate that, for some grape varieties, an intense electroporation of skin cells may engender a very intense release of phenolic compounds that precipitate rather than maintain themselves in the wine. The varying effect of PEF on the extraction of polyphenols for different grape varieties has been explained in terms of polyphenol extractability, which depends, in turn, on grape maturity, cell morphology, and composition of the skin cell wall (López et al. 2008a). Results suggest that the electroporation of grape skin cells is more useful in situations in which the extraction of phenolic compounds from the grape skins is more difficult (López et al. 2008a).

The implementation of PEF technology for improving polyphenolic extraction requires it to be applied in continuous flow, in order to be able to treat the hundreds or thousands of tons of grapes that have to be processed annually in a winery. The first studies conducted in continuous flow at pilot plant scale (120 kg.h^{-1}) permitted researchers to validate the results obtained in batch. More importantly, however, they were able to produce sufficient amounts in order to evaluate the sensory characteristics of wine obtained from PEF-treated grapes and its evolution in the course of aging (Puértolas et al. 2010a).

Table 2 provides an overview of or main results reported on the benefits of applying PEF in continuous flow in terms of improvement CI, TAC, and TPI. It shows results obtained at pilot plant scale in which hundreds of kg.h^{-1} were processed, but also others obtained on a semi-industrial scale with flows amounting to tons per hour. Studies obtained in continuous flow confirm those previously obtained at lab scale in batch with relatively low amounts. Differences observed in terms of global effect were unimportant despite the varying treatment chamber configurations used in continuous flow (while batch treatments were conducted with a parallel electrode treatment chamber, the configuration of the treatment chamber in continuous flow was colinear).

As compared with a parallel electrode treatment chamber configuration, the colinear configuration has lower energetic requirements thanks to its higher load resistance, and the chamber's circular section facilitates the flow of crushed grapes (van den Bosch 2007). However, the inhomogeneity in the distribution of the electric field in this configuration could cause a proportion of the grape skin cells either to remain unaffected or insufficiently affected by an electric field strength too low to cause electroporation.

López-Giral et al. (2015) studied the effect of PEF applied in continuous flow (400 kg.h^{-1}) on phenolic compound extraction from three grape varieties (*Graciano*, *Tempranillo*, *Grenache*) in the course of two vintages. Results showed that the effect

Table 2 Improvements in color intensity (CI), anthocyanin content (TAC), and total polyphenol index (TPI) in wines obtained from grapes treated by PEF in continuous flow

Grape variety	PEF treatment conditions	Δ CI%	Δ TAC%	Δ TPI%	Refs.
Cabernet Sauvignon	5 kV.Cm^{-1}, 3.7 kJ.Kg^{-1}	34	38	23	Puértolas et al. (2010a)
Grenache	4 kV.Cm^{-1}, 1.5 kJ.Kg^{-1}	13	25	24	Luengo et al. (2014)
Graciano	4.6 kV.Cm^{-1}, 720μs	103	84	51	López-Giral et al. (2015)
Tempranillo	4.6 kV.Cm^{-1}, 720μs	184	184	93	
Grenache	4.6 kV.Cm^{-1}, 720μs	230	458	98	
Grenache	4.0 kV.Cm^{-1}, 370μs	52	18	29	Maza et al. (2019)

of PEF depends on the grape variety, but they also showed that the different physicochemical composition of grapes, varying from one vintage to the next, exerted a significant influence on the effect of PEF. The improvement in the extraction of polyphenols from the PEF-treated grapes was more pronounced when the concentration of phenolic compounds in the skins was lower. Therefore, the application of PEF technology to improve the extraction of polyphenols would be of particular interest for those vintages in which the concentration of those compounds in the grape skins is relatively low.

The potential of PEF for reducing maceration time has also been investigated in studies conducted in continuous flow at pilot plant scale and at semi-industrial scale. *Cabernet Sauvignon* grapes treated by PEF at a flow of 120 kg.h^{-1} were vinified in tanks of 80 L capacity (Puértolas et al. 2010a). Grape pomace was separated from the fermenting must after 96 h, whereas in control the grape skins remained in contact with the fermenting must until the end of fermentation (144 h). Although the maceration time for wine obtained from *Cabernet Sauvignon* grapes treated by PEF (5 kV.cm^{-1}, 150μs, 3.7 kJ.kg^{-1}) was 48 h shorter, by the end of alcoholic fermentation, the resulting wine presented higher CI, TAC, and TPI values than the control wine.

Trials conducted on a semi-industrial scale in wineries confirmed results obtained at laboratory and pilot plant scale. Luengo et al. (2014) processed 3000 kg of *Grenache* grapes subjected to a PEF treatment (4.3 kV.cm^{-1}, 60μs) using a colinear treatment chamber at a flow of 1900 kg.h^{-1}. The wine obtained from PEF-treated grapes with a maceration time of 7 days was compared with wine obtained from untreated and PEF-treated grapes with the current maceration time used in the winery (14 days). After 7 days of maceration, the CI, TAC, and TPI of the wine obtained from grapes treated by PEF were 13%, 25%, and 24% higher, respectively, than in wine obtained from untreated grapes after 14 days of maceration. These results confirm the potential of PEF to obtain wine with a sufficient concentration of polyphenolic compounds while still applying moderate maceration times.

More recently, 12 tons of *Grenache* grapes were processed at a flow rate of 2500 kg.h^{-1}, and wines obtained from untreated and PEF-treated (4.0 kV.cm^{-1}, 370μs) grapes after 3 and 6 days of maceration were compared (Maza et al. 2019). The pre-treatment of grapes by PEF permitted to reduce maceration time from 6 to 3 days without causing a significant decrease in red wine color or in the concentration of polyphenolic compounds; alternatively, it allowed to increase those parameters depending on polyohenol extraction when maceration time was extended to 6 days. The more elevated tannin concentration observed in PEF-treated wines was a consequence of a higher degree of extraction of tannins located in the grape skins rather than in the seeds.

Evolution of Wine Obtained with Grapes Treated by PEF During Aging in Bottles and Oak Barrels

Wine obtained after concluding the maceration- fermentation step requires a stabilization stage followed by an aging period in bottle or in oak barrels prior to consumption (see Fig. 1). During the aging stage, polyphenols participate in successive reactions that exert a considerable influence on a wine's overall sensory quality (Setford et al. 2017). These modifications are a consequence of precipitations, degradation reactions, or polymerization reactions that lead to the formation of new stable compounds, and they help bring about important changes in the wine's sensory properties (Guadalupe and Ayestarán 2008). Since these reactions depend to a great extent on the type and degree of concentration of polyphenols obtained during fermentation, it is important to have knowledge about the evolution of wines produced from grapes treated with PEF during aging.

The evolution of wine obtained from PEF-treated grapes during aging in bottle and oak barrels was investigated for the first time by Puértolas et al. (2010b). The evolution of the main wine indexes depending on polyphenolic extraction along aging of *Cabernet Sauvignon* wine obtained from PEF-treated grapes follows the same trend as that of control wine in terms of evolution of color intensity, anthocyanin content, total phenolic content, and tannin content during wine aging in bottle or in oak barrels (6 months) followed by bottling (8 months). The preferable chromatic characteristics and higher polyphenolic content obtained from the PEF treatment after the fermentation process were maintained after aging.

Grenache wines obtained from PEF-treated grapes were aged during 24 months in bottles, as well as during 6 months in oak barrels followed by 18 months of aging in bottle (Maza et al. 2019). Similar conclusions on the evolution of indexes depending on polyphenolic extraction were obtained after even longer aging periods. In terms of color intensity, phenolic families (anthocyanins, hydroxycinnamic acids, flavonols, and flavanols), and individual phenolics, the *Grenache* wine obtained with PEF-treated grapes with different maceration times evolved similarly to the wine obtained with untreated grapes during aging in bottle or in oak barrel followed by bottling. In all cases, a decrease in the concentration of those compounds was observed across time, and the differences between the wine obtained from untreated and PEF-treated grapes persisted at the end of the aging period.

Sensory Properties of Wines Obtained with Grapes Treated by PEF

Aromatic molecules also play a highly significant role in red winemaking, owing to their contribution to sensory properties. Several investigations on the effect of PEF treatments on wine aroma compounds have been conducted.

The effect of PEF on the volatile composition of wine was investigated for the first time by Garde-Cerdán et al. (2013). The volatile composition of *Tempranillo* and *Graciano* wines was not improved by treating the grapes with PEF; however, the aromatic characteristics of *Grenache* wine were improved by increasing the concentration of monoterpenoids, β-ionone, total esters, and volatile phenol compounds.

With the same grape variety, another study was conducted in which control wine was compared with wine obtained from grapes treated by PEF after 3 and 6 days of maceration. The content of all major aromatic compounds was greater in the wines that had undergone a longer maceration period (Maza et al. 2019). Slight differences were observed among the wines obtained with untreated and PEF-treated grapes and subjected to the two maceration times investigated. In the case of aroma compounds liberated from precursors in the grapes and present in trace concentrations ($\mu g.L^{-1}$) in wine, the concentration of β-ionone associated with the floral aroma of "violets" that went undetected in control exceeded the odor threshold in the wines obtained from grapes treated by PEF.

PEF could therefore help extract certain individual aroma precursors present in grapes, thereby increasing the concentration of certain molecules which, once liberated from precursors, can exert a positive effect on wine aroma.

The effect of PEF on the organoleptic characteristics of wine has been also assessed by sensory evaluation. A panel of professional testers preferred *Merlot* wine obtained by grapes treated by PEF at 0.5 and 0.7 $kV.cm^{-1}$ for 40 ms than control wine or wine obtained by grapes subjected to a more intense PEF treatment (0.7 $kV.cm^{-1}$, 100 ms) (Delsart et al. 2012). While the diffusion of tannins to the wine after applying the most intense treatment was probably excessive, their diffusion was lower in the case of less intense PEF treatments, and the resulting wines were more aromatic and fruity, thus suggesting that PEF treatments promote an additional diffusion of aromatic compounds from the cell skins.

Results of a triangle and preference test showed significant sensory differences between the wines obtained with untreated or PEF-treated grapes after 3 and 6 days of maceration when aged either in bottle or in oak barrels. All panelists were able to differentiate the wines obtained with untreated or PEF-treated grapes from one another, and a majority of panelists (86%) preferred wines elaborated from grapes treated with PEF. These results indicate that the improvement in polyphenolic extraction brought about by the application of a PEF treatment prior to maceration permits to obtain wines that are sensorially different and that were preferred by panelists.

Application of PEF in Wineries for Microbial Inactivation

Microbial processes play a fundamental role in the final quality flavor and aroma characteristics of wine. While the yeast *Saccharomyces cerevisiae* conducts alcoholic fermentation, malolactic fermentation is conducted by lactic acid bacteria.

Microbial control in wineries is necessary in order to prevent wine spoilage in view of the microorganisms involved in both types of fermentation once the process is finished, but also in order to prevent the growth of other microorganisms that are not involved in the fermentation process, such as *non-Saccharomyces* yeast or acetic acid bacteria (Drysdale and Fleet 1988).

The development of spoiling microroganisms in wine may lead to great economic losses in wineries (Zuehlke et al. 2013). The growth of undesirable yeast during alcoholic fermentation may impair the development of the inoculated yeast and may likewise lead to important sensorial changes in the wine just after fermentation. On the other hand, lactic acid bacteria are responsible for the alteration known as "piqûre lactique": acetic acid bacteria are able to spoil the wine via ethanol acidification, and yeast from the genus *Brettanomyces* are involved in the formation of unpleasant odors described as "leather," "animal," and "horse sweat."

To avoid microbial spoilage, the most common practice applied in wineries is the addition of sulfur dioxide (SO_2), due to its effectivity in inhibiting the development of *non-Saccharomyces* yeast and spoilage bacteria. Although SO_2 is widely regarded as indispensable due to its combined antimicrobial and antioxidant activities, certain drawbacks are associated with it. It has been reported that sulfur dioxide may cause organoleptic deviations and unpleasant flavors when it is added in excess; furthermore, it can constitute a health hazard by producing allergic reactions in sensitive consumers (Lester 1995; Yang and Purchase 1985). One of the oldest challenges for wineries has thus been to find alternatives that permit the reduction or the elimination of the use of SO_2 in winemaking.

PEF was recently evaluated as one of the alternatives for obtaining additive-free wine, due to the treatment's ability to inactivate vegetative cells of microorganisms at lower temperatures that those used in thermal processing.

The application of PEF to must of the *Parellada* white grape variety prior to fermentation as an alternative to the addition of SO_2 for controlling the development of spoilage microorganisms was investigated for the first time by Garde-Cerdán et al. (2008). They carried out a reproducible fermentation of PEF-treated must without significantly modifying the composition of volatile compounds responsible for the typical white wine flavor stemming from alcoholic fermentation and aging. The concentration of SO_2 in wine can therefore be reduced to safer levels or even eliminated.

Puértolas et al. (2009) investigated the PEF resistance of wine spoilage microbiota such as *Dekkera anomala*, *Dekkera bruxellensis*, *Lactobacillus plantarum*, and *Lactobacillus hilgardii*. PEF treatments led to substantial inactivation of all investigated microorganisms, whereby yeasts were more sensitive to PEF than bacteria.

The most complete study to date of the inactivation of wine-associated microbiota by PEF was conducted by González-Arenzana et al. (2015). Four different PEF treatments in continuous process were tested in a continuous-flow system with the purpose of inactivating 25 species of yeast, lactic acid bacteria, and acetic acid bacteria, all associated with winemaking. The obtained inactivation was highly dependent on the type of microorganism. Overall, the level of inactivation varied from

0.64 to 4.94 log_{10} cycles. The results from that study were subsequently validated upon different winemaking stages of red *Tempranillo* Rioja wines: after alcoholic fermentation, after malolactic fermentation, and during aging in oak barrels.

The application of PEF to wine after alcoholic fermentation aimed to reduce the microbial population in order to facilitate the growth of a commercial strain *Oenococcus oeni* for subsequente malolactic fermentation (González-Arenzana et al. 2019a). The applied PEF treatment (33 $kV.cm^{-1}$, 105 ms, 158 $kJ.kg^{-1}$) led to a significant reduction in the yeast population and completely eliminated acetic acid bacteria; however, a lower level of inactivation was observed in the case of lactic acid bacteria in the four different wines assayed. In three of them, a slight shortening of the malolactic fermentation period duration monitored through malic lactic acid consumption after inoculating *O. oeni* and an improved sensorial quality after the PEF treatment were observed.

Wineries commonly add SO_2 to wine after malolactic fermentation to prevent spoilage and to reduce the activity of enzymes that cause wine oxidation. The effect of the stabilization process currently used in a winery (the addition of 30 $mg.L^{-1}$ SO_2) was compared with the effect of a PEF treatment (33 $kV.cm^{-1}$, 105 ms, 158 $kJ.kg^{-1}$), as well as with the combination of the addition of 15 $mg.L^{-1}$ SO_2 and a PEF treatment (González-Arenzana et al. 2019a). Four days after treatment, it was observed that the PEF treatment was more effective in reducing the microbial population of the three investigated wines than the addition of SO_2. After 6 months of aging, neither yeast nor acetic bacteria were detected in any of the wines, independently of the treatment previously applied; moreover, some of the lactic acid bacteria were still viable in some of the wines. After these observations and the evaluation of the wines' physicochemical and sensory properties, the authors concluded that PEF constitutes an alternative to SO_2 in terms of microbial stabilization, and it can preserve and/or enhance physicochemical and sensory qualities.

One of the main problems involved in aging wine in oak barrels is the difficulty of disinfecting the barrels, since the wood has a porous structure. As a consequence, certain microorganisms surviving in the wood may spoil the wine in the barrels. The main agent responsible for wine spoilage during aging in oak barrels is *Brettanomyces/Dekkera* yeast. These yeasts that produce ethyl-phenols responsible for unpleasant odors in the wine are known to tolerate wine ethanol content and commonly applied doses of SO_2. The capability of PEF to decontaminate wines containing *Brettanomyces/Dekkera* yeast was investigated by González-Arenzana et al. (2019b). After the PEF treatment, they observed a significant microbial inactivation of the most important spoilage microorganisms active in aged wines, including *Brettanomyces/Dekkera*, thereby leading to a significant reduction of volatile phenols as compared with the untreated wine.

In conclusion, studies conducted to demonstrate the ability of PEF to inactivate wine spoilage microorganisms, along with validation of the technique for microbial decontamination in different stages of winemaking and with the demonstration that PEF did not negatively affect physicochemical and sensory properties, all support PEF's potential as an alternative that can permit the reduction or elimination of the use of SO_2 in wineries.

Application of PEF in Wineries for Accelerating Aging on the Lees

Aging on the lees is a technique traditionally used in the production of sparkling wines but has more recently been extended to white wine and even red wine (Pérez-Serradilla et al. 2008). The technique consists in maintaining the wine in contact with the lees for a period of time after fermentation. The lees are the residues that remain in the bottom of the fermentation tanks after alcoholic fermentation. Although their makeup is variable, they are mainly composed of yeast (Guilloux-Benatier and Chassagne 2003).

During the aging of wine on the lees, yeast autolysis takes place, i.e., a lytic event in the cells caused by intracellular yeast enzymes. As a consequence of this yeast degradation, certain yeast compounds such as proteins, nucleic acids, lipids, and polysaccharides are transferred to the wine (Charpentier et al. 2004). One of the transferred compounds exerting an effect on physicochemical and sensory properties of wine is mannoproteins: highly glycosylated proteins which constitute the major component of the cell wall in yeast. Their presence in wine produces positive effects such as the reduction of haze formation, the prevention of tartaric salt precipitation, diminution of astringency, and notable improvement in terms of mouthfeel, aroma intensity, and color stability (Pérez-Serradilla and de Castro 2008).

Traditional aging on the lees is an extended process that lasts from a few months to years; it requires considerable investment in equipment (tanks, barrels), entails elevated labor costs (periodic stirring, bâtonnage, and sensorial analyses), and implies immobilization of winery stocks. Furthermore, this practice may negatively affect wine quality, thereby increasing the risk of wine oxidation and microbial contamination. Due to the drawbacks associated with aging on the lees, wineries have a great interest in reducing the duration of the process.

It has been recently proven that PEF treatment triggers the autolysis of *S. cerevisiae* and accelerates the release of mannoproteins (Martínez et al. 2016). Several mechanisms associated with electroporation could be involved in the induction of autolysis by PEF. On the one hand, electroporation would permit water from the surrounding media to enter into the cytoplasm, where there is a higher solute concentration. The decrease of osmotic pressure within the cytoplasm as a consequence of the water inlet could lead to plasmolysis of the organelles and the release of the enzymes involved in cell wall degradation (proteases and β-glucanases). On the other hand, the electroporation of the cytoplasmic membrane by PEF could facilitate the contact of those released enzymes with the outermost layer of the yeast cell wall where the mannoproteins are located (Fig. 2). This effect, initially observed in yeast suspended in buffer, has also been validated in aging on the lees of white *Chardonnay* wine (Martínez et al. 2019).

Chardonnay aging in the presence of *S. cerevisiae* yeast treated with PEF was compared with traditional aging on the lees (Martínez et al. 2019). Mannoprotein release increased drastically in *Chardonnay* wine containing PEF-treated (5 and 10 kV.cm^{-1}, 75μs) yeasts.

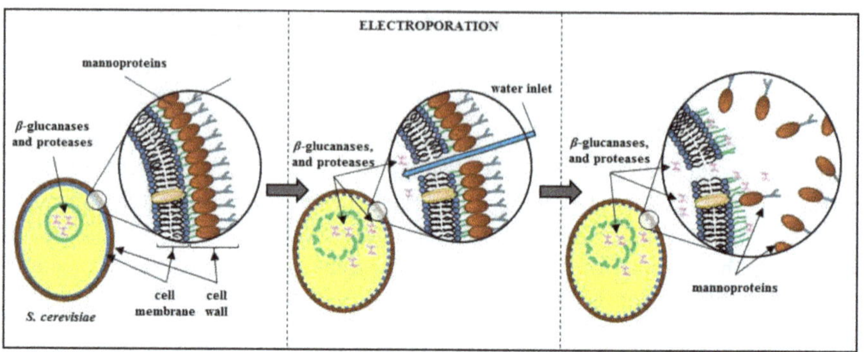

Fig. 2 Release of mannoproteins from *Saccharomyces cerevisiae* cells subjected to PEF treatments

While no mannoprotein release was observed in the first 10 days of aging on the lees in wine containing untreated yeast, mannoprotein concentration increased by 40 and 60% in wines containing yeast treated by PEF at 5 and 10 kV.cm^{-1}, respectively. After 30 days of incubation, the mannoprotein concentration in wines containing yeast treated under the most intense PEF treatment conditions reached the maximum value.

Control cells, on the other hand, required 6 months to reach the same maximum level. Chromatic characteristics, total polyphenol index, total volatile acidity, pH, ethanol, and CIELAB parameters of the wine were not affected during aging on the lees with PEF-treated yeast. On the other hand, mannoproteins released from PEF-treated yeast decreased wine turbidity after centrifugation and showed more pronounced foaming properties than mannoproteins released during traditional aging on the lees (Fig. 3).

Conclusions

PEF can be regarded as a technology with multiple applications for wineries in view of its ability to improve different operations conducted therein. The effects of PEF on different winemaking stages respond to some of the food industry's most important current demands, such as the improvement of energy efficiency and sustainability, as well as the reduction of the use of chemicals.

The physical techniques currently applied to reduce the duration of the maceration-fermentation stage in red winemaking such as thermovinification and flash expansion are based on heating (Morel-Salmi et al. 2006). Considerable efforts have been conducted to optimize energy consumption and heat recovery in processes based on heating. However PEF may yet offer further potential to help reduce energy consumption and operational costs while improving the overall sustainability of food production. Table 3 compares the energy requirements for grape

Fig. 3 Turbidity and foam measurements of *Chardonnay* wine containing *S. cerevisiae* cells either untreated (**a**), PEF treated at 5 kV.cm^{-1} for 75μs (**b**), or PEF treated at 10 kV.cm^{-1} 75μs (**c**) after 30 days of storage

Table 3 Estimation of the energetic requirements of thermovinification, flash expansion, and PEF for the improvement of polyphenol extraction during winemaking

Technology	Specific energy delivered to the grape (kJ.kg^{-1})	Additional specific energy[a] (kJ.kg^{-1})	Total specific energy (kJ.kg^{-1})
Thermovinification	161.92	40.68	202.6
Flash-release	251.88	45.86	297.7
PEF	6.70	–	6.70

[a]The energy required for the complete operation of the thermal system (pumps, refrigeration, and condensation systems)

processing via thermovinification, flash expansion, and PEF to obtain an equivalent effect in terms of polyphenol extraction in red winemaking.

The energy required to increase the temperature of grapes is much higher than the energy required to electroporate grape skin cells by PEF. As the PEF treatment only causes a temperature increment of a few degrees, one of its additional advantages is that it is not necessary to cool the grapes after heating to initiate fermentation. Another important issue when comparing thermal technologies and PEF from an energetic point of view is that PEF delivers energy directly to the product, thus making it much more efficient than heating techniques that transfer thermal energy through an intermediate medium such as water, water vapor, or oil. Whereas thermal techniques require water, PEF can help winemakers to obtain similar objectives without increasing water consumption. PEF is thus considerably more sustainable, since it reduces resource consumption as well as CO_2 emissions.

Regarding the implementation of PEF, another issue to be considered is the ease of installation of the PEF unit in the winery and the possibility of using the same generator for different applications (Fig. 4). The overall space required for the installation of thermovinification or flash expansion is much larger; generally, moreover, a winery has to undergo renovation in order to install such units along with their associated auxiliary units. PEF technology differs from others in view of its portability. The pulse generator unit can be separate from the treatment chamber, thereby allowing a rapid adaptation of the process, depending on the product to be treated. This can allow the PEF generator used to electroporate of the grape skins to be assigned to further uses, such as microbial inactivation or treatment of lees.

Table 4 shows the estimated costs for the introduction of a PEF unit in a winery according to production capacity. Although the same unit can be used for different applications, the total cost per liter of produced wine has been calculated only in view of grape treatment prior to maceration-fermentation. As observed, the application of PEF slightly would increase the cost per liter of wine.

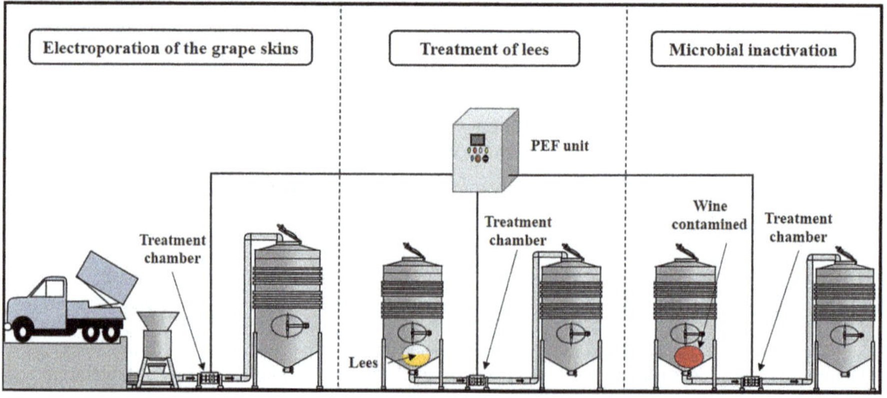

Fig. 4 Use of pulsed electric field for different applications in a winery

Table 4 Total costs of the implementation of a PEF unit in a winery according to the production capacity

	Small winery	Medium winery	Large winery
Processed grapes per year (tons)	500	3000	15,000
Voltage input (kV)	10	15	40
Production capacity (ton.h^{-1})	5	10	40
Investment (€)	60,000	80,000	250,000
Depreciation range (years)	10	10	10
Installation (€)	5500	6500	20,000
Replacement value (€)	72,000	96,000	300,000
Interest (%)	6	6	6
Depreciation range (€.annum^{-1})	7200	9600	30,000
Interest (€.annum^{-1})	3600	5760	18,000
Maintenance (€.annum^{-1})	500	1000	2500
Fixed costs (€.annum^{-1})	**11,300**	**16,360**	**50,500**
Total specific energy (kJ.kg^{-1})	8–10	8–10	8–10
Specific energy imput (kWh)	4.5	13.8	69.0
	0.90 kWh.ton^{-1}	1.38 kWh.ton^{-1}	1.73 kWh.ton^{-1}
Power price	0.13 €.kWh^{-1}	0.13 €.kWh^{-1}	0.13 €.kWh^{-1}
	0.58 €.h^{-1}	1.78 €.h^{-1}	8.90 €.h^{-1}
	0.12 €.ton^{-1}	0.18 €.ton^{-1}	0.22 €.ton^{-1}
Variable costs (€.annum^{-1})	**58**	**534**	**3338**
Total costs	11,358 €.annum^{-1}	16,894 €.annum^{-1}	53,838 €.annum^{-1}
Total costs per ton (€)	**22.72**	**5.63**	**3.59**
Total costs per liter of wine (€)	**0.035**	**0.009**	**0.005**

In conclusion, the development of pulse power systems capable of responding to the processing capacity demands of wineries, the low energy consumption involved in PEF's different applications in winemaking, and the easy incorporation of PEF treatment chambers into existing processing lines all constitute solid arguments in favor of a successful transfer of PEF technology to wineries in a near future.

References

Amrani-Joutei K, Glories Y (1995) Tannins and anthocyanins of grape berries: localization and extraction technique. Rev Fr Oenol (153):28–31

Bautista-Ortín AB, Martínez-Cutillas A, Ros-García JM, López-Roca JM, Gómez-Plaza E (2005) Improving colour extraction and stability in red wines: the use of maceration enzymes and enological tannins. Int J Food Sci Technol 40(8):867–878. https://doi.org/10.1111/j.1365-2621.2005.01014.x

Charpentier C, Dos Santos AM, Feuillat M (2004) Release of macromolecules by *Saccharomyces cerevisiae* during ageing of French flor sherry wine "Vin jaune". Int J Food Microbiol 96(3):253–262. https://doi.org/10.1016/j.ijfoodmicro.2004.03.019

Cheynier V, Dueñas-Paton M, Salas E, Maury C, Souquet J-M, Sarni-Manchado P, Fulcrand H (2006) Structure and properties of wine pigments and tannins. Am J Enol Vitic 57:298–305
Delfini C, Cocito C, Bonino M, Schellino R, Gaia P, Baiocchi C (2001) Definitive evidence for the actual contribution of yeast in the transformation of neutral precursors of grape aromas. J Agric Food Chem 49(11):5397–5408. https://doi.org/10.1021/jf010613a
Delsart C, Ghidossi R, Poupot C, Cholet C, Grimi N, Vorobiev E, Peuchot MM (2012) Enhanced extraction of phenolic compounds from Merlot grapes by pulsed electric field treatment. Am J Enol Vitic 63(2):205–211. https://doi.org/10.5344/ajev.2012.11088
Donsì F, Ferrari G, Fruilo M, Pataro G (2010) Pulsed electric field-assisted vinification of Aglianico and Piedirosso grapes. J Agric Food Chem 58(22):11606–11615. https://doi.org/10.1021/jf102065v
Drysdale GS, GH Fleet (1988) Acetic Acid Bacteria in Winemaking: A Review. American J Enology and Viticulture 39(2):143–54
El Darra N, Grimi N, Maroun RG, Louka N, Vorobiev E (2013) Pulsed electric field, ultrasound, and thermal pretreatments for better phenolic extraction during red fermentation. Eur Food Res Technol 236(1):47–56. https://doi.org/10.1007/s00217-012-1858-9
Garde-Cerdán, Teresa, A. Robert Marsellés-Fontanet, Margaluz Arias-Gil, Carmen Ancín-Azpilicueta, Olga Martín-Belloso (2008) Influence of SO2 on the Evolution of Volatile Compounds through Alcoholic Fermentation of Must Stabilized by Pulsed Electric Fields. European Food Research and Technology 227(2):401–8. https://doi.org/10.1007/s00217-007-0734-5
Garde-Cerdán T, González-Arenzana L, López N, López R, Santamaría P, López-Alfaro I (2013) Effect of different pulsed electric field treatments on the volatile composition of Graciano, Tempranillo and Grenache grape varieties. Innov Food Sci Emerg Technol 20:91–99. https://doi.org/10.1016/j.ifset.2013.08.008
Geffroy O, Lopez R, Serrano E, Dufourcq T, Gracia-Moreno E, Cacho J, Ferreira V (2015) Changes in analytical and volatile compositions of red wines induced by pre-fermentation heat treatment of grapes. Food Chem 187:243–253. https://doi.org/10.1016/j.foodchem.2015.04.105
Guilloux-Benatier M, Chassagne D (2003) Comparison of components released by fermented or active dried yeasts after aging on lees in a model wine. J Agr Food Chem 51(3):746–751. https://doi.org/10.1021/jf020135j
González-Arenzana L, Portu J, López R, López N, Santamaría P, Garde-Cerdán T, López-Alfaro I (2015) Inactivation of wine-associated microbiota by continuous pulsed electric field treatments. Innov Food Sci Emerg Technol 29:187–192. https://doi.org/10.1016/j.ifset.2015.03.009
González-Arenzana L, Portu J, López N, Santamaría P, Gutiérrez AR, López R, López-Alfaro I (2019a) Pulsed electric field treatment after malolactic fermentation of Tempranillo Rioja wines: influence on microbial, physicochemical and sensorial quality. Innov Food Sci Emerg Technol 51:57–63. https://doi.org/10.1016/j.ifset.2018.05.019
González-Arenzana L, López-Alfaro I, Gutiérrez AR, López N, Santamaría P, López R (2019b) Continuous pulsed electric field treatments' impact on the microbiota of red Tempranillo wines aged in oak barrels. Food Biosci 27:54–59. https://doi.org/10.1016/j.fbio.2018.10.012
Guadalupe Z, Ayestarán B (2008) Changes in the color components and phenolic content of red wines from *Vitis vinifera* L. Cv. "Tempranillo" during vinification and aging. Eur Food Res Technol 228(1):29–38. https://doi.org/10.1007/s00217-008-0902-2
Lester MR (1995) Sulfite sensitivity: significance in human health. J Am Coll Nutr 14(3):229–232. https://doi.org/10.1080/07315724.1995.10718500
López N, Puértolas E, Condón S, Álvarez I, Raso J (2008a) Application of pulsed electric fields for improving the maceration process during vinification of red wine: influence of grape variety. Eur Food Res Technol 227(4):1099. https://doi.org/10.1007/s00217-008-0825-y
López N, Puértolas E, Condón S, Álvarez I, Raso J (2008b) Effects of pulsed electric fields on the extraction of phenolic compounds during the fermentation of must of Tempranillo grapes. Innov Food Sci Emerg Technol 9(4):477–482. https://doi.org/10.1016/j.ifset.2007.11.001

López N, Puértolas E, Hernández-Orte P, Álvarez I, Raso J (2009) Effect of a pulsed electric field treatment on the anthocyanins composition and other quality parameters of Cabernet Sauvignon freshly fermented model wines obtained after different maceration times. LWT Food Sci Technol 42(7):1225–1231. https://doi.org/10.1016/j.lwt.2009.03.009

López-Giral N, González-Arenzana L, González-Ferrero C, López R, Santamaría P, López-Alfaro I, Garde-Cerdán T (2015) Pulsed electric field treatment to improve the phenolic compound extraction from Graciano, Tempranillo and Grenache grape varieties during two vintages. Innov Food Sci Emerg Technol 28:31–39. https://doi.org/10.1016/j.ifset.2015.01.003

Loscos N, Hernández-Orte P, Cacho J, Ferreira V (2007) Release and formation of varietal aroma compounds during alcoholic fermentation from nonfloral grape odorless flavor precursors fractions. J Agric Food Chem 55(16):6674–6684. https://doi.org/10.1021/jf0702343

Luengo E, Franco E, Ballesteros F, Álvarez I, Raso J (2014) Winery trial on application of pulsed electric fields for improving vinification of Garnacha grapes. Food Bioprocess Technol 7(5):1457–1464. https://doi.org/10.1007/s11947-013-1209-2

Martínez JM, Cebrián G, Álvarez I, Raso J (2016) Release of mannoproteins during *Saccharomyces cerevisiae* autolysis induced by pulsed electric field. Front Microbiol 7. https://doi.org/10.3389/fmicb.2016.01435

Martínez JM, Delso C, Maza MA, Álvarez I, Raso J (2019) Pulsed electric fields accelerate release of mannoproteins from *Saccharomyces cerevisiae* during aging on the lees of Chardonnay wine. Food Res Int 116:795–801. https://doi.org/10.1016/j.foodres.2018.09.013

Maza MA, Martínez JM, Hernández-Orte P, Cebrián G, Sánchez-Gimeno AC, Álvarez I, Raso J (2019) Influence of pulsed electric fields on aroma and polyphenolic compounds of Garnacha wine. Food Bioprod Process. https://doi.org/10.1016/j.fbp.2019.06.005

Morel-Salmi C, Souquet J-M, Bes M, Cheynier V (2006) Effect of flash release treatment on phenolic extraction and wine composition. J Agric Food Chem 54(12):4270–4276. https://doi.org/10.1021/jf053153k

Nichenametla SN, Taruscio TG, Barney DL, Exon JH (2006) A review of the effects and mechanisms of polyphenolics in Cancer. Crit Rev Food Sci Nutr 46(2):161–183. https://doi.org/10.1080/10408390591000541

OIV (2018) Report on the world vitivinicultural situation. Retrieved July 26, 2019, from Oiv.int website: http://www.oiv.int/en/oiv-life/oiv-2018-report-on-the-world-vitivinicultural-situation

Pérez-Serradilla JA, de Castro MDL (2008) Role of lees in wine production: a review. Food Chem 111(2):447–456. https://doi.org/10.1016/j.foodchem.2008.04.019

Puértolas E, López N, Condón S, Raso J, Álvarez I (2009) Pulsed electric fields inactivation of wine spoilage yeast and bacteria. Int J Food Microbiol 130(1):49–55. https://doi.org/10.1016/j.ijfoodmicro.2008.12.035

Puértolas E, Saldaña G, Álvarez I, Raso J (2010b) Effect of pulsed electric field processing of red grapes on wine chromatic and phenolic characteristics during aging in oak barrels. J Agric Food Chem 58(4):2351–2357. https://doi.org/10.1021/jf904035v

Puértolas E, Hernández-Orte P, Saldaña G, Álvarez I, Raso J (2010a) Improvement of winemaking process using pulsed electric fields at pilot-plant scale. Evolution of chromatic parameters and phenolic content of Cabernet Sauvignon red wines. Food Res Int 43(3):761–766. https://doi.org/10.1016/j.foodres.2009.11.005

Salas E, Fulcrand H, Meudec E, Cheynier V (2003) Reactions of anthocyanins and tannins in model solutions. J Agric Food Chem 51:7951–7961. https://doi.org/10.1021/jf0345402

Samoticha J, Wojdyło A, Chmielewska J, Oszmiański J (2017) The effects of flash release conditions on the phenolic compounds and antioxidant activity of Pinot noir red wine. Eur Food Res Technol 243(6):999–1007. https://doi.org/10.1007/s00217-016-2817-7

Setford PC, Jeffery DW, Grbin PR, Muhlack RA (2017) Factors affecting extraction and evolution of phenolic compounds during red wine maceration and the role of process modelling. Trends Food Sci Tech 69:106–117. https://doi.org/10.1016/j.tifs.2017.09.005

Togores JH (2011) Tratado de Enología. ISBN: 8484764141 Mundi-Prensa, España

Tomera JF (1999) Current knowledge of the health benefits and disadvantages of wine consumption. Trends Food Sci Technol 10(4):129–138. https://doi.org/10.1016/S0924-2244(99)00035-7
van den Bosch HFM (2007) 4 chamber design and process conditions for pulsed electric field treatment of food. In: Lelieveld HLM, Notermans S, de Haan SWH (eds) Food preservation by pulsed electric fields, pp 70–93. https://doi.org/10.1533/9781845693831.1.70
Yang WH, Purchase ECR (1985) Adverse reactions to sulfites. Can Med 133(9):865–880. Retrieved from https://www.ncbi.nlm.nih.gov/pmc/articles/PMC1346296/
Zuehlke JM, B Petrova CG Edwards (2013) Advances in the Control of Wine Spoilage by Zygosaccharomyces and Dekkera/Brettanomyces. Annual Review of Food Science and Technology 4(1):57–78. https://doi.org/10.1146/annurev-food-030212-182533.

Applying Pulsed Electric Fields to Improve Olive Oil Extraction

Ana Cristina Sánchez-Gimeno, Ignacio Álvarez, and Javier Raso

Abbreviations

EVOO	Extra virgin olive oil
EY	Extraction yield
OE	Oil extractability
PEF	Pulsed electric fields

Introduction

Olive oil is an edible oil widely appreciated for its sensorial characteristics, as well as for its high nutritional value (Nocella et al. 2018). A series of health benefits have been associated with olive oil consumption, including prevention of cardiovascular diseases and of several types of cancers. These health benefits are attributed to the presence of monounsaturated fatty acids (mainly oleic acid) and of minor components with biological properties such as sterols, tocopherols, phenols, and carotenoids (Schwingshackl and Hoffmann 2017). Although olive oil is regarded as an essential component of the Mediterranean diet, large amounts of olive oil are currently produced and consumed worldwide (Uylaser and Yildiz 2014).

Virgin olive oil is olive juice obtained without the use of chemicals or industrial refining processes. According to the International Olive Council, virgin olive oil is obtained from the fruit of the olive tree (*Olea europaea* L.) solely by mechanical or other physical means under certain conditions, particularly thermal conditions, that do not lead to alterations in the oil and in which the olives have not undergone any treatment other than washing, decantation, centrifugation, and filtration. The following parameters determine the quality of olive oil: free acidity, peroxide value, absorption coefficients in the ultraviolet region (K_{232} and K_{270}), fatty acid ethyl esters, and sensory analysis (characteristic aroma, flavour, and zero negative attributes) (European Comission Regulation, 1991). Based on these physico-chemical

A. C. Sánchez-Gimeno · I. Álvarez · J. Raso (✉)
Tecnología de los Alimentos, Facultad de Veterinaria, Instituto Agroalimentario de Aragón-IA2, Universidad de Zaragoza, Zaragoza, Spain
e-mail: jraso@unizar.es

J. Raso et al. (eds.), *Pulsed Electric Fields Technology for the Food Industry*, Food Engineering Series, https://doi.org/10.1007/978-3-030-70586-2_11

parameters and sensory properties, EU legislation classifies olive oil into different categories: extra virgin olive oil (EVOO), virgin olive oil, and lampant olive oil (European Comission Regulation, 1991). EVOO is the olive oil of the highest quality.

It is estimated that current industrial systems extract no more than 80% of the oil contained in the olive fruit: this represents a considerable monetary loss for olive oil firms (Chiacherini et al. 2007). Although the residual oil in the pomace is normally extracted using organic solvents, its quality is much lower because most of its nutritional and sensorial attributes are lost during refining.

The aim of this chapter is to evaluate the ability of pulsed electric field (PEF) technology to improve the degree of extraction and enhance oil quality in the industrial olive oil extraction process.

Olive Oil Extraction

Olive oil is extracted from the olive fruits using only mechanical or other physical means. In the traditional system, olives were crushed by a grindstone consisting of several stone wheels that revolved on a large stone bowl where the olive fruits were located. The oil was separated from the obtained olive paste in a hydraulic press in which the paste was spread on disks made of different fibres and stacked on top of one another. However, due to the discontinuity of the process, to the pressing systems' low capacity, to the high requirements in manpower, and to hygienic problems, this traditional system has been replaced by higher-capacity extraction systems where the machines work in flow (Fig. 1) (Kalogianni et al. 2019).

The first step in olive oil extraction consists in removal of leaves and olive washing. Once the leaves and other debris have been removed by passing the olives over a vibrating screen with a blower, the olives are washed. Washing aims to remove dirt and other impurities; during this step, the olives are shaken in metal nets to remove excess water.

After washing, the next step is olive crushing. In modern facilities, the traditional stone mill has been replaced by a hammer mill or disk crusher that works in continuous flow. During crushing, the olive fruit tissues are broken, thereby facilitating the release of the oil droplets from the vacuoles of the olive pulp cells. Recently, in some processing lines, a system called destoning or depitting has been introduced:

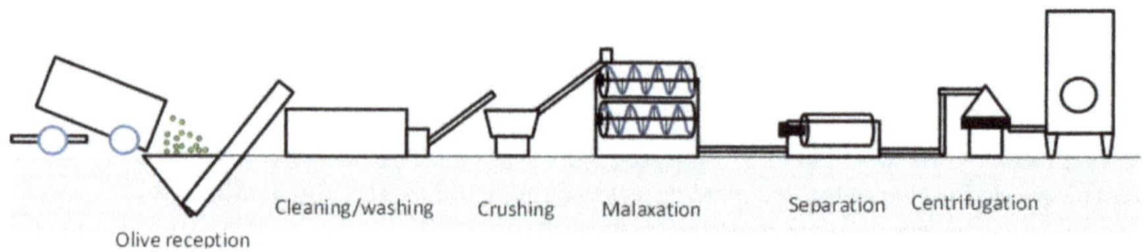

Fig. 1 Scheme of the olive oil processing line

it removes the olive fruit stones and ensures that the olive paste consists only of the fleshy part (Luaces et al. 2007). This prevents the transfer of oxidative enzymes such as phenoloxidase and lipoxygenase from the seeds (Servili et al. 2007).

The only softening operation in current olive oil extraction lines is malaxation, the step after crushing. During malaxation, the paste is slowly stirred (20–30 rpm) in a semi-cylindrical tank equipped with a heating jacket for temperature regulation. Malaxation is conducted at moderate temperatures (20–35 °C) for a period ranging from 30 to 90 min (Clodoveo 2012). This process encourages the coalescence of the oil droplets released from the olive pulp after crushing, thus preparing for separation of the oil from the olive paste. Some coadjuvants added during malaxation, such as talc, calcium carbonate, or sodium chloride, help to break down the oil/water emulsion, thereby facilitating coalescence of oil droplets and subsequent separation of the oil (Kalogianni et al. 2019). The only coadjutant currently authorized for this purpose is talc (European Comission Regulation, 1991).

The malaxation step has an important impact on oil extraction yield, but also chemical and enzymatic reactions such as oxidation and hydrolysis, which take place during this step, may affect the phenol content of olive oil and, as a consequence, modify its nutritional and sensorial properties. Malaxation time and temperature, along with the composition of the atmosphere in the malaxer, are the parameters that have the greatest impact on extraction yield as well as on final olive oil quality (Clodoveo 2012).

Traditional techniques used for olive oil separation such as presses or percolation systems have been replaced by horizontal continuous centrifugal separation systems called decanters. The degree of separation of the oil in the decanter depends on the differences in density among the constituents of the olive paste (insoluble solids, water, and oil) (Uceda et al. 2006). Current available continuous separation systems used in olive oil extraction can be classified as two-phase or three-phase, depending on the product obtained at the end of the process. A three-phase decanter requires the addition of water, and the outcome consists of oil, water, and solids (olive pomace). On the other hand, separation of oil in a two-phase decanter occurs without addition of water, or only a small amount thereof, and the outcome consists of oils and olive pomace that contain the vegetative water of the olives. As the extraction yield obtained with two-phase decanters is similar, yet with lower water consumption and less generation of wastewater, this system is increasingly replacing three-phase decanters (Di Giovachino et al. 2001).

In order to remove oil turbidity caused by the presence of solids in suspension and the dispersion of microscopic water droplets, the oil extracted from the decanter is clarified in an automated discharge vertical centrifuge. Before bottling, the oil is kept in tanks to promote the sedimentation of water droplets and solids that might still be present in the oil after vertical centrifugation.

In summary, olive oil extraction involves several steps that aim to obtain the highest yield from olive fruits and the best resulting oil quality. This objective requires an optimization of processing parameters, because conditions that encourage the greatest degree of oil recovery generally tend to have a detrimental effect on olive oil quality.

The Application of PEF to Improve the Olive Oil Extraction Process

The oil is located in the cells of the pulp, which is the mesocarp of the olive fruit (Ranalli et al. 2001). Within the cells, the oil is partly located in the vacuoles (approximately 76%), where it is free, and the other portion is in the cytoplasm (approximately 24%), where it is dispersed in the form of minute droplets bound to colloids. In view of these locations, and considering that PEF technology improves mass transfer processes through cell envelopes by causing electroporation of the cytoplasmic membranes, a series of studies have been conducted to evaluate whether PEF can exert positive effects on olive oil extraction yield (Table 1). These studies have been conducted not only at laboratory scale but also on an industrial scale.

Different location alternatives for the installation of a PEF treatment chamber are shown in Fig. 2, according to the type of processing line. In order to evaluate the benefits of PEF technology for olive oil extraction and for enhancement of oil quality, the effects of a PEF treatment have been evaluated at three locations: upstream of the crushing stage; downstream of the crushing stage and upstream of malaxation; and, finally, downstream of malaxation and upstream of the decanter.

Table 1 Benefits of the application of PEF at different moments of the olive oil extraction process

Olive cultivar	Location of the PEF treatment	PEF treatment conditions	Malaxation variables	Decanter type	Improvement in olive oil extraction	References
Tsounati, Amfissis Manaki	Before crushing	Batch 1.8 kV/cm 1.6–70.0 kJ/kg	30 °C for 20 min	Laboratory centrifuge	1–18% EY increment	Andreou et al. (2017)
Arbequina	Before malaxation	Batch 1 and 2 kV/cm 1.4 and 5.2 kJ/kg	15 and 26 °C for 30 min	Laboratory centrifuge(2-phase)	Reduction of malaxation temperature	Abenoza et al. (2013)
Noceralla del Belice	Before malaxation	2300 kg/h 2 kV/cm 7.8 kJ/kg	27 °C for 30 min	3-phase decanter	6% OE increment	Tamborrino et al. (2020)
Arroniz	After malaxation	520 kg/h 2 kV/cm 11.25 kJ/kg	24 °C for 60 min	2-phase decanter	13.3% EY increment	Puértolas and Martínez de Marañón (2015)
Carolea, Ottobratica and *Coratina*	After malaxation	200 kg/h 1.7 kV/cm 17 kJ/kg	25 °C	2-phase decanter	2.3–6% EY increment	Veneziani et al. (2019)

EY extraction yield

OE Oil extractability (ratio between % of oil extracted from the olives with respect to the % of oil content in olives)

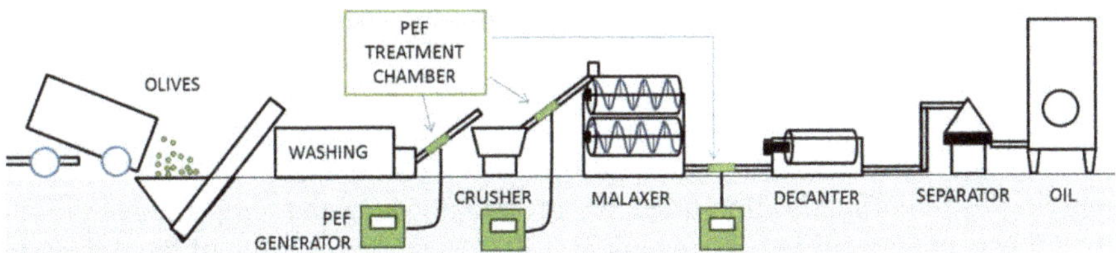

Fig. 2 Scheme of the olive oil processing line with the possible locations of the PEF treatment chamber

Effects of the Application of PEF Upstream of the Crushing Stage

The benefit on oil yield from applying a PEF treatment to olive oil before crushing has been investigated at laboratory scale (Andreou et al. 2017). Olive fruits from three different varieties (*Tsounati*, *Amfissis*, and *Manaki*) were treated in a parallel plate batch treatment chamber of 400 mL volume at an electric field strength of 1.8 kV/cm and a total specific energy ranging from 1.6 to 70.0 kJ/kg. After PEF treatments, olives were crushed and malaxation was conducted at 30 °C for 20 min.

The effect of PEF depended on processing conditions, as well as on the olive varieties investigated. The highest benefit was obtained for the *Manaki* variety, for which the extraction yield increased by 18%, thereby representing an increment of 2.4 kg/100 kg of processed olives in the amount of oil extracted compared to control samples. In the case of the *Tsounati* variety, the extraction yield increment was 3%, whereas for the *Amfissis* variety, the increase ranged from 1 to 9% according to treatment conditions. In view of the results obtained at different specific energies, the authors concluded that the most efficient treatment in terms of energy costs and yield benefits was 1.8 kV/cm at a specific energy of 20 kJ/kg.

Although the oils obtained from olives treated by PEF displayed an increment in free acidity, the physicochemical parameters of all oils remained below the EU legislation maximum limit for EVOO (<0.80% oleic acid). Both more prolonged shelf life based on the determination of peroxide values and greater oxidative stability observed in the oil obtained from PEF pre-treated olives were attributed to the fact that the PEF treatment also increases phenolic extraction (although the phenolic compositions of the oils were not determined).

It is technologically feasible to apply a PEF treatment to olive fruits before crushing by using belt PEF systems similar to those used to process whole pieces such as potatoes. However, the installation of the required PEF treatment chamber at that location would require considerable modifications in the processing line of olive processing plants. As a consequence, in industrial tests conducted to evaluate the benefits of PEF technology, the treatment has been applied to olive paste after crushing, as well as before or after malaxation.

PEF Effects Downstream of the Crushing Stage and Upstream of Malaxation

A study aiming to evaluate the effect of a PEF treatment on oil extraction yield applying different malaxation times (0, 15, and 30 min) and temperatures (15 and 26 °C) was conducted by Abenoza et al. (2013) with olive fruits of the *Arbequina* variety. Laboratory-scale equipment (Abencor System, MC2 Ingenieria y Sistemas, Seville, Spain) was used for olive oil extraction. PEF treatments at two different intensities (1 kV/cm–1.5 kJ/kg and 2 kV/cm–5.2 kJ/Kg) were applied to the olive paste before malaxation using a collinear treatment chamber at a flow of 120 kg/h and with a mean residence time in the treatment zone of 0.42 s.

The greatest improvement in extraction yield when the PEF treatment was applied to olive paste was obtained when the oil was separated from the olive paste by centrifugation without malaxation. Under these conditions, extraction yield improved by 23 and 54% when the olive paste was treated with PEF at 1 and 2 kV/cm, respectively. However, the amount of oil obtained when the olive paste was not malaxated was half of that obtained after a malaxation of 30 min. Therefore, despite the application of PEF, the malaxation step could not be eliminated in order to improve the oil quality and the efficiency of extraction process lines in industrial olive mills.

When the olive paste was malaxated at 26 °C, PEF treatment did not increase the extraction yield. However, the application of a PEF treatment to the olive paste permitted a reduction in malaxation temperature from 15 to 26 °C without impairing the extraction yield.

Physico-chemical, nutritional, and sensorial properties of control oil (malaxated at 26 °C for 30 min) were compared with the oil obtained from olive paste treated by PEF (2 kV/cm) and malaxated at 15 °C for 30 min. Values of analytical parameters established by the Regulation (European Comission Regulation, 1991 did not exceed the established limits for EVOO in the oil obtained from olive paste treated by PEF. Although the concentration of α-tocopherol was slightly higher in the oil obtained with paste treated by PEF, higher concentrations of pigments (chlorophylls and carotenoids) and phenolic compounds were detected in the control oil. Although it is well known that PEF promotes the extraction of compounds located inside the cells (Puertolas et al. 2012), the lower concentration of these compounds was explained by the lower temperature of malaxation used for the olive paste treated by PEF. The PEF treatment did not affect the fatty acid content including oleic acid, and a similar oxidative stability was estimated for both oils.

According to sensory analysis, the lower malaxation temperature for the oil obtained with olive paste treated by PEF resulted in a more fruity, less bitter, and less pungent oil than the control oil.

Therefore, results obtained in that study indicate that a PEF treatment applied to olive paste after crushing could help reduce malaxation temperature: this, in turn, would improve sensory quality of olive oil and lead to beneficial energy savings for the olive oil industry.

A study aiming to evaluate if the application of a PEF treatment to olive paste before malaxation could improve extractability and enhance oil quality in a commercial olive mill was conducted by Tamborrino et al. (2020). A PEF treatment of 2 kV/cm and 7.8 kJ/kg was applied to the olive paste of *Nocerella del Belice* cultivar at a flow of 2300 kg/h. After the treatment, the olive paste was malaxated for 30 min at 27 °C, and then the oil was separated using a three-phase decanter.

Oil extractability, representing the ratio between the percentage of oil extracted from olives and the percentage of oil content in the olives, increased from 79.5% to 85.5% when the olive paste was PEF-treated before malaxation. This increment in extractability resulted in a significant reduction in oil loss in the pomace of about 40%. As previously observed, PEF treatment did not affect the analytical parameters established by Regulation EC/2568/91. In this case, in which malaxation was extended to 30 min as in the control samples, the total phenolic content of the olive oil obtained from paste treated by PEF was similar to that of control oil.

Although sensory analysis was not performed in this study, the volatile molecules of the oil obtained from olive paste treated by PEF were evaluated. The most notable effect of the PEF treatment was a reduction of 25% in the concentration of total alcohols that are responsible for fruity and ripe fruit aromas.

Effects of PEF Treatment Downstream of the Malaxation Stage and Upstream of the Separation Stage

The effect of the application of PEF on olive paste after malaxation in terms of extraction yield, as well as in terms of chemical and sensorial quality, was investigated for the first time by Puértolas and Martínez-Marañon (2015). The study was conducted in an industrial oil mill equipped with a two-phase decanter at a flow of 520 kg/h with olive fruits of the *Arroniz* variety. After malaxation at 24 °C for 60 min, the olive paste was PEF-treated at 2 kV/cm and 11.3 kJ/kg.

The applied PEF treatment yielded an additional 2.7 kg of olive oil for each 100 kg of processed olives. The effect of PEF in improving the oil extraction could not have been attributed to electroporation of the cell membranes of olive mesocarp cells, because after malaxation the loss of the integrity of the cytoplasmic membrane of all cells is assumed. Since part of the oil that is lost with the pomace at the decanter exit corresponds to oil droplets surrounded by a lipoprotein membrane emulsified in the vegetable water, the authors hypothesized that the PEF treatment could have disrupted the lipoprotein membranes, thereby facilitating the release of oil.

Similarly to results obtained when the PEF treatment was applied before crushing or malaxation, the values of the analytical parameters established by European Comission Regulation, 1991 for the oil obtained when the PEF treatment was applied after malaxation did not exceed the established limits for EVOO. On the other hand, the PEF treatment increased total phenolic content by 11.5%, total phytosterol content by 9.9%, and total tocopherol content by 15.0%, whereby the increment in α-tocopherol corresponded to 25%.

Regarding sensory attributes, panellists did not detect any specific off-flavour or taste associated with PEF treatment, thereby showing that oil obtained from paste treated by PEF has a sensory profile that fits within the requirements of the EVOO category.

The influence of the type of olive cultivar (*Carolea*, *Ottobratica*, or *Coratina*) on olive oil extraction yield and olive oil quality obtained from olive paste treated by PEF after malaxation was investigated by Veneziani et al. (2019). The three Italian olive cultivars had different geographical origins and morphological characteristics, and they resulted from divergent agronomical practices. Oil extraction was conducted in a mill with a working capacity of 200 kg of olives per hour. Olive oil was separated from the olive paste using a two-phase decanter, and a vertical centrifuge was used to separate the olive oil from the residual water phase. Similarly to the previous study conducted by Puértolas and Martínez Marañón (2015), a PEF treatment at an electric field strength of 1.7 kV/cm and a specific energy of 17 kJ/kg was applied after the malaxation phase.

PEF treatment of olive paste prior to oil separation in the decanter increased oil yield for the three investigated cultivars with degrees of improvement ranging from 2.3 to 6.0%. These enhancement values represented an increment of 0.9, 0.4, and 0.8 kg of oil extracted per 100 kg of processed olive for *Carolea*, *Ottobratica*, and *Coratina* cultivars, respectively. As previously observed in other studies, the PEF treatment did not negatively affect the analytical parameters established by European Comission Regulation, 1991 for EVOO. Results obtained in this study confirmed the previous conclusions obtained by Puertolas and Martínez Marañón (2015), namely, that the application of a PEF treatment after malaxation increased the amount of phenolic compounds in the resulting olive oils. A significant increment of total phenols of 14.3, 7.05, and 3.2% for *Carolea*, *Ottobratica*, and *Coratina* cultivars was observed, respectively. A more elevated concentration of phenolic compounds in the oil obtained with olive paste treated by PEF correlated with higher oxidative stability as analysed by the Rancimat test. However, the PEF treatment did not affect α-tocoferol concentration, nor that of main aldehydes, alcohols, and esters involved in flavour.

Cost Analysis

Studies conducted at laboratory scale, but also tests performed in industrial olive mills, have demonstrated that the application of PEF treatment to olives before crushing or to olive paste before or after malaxation leads to an increase in oil extraction yield without negatively impairing sensorial attributes and maintaining the legal EU standards for the highest-quality olive oil (EVOO). Other additional advantages observed after the application of a PEF treatment are the possibility of decreasing malaxation temperature without affecting extraction yield and an increment in the concentration of bioactive compounds such as phenols, phytosterols, and tocopherols.

Thanks to the low energy consumption of the PEF treatments used in the studies reviewed above (≤ 20 kJ/kg), as well as to the easy integration of the treatment chambers into existing processing lines, PEF should soon establish itself as a viable technology to help olive oil factories improve their profit margins.

The implementation of PEF technology in an olive mill requires an investment for equipment purchase, along with the operational cost mainly represented by electricity consumption. Table 2 shows a draft estimation of total cost of introduction of the PEF technology in an industrial olive oil plant for production capacities of 1500 and 3000 tons/year. Processing conditions and power requirements of a PEF unit have been estimated from results published in the literature.

The seasonality of the olive crop may represent an obstacle for the introduction of PEF technology in olive oil processing plants, because the PEF unit would only be used during a short period of the year. According to the results obtained in a series of studies conducted by different authors with different olive varieties, PEF treatment could help recover up to 50% of the oil that currently remains in the olive pomace. This increment corresponds to a yield improvement in the range of 2–13%.

Table 2 Estimation of total cost of the introduction of a PEF modulator in an olive oil processing plant for a production capacity of 1500 and 3000 tons/year

	Small	**Medium**
	1500 tons/year	**3000 tons/year**
	Chamber Ø4 X 2.5 G	**Chamber Ø4 X 2.5 G**
Input voltage	10 kV	10 kV
Production capacity	2000 kg/h	4000 kg/h
Investment	60.000 €	60.000 €
Depreciation range	10 years	10 years
Installation	5.500 €	6.500 €
Residual value	–	–
Replacement value	72.000 €	72.000 €
Interest	6%	6%
Depreciation range	7.200 €/annum	7.200 €/annum
Interest	3.600 €/annum	3.600 €/annum
Maintenance	500 €/annum	1.000 €/annum
Fixed costs	**11.300 €/annum**	**11.800 €/annum**
Total specific energy	10 kJ/kg	10 kJ/kg
Specific energy input	5,3 kWh	10,7 kWh
	0,0027 kWh/ton	0,0027 kWh/ton
Power price	0,13 €/kWh	0,13 €/kWh
	0,68 €/h	1,38 €/h
	0,00034 €/ton	0,00035 €/ton
Variable costs	**513 €/annum**	**1.035 €/annum**
Total costs	11.813 €/annum	12.835 €/annum
Total costs x kg	**0,008 €/kg**	**0,004 €/kg**
Total costs x liter (oil)	**0,039 €/L**	**0,021 €/L**

The benefit derived from the significant increment in the oil obtained each day in an industrial oil mill could represent a relatively high return on investment. On the other hand, olive crop seasonality could be transformed into an opportunity to create companies that rent out PEF modulators. Such companies could propose the rental of PEF modulators to different sectors whose activity depends on the seasonality of crops in different periods of the year.

Conclusions

After the introduction of continuous-type extraction systems in most olive oil plants, the process of olive oil extraction has changed very little over the last 25 years (Clodoveo and Hachicha Hbaieb 2013). In recent years, certain advances have been made in the potential application of emerging technologies such as ultrasound and microwaves in addition to PEF (Tamborrino et al. 2014, Leone et al. 2015, Clodoveo et al. 2017). In studies conducted in processing plants, ultrasound and microwaves have been applied to olive paste after crushing with the objective of reducing the time required for the olive paste to achieve malaxation temperature, thereby reducing malaxation time. These treatments improve process efficiency and increase the oil's polyphenol content; however, an increment in extraction yield is not observed. As compared with these technologies, the integration of PEF into the olive oil mechanical extraction process has been shown to lead to significant increments in oil yield while exerting a positive impact on olive oil quality by enhancing bioactive compounds known to be beneficial to human health.

Although reported studies on the application of PEF in the olive oil industry have been conducted on a semi-industrial and on an industrial scale, an improved understanding of the mechanism involved in the observed benefits could encourage the successful transfer of this technology to industrial olive mills. Further investigations are necessary to assess the influence of olive maturity indexes on the effect of PEF, the effect of varying process parameters in the extraction plant machines, and the effect of the application of PEF before or after malaxation on the olive paste's rheological characteristics. Cell lysis by PEF results in the release of intracellular components that may lead to modification of those rheological characteristics. This information is essential in order to optimize processing parameters of the most critical points in the mechanical extraction process (malaxation and separation) when the paste is treated by PEF, since the olive paste's viscosity exerts a considerable influence on the performance of the malaxer and the decanter.

References

Abenoza M, Benito M, Saldaña G, Álvarez I, Raso J, Sánchez-Gimeno AC (2013) Effects of pulsed electric field on yield extraction and quality of olive oil. Food Bioprocess Technol 6:1367–1373

Andreou V, Alexandrakis GDZ, Katsaros G, Oikonomou D, Toepfl S, Heinz V, Taoukis P (2017) Shelf-life evaluation of virgin olive oil extracted from olives subjected to nonthermal pretreatments for yield increase. Innov Food Sci Emerg Technol 40:52–57

Chiacchierini E, Mele G, Restuccia D, Vinci G (2007) Impact of innovative and sustainable extraction technologies on olive oil quality. Trends Food Sci Technol 18:299–305

Clodoveo ML (2012) Malaxation: influence on virgin olive oil quality. Past, present and future—an overview. Trends Food Sci Technol 25:13–23

Clodoveo ML, Hachicha Hbaieb R (2013) Beyond the traditional virgin olive oil extraction systems: searching innovative and sustainable plant engineering solutions. Food Res Int 54:1926–1933

Clodoveo ML, Paduano A, Di Palmo T, Crupi P, Moramarco V, Distaso E, Pesce V (2017) Engineering design and prototype development of a full scale ultrasound system for virgin olive oil by means of numerical and experimental analysis. Ultrason Sonochem 37:169–181

Di Giovacchino L, Costantini N, Serraiocco A, Surricchio G, Basti C (2001) Natural antioxidants and volatile compounds of virgin olive oils obtained by two or three-phases centrifugal decanters. Eur J Lipid Sci Technol 103:279–285

European Comission Regulation (EC) N° 2568/1991 of 1 July of 1991 on the characteristics of olive oil and olive-residue oil and on the relevant methods of analysis. Off J Eur Communities L248: 1–114

European Comission Regulation (EC) N° 1513/2001 of 23 July 2001 amending Regulations No 136/66/EEC and (EC) No 1638/98 as regards the extension of the period of validity of the aid scheme and the quality strategy for olive oil. Offi J L201: 4–7

Kalogianni EP, Despoina Georgiou D, Hasanov JH (2019) Olive oil processing: current knowledge, literature gaps, and future perspectives. J Am Oil Chem Soc 96:481–507

Leone A, Tamborrino A, Zagaria R, Sabella E, Romaniello R (2015) Plant innovation in the olive oil extraction process: a comparison of efficiency and energy consumption between microwave treatment and traditional malaxation of olive pastes. J Food Eng 146:44–52

Luaces P, Romero C, Gutierrez F, Sanz C, Perez AG (2007) Contribution of olive seed to the phenolic profile and related quality parameters of virgin olive oil. J Sci Food Agric 87:2721–2727

Nocella C, Cammisotto V, Fianchini L, D'Amico A, Novo M, Castellani V, Stefanini L, Violi F, Carnevale R (2018) Extra virgin olive oil and cardiovascular diseases: benefits for human health. Endocr Metab Immune Disord Drug Targets 18(1):4–13

Puértolas E, Martínez de Marañón I (2015) Olive oil pilot production assisted by pulsed electric field: impact on extraction yield, chemical parameters and sensory properties. Food Chem 167:497–502

Puértolas E, Luengo E, Álvarez I, Raso J (2012) Improving mass transfer to soften tissues by pulsed electric fields: fundamentals and applications. Annu Rev Food Sci Technol 3:263–282

Ranalli A, Contento S, Schiavone C, Simone N (2001) Malaxing temperature affects volatile and phenol composition as well as other analytical features of virgin olive oil. Eur J Lipid Sci Technol 103:228–238

Schwingshackl L, Hoffmann G (2012) Monounsaturated fatty acids and risk of cardiovascular disease: synopsis of the evidence available from systematic reviews and meta-analyses. Nutrients 4(12):1989–2007

Servili M, Taticchi A, Esposto S, Urbani S, Selvaggini R, Montedoro G (2007) Effect of olive stoning on the volatile and phenolic composition of virgin olive oil. J Agric Food Chem 55:7028–7035

Tamborrino A, Romaniello R, Zagaria R, Leone A (2014) Microwave-assisted treatment for continuous olive paste conditioning: impact on olive oil quality and yield. Biosyst Eng 127:92–102

Tamborrino A, Urbani S, Servili M, Romaniello R, Perone C, Leone A (2020) Pulsed electric fields for the treatment of olive pastes in the oil extraction process. Appl Sci 10:114–124

Uceda M, Jimenez A, Beltran G (2006) Olive oil extraction and quality. Grasas Aceites 57:25–31

Uylaşer V, Yildiz G (2014) The historical development and nutritional importance of olive and olive oil constituted an important part of the Mediterranean diet. Crit Rev Food Sci Nutr 54:1092–1101

Veneziani G, Esposto S, Taticchi A, Selvaggini R, Sordini B, Lorefice A, Daidone L, Pagano M, Tomasone R, Servili M (2019) Extra-virgin olive oil extracted using pulsed electric field technology: cultivar impact on oil yield and quality. Front Nutr 6:134–141

Integration of Pulsed Electric Fields in the Biorefinery Concept to Extract Microalgae Components of Interest for Food Industry

Christian Adrian Gusbeth and Wolfgang Frey

Abbreviations

PEF	Pulsed electric field
HPH	High-pressure homogenization
HSH	High shear homogenization
DW	Dry weight
BDW	Biomass dry weight
SUS	Suspension

Introduction

"One of the greatest global challenges of the 21st century is to provide a growing world population with sustainable food, raw materials and energy in times of climate change." This is not only the strategy of a national German research programs according to *Bioeconomy 2030*, BMBF (2010), but also part of the next EU research and innovation investment program *Horizon Europe*.

Microalgae can help to tackle this important challenge as these autotrophic growing microorganisms are able to bind CO_2 and thereby to reduce greenhouse gas emission and, in addition, to produce valuable components such as proteins, carbohydrates, and lipids. Some algae species can produce up to 60% of their body weight in the form of triacylglycerols (Metzger and Largeau (2005)), which are lipids consisting of three long chains of fatty acids attached to a glycerol backbone. These lipids are similar to triacylglycerols found in large quantities in natural oil from oilseed crops that are suitable as biodiesel. An important feature of microalgae production is that they do not have to compete with other biomass alternatives but have the capacity to use water and land resources that are not considered for crop

C. A. Gusbeth (✉) · W. Frey
Karlsruhe Institute of Technology, Karlsruhe, Germany
e-mail: christian.gusbeth@kit.edu; wolfgang.frey@kit.edu

J. Raso et al. (eds.), *Pulsed Electric Fields Technology for the Food Industry*, Food Engineering Series, https://doi.org/10.1007/978-3-030-70586-2_12

production. As a result, over the past three decades, algae have been widely discussed as an alternative source of biofuel that does not compete with the existing oilseed market.

Nowadays, an energy neutral production of microalgae biomass and their subsequent utilization for production of biofuel are unfeasible. This is paid, on one hand, to the high costs for microalgae cultivation (e.g., circulation and mixing) and, on the other hand, to the intensive downstream processing, which requires harvesting, drying, cell disruption, and finally purification of the product. These costs can be attributed up to 60% of the total production costs (Coons et al. (2014), Delrue et al. (2012), Molina Grima et al. (2003)). One solution to increase the efficiency is to use the residual biomass after obtaining high- and middle-value products (e.g., pigments, polysaccharides, or proteins) for biofuel production. This idea of splitting an educt into several products of different chemical composition and value is called biorefinery (Posten & Walter (2012)). The International Energy Agency (IEA) (2009) defines biorefiney as "the sustainable processing of biomass into a spectrum of marketable products and energy." Following this concept, a cascade processing of the microalgae has to be used in order to enable multiple component recovery, which improves the economics (Ruiz et al. (2016), Vanthoor-Koopmans et al. (2013), Wijffels et al. (2010)).

Valuable algal components are generally stored intracellularly either in organelles such as oil droplets or bound to membranes, which are enclosed by robust polysaccharide cell walls. Therefore, effective pre-treatment methods are required to break the cell envelopes, especially the cell wall, while the quality of the target components is not impaired. Conventional cell disruption technologies for component release, e.g., bead milling or high-pressure homogenization (HPH), can break this barrier but are energy-intensive and provide poor product quality since component fractions are either mixed or emulsified. Cell disruption methods must ensure that the products are not impaired in terms of quality and functionality by avoiding exposure to caustic agents and undesirable heating effects due to shear forces and high pressures. For this application, PEF treatment can be used as a mild and effective method of cell disruption, facilitating recovery of unaltered constituents at low energy costs (Vanthoor-Koopmans et al. (2013), Grimi et al. (2014)). PEF technology allows cascade processing for multiple component recovery, since it does not destroy cell shape and maintains gravimetrical biomass separability after single extraction step (Eing et al. (2013), Goettel et al. (2013), Kotnik et al. (2015), Silve (2018a, b), Jaeschke et al. (2019), Scherer et al. (2019), Akaberi et al. (2019, 2020)). In our approach, water-soluble proteins and carbohydrates are recovered first, followed by a solvent extraction of lipids, and all this in a wet process without the necessity to dry the biomass. This would contribute to significant energy savings in industrial-scale processes and also to a full valorization of the algae biomass.

In the last years, the focus of research is shifting toward using microalgal extracts as nutritional supplements, as fertilizer, or as raw material for cosmetic and pharmaceutical products (IGV Planttech (2019), Biorizon biotech (2020)). Therefore, this chapter focuses on the first stage of the biorefinery cascade: the PEF-assisted recovery of proteins and pigments of *C. vulgaris* and *A. platensis* as representatives for green algae and cyanobacteria, respectively. Improving the treatment procedure and subsequent downstream processing of the biomass are other aspects that are discussed in this chapter. Using the examples of these two microorganisms, the requirements for upscaling this process to high biomass throughputs and the mechanisms of protein release are also discussed.

Cell Disintegration Methods

Conventional cell disruption technologies for component release, e.g., bead milling, HPH, high shear homogenization (HSH), ultra-sonication, and thermal and chemical treatment, can break the cell envelope (consisting of cell membrane and cell wall) and are therefore very efficient in recovering individual components. A drawback of these methods is that the processing of the residual biomass for further valorization becomes difficult due to various factors, such as chemical and thermal alteration of the target product or emulsification of all component fractions. Some of these disruption methods cannot be scaled up properly to process large amounts of microalgae because they create cell debris that is hard to separate in industrial settings and they can also generate a fair amount of heat that might be detrimental to the compound of interest Carullo et al. (2018), Kapoore et al. (2018). Among the abovementioned methods, HPH and HSH are well-established techniques for algae biomass processing, which have been proven for processing of yeast and bacteria suspensions. Nevertheless, a comparison of the energy consumption by different disruption techniques is a difficult task since each application has specific and individual evaluations.

The specific energetic consumption for pre-treatment of microalgae biomass is crucial for the final process economics and sustainability. In the case of bead milling, the specific energy consumption for disruption of *C. vulgaris* with a cell disintegration degree higher than 80% is in the range of 7.5–10.0 $kWh \cdot kg_{DW}^{-1}$ (DW, dry weight), as reported by Doucha and Livansky (2008). Postma (2017) reported that a selective protein extraction can reduce energy consumption while maintaining the protein yield over 30% (related to dry biomass). In this case, only water-soluble proteins were obtained. In this case, the specific energy consumption was higher than 0.8 $kWh \cdot kg_{DW}^{-1}$. It is also difficult to estimate the energy consumption of HPH, because it depends on various factors such as algal species, the type of the equipment, required degree of disintegration, biomass concentration, etc. According to Samarasinghe et al. (2012), the energy requirement for processing 1 m^3 of algal slurry (100 $g_{DW} \cdot kg_{sus}^{-1}$) in a single pass at low pressure (690 bar) is 69 MJ, which corresponds to specific energy consumption of 0.2 $kWh \cdot kg_{DW}^{-1}$. This treatment resulted in a degree of disintegration lower than 12%, for *Nannochloris oculata*. In order to achieve a higher degree of disintegration (>90%), at least three passes and higher pressures (>2000 bar) are required, which results in a specific energy consumption of 1.7 $kWh \cdot kg_{DW}^{-1}$. In our lab, we used the Avestin EmulsiFlex-C3 HPH with a pressure of 2 kbar as a benchmark procedure for cell disintegration. The specific energy consumption of this HPH can be calculated by taking into account the electric power (0.75 kW) at 2 kbar pressure. Accordingly, the specific energy demand for a single pass at a flow rate of 0.83 $ml \cdot s^{-1}$ and a biomass concentration of 100 $g_{DW} \cdot kg_{SUS}^{-1}$ is 2.5 $kWh \cdot kg_{DW}^{-1}$. On industrial scale at higher throughput (1000 $l \cdot h^{-1}$), the energy requirement drops significantly when compared with laboratory equipment, such as in the case of the HPH equipment from GEA Westfalia (see Table 1), with a specific energy consumption in the range of 1.1–2.2 $kWh \cdot kg_{DW}^{-1}$.

In contrast to the disintegration methods mentioned before, the PEF treatment is not breaking the cell envelope, and therefore it is considered as a mild cell disruption method. Nevertheless, recent publications have shown the strong potential of PEF as a pre-treatment method to extract intracellular components, such as lipids, proteins, and pigments from microalgae (Goettel et al. (2013), Eing et al. (2013), Luengo et al. (2014), Pataro et al. (2017), Silve (2018a, b), Jaeschke et al. (2019), Scherer et al. (2019), Akaberi et al. (2020)). In these studies, the algae suspension was treated with a specific energy in the range of 50–150 $kJ \cdot kg_{SUS}^{-1}$. As a result, for a biomass concentration of the algae slurry between 10–100 $g_{DW} \cdot kg_{SUS}^{-1}$, the specific energy consumption for PEF treatment is in the range of 0.14–4.2 $kWh \cdot kg_{DW}^{-1}$. Table 1 gives an overview of the specific energy consumption for pre-treatment of microalgae, especially *C. vulgaris*, using various HPH equipment, bead milling, and PEF treatment. In view of these findings, the specific energy consumption does not seem to be decisive reason for the choice of the disintegration method. Furthermore,

Table 1 Specific energy demand of various cell disintegration methods using different equipment for microalgae biomass

Method/ biomass concentration in $g_{DW} \cdot kg_{SUS}^{-1}$	Equipment	Algae species	Degree of cell disintegration in %	Specific energy consumption in $kWh \cdot kg_{DW}^{-1}$	References
HPH 100	*NanoDeBEE* 0.69 kbar, 3 passes	*N. oculata*	> 90	0.6–1.7	Samarasinghe et al. (2012)
	Avestin EmulsiFlex –C3 2 kbar, 5 passes	*C. vulgaris*	–	12.5	–
	GEA Ariete NS3015H 0.4–1.0 kbar, 3 passes	*C. vulgaris*	–	0.3–0.4	SABANA e-bulletin No. 4, 2019
Bead milling dry biomass	Dyno-Mill	*C. vulgaris*	>80 –	7.5–10.0 0.8	Postma (2017)
PEF 5	Transmission line pulse generator: 40 $kV \cdot cm^{-1}$, 1 μs, square pulses	*A. protothecoides*	–	0.3–0.8	Goettel et al. (2013)

it gives priority to select the method, which optimizes the utilization of the entire algal biomass by integration in a biorefinery concept.

Biorefinery Concept Based on PEF Technology

PEF technology opens the possibility for a cascade biorefinery in which all fractions can be successfully recovered, as reported in several studies (Wijffels et al. (2010), Kotnik et al. (2015), Goettel et al. (2013), Eing et al. (2013)). According to the biorefinery concept, the key feature of an alternative pre-treatment method should be the ability to extract high valuable intracellular components from a wet biomass prior to the recovery of lipids by organic solvents. Thus, the degradation of intracellular components is avoided, and the extremely costly drying step is removed from the production line (Xu et al. (2011), Lardon et al. (2009)). Figure 1 shows the microalgae valorization concept based on PEF technology, in which the cell permeabilization is followed by two stages of biorefinery. During the first stage, high-value water-soluble compounds, such as proteins, polysaccharides, and pigments, are recovered, while in the second stage, the lipids are extracted via a green organic solvent (e.g., ethanol). Finally, the residual biomass can be used in energetic processes (e.g., thermochemical conversion (Guo et al. (2019))) or further valorized (e.g., anaerobic digestion).

Although the use of PEF technology for lipid extraction has been demonstrated in various studies (Eing et al. (2013), Goettel et al. (2013), Silve (2018a, b)), an efficient extraction of proteins from microalgae using PEF treatment has not been shown. In fact, some studies have claimed that protein extraction yield is usually too low for PEF treatment to be a feasible option for industrial-scale applications (Postma (2016), Safi (2017), Zocher (2016)). Therefore, in the last years, many efforts have been done to identify the most economical pathway – e.g., suitable PFE treatment and extraction methods for protein recovery.

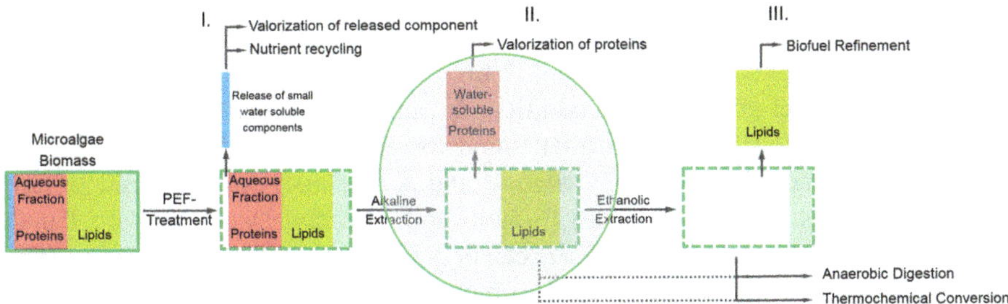

Fig. 1 Biorefinery concept based on PEF technology in which all biomass fractions are used efficiently

Impact of Various Processing Conditions on Recovery of Proteins and Pigments

The main advantage of PEF-assisted extraction is the possibility to get higher degree of extraction selectivity toward certain valuable fractions such as proteins and lipids. In general, PEF treatment of pre-concentrated biomass has to be performed with pulses of a duration of less than 10 μs. Longer pulses cause biomass deposition at the electrodes and may result in clogging of treatment chambers (Sträßner et al. (2016)). In the case of lipid extraction for biofuel production, PEF treatment has shown promising results for reducing the environmental and economic costs of the process (Silve et al. (2018a)). For example, up to 92% of the total lipid content was recovered after PEF treatment and ethanol/hexane extraction of mixotrophically grown *Auxenochlorella protothecoides*. Moreover, it was shown that the lipid yield obtained after PEF treatment with low treatment energy of 0.07 $kWh \cdot kg_{DW}^{-1}$ and 20 h of incubation period was the same as the lipid yield as right after the treatment with high specific energy of 0.42 $kWh \cdot kg_{DW}^{-1}$ without any incubation time (Silve et al. (2018b)). Even though PEF treatment did not cause a spontaneous release of lipids, the efficiency of lipid recovery is much higher than that of proteins, probably due to the combination of the extraction with organic solvents that can penetrate the cells and dissolve the lipids. In contrast to lipids, which are stored intracellularly mainly in oil droplets, proteins are located either in cytoplasm or linked to membranes. Therefore, some authors (Coustets et al. (2014)) proposed that PEF treatment only leads to the release of water-soluble cytosolic proteins without affecting vacuole membrane integrity. They also found that an incubation step in a salty buffer is necessary for an effective recovery of proteins, but they have not identified the parameters, which might influence the protein release. Therefore, the following section pays particular attention to the impact of various treatment parameters on the efficiency of protein recovery from *C. vulgaris* and *A. platensis*, as examples.

Both microorganisms are certified for food and feed application as they are generally regarded as safe (GRAS status) and are being marketed as food additives since many years. The idea to use microalgae as dietary supplement is not new (Morimura and Tamiya (1953)) and is owing to the high protein content of 42–58$\%_{BDW}$ in *C. vulgaris* and up to 70$\%_{BDW}$ in *A. platensis* (Morris et al. (2008), Seyfabadi et al. (2011), Servaites et al. (2012), Safi et al. (2013)). Among various proteins, phycobiliproteins like C-phycocyanins are supplementary light-absorbing complexes (pigments) that are present in high concentration in *A. platensis* (up to 20$\%_{BDW}$) (Safaei et al. (2019)). In addition, the amino acid profile of these proteins has been shown to be similar to that found for egg proteins (Safi et al. (2014)). Dietary supplementation with *C. vulgaris* and *A. platensis* is considered to have positive benefits for human health (Soheili and Khosravi-Darani (2012), Panahi et al. (2015)). For instance, *C. vulgaris* can help in lowering serum cholesterol (Ryu et al. (2014)) and thus is considered to have preventive effects in cardiovascular diseases. Some authors supposed that consumption of *C. vulgaris* has a preventive effect on diabetes as it lowers blood sugar level and mitigates insulin resistance

(Cherng and Shih (2005), Lee and Kim (2009)). Phycocyanins from *A. platensis* are mostly used as color supplements in food and cosmetics (Fernández-Rojas et al. (2014)) and have shown to exhibit antioxidant activity, being associated to the decrease of the risk of degenerative, neuro-, and renal diseases (Li et al. (2015), Memije-Lazaro et al. (2018), Park et al. (2015), Raja et al. (2015)). In the following, the impact of incubation time after PEF treatment, temperature, pH value, and biomass concentration on phenomena that underlie the process of protein release is discussed.

Impact of Incubation Time and Temperature

PEF treatment enables protein release via diffusion, which is strongly time and temperature dependent. Up to 50% of the protein content from *C. vulgaris* can be recovered after PEF treatment with specific energy of 150 $kJ \cdot kg_{SUS}^{-1}$ (40 $kVcm^{-1}$, 1 μs square pulses, 5–10 $g_{DW} \cdot kg_{SUS}^{-1}$) and 24 h incubation period as reported by Scherer et al. (2019). The incubation temperature had an unexpected impact on the release kinetics of proteins from *C. vulgaris* (Fig. 2). According to this study, the optimal temperature for protein release was in the physiological temperature range between 20 °C and 40 °C. The incubation at higher temperatures, above 40 °C, does not lead to faster release, as would be expected for a diffusion-driven process. At 50 °C, the release again was limited, and the yield over the 24 h incubation period was only half of what was obtained under physiological conditions. The authors of

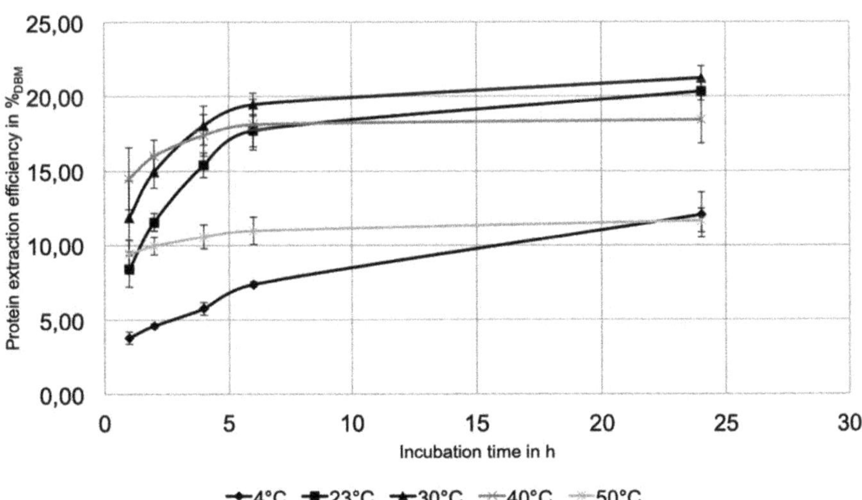

Fig. 2 Time course of protein recovery efficiency after PEF treatment of *C. vulgaris* in dependence of incubation temperature. The microalgae suspension (5 $g_{DW} \cdot kg_{SUS}^{-1}$) was treated with a specific treatment energy of 150 $kJ \cdot kg_{SUS}^{-1}$ and then incubated at different temperatures (Figure: Scherer et al. (2019))

this study considered that besides diffusion, the protein release is enzyme-mediated. In this model enzyme-driven protein release, PEF treatment induces irreversible membrane electroporation in *C. vulgaris* and consequently cell death, followed by self-digestion (autolysis), which results in protein liberations. Similar mechanisms, in which the release of proteins and pigments is facilitated through an autolytic process associated with programed cell death, have been proposed and described for yeast (Simonis et al. (2017), Martinez et al. (2018a)) and red algae *Rhodotorula glutinis* (Martinez et al. (2018b)). The overall results suggest that proteins are released by a proteolytic activity after triggering cell death via PEF treatment. These outcomes are supported by the observed DNA laddering, which is one of the indications for programmed cell death. In *C. vulgaris,* DNA laddering begins within 1 h after PEF treatment and progresses over time, with the genomic DNA completely fragmenting within 24 h, as reported by Scherer et al. (2019). This is in line with protein release, which also begins within 1 h after PEF treatment. In addition, Western blot analysis of water-soluble protein fractions obtained after PEF treatment reveals that proteins from all the organelles are released in the supernatant following treatment. A signal for RuBisCo, histone H3, and actin can be detected within 1 h after PEF treatment (Scherer et al. (2019)), while the signal for COXII, a protein present in mitochondrion, appears after an extended period of 6 h post-PEF treatment. These results show that all intracellular compartments are digested and proteins and pigments can be released from the entire cell.

For *A. platensis* the kinetics of released proteins and C-phycocyanin into the external medium was also strongly influenced by incubation temperature after PEF treatment with 56–114 $kJ \cdot kg_{SUS}^{-1}$ (40 $kV \cdot cm^{-1}$, 1 μs square pulses, 2.6 $g_{DW} \cdot kg_{SUS}^{-1}$), as reported by Akaberi et al. (2020). In contrast to *C. vulgaris*, the release of intracellular components from *A. platensis* is dominated by diffusion. In the first hour after treatment, over 70% of the total protein content was recovered. After 4 h of incubation at either 23 or 40 °C, the maximum protein yield (~97%) was reached. For these temperatures, at least 2 h of incubation were required to obtain the comparable amount of proteins as obtained via HPH treatment ($60\%_{BDW}$). The biomass incubated at 4 °C showed a drastically lower protein release, which cannot be compensated even after a longer incubation period of 24 h. This might be an indication of inhibited autolysis in *A. platensis* after cell death.

C-phycocyanin release after PEF treatment was also strongly time and temperature dependent. The amount of C-phycocyanin released after 1 h of incubation at 23 and 40 °C ($5.5\%_{BDW}$ and $4.9\%_{BDW}$, respectively) was comparable. After 4 h of incubation, C-phycocyanin concentration reached the same level as those obtained after HPH treatment (up to $10\%_{BDW}$). A drawback of a long incubation time is that at high temperatures (> 40 °C) and unfavorable pH changes in the incubation medium, molecular degradation of C-phycocyanin can happen.

In short, post-PEF incubation temperature strongly influences the protein release from *C. vulgaris* as well as *A. platensis*. A recommended incubation temperature for both microorganisms is in the physiological range between 23 and 30 °C, which can be applied to other microorganisms in a first approach. Significant protein recovery can be achieved for both microorganisms by an incubation period of at least 5 h for

C. vulgaris (50% of total proteins) and 2 h for *A. platensis* (98% of total proteins) after PEF treatment.

Impact of Post-PEF Incubation pH

The protein conformation changes between different thermodynamic states when the incubation conditions (temperature, pH) are altered. By changing the pH of the medium, the hydrogen bonds and the salt bridges dissolve, which change the folding and the solubility of proteins. Accordingly, the pH of the incubation medium and the stability of pH over time are important factors, which can affect the release and conformation of proteins during the incubation period.

Protein recovery efficiency from *C. vulgaris* at various pH values shows that the extraction yield in incubation medium is low at pH 7, but it has an extraction optimum at pH 8–9 and it decreases slightly with further increasing of the pH (Fig. 3). The initial pH of the medium changes during longer incubation, due to the permanent release of cytosolic components and to the ongoing autolytic process. Therefore, for longer incubation times, the pH of the medium must be stabilized using a buffer system, such as Tris.

The efficiency of protein recovery from *A. platensis* suspension (3.6 $g_{DW}\cdot kg_{SUS}^{-1}$) after PEF treatment (40 $kV\cdot cm^{-1}$, 1 μs square pulses, 56–114 $kJ\cdot kg_{SUS}^{-1}$) was also pH dependent, as reported by Akaberi et al. (2020). Incubation in pH 6 and pH 8

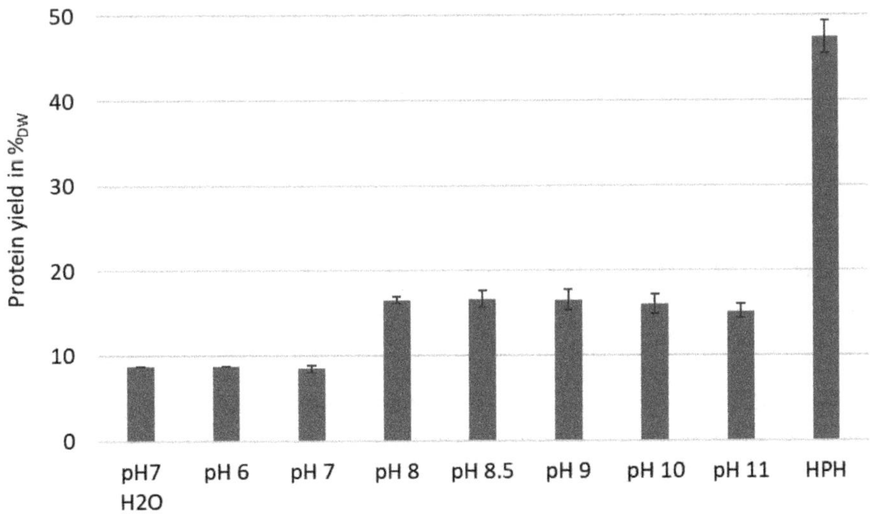

Fig. 3 Protein recovery efficiency from *C. vulgaris* at various pH values. The algae suspension (5 $g_{DW}\cdot kg_{SUS}^{-1}$) was PEF treated (40 $kV\cdot cm^{-1}$, 1 μs, spec. energy of 150 $kJ\cdot kg_{SUS}^{-1}$ means 8.4 $kWh\cdot kg_{DW}^{-1}$) and afterward incubated for 24 h at room temperature under the influence of different pH values in water or leftover medium conditioned to defined pH values

buffer showed a gradual increase of released proteins within 24 h, with $40.5\%_{BDW}$ and $47.6\%_{BDW}$, respectively. In comparison, the total protein content, obtained by HPH treatment of *A. platensis* suspension, was $56.7\%_{BDW}$. When using initial buffer as an incubation medium, the kinetics of protein release was fast at the beginning and decreased gradually after a certain time. At low biomass concentration (3.6 $g_{DW}{\cdot}kg_{SUS}^{-1}$), a drastic shift in pH from 10.5 to 7.0 within 24 h was observed during incubation in initial medium. This was due to the low buffering capacity of the initial buffer and the tremendous release of intracellular compounds into the medium. Regarding C-phycocyanin recovery, incubation in a buffered medium with a pH of 6 or 8 prevents the molecular degradation of protein complex. At pH 8, more than 97% C-phycocyanin ($10.5\%_{BDW}$) was recovered within 3 h after PEF treatment (Akaberi et al. (2020)). The blue color of C-phycocyanin fades drastically when the *A. platensis* suspension was incubated in initial buffer without buffering capacity.

In summary, the monitoring and adjustment of the pH are required in the downstream processes, especially when an incubation step is carried out, in order to maintain the quality of the extracted component and to ensure efficiency of extraction. For *C. vulgaris* suspensions, the protein release is effective at alkaline pH of around 8.5–9 in a buffered environment, whereas a constant pH of 8 is required for *A. platensis*, when high biomass concentrations are used.

Impact of Biomass Concentration

Similarly to HPH, concentrating the biomass before the treatment enables to reduce considerably the energy input per kg of dry biomass. However, the biomass concentration of the microalgae slurry can affect both the PEF treatment and the diffusion during post-PEF incubation period. Due to the high biomass concentration before PEF treatment, cells can aggregate into cluster, which leads to electrical shielding of the cells inside the cluster and thus to a lower membrane electroporation. In addition, the high amounts of externalized components lead to molecular degradation due to redox reactions or enzymatic digestion, which can affect the targeted product.

Experiments with *C. vulgaris* suspensions of different biomass concentrations (36–167 $g_{DW}{\cdot}kg_{SUS}^{-1}$) revealed that the suspension's biomass content had no negative influence on the efficiency of the PEF treatment as reported by Goettel et al. (2013). It seems that at these biomass concentrations and an electric field intensity of 40 $kV{\cdot}cm^{-1}$, the so-called percolation threshold is not exceeded (El Zakhem et al. (2006)), and therefore mutual electric shielding is negligible. However, this is not the case when dense *A. platensis* suspension is PEF treated. The kinetics of protein/C-phycocyanin release is considerably influenced by the homogeneity of the suspension (Akaberi et al. (2020)). The long filamentous structure of *A. platensis* is prone to clump together and form large aggregates of cells. Homogeneous suspensions have a faster release kinetics when compared with heterogeneous suspensions. The slow release of proteins and C-phycocyanin in the case of a heterogeneous suspension (containing clusters of cells) is attributed to the mutual electric shielding of cells, which causes a decrease in the amplitude of induced transmembrane voltage (Guittet et al. (2017)). Consequently, higher specific treatment energies are required for these suspensions, in order to maintain the extraction efficiency. For these reasons, a gentle homogenization of cell suspension prior to the PEF application is recommended in order to avoid cluster formation.

Besides the mutual shielding effect at high biomass concentrations, the release of components and the quality of the extract are also affected. For instance, with increasing the biomass concentration of *C. vulgaris* suspension from 2.5 to 12.5 $g_{DW} \cdot kg_{SUS}^{-1}$, the protein yield after an incubation step of 24 h decreased significantly (from 50% to 30%) (Fig. 4). This can be explained by the fact that biomass concentration influences the diffusion gradient of the intracellular proteins into the medium. With regard to total protein release and C-phycocyanin release from *A. platensis* at higher biomass concentrations (~ 10 $g_{DW} \cdot kg_{SUS}^{-1}$), the following can be stated: In both cases at high biomass concentration, the release kinetics is slower, and the degradation of C-phycocyanin in the medium no longer occurs. Another observation is that the maximum protein and C-phycocyanin yield can only be obtained after an incubation time of 5 h in minimum. In contrast, comparable yields can be achieved after 2 h of incubation when low biomass concentrations were processed. This prolonged incubation time at higher biomass concentrations can be explained by a lower contribution of diffusion to component externalization at higher cell density.

The impairment of the component release and quality with increased cell density represent a new challenge for upscaling in industrial applications and must be clarified in order to maintain the extraction efficiency.

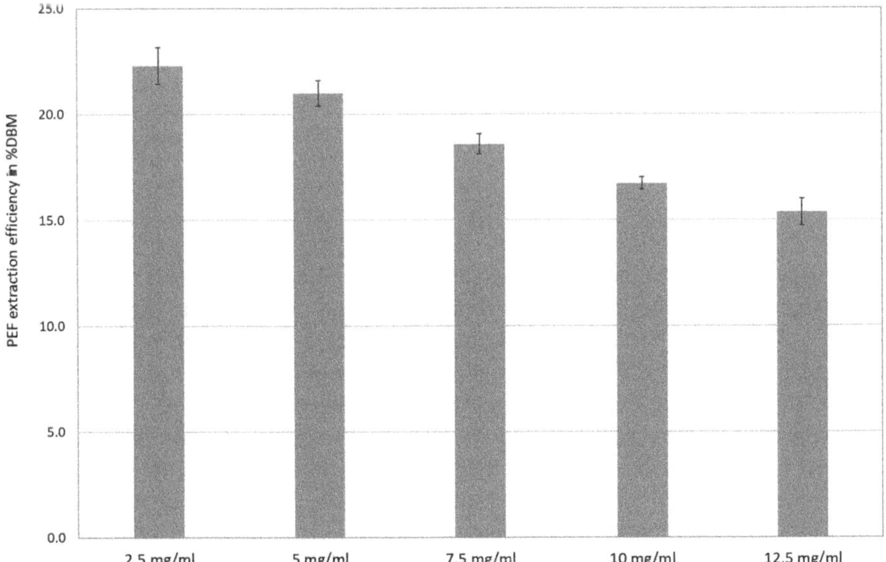

Fig. 4 Protein recovery efficiency from *C. vulgaris* in dependence of biomass concentration. *C. vulgaris* suspensions were PEF treated (40 $kV \cdot cm^{-1}$, 1 μs square pulses, 150 kJ $\cdot kg_{SUS}^{-1}$) and incubated for 24 h at room temperature (Figure: Scherer et al. (2019))

Specific Treatment Energy

The specific treatment energy is a decisive factor for the successful implementation of PEF technology in the large-scale processing of microalgae. The energy required for efficient PEF treatment of microalgae is directly related to the volume of suspension and its properties such as conductivity and biomass concentration but also to the microalgae species and the target product.

As discussed in the previous section, the reduction in energy requirements can be achieved by an additional incubation step, which fosters the release of valuable components after PEF treatment. In this approach, PEF treatment induces cell death, which triggers an autolytic process. Accordingly, the lowest energy demand required for inducing autolytic processes in *C. vulgaris* equals to the specific lethal energy dose. Under this conditions the required PEF treatment energy could be insignificantly low, since the lethal dose for *C. vulgaris* is in the range of 1.4 $kJ \cdot kg^{-1}$ (for 1 μs square pulses of 40 $kV \cdot cm^{-1}$) (Gusbeth et al. (2013)), which corresponds to an energy consumption of only 0.004 $kWh \cdot kg_{DW}^{-1}$, for an algae suspension of 100 $g_{DW} \cdot kg_{SUS}^{-1}$.

Since in the case of *A. platensis*, the autolytic process after PEF treatment can be neglected and diffusion predominates, the lowest treatment energy required for an efficient protein release does not depend on the lethal dose. The kinetic of the protein/C-phycocyanin release is not only time and temperature dependent but also dose dependent. With a high specific treatment energy 114 $kJ \cdot kg_{sus}^{-1}$ (corresponds to 3.2 $kWh \cdot kg_{DW}^{-1}$, for 10 $g_{DW} \cdot kg_{SUS}^{-1}$), the maximum protein yield is reached within 3–5 h, while with 56 $kJ \cdot kg_{sus}^{-1}$ it takes more than 20 h.

In summary, these two examples clearly show that the energy demand of efficient PEF pre-treatment to obtain valuable intracellular components at high quality and in large quantities must be identified for each microorganism and individual purposes.

Conclusion

This approach uses an incubation step after PEF treatment to increase the efficiency of extracting valuable intracellular components. As example, the time-dependent release of proteins and pigments is affected by process factors such as biomass concentration, post-PEF incubation temperature, and pH. Using *A. platensis*, the most important benefit that PEF treatment offered by subsequent incubation in pH 8 buffer is the enhancement of the purity ratio of C-phycocyanin. For some algae species, such as *C. vulgaris*, the induced cell death is sufficient to trigger an autolytic process, which boosts the protein release. By incubating the microalgae suspension after PEF treatment under appropriate conditions (temperature between 20 °C and 40 °C and pH in the range from 8.5 to 9.5), up to 50% of the total protein content can be obtained. This allows a significant reduction of energy consumption and hence reduction of operation costs. From a bioprocess engineering point of view, this approach opens the opportunity to use PEF treatment in a biorefinery concept, where water-soluble ingredients are extracted before solvent extraction of lipids.

References

Akaberi S, Gusbeth C, Silve A, Senthil Senthilnathan D, Navarro-López E, Molina-Grima E, Müller G, Frey W (2019) Effect of pulsed electric field treatment on enzymatic hydrolysis of proteins of Scenedesmus almeriensis. Algal Res 43. https://doi.org/10.1016/j.algal.2019.101656

Akaberi S, Krust D, Müller G, Frey W, Gusbeth C (2020) Impact of incubation conditions on protein and C-Phycocyanin recovery from *Arthrospira platensis* post- pulsed electric field treatment. Bioresour Technol 306. https://doi.org/10.1016/j.biortech.2020.123099

Biorizon biotech, Catalogo Biorizon (2020), https://www.biorizon.es

Bundesministerium für Bildung und Forschung (BMBF) (2010) Nationale Forschungsstrategie BioÖkonomie 2030 Unser Weg zu einer bio-basierten Wirtschaft. DruckVogt, Berlin. http://www.bmbf.de

Carullo D et al (2018) Effect of pulsed electric fields and high pressure homogenization on the aqueous extraction of intracellular compounds from the microalgae *Chlorella vulgaris*. Algal Res Elsevier:60–69. https://doi.org/10.1016/j.algal.2018.01.017

Cherng J-Y, Shih M-F (2005) Potential hypoglycemic effects of Chlorella in streptozotocin-induced diabetic mice. Life Sci 77(9):980–990. https://doi.org/10.1016/j.lfs.2004.12.036

Coons J E, Kalb D M, Dale T, Marrone B (L 2014) Getting to low-cost algal biofuels: a monograph on conventional and cutting-edge harvesting and extraction technologies. Algal Res 6:250–270

Coustets M, Joubert-Durigneux V, Hérault J, Schoefs B, Blanckaert V, Garnier J-P, Teissié J (2014) Optimization of proteins electroextraction from microalgae by a flow process. Bioelectrochemistry 103:74–81. https://doi.org/10.1016/j.bioelechem.2014.08.022

Delrue F, Setier P-A, Sahut C, Cournac L, Roubaud A, Peltier G, Froment A-K (2012) An economic, sustainability, and energetic model of biodiesel production from microalgae. Bioresour Technol 111:191–200

Doucha J, Lívanský K (2008) Influence of processing parameters on disintegration of *Chlorella* cells in various types of homogenizers. Appl Microbiol Biotechnol 81:431–440. https://doi.org/10.1007/s00253-008-1660-6

Eing C, Goettel M, Straeßner R, Gusbeth C, Frey W (2013) Pulsed electric field treatment of microalgae-benefits for microalgae biomass processing. IEEE Trans Plasma Sci 41:2901–2907

El Zakhem H et al (2006) Behavior of yeast cells in aqueous suspension affected by pulsed electric field. J Colloid Interface Sci 300:553–563

Fernández-Rojas B, Hernández-Juárez J, Pedraza-Chaverri J (2014) Nutraceutical properties of phycocyanin. J Funct Foods 11:375–392

Goettel M, Eing C, Gusbeth C, Straeßner R, Frey W (2013) Pulsed electric field assisted extraction of intracellular valuables from microalgae. Algal Res 2:401–408

Grimi N, Dubois A, Marchal L, Jubeau S, Lebovka NI, Vorobiev E (2014) Selective extraction from microalgae *Nannochloropsis sp.* using different methods of cell disruption. Bioresour Technol 153:254–259

Guittet A, Poignard C, Gibou F (2017) A Voronoi Interface approach to cell aggregate electropermeabilization. J Comput Phys 332:143–159

Guo B, Yang B, Silve A, Akaberi S, Scherer D, Papachristou I, Frey W, Hornung U, Dahmen N (2019) Hydrothermal liquefaction of residual microalgae biomass after pulsed electric field-assisted valuables extraction. Algal Res 43. https://doi.org/10.1016/j.algal.2019.101650

Gusbeth C, Eing C, Göttel M, Frey M (2013) Boost of algae growth by ultrashort pulsed electric field treatment. IEEE Int Conf Plasma Scienc (ICOPS) SF, CA, 1-1

IGV PLANTTECH Product Range and Services, Extraction and Cultivation, 2. Edition (2019) Issued August, https://www.igv-gmbh.com

International Energy Agency (IEA) (2009) Bioenergy Task 42 Biorefinery, https://www.iea-bioenergy.task42-biorefineries.com

Jaeschke DP, Mercali GD, Marczak LDF, Müller G, Frey W, Gusbeth C (2019) Extraction of valuable compounds from *Arthrospira platensis* using pulsed electric field treatment. Bioresour Technol 283:207–212

Kapoore R et al (2018) Microwave-assisted extraction for microalgae: from biofuels to biorefinery. Biology:7. https://doi.org/10.3390/biology7010018

Kotnik T, Frey W, Sack M, Haberl Meglicˇ S, Peterka M, Miklavcˇicˇ D (2015) Electroporation-based applications in biotechnology. Trends Biotechnol 33:480–488

Lardon L, Hélias A, Sialve B, Steyer J-P, Bernard O (2009) Life-cycle assessment of biodiesel production from microalgae. Environ Sci Technol 43:6475–6481

Lee HS, Kim MK (2009) Effect of Chlorella vulgaris on glucose metabolism in Wistar rats fed high fat diet. J Med Food 5:1029–1037. https://doi.org/10.4162/nrp.2008.2.4.204

Li B, Gao M-H, Chu X-M, Teng L, Lv C-Y, Yang P, Yin Q-F (2015) The synergistic antitumor effects of all-trans retinoic acid and C-phycocyanin on the lung cancer A549 cells in vitro and in vivo. Eur J Pharmacol 749:107–114

Luengo E, Condón-Abanto S, Álvarez I et al (2014) Effect of pulsed electric field treatments on permeabilization and extraction of pigments from *Chlorella vulgaris*. J Membr Biol 247:1269–1277. https://doi.org/10.1007/s00232-014-9688-2

Martínez JM, Delso C, Aguilar D et al (2018a) Factors influencing autolysis of *Saccharomyces cerevisiae* cells induced by pulsed electric fields. Food Microbiol 73:67–72

Martínez JM, Delso C, Angulo J et al (2018b) Pulsed electric field-assisted extraction of carotenoids from fresh biomass of *Rhodotorula glutinis*. Innov Food Sci Emerg Technol 47:421–427

Memije-Lazaro IN, Blas-Valdivia V, Franco-Colín M, Cano-Europa E (2018) *Arthrospira maxima* (Spirulina) and C-phycocyanin prevent the progression of chronic kidney disease and its cardiovascular complications. J Funct Foods 43:37–43

Metzger P, Largeau C (2005) Botryococcus braunii: a rich source for hydrocarbons and related ether lipids. Appl Microbiol Biotechnol 66:486–496. https://doi.org/10.1007/s00253-004-1779-z

Molina Grima E, Belarbi E-H, Acién Fernández FG, Robles Medina A, Chisti Y (2003) Recovery of microalgal biomass and metabolites: process options and economics. Biotechnol Adv 20:491–515

Morimura Y, Tamiya N (1953) Preliminary experiments in the use of chlorella as human food. Food Technol 8:179–182

Morris HJ et al (2008) Utilisation of Chlorella vulgaris cell biomass for the production of enzymatic protein hydrolysates. Bioresour Technol 99:7723–7729. https://doi.org/10.1016/j.biortech.2008.01.080

Panahi Y, Darvishi B, Jowzi N, Nejat S, Sahebkar A (2015) Chlorella vulgaris: a multifunctional dietary supplement with diverse medicinal properties. Curr Pharm Des. https://doi.org/10.2174/1381612822666151112145226

Park K-H, Kim J-R, Choi I, Kim J-R, Cho K-H (2015) ω-6 (18:2) and ω-3 (18,3) fatty acids in reconstituted high-density lipoproteins show different functionality of anti-atherosclerotic properties and embryo toxicity. J Nutr Biochem 26:1613–1621

Pataro G, Goettel M, Straeßner R, Gusbeth C, Ferrari G, Frey W (2017) Effect of PEF Treatment on Extraction of Valuable Compounds from Microalgae C. Vulgaris. Chem Eng Transac 57:67–72

Posten C, Walter C (eds) (2012) Microalgal biotechnology: potential and production. De Gruyter, Berlin, Boston

Postma PR (2016) Selective extraction of intracellular components from the microalga *Chlorella vulgaris* by combined pulsed electric field-temperature treatment. Bioresour Technol 203:80–88. https://doi.org/10.1016/j.biortech.2015.12.012

Postma PR et al (2017) Energy efficient bead milling of microalgae: effect of bead size on disintegration and release of proteins and carbohydrates. Bioresour Technol 224:670–679. https://doi.org/10.1016/j.biortech.2016.11.071
Raja R, Hemaiswarya S, Ganesan V, Carvalho IS (2015) Recent developments in therapeutic applications of cyanobacteria. Crit Rev Microbiol 1:1–12
Ruiz J, Olivieri G, de Vree J, Bosma R, Willems P, Reith JH, Eppink MH, Kleinegris DM, Wijffels RH, Barbosa MJ (2016) Towards industrial products from microalgae. Energ Environ Sci 9:3036–3043
Ryu NH et al (2014) Impact of daily Chlorella consumption on serum lipid and carotenoid profiles in mildly hypercholesterolemic adults: a double-blinded, randomized, placebo-controlled study. Nutr J 13:1–8
SABANA e-bulletine No.4, May (2019) Sustainable Algae Biorefinery for Agriculture and Aquaculture, SABANA project H2020-BG-2016-2017, https://www.eusabana.eu
Safaei M, Maleki H, Soleimanpour H, Norouzy A, Zahiri HS, Vali H, Noghabi KA (2019) Development of a novel method for the purification of C-phycocyanin pigment from a local cyanobacterial strain *Limnothrix sp.* NS01 and evaluation of its anticancer properties. Sci Rep 9:1–16
Safi C et al (2013) Influence of microalgae cell wall characteristics on protein extractability and determination of nitrogen-to-protein conversion factors. J Appl Phycol 25:523–529. https://doi.org/10.1007/s10811-012-9886-1
Safi C et al (2014) Morphology, composition, production, processing and applications of *Chlorella vulgaris*: a review. Renew Sustain Energy Rev 35:265–278. https://doi.org/10.1016/j.rser.2014.04.007
Safi C et al. (2017) Energy consumption and water-soluble protein release by cell wall disruption of Nannochloropsis gaditana, Bioresource Technology. Elsevier Ltd, 239, pp. 204–210. https://doi.org/10.1016/j.biortech.2017.05.012
Samarasinghe N, Fernando S, Lacey R, Faulkner WB (2012) Algal cell rupture using high pressure homogenization as a prelude to oil extraction. Renew Energy 48:300–308
Scherer D, Krust D, Frey W, Mueller G, Nick P, Gusbeth C (2019) Pulsed electric field (PEF)-assisted protein recovery from *Chlorella vulgaris* is mediated by an enzymatic process after cell death. Algal Res 41:101536
Servaites JC, Faeth JL, Sidhu SS (2012) A dye binding method for measurement of total protein in microalgae. Anal Biochem 421:75–80. https://doi.org/10.1016/j.ab.2011.10.047
Seyfabadi J, Ramezanpour Z, Khoeyi ZA (2011) Protein, fatty acid, and pigment content of Chlorella vulgaris under different light regimes. J Appl Phycol 23:721–726. https://doi.org/10.1007/s10811-010-9569-8
Silve A et al (2018a) Extraction of lipids from wet microalga *Auxenochlorella protothecoides* using pulsed electric field treatment and ethanol-hexane blends. Algal Res 29:212–222. https://doi.org/10.1016/j.algal.2017.11.016
Silve A et al (2018b) Incubation time after pulsed electric field treatment of microalgae enhances the efficiency of extraction processes and enables the reduction of specific treatment energy. Bioresour Technol 269:179–187. https://doi.org/10.1016/j.biortech.2018.08.060
Simonis P et al (2017) Caspase dependent apoptosis induced in yeast cells by nanosecond pulsed electric fields. Bioelectrochemistry 115:19–25
Soheili M, Khosravi-Darani K (2012) The potential health benefits of algae and micro algae in medicine: a review on *Spirulina platensis*. Curr Nutr Food Sci 7:279–285. https://doi.org/10.2174/157340111804586457
Straeßner R, Silve A, Eing C, Rocke S, Wuestner R, Leber K, Mueller G, Frey W (2016) Microalgae precipitation in treatment chambers during pulsed electric field (PEF) processing. Innov Food Sci Emerg Technol 37:391–399
Vanthoor-Koopmans M, Wijffels RH, Barbosa MJ, Eppink MHM (2013) Biorefinery of microalgae for food and fuel. Bioresour Technol 135:142–149

Wijffels RH, Barbosa MJ, Eppink MHM (2010) Microalgae for the production of bulk chemicals and biofuels. Biofuels Bioprod Biorefin 4:287–295

Xu L, Brilman DWF, Withag JAM, Brem G, Kersten S (2011) Assessment of a dry and a wet route for the production of biofuels from microalgae: energy balance analysis. Bioresour Technol 102:5113–5122

Zocher K, Banaschik R, Schulze C, Schulz T, Kredl J, Miron C, Frey W, Kolb J (2016) Comparwison of extraction of valuable compounds from microalgae by atmospheric pressure plasmas awnd pulsed electric fields. Plasma Med 4:273–302. https://doi.org/10.1615/PlasmaMed.2017019104

Drying Improving by Pulsed Electric Fields

Artur Wiktor, Oleksii Parniakov, and Dorota Witrowa-Rajchert

Abbreviations

CDI Cell disintegration index
L* Lightness in CIE L*a*b* colour measurement system
a* Share of red-green colour in CIE L*a*b* colour measurement system
b* Share of blue-yellow colour in CIE L*a*b* colour measurement system
MR Moisture ratio
OD Osmotic dehydration
PEF Pulsed electric field

Introduction

Population growth is one of the most important factors that impacts on changes and dynamics of global food market. In 2019 the worldwide population reached 7.7 billion, and it is estimated that in 2050 it will reach 9.7 billion and in 2100 will peak to nearly 11 billion (United Nations 2020). However, growth population is not the only determinant of food production, as it appears that between 1961 and 2013, the food consumption growth outpaced the population growth (European Union 2015). Other factors that influence the global food market size and value are related to growth of income that shifts the consumption towards high-value products and changes of consumers' habits and preferences. These changes, in turn, are linked to many different drivers as, for instance, climate change or health concerns (European Union 2019). In the report entitled "50 trends influencing Europe's food sector by 2035" (Fraunhofer Institute For Systems and Innovations Research ISI 2020), the sustainable production and value chains, food losses and waste and climate changes are

A. Wiktor (✉) · D. Witrowa-Rajchert
Department of Food Engineering and Process Management, Institute of Food Sciences, Warsaw University of Life Sciences (WULS-SGGW), Warsaw, Poland
e-mail: artur_wiktor@sggw.edu.pl

O. Parniakov
Elea Vertriebs- und Vermarktungsgesellschaft GmbH, Quakenbrück, Germany

J. Raso et al. (eds.), *Pulsed Electric Fields Technology for the Food Industry*, Food Engineering Series, https://doi.org/10.1007/978-3-030-70586-2_13

listed as one of the most important drivers for the development of food production sector. It means that food industry needs new solutions that can fulfil needs of consumers that are more and more aware about social, environmental and health impact of food and its consumption. Pulsed electric field (PEF) is one of the most promising non-thermal technologies that has a potential either to replace some traditional techniques of food processing or to enhance the kinetics of many unit operations applied in food technology and improve the quality of food. The scientific literature shows that PEF can be applied to improve extraction (Liu et al. 2018a, b; Nowacka et al. 2019; Ricci and Parpinello 2018), juice pressing (Jaeger et al. 2012; Kantar et al. 2020), freezing (Parniakov et al. 2016; Wiktor et al. 2015a, b) or osmotic dehydration (Tylewicz et al. 2017; Wiktor et al. 2014). There are also some examples justifying that PEF can be successfully applied to extend the shelf life of liquid food (Rivas et al. 2006; Timmermans et al. 2019). One of the most prominent examples of PEF utilization, already commercialized, is potato and snack industry (Fauster et al. 2018). Such a broad range of possible application of PEF is associated with its mechanism which is called electroporation and, in simplification, depends on electrically induced perforation of the cell membrane. The detailed description of the electroporation, and all the processes and phenomena that are related to it, is presented in some of previous chapters of this book. However, the disintegration of the cellular structure, as it can be caused by PEF, may be also profitable when it comes to the intensification of drying or freeze-drying. This is because cellular structure is one of the main limiting factors of mass and heat transfer that occurs during water removal. This chapter presents the effects of PEF, applied as pre-treatment before drying and freeze-drying, on process kinetics and quality of food products.

Drying Progress

The technological aim of food drying is linked mainly to the removal of water and reduction of product's water activity below some critical value to limit the microorganism growth and chemical or enzymatic reactions (Ratti 2001). Some other consequences associated with drying are a reduction of mass and volume of the products, preservation of surplus or preparation for subsequent processing or applications (Wray et al. 2015). As one of the most popular unit operations in food processing, drying is also one of the oldest and most energy consuming. According to some estimations, drying accounts for 12–25% of energy used by whole food industry (Kumar et al. 2014; Pirasteh et al. 2014). Considering the fact that in the cases of energy-intensive processes, reduction of energy consumption by only 1% may lead to 10 times higher increment of the profits (Kumar et al. 2014). It means that all activities aimed towards intensification of drying process are of paramount importance. Although drying is used to preserve food or produce new type of food, the quality of such processed material is usually lower when it is compared to the raw material. This is because during conventional drying processes (e.g. convection drying), food is exposed to elevated temperature for a long time. This drawback can be compensated by application of other drying methods that either last for shorter time (e.g. microwave assisted air drying) or are performed in lower temperature and/or under vacuum (e.g. vacuum drying and freeze-drying). It has been reported that carotene and vitamin C retention was higher for carrot slices dried by vacuum microwave and freeze-dried method than for air-dried material (Lin et al. 1998). It means that reduction of drying time not only contributes to better energy efficiency but also prospectively can improve the quality of final product.

Drying progress, in general, can be divided into two periods (Fig. 1). During initial period of drying, material is warmed up, and the surface and capillary water is being removed. During this phase water is removed with constant rate, and product temperature does not change a lot. This is why this phase is called constant drying rate period. When water content of processed material reaches critical water content, which is specific for each raw material and each process, the second phase of the processes starts. This period is characterized be the falling drying rate since there is no water on the material's surface anymore and water needs to diffuse from deeper regions. In this period, the internal mass transfer resistance is higher than convective mass transfer resistance. Moreover, from this point, the properties of material, and especially cellular structure of the product, which hinders diffusion of water, start to play more important role than processing parameters such as drying temperature, air flow or humidity, which are very significant during the first drying period. What more, as already mentioned, the elevation of temperature or increment of air flow may cause lowering of product quality by degradation of bioactive compounds which are sensitive for high temperature and/or oxygen (Kaya et al. 2008; Frias et al. 2010).

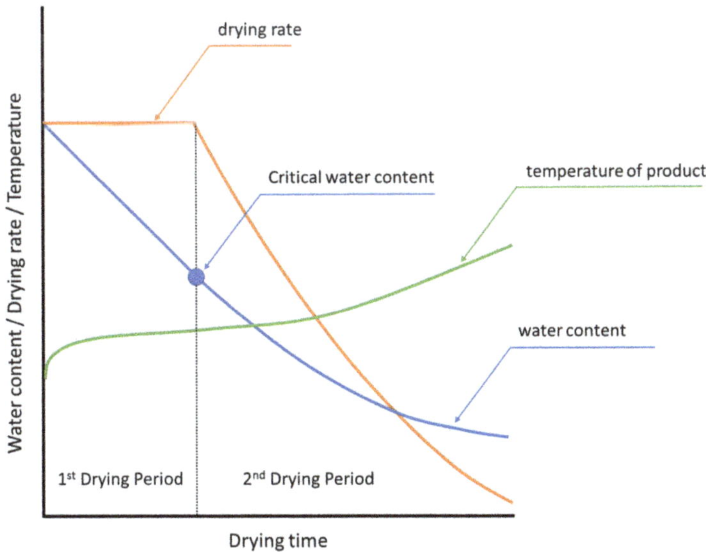

Fig. 1 Drying process progress presented as changes of water content or product, temperature of the product and drying rate during progress of drying (own elaboration)

The degradation of cellular structure and thus intensification of drying may be done by thermal, mechanical or chemical methods. However, application of these methods is associated with many drawbacks like high consumption of water and energy (e.g. blanching), changes of physical properties of material (e.g. cutting/grinding) or consumers' concerns (chemical treatment) (Doymaz and Ismail 2010; Witrowa-Rajchert et al. 2014). Since PEF application, due to electroporation phenomenon, induces rupture of the cellular structure without significant increase of the temperature of the product, it exhibits high potential in enhancement of drying process and improvement of dried product quality (Barba et al. 2015).

The Effect of PEF Pre-treatment on Drying

Air Drying

The reduction of drying time by application of PEF as a pre-treatment before drying has been evidenced in many different scientific papers using different food matrices as examples. For instance, Alam et al. (2018) demonstrated that air drying of carrot supported by PEF pre-treatment can last 13–27% shorter in comparison to the reference process without any pre-treatment. Similar findings have been reported by Wiktor et al. (2016a, b) (reduction of drying time by ca. 8%) and by Gachovska, Simpson, Ngadi, and Raghavan (2009) (drying time reduced by 11–30%) depending on parameters of PEF treatment. The effect of PEF, expressed by the reduction of drying time, strongly depends on processing parameters which is why the results reported by different authors range from 8% to 30% for the same matrix. One of the most important factors that affects the drying course of PEF pre-treated material is cell disintegration index (CDI), which measures the effectiveness of electroporation. In general, the higher the effectiveness of PEF treatment, the greater the benefit that can be achieved by PEF. Figure 2 represents the results of meta-analysis that has been performed based on existing publications that reported CDI and that were related to drying enhancement. Obtained results indicate that there is statistically significant positive correlation between CDI and drying time reduction. However, as it was reported by Ostermeier et al. (2018), higher effectiveness of electroporation does not have to intensify drying process in all cases. Hence, it confirms that optimization studies need to be done each time before implementation of the technology. The effect of PEF on drying is also linked to the matrix itself. First, the effectiveness of PEF is related to many different properties of treated food such as electrical conductivity, chemical composition or porosity. Furthermore, also drying process depends on physical and chemical properties of food. Therefore, drying time reduction of air drying of apple tissue preceded by PEF treatment as presented in the literature ranges from 0% to 30% (Arevalo et al. 2004; Wiktor et al. 2013; Chauhan et al. 2018). The most of reported research focuses on the effect of PEF on the kinetics of air drying which is one of the most popular and easiest drying

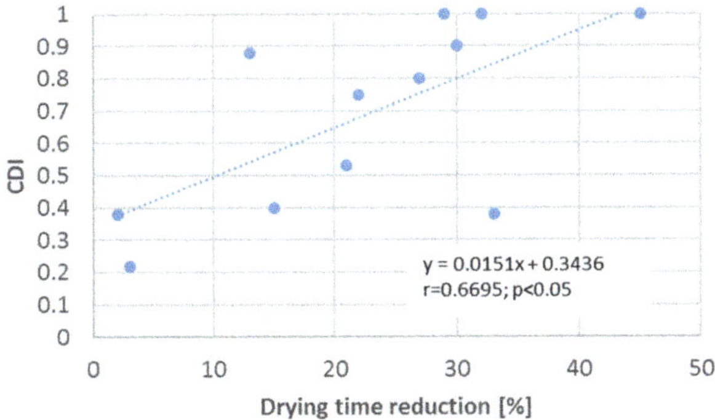

Fig. 2 The relationship between the efficiency of PEF treatment expressed as CDI (cell disintegration index) and the reduction of air drying time as reported in the literature for different matrices (Ade-Omowaye et al. 2000; Lebovka et al. 2007; Shynkaryk et al. 2008; Wiktor et al. 2013; Alam et al. 2018; Chauhan et al. 2018; Dermesonlouoglou et al. 2018; Ostermeier et al. 2018)

methods alike. However, based on existing data, it can be stated that drying method is another factor that affects the effect of PEF measured by the means of drying time reduction. For instance, vacuum drying and air drying of basil leaves preceded by PEF treatment performed with 65 pulses with width of 150 μs and at electric field intensity of 0.65 kV/cm lasted by 57 and 33% shorter, respectively, in comparison to trials with untreated material (Telfser and Gómez Galindo 2019).

During drying the conditions are favourable for many of deteriorative reactions. High temperature usually accelerates degradation of thermolabile compounds and together with air flow can also prompt oxidation of many different compounds. Since PEF reduces drying time, it is anticipated that it positively impacts retention of bioactive, sensitive compounds. On contrary, during PEF treatment some reactive chemical species such as free radicals can be formed. Disintegrated by electroporation cellular structure promotes leakage of polar molecules, which can also lead to decline of many of bioactive compounds. Therefore, the impact of PEF on quality of dried food is a resultant of all phenomena that occur during drying and PEF pretreatment alike. Some of published articles demonstrate that it is possible to improve retention of anthocyanin in dried food as exemplified by blueberry but only when air drying was performed at 60–75 °C (Yu et al. 2017). However, at the same time, PEF pre-treated air-dried material exhibited lower antioxidant activity. Vacuum-dried material exhibited the same, from statistical point of view, free radical scavenging properties regardless it was untreated or PEF treated. In the case of hydrophobic bioactive compounds, it has been demonstrated that combination of PEF with ultrasound significantly improved the retention of carotenoids in carrot subjected to air drying (Wiktor et al. 2019a).

Moreover, chemical properties of food are not the only that are altered during drying. Evaporation of water changes the physical properties of food – mainly those that are linked to the cellular structure such as reconstitution properties or hygroscopicity. What more, degradation of bioactive compounds is accompanied very often by colour change, which further influences consumers' behaviour. Published data shows that the effect of PEF on colour of dried food strongly depends on matrix, which is related to the chemical composition of food, drying method and PEF parameters. For instance, no changes of lightness (L*), lower a* and higher b* were stated for PEF pre-treated air-dried basil leaves (Kwao et al. 2016). In turn, Alam et al. (2018) found out that carrot dried by convective method preceded by PEF treatment exhibited lower L* but no changes of a* or b*. Research performed by Wiktor et al. (2016b) evidenced that air-dried carrot pre-treated by PEF had the same (from statistical point of view) lightness but different a* and b* values. No changes of colour were found for the PEF-treated blueberries regardless of the drying method that has been used (Yu et al. 2017). However, as aforementioned the type of matrix and the bioactive compounds or pigments that it contains play very important role when it comes to the response of tissue and colour change on PEF pre-treatment. For instance, tomatoes and mangoes dried by the convection method were brighter in colour when drying was performed without PEF pre-treatment (Figs. 3, 4). At the same time, the colour of PEF pre-treated material was more saturated and vivid. Such situation may be either related to better retention of pigments (carotenoids) in the tissue subjected to PEF pre-treatment prior to drying or to modification of enzymatic activity of the tissue, which implied some colour changes. Indeed, it has been reported that PEF application can modify enzyme activity – either it can inhibit their activity or it can enhance it by making the substrates more available (Ohshima et al. 2007; Wiktor et al. 2015b; Mannozzi et al. 2019). However, as mentioned before there is no universal pre-treatment protocol for all raw materials. Parameters that can lead to beneficial results in the case of one material can cause completely different and unexpected results when applied to another tissue. For instance, Fig. 5 shows the appearance and colour of untreated and PEF-treated air-dried spring onion. Samples subjected to PEF pre-treatment after drying were darker, browner and more shrunken when compared to untreated spring onion.

It is worth emphasizing that by PEF treatment rehydration properties of some food matrices can be also modified. Once again, in this case, the literature is not consistent. Some of researchers observed improved rehydration properties of PEF-pre-treated dried material (Adedeji et al. 2008; Telfser and Gómez Galindo 2019), whereas some others reported that reconstitution lasted longer (Amami et al. 2007; Shynkaryk et al. 2008). In turn, the combination of sonication (US) and PEF as

A)

B)

Fig. 3 The images of (**a**) untreated and (**b**) PEF (1 kV/cm; 1 kJ/kg)-pre-treated tomato slices air dried at 70 °C (own elaboration, unpublished data)

Fig. 4 The images of (**a**) untreated and (**b**) PEF (1 kV/cm; 1 kJ/kg)-pre-treated mango sticks air dried at 70 °C (own elaboration, unpublished data)

Fig. 5 The images of (**a**) untreated and (**b**) PEF (1 kV/cm; 1 kJ/kg)-pre-treated spring onion air dried at 70 °C (own elaboration, unpublished data)

pre-treatment did not change significantly rehydration properties of microwave-assisted air-dried carrots. However, what is worth emphasizing is that the hygroscopicity of samples pre-treated by combination of PEF and ultrasounds was higher (Wiktor and Witrowa-Rajchert 2020). What is interesting is the same matrix pre-treated using exactly the same parameters of PEF but air dried exhibited lower hygroscopic properties (Wiktor et al. 2019a).

Textural properties, next to the rehydration and hygroscopicity, belong to the most important properties of dried material, especially the one that should serve as a snack. When it comes to texture of dried material, two important components seem to play an important role: mechanical and acoustic properties (Gondek et al. 2006; Costa et al. 2011). Although the effect of PEF on mechanical and acoustic properties has been demonstrated by different researchers (Lebovka et al. 2004; Wiktor et al. 2016a), the amount of data about impact of PEF on texture of dried material is limited. The vast majority of research in this field was carried out using rehydrated material. However, most of articles suggest that application of PEF before drying either does not change mechanical properties of reconstituted material or makes the structure firmer (Shynkaryk et al. 2008; Gachovska et al. 2009; Alam et al. 2018; Liu et al. 2018a, b). Nevertheless, the course of the changes depended strongly also on the drying parameters, for instance, temperature.

Freeze-Drying

Some articles report the possibility of utilization of PEF pre-treatment to improve freeze-drying process. For instance, research conducted by Lammerskitten et al. (2019a) showed that it is possible to obtain high-quality freeze-dried apples by performing freezing inside the drying chamber. What more, it was found that using such approach drying time (up to MR = 0.004) was reduced from 840 to 368 minutes. At the same time, water uptake during rehydration of PEF-pre-treated apples was improved without any significant losses of soluble solids or significant changes of hygroscopicity. PEF-pre-treated freeze-dried apples exhibited also better crunchiness and lower shrinkage in comparison to control samples produced using the same technology but without any pre-treatment (Lammerskitten et al. 2019b). Research related to PEF-assisted freeze-drying using strawberry as raw material showed that it is possible to modify some physical properties of dried product, for instance, its shape and porosity (Fig. 6), which, as aforementioned, has serious impact on its textural properties. Lower shrinkage and volume loss of PEF-pre-treated freeze-dried strawberries and bell pepper were also reported by Fauster et al. (2020).

The utilization of PEF prior to freezing in freeze-drying process may also lead to better preservation of colour, as it has been reported for apples (Lammerskitten et al. 2019b) or can be seen in Fig. 7 for spring onion. The reason that stays behind such behaviour is most probably related to faster processing time and modification of enzymatic activity.

Osmotic Dehydration

Existing literature demonstrates that PEF can be also utilized to modify the course of osmotic dehydration (OD). OD is an operation that aims to remove some of the water from the product by placing it in a concentrated solution of the osmo-active substance. Although osmotic dehydration is not an independent process, but rather a pre-treatment before drying, a lot of research is done because of the impact of this process on nutritional value of the products. In the case of OD, two counter-current flows are observed: water removal from the processed material into the solution and

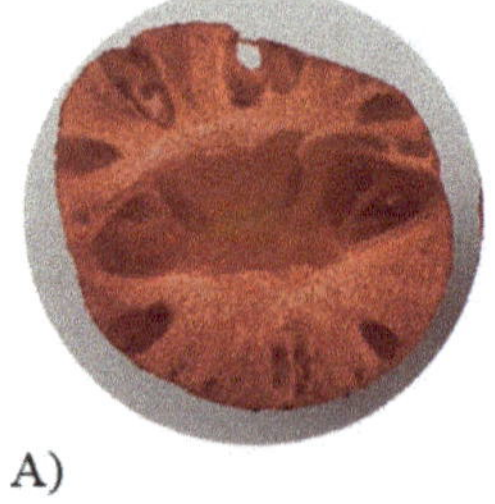

A)

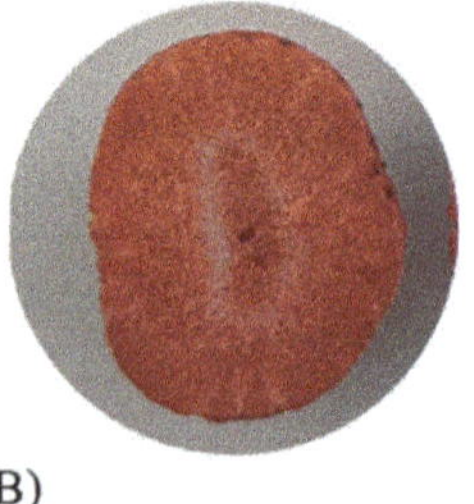

B)

Fig. 6 The cross section of freeze-dried strawberries: (**a**) untreated and (**b**) PEF (1 kV/cm; 1 kJ/kg) pre-treated (own elaboration, unpublished data)

Fig. 7 The images of (**a**) untreated and (**b**) PEF (1 kV/cm; 1 kJ/kg)-pre-treated freeze-dried spring onion (own elaboration, unpublished data)

solid uptake from osmotic medium into material. Solid gain, when OD is performed in sugar solutions, can be considered as both positive and negative phenomena depending on the final application of osmotic dehydrated product. Some of published articles demonstrate that PEF can intensify OD in a way that enhances water loss but does not influence the solid uptake when compared to the control, untreated fruits and vegetables (Ade-Omowaye et al. 2003; Wiktor et al. 2014). Such results can be interesting for production of sweet dried snacks that fit to current nutritional trends aimed towards reduction of simple sugar consumption. The possibility of utilization of PEF in combination with blanching and ultrasound applied prior to drying and OD performed in ternary osmotic solution composed of sugar and steviol glycosides was demonstrated using cranberries. Drying of PEF-pre-treated cranberries lasted shorter than drying of cut material. What more, such processed material was characterized by lower sucrose content than fruits processed by traditional method (Wiktor et al. 2019b). These results prove that PEF enables the production of food with reduced water content and at the same time lower content of sucrose, which is often used as an osmotic factor. Products of this type, subjected to further processing, e.g., drying, could be dedicated to people who because of health reasons should limit the consumption of sugars but would not want to give up eating certain dried fruits.

Conclusion

Pulsed electric field (PEF) technology is a useful strategy to assist dehydration of plant material. Overall, enhanced drying kinetics and improved final product quality after the application of PEF were described by the numerous researchers, thus showing the high added value of PEF for the food industry. Additionally, the application of PEF-assisted drying was interesting to preserve antioxidant thermolabile compounds such as anthocyanins, vitamin C, etc. However, each raw material shows different response to the PEF pre-treatment. Therefore, the investigation and optimization of treatment protocols are required. Moreover, the adaptation of suitable drying conditions for the PEF-treated material is necessary in order to achieve all the benefits of this technology.

References

Adedeji AA, Gachovska TK, Ngadi MO, Raghavan GSV (2008) Effect of pretreatments on drying characteristics of okra. Drying Technol 26:1251–1256. https://doi.org/10.1080/07373930802307209

Ade-Omowaye BIO, Angersbach A, Eshtiaghi NM, Knorr D (2000) Impact of high intensity electric field pulses on cell permeabilisation and as pre-processing step in coconut processing. Innov Food Sci Emerg Technol 1:203–209. https://doi.org/10.1016/S1466-8564(00)00014-X

Ade-Omowaye BIO, Taiwo KA, Eshtiaghi NM, Angersbach A, Knorr D (2003) Comparative evaluation of the effects of pulsed electric field and freezing on cell membrane permeabilisation and mass transfer during dehydration of red bell peppers. Innov Food Sci Emerg Technol 4:177–188. https://doi.org/10.1016/S1466-8564(03)00020-1

Alam MR, Lyng JG, Frontuto D, Marra F, Cinquanta L (2018) Effect of pulsed electric field pretreatment on drying kinetics, color, and texture of parsnip and carrot. J Food Sci 83:2159–2166. https://doi.org/10.1111/1750-3841.14216

Amami E, Fersi A, Khezami L, Vorobiev E, Kechaou N (2007) Centrifugal osmotic dehydration and rehydration of carrot tissue pre-treated by pulsed electric field. LWT- Food Sci Technol 40:1156–1166. https://doi.org/10.1016/j.lwt.2006.08.018

Arevalo P, Ngadi MO, Bazhal MI, Raghavan GSV (2004) Impact of pulsed electric fields on the dehydration and physical properties of apple and potato slices. Drying Technol 22:1233–1246. https://doi.org/10.1081/DRT-120038589

Barba FJ, Parniakov O, Pereira SA, Wiktor A, Grimi N, Boussetta N, Saraiva JA, Raso J, Martin-Belloso O, Witrowa-Rajchert D, Lebovka N, Vorobiev E (2015) Current applications and new opportunities for the use of pulsed electric fields in food science and industry. Food Res Int:77. https://doi.org/10.1016/j.foodres.2015.09.015

Chauhan OP, Sayanfar S, Toepfl S (2018) Effect of pulsed electric field on texture and drying time of apple slices. J Food Sci Technol 55:2251–2258. https://doi.org/10.1007/s13197-018-3142-x

Costa F, Cappellin L, Longhi S, Guerra W, Magnago P, Porro D, Soukoulis C, Salvi S, Velasco R, Biasioli F, Gasperi F (2011) Postharvest biology and technology assessment of apple (Malus × domestica Borkh.) fruit texture by a combined acoustic-mechanical profiling strategy. Postharvest Biol Technol 61:21–28. https://doi.org/10.1016/j.postharvbio.2011.02.006

Dermesonlouoglou E, Chalkia A, Dimopoulos G, Taoukis P (2018) Combined effect of pulsed electric field and osmotic dehydration pre-treatments on mass transfer and quality of air dried goji berry. Innov Food Sci Emerg Technol 49:106–115. https://doi.org/10.1016/j.ifset.2018.08.003

Doymaz I, Ismail O (2010) Processing drying characteristics of sweet cherry. Food Bioprod Process 89:31–38. https://doi.org/10.1016/j.fbp.2010.03.006

European Comission (2015) World food consumption patterns – trends and drivers. EU Agricultural Markets Briefs 6:1–11. https://ec.europa.eu/info/sites/info/files/food-farming-fisheries/trade/documents/agri-market-brief-06_en.pdf. Accessed 19.04.2020

European Comission (2019) Global food supply and demand. Consumer trends and trade challenges 16: 1–12, https://ec.europa.eu/info/sites/info/files/food-farming-fisheries/farming/documents/market-brief-food-challenges-sep2019_en.pdf. Accessed 19.04.2020

Fauster T, Schlossnikl D, Rath F, Ostermeier R, Teufel F, Toepfl S, Jaeger H (2018) Impact of pulsed electric field (PEF) pretreatment on process performance of industrial French fries production. J Food Eng 235:16–22. https://doi.org/10.1016/j.jfoodeng.2018.04.023

Fauster T, Giancaterino M, Pittia P, Jaeger H (2020) Effect of pulsed electric field pretreatment on shrinkage, rehydration capacity and texture of freeze-dried plant materials. LWT- Food Sci Technol 121:108937. https://doi.org/10.1016/j.lwt.2019.108937

Fraunhofer Institute for Systems and Innovations Research ISI (2020) 50 trends influencing Europe's food sector by 2035, https://www.isi.fraunhofer.de/content/dam/isi/dokumente/ccv/2019/50-trends-influencing-Europes-food-sector.pdf, Accessed 19.04.2020

Frias J, Penas E, Ullate M, Vidal-Valverde C (2010) Influence of drying by convective air dryer or power ultrasound on the vitamin C and β -carotene content of carrots. J Agric Food Chem 58:10539–10544. https://doi.org/10.1021/jf102797y

Gachovska TK, Simpson MV, Ngadi MO, Raghavan GSV (2009) Pulsed electric field treatment of carrots before drying and rehydration. J Sci Food Agric 89:2372–2376. https://doi.org/10.1002/jsfa.3730

Gondek E, Lewicki PP, Ranachowski Z (2006) Influence of water activity on the acoustic properties of breakfast cereals. J Texture Stud 37:497–515

Jaeger H, Schulz M, Lu P, Knorr D (2012) Adjustment of milling, mash electroporation and pressing for the development of a PEF assisted juice production in industrial scale. Innov Food Sci Emerg Technol 14:46–60. https://doi.org/10.1016/j.ifset.2011.11.008

Kantar S, Boussetta N, Lebovka N, Foucart F, Rajha HN, Maroun RG, Louka N, Vorobiev E (2020) Pulsed electric field treatment of citrus fruits: improvement of juice and polyphenols extraction. Innov Food Sci Emerg Technol 46:153–161. https://doi.org/10.1016/j.ifset.2017.09.024

Kaya A, Aydın O, Kolaylı S (2008) Food and bioproducts processing effect of different drying conditions on the vitamin C (ascorbic acid) content of Hayward kiwifruits (Actinidia deliciosa Planch). Food Bioprod Process 88:165–173. https://doi.org/10.1016/j.fbp.2008.12.001

Kumar C, Karim MA, Joardder MUH (2014) Intermittent drying of food products: a critical review. J Food Eng 121:48–57. https://doi.org/10.1016/j.jfoodeng.2013.08.014

Kwao S, Al-Hamimi S, Damas MEV, Rasmusson AG, Gómez Galindo F (2016) Effect of guard cells electroporation on drying kinetics and aroma compounds of Genovese basil (Ocimum basilicum L.) leaves. Innov Food Sci Emerg Technol 38:15–23. https://doi.org/10.1016/j.ifset.2016.09.011

Lammerskitten A, Mykhailyk V, Wiktor A, Toepfl S, Nowacka M, Bialik M, Czyżewski J, Witrowa-Rajchert D (2019a) Impact of pulsed electric fields on physical properties of freeze-dried apple tissue. Innov Food Sci Emerg Technol 57:102211. https://doi.org/10.1016/j.ifset.2019.102211

Lammerskitten A, Wiktor A, Siemer C, Toepfl S, Mykhailyk V, Gondek E, Rybak K, Witrowa-Rajchert D, Parniakov O (2019b) The effects of pulsed electric fields on the quality parameters of freeze-dried apples. J Food Eng 252:36–43. https://doi.org/10.1016/j.jfoodeng.2019.02.006

Lebovka NI, Praporscic I, Vorobiev E (2004) Effect of moderate thermal and pulsed electric field treatments on textural properties of carrots, potatoes and apples. Innov Food Sci Emerg Technol 5:9–16. https://doi.org/10.1016/j.ifset.2003.12.001

Lebovka NI, Shynkaryk NV, Vorobiev E (2007) Pulsed electric field enhanced drying of potato tissue. J Food Eng 78:606–613. https://doi.org/10.1016/j.jfoodeng.2005.10.032

Lin TM, Durance TD, Scaman CH (1998) Characterization of vacuum microwave, air and freeze dried carrot slices. Food Res Int 31:111–117. https://doi.org/10.1016/S0963-9969(98)00070-2

Liu C, Grimi N, Lebovka N, Vorobiev E (2018a) Effects of pulsed electric fields treatment on vacuum drying of potato tissue. Lwt 95:289–294. https://doi.org/10.1016/j.lwt.2018.04.090

Liu ZW, Zeng XA, Ngadi M (2018b) Enhanced extraction of phenolic compounds from onion by pulsed electric field (PEF). J Food Process Preserv 42:e13755. https://doi.org/10.1111/jfpp.13755

Mannozzi C, Rompoonpol K, Fauster T, Tylewicz U (2019) Influence of pulsed electric field and ohmic heating pretreatments on enzyme and antioxidant activity of fruit and vegetable juices. Foods 8:247

Nowacka M, Tappi S, Wiktor A, Rybak K, Miszczykowska A, Czyzewski J, Drozdzal K, Witrowa-Rajchert D, Tylewicz U (2019) The impact of pulsed electric field on the extraction of bioactive compounds from beetroot. Foods:8. https://doi.org/10.3390/foods8070244

Ohshima T, Tamura T, Sato M (2007) Influence of pulsed electric field on various enzyme activities. J Electrostat 65:156–161. https://doi.org/10.1016/j.elstat.2006.07.005

Ostermeier R, Giersemehl P, Siemer C, Töpfl S, Jäger H (2018) Influence of pulsed electric field (PEF) pre-treatment on the convective drying kinetics of onions. J Food Eng 237:110–117. https://doi.org/10.1016/j.jfoodeng.2018.05.010

Parniakov O, Bals O, Lebovka N, Vorobiev E (2016) Effects of pulsed electric fields assisted osmotic dehydration on freezing-thawing and texture of apple tissue. J Food Eng 183:32–38. https://doi.org/10.1016/j.jfoodeng.2016.03.013

Pirasteh G, Saidur R, Rahman SMA, Rahim NA (2014) A review on development of solar drying applications. Renew Sustain Energy Rev 31:133–148. https://doi.org/10.1016/j.rser.2013.11.052

Ratti C (2001) Hot air and freeze-drying of high-value foods: a review. J Food Eng 49:311–319. https://doi.org/10.1016/S0260-8774(00)00228-4

Ricci A, Parpinello GP (2018) Recent advances and applications of pulsed electric fields (PEF) to improve polyphenol extraction and color release during red winemaking. Beverages 4:18. https://doi.org/10.3390/beverages4010018

Rivas A, Rodrigo M, Martinez A, Barbosa-Cánovas GV, Rodrigo M (2006) Effect of PEF and heat pasteurization on the physical – chemical characteristics of blended orange and carrot juice. LWT- Food Sci Technol 39:1163–1170. https://doi.org/10.1016/j.lwt.2005.07.002

Shynkaryk MV, Lebovka NI, Vorobiev E (2008) Pulsed electric fields and temperature effects on drying and rehydration of red beetroots. Drying Technol 26:695–704. https://doi.org/10.1080/07373930802046260

Telfser A, Gómez Galindo F (2019) Effect of reversible permeabilization in combination with different drying methods on the structure and sensorial quality of dried basil (Ocimum basilicum L.) leaves. Lwt 99:148–155. https://doi.org/10.1016/j.lwt.2018.09.062

Timmermans RAH, Mastwijk HC, Berendsen LBJM, Nederhoff AL, Matser AM, Van Boekel MAJS, Groot MNN (2019) International journal of food microbiology moderate intensity pulsed electric fields (PEF) as alternative mild preservation technology for fruit juice. Int J Food Microbiol 298:63–73. https://doi.org/10.1016/j.ijfoodmicro.2019.02.015

Tylewicz U, Tappi S, Mannozzi C, Romani S, Laghi L, Ragni L, Rocculi P, Rosa MD (2017) Accepted manuscript. J Food Eng 213:2–9. https://doi.org/10.1016/j.jfoodeng.2017.04.028

United Nations (2020) Growing at a slower pace, world population is expected to reach 9.7 billion in 2050 and could peak at nearly 11 billion around 2100. https://www.un.org/development/desa/en/news/population/world-population-prospects-2019.html. Accessed 19.04.2020

Wiktor A, Witrowa-Rajchert D (2020) Drying kinetics and quality of carrots subjected to microwave-assisted drying preceded by combined pulsed electric field and ultrasound treatment. Drying Technol 38:176–188. https://doi.org/10.1080/07373937.2019.1642347

Wiktor A, Iwaniuk M, Śledź M, Nowacka M, Chudoba T, Witrowa-Rajchert D (2013) Drying kinetics of apple tissue treated by pulsed electric field. Drying Technol 31. https://doi.org/10.1080/07373937.2012.724128

Wiktor A, Śledź M, Nowacka M, Chudoba T, Witrowa-Rajchert D (2014) Pulsed electric field pretreatment for osmotic dehydration of apple tissue: experimental and mathematical modeling studies. Drying Technol 32. https://doi.org/10.1080/07373937.2013.834926

Wiktor A, Schulz M, Voigt E, Knorr D, Witrowa-Rajchert D (2015a) Impact of pulsed electric field on kinetics of immersion freezing, thawing, and on mechanical properties of carrot. Zywn Nauk Technol Jakosc/Food Sci Technol Qual 22. https://doi.org/10.15193/zntj/2015/99/027

Wiktor A, Sledz M, Nowacka M, Rybak K, Chudoba T, Lojkowski W, Witrowa-Rajchert D (2015b) The impact of pulsed electric field treatment on selected bioactive compound content and color of plant tissue. Innov Food Sci Emerg Technol 30:69–78. https://doi.org/10.1016/j.ifset.2015.04.004

Wiktor A, Gondek E, Jakubczyk E, Nowacka M, Dadan M, Fijalkowska A, Witrowa-Rajchert D (2016a) Acoustic emission as a tool to assess the changes induced by pulsed electric field in apple tissue. Innov Food Sci Emerg Technol 37. https://doi.org/10.1016/j.ifset.2016.04.008

Wiktor A, Nowacka M, Dadan M, Rybak K, Lojkowski W, Chudoba T, Witrowa-Rajchert D (2016b) The effect of pulsed electric field on drying kinetics, color, and microstructure of carrot. Drying Technol 34. https://doi.org/10.1080/07373937.2015.1105813

Wiktor A, Dadan M, Nowacka M, Rybak K, Witrowa-Rajchert D (2019a) The impact of combination of pulsed electric field and ultrasound treatment on air drying kinetics and quality of carrot tissue. LWT- Food Sci Technol 110:71–79. https://doi.org/10.1016/j.lwt.2019.04.060

Wiktor A, Nowacka M, Anuszewska A, Rybak K, Dadan M (2019b) Drying kinetics and quality of dehydrated cranberries pretreated by traditional and innovative techniques. J Food Sci 84:1820–1828. https://doi.org/10.1111/1750-3841.14651

Witrowa-Rajchert D, Wiktor A, Sledz M, Nowacka M (2014) Selected emerging technologies to enhance the drying process: a review. Drying Technol 32. https://doi.org/10.1080/07373937.2014.903412

Wray D, Ramaswamy HS, Wray D, Ramaswamy HS (2015) Novel concepts in microwave drying of foods novel concepts in microwave drying of foods. Drying 33:769–783. https://doi.org/10.1080/07373937.2014.985793

Yu Y, Jin TZ, Xiao G (2017) Effects of pulsed electric fields pretreatment and drying method on drying characteristics and nutritive quality of blueberries. J Food Process Preserv:41. https://doi.org/10.1111/jfpp.13303

Pulsed Electric Fields Application in Meat Processing

Roman Karki, Indrawati Oey, Phil Bremer, and Pat Silcock

Abbreviations

a^*	Redness
b^*	Yellowness
BF	*Biceps femoris* (outside flat) muscle
C	Chroma ($C = \sqrt{a^{*2} + b^{*2}}$)
C16:0	Palmitic acid
C18:0	Stearic acid
C18:1 $n-6$	Oleic acid
C18:2 $n-6$	Linoleic acid
C18:3 $n-6$	Linolenic acid
C20:4 $n-6$	Arachidonic acids
C22:6 $n-3$	Docosadienoic acid
CIE	Commission Internationale d'Eclairage
Cryo-SEM	Cryo-scanning electron microscopy
DP	*Deep pectoralis* (brisket) muscle
DSC	Differential scanning calorimetry
EFS	Electric field strength
Fe	Iron
h	Hue angle ($h = \tan^{-1} \frac{b^*}{a^*}$)
HPEF	High-intensity PEF
K	Potassium
L^*	Lightness
LD	*Longissimus dorsi* muscle

R. Karki · P. Bremer · P. Silcock
Department of Food Science, University of Otago, Dunedin, New Zealand
e-mail: roman.karki@postgrad.otago.ac.nz; phil.bremer@otago.ac.nz; pat.silcock@otago.ac.nz

I. Oey (✉)
Department of Food Science, University of Otago, Dunedin, New Zealand

Riddet Institute, Palmerston North, New Zealand
e-mail: indrawati.oey@otago.ac.nz

J. Raso et al. (eds.), *Pulsed Electric Fields Technology for the Food Industry*, Food Engineering Series, https://doi.org/10.1007/978-3-030-70586-2_14

LL	*Longissimus lumborum* (beef loin) muscle
LPEF	Low-intensity PEF
LT	*Longissimus thoracis* (rib-eye) muscle
MDA	Malondialdehyde
NMR	Nuclear magnetic resonance spectroscopy
P	Phosphorus
PEF	Pulsed electric fields
PF	Pulse frequency
PN	Pulse number
PUFA	Polyunsaturated fatty acids
PW	Pulse width
SE	Specific energy
SEM	Standard error of mean
SFA	Saturated fatty acids
SM	*Semimembranosus* (inside round) muscle
TBARS	Thiobarbituric acid reactive substances
TDS	Temporal dominance of sensations
TEM	Transmission electron microscopy
WHC	Water holding capacity
Zn	Zinc

Introduction

PEF processing applies high-voltage electric energy to meat for a quick time (μs to ms) in pulses to increase the permeability of cell membrane or to form permanent pores (Bhat et al. 2019b; Gomez et al. 2019; Ostermeier et al. 2018). The effectiveness of PEF treatment depends on processing variables including the design of the treatment chamber and the intrinsic properties of the meat. PEF technology provides many benefits to meat processing including facilitating enzyme reaction, enhancing mass and heat transfer, and improving the operation efficiency of subsequent unit operations such as sous vide, ageing, curing, and drying with minimal energy consumption (Alahakoon et al. 2018, 2019a, b; Arroyo et al. 2015b; Bekhit et al. 2016; Bhat et al. 2019b, c; Blahovec et al. 2017; Faridnia et al. 2014; McDonnell et al. 2014; Puértolas et al. 2016; Roldan et al. 2013; Suwandy et al. 2015d; Toepfl and Heinz 2007, 2008).

In this book chapter, both the beneficial and adverse effects of PEF processing on meat quality will be outlined, and the potential of PEF pretreatment for various meat processing application will be discussed.

Effect of PEF on Meat Quality

The most important meat quality parameters are colour, flavour, juiciness, and tenderness (Listrat et al. 2016). Consumers judge meat freshness based on colour, whereas tenderness influences their repurchasing decision (Bekhit et al. 2016; Suwandy et al. 2015c). Therefore, a processing technology which can improve tenderness and flavour, while preserving colour and minimising lipid oxidation, will be of significant value (Kantono et al. 2019).

Several physical, chemical, and enzymatic methods have been used to tenderise meat (Bekhit et al. 2014a; Devine et al. 2002; Honikel et al. 1983; Koohmaraie and Geesink 2006; Liu et al. 2006; Nuss and Wolfe 1981); but many of these processes also produce adverse effects such as causing a mushy texture due to severe structural changes, the development of off-flavours, or increases in microbial numbers (Bekhit et al. 2016). PEF can induce structural changes, improve meat tenderness, enhance meat processing, and contribute to the development of novel meat products (Arroyo et al. 2015a; Bekhit et al. 2014b; Faridnia et al. 2015; Smetana et al. 2019). The subsequent increase in permeability due to PEF can accelerate proteolysis (Bekhit et al. 2014b; Suwandy et al. 2015b).

The tenderising effect of PEF on meat has been reported for beef *semitendinosus* (ST, outside round), *semimembranosus* (SM, inside round), *longissimus thoracis* (LT, rib-eye), *longissimus lumborum* (LL, beef loin), *biceps femoris* (BF, outside flat), and *deep pectoralis* (DP, brisket) meat cuts (Alahakoon et al. 2019a; Bekhit et al. 2014b, 2016; Faridnia et al. 2015, 2016; Suwandy et al. 2015b, c, d). Moderate-intensity PEF treatment (electric field strength of 0.27–1.4 kV/cm) has a minimal effect on cooking loss, lipid oxidation, colour, fatty acid profile, or the ratio of polyunsaturated/saturated fatty acids and omega 6/omega 3 fatty acids of different meat muscles.

The different and sometimes contradictory reported effects of PEF treatment on meat quality may be because of the different experimental conditions used, including PEF processing parameters and intensity, treatment chamber, sample size, muscle types, and the sample preparation prior to PEF treatment such as freezing and thawing (Faridnia et al. 2015). The heterogeneity of meat cuts; variation in quantity and composition of protein, fat, and collagen content; different electrical and topological properties; and muscle fibre orientation can also affect the distribution of electric field and the outcome of PEF treatment (Alahakoon et al. 2017a; Suwandy et al. 2015c).

Effect of PEF on Meat Tenderness

The tenderness of meat and meat products has a dramatic impact on consumer's repurchasing decisions (Bolumar et al. 2013). The tenderness of a meat cut depends on the amount of connective tissue presence, the extent of proteolysis of the myofibrillar proteins, as well as animal (age, breed), environmental, and genetic factors (Anderson et al. 2012).

Depending on the intensity, PEF treatment causes the formation of pores in the muscle fibres that enhances meat tenderness through an increased release of Ca^{2+} and μ-calpain enzymes which stimulates the glycolytic processes required for early proteolysis (Bekhit et al. 2014b; Bhat et al. 2019b; Suwandy et al. 2015a); increases myofibrillar fragmentation and degradation of myosin heavy chains (Faridnia et al. 2014); reduces the muscle size and increases the number of visible gaps between muscle cells (Gudmundsson and Hafsteinsson 2001); and increases the porosity and thermal solubility of connective tissue (Alahakoon et al. 2017a).

The effect of PEF on meat quality depends on various process parameters such as electric field strength (EFS), specific energy (SE), pulse width (PW), pulse frequency (PF), and pulse number (PN). Moderate-intensity PEF treatments using a range of process parameters such as EFS, 0.27–1.4 kV/cm; PW, 20μs; PF, 20–90 Hz; and SE, 16.2–250 kJ/kg on different beef muscles have been reported to increase tenderness in various beef muscles without having a significant adverse affect on colour, water holding capacity, or lipid oxidation (Table 1) (Bekhit et al. 2014b, 2016; Faridnia et al. 2015, 2016; Suwandy et al. 2015b, c, d). Alahakoon et al. (2017b) confirmed that PEF, even with moderate intensity (EFS: 1.0–1.5 kV/cm), increased the porosity and solubility of connective tissue isolated from beef DP muscle highlighting the ability of PEF to enhance tenderness and lower cooking time of even low-value tough meat cuts.

The increase in porosity in meat muscles has been confirmed by several microscopic studies. Faridnia et al. (2014) observed that the meat structure becomes more porous as the EFS increases. Cryo-scanning electron microscopy (cryo-SEM) images of beef LT muscles showed that EFS of 0.3 kV/cm resulted in the myofibrils having a porous structure compared to the intact surface of the untreated muscle (Fig. 1).

Chian et al. (2019) observed that PEF-treated (EFS, 1.00–1.25 kV/cm; PN, 500 and 2000; SE, 48 ± 5 kJ/kg and 178 ± 11 kJ/kg) beef ribeye (LT) muscles showed significantly longer sarcomere length than in control ($p < 0.001$) and extent of elongation was higher at high treatment intensity. This showed that PEF caused the physical disruption of the muscle fibres and weakening of Z-disk and I-band junctions resulting in the elongation of sarcomeres, which was positively correlated to increased tenderness.

Pretreatment such as freezing–thawing prior to PEF increases tenderness because of myofibrillar disintegration. Faridnia et al. (2015) demonstrated that a freezing–thawing cycle prior to the PEF treatment led to additional physical damage of the muscle as seen in transmission electron micrographs (TEM) of beef ST muscles (Fig. 2). Frozen–thawed samples had clear district A-bands, I-bands, and Z-lines, and the ultrastructure of samples had intact myofibrils as shown in Fig. 2*A*, whereas myofibrils were ruptured along the Z-lines due to PEF treatment (Fig. 2*B*).

High-intensity PEF can have a negative impact on meat quality owing to the shrinkage and denaturation of myofibrillar proteins and enzymes (Table 1). PEF treatment (1.1–2.8 kV/cm and 12.7–226 kJ/kg) raised the temperature of post-rigour

Table 1 Effect of PEF on meat tenderness

Muscles type	PEF operating parameters	Effect on meat tenderness	References
Triceps brachii (TB) muscles	EFS, 3.5 kV/cm; PF, 20 Hz; treatment time, 5 s	Decreased instrumental maximum shear force by 21.5%	Lopp and Weber (2005)
Post-rigour beef *semitendinosus* (ST) muscle	Batch PEF treatment with EFS, 1.1–2.8 kV/cm; SE, 12.7–226 kJ/kg; PF, 5–200 Hz; PN, 152–300	Increase in muscle temperature up to 30 °C which significantly increases the water loss from myofibrils; insignificant decrease in instrumental texture profile indicating minimal increase in tenderness	O'Dowd et al. (2013)
Beef loin *M. longissimus lumborum* (LL) and topside *M. semimembranosus* (SM) muscles	Batch PEF treatment with parallel fibre orientation to applied electric field; with EFS, 0.27–0.56 kV/cm; PW, 20μs; PF, 20, 50, and 90 Hz; variable PN, 607–2726 SE, 3.1–73.2 kJ/kg	LL muscle becomes more tender (19.5% reduction in the shear force); degree of tenderness of SM increased with an increase in pulse frequency of treatment (4.1, 10.4, and 19.1% reduction in the shear force at 20, 50, and 90 Hz, respectively)	Bekhit et al. (2014b)
Beef ST muscle	Batch PEF treatment using square-wave bipolar pulses at EFS, 1.4 kV/cm; PW, 20μs; PF, 50 Hz; and SE, 250 kJ/kg	Cause significant microstructural changes, which in combination with freezing–thawing treatment increased meat tenderness as shown by reduced shear force values; no significant increase in cooking loss without affecting free fatty acid profile and the ratio of polyunsaturated/saturated fatty acids and omega 6/omega 3	Faridnia et al. (2015)
Turkey breast	EFS, up to 3 kV/cm; PW, 20μs; PN, 100, 200, and 300; PF, 10, 55, and 110 Hz; SE, 11–94 kJ/kg	Lipid oxidation in all PEF-treated samples not significantly different to the control. No difference in weight loss, cook loss, lipid oxidation, texture, and colour (raw and cooked) either on fresh or frozen samples: differences between the PEF-treated samples and the controls in terms of texture and odour during sensory evaluation	Arroyo et al. (2015a)

(continued)

Table 1 (continued)

Muscles type	PEF operating parameters	Effect on meat tenderness	References
Beef LL and SM muscles	Batch PEF treatment of EFS, 0.28–0.56 kV/cm; PPP, 99.6–471 kW; PPC, 43.7–112.2 A; PW, 20μs; PF, 20–90 Hz; SE, 3.2–70.1 kJ/kg	Beef LL becomes tougher with increasing treatment frequency, whereas beef SM becomes tender (21.6% reduction in measured shear force). At low-intensity treatment (20 Hz) increased tenderness because of myofibrillar proteolysis shown by increased troponin and desmin degradation	Suwandy et al. (2015b)
Beef LL muscle	Batch PEF treatment with EFS, 0.58–0.73 kV/cm; PPP, 486.3–507.8 kW; PPC, 97–104.3 A; PW, 20μs; PF, 390 Hz; and SE, 16.2–19.3 kJ/kg	Insignificant effect of muscle fibre orientation or initial pH on measured shear force, total water loss, meat colour, and lipid stability; increased proteolysis demonstrated by an increase in troponin T and desmin degradation. Decreased shear force value due to increased proteolysis at lower muscle pH (5.5–5.8) than at higher pH (>6.1) indicating tenderisation effect	Suwandy et al. (2015c)
Cold deboned beef SM and LL muscles	Batch PEF treatment with 0, 1, 2, and 3 times repeated application parallel to fibre orientation, EFS, 0.50–0.58 kV/cm; PF, 90 HZ; PW, 20μs; treatment time, 30 s	No tenderising effect on SS muscle but LL muscle becomes more tender with each repeat, with increased proteolysis shown by increased troponin and desmin degradation; no significant effect on purge loss; lipid oxidation by PEF	Suwandy et al. (2015d)
Hot deboned beef loin (LL) and topside (SM) muscles	Batch PEF treatment with 1, 2, and 3 repeats, EFS, 0.44–0.46 kV/cm, and SE, 633.2–70.8 kJ/kg on LL muscle and 0.32–0.35 kV/cm, and SE, 17.5–19 kJ/kg on SM muscle; PF, 90 Hz; and PW, 20μs	Three times repeated treatment on LL muscle reduced the tenderness whereas in MM muscle reduced shear force value at 3 days post-treatment is reported; lower cooking loss; increase in proteolysis in single PEF treatment and decrease in proteolysis in 3× PEF-treated samples due to protein denaturation as suggested by results from Western blotting	Bekhit et al. (2016)

(continued)

Table 1 (continued)

Muscles type	PEF operating parameters	Effect on meat tenderness	References
Beef outside flat (*biceps femoris*, BF) muscle	Batch PEF treatment of EFS, 1.7–2.0 kV/cm, and SE, 185 kJ/kg	Increase in muscle tenderness indicated by significantly decreased shear force value; no effect on cooking loss but significant increase ($P < 0.05$) in purge loss; dramatic increase in Z-lines fracture in myofibrils compared to the untreated samples during ageing	Faridnia et al. (2016)
Isolated connective tissue obtained from beef deep pectoralis muscle (brisket)	Batch PEF treatment in treatment chamber of dimension (4 cm width, length 6 cm, and depth 6 cm) and EFS, 1.0 and 1.5 kV/cm, and SE, 40–50 and 90–100 kJ/kg, within an agar matrix at electrical conductivities from 1 to 7 mS/cm	Significant decrease ($p < 0.05$) in denaturation temperature, increased ringer soluble collagen fraction, and porosity of connective tissue which indicates the PEF ability to improve tenderness and to decrease cooking time of tough meat cuts	Alahakoon et al. (2017b)
Cold deboned beef LL muscle	Low intensity PEF (LPEF, 0.5 kV/cm, 200 Hz, 20μs, and 12.4 kJ/kg) and high intensity PEF (HPEF, 2 kV/cm, 200 Hz, 20μs, and 149.8 kJ/kg) treatment	HPEF treatment reduces the quality of meat including tenderness	Khan et al. (2017)
Fresh and frozen beef BF and ST muscles	EFS, 0.8–1.1 kV/cm; PW, 20μs; PF, 50 Hz; SE, 150 kJ/kg	Improved meat tenderness and colour in both fresh and frozen samples but increased fat oxidation and saturated fatty acids in frozen samples	Kantono et al. (2019)
Beef topside (SM) muscle from culled dairy cows of 9 years of age	Two PEF treatments: EFS, 0.36 kV/cm; PW, 20μs; PF, 90 Hz; SE, 18.63 kJ/kg (T1), and EFS, 0.60 kV/cm; PF, 20 Hz; PW, 20μs; SE, 73.28 kJ/kg (T2)	The mean values for shear force for samples treated with PEF were lower than control but not significant ($p > 0.05$)	Bhat et al. (2019a)

EFS Electric field strength, *PPP* pulse peak power, *PPC* pulse peak current, *PW* pulse width, *PF* pulse frequency, *PN* pulse number, *SE* specific energy

beef ST muscle by 30 °C, which causes denaturation of myofibrillar protein, reduced the water holding capacity, and increased drip loss with no significant increase in meat tenderness (O'Dowd et al. 2013). The adverse effect on meat quality will increase as the treatment intensity is increased. This is supported by Khan et al.

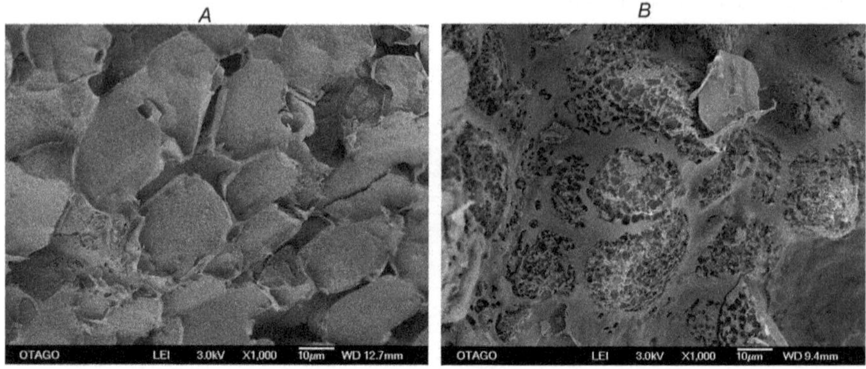

Fig. 1 Cryo-SEM micrographs of *longissimus thoracis* muscles (×1000 magnification): (**A**) untreated; (**B**) after a PEF treatment (0.3 kV/cm, 1 Hz, 20μs, and 0.16 ± 0.02 kJ/kg) (Images provided by Dr. Farnaz Faridnia)

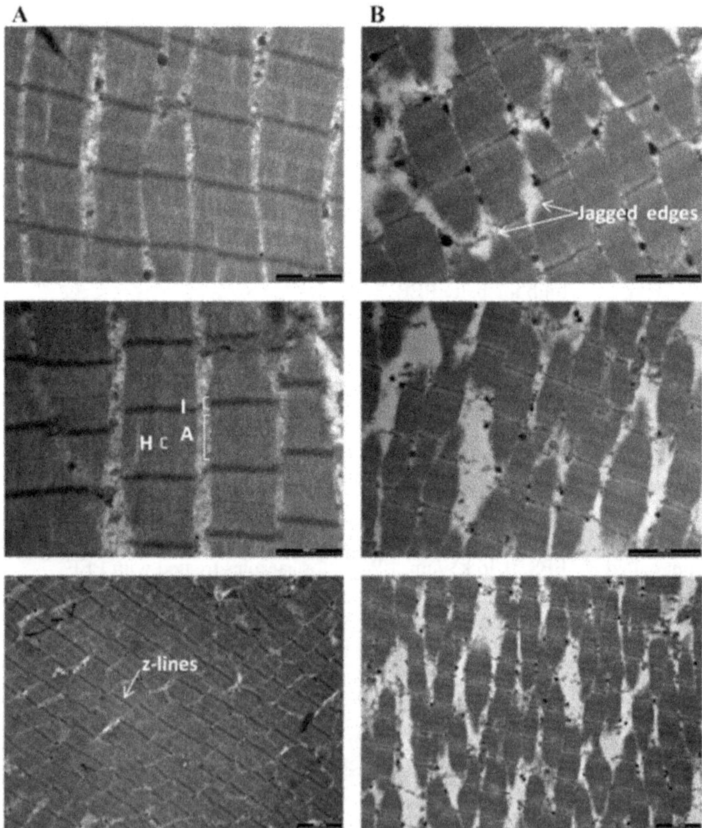

Fig. 2 TEM of beef semitendinosus (ST) muscles showing rupture of myofibrils along with the Z-lines. (**a**) Control: magnification 17,500×. Frozen–thawed beef ST muscles before PEF processing; intact myofibrils. (**b**) Pulsed electric fields (PEF) treated: magnification 17,500×. Frozen–thawed beef ST muscles after PEF processing. Visible white areas in frozen–thawed PEF-treated samples are connective tissues around muscle cells indicating jagged edges and myofibril separation from Z-lines (Reproduced from Faridnia et al. 2015)

(2017) who observed that high-intensity PEF treatment (2 kV/cm, 200 Hz, 20μs, and 149.8 kJ/kg) causes a decrease in tenderness of beef LL muscle. Some studies found that the ability of PEF to tenderise the meat from these animals is reduced when the animal gets older. Bhat et al. (2019a) observed that PEF treatments (0.36 kV/cm and 18.63 kJ/kg or 0.60 kV/cm and 73.28 kJ/kg) did not improve tenderness in beef topsides (*SM*) obtained from 9-year-old dairy cows.

Therefore, the ability of PEF to enhance meat tenderness depends on several factors such as PEF processing parameters as well as the breed, age, and sex of an animal, and the composition and physicochemical characteristics of meat including pretreatment and post-treatment conditions of meat. It is noted that PEF treatment intensity should occur at a certain critical value that can disintegrate cell membranes (0.5 kV/cm for sarcolemma (Toepfl et al. 2006)) and accelerate biochemical reactions within muscles, but not so high as to cause detrimental effects. As evident from several experiments, PEF efficiency to enhance meat tenderness depends upon electrical conductivity, fibre types, and composition and orientation of muscle fibre, which, in turn, affects the distribution of the electric field in the treatment chamber. Pretreatment and post-treatment chilling and freezing–thawing, post-mortem status, and post-slaughter hot or cold boning also provide tenderising effect besides PEF treatment.

Effect of PEF on Water Holding Capacity of Meat

The water holding capacity (WHC) of meat affects its texture, tenderness, and flavour (Arroyo et al. 2015a). As myofibril proteins hold the major proportion of water in meat, their denaturation, especially of myosin, reduces the meat's water holding capacity (Wiklund et al. 2001). Moreover, myofibrillar shrinkage and contraction reduces the amount of water that can be contained in the muscle fibres (Micklander et al. 2005; Suwandy et al. 2015b). PEF causes a change in the myofibrillar structure, the fragmentation of the myofibrillar proteins to small particles, and the formation of pores, which may lead to a loss of water (O'Dowd et al. 2013; Suwandy et al. 2015b).

There are contradictory conclusions regarding the effect of PEF on water holding capacity of meat. Some studies found that the weight loss in various meat cuts after PEF treatment increases from 0.2 to 3.6% with elevating PEF treatment intensity (0.28–0.54 kV/cm) (Bekhit et al. 2014b; Suwandy et al. 2015b), while other authors have reported that PEF treatments of EFS 1.1–2.8 kV/cm and SE ranging from 12.7 to 226 kJ/kg do not have an adverse effect on the WHC of beef ST muscle (O'Dowd et al. 2013).

It was also concluded that PEF did not cause significant loss of water held within muscle fibre, when measured by nuclear magnetic resonance (NMR) T_2 relaxometry study. For example, there were no significant changes in the NMR $T_{2\text{-}1}$ and $T_{2\text{-}2}$ relaxation times (42.59 and 202.69 ms, respectively) between control, PEF-treated, and water bath heated samples (Fig. 3) (O'Dowd et al. 2013). $T_{2\text{-}1}$ relaxation time

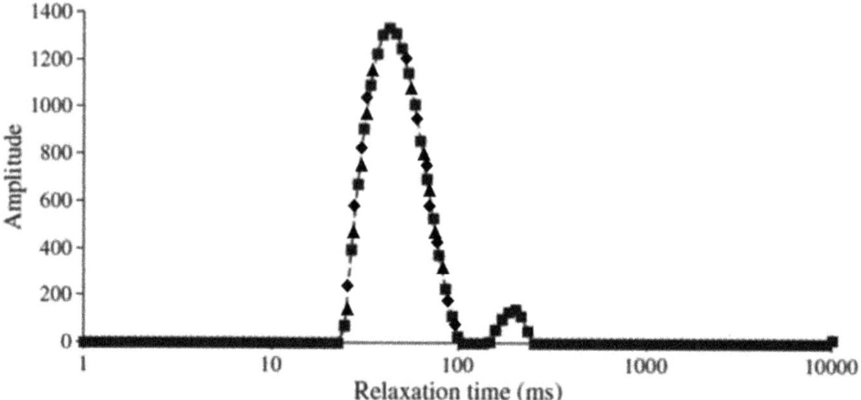

Fig. 3 NMR distribution of T_2 relaxation times (ms) for control (···▲···), PEF-treated (-■-), and water bath-treated (···◆···) beef (coincided with each other) (Reproduced from O'Dowd et al. 2013)

represents the portion of water inside the myofibrils (96.78% of total water); this water is tightly bound but known to decrease with post-mortem time as the breakdown of the cell membrane increases. $T_{2\text{-}2}$ relaxation time represents the more mobile or loosely bound water in the extracellular spaces (3.22% of the total water) of the muscle (Bertram et al. 2001; Tornberg et al. 2000). No significant changes in the $T_{2\text{-}1}$ and $T_{2\text{-}2}$ relaxation times indicate that there were only minor changes in the muscle cell membrane and observed loss in weight is because of expulsion of fluid held in an extracellular space of muscle fibre. More research needs to be carried to evaluate the effect of different PEF treatments on the WHC of meat in order to elucidate under which conditions PEF can affect this parameter. This is a key point mainly when meat is being dried or subjected to a process where water mass transfer occurs.

Effect of PEF on Purge Loss of Meat

Purge loss is the amount (%) of water lost from meat during post-rigour storage in trays or modified atmospheric packaging on retail display or during ageing of meat under atmospheric or vacuum conditions. PEF treatments can increase purge loss with higher amount occurring in thawed samples and with post-PEF ageing (Table 2). Thawing before PEF disrupts the myofibril structure which reduces their ability to hold moisture (Faridnia et al. 2015; Huff-Lonergan and Lonergan 2005; Pearce et al. 2011). Purge loss was significantly affected by PEF processing parameters such as frequency and voltage (Bekhit et al. 2014b). Microscopic examination of transverse sections of LL muscle by Khan et al. (2017) showed that, at 40× magnification, the PEF-treated (0.5 kV/cm, 200 Hz, 20μs, and 12.4 kJ/kg) muscle fibre bundles were elongated and had higher purge loss (%) than control samples. However, Suwandy et al. (2015d) have found that purge loss for beef LL muscle was not significantly affected by repeated PEF treatments or by ageing. A similar observation was recorded by Bhat et al. (2019a) in SM muscle of 9-year-old dairy cow in which purge loss after post-PEF (EFS, 0.36–0.6 kV/cm, and SE, 18.63–73.28 KJ/kg) ageing did not significantly increase. This result was speculated to be due to the temperature increase in the meat during PEF not being of a sufficient magnitude to cause severe protein denaturation.

Table 2 Effect of PEF on purge loss of meat

Muscles studied	PEF operating parameters	Effect on meat tenderness	References
Beef ST muscle	Batch PEF treatment using square-wave bipolar pulses at EFS, 1.4 kV/cm; PW, 20μs; PF, 50 Hz; and SE, 250 kJ/kg	Purge loss significantly ($p < 0.001$) increased after 7 days of vacuum ageing for both fresh and frozen–thawed samples	Faridnia et al. (2015)
Beef LL and SM muscles	Batch PEF treatment of input EFS, 0.28–0.56 kV/cm; PPP, 99.6–471 kW; PPC, 43.7–112.2 A; PW, 20μs; PF, 20–90 Hz; SE, 3.2–70.1 kJ/kg	Purge loss (%) of hot-boned beef LL muscles was affected by PEF treatments ($p = 0.028$) and ageing ($p < 0.01$) which increases with ageing period of 3, 7, and 14 days	Suwandy et al. (2015b)
Beef LL muscle	Batch PEF treatment with EFS, 0.58–0.73 kV/cm; PPP, 486.3–507.8 kW; PPC, 97–104.3 A; PW, 20μs; PF, 390 Hz; and SE, 16.2–19.3 kJ/kg	Purge loss (%) of cold-boned beef LL muscles increased with increased post-mortem ageing	Suwandy et al. (2015c)
Cold deboned beef SM and LL muscles	Batch PEF treatment with 0, 1, 2, and 3 repeated application parallel to fibre orientation, EFS, 0.50–0.58 kV/cm; PF, 90 Hz; PW, 20μs; treatment time, 30s	Purge loss for beef LL muscle was not affected by the repeated PEF applications or by ageing, but increased by 0.13% for every extra day of ageing, which they concluded insignificant due to limited power of statistical test	Suwandy et al. (2015d)
Beef topside (SM) muscle from culled dairy cows of 9 years of age	Two PEF treatments: EFS, 0.36 kV/cm; PW, 20μs; PF, 90 Hz; SE, 18.63 kJ/kg (T_1), and EFS, 0.60 kV/cm; PF, 20 Hz; PW, 20μs; SE, 73.28 kJ/kg (T_2)	No effect ($P > 0.05$) on the purge loss during post-treatment ageing	Bhat et al. (2019a)

EFS Electric field strength, *PPP* pulse peak power, *PPC* pulse peak current, *PW* pulse width, *PF* pulse frequency, *PN* pulse number, *SE* specific energy

Effect of PEF on Cooking Loss of Meat

The major reason for water loss during cooking is the thermal denaturation of myofibrillar proteins and the shrinkage and volume reduction of the myofibrils (Cheng and Sun 2008). Higher cooking loss is associated with a decrease in juiciness and succulence (Arroyo et al. 2015a). Several studies on different beef muscles or turkey breast have indicated that PEF treatments, even with prior freezing–thawing or in old animal, have a lower or insignificant increase in % cooking loss compared to control (Table 3) (Arroyo et al. 2015a; Bekhit et al. 2014b; Bhat et al. 2019a; Faridnia et al. 2015; Khan et al. 2017). In contrast, it has been reported that PEF-treated hot-boned LL muscle (Suwandy et al. 2015b), hot-boned beef SM muscle post-PEF ageing for 7 days (Bekhit et al. 2016), and LL muscle after repeated PEF treatments (Suwandy et al. 2015d) had higher % cooking loss compared to untreated control samples or cold-boned meat. These differences were postulated to be due to different factors during PEF treatments such as treatment intensity, higher degree of protein denaturation after the treatment, as well as differences in the intrinsic moisture and the type and amount of protein.

Effect of PEF on Colour of Meat

Meat colour is a critical quality attribute as the visual appearance of raw meat influences the consumer's decision to purchase and the acceptance of the cooked meat (Arroyo et al. 2015b). The colour of fresh meat depends on the content and physio-chemical state of myoglobin, i.e. purple (reduced myoglobin), red (oxymyoglobin), and brown (metmyoglobin) (Abril et al. 2001). When the concentration of brown pigment "metmyoglobin" reaches at least 60% of the maximum possible, the meat is unacceptable to consumers, and they will not buy it (Bekhit et al. 2016).

Generally, the colour of meat is measured using an instrumental colorimetric method, which provides an output expressed as Commission Internationale d'Eclairage (CIE) L^*, a^*, and b^* values, which represent lightness, redness, and yellowness, respectively. Several studies on different beef muscles have indicated that lightness values (L^*) decrease after PEF treatment and post-treatment storage or display time (Table 4) (Kantono et al. 2019; O'Dowd et al. 2013; Suwandy et al. 2015d). However, Khan et al. (2017) reported a significant ($p < 0.001$) increase in L^* of beef LL muscles treated at high PEF intensity (HPEF, 2 kV/cm, 200 Hz, 20μs, and 149.8 kJ/kg) compared to untreated controls or samples treated at low PEF (LPEF 0.5 kV/cm, 200 Hz, 20μs, 12.4 kJ/kg) after 24 h of blooming. This may have been because of an increase in water on the surface of the meat (due to its lower water holding capacity) that resulted in more light being reflected from its surface (Bekhit et al. 2001). This theory is supported by the higher purge loss observed due to PEF at high intensity.

Table 3 Effect of PEF on cooking loss of meat

Muscles studied	PEF operating parameters	Effect on meat tenderness	References
Beef loin *M. longissimus lumborum* (LL) and topside *M. semimembranosus* (SM) muscles	Batch PEF treatment with parallel fibre orientation to applied electric field, with EFS, 0.27–0.56 kV/cm; PW, 20μs; PF, 20, 50, and 90 Hz; variable pulse number, 607–2726 and SE, 3.1–73.2 kJ/kg	LL muscle had significantly lower ($p < 0.001$) % cooking loss than untreated LL muscle	Bekhit et al. (2014b)
Beef ST muscle	Batch PEF treatment using square-wave bipolar pulses at EFS, 1.4 kV/cm; PW, 20μs; PF, 50 Hz; and SE, 250 kJ/kg	No significant increase in cooking loss; freezing and thawing prior to PEF did not significantly ($p < 0.05$) affect the % cooking loss	Faridnia et al. (2015)
Turkey breast	EFS, up to 3 kV/cm; PW, 20μs; PN, 100, 200, and 300; PF, 10, 55, and 110 Hz; SE, 11–94 kJ/kg	No significant difference ($p > 0.05$) for fresh (averaged cook loss = 11.9%) and frozen meat samples (averaged cook loss = 14.3%) after PEF treatment and control	Arroyo et al. (2015a)
Beef LL and SM muscles	Batch PEF treatment of input V, 5–10 kV; EFS, 0.28–0.56 kV/cm; PPP, 99.6–471 kW; PPC, 43.7–112.2 A; PW, 20μs; PF, 20–90 Hz; SE, 3.2–70.1 kJ/kg	Cooking loss (%) of PEF-treated hot-boned LL muscle was higher than non-treated samples	Suwandy et al. (2015b)
Cold deboned beef SM and LL muscles	Batch PEF treatment with 0, 1, 2, and 3 repeated application parallel to fibre orientation, EFS, 0.50–0.58 kV/cm; PF, 90 HZ; PW, 20μs; treatment time, 30 s	Significant increase ($p = 0.03$) in % cooking loss of LL muscle regardless of the number of PEF treatment repeats, but in beef SM muscle not significantly affected by PEF treatment ($p > 0.05$)	Suwandy et al. (2015d)
Hot deboned beef loin (LL) and topside (SM) muscles	Batch PEF treatment with 1, 2, and 3 repeats, EFS, 0.44–0.46 kV/cm, and SE, 633.2–70.8 kJ/kg on LL muscle and 0.32–0.35 kV/cm, and SE, 17.5–19 kJ/kg on SM muscle; PF, 90 Hz; and PW, 20μs	The cooking loss (%) of hot-boned beef SM muscle was affected by ageing ($p < 0.01$) with samples aged for 7 days having the highest cooking loss compared to those aged for 3, 14, and 21 days	Bekhit et al. (2016)

(continued)

Table 3 (continued)

Muscles studied	PEF operating parameters	Effect on meat tenderness	References
Cold deboned beef LL muscle	Low PEF (LPEF, 0.5 kV/cm, 200 Hz, 20μs, and 12.4 kJ/kg) and high PEF (HPEF, 2 kV/cm, 200 Hz, 20μs, and 149.8 kJ/kg) treatment	No significant difference in % cooking loss between treated and untreated meat	Khan et al. (2017)
Beef topside (SM) muscle from culled dairy cows of 9 years of age	Two PEF treatments: EFS, 0.36 kV/cm; PW, 20μs; PF, 90 Hz; SE, 18.63 kJ/kg (T1), and V, 10 kV; EFS, 0.60 kV/cm; PF, 20 Hz; PW, 20μs; SE, 73.28 kJ/kg (T2)	No effect ($p > 0.05$) on observed cooking loss (%)	Bhat et al. (2019a)

EFS Electric field strength, *PPP* pulse peak power, *PPC* pulse peak current, *PW* pulse width, *PF* pulse frequency, *PN* pulse number, *SE* specific energy

The redness of meat is highly dependent on the meat structure, the concentration, and the oxidation state of myoglobin present in the muscle (de Huidobro et al. 2005). The redness (a^*) and yellowness (b^*) values were lower in PEF-treated beef muscles than in controls, and these values were dependent on treatment intensity, muscle type, ageing, and display time (time the samples are kept under conditions similar to retail display). The decrease in a^* values is due to an increase in the level of oxidation of myoglobin to metmyoglobin (Faridnia et al. 2016; Khan et al. 2017; O'Dowd et al. 2013; Suwandy et al. 2015d). It is, however, important to note that Kantono et al. (2019) found that PEF treatment caused an immediate increase in a^* values, which shows that PEF may not directly be having a detrimental effect on redness of meat. Chian et al. (2019) also did not observe significant difference ($p > 0.05$) in instrumental colour values between control and PEF-treated beef LT muscles.

Similarly, chroma ($C = \sqrt{a^{*2} + b^{*2}}$) values, browning indexes (630 nm/580 nm and 630–580 nm, which are a good indicator of metmyoglobin formation), and Hue angle ($h = \tan^{-1} \frac{b^*}{a^*}$) values were affected by display time and repeated PEF treatments ($p < 0.05$) (Khan et al. 2017; Suwandy et al. 2015d). The h value generally increases during ageing, as a result of myoglobin and oxymyoglobin oxidation (Bekhit and Faustman 2005). The oxidation of myoglobin also depends on temperature, oxygen pressure, and lipid oxidation, with the rate of myoglobin oxidation increasing as the temperature is increased (Faustman et al. 2010). Therefore, the increased discolouration and modification of colour in cold-boned beef with repeated PEF treatments could be due in part to a variety of processes including PEF-induced electroporation, an increase in temperature, and lipid oxidation.

For white meat such as fresh turkey breast, PEF treatment does not appear to significantly impact on colour of the meat when raw or cooked ($p \geq 0.05$) (Arroyo et al. 2015a).

Table 4 Effect of PEF on meat colour

Muscles studied	PEF operating parameters	Effect on meat tenderness	References
Post-rigour beef *semitendinosus* (ST) muscle	Batch PEF treatment with EFS, 1.1–2.8 kV/cm; SE, 12.7–226 kJ/kg; PF, 5–200 Hz; and PN, 152–300	Decrease in lightness (L^*) value than samples held at 5 °C; the redness (a^*) and yellowness (b^*) values were lower in the treated samples than in the controls	O'Dowd et al. (2013)
Turkey breast	EFS, up to 3 kV/cm; PW, 20μs; PN, 100, 200, and 300; PF, 10, 55, and 110 Hz; and SE, 11–94 kJ/kg	No significant ($p > 0.05$) effect on the colour after PEF but significant increase in L^* and b^* values and decrease in a^* after cooking	Arroyo et al. (2015a)
Cold deboned beef SM and LL muscles	Batch PEF treatment with 0, 1, 2, and 3 repeated application parallel to fibre orientation, EFS, 0.50–0.58 kV/cm; PF, 90 HZ; PW, 20μs; treatment time, 30 s	L^* value was only affected by display time ($p < 0.01$) after PEF; a^* decreased with number of PEF repeats ($p = 0.04$); b^* was affected by display time ($p < 0.05$) and the interactions between display time, muscle type, and ageing ($p < 0.05$); chroma (C) values were affected by display time ($p < 0.05$) and the interactions between display time, muscle type, and ageing ($p < 0.05$). Browning indexes affected by display time ($p < 0.05$) and the interactions between display time, muscle type, and ageing ($p < 0.05$); Hue (h) values affected by muscle type, display time, and the interactions between the three factors (at all $p < 0.05$) and increased with number of PEF repeat	Suwandy et al. (2015d)
Beef outside flat (*biceps femoris*, BF) muscle	Batch PEF treatment of EFS, 1.7–2.0 kV/cm, and SE, 185 kJ/kg	Display time has a significant effect on the meat colour values ($P < 0.001$) of PEF-treated meat, a^* decreased for all treatments over the days of light exposure, but PEF treatment can be used without adverse effects on meat instrumental colour parameters	Faridnia et al. (2016)

(continued)

Table 4 (continued)

Muscles studied	PEF operating parameters	Effect on meat tenderness	References
Cold deboned beef LL muscle	Low PEF (LPEF, 0.5 kV/cm, 200 Hz, 20μs, and 12.4 kJ/kg) and high PEF (HPEF, 2 kV/cm, 200 Hz, 20μs, and 149.8 kJ/kg) treatment	After 24 h of blooming, L^* values of HPEF-treated meat were significantly ($p < 0.001$) higher than that of control and LPEF-treated samples. LPEF-treated and non-treated control samples have higher a^* values compared to HPEF, b^* values were affected by the different treatments and the display time, C value of the meat samples was significantly ($p < 0.001$) affected by display time and treatments which decreased with increasing display time, h value significantly ($p < 0.01$) increased as the display time increased and significantly ($p < 0.001$) in HPEF-treated samples. The results obtained for a^* and h values were reflected in the browning index which was higher in HPEF-treated samples	Khan et al. (2017)
Fresh and frozen beef BF and ST muscles	EFS, 0.8–1.1 kV/cm; PW, 20μs; PF, 50 Hz; SE, 150 kJ/kg	Decreased L value of ST muscle immediately after PEF even without storage but increased after 7 days of storage; L value of BF increased immediately after PEF without storage and then decreased after 7 days of storage; a^* increased immediately after PEF	Kantono et al. (2019)
Beef rib-eye (*longissimus thoracis*) muscle	EFS, 1.00–1.25 kV/cm; PN, 500 and 2000; SE, 48 ± 5 kJ/kg and 178 ± 11 kJ/kg	No significant difference ($p > 0.05$) in instrumental colour values between control and PEF treated	Chian et al. (2019)

EFS Electric field strength, *PPP* pulse peak power, *PPC* pulse peak current, *PW* pulse width, *PF* pulse frequency, *PN* pulse number, *SE* specific energy

Effect of PEF on Sensory Properties of Meat

Sensory analysis can be used to ascertain the quality differences due to PEF processing detected by instrumental analysis. Even though the physical characteristics of meat can be assessed instrumentally, the use of sensory evaluation has an advantage over instrumental evaluation as it reflects the actual eating experience. The effect of PEF treatment on the sensory properties treated of meat has been assessed using affective and objective methods.

Arroyo et al. (2015a) used triangle tests and an acceptance test with 40 panellists to assess whether sensory differences existed between PEF-treated (EFS: up to 3 kV/cm) and untreated turkey samples. Triangle tests determined that the odour and texture were significantly different between the PEF-treated turkey and untreated control. The panellists significantly ($p < 0.05$) liked the odour of the control sample more than the PEF-treated sample, with average scores of 6.15 (SEM = 0.20) and 5.58 (SEM = 0.19), respectively. There was no detailed information about the method used for triangle test.

Kantono et al. (2019) examined the effect of freezing and PEF treatments (EFS, 0.8–1.1 kV/cm; PW, 20μs; PF, 50 Hz; and SE, 130 kJ/kg) on the temporal sensory properties of beef BF and ST muscles. Temporal dominance of sensations (TDS) was used to assess the temporal changes in brothy, browned, juicy, livery, and oxidised attributes during consumption of beef ST and BF muscles before and after PEF treatments, as well as after 7 days of storage. The dominant attributes across all samples were brothy, browned, and oxidised. The panel found that brothy was the most dominant attribute for the first 8 s of consumption followed by oxidised attribute from 8 to 80 s. Browned attribute reached significance as brothy attribute dominance rate was decreasing but before the oxidised attribute. The livery and juicy attributes did not reach significance and only occasionally reached above chance level.

Effect of PEF on Thermal Properties and Solubility of Meat Proteins

Differential scanning calorimetry (DSC) has been used to study the changes in thermal properties and denaturation behaviour of meat proteins. It has been reported that PEF does not result in the denaturation of myofibrillar proteins (Chian et al. 2019). There was no significant difference ($p > 0.05$) in the first, second, and third peak temperatures (which correspond to the thermal denaturation temperature of myosin, collagen and sarcoplasmic proteins, and actin, respectively) during DSC studies between control and PEF-treated (EFS, 1.00–1.25 kV/cm; PN, 500 and 2000; SE, 48 ± 5 kJ/kg and 178 ± 11 kJ/kg) beef rib-eye (LT) muscles. However, PEF can only induce electroporation and do not cause extensive myofibrillar fragmentation as evidenced by similar protein profile of PEF-treated and control beef LT muscle on SDS-PAGE (Faridnia et al. 2014).

Sionkowska (2005) reported that with DSC the observed denaturation temperature of proteins decreased as the polypeptide chain length decreased. A DSC study of PEF-treated dehydrated connective tissue isolated from beef brisket by Alahakoon et al. (2017b) showed that PEF causes degradation of the connective tissue structure into separate, heterogeneous fractions or sections of varying thermal stability. PEF treatment induced the broadening of endotherm and decreased the onset peak and endpoint denaturation temperatures (Fig. 4). Broadening of the endotherm was postulated to be due to the reduction in the number of hydrogen bonds in the collagen molecules, which reduces the energy needed for the activation of the denaturation process (Alahakoon et al. 2017b).

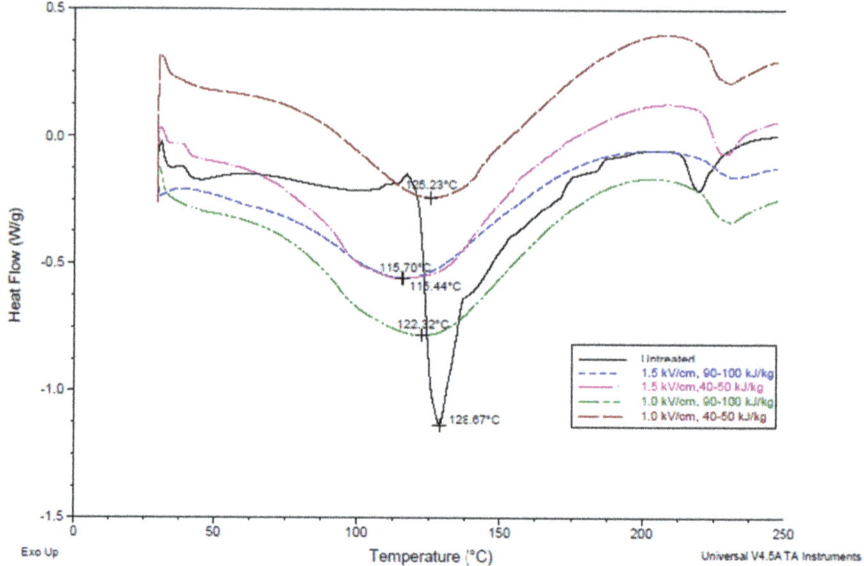

Fig. 4 Representative thermogram of differential scanning calorimetry (DSC) of untreated and PEF-treated (treated under 3 mS/cm electrical conductivity) intramuscular connective tissue (Reproduced from Alahakoon et al. 2017b)

The decrease in denaturation temperature because of PEF, as seen from the DSC data, was attributed to the cleavage of polypeptide chain and molecular destabilisation of collagen, which accelerates the conversion of collagen into gelatin during heating. The extent of connective tissue denaturation increased with a gradual increase in EFS and SE; the most significant reduction in the onset, peak, and end-point temperatures of denaturation was at EFS of 1.5 kV/cm and SE of 90–100 kJ/kg. The denaturation enthalpy values of PEF-treated connective tissue decreased slightly with increases in the EFS and SE. Overall, these changes in onset denaturation and peak temperatures reflected the degradation of the connective tissue, an increase in the amount of soluble collagen, and the rate of collagen degradation during heating (Alahakoon et al. 2017b).

It has been observed that PEF weakens the structure of intramuscular connective tissue, decreases the cross-links of collagen molecules increasing their solubility, causes the breakage of hydrogen bonds, and encourages other interactions that destabilise the triple helix collagen leading to exposure of the hydroxyl groups to reaction and causing subsequent hydration of the proteins and ultimately an increase in meat tenderness (Alahakoon et al. 2017b; Zhao et al. 2012). Alahakoon et al. (2017b) observed that PEF-treated connective tissue isolated from beef brisket had a faster rate of solubilisation compared to the untreated sample at 60 and 70 °C. At 60 °C, 50% of the collagen has solubilised within 36 h for PEF-treated connective tissue compared to 48 h for control. Thus, a disruption of the physical structure of

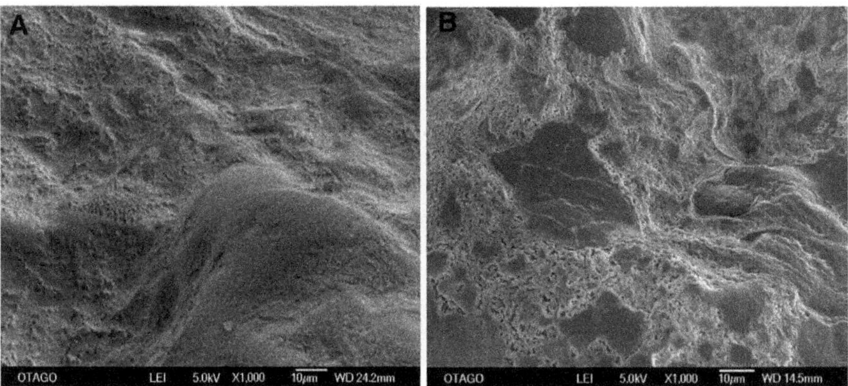

Fig. 5 Cryo-SEM micrographs (1000× magnification) of surface changes of untreated (**a**) and PEF-treated (**b**) isolated connective tissue (Reproduced from Alahakoon et al. 2017b)

collagen owing to electroporation was clearly visible in the images obtained using cryo-SEM (Fig. 5) suggesting an increase in its porosity, which decreases the thermal stability of collagen while increasing collagen solubility.

Effect of PEF on Lipid Oxidation in Meat

Fatty acids in muscle tissue are not only determinant of the nutritional value of meat but also affect the meat quality, including its tenderness, colour, lipid stability, odour, and flavour (Faridnia et al. 2015; Wood et al. 2004). The rate and extent of lipid oxidation depends on many factors: meat composition, fatty acid content, level of unsaturation, presence of antioxidants, post-mortem factors such as hot boning, post-mortem pH decrease, freezing–thawing, heat, post-mortem temperature, and other processing conditions and tenderising processes such as mechanical separation, grinding, electrical stimulation, and PEF (Arroyo et al. 2015a). As PEF causes electroporation in sarcolemma, it might also accelerate lipid oxidation due to leaching of intracellular prooxidants. This may make lipids more prone to oxidation because of increased interaction between prooxidants (haem/non-haem iron) and unsaturated fatty acids generating free radicals and lipid hydroperoxides. Hydroperoxides break down into secondary products such as acids, alcohols, aldehydes, and ketones which are responsible for oxidised and rancid off-flavour known as warmed-over flavour. This may alter the fatty acid composition and volatile profile of meat. In severe cases it may also lead to rancid flavour. Lipid oxidation not only leads to off-flavour but also causes loss of fresh meat colour and production of potentially harmful reactive compounds (e.g. peroxides) (Suwandy et al. 2015d).

Several studies on beef muscles and turkey breast, either fresh or frozen, concluded that PEF treatment did not significantly increase lipid oxidation (Table 5)

Table 5 Effect of PEF on lipid oxidation of meat

Muscles studied	PEF operating parameters	Effect on meat tenderness	References
Beef ST muscle	Batch PEF treatment using square-wave bipolar pulses at EFS, 1.4 kV/cm; PW, 20µs; PF, 50 Hz; and SE, 250 kJ/kg	Extent of lipid oxidation at the end of storage period was lower than that can be detected by sensory evaluation	Faridnia et al. (2015)
Turkey breast	EFS, 1.25–3 kV/cm; PW, 20µs; PN, 100, 200, and 300; PF, 10, 55, and 110 Hz; and SE, 11–94 kJ/kg	Lipid oxidation in all PEF-treated samples not significantly different to the control	Arroyo et al. (2015a)
Beef LL muscle	Batch PEF treatment with EFS, 0.58–0.73 kV/cm; pulse peak power (PPP), 486.3–507.8 kW; pulse peak current (PPC), 97–104.3 A; PW, 20µs; PF, 390 Hz; and SE, 16.2–19.3 kJ/kg	No significant effect on the TBARS value	Suwandy et al. (2015c)
Cold deboned beef SM and LL muscles	Batch PEF treatment with 0, 1, 2, and 3 repeated application parallel to fibre orientation, EFS, 0.50–0.58 kV/cm; PF, 90 HZ; PW, 20µs; treatment time, 30 s	No significant effect on lipid oxidation	Suwandy et al. (2015d)
Cold deboned beef LL muscle	Low PEF (LPEF, EFS, 0.5 kV/cm; PF, 200 Hz; PW, 20µs; and SE, 12.4 kJ/kg) and high PEF (HPEF, EFS, 2 kV/cm; PF, 200 Hz; PW, 20µs; and SE, 149.8 kJ/kg) treatments	HPEF samples had higher lipid oxidation at the end of display time compared to LPEF; but both of these groups of samples were not different from the non-treated control samples	Khan et al. (2017)
Fresh and frozen beef BF and ST muscles	EFS, 0.8–1.1 kV/cm; PW, 20µs; PF, 50 Hz; SE, 150 kJ/kg	TBARS values significantly increased in all beef samples immediately after PEF. Lipid oxidation of BF muscles was significantly higher ($p < 0.05$) than ST muscles at 0 and 7 days of storage	Kantono et al. (2019)

EFS Electric field strength, *PPP* pulse peak power, *PPC* pulse peak current, *PW* pulse width, *PF* pulse frequency, *PN* pulse number, *SE* specific energy

(Arroyo et al. 2015a; Khan et al. 2017; Suwandy et al. 2015c, d). However, Faridnia et al. (2015) found that when PEF treatment was applied to frozen–thawed beef samples, TBARS values were significantly ($p < 0.05$) higher (0.96 mg malondialdehyde (MDA)/kg meat) than untreated samples (about 0.64 mg MDA/kg meat) 18 days post-PEF storage at 4 °C. However, the degree of lipid oxidation, as determined by thiobarbituric acid reactive substances (TBARS), was lower than what can be typically perceived by sensory panellists. A TBARS value of >1.0 mg MDA/kg meat has been reported as threshold MDA concentration to begin to perceive rancid flavours and odours (Zakrys et al. 2008). Similarly, Kantono et al. (2019) also observed that TBARS values significantly increased in all beef samples immediately after PEF and after 7 days of storage (0.06–0.12 mg MDA/kg meat) but were still below the critical level.

Effect of PEF on Fatty Acid Composition of Meat

The most abundant fatty acids in beef are palmitic acid (C16:0), stearic acid (C18:0), and oleic acid (C18:1 $n-6$). There are lower amounts of linoleic (C18:2 $n-6$) and linolenic (C18:3 $n-6$) acids. Nutritionally, the ratios of polyunsaturated fatty acids (PUFAs) and saturated fatty acids (SFA) and omega 6/omega 3 ($n-6/n-3$) are important (Faridnia et al. 2015). Nutritional guidelines recommend that the PUFA/SFA ratio should be above 0.4–0.5 (Wood et al. 2008) and the n6/n3 ratio should be <4 (McArdle et al. 2010). PEF does not appear to have a detrimental effect on the nutritional value of fatty acids in beef meat. Faridnia et al. (2015) found that in PEF-treated meat, the PUFA/SFA ratios were in the range of 0.31–0.33 and the n6/n3 ratios in all treated samples remained within the recommended levels.

Similarly, Kantono et al. (2019) observed that PEF did not cause a significant immediate reduction of PUFAs such as linoleic and arachidonic (C20:4 $n-6$) acids in beef BF muscle, except docosadienoic (C22:6 $n-3$) acid. However, most fatty acids significantly decreased in PEF-treated BF samples after 7 days of storage, which could result from fatty acid oxidation, as indicated by the significantly higher TBARS value compared to the non-PEF samples.

Effect of PEF on Volatile Profile of Meat

Lipid oxidation is the major pathway leading to the formation of off-flavours (due to carbonyls, alcohols, and acids) as well as toxic compounds, such as MDA and cholesterol oxidation products (Utrera and Estévez 2013). Apart from lipid oxidation, other biochemical reactions post-mortem such as enzymatic oxidation of unsaturated fatty acids and further interactions with proteins, peptides, and amino acids generate volatile compounds (Faridnia et al. 2015). Hexanal, which originates mainly from linoleic and arachidonic acids, can lead to a faintly rancid aroma, and it is considered to be a good indicator for oxidation (Martín et al. 2000). The concentration of hexanal levels is positively correlated to TBARS values and negatively correlated to acceptance of meat flavour (Calkins and Hodgen 2007). It is also the reason for development of fruity or fatty aromas (Kang et al. 2013). Other volatile aldehydes, such as heptanal, octanal, and nonanal, are the product of oleic acid oxidation, with oleic acid being the principal MUFA in beef (Machiels et al. 2004).

Faridnia et al. (2015) concluded that PEF and freezing pretreatment affected the volatile profile of meat due to lipid and protein oxidation. The authors reported the amounts of pentanal, hexanal, heptanal, benzaldehyde, octanal, and nonanal were significantly ($p < 0.05$) higher in PEF-treated beef ST muscles frozen prior to PEF than untreated samples. Also, 3-methylbutanal (compounds responsible for the roasted beef flavour (Machiels et al. 2004)) and 2-methylbutanal were significantly higher ($p < 0.05$) in all frozen PEF-treated samples than untreated samples, which indicates the formation of these aldehydes from Strecker degradation of leucine and isoleucine during cooking.

Dimethyl disulphide was also found at higher levels in frozen PEF-treated meat samples, which may have been derived from the degradation of sulphur-containing amino acid and a reaction between cysteine and ribose. In addition, the 2,3-octanedione content of both fresh and frozen PEF-treated meat were significantly higher ($p < 0.05$) than that of the control. Stetzer et al. (2008) have reported that 2,3-octanedione is derived from lipid oxidation and responsible for oxidised fat flavour and an influencer of overall flavour.

Effect of PEF on Mineral Profile of Meat

Despite red meat being an excellent source of minerals such as iron (Fe), zinc (Zn), and phosphorus (P), the increased permeability of the myofibril cell membrane after PEF treatment may lead to the potential loss of these minerals during storage or cooking (Khan et al. 2017). Low-intensity PEF treatment (EFS, 0.5 kV/cm; PF, 200 Hz; PW, 20μs; and SE, 12.4 kJ/kg) did not result in a significant decrease in P, K, Fe, and Zn content in cold-boned beef loins than high-intensity PEF treatment (Khan et al. 2018). In fact it was observed that low-intensity PEF-treated (LPEF, 0.5 kV/cm, 200 Hz, 20μs, and 12.4 kJ/kg) had a higher P content (6893.3 mg/kg dry basis) than high-intensity PEF-treated (HPEF, 2 kV/cm, 200 Hz, 20μs, and 149.8 kJ/kg) samples (6551.9 mg/kg dry basis, respectively) (Khan et al. 2017). The lower P content in HPEF-treated meat is reasoned to be due to increased electroporation and more leaching of P from the meat than LPEF.

Similar trends were found with the concentrations of K; LPEF-treated beef had higher K concentrations (14.101.8 mg/kg dry weight) than HPEF beef samples (2.983.1 mg/kg dry weight) due to lower degree of electroporation and lesser leaching out from meat samples (Khan et al. 2017). Similarly, the lower concentrations of K and P in cooked meat and HPEF suggest that the proportion of these minerals present in the sarcoplasmic fraction is lost during cooking and after PEF at high treatment intensity, which led to higher purge loss. The concentration of Fe was more significantly affected by HPEF (EFS, 2 kV/cm; PF, 200 Hz; PW, 20μs; and SE, 149.8 kJ/kg) treatment ($p = 0.011$) (5.6 mg/kg dry basis) than LPEF (EFS, 0.5 kV/cm; PF, 200 Hz; PW, 20μs; and SE, 12.4 kJ/kg) and control samples (95.0 and 94.8 mg/kg dry basis, respectively). However, the concentration of Zn was not significantly affected by PEF treatments.

Effect of PEF on Microorganisms in Meat

There are only a few studies on the inactivation of microorganisms in meat using PEF (Table 6). Reported $\log_{10}$ reduction of microorganisms is in the range of 1.44–2.77 only, which is inadequate for ensuring microbial safety. The electric field strengths applied were in the range of 0.25–15 kV/cm, which is far below the level required to inactivate microorganisms and ensure microbial safety found in other foods (i.e. 20–60 kV/cm). This electric field strength level, if applied, may cause a significant ohmic heating and an increase in temperature resulting in denaturation of meat proteins affecting meat quality. Therefore, PEF treatments applied in meat will not kill microorganisms but may help to limit the growth of the initial microbial population.

Application of PEF as Pretreatment to Sous-Vide Processing

Low-temperature-long-time cooking like sous vide can make tough meat tender (Baldwin 2012). Sous-vide processing at temperature–time combinations of 58–63 °C for 10–48 h is recommended for tender beef, pork, or lamb meat (Ruiz et al. 2013). Tough meat cuts with higher levels of connective tissue need longer sous-vide times compared to tender meat cuts. The recommended temperature–time for tough meat cuts like beef chuck, brisket, and pork shoulders is between 55 and 60 °C for 24–48 h (Baldwin 2012).

The sensory characteristics of meat change during cooking as a function of temperature and time as different physiochemical processes depend on the rate of change in temperature. The challenge is to reach a desired endpoint for optimal sensory characteristics without generating adverse effects on other meat quality attributes (Alahakoon et al. 2019a).

Sous-vide processing is relatively uneconomical because of high energy requirements and long processing time. The process may also have detrimental effects on other quality parameters such as meat colour, juiciness, and doneness (Alahakoon et al. 2018). Cooking causes the denaturation of myofibrillar protein, lowering their ability to contain free water within the myofibrils (Tornberg 2005). This changes the water distribution within myofibrillar structure, and free water is expelled out, reducing juiciness (Bertram et al. 2006). The injection of exogenous proteolytic enzymes, prior to sous vide, is used to reduce sous-vide time and temperature. However, owing to the high inactivation temperature of these enzymes, they may not be inactivated during the initial phase of sous-vide processing causing over-tenderisation and loss of quality (Alahakoon et al. 2019a).

PEF is one of the novel technologies being explored as an alternative to the use of enzymes. It has been found that mild PEF treatment did not denature muscle proteins to the extent that it can have a negative effect on meat quality (Arroyo et al. 2015b). Some studies on the use of PEF as pretreatment (EFS, 0.7–1.5 kV/cm; SE,

Table 6 Effect of PEF treatment on pathogenic and spoilage microorganisms in food

Food types	Treatment variables	Target microorganisms	Effect	References
Skinless chicken breast	EFS, 3.75 or 15 or kV/cm; PW, 10 m; PF, 5 Hz; treatment time, 10, 15, 20, 25, and 30 s	*Campylobacter jejuni* *Campylobacter coli* *Escherichia coli* (ATCC 25922) *Salmonella enteritidis* (ATCC 13076)	PEF did not result in any significant reductions in total viable counts of all organisms	Haughton et al. (2012)
Beef *semitendinosus* muscles	Combined freezing and PEF treatment of EFS, 1.4 kV/cm; square-wave bipolar pulses; pulse width, 20μs; frequency, 50 Hz; specific energy, 250 kJ/kg	Aerobic bacteria	TVC immediately after treatment, 2.69–2.77 log CFU/g; no significant increase in TVC up to 7 days; bacterial count increased rapidly after 8 days of storage at 4 °C	Faridnia et al. (2015)
Porcine blood plasma	PEF application of EFS, 12 kV/cm; pulse frequency, 169 Hz; specific energy, 113 kJ/kg; treatment time, 130μs; and at initial temperature of 30 °C	*Pseudomonas* spp. and *Enterobacteriaceae*	1.44 log_{10} CFU/mL reduction in total aerobic plate count (TPC); count of *Pseudomonas* spp. and *Enterobacteriaceae* below the detection limit; resulted in extended shelf life (up to 14 days) at 3 °C based on strictest criteria for microbiological acceptability (5×10^4 CFU/mL for TPC, 5×10^3 CFU/mL for *Pseudomonas* spp., 5×10^2 CFU/mL for *Enterobacteriaceae*)	Boulaaba et al. (2018)
Beef briskets	PEF treatment of EFS, 0.7 and 1.5 kV/cm, specific energy (90–100 kJ/kg), and sous-vide processing (60 °C/12 or 24 h)	Aerobic and lactic acid bacteria	Sous-vide processing (12 or 24 h) after PEF treatment reduced both aerobic and lactic acid bacteria numbers to below the detection limit	Alahakoon et al. (2019b)

(continued)

Table 6 (continued)

Food types	Treatment variables	Target microorganisms	Effect	References
Raw chicken	Batch system with 8 cm electrode gap and electrode surface area of 2.76 cm^2: rectangular bipolar pulse with EFS, 0.25–1 kV/cm; PW, 20μs; PF, 1 Hz; and PN, 50 followed by resuspension in ¼ MIC of oregano	*Campylobacter jejuni* strains. *Campylobacter. jejuni* (1146DF)	1.5 $\log_{10}$ reduction at 1 kV/cm followed by resuspension	Clemente et al. (2020)

EFS Electric field strength, *PPP* pulse peak power, *PPC* pulse peak current, *PW* pulse width, *PF* pulse frequency, *PN* pulse number, *SE* specific energy

50–100 kJ/kg) prior to sous vide (60–70 °C/12–72 h) of tough beef brisket muscles have reported an increase in tenderness without affecting colour stability, lipid oxidation, or cooking loss and a reduction in the biological variation in meat quality (Table 7) (Alahakoon et al. 2018, 2019a).

Alahakoon et al. (2019a) optimised a PEF and sous-vide treatment combination for beef brisket; PEF pretreatment at 0.7 kV/cm followed by sous vide at 60 °C for 24 h was the best conditions to reduce both the shear force and hardness without it having a significant effect on lipid oxidation, colour stability, cooking loss, or in vitro protein digestibility. Furthermore, the PEF-sous vide processed brisket samples more closely resembled the steak than traditional sous vide processed tough meat cuts (Alahakoon et al. 2019a).

Application of PEF as Pretreatment to Meat Ageing

Post-mortem ageing is a conventional commercial method used to increase meat tenderness (Sitz et al. 2006). The endpoint tenderness achieved and the rate of ageing depend on the size and location of muscle, muscle types (there is a faster rate in fast-twitch glycolytic type II muscles (e.g. *longissimus dorsi*) than slow-twitch oxidative type I muscles (e.g. *triceps brachii*), and duration of ageing; but the method has a poor reproducibility (Faridnia et al. 2014). Moreover, the process is slow and costly as refrigerated storage is needed for a long (≥28 days) period (Herrera-Mendez et al. 2006). The process may be accelerated by using various interventions: physical disruption techniques such as hanging, blade/needle tenderisation,

Table 7 Effect of PEF as pretreatment to sous vide on meat quality

Meat types	Treatment parameters	Properties studied	Effects	References
Lamb loins	Sous vide at 60 °C for 24 h	Tenderness	Lower shear force values due to solubilisation and partial denaturation of the connective tissue and sarcoplasmic protein	Roldan et al. (2013)
Beef brisket	PEF of EFS, 1.5 kV/cm; SE, 90–100 kJ/kg cooked by sous vide at 60, 65, or 70 °C/24, 48, or 72 h	Tenderness	The hardness was lowest for both non-PEF-treated and PEF-treated meat sous vide processed for 72 h at 60 or 65 °C	Alahakoon et al. (2018)
Beef brisket	PEF of EFS, 0.7, 1.0, and 1.5 kV/cm; SE, 90–100 kJ/kg followed by sous vide at 60 °C for 12, 18, and 24 h	Tenderness	PEF with EFS of 0.7 kV/cm and sous vide for all time, and all PEF treatments that are sous vide processed for 24 h had significantly lower shear force values compared to non-PEF-treated sous-vide processed beef brisket. The lowest values for hardness (<20.2 N), shear force (100.5 N/mms), and cooking loss (<17.4%) were observed in PEF with EFS 0.7 kV/cm and sous-vide processing for 12–20.2 h or 1.5 kV/cm PEF and sous-vide processing for 20.8–23.7 h	Alahakoon et al. (2019a)
Isolated connective tissue from beef *deep pectoralis* muscle (brisket) on agar matrix	PEF treatment EFS, 1.0 and 1.5 kV/cm; SE, 50 and 100 kJ/kg	Denaturation; collagen solubilisation; and ultrastructure changes	Differential scanning calorimetry showed that PEF treatment significantly ($p < 0.05$) decreased the denaturation temperature compared to untreated samples: ringer soluble collagen fraction was augmented with increasing electric field strength and the specific energy. SEM images showed increased porosity	Alahakoon et al. (2017b)

(continued)

Table 7 (continued)

Meat types	Treatment parameters	Properties studied	Effects	References
Beef brisket	PEF of EFS, 1.5 kV/cm; SE, 90–100 kJ/kg cooked by sous vide at 60, 65, or 70 °C/24, 48, or 72 h	Collagen solubility and porosity	Collagen solubility of PEF-treated and sous-vide processed beef brisket at 60 °C for 24 or 72 h or 65 °C for 48 h was higher than that of the non-PEF. The isolated connective tissue after PEF has a more porous surface than non-PEF-treated samples. PS-OCT images show that PEF caused a breakdown in the collagen	Alahakoon et al. (2018)
Beef brisket	PEF of EFS, 1.5 kV/cm; SE, 90–100 kJ/kg cooked by sous vide at 60, 65, or 70 °C/24, 48, or 72 h	Water holding capacity	Significant negative regression coefficients of linear terms of sous-vide temperature and time of PEF-treated meat on cooking loss; in non-PEF-treated samples, similar effect was not observed which indicates the ability of PEF treatment to increase water holding capacity	Alahakoon et al. (2018)
Beef brisket	PEF of EFS, 0.7, 1.0, and 1.5 kV/cm; SE, 90–100 kJ/kg followed by sous vide at 60 °C for 12, 18, and 24 h	Water holding capacity	Cooking loss of PEF of 0.7, 1.0, and 1.5 kV/cm and sous-vide processed beef brisket was 0.79 ± 0.01, 0.91 ± 0.01, and 0.98 ± 0.01%, respectively (very low). PEF treatments of 0.7 kV/cm and 1.0 kV/cm and sous vide at 60 °C for 12 h had a significantly lower cooking loss compared to non-PEF meat	Alahakoon et al. (2019a)
Beef brisket	PEF of EFS, 1.5 kV/cm; SE, 90–100 kJ/kg cooked by sous vide at 60, 65, or 70 °C/24, 48, or 72 h	Instrumental colour (L^*, a^*, b^* values)	The increase in sous-vide temperature and time significantly increases the lightness (L^*) of PEF-treated meat: the instrumental redness (a^*) and yellowness (b^*) did not decrease significantly	Alahakoon et al. (2018)

(continued)

Table 7 (continued)

Meat types	Treatment parameters	Properties studied	Effects	References
Beef brisket	PEF of EFS, 0.7, 1.0, and 1.5 kV/cm; SE, 90–100 kJ/kg followed by sous vide at 60 °C for 12, 18, and 24 h	Instrumental colour (*L**, *a**, *b** values)	PEF did not show any significant effect on *a** at all temperature–time combinations; *b** increased with increasing sous-vide time at each temperature due to the formation of metmyoglobin	Alahakoon et al. (2019a)
	PEF of EFS, 0.7, 1.0, and 1.5 kV/cm; SE, 90–100 kJ/kg followed by sous vide at 60 °C for 12, 18, and 24 h	Lipid oxidation	PEF and sous-vide processing did not have a significant effect on the TBARS values	Alahakoon et al. (2019a)
		Peptic and pancreatic digestibility	PEF and sous vide do not have any negative impact on protein digestibility	Alahakoon et al. (2019a)

EFS Electric field strength, *PPP* pulse peak power, *PPC* pulse peak current, *PW* pulse width, *PF* pulse frequency, *PN* pulse number, *SE* specific energy

electrical stimulation, and control of pH and chilling regimes (Bolumar et al. 2013), injection or infusion of calcium chloride for activation of calpain enzyme system (Koohmaraie and Geesink 2006), and injection of proteolytic enzymes of plant origin (papain and bromelain) or of animal origin (porcine) (Pietrasik et al. 2010). However, these technologies were uneconomical and inconsistent, caused under- or over-tenderisation, or resulted in the development of off-flavours (Bekhit et al. 2016).

PEF is an economical, energy-saving, and fast meat tenderisation technology which is being explored for commercial applications (Arroyo et al. 2015b). Some studies on using PEF (voltage, 5–10 kV; EFS, 0.2–0.614 kV/cm; PW, 20–70.8μs; PF, 1–90 Hz; PN, 30–1528; SE, 3.1–64.7 kJ/kg; and treatment time, 600–30,560μs) as a pretreatment for ageing beef muscles (1–28 days at 4 °C) have reported an increase in tenderness without affecting other quality parameters (Table 8) (Arroyo et al. 2015b; Bekhit et al. 2014b, 2016; Bhat et al. 2019b; Suwandy et al. 2015d).

The increase in tenderness owing to a PEF pretreatment could be because of the electroporation of biomembranes and an increased permeability facilitating the movement and release of intracellular and extracellular components principally intracellular lysosomal cathepsins, multicatalytic proteinase complex, calpains, and calcium ions. Calcium ions induce the early activation of calpain enzyme system which increases tenderness and shortens ageing. The calpain enzyme system is the two calcium-dependent cysteine proteases, calpains 1 and 2 (Gomez et al. 2019; Ouali et al. 2013), which are associated with fragmentation of myofibrillar proteins. Calpain 1 has been reported to lose most of its activity within 3 h at 25 °C, whereas

Table 8 Effect of PEF pretreatment and ageing on meat quality

Meat types	Treatment parameters	Effects	References
Beef *longissimus thoracis* (LT) muscles	PEF of EFS, 0.2–0.6 kV/cm; PW, 20µs; PF, 1–50 Hz; PN, 30–1528; treatment time, 600–30,560µs; SE, 0.05–34.33 kJ/kg; and ageing time, 1 and 3 days at 4 °C; 95% relative humidity; air speed, 2–3 m/s. under vacuum packaging	PEF treatment had no additional significant impact on the mean shear force values ($p > 0.05$), while ageing resulted in a decrease ($p < 0.001$) in shear force values regardless of treatments; PEF treatment increased the purge loss but did not cause significant ($p > 0.05$) cooking loss	Faridnia et al. (2014)
Beef LL and SM muscles	PEF of EFS, 0.31–0.56 kV/cm; PW, 20 µS; PF, 20, 50, and 90 Hz; SE, 3.1–64.7 kJ/kg followed by ageing for 3, 7, 14, and 21 days at 4 °C after vacuum packaging	Shear force values for 1-day post-mortem in LL muscle decreased with PEF regardless of the post-treatment ageing, and WBSF values for 1- or 3-day post-mortem SM muscle decreased as ageing storage time increased; sensory analysis has concluded that 60% panellist scored PEF treated as tender; only 27.5% did for untreated samples	Bekhit et al. (2014b)
Beef *longissimus thoracis et lumborum* (LTL) muscles	PEF of EFS, 1.4 kV/cm; PW, 20µs; PF, 10 Hz; PN, 300 and 600 using square waveform pulses; SE, 25 kJ/kg for 300 PN and 50 kJ/kg for 600 PN followed by 2, 10, 18, and 28 days of ageing at 4 °C or ageing for that times followed by PEF treatments	PEF treatment has no direct relationship between ageing time and weight loss which increases with higher treatment intensity (greater at specific energy of 50 kJ/kg than at 25 kJ/kg); colorimetric redness (a^*) and yellowness (b^*) values of control sample decrease with increasing ageing period; lightness was not affected by ageing. PEF-treated sample a^* and b^* values significantly increased with time. PEF after 2, 10, 18, and 26 days showed no significant differences in % water loss; L^*, a^*, and b^* values; and cooking loss between untreated and PEF treated; gradual decrease in WBSF for PEF-treated samples with ageing	Arroyo et al. (2015b)

(continued)

Table 8 (continued)

Meat types	Treatment parameters	Effects	References
Beef LL and SM muscles	PEF of EFS, 0.50–0.58 kV/cm; PF, 90 Hz; PW, 20μs; treatment time, 30 s followed by ageing for 3, 7, 14, and 21 days at 4 °C	Shear force of PEF-treated LL muscles was lower than control samples at all ageing times; decreased by 2.5 N for each additional application of PEF over the four PEF repeats used (0×, 1×, 2×, and 3×). The final temperatures were 31.3 ± 3.4 °C, 33.3 ± 1.3 °C, and 38.3 ± 5.23 °C for 1×, 2×, and 3× treated samples, respectively	Suwandy et al. (2015d)
Hot deboned beef loins (LL) and topsides (SM)	1×, 2×, and 3× repeated PEF EFS, 0.33–0.48 kV/cm; PW, 20μs; PF, 90 Hz; SE, 17.5–70.8 kJ/kg followed by ageing for 3, 7, 14, and 21 days at 4 °C	Increase in conductivity, ranged from 1.3 to 5.0 mS/cm for LL muscles and 1.5 to 3.0 mS/cm for SM muscles; increase in purge loss LL with increasing number of PEF repeats ($p \leq 0.001$); significant increase ($p \leq 0.001$) in the purge loss with increase in ageing period; toughness of LL muscles was highest in 3× repeated PEF treatment, no differences among samples from control, 1× and 2× repeated PEF treatment repeats	Bekhit et al. (2016)
Beef *biceps femoris* muscles	PEF treatments of EFS, 0.38 ± 0.01 kV /cm; PPP, 144.8 ± 1.10 kW; PPC, 46.8 ± 1.38; PW, 20μs; PN, 2725; SE, 21.17 ± 0.72; temperature, 4.39 ± 0.85 °C (T_1), and EFS, 0.61 ± 0.01 kV/cm; PPP, 553.8 ± 9.68 kW; PPC, 111.3 ± 1.68; PW, 20μs; PN, 2723; SE, 74.24 ± 2.97; temperature, 4.78 ± 1.20 °C (T_2); vacuum packed at ages 1, 7, and 14 days at 4 °C	The average rise in the temperature in T_1, 8.18 °C, and T2, 11.08 °C; Western blotting, casein zymography, and myofibrillar protein profile depict increase calpain activity and proteolysis during ageing in PET-treated muscles; higher rate of tenderisation of PEF-treated sample during first week of ageing	(Bhat et al. 2019b)

(continued)

Table 8 (continued)

Meat types	Treatment parameters	Effects	References
Cold-boned *longissimus dorsi* obtained from red deer (*Cervus elaphus*)	2 PEF treatments: EFS, 0.2 kV/cm; PF, 50 Hz; PW, 20μs; SE, 1.93 kJ/kg, and EFS, 0.5 kV/cm; PF, 90 Hz; PW, 20μs; SE, 70.2 kJ/kg followed by ageing for 21 days at 4 ± 1 °C	No significant impact of PEF was observed on the shear force and Myofibrillar Fragmentation Index (MFI); slight increase in the calpain activity and proteolysis of troponin T; casein zymography, Western blotting, and myofibrillar protein profile suggest that PEF did not produce any significant impact on the tenderisation process of venison	Bhat et al. (2019c)

EFS Electric field strength, *PPP* pulse peak power, *PPC* pulse peak current, *PW* pulse width, *PF* pulse frequency, *PN* pulse number, *SE* specific energy

calpain 2 has been observed to be more stable (Ertbjerg et al. 2012). Western blotting, casein zymography, and analysis of myofibrillar protein indicated an increased calpain activity and proteolysis during ageing of PEF-treated beef BF muscles (Bhat et al. 2019b).

In contrast, Faridnia et al. (2014) reported that PEF treatment did not have a significant impact on the mean shear force values in beef LT muscles, while ageing resulted in a decrease in shear force values even without PEF treatment. Bekhit et al. (2016) also observed the toughness of hot-boned LL muscles was at the highest in repeated (three times) PEF treatment, and a study conducted by Bhat et al. (2019c) on cold-boned *longissimus dorsi* (LD) muscle of red deer (*Cervus elaphus*) also reported that PEF did not significantly change the instrumental shear force, Myofibrillar Fragmentation Index (MFI), casein zymography, Western blotting, and myofibrillar protein profile of venison LD muscles during ageing.

Application of PEF in Meat Salting, Brining, and Drying

PEF treatment induces cell membrane electroporation and increases the cell permeability, myofibrillar fragmentation, and the size of muscle cell gaps, all changes which enhance the mass transport within meat and therefore the efficiency of diffusion of solutes and water in meat processing unit operations (Gudmundsson and Hafsteinsson 2001; Smetana et al. 2019; Toepfl and Heinz 2007). Furthermore, PEF treatment is rapid (within seconds) and cost-effective (energy requirement: 10 kJ/kg) method for improvement of mass transfer operations and preservation in dry cured meat products (Toepfl and Heinz 2007). Klonowski et al. (2006) also mentioned unpublished results regarding the PEF treatment of ham (done by the Berlin University of Technology) that resulted in porous, swamp-like tissue structure with increased weight after brine injection, higher water holding capacity, and less loss during cooking, and the PEF-treated ham was significantly softer and tender than untreated ham.

Several studies have concluded that PEF treatment of chicken, pork, and beef muscles resulted in an increase in the electrical conductivity, diffusion of curing salt, and release of intracellular enzymes resulting in faster maturation and ripening of cured meat, a faster drying rate, and an increase in proteolytic activity, enhanced fermentation, and a decrease in the time required for pH drop, all without affecting sensory properties or appearance of cured sausage meat (Table 9) (Astrain-Redin et al. 2019; Levkov et al. 2019; McDonnell et al. 2014; Toepfl and Heinz 2007, 2008). The increase in the effective diffusivity of water and enhanced mass transport by PEF can also be utilised to design PEF-assisted industrial-scale deep-fried meat products. However, at higher treatment intensity (4 kV/cm, 20 kJ/kg) and subsequent hand salting, the surface of meat was denatured, which inhibited the movement of moisture from interior to the surface (Toepfl and Heinz 2007).

PEF can also provide a novel method to produce healthier reduced-sodium processed meat products. Due to better diffusivity of salt after PEF treatment, the salt concentration can be lowered in cured meat. Bhat et al. (2020) investigated the use of PEF for sodium reduction in beef jerky made from beef topside muscles. PEF-treated samples had a significantly lower sodium content than the control, without affecting sensory scores, instrumental shear force (N), and toughness (N/mms) but with better scores during chewing, and did not affect colour, yield (%), and oxidative and microbial stability of the products.

Conclusions

PEF processing has been shown to offer potential benefits in the processing of meat. PEF treatments are energy efficient and are of very short duration. The efficiency of PEF treatments for meat application depends on PEF process parameters such as EFS, treatment time, and specific energy as well as the intrinsic meat properties. Furthermore, PEF pretreatment before ageing improves tenderness and shortens ageing times which otherwise are very slow and costly with poor reproducibility. PEF can also enhance mass transfer of water and solutes in or out of the meat which results in more cost-effective meat curing, maturation, fermentation, and subsequent drying and can decrease the use of sodium producing healthier cured meat products.

The positive effects of PEF on meat structure modifications, tenderisation, shortening ageing time, and enhanced mass transfer are of great commercial value. The latest application of PEF pretreatment prior to sous-vide processing also shows promising results that the tenderness of low-value tough meat cuts increases without adverse effects on colour stability, lipid oxidation, or cooking loss. Since, in multicellular tissues like meat, the occurrence of PEF-induced changes is complex because of the presence of components with various electrical and morphological

Table 9 Application of PEF in meat curing, smoking, and drying

Meat types	Treatment parameters	Effects	References
Dry and wet cured pork shoulder	Pilot scale and batch treatment; treatment chamber size, 50 cm × 30 cm × 15 cm; stainless steel electrode with gap of 8–15 cm; EFS, 3–4 kV/cm; PF, 1–20 Hz; SE, 5–20 kJ/kg; followed by dry curing by 8% nitrite salt on the product surface or by injection of 10% of saturated brine and subsequent drying by air at 8 °C and 95% relative humidity (RH)	Increase in tissue conductivity; enhanced drying rate depending on treatment intensity and salting methods; the shortest drying time at 3 kV/cm, 5 kJ/kg followed by brine injection; treatment at 4 kV/cm and 20 kJ/kg followed by hand salting formed the dry, denatured meat surface inhibiting mass transfer from the centre to the surface	Toepfl and Heinz (2007)
Whole pork haunches	PEF treatment at 3 kV/cm and 5 kJ/kg followed by dry curing with 8% nitrite salt	Enhanced diffusion of salt and nitrite into to the product core after storage time of a few weeks	Toepfl and Heinz (2007)
Minced sausage meat	Continuous treatment chamber with a diameter of 6 cm with central high-voltage electrode and grounded counterparts, separated by polyethylene insulators; electrode gap, 6 cm; peak voltage, 24 kV; EFS, 2 kV/cm; PF, 20 Hz; SE, 2–10 kJ/kg; flow rate, 2 t/h; and a specific energy input of 2 kJ/kg followed by fermentation	Decrease in time required to drop pH to a value of 5 by 30%; no detrimental effect on sensorial properties and appearance	Toepfl and Heinz (2007)
Cured pork shoulder	Lab-scale PEF treatment, EFS, 3 kV/cm, followed by curing and drying at 8 °C and 95% RH	Accelerated rate of water release during drying at 8 °C and 95% RH and increased release of intracellular enzymes that fastened the ripening	Toepfl and Heinz (2008)
Pork *M. longissimus thoracis et lumborum* (LTL) muscle	PEF treatment of EFS, 1.2 or 2.3 kV/cm; PF, 100 or 200 Hz; PN, 150 or 300; SE, 20–185 kJ/kg	Significant increase in NaCl content in PEF-treated sample than control; treatment at 1.2 kV/cm and 67.6 kJ/kg, 100 Hz, and PN 300 gave the best NaCl uptake (13% above the control); no effect on cook loss and water binding capacity	McDonnell et al. (2014)

(continued)

Table 9 (continued)

Meat types	Treatment parameters	Effects	References
Chicken breast	PEF generator, based on insulated-gate monopolar transistor switching capable of peak voltage, 1 kV, current up to 160 A; PW, 5–100μs; PF, 1–16 Hz with energy conversion efficiency of 88%. PEF treatment with EFS of 0.5 kV/cm; PW, 50μs; PF, 1 Hz; and PN, 120	Increased the effective diffusivity of water by 13–24% and reduced convective air drying time by 6.4–15.3%	Levkov et al. (2019)
Pork loin	Cylindrical-parallel-plate-electrode treatment chamber with a diameter of 4 cm and a gap of 5 cm; EFS, 0.1–5.0 kV/cm; PW, 10–200μs; PN, 0–200; SE, 0–56 kJ/kg	PEF treatment of 1 kV/cm, 200μs, and 28 kJ/kg was optimal that was the most suitable with water content reduction of 60.4% in treated meat samples dried at 4 °C; higher drying rates in minced pork meat with a particle size of 4.0 mm; the application of PEF with above-mentioned intensity to Spanish cured sausages "longaniza" reduced the drying time from 17 to 9–10 days (a reduction of 41–47%)	Astrain-Redin et al. (2019)
Beef topsides	PEF assembly with treatment chamber of 13 × 8 × 5 cm dimension with two parallel stainless steel electrodes held 8 cm apart; EFS, 0.52 kV/cm; PW, 20μs; PF, 20 Hz; SE, 13.79 ± 0.36 kJ/kg followed by curing with 1.2% NaCl in curing solution and 2% NaCl in control	PEF-treated samples had significantly ($p < 0.05$) lower sodium content than the control; however, the sensory scores were comparable ($p > 0.05$) with control, and > 84% of the panellists preferred PEF-treated samples over control for saltiness; no significant ($p > 0.05$) effect on sensory scores, instrumental shear force (N) and toughness (N/mm s), colour, yield (%), and oxidative and microbial stability of the products	Bhat et al. (2020)

EFS Electric field strength, *PW* pulse width, *PF* pulse frequency, *PN* pulse number, *SE* specific energy properties, further research is needed for optimisation and commercial PEF application in meat processing.

Research to date suggests that PEF is not a feasible approach to attain commercial sterility (or even the pasteurisation of meat products) because of higher treatment intensity needed to kill the microorganisms present and the resistance of bacterial spores to PEF. Therefore, other hurdles or preservation techniques such as cooking, refrigeration, freezing, the lowering of pH, or atmospheric modification should be used in combination with PEF to ensure food safety and the prevention of meat spoilage.

Future Directions

Despite the potential uses for PEF in meat processing, there are still challenges to develop a successful commercial PEF application. Published studies on the effect of PEF on meat quality can be conflicting due to various factors such as the use of different animal species, breed, age, sex, muscle types, genetic and environment factors, and differences in PEF operating variables. The use of high PEF intensity treatments may enhance tenderness but may have detrimental effects on other quality parameters such as water holding capacity, meat colour, juiciness, and doneness. At high-intensity treatments, the process may be uneconomical because of high energy requirements. Considering the heterogeneity in meat composition even in the same muscle, there is a need for further investigations studying the relationship between quality parameters during and after PEF treatment and the optimisation of process variables and processing conditions for specific meat and muscle types prior to commercial application. Furthermore, as research in PEF application for meat processing is still at a laboratory or pilot scale, the development of reliable, economical, and user-friendly industrial-scale equipment is necessary.

Acknowledgements The authors acknowledge funding from the New Zealand Ministry of Business, Innovation and Employment (contract MAUX1402) for the Food Industry Enabling Technologies (FIET) programme.

References

Abril M, Campo MM, Onenc A, Sanudo C, Alberti P, Negueruela AI (2001) Beef colour evolution as a function of ultimate pH. Meat Sci 58:69–78. https://doi.org/10.1016/s0309-1740(00)00133-9

Alahakoon AU, Faridnia F, Bremer P, Silcock P, Oey I (2017a) Pulsed electric fields effects on meat tissue quality and functionality. In: Handbook of Electroporation, pp 2455–2475. https://doi.org/10.1007/978-3-319-32886-7_179

Alahakoon AU, Oey I, Silcock P, Bremer P (2017b) Understanding the effect of pulsed electric fields on thermostability of connective tissue isolated from beef pectoralis muscle using a model system. Food Res Int 100:261–267. https://doi.org/10.1016/j.foodres.2017.08.025

Alahakoon AU, Oey I, Bremer P, Silcock P (2018) Optimisation of sous vide processing parameters for pulsed electric fields treated beef briskets. Food Bioprocess Technol 11:2055–2066. https://doi.org/10.1007/s11947-018-2155-9

Alahakoon AU, Oey I, Bremer P, Silcock P (2019a) Process optimisation of pulsed electric fields pre-treatment to reduce the sous vide processing time of beef briskets. Int J Food Sci Technol 54:823–834. https://doi.org/10.1111/ijfs.14002

Alahakoon AU, Oey I, Bremer P, Silcock P (2019b) Quality and safety considerations of incorporating post-PEF ageing into the pulsed electric fields and sous vide processing Chain. Food Bioprocess Technol 12:852–864. https://doi.org/10.1007/s11947-019-02254-6

Anderson MJ, Lonergan SM, Fedler CA, Prusa KJ, Binning JM, Huff-Lonergan E (2012) Profile of biochemical traits influencing tenderness of muscles from the beef round. Meat Sci 91:247–254. https://doi.org/10.1016/j.meatsci.2012.01.022

Arroyo C, Eslami S, Brunton NP, Arimi JM, Noci F, Lyng JG (2015a) An assessment of the impact of pulsed electric fields processing factors on oxidation, color, texture, and sensory attributes of Turkey breast meat. Poult Sci 94:1088–1095. https://doi.org/10.3382/ps/pev097

Arroyo C, Lascorz D, O'Dowd L, Noci F, Arimi J, Lyng JG (2015b) Effect of pulsed electric field treatments at various stages during conditioning on quality attributes of beef longissimus thoracis et lumborum muscle. Meat Sci 99:52–59. https://doi.org/10.1016/j.meatsci.2014.08.004

Astrain-Redin L, Raso J, Cebrian G, Alvarez I (2019) Potential of pulsed electric fields for the preparation of Spanish dry-cured sausages. Sci Rep 9(11). https://doi.org/10.1038/s41598-019-52464-3

Baldwin DE (2012) Sous vide cooking: a review. Int J Gastron Food Sci 1:15–30. https://doi.org/10.1016/j.ijgfs.2011.11.002

Bekhit AE-DA, Faustman C (2005) Metmyoglobin reducing activity. Meat Sci 71:407–439. https://doi.org/10.1016/j.meatsci.2005.04.032

Bekhit AE-DA, Geesink GH, Morton JD, Bickerstaffe R (2001) Metmyoglobin reducing activity and colour stability of ovine longissimus muscle. Meat Sci 57:427–435. https://doi.org/10.1016/S0309-1740(00)00121-2

Bekhit AE-DA, Hopkins DL, Geesink G, Bekhit AA, Franks P (2014a) Exogenous proteases for meat tenderization. Crit Rev Food Sci Nutr 54:1012–1031. https://doi.org/10.1080/10408398.2011.623247

Bekhit AE-DA, van de Ven R, Suwandy V, Fahri F, Hopkins DL (2014b) Effect of pulsed electric field treatment on cold-boned muscles of different potential tenderness. Food Bioprocess Technol 7:3136–3146. https://doi.org/10.1007/s11947-014-1324-8

Bekhit AE-DA, Suwandy V, Carne A, van de Ven R, Hopkins DL (2016) Effect of repeated pulsed electric field treatment on the quality of hot-boned beef loins and topsides. Meat Sci 111:139–146. https://doi.org/10.1016/j.meatsci.2015.09.001

Bertram HC, Karlsson AH, Rasmussen M, Pedersen OD, Dønstrup S, Andersen HJ (2001) Origin of multiexponential T2 relaxation in muscle myowater. J Agric Food Chem 49:3092–3100

Bertram HC, Kohler A, Bocker U, Ofstad R, Andersen HJ (2006) Heat-induced changes in myofibrillar protein structures and myowater of two pork qualities. A combined FT-IR spectroscopy and low-field NMR relaxometry study. J Agric Food Chem 54:1740–1746. https://doi.org/10.1021/jf0514726

Bhat ZF, Morton JD, Mason SL, Bekhit AE-DA (2019a) Does pulsed electric field have a potential to improve the quality of beef from older animals and how? Innov Food Sci Emerg Technol 56:102194. https://doi.org/10.1016/j.ifset.2019.102194

Bhat ZF, Morton JD, Mason SL, Bekhit AE-DA (2019b) Pulsed electric field operates enzymatically by causing early activation of calpains in beef during ageing. Meat Sci 153:144–151. https://doi.org/10.1016/j.meatsci.2019.03.018

Bhat ZF, Morton JD, Mason SL, Mungure TE, Jayawardena SR, Bekhit AEA (2019c) Effect of pulsed electric field on calpain activity and proteolysis of venison. Innov Food Sci Emerg Technol 52:131–135. https://doi.org/10.1016/j.ifset.2018.11.006

Bhat ZF, Morton JD, Mason SL, Bekhit AE-DA (2020) The application of pulsed electric field as a sodium reducing strategy for meat products. Food Chem 306:125622. https://doi.org/10.1016/j.foodchem.2019.125622

Blahovec J, Vorobiev E, Lebovka N (2017) Pulsed electric fields pretreatments for the cooking of foods. Food Eng Rev 9:226–236. https://doi.org/10.1007/s12393-017-9170-x

Bolumar T, Enneking M, Toepfl S, Heinz V (2013) New developments in shockwave technology intended for meat tenderization: opportunities and challenges. A review. Meat Sci 95:931–939. https://doi.org/10.1016/j.meatsci.2013.04.039

Boulaaba A, Kiessling M, Egen N, Klein G, Töpfl S (2018) Effect of pulsed electric fields on the endogenous microflora and physico-chemical properties of porcine blood plasma. J Food Saf Food Qual 69:164–170. https://doi.org/10.2376/0003-925X-69-164

Calkins CR, Hodgen JM (2007) A fresh look at meat flavor. Meat Sci 77:63–80. https://doi.org/10.1016/j.meatsci.2007.04.016

Cheng Q, Sun DW (2008) Factors affecting the water holding capacity of red meat products: A review of recent research advances. Crit Rev Food Sci Nutr 48:137–159. https://doi.org/10.1080/10408390601177647

Chian FM, Kaur L, Oey I, Astruc T, Hodgkinson S, Boland M (2019) Effect of pulsed electric fields (PEF) on the ultrastructure and in vitro protein digestibility of bovine longissimus thoracis. LWT-Food Sci Technol 103:253–259. https://doi.org/10.1016/j.lwt.2019.01.005

Clemente I, Condón-Abanto S, Pedrós-Garrido S, Whyte P, Lyng JG (2020) Efficacy of pulsed electric fields and antimicrobial compounds used alone and in combination for the inactivation of Campylobacter jejuni in liquids and raw chicken. Food Control 107. https://doi.org/10.1016/j.foodcont.2019.01.017

de Huidobro FR, Miguel E, Blazquez B, Onega E (2005) A comparison between two methods (Warner-Bratzler and texture profile analysis) for testing either raw meat or cooked meat. Meat Sci 69:527–536. https://doi.org/10.1016/j.meatsci.2004.09.008

Devine CE, Payne SR, Peachey BM, Lowe TE, Ingram JR, Cook CJ (2002) High and low rigor temperature effects on sheep meat tenderness and ageing. Meat Sci 60:141–146. https://doi.org/10.1016/S0309-1740(01)00115-2

Ertbjerg P, Christiansen LS, Pedersen AB, Kristensen L (2012) The effect of temperature and time on activity of calpain and lysosomal enzymes and degradation of desmin in porcine longissimus muscle. In: Proceedings of the 58th international congress of meat science and technology, Montreal, QC, Canada

Faridnia F, Bekhit AE-DA, Niven B, Oey I (2014) Impact of pulsed electric fields and post-mortem vacuum ageing on beeflongissimus thoracismuscles. Int J Food Sci Technol 49:2339–2347. https://doi.org/10.1111/ijfs.12532

Faridnia F, Ma QL, Bremer PJ, Burritt DJ, Hamid N, Oey I (2015) Effect of freezing as pre-treatment prior to pulsed electric field processing on quality traits of beef muscles. Innov Food Sci Emerg Technol 29:31–40. https://doi.org/10.1016/j.ifset.2014.09.007

Faridnia F, Bremer P, Burritt DJ, Oey I (2016) Effects of pulsed electric fields on selected quality attributes of beef outside flat (Biceps femoris). In: Jarm T, Kramar P (eds) 1st world congress on electroporation and pulsed electric fields in biology, medicine and food & environmental technologies, Singapore, 2016. Springer, Singapore, pp 51–54

Faustman C, Sun Q, Mancini R, Suman SP (2010) Myoglobin and lipid oxidation interactions: mechanistic bases and control. Meat Sci 86:86–94. https://doi.org/10.1016/j.meatsci.2010.04.025

Gomez B et al (2019) Application of pulsed electric fields in meat and fish processing industries: an overview. Food Res Int 123:95–105. https://doi.org/10.1016/j.foodres.2019.04.047

Gudmundsson M, Hafsteinsson H (2001) Effect of electric field pulses on microstructure of muscle foods and roes. Trends Food Sci Technol 12:122–128. https://doi.org/10.1016/S0924-2244(01)00068-1

Haughton PN, Lyng JG, Cronin DA, Morgan DJ, Fanning S, Whyte P (2012) Efficacy of pulsed electric fields for the inactivation of indicator microorganisms and foodborne pathogens in liquids and raw chicken. Food Control 25:131–135. https://doi.org/10.1016/j.foodcont.2011.10.030

Herrera-Mendez CH, Becila S, Boudjellal A, Ouali A (2006) Meat ageing: reconsideration of the current concept. Trends Food Sci Technol 17:394–405. https://doi.org/10.1016/j.tifs.2006.01.011

Honikel KO, Roncales P, Hamm R (1983) The influence of temperature on shortening and rigor onset in beef muscle. Meat Sci 8:221–241. https://doi.org/10.1016/0309-1740(83)90046-3

Huff-Lonergan E, Lonergan SM (2005) Mechanisms of water-holding capacity of meat: the role of postmortem biochemical and structural changes. Meat Sci 71:194–204. https://doi.org/10.1016/j.meatsci.2005.04.022

Kang G et al (2013) Effects of high pressure processing on fatty acid composition and volatile compounds in Korean native black goat meat. Meat Sci 94:495–499. https://doi.org/10.1016/j.meatsci.2013.03.034

Kantono K et al (2019) Physicochemical and sensory properties of beef muscles after pulsed electric field processing. Food Res Int 121:1–11. https://doi.org/10.1016/j.foodres.2019.03.020

Khan AA, Randhawa MA, Carne A, Mohamed Ahmed IA, Barr D, Reid M, Bekhit AE-DA (2017) Effect of low and high pulsed electric field on the quality and nutritional minerals in cold boned beef M. longissimus et lumborum. Innov Food Sci Emerg Technol 41:135–143. https://doi.org/10.1016/j.ifset.2017.03.002

Khan AA, Randhawa MA, Carne A, Mohamed Ahmed IA, Barr D, Reid M, Bekhit AE-DA (2018) Effect of low and high pulsed electric field processing on macro and micro minerals in beef and chicken. Innov Food Sci Emerg Technol 45:273–279. https://doi.org/10.1016/j.ifset.2017.11.012

Klonowski I, Heinz V, Toepfl S, Gunnarsson G, Þorkelsson GJ (2006) Applications of pulsed electric field technology for the food industry Iceland fishes laboratory report 06-06 project no 1677:1-10

Koohmaraie M, Geesink GH (2006) Contribution of postmortem muscle biochemistry to the delivery of consistent meat quality with particular focus on the calpain system. Meat Sci 74:34–43. https://doi.org/10.1016/j.meatsci.2006.04.025

Levkov K, Vitkin E, González CA, Golberg A (2019) A laboratory IGBT-based high-voltage pulsed electric field generator for effective water diffusivity enhancement in chicken meat food bioprocess tech doi:https://doi.org/10.1007/s11947-019-02360-5

Listrat A et al (2016) How muscle structure and composition influence meat and flesh quality. Sci World J 3182746:2016. https://doi.org/10.1155/2016/3182746

Liu MN, Solomon MB, Vinyard B, Callahan JA, Patel JR, West RL, Chase CC (2006) Use of hydrodynamic pressure processing and blade tenderization to tenderize top rounds from Brahman cattle. J Muscle Foods 17:79–91. https://doi.org/10.1111/j.1745-4573.2006.00037.x

Lopp A, Weber H (2005) Research into the optimising the tenderness of beef from: parts of the forequarter. Fleischwirtschaft 85:111–116

Machiels D, Istasse L, van Ruth SM (2004) Gas chromatography-olfactometry analysis of beef meat originating from differently fed Belgian Blue, Limousin and Aberdeen Angus bulls. Food Chem 86:377–383. https://doi.org/10.1016/j.foodchem.2003.09.011

Martín L, Timón ML, Petrón MJ, Ventanas J, Antequera T (2000) Evolution of volatile aldehydes in Iberian ham matured under different processing conditions. Meat Sci 54:333–337. https://doi.org/10.1016/S0309-1740(99)00107-2

McArdle R, Marcos B, Kerry JP, Mullen A (2010) Monitoring the effects of high pressure processing and temperature on selected beef quality attributes. Meat Sci 86:629–634. https://doi.org/10.1016/j.meatsci.2010.05.001

McDonnell CK, Allen P, Chardonnereau FS, Arimi JM, Lyng JG (2014) The use of pulsed electric fields for accelerating the salting of pork. LWT-Food Sci Technol 59:1054–1060. https://doi.org/10.1016/j.lwt.2014.05.053

Micklander E, Bertram HC, Marno H, Bak LS, Andersen HJ, Engelsen SB, Norgaard L (2005) Early post-mortem discrimination of water-holding capacity in pig longissimus muscle using new ultrasound method. LWT-Food Sci Technol 38:437–445. https://doi.org/10.1016/j.lwt.2004.07.022

Nuss JI, Wolfe FH (1981) Effect of postmortme storage temperatures on isometric tension, PH, ATP, clycogen and Glucose-6-phosphate for selected bovine muscles. Meat Sci 5:201–213. https://doi.org/10.1016/0309-1740(81)90003-6

O'Dowd LP, Arimi JM, Noci F, Cronin DA, Lyng JG (2013) An assessment of the effect of pulsed electrical fields on tenderness and selected quality attributes of post rigour beef muscle. Meat Sci 93:303–309. https://doi.org/10.1016/j.meatsci.2012.09.010

Ostermeier R, Giersemehl P, Siemer C, Topfl S, Jager H (2018) Influence of pulsed electric field (PEF) pre-treatment on the convective drying kinetics of onions. J Food Eng 237:110–117. https://doi.org/10.1016/j.jfoodeng.2018.05.010

Ouali A, Gagaoua M, Boudida Y, Becila S, Boudjellal A, Herrera-Mendez CH, Sentandreu MA (2013) Biomarkers of meat tenderness: present knowledge and perspectives in regards to our current understanding of the mechanisms involved. Meat Sci 95:854–870. https://doi.org/10.1016/j.meatsci.2013.05.010

Pearce KL, Rosenvold K, Andersen HJ, Hopkins DL (2011) Water distribution and mobility in meat during the conversion of muscle to meat and ageing and the impacts on fresh meat quality attributes—A review. Meat Sci 89:111–124. https://doi.org/10.1016/j.meatsci.2011.04.007

Pietrasik Z, Aalhus JL, Gibson LL, Shand PJ (2010) Influence of blade tenderization, moisture enhancement and pancreatin enzyme treatment on the processing characteristics and tenderness of beef semitendinosus muscle. Meat Sci 84:512–517. https://doi.org/10.1016/j.meatsci.2009.10.006

Puértolas E, Koubaa M, Barba FJ (2016) An overview of the impact of electrotechnologies for the recovery of oil and high-value compounds from vegetable oil industry: Energy and economic cost implications. Food Res Int 80:19–26. https://doi.org/10.1016/j.foodres.2015.12.009

Roldan M, Antequera T, Martin A, Mayoral AI, Ruiz J (2013) Effect of different temperature-time combinations on physicochemical, microbiological, textural and structural features of sous-vide cooked lamb loins. Meat Sci 93:572–578. https://doi.org/10.1016/j.meatsci.2012.11.014

Ruiz J, Calvarro J, Sánchez del Pulgar J, Roldán M (2013) Science and technology for new culinary techniques. J Culinary Sci Technol 11:66–79. https://doi.org/10.1080/15428052.2013.755422

Sionkowska A (2005) Thermal denaturation of UV irradiated wet rat tail tendon collagen. Int J Biol Macromol 35:145–149. https://doi.org/10.1016/j.ijbiomac.2005.01.009

Sitz BM, Calkins CR, Feuz DM, Umberger WJ, Eskridge KM (2006) Consumer sensory acceptance and value of wet-aged and dry-aged beef steaks. J Anim Sci 84:1221–1226. https://doi.org/10.2527/2006.8451221x

Smetana S, Terjung N, Aganovic K, Alahakoon AU, Oey I, Heinz V (2019) Chapter 10 – Emerging technologies of meat processing. In: Galanakis CM (ed) Sustainable meat production and processing. Academic Press, pp 181–205. https://doi.org/10.1016/B978-0-12-814874-7.00010-9

Stetzer AJ, Cadwallader K, Singh TK, McKeith FK, Brewer MS (2008) Effect of enhancement and ageing on flavor and volatile compounds in various beef muscles. Meat Sci 79:13–19. https://doi.org/10.1016/j.meatsci.2007.07.025

Suwandy V, Carne A, van de Ven R, Bekhit AE-DA, Hopkins DL (2015a) Effect of pulsed electric field on the proteolysis of cold boned beef M. Longissimus lumborum and M. Semimembranosus. Meat Sci 100:222–226. https://doi.org/10.1016/j.meatsci.2014.10.011

Suwandy V, Carne A, van de Ven R, Bekhit AE-DA, Hopkins DL (2015b) Effect of pulsed electric field treatment on hot-boned muscles of different potential tenderness. Meat Sci 105:25–31. https://doi.org/10.1016/j.meatsci.2015.02.009

Suwandy V, Carne A, van de Ven R, Bekhit AE-DA, Hopkins DL (2015c) Effect of pulsed electric field treatment on the eating and keeping qualities of cold-boned beef loins: impact of initial pH and fibre orientation. Food Bioprocess Technol 8:1355–1365. https://doi.org/10.1007/s11947-015-1498-8

Suwandy V, Carne A, van de Ven R, Bekhit AE-DA, Hopkins DL (2015d) Effect of repeated pulsed electric field treatment on the quality of cold-boned beef loins and topsides. Food Bioprocess Technol 8:1218–1228. https://doi.org/10.1007/s11947-015-1485-0

Toepfl S, Heinz V (2007) Application of pulsed electric fields to improve mass transfer in dry cured meat products. Fleischwirtschaft Int 22:2007

Toepfl S, Heinz V (2008) The stability of uncooked cured products use of pulsed electric fields to accelerate mass transport operations in uncooked cured products. Fleischwirtschaft 88:127–130

Toepfl S, Heinz V, Knorr D (2006) Applications of pulsed electric fields technology for the food industry. doi:https://doi.org/10.1007/978-0-387-31122-7_7

Tornberg E (2005) Effects of heat on meat proteins – implications on structure and quality of meat products. Meat Sci 70:493–508. https://doi.org/10.1016/j.meatsci.2004.11.021

Tornberg E, Wahlgren M, Brondum J, Engelsen SB (2000) Pre-rigor conditions in beef under varying temperature- and pH-falls studied with rigometer, NMR and NIR. Food Chem 69:407–418. https://doi.org/10.1016/s0308-8146(00)00053-4

Utrera M, Estévez M (2013) Oxidative damage to poultry, pork, and beef during frozen storage through the analysis of novel protein oxidation markers. J Agric Food Chem 61:7987–7993. https://doi.org/10.1021/jf402220q

Wiklund E, Stevenson-Barry JM, Duncan SJ, Littlejohn RP (2001) Electrical stimulation of red deer (Cervus elaphus) carcasses—effects on rate of pH-decline, meat tenderness, colour stability and water-holding capacity. Meat Sci 59:211–220. https://doi.org/10.1016/S0309-1740(01)00077-8

Wood JD et al (2004) Effects of fatty acids on meat quality: a review. Meat Sci 66:21–32. https://doi.org/10.1016/S0309-1740(03)00022-6

Wood JD et al (2008) Fat deposition, fatty acid composition and meat quality: a review. Meat Sci 78:343–358. https://doi.org/10.1016/j.meatsci.2007.07.019

Zakrys PI, Hogan SA, O'Sullivan MG, Allen P, Kerry JP (2008) Effects of oxygen concentration on the sensory evaluation and quality indicators of beef muscle packed under modified atmosphere. Meat Sci 79:648–655. https://doi.org/10.1016/j.meatsci.2007.10.030

Zhao G-Y et al (2012) Tenderization effect of cold-adapted collagenolytic protease MCP-01 on beef meat at low temperature and its mechanism. Food Chem 134:1738–1744. https://doi.org/10.1016/j.foodchem.2012.03.118

Other Applications of Pulsed Electric Fields Technology for the Food Industry

Diederich Aguilar-Machado, Julio Montañez, Javier Raso, and Juan Manuel Martínez

Abbreviations

PEF	Pulsed electric field
Zp	Disintegration index
fw	Fresh weight matter
dw	Dried weight matter
PC	Phenolic compounds
FC	Flavonoids compounds
GAE	Gallic acid equivalent
SB	Steam blanching
US	Ultrasound
MW	Microwave
SFE	Supercritical fluid extraction
PLE	Pressurized liquid extraction
°C	Grade Celsius
h	Hour
μs	Microseconds
min	Minutes
W	Watts
ms	Milliseconds
BPE	β-Phycoerythrin
TPEY	Total polyphenol extraction yield
SAE	Supplementary aqueous extraction

D. Aguilar-Machado
Tecnología de los Alimentos, Facultad de Veterinaria, Instituto Agroalimentario de Aragón-IA2, Universidad de Zaragoza, Zaragoza, Spain

Department of Chemical Engineering, Facultad de Ciencias Químicas, Universidad Autónoma de Coahuila, Saltillo, Mexico

J. Montañez
Department of Chemical Engineering, Facultad de Ciencias Químicas, Universidad Autónoma de Coahuila, Saltillo, Mexico

J. Raso · J. M. Martínez (✉)
Tecnología de los Alimentos, Facultad de Veterinaria, Instituto Agroalimentario de Aragón-IA2, Universidad de Zaragoza, Zaragoza, Spain

J. Raso et al. (eds.), *Pulsed Electric Fields Technology for the Food Industry*, Food Engineering Series, https://doi.org/10.1007/978-3-030-70586-2_15

AEY	Anthocyanin extraction yield
QE	Quercetin
ATCC	American Type Culture Collection
kV/cm	Kilovolt/centimeter (electric field strength)
kJ/kg	Kilojoule/kilogram (specific energy)
Hz	Hertz

Introduction

The feasibility of pulsed electric fields (PEF) for a wide pool of different applications in the food industry has been demonstrated in many diverse studies. Regardless of some well-known uses in different sectors such as fruit juices, potato processing, wineries, olive oil extraction, drying, or meat processing, other PEF emerging applications are recently taking importance showing promising benefits and results. PEF technology is a valuable tool that can improve extractability and recovery of functional and high valuable compounds (Fig. 1) or modify the food structure with a diverse variety of practical applications. The use of PEF has been expanded in different areas within food science, food engineering, and food biotechnology. The growing demand for products derived from natural sources has encouraged in the last years the study of different applications of PEF where biotechnological applications are increasingly relevant. Other important approaches such as the reduction of energy expenditure and the development of more efficient processes have been other areas of opportunity for PEF. The benefits of this technology are reflected not only in the quality of the products obtained but also in energy savings and the lower environmental impact that this represents.

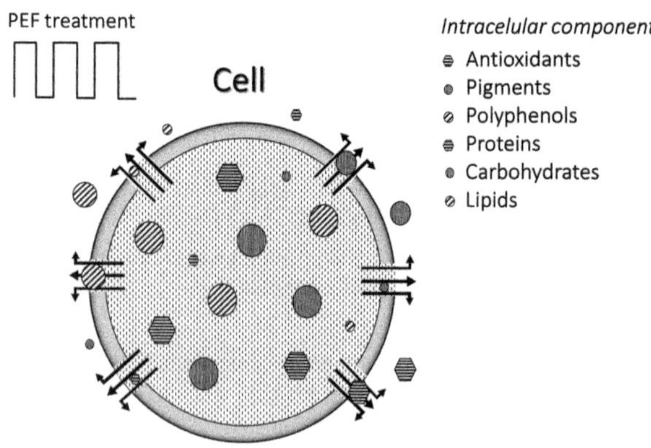

Fig. 1 PEF-assisted extraction of intracellular compounds from plant and microbial cells

Extraction of Pigments by PEF

Pigments are compounds that change the color of reflected or transmitted light as the result of wavelength-selective absorption. Carotenoids, anthocyanins, betanin, and chlorophylls, among others, are the main pigments that provide the color to food. These coloring agents have drawn renewed interest because these not only make food attractive but also have potential health benefits (Rodríguez-Amaya 2016; Saini and Keum 2018). While some plants and microorganisms can synthesize pigments, other organisms such as animals must obtain these molecules through the diet. In the last years, the market for food pigments has rapidly increased and it is expected to continue growing (Institute of Food Technologist 2016). Pigments are nowadays mostly chemically produced; however, the growing awareness of the environmental hazards and the negative assessment of synthetic food dyes from modern consumers have given a strong interest in natural coloring alternatives (Carocho et al. 2014). Suiting this trend, major food and beverage companies have committed to removing artificial colorants from their products substituting those by natural ones (Cortez et al. 2017). Besides, the consumption of natural pigments has been associated with a reduction of the risk of suffering diseases such as cancer, diabetes, and obesity in comparison to synthetic ones (Li et al. 2016; Cooperstone and Schwartz 2016; Fakhri et al. 2018). Therefore, it makes sense to seek feasible sources of natural pigments from microorganisms or plants.

To incorporate natural pigments in food with sufficient concentration and with increased bioavailability, the pigments have to be extracted and isolated from the natural source and subsequently added to the food or beverage. In the case of unicellular organisms, some of the pigments are released spontaneously to the extracellular medium, but in most of the cases, pigments remain inside the microbial cells. The structure of the cells and the location of the pigments are specific for each type of cell in which pigments are contained. Therefore, the design of effective strategies for their extraction and subsequent purification requires exhaustive research of these features. Both in vegetal or microbial cells, intracellular compounds have to cross the cell envelopes to be recovered. The presence of an intact cytoplasmic membrane which acts as a semipermeable barrier, or also membranes of organelles and other envelopes such as thick cellular walls, makes the release of valuable compounds difficult. The recovery of these compounds, which are intracellularly locked, should occur through a sustainable process in which a crucial role is played by the cell disintegration technique used to improve the efficiency of the extraction step. Figure 2 shows a process flow diagram of pigment production from vegetable matrices or microorganisms. The process begins with the cultivation of strains of microorganisms or vegetable species that produce the pigment of interest and its harvesting. After the concentration of the microbial suspension or the size reduction of vegetable matrices, the extraction process has been traditionally carried out after dehydrating the biomass. This approach allows the extraction of hydrophobic pigments using organic solvents and the storage of the biomass, thereby providing the option of delaying the extraction of the target compound. Nevertheless, dehydrating

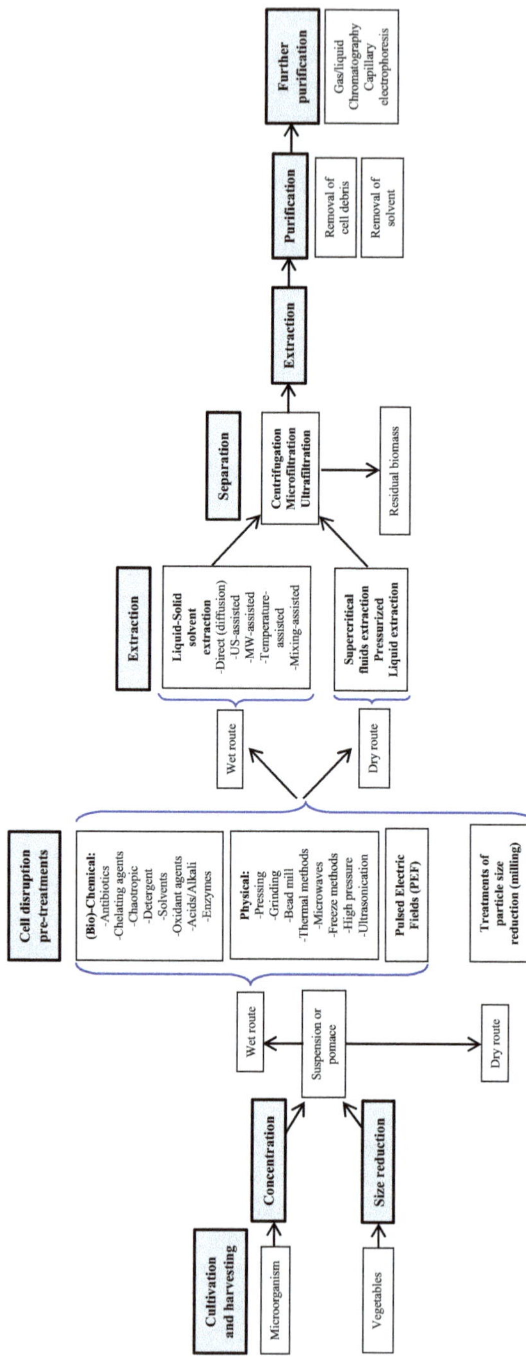

Fig. 2 Process diagram of pigment production from vegetable and microorganisms' matrices

the biomass previously to the extraction stage consumes a considerable amount of energy and may provoke the degradation of high-value compounds due to oxidation. Thus, it should be preferable to use moist biomass, where fewer unit operations are required. Either following the dry or wet route, cell disruption pre-treatments must be applied to pierce the cell envelopes and facilitate the subsequent extraction. The selected technique should allow a mild cell disruption of biomass to selectively improve the extraction efficiency, reducing processing time, temperature, and amount of organic solvents but maintaining the integrity of the extracted molecules with no losses of functionality. The cell disruption pre-treatments traditionally applied are based on chemicals, enzymes, or physical methods. Chemical treatments are effective, but for food applications, the contamination of the suspensions by the active chemical, which often is a nonfood grade solvent, results in the complication of downstream purification (Lam and Lee 2015). Enzymes require long incubation times, are expensive, and are limited to the use under optimal conditions (Lam and Lee 2015; Middelberg 1995). Concerning the application of mechanical treatments, they usually produce an increment of temperature, and biomass must be then cooled down to avoid undesirable heating effects. Besides, these techniques generally prevent the extraction of thermolabile compounds and produce excessive denaturalization of cell envelopes that lead to the release of cell debris, and subsequent expensive purification processes are required (McMillan et al. 2013; Skorupskaite et al. 2017; Walther et al. 2017). Other nonconventional technologies including ultrasound (US), microwave (MW), mixing, supercritical fluid extraction (SFE), pressurized liquid extraction (PLE), or heating methods have also been tested to improve the yields from direct solvent diffusion (Macías-Sanchez et al. 2009; Jaime et al. 2010; Pasquet et al. 2011; Plaza et al. 2012; Chemat et al. 2017).

Regardless of the method of the cell disruption used, biomass can be either dried (dry route), or the process follows up using the moist biomass (wet route). The extraction is usually carried out by dissolving the pigments in a liquid solvent to later recover them from the mixture. The solvent (organic or aqueous, depending on the polarity of the pigment to be extracted) is blended with the solid, and contact is kept throughout the required time interval. The release of the compound that is located inside a cell requires the solvent penetration and then the dissolution of the compound, followed by the diffusion to the surface of the cell and, finally, to the extracellular solvent. After the extraction stage, the extract contained in the solvent is purified to different degrees depending on the final application of the extracted pigment. For example, the production of phycobiliproteins as biomarkers for immunology assays requires highly purified molecules, while the extracts used as pigments for foods or beverages usually contain a mixture of molecules (Martínez et al. 2020b).

Pulsed electric field (PEF) is a nonthermal technology able to enhance the migration of intracellular compounds located in the cytoplasm of cells through the cytoplasmic membrane. The loss of the selective permeability of the membrane improves the mass transfer and enhances the release of intracellular compounds to the surrounding media. This permeabilization caused by PEF has been mainly evaluated through cell disintegration index and percentage of irreversible cells electroporated for vegetables-based matrices and microbial cells, respectively. The efficacy of electroporation required for extraction depends on PEF treatment intensity and the type of sample studied, as well as other parameters such as temperature, conductivity pH, sample size, and form (Puértolas and Barba 2016).

This section provides insights into the potential of electroporation by PEF as a pre-treatment to improve the subsequent extraction of pigments as well as other bioactive compounds from plants and microbial cells.

Extraction of Pigments from Plant-Based Matrices

Plant tissues are rich sources of pigments that can be used as natural food colorants but also provide a supportive role in disease prevention, therapy, and health promotion. In general, anthocyanins are associated with the prevention of vision disorders, cancer, and cardiovascular and neurological diseases, while carotenoids are thought to prevent cancer and photooxidative damage and improve immune system, and betalains and chlorophylls are associated as well with anticancer activity (Leong et al. 2018). Also, within the same group of pigments, there are numerous specific compounds; each one can also bear particular bioactive potential.

Commonly, the procedures of pigment extraction from vegetables are based on using high temperatures and/or polluting solvents, with the drawbacks mentioned in the previous section. It must be highlighted that plant pigments are highly susceptible to degradation or lose their color attributes. For example, the chlorophyll of green vegetables exposed to mild heat treatment and light can be quickly degraded into its derivatives like pheophytin turning into dark yellowish. Anthocyanin color varies with the pH of the medium and is very unstable depending on environmental conditions (heat, acid, peroxides, enzymes, oxygen, and light). Carotenoids can readily be oxidized by the presence of oxygen depending on temperature and the light itself also provokes degradation. With the total disruption of cells by unspecific treatments, various substances retained in different cellular compartments come into contact with each other (enzymes, oxygen, pigments), and the risk of degradation rises. Different organelles inside the plant cells may contain the specific types of pigments but usually are inside vacuoles. Permeabilization of cytoplasmic membrane by PEF can lead to destabilization of vacuoles due to osmotic imbalance contributing to the selective release of pigments (Luengo et al. 2014a). Therefore, PEF treatment can enhance the selective diffusivity of components using the appropriate solvents at ambient temperatures, preventing thermal degradation of vegetal cell walls and the release of debris.

Several detailed studies of pigment extraction from different plant materials after PEF have been conducted. Typically, before PEF treatment, the tissue is processed to reduce the size into pieces or to produce a pomace (Fig. 2). The sample can be subjected directly to the PEF treatment or embedded in a solution. Most of the studies reported in literature deal with the extractability of betanin, anthocyanin, and carotenoids following the PEF treatment of red beetroot, cabbage, grape, tomato, carrot, or paprika (Brianceau et al. 2015; Grimi et al. 2009; Luengo et al. 2016). In these studies, the treatment parameters (electric field strength, treatment time, pulse width, temperature of application, pH, and specific energy applied) have been optimized to maximize the extraction yields by producing irreversible electroporation. Generally, due to the larger size of vegetal cells, the electric field strengths required to electroporate the cells are lower than in the case of microbial cells. Total extraction has been achieved using moderate electric field strengths (0.5–5.0 kV/cm) both in the range of microseconds or milliseconds, with low energy consumption. Generally, longer pulses in the order of milliseconds result in larger pores (Saulis 2010; Saulis and Saule 2012), and the same occurs with the extension of treatment time or the number of pulses (Silve et al. 2014). Likewise, temperature, pH, and electrical conductivity of the medium have been described as important parameters also affecting extraction yields. The application of PEF to a highly conductive solution results in a higher current across the electrodes and a more ohmic heating effect, thus increasing temperature in comparison to a lower conductivity solution. Also, the variables involved during extraction such as temperature and type of solvent affect the extraction yields and thus modulate the effect achieved after PEF pre-treatment.

Ade-Omowaye et al. (2001) investigated the impact of PEF treatment (1.7 kV/cm; 0.5 kJ/kg) before high-pressure extraction on yield and some quality parameters of juice from paprika and compared the results with those of juice obtained from enzymatically treated or untreated paprika mash. In addition to pH, soluble solids (°Brix), total dry matter and color, total carotenoids, and β-carotene were determined in the resultant juice samples. PEF treatments resulted in about 10% increase in juice yield, better redness color, and 60% more β-carotene extraction being the remaining juice quality parameters similar to those obtained from untreated and enzymatically treated paprika.

Regarding the extraction of betanin from red beetroot, López et al. (2009) reported the use of PEF for the treatment of thin disks at different intensities (number of pulses and field strengths) for the release of this pigment into media of different pH and temperature. The use of extracting media having pH 3.5 and a temperature of 30 °C resulted in the highest yields. When samples were subjected to 5 pulses of 2μs at 7 kV/cm, a release about 90% of total betanin in 300 min was observed; this extraction time was five times faster than the non-PEF-treated samples. Pressing the samples during the extraction process also accelerated the extraction. Loginova et al. (2011a) studied the effect of temperature (30–80 °C) on the degradation of colorants and kinetics of their extraction from the red beet by using PEF. The temperature increased the extraction yields, but degradation became important above 60 °C. PEF treatment removed the membrane barriers, being extraction ruled by unrestricted diffusion from PEF-treated samples. More recently, Luengo et al. (2016) compared the efficiency of PEF treatments in the range of milliseconds (0.4 to 0.6 kV/cm; 10 to 60 ms) and microseconds (4 to 6 kV/cm; 30 to 150μs) during the extraction of betanin from red beet. The most intense treatment conditions applied in the milliseconds range (0.6 kV/cm; 40 ms) and the microseconds range (6 kV/cm; 150μs) increased the yields 6.7 and 7.2 times, respectively, compared with untreated samples. However, lower specific energy was consumed in the PEF treatments applied in the range of µs (28.8 kJ/kg) than in the PEF treatments applied in the ms range (43.2 kJ/kg).

Improvement of anthocyanin extraction from different plant materials after PEF treatment has also been investigated. Gachovska et al. (2010) evaluated the effect of PEF on anthocyanin extraction from red cabbage mash using water as a solvent. PEF treatments (2.5 kV/cm; 750µs) enhanced total anthocyanin extraction by 2.15 times in comparison to the untreated one. Comparing heat and light stability of PEF-treated and untreated samples, not significantly different was observed ($P > 0.05$). Puértolas et al. (2013) studied the influence of PEF parameters on the anthocyanin extraction yield (AEY) from purple-fleshed potato at different extraction times (60–480 min) and temperatures (10–40 °C) using water and ethanol (48% and 96%) as solvents. A PEF treatment of 3.4 kV/cm and 105µs (35 pulses of 3µs) resulted in the highest cell disintegration index ($Zp = 1$) at the lowest specific energy requirements (8.92 kJ/kg). After 480 min at 40 °C, the AEY obtained for the untreated sample using 96% ethanol as the solvent (63.9 mg/100 g fw) was similar to that obtained in the PEF-treated sample using water (65.8 mg/100 g fw). Therefore, PEF could allow substituting the use of an organic solvent such as ethanol by water without impairing extraction yield. Liu et al. (2018) reported PEF-assisted aqueous extraction as a selective and eco-friendly method to recover water-soluble polyphenols (PC) and pigments flavonoids (FC) from fresh red onion. Optimal PEF conditions (2.5 kV/cm, 90 pulses, and 45 °C extraction temperature) resulted in an enhancement of the extraction yield of PC and FC up to 102.86 mg GAE 100/g fw and 37.58 mg QE 100/g fw, respectively. These yields suppose an increment by 2.2 and 2.7 times for PC and FC regarding non-PEF-treated samples.

Roohinejad et al. (2014) studied the effect of electric field strength (0.1–1.0 kV/cm) and frequency (5–75 Hz) of PEF processing on the extractability of carotenoids from carrot pomace using different vegetable oils. PEF improved the yields in all cases in comparison to untreated samples, but the effect of solvent depended on the PEF treatment conditions. Sunflower had the highest carotenoid extractability with 45.50 and 39.42µg/g for β- and α-carotene, respectively. The influence of the application of PEF (3–7 kV/cm and 0–300µs) on the extraction of lycopene from tomato waste was investigated by Luengo et al. (2014a). A PEF treatment at 5 kV/cm for 90µs improved the carotenoid extraction from tomato peel by 39% as compared with the control in a mixture of hexane/ethanol/acetone (50:25:25). On the other hand, this PEF treatment allowed reducing the hexane content in the solvent combination from 45% to 30% without affecting the carotenoid extraction yield. The antioxidant capacity of the extracts obtained from tomato peel was correlated with the carotenoid concentration, and it was not affected by the PEF treatment. Pataro et al. (2018) demonstrated that the combination of steam blanching (SB) (1 min at 50–70 °C), with PEF treatments (0.25–0.75 kV/cm, 1 kJ/kg) of whole tomatoes, in addition to reducing the energy required for tomato peeling, can significantly contribute to the recovery of carotenoids from the peels in acetone. PEF and SB caused the increase of the yield in total carotenoids (up to 188% for PEF and 189% for SB) and antioxidant power (up to 372% for PEF and 305% for SB) concerning the peels from untreated tomatoes. Furthermore, the application of a combined treatment (PEF + SB) showed a synergistic effect in extraction, being lycopene, the main carotenoid extracted, and neither PEF nor SB caused any degradation. These results demonstrated that the integration of PEF in the processing line of tomato fruits before SB contributes to the valorization of tomato processing by-products.

Extraction of Valuable Pigments from Microbial Cells

Aside from plants, microorganisms have been considered potential sources of natural pigments because they are able of producing high yields while growing in low-cost substrates such as agro-industrial wastes (Buzzini and Martini 2000). Some of the natural isomers of the pigments that microorganisms produce have proven superior in terms of nutritional and therapeutic value and thus are advantageous despite being more expensive than synthetic forms (Becker 2004). For example, the carotenoid astaxanthin provides the orange-red color to crustaceans and salmonids which plays an important role in consumer appeal, but animals cannot synthesize the pigment and need to incorporate it in the diet. *Xanthophyllomyces dendrorhous* yeast and *Haematococcus pluvialis* microalgae naturally accumulate this pigment, and a lot of scientific papers and patents are devoted to microbial astaxanthin production (Johnson 2003; Schmidt et al. 2011). Besides, the modification of strains to obtain overproducing mutants enables significant improvement of pigment yields (Jacobson et al. 2003). On the other hand, the yeast of the genus *Rhodotorula* produces the carotenoids torulene and torularhodin, apart from a small quantity of β-carotene. Torularhodin is a singular red carotenoid with potential applications ranging from food technology to therapeutics thanks to its strong antioxidant activity, being a vitamin A precursor, and anticancer potential (Moliné et al. 2012; Zoz et al. 2015; Du et al. 2016).

The permeabilization of the cytoplasmic membrane after PEF treatment is generally described as the cause of the improved release of the compounds of interest. The PEF improvements in the recovery of pigments from microorganisms such as yeast and microalgae are summarized in Table 1. It has been recently reported that PEF treatments also activate yeast autolysis, thereby leading to the self-degradation of the constituents of yeast cells by their own hydrolytic enzymes (Martínez et al. 2016). In this sense, the potential of PEF for inducing autolysis has been studied in the red yeast *Rhodotorula glutinis* to improve carotenoid extraction. Martínez et al. (2018) designed an efficient and eco-friendly process of carotenoid extraction from fresh biomass. While an extended contact of *R. glutinis* biomass in ethanol after PEF treatment resulted in negligible extraction, 24 h of incubation of the treated cells in aqueous medium allowed achieving subsequent carotenoid extraction in ethanol. This fact was associated with the trigger effect of PEF on enzymatic reactions of autolysis, and it was confirmed with flow cytometer which evidenced morphological changes in PEF-treated *R. glutinis* cells during aqueous incubation. This improvement in the release of carotenoids was explained by the lysis of the bonds

Table 1 Recovery of pigments from microorganisms by pulsed electric fields (PEF) technology

Microorganisms	Pigment	Improvement	Electric field strength (kV cm^{-1})	Total specific energy (kJ/kg)	References
Yeast					
R. glutinis	Carotenoids[a]	3 times greater extraction than from untreated (375μg/g $_{dw}$)	15 kV/cm	68 kJ/kg	Martínez et al. (2018)
R. glutinis	Carotenoids[b]	80% of total carotenoid (267μg/g $_{dw}$)	15 kV/cm	84 kJ/kg	Martínez et al. (2020a)
X. dendrorhous	Astaxanthin[a]	70% of total carotenoid (2.4 mg/g $_{dw}$) Up to 84% corresponded to trans-astaxanthin	20 kV/cm	54 kJ/kg	Aguilar-Machado et al. (2020)
Microalgae					
C. vulgaris	Lutein	3.5–4.2-fold in comparison to untreated	25 kV/cm	61 kJ/kg	Luengo et al. (2015a, b)
A. platensis	C-Phycocyanin	Total content extraction after 6 h	25 kV/cm	93 kJ/kg	Martínez et al. (2017)
Nannochloropsis	Proteins, pigments	Efficient extraction of proteins in water at the first step and improved extraction in DMSO/ethanol of pigments (second step)	20 kV/cm	96 kJ/kg	Parniakov et al. (2015)
P. cruentum	β-Phycoerythrin	Total content extraction after 24 h	8 kV/cm	15 kJ/kg	Martínez et al. (2019a, b)
H. pluvialis	Astaxanthin	96% of the total carotenoid content (18.3 mg/g $_{dw}$)	1 kV/cm	50 kJ/kg	Martínez et al. (2019a, b)

Data regarding total specific energy depends on the conductivity of the media in which treatments were applied, a crucial parameter
[a]Fresh biomass
[b]Dried biomass

of carotenoids with other components of the cytoplasm by enzymes released from the organelles, as a result of the distressing osmotic conditions of the cytoplasm after electroporation (Martínez et al. 2018). Likewise, the potential of PEF to improve carotenoid extraction from dry biomass of *R. glutinis* was also demonstrated (Martínez et al. 2020a). The extraction from untreated or PEF-treated cells that were immediately freeze-dried after the pre-treatment using acetone, hexane, and ethanol as solvent was not very effective (extraction yield <20% total content).

Conversely, PEF treatment and subsequent intermediate incubation in an aqueous buffer for 24 h, followed by freeze-drying and extraction, led to a large improvement in the extraction yield with the three solvents assayed. Ethanol was the most efficient, reaching an extraction yield of 80% of total carotenoid, which represents a recovery of 267μg/g dw being torularhodin esters the main carotenoid found in the extracts. On the other hand, Aguilar-Machado et al. (2020) recently demonstrated that the application of PEF treatment and subsequent aqueous incubation of fresh biomass of mutant *X. dendrorhous* ATCC 74219 allowed the ethanolic extraction of 70% of the total carotenoids. From total carotenoid extracted, around 84% corresponded to all-trans-astaxanthin. The effective extraction was attributed to the irreversible electroporation caused by PEF and the hydrolytic activity of the enzyme esterase triggered by PEF during the incubation time. The application of PEF treatments to the fresh biomass with an immediate suspension in ethanol was ineffective for the extraction of carotenoids. However, after 12 h of aqueous incubation and subsequent resuspension in ethanol, it resulted in an effective astaxanthin extraction. The esterase activity triggered by PEF seems to play an important role by the effective extraction achieved in certain species. These esterases could hydrolyze the triacylglycerols of lipid droplets that contain the astaxanthin resulting in the loss of their structure and the subsequent carotenoids release by diffusion into the extraction media.

Despite the potential of PEF-assisted extraction of pigments from yeasts, the most extensive group of microorganisms exploited for pigment production is microalgae. These photosynthetic organisms contain basically three types of pigments: carotenoids, chlorophylls, and phycobiliproteins. Among carotenoids, apart from α- and ß-carotenes, microalgae such as *Chlorella* sp. contain lutein, which is approved as a food colorant by the European Union (E-161 b) and has also a potential role in preventing diseases due to its antioxidant capabilities (Carpentier et al. 2009). Like carotenoids, chlorophylls are lipid-soluble compounds with low polarity, and they are used commercially as important natural green pigments (Henriques et al. 2007). Chlorophyll has also been associated with health benefits as a nutraceutical agent with antioxidant, anti-inflammatory, antimutagenic, and antimicrobial properties (da Silva Ferreira and Sant'Anna 2017). Another type of photosynthetic pigment is the phycobiliprotein, a water-soluble protein assembled in the thylakoid membranes of chloroplasts. The cyanobacterium *Arthrospira* and the rhodophyte *Porphyridium* are already commercially used for its production (Viskari and Colyer 2003; Román et al. 2002). The main potential of these molecules is to serve as natural dyes. However, their health-promoting properties have been also proven along with a broad range of pharmaceutical applications and as fluorescent biomarkers in immunology (Qiu et al. 2004).

The potential of PEF technology as a pre-treatment to improve the extraction of microalgae pigments has been widely reported (Martínez et al. 2020b). In general, PEF pre-treatment improved the subsequent release of water-soluble as well as nonpolar pigments utilizing an appropriate solvent. Electric field strength, treatment time, and specific energy influence the effectiveness of PEF treatment on microalgae. The effective permeabilization after PEF treatment of enough intensity has been evidenced by a strong increase in the electrical conductivity of the suspension media and the uptake of propidium iodide (Kempkes 2016). Beyond the critical electric field strength, permeabilization commonly increases along with more intense electric field strength and longer treatment durations. However, the shortness of treatment duration from milliseconds to microseconds, combined with a small increase of electric field strength, allows maintaining or even increasing the extraction yield while reducing the energy consumption (Luengo et al. 2015a). Another crucial factor influencing the electroporation of microalgae is the temperature during PEF treatment. The rise of temperature within levels that do not cause degradation of the pigments by heat (below 40 °C) allows a reduction of the required electric field strength and treatment time to achieve a certain extraction yield and, as a result, a reduction in the total specific energy supplied by the treatment (Luengo et al. 2015b; Martínez et al. 2017).

Luengo et al. (2014b; 2015a, b) studied and optimized the extraction of chlorophylls and carotenoids (lutein) from the microalgae *Chlorella vulgaris*, regarding the influence of electrical parameters and temperature. A PEF treatment of 20 kV/cm for 75μs increased extraction yields for carotenoids and chlorophylls *a* and *b* by 1.2, 1.6, and 2.1 times, respectively (Luengo et al. 2014b). A high correlation was observed between irreversible electroporation and the percentage of yield increase when the extraction was conducted 1 h after the application of PEF treatment, but not when the extraction was conducted just after PEF treatment. The authors compared the influence of microsecond against millisecond pulses in conjunction with electric field strength (Luengo et al. 2015a). To achieve a maximum extraction yield, the energy of treatments required in the milliseconds range was 150 kJ/L, while for treatments in the microsecond order, only 30 kJ/L was required. Regarding the influence of treatment medium temperature, a treatment of 25 kV/cm–100μs at 25 to 30 °C increased the lutein extraction yield 3.5- to 4.2-fold in comparison with control, resulting in the most suitable treatment conditions for maximizing lutein extraction from *Chlorella vulgaris* at the lowest energy cost (Luengo et al. 2015b).

Martínez et al. (2017) applied PEF to fresh biomass of *Artrosphira platensis* to enhance the selective extraction of the water-soluble protein phycocyanin in aqueous media. Electric field strength (15 to 25 kV/cm), treatment time (60 to 150μs), and temperature of application (10 to 40 °C) were found to influence extraction yields, but a delay of 150 min at the onset of extraction was observed for all conditions. This delay was attributed to the fact that low molecular weight molecules can pass through the cytoplasmic membrane directly after electroporation, while the leakage of components of larger molecular weight might need that the pores generated by PEF treatment enlarge over the progression of time. Likewise, Jaeschke et al. (2019) achieved high extraction yields of phycocyanin from *A. platensis* after PEF treatments of 40 kV/cm using 1 μs wide pulses (112 kJ/kg), being observed that the yields increased by increasing incubation time after PEF treatment.

More recently, Martínez et al. (2019a) studied the extraction of another water-soluble phycobiliprotein, β-phycoerythrin (BPE), into aqueous media by the application of PEF to fresh *Porphyridium cruentum*. While the release of this water-soluble protein was undetectable in the untreated cells even after long incubation times, the entire content was released from PEF-treated cells after 24 hr of extraction. The protein was not released immediately, nevertheless; a delay time of over 6 h was needed till the pigment could be detected in the extraction buffer. This kinetics pointed out that BPE release requires not only the diffusion of the pigment through the cell membrane but also the disassembling of the molecule from the cell organization. In this respect, it was postulated that PEF could trigger the release of hydrolytic enzymes from the *P. cruentum* organelles that would lyse the bonds between the pigment and other compounds of the cell; thus, the water-BPE complex could diffuse through the membrane, carried by a concentration gradient.

In this sense, the approach of triggering the autolysis of microalgae using PEF was afterward evaluated for the extraction of carotenoid astaxanthin from the Nordic strain *Haematococcus pluvialis* (Martínez et al. 2019b). The efficiency of PEF pretreatment of fresh biomass of *H. pluvialis* followed by incubation in the growth medium was compared to classical disruption methods (bead-beating, freezing-thawing, thermal treatment or ultrasound) for the subsequent extraction of astaxanthin in ethanol. *H. pluvialis* nitrogen-starved cells treated with PEF followed by aqueous incubation for 6 h resulted in an extraction of 96% (18.3 mg $_{car}$/g $_{dw}$) of the total carotenoid content compared to 80% (15.3mg $_{car}$/g $_{dw}$) using the other methods. The higher proportion of free astaxanthin than astaxanthin forming esters in the extracts from PEF-treated samples was associated with the triggering esterase activity by PEF. This PEF autolysis induction during aqueous incubation permitted to substitute the injurious organic chemicals such as acetone or hexane by ethanol, a safer and eco-friendly solvent. The mediation by an enzymatic process after cell death of the extraction assisted by PEF of other compounds, like proteins, has been confirmed as well by other authors (Scherer et al. 2019). However, it is required to go deeper into the understanding of this mechanism and identify the enzymes involved in microalgae autolysis triggered by PEF in other species and pigments of interest. This would allow the application of this approach for industrial implementation.

Generally, PEF is a very promising technique useful for microbial cell piercing on a large scale. The efficiency of PEF treatments is high enough to achieve significant pigment extraction yields. The treatment improves extraction selectively, and thus the extracts obtained by PEF-assisted method are purer than those obtained using other methodologies that consist of total cell destruction such as bead-beating, crushing, ultrasound, or high-pressure homogenization producing the release of impurities; thus the addition of more unit operations in downstream processing is needed. Furthermore, PEF treatment consumes much lower energy than these conventional techniques (Carullo et al. 2018; de Boer et al. 2012) and presents the opportunity of performing the extraction with fresh biomass instead of dried biomass, leading to the replacement of detrimental chemicals with greener solvents (Martínez et al. 2020b). PEF can be implemented in a continuous flow for the extraction of pigments from microorganisms, allowing for a feasible scale-up to processing capacities in the magnitudes of thousands of liters per hour. The simplicity, speed, and viability of adaptation of PEF to industrial equipment harbor the possibility of combining it with other methods. However, an important aspect is that

PEF parameters should be tailored to each microbial species, considering their structure, size, and other factors affecting efficiency. On the other hand, the recent discovery of the triggering effect of enzymatic activity of cells after electroporation and incubation opens up the possibility of new applications of PEF for the facilitation of extraction of pigments that are bounded or assembled in structures. PEF parameters, along with suspension storage conditions, must be optimized to reach the desired effect. However, this technology also has some drawbacks and limitations for this application. The power required to apply the treatments is highly dependent on the conductivity of the treatment medium. Therefore, when microbial cells are cultivated in media highly conductive, it is not possible to directly treat the microbial biomass, thereby requiring decreasing the value of conductivity by diluting or resuspending microbial cells in another medium.

In addition to the extraction of these pigments already proven, PEF could result useful for other yeasts and bacteria that accumulate pigments of interest with promising therapeutic and colorant properties. Bacterial species of genera *Flavobacterium* and *Paracoccus* accumulate carotenoid zeaxanthin (Manikandan et al. 2016), which is also found in genetically modified *E. coli* (Zhang et al. 2018). Other bacteria, such as the photosynthetic bacterium *Bradyrhizobium* or *Halobacterium*, synthesize canthaxanthin, and the widely used carotenoid astaxanthin is also produced by *Agrobacterium aurantiacum* (Dufossé 2006). The hydrosoluble vitamin riboflavin (vitamin B2) that has a variety of applications as a yellow food colorant is accumulated by yeast such as *Candida guilliermondii* or *Debaryomyces subglobosus* (Dufossé 2006). Anthocyanins are currently obtained from plants, although its microbial production using genetically engineered bacteria is a promising strategy for future therapeutic use (Smeriglio et al. 2016; Zha and Koffas 2017).

Enhancing Expression of Juices by PEF

Extraction by pressure is a common procedure to obtain fruit and vegetable juices. Different equipment for mechanical extraction (screw, belt, hydraulic presses, and filter presses) are conventionally employed at the industrial level. In these processes, the extraction yields are increased by applying a physical (heating, milling), chemical (alkalis), or enzymatic (maceration) pre-treatment that facilitates the leaching of the intracellular components during the pressing step. Nevertheless, an intense pre-treatment may affect the sensorial and nutritional properties of juices, and cause a further release of undesirable components, demanding multistage juice clarification (Vorobiev and Lebovka 2017).

Pulsed electric fields (PEF) have been considered as a valuable alternative to reduce the intensity or even replace conventional pre-treatments for juice extraction. PEF can improve the extraction efficiency by increasing yield, decreasing processing time, and reducing energy requirements in comparison to other pre-treatments, besides to decrease the intensity of subsequent processing steps (Barba et al. 2015).

Since early 2000, PEF-assisted pressing for juice extraction from different tissues including carrots, apples, white grapes, blueberries, and potatoes has been reported (Bazhal and Vorobiev 2000; Lebovka et al. 2004; Praporscic et al. 2007a; Grimi et al. 2011; Jaeger et al. 2012; Lamanauskas et al. 2015). These studies have demonstrated the positive effect of PEF in both quality and yield of juices. PEF application allowed to express a high quality must from three varieties of white grape (Sauvignon, Muscadelle, and Semillon), with an increase in extraction yield from 49–54% to 76–78% after 45 min of constant pressing (5 bars). Optimal energy applied of 20 kJ/kg at electric field strength E = 750 V/cm accelerated selective damage of cell membrane without major damage in the cell wall (Praporscic et al. 2007a). This effect was also verified in other plant tissues: an enhancing of juice expression after PEF application noticeably decreased the absorbance (color) and increased the °Brix value of apple and carrot juice. PEF treatment enhanced the juice yields and regulates the qualitative characteristics of the juice (Praporscic et al. 2007b).

Generally, fruits and vegetables are previously cut into small pieces to increase the pressing efficacy. Grimi et al. (2011) conducted an interesting study about the impact of apple processing modes on extracted juice quality. They reported that PEF as a pre-treatment accelerates the kinetics of apple juice extraction from both whole or sliced apples during pressing. Nevertheless, better intensification of juice yield after PEF treatment of slices compared to the treatment of whole samples was observed. Conversely, whole apples presented improved juice characteristics, less turbidity, and higher antioxidant capacity. While sliced samples resulted in better yield, whole samples resulted in higher quality juice. According to Jaeger et al. (2012), the increase of juice yield after PEF treatment is strongly dependent on the mash structure and de-juicing system. They reported an increase of juice yield after PEF treatment which ranged from 0% to 11% for carrot and 8% to 31% for apple mash. PEF allowed achieving a higher level of disintegration during tissue pressing with yield improvements dependent on the type of press used. Carbonell-Capella et al. (2016) studied the effect of PEF on apple juice quality obtained by pressing at subzero temperatures. The combination of PEF + pressing (5 bars) at a subzero temperature increases the nutritional content of apple juice with high levels of carbohydrates, antioxidants, and bioactive compounds compared with non-PEF-treated samples. Another benefit of PEF was the reduction of both freezing and thawing time during extraction.

Data available to date demonstrated the benefits of applying PEF in the juice extraction from fruit and vegetables. On the other hand, the low energy consumption (typically 1–15 kJ/kg) of PEF as a pre-treatment for plant tissues is substantially low compared with other methods (Vorobiev and Lebovka 2010) such as mechanical (W = 20–40 kJ/kg), enzymatic (W = 60–100 kJ/kg), and heating or freezing/thawing (W > 100 kJ/kg) (Töepfl 2006).

Improvement of Sugar Beet Processing by PEF

According to the worldwide production of sugar from sugar beet in 2016–2017 reached 39.7 million tons, representing 22.5% of global sugar production. The manufacturing process of sugar beet consists of a multistage complex process (slicing, extraction, purification, evaporation, crystallization, and drying) and represents one of the most energy-intensive processes in the food industry (Asadi 2006; Rajaeifar et al. 2019). The extraction is a key unit operation influencing all the subsequent transformation process. Conventionally, beets are firstly sliced (cossettes) and preheated for 10 min at high temperatures (85–95 °C) to disintegrate the cell membranes. Then, the cossettes are heated for up to 90 min between 70 and 75 °C in a countercurrent process where sucrose is extracted by diffusion. This process results in significant thermal degradation of the cell wall and the release of non-desired components such as pectins and proteins. These impurities decrease the quality of extracts and force a complex multistage purification process. Such high extraction temperature also accelerates chemical reactions such as the Maillard reaction that lead to the undesired color changes (Van der Poel 1998; Asadi 2006; Loginova et al. 2011a). The reduction of extraction temperatures to avoid these problems would represent prolonged times to achieve the required yield and would lead to the growth of thermophile microorganisms that consume sugar. During the last decades, research has focused on the development of process improvements to reduce the high energy consumption in order to obtain high-quality extracted juice (Almohammed et al. 2016). In this context, researchers have explored different schemes to improve the beet sugar processing using pulsed electric fields (PEF) such as (i) conventional extraction from PEF-treated tissues at low or moderate temperatures (cold diffusion), (ii) PEF-assisted pressing of tissues without heating (cold pressure), and (iii) combination of pressure and aqueous diffusion.

The efficacy of PEF to electroporate the beet cells and enhance the solid-liquid extraction has been widely demonstrated. PEF treatment allows similar cell disintegration index with low energy consumption (8 kJ/kg) compared to thermal treatment at 80 °C for 10 min (195 kJ/kg). The application of 1 kV/cm for 25 ms remarkably facilitated the release of sugar from sugar beet cells at cold temperatures (El Belghiti and Vorobiev 2004; Lebovka et al. 2007; Maskooki and Eshtiaghi 2012). Conversely to the conventional method, sugar extraction of PEF-treated cossettes can be carried out by aqueous diffusion at moderate (40–50 °C) or even lower temperatures (20–30 °C) with high extraction yields (El-Belghiti et al. 2005; Lebovka et al. 2007; Loginova et al. 2011a). The low extraction temperature reduces the energy input and the release of undesirable components, opening new possibilities for the use of ultrafiltration as a purification methodology. Furthermore, the extracts obtained from cold diffusion of PEF-treated sugar beets have lower quantities of protein, pectin, and color, as well as higher purity than extracts obtained after thermal diffusion (70 °C) (Loginova et al. 2011b).

López et al. (2009) demonstrated the possibility of decreasing the extraction temperature from 70 to 40 °C applying 7 kV/cm for 40μs (3.9 kJ/kg), showing that

80% of sucrose was extracted in 60 min, reducing the thermal energy consumption by more than 50%. The combination of mild heating and PEF shortened the diffusion time by increasing the coefficient of diffusion and enhancing the kinetics of extraction (El-Belghiti et al. 2005; Lebovka et al. 2007). Pilot-scale studies in a countercurrent extractor confirmed the results previously obtained at laboratory scale. PEF treatment of 0.26 kV/cm for 50 ms followed by aqueous extraction at 30 °C reduces the draft (ratio of the mass of diffusion juice and mass of cossettes) from 120% to 100%; therefore the water consumption and energy required for evaporation also decreased. The sugar beet pulp pre-treated by PEF can also be well extracted by low or mild thermal extraction. PEF-assisted cold extraction represents a promising alternative for preparing high-quality sugar beet juices with low concentration in impurities (Loginova et al. 2011a; Loginova et al. 2011b).

PEF-assisted pressing is another approach proposed to reduce the energy demand and improve the quality of the raw juice extracted (Bouzrara and Vorobiev 2000, 2001; Eshtiaghi and Knorr 2002). This methodology eliminates the aqueous sugar diffusion step and replaces it with PEF-assisted pressing. Jemai and Vorobiev (2006) evaluated the use of PEF as an intermediate treatment for cold juice extraction at pilot scale. The cossettes were firstly compressed twice with an intermediated PEF treatment followed by washing and finally compressing. The purity of PEF-treated samples reached 98% and was higher in comparison to untreated ones. On the other hand, the preheating of cossettes at 50 °C after PEF treatment (0.6 kV/cm, 10 ms at 20 °C) produced an additional softening of sugar beet, increasing the juice yield between 5% and 10%; however the purity was reduced to 92.9% (Mhemdi et al. 2012).

Almohammed et al. (2015) proposed a new strategy combining PEF, pressing, and liming to enhance the cold pressing process. The cossettes were firstly electroporated and then subjected to a double pressing with intermediate limed juice soaking (0.6 g CaO/100 g sugar beet). The liming accelerated the extraction kinetics and improved the raw juice quality. This methodology was then optimized to minimize the sucrose losses during the final pressed pulp. The implementation of a double pressing (5–15 bars) of cossettes with intermediate lime milk soaking and a double pressing of limed pulp with intermediate water soakings resulted in a final yield of solutes of 99.53%. The expressed juice resulted in being more concentrated, with higher purity and lower coloration in comparison with the juice obtained by conventional hot water extraction (Almohammed et al. 2016).

Finally, a combination of these innovative methods described above was proposed by Mhemdi, Bals, and Vorobiev (2016). The electroporated cossettes are firstly pressed at 5 bars for 4 min to obtain a juice yield of 50% and then introduced in a countercurrent heat diffuser at 30 or 70 °C for extraction of remaining sucrose in the pressed cake. This methodology allows a draft reduction from 120% to 80% leading to substantial water and energy saving. The pressing-diffusion juice was more concentrated in sucrose (16.2 °Bx vs. 14.6 °Bx), less colored (7000 vs. 10,000), and purer (92.8% vs. 91.8%) than conventional diffusion juice.

All methodologies proposed represent an alternative with considerable advantages, such as energy saving and quality improvement of sugar beet juice extraction.

These benefits have very positive repercussions in the subsequent stages of beet sugar manufacture. The implementation of any alternative to improve sugar beet processing will depend on the installed capacity of the factory and the available equipment.

Improving Tomato Peeling by PEF

Tomato peeling is the first unit operation performed during manufacturing most of the derived tomato products (Das and Barringer 2006). Traditionally, tomato peeling involves some biochemical, thermal, and physic mechanisms that allow loosening the skin and facilitating its removal (Smith et al. 1997). Nowadays, the lye peeling is the most applied method for tomato peeling, performed by the immersion in a highly concentrated sodium hydroxide solution (up to 25 g/L) at high temperatures (60–100 °C). However, this method generates high quantities of waste streams extremely alkaline and polluted (Floros and Chinnan 1990; Rock et al. 2012). As an alternative, the tomato industry has implemented the pressurized steam peeling process. This consists of heating the tomatoes for a few seconds with pressurized steam and then vacuum cooling to prevent overcooking (Rock et al. 2012). This method reduces environmental pollution; however, it results in a loss in product firmness, low yields, and greater water and energy consumption compared to traditional lye peeling (Smith et al. 1997).

PEF has been recently evaluated by several authors as a technological alternative to improving tomato peeling. Arnal et al. (2018) assessed the potential of PEF at the industrial scale (35 t/h) to facilitate the steam peeling process. PEF treatment before the steam peeling with an electric field strength of 0.45 kV/cm and energy inputs of 0.40 kJ/kg allowed a reduction of the hot steam required for the thermophysical peeling stage. The application of PEF also allowed an easier detachment and removal of tomato skin during the steam peeling process. Compared with the conventional steam process, energy consumption and peeling time were reduced by 20%. Pataro et al. (2018) have also addressed the pulsed electric fields as a pre-treatment on whole tomatoes combined with steam blanching to improve the tomato peeling process with less energy consumption.

The improvement in tomato peeling and the reduction of energy consumption were related to the electroporation phenomenon increasing the membrane permeability and also modifying the tissues. The exposure to a moderate electric field (0.2–0.5 kV/cm) promotes the mass transfer of water inside the tomato, thus enhancing the amount of water available under the tomato skin compared with the untreated sample. As more water was vaporized when heating the PEF-treated tomatoes during the steam peeling process, a greater pressure difference across the tomato skin occurs facilitating the formation of more cracks in the tomato peels, which are then easily removed by the pinch roller system (Pataro et al. (2018); Arnal et al. 2018).

The applicability of PEF in the tomato peeling process without the combination with any other techniques has been also investigated. Recently, Andreou et al.

(2020) reported an improvement in the peeling of whole tomatoes applying low electric field strength with a large treatment duration. The application of PEF treatment (1.5 kV/cm and 2000 pulses of 15μs) led to up to 92% decrease in the effort for tomato peel detachment with a decrease of firmness of 21.6% compared with the untreated tomato. The relationship between the cell disintegration index (*Zp*) and the peel tomato detachment seems to indicate that a major cell disintegration index facilitating the tomato peeling. Both the shape and texture of the final tomato peeled treated by PEF were better preserved in comparison with blanched and steam peeled samples since both temperature and treatment duration are substantially lower.

PEF technology can be incorporated easily into an established process line or be taken into account in the design of new process plants. This methodology improves the quality and functionality of tomato compared to the traditional tomato peeling techniques leading to a decrease in energy consumption.

Application of PEF for By-products Revalorization

Food wastes and by-products are produced by a variety of sources through the food chain supply. The disposal of large amounts of food waste or by-products results in significant social, economic, and environmental problems (Baiano 2014; Xu et al. 2018). During the last decades, the scientific community has been interested in revalorizing these materials for their commercialization or incorporation in food and cosmetics products. Most studies have been focused on the extraction of high-valuable compounds, mainly polyphenols present in by-products/waste cellular tissues. Polyphenols are secondary plant metabolites present in a majority of fruits, herbs, and vegetables. These compounds can be classified as flavonoids, phenolic acids, stilbenes, lignans, and tannins (Brglez Mojzer et al. 2016). Many scientific reports point out the vital role of polyphenols in the regulation of metabolism, cell proliferation, weight, and chronic diseases. The studies carried out support its ability to prevent or serve as a treatment against certain cancer types, neurodegenerative diseases, cardiovascular disease, and diabetes type 2 among others. The conventional methods for these bioactive compounds extraction are based on solid-liquid and liquid-liquid solvent extraction. The methods mostly implemented are maceration, Soxhlet extraction, and distillation. However, the main disadvantages of the abovementioned extraction technologies are the prolonged extraction time, low selectivity, decomposition of thermolabile compounds, and high risk of environmental contamination due to the extensive use of non-environmentally friendly solvents (Brglez Mojzer et al. 2016). Despite conventional extraction methods usually reaching high yields, considerable quantities of organic solvents, extended extraction times, and high energy input are needed. These high costs associated are generally not acceptable for waste revalorizing. Based on that alternative methods including PEF are being developed and evaluated for by-products revalorization (Baiano 2014; Poojary et al. 2016).

Some examples of PEF-assisted extraction to recover high-added value compounds from different areas in the food industry are summarized in Table 2. PEF has been applied as a pre-treatment before the extraction processes such as cold pressing, solvent or aqueous extraction at low or moderate temperatures. The permeabilization of the tissues substantially increases the mass transfer processes and the release of the intracellular components from cells. In this sense, the extraction of bioactive compounds from fruit and vegetable peels and seeds from apple, orange, mango, tomato, and olive, among others, has been studied.

Luengo et al. (2013) investigated the influence of PEF treatment on the extraction by pressing of total polyphenols and flavonoids (naringin and hesperidin) from orange peel by-product. They reported that after 30 min of pressurization at 5 bars, the total polyphenol extraction yield (TPEY) increased by 20%, 129%, 153%, and

Table 2 Reports on the application of pulsed electric fields for wastes/by-products revalorization of different areas of the food industry

Food group	Food waste/ by-product source	High-added value compound	Improvements	Treatment conditions	**References**
Fruits	Apple peels	Polyphenols	3.15 mg GAE/100 g dw	1.2 kV/cm; 0.75 s	Wang et al. (2020)
	Blueberry waste	Anthocyanins	223.13 mg/L	18.71 kV/cm; 60μs	Zhou et al. (2015)
	Orange peel	Polyphenols	34.80 mg GAE/100 g fw	7 kV/cm; 60μs + pressing	Luengo et al. (2013)
	Mango peel	Polyphenols	2169 mg/kg dw	Initial 13.3 kV/cm, 20 ms + 3 h SAE	Parniakov et al. (2016)
	Papaya peel	Proteins	20 mg/L	13.3 kV/cm; 20 ms + 3 h SAE	Parniakov et al. (2014)
Vegetables	Tomato peel	Lycopene	14.31 mg/100 g dw	1.0 kV/cm; 7.5 ms	Andreou et al. (2020)
	Potato peel	Steroidal alkaloids	1856.2μg/g dw	0.75 kV/cm; 600μs	Hossain et al. (2015)
	Olive seed	Polyphenols	146 mg GAE/L	Initial 13.3 kV/cm, 3 ms	Roselló-Soto et al. (2015)
	Lemon peel	Hesperidin Eriocitrin	84 mg/100 g fw 176 mg100 g fw	7 kV/cm; 90μs	Peiró et al. (2019)
Winery	Fermented grape pomace	Selective anthocyanins	1.2 g/100 g r.m.	1.2 kV/cm; 100 ms	Brianceau et al. (2015)
Poultry	Chicken breast muscle	Proteins	78 mg mL^{-1}	First step 1 kV; 1.25 ms Second step 0.5 kV; 750 ms	Ghosh et al. (2019)
Fish	Fish gills, bones, and heads	Antioxidant extract	Biological active peptides	1.40 kV/cm; 2 ms	Franco et al. (2020)

SEA, Supplementary aqueous extraction

159% for orange peel PEF treated at 1, 3, 5, and 7 kV/cm, respectively, for 60μs. Likewise, the antioxidant activity of the extract increased 51%, 94%, 148%, and 192% respectively after these treatments, having the main advantage that this procedure reduces extraction time and dispenses with using organic solvents. In the same way, PEF treatment of 7 kV/cm for 900μs increased by 300% the extraction yield from lemon peels (Peiró et al. 2019). PEF treatment also facilitated the aqueous extraction of polyphenols and proteins at moderated temperatures (50 °C) from mango and papaya peels. PEF treatment followed by supplementary aqueous extraction (SAE) increases in 400% the total phenolic compounds from mango peels, while a significant increase of proteins and antioxidant capacities from papaya peels was also reported (Parniakov et al. 2014; Parniakov et al. 2016). The rigid cell structure of lignocellulosic waste and the low moisture content of some vegetal waste, such as seeds, leaf, shoot, and branches, hinder the extraction of compounds; however, PEF-assisted extraction has been successfully evaluated to recover biocompounds from these matrices using higher electric field strengths and pulse duration (above 10 kV/cm) (Roselló-Soto et al. 2015; Yu et al. 2016).

Different studies have also demonstrated that pre-treatment with PEF allows the aqueous extraction of polyphenols from many leaves avoiding the extensive use of organic solvent. Improvements in the aqueous extraction of polyphenols and antioxidants from borage leaves applying PEF as a pre-treatment were carried out by Segovia et al. (2015). The electroporation of borage leaves with 5 kV/cm for 60μs and extraction temperature of 10 °C increased the polyphenol extraction yield by 6.6 times compared to control, while the antioxidant activity was increased by 13.7 times in comparison with non-PEF-treated samples. Bozinou et al. (2019) compared the efficiency of PEF with other technologies (microwave, ultrasound, and boiling-assisted extraction) to extract polyphenols from freeze-dried *Moringa oleifera* leaves using water as solvent. The highest extraction yield of polyphenols (40.24 mg GAE/g of dw) was achieved by PEF with a treatment of 7 kV/cm using 1 pulse of 20 ms every 100μs for 40 min. Unlike the other methodologies evaluated, the temperature during the PEF-assisted extraction process never exceeded 30 °C. PEF technology can also be used to remove the conventional thermal dehydration procedure before polyphenol extraction from tea leaves. It was demonstrated that using PEF as a pre-treatment previous to solid-liquid solvent extraction (50% acetone/water), polyphenols extracted rate was 3.6 times faster than untreated leaves, with lower energy input and without significantly altering the phenolic profile. PEF pre-treatment (1 kV/cm, 10 ms) with subsequent extraction enabled the extraction of 398 mg/L of polyphenols in 2 h, which corresponded to 77% of total polyphenols approximately (Liu et al. 2019).

As described above, extraction studies have been carried out using different kinds of samples including fresh whole leaves (tea leaves), fresh ground leaves in small sizes (borage leaves, onion), lyophilized and milled samples (*Moringa*), and even fine powders (cinnamon) with promising results in all cases. PEF resulted effective for polyphenol extraction in ethanol solution (ethanol 1:10 w/v) from cinnamon powder. As in most cases, extraction is influenced by the electric field intensity and time of treatment (Pashazade et al. 2020). In any case, the different studies showed a partial or even total removal of organic solvents for polyphenol extraction by replacing them with water. The main advantage of using PEF is that similar yields of bioactive compounds are reached compared to traditional extraction; however there is a great significance of lower energy requirements, shorter extraction times, and lower processing temperatures to protect thermolabile compounds. Depending on the case study, all these advantages or at least one of them can be obtained during the PEF-assisted extraction of bioactive compounds from fruit and vegetable by-products.

Valorization of winery waste and by-products for recovery of antioxidants has also been studied (Barba et al. 2016). The combination of densification and PEF treatment at optimal conditions (1.2 kV/cm; 18 kJ/kg) increased by 12.9% the content of total polyphenols from fermented grape pomace compared to non-treated samples. Results also demonstrated the selective nature of PEF treatment in anthocyanin extraction with a reduced solvent amount and extraction time (Brianceau et al. 2015).

The meat industry produces significant amount of wastes with high economic and environmental costs in the food supply chain. The revalorization of the chicken breast muscle was studied by Ghosh et al. (2019). PEF treatment combined with mechanical pressing can be used for protein extraction with a chemical-free process. A two-step-based protocol, which consists of PEF treatment at high voltage and short pulses followed by low voltage and long pulses before pressing, allowed a protein extraction of 78 mg/mL.

Pulsed electric fields have proven to be a real tool to help for adequate revalorization of agro-industrial wastes. In any case, any improvement during the extraction process in yields will be depending on the matrix and the PEF processing conditions tested.

Conclusions

As has been demonstrated in the studies presented throughout this chapter, PEF is a versatile technology that can be implemented to improve the traditional processes, or it can even be considered for the design of new and innovative industrial processes. The main advantages of PEF against traditional processes have been evidenced along the different studies presented. The electroporation phenomenon results in the improvement of mass transfer and, therefore, in the extraction yields of intracellular compounds. Other benefits such as the reduction of the extraction time and temperature, the reduction of the amount of organic solvents used in the extraction processes, and the smaller presence of cell debris in the extraction medium are also attributed to the use of PEF. On the other hand, the number of applications of this technology is increasing rapidly. It has been reported PEF allows the selective extraction of different intracellular compounds, reducing the subsequent purification steps. More recently, it has been demonstrated that beyond mass transfer improvement, the application of pulsed electric fields triggers and accelerates some enzymatic reactions closely related to the extraction of intracellular components of interest to the food and beverage industry from microorganisms. In conclusion, PEF is a technology that can be relatively easily implemented, has a low energy consumption and minimal environmental impact, and thus has become a very attractive alternative for processes improvement into the food and biotechnological industries.

References

Ade-Omowaye BIO, Angersbach A, Taiwo KA, Knorr D (2001) The use of pulsed electric fields in producing juice from paprika (Capsicum annuum L.). J Food Process Pres 25(5):353–365

Aguilar-Machado D, Delso C, Martínez JM et al (2020) Enzymatic processes triggered by PEF for astaxanthin extraction from Xanthophyllomyces dendrorhous. Front Bioeng Biotechnol 8:857

Almohammed F, Mhemdi H, Grimi N, Vorobiev E (2015) Alkaline pressing of electroporated sugar beet tissue: process behavior and qualitative characteristics of raw juice. Food Bioproc Tech 8(9):1947–1957

Almohammed F, Mhemdi H, Vorobiev E (2016) Several-staged alkaline pressing-soaking of electroporated sugar beet slices for minimization of sucrose loss. Innov Food Sci Emerg 36:18–25

Andreou V, Dimopoulos G, Dermesonlouoglou E, Taoukis P (2020) Application of pulsed electric fields to improve product yield and waste valorization in industrial tomato processing. J Food Eng 270:109778

Arnal ÁJ, Royo P, Pataro G et al (2018) Implementation of PEF treatment at real-scale tomatoes processing considering LCA methodology as an innovation strategy in the agri-food sector. Sustainability 10(4):979

Asadi M (2006) Beet-sugar handbook. John Wiley & Sons., Inc., Publication, Hoboken

Baiano A (2014) Recovery of biomolecules from food wastes – a review. Molecules 19(9):14821–14842

Barba FJ, Parniakov O, Pereira SA et al (2015) Current applications and new opportunities for the use of pulsed electric fields in food science and industry. Food Res Int 77:773–798

Barba FJ, Zhu Z, Koubaa M et al (2016) Green alternative methods for the extraction of antioxidant bioactive compounds from winery wastes and by-products: a review. Trends Food Sci Technol 49:96–109

Bazhal M, Vorobiev E (2000) Electrical treatment of apple cossettes for intensifying juice pressing. J Sci Food Agric 80(11):1668–1674

Becker W (2004) 18 microalgae in human and animal nutrition. In: Handbook of microalgal culture: biotechnology and applied phycology, vol 312. Wiley Online Library, p 93

Bouzrara H, Vorobiev E (2000) Beet juice extraction by pressing and pulsed electric fields. Int Sugar J 102(1216):194–200

Bouzrara H, Vorobiev E (2001) Non-thermal pressing and washing of fresh sugarbeet cossettes combined with a pulsed electrical field. Zuckerindustrie 126(6):463–466

Bozinou E, Karageorgou I, Batra G et al (2019) Pulsed electric field extraction and antioxidant activity determination of Moringa oleifera dry leaves: a comparative study with other extraction techniques. Beverages 5(1):8

Brglez Mojzer E, Knez Hrnčič M, Škerget M et al (2016) Polyphenols: extraction methods, antioxidative action, bioavailability and anticarcinogenic effects. Molecules 21(7):901

Brianceau S, Turk M, Vitrac X, Vorobiev E (2015) Combined densification and pulsed electric field treatment for selective polyphenols recovery from fermented grape pomace. Innov Food Sci Emerg Technol 29:2–8

Buzzini P, Martini A (2000) Production of carotenoids by strains of Rhodotorula glutinis cultured in raw materials of agro-industrial origin. Bioresour Technol 71(1):41–44
Carbonell-Capella JM, Parniakov O, Barba FJ et al (2016) "Ice" juice from apples obtained by pressing at subzero temperatures of apples pretreated by pulsed electric fields. Innov Food Sci Emerg 33:187–194
Carocho M, Barreiro MF, Morales P, Ferreira IC (2014) Adding molecules to food, pros and cons: a review on synthetic and natural food additives. Compr Rev Food Sci Food Saf 13(4):377–399
Carpentier S, Knaus M, Suh M (2009) Associations between lutein, zeaxanthin, and age-related macular degeneration: an overview. Crit Rev Food Sci 49(4):313–326
Carullo D, Abera BD, Casazza AA et al (2018) Effect of pulsed electric fields and high pressure homogenization on the aqueous extraction of intracellular compounds from the microalgae Chlorella vulgaris. Algal Res 31:60–69
Chemat F, Rombaut N, Sicaire AG et al (2017) Ultrasound assisted extraction of food and natural products. Mechanisms, techniques, combinations, protocols and applications. A review. Ultrason Sonochem 34:540–560
Cooperstone JL, Schwartz SJ (2016) Recent insights into health benefits of carotenoids. In: Handbook on natural pigments in food and beverages. Woodhead Publishing, pp 473–497
Cortez R, Luna-Vital DA, Margulis D, González de Mejía E (2017) Natural pigments: stabilization methods of anthocyanins for food applications. Compr Rev Food Sci Food Saf 16(1):180–198
da Silva FV, Sant'Anna C (2017) Impact of culture conditions on the chlorophyll content of microalgae for biotechnological applications. World J Microbiol Biotechnol 33(1):20
Das DJ, Barringer SA (2006) Potassium hydroxide replacement for lye (sodium hydroxide) in tomato peeling. J Food Process Pres 30(1):15–19
de Boer K, Moheimani NR, Borowitzka MA, Bahri PA (2012) Extraction and conversion pathways for microalgae to biodiesel: a review focused on energy consumption. J Appl Phycol 24(6):1681–1698
Du C, Li Y, Guo Y et al (2016) Correction: Torularhodin, isolated from Sporidiobolus pararoseus, inhibits human prostate cancer LNCaP and PC-3 cell growth through Bcl-2/Bax mediated apoptosis and AR down-regulation. RSC Adv 6(13):10249–10249
Dufossé L (2006) Microbial production of food grade pigments. Food Technol Biotech 44(3):313–323
El-Belghiti K, Vorobiev E (2004) Mass transfer of sugar from beets enhanced by pulsed electric field. Food Bioprod Process 82(3):226–230
El-Belghiti K, Rabhi Z, Vorobiev E (2005) Kinetic model of sugar diffusion from sugar beet tissue treated by pulsed electric field. J Sci Food Agric 85(2):213–218
Eshtiaghi MN, Knorr D (2002) High electric field pulse pretreatment: potential for sugar beet processing. J Food Eng 52(3):265–272
Fakhri S, Abbaszadeh F, Dargahi L, Jorjani M (2018) Astaxanthin: a mechanistic review on its biological activities and health benefits. Pharmacol Res 136:1–20
Floros JD, Chinnan MS (1990) Diffusion phenomena during chemical (NaOH) peeling of tomatoes. J Food Sci 55(2):552–553
Franco D, Munekata PE, Agregán R et al (2020) Application of pulsed electric fields for obtaining antioxidant extracts from fish residues. Antioxidants 9(2):90
Gachovska T, Cassada D, Subbiah J et al (2010) Enhanced anthocyanin extraction from red cabbage using pulsed electric field processing. J Food Sci 75(6):E323–E329
Ghosh S, Gillis A, Sheviryov J et al (2019) Towards waste meat biorefinery: extraction of proteins from waste chicken meat with non-thermal pulsed electric fields and mechanical pressing. J Clean Prod 208:220–231
Grimi N, Lebovka NI, Vorobiev E, Vaxelaire J (2009) Effect of a pulsed electric field treatment on expression behavior and juice quality of chardonnay grape. Food Biophys 4(3):191–198
Grimi N, Mamouni F, Lebovka N et al (2011) Impact of apple processing modes on extracted juice quality: pressing assisted by pulsed electric fields. J Food Eng 103(1):52–61

Henriques M, Silva A, Rocha J (2007) Extraction and quantification of pigments from a marine microalga: a simple and reproducible method. In: Communicating current research and educational topics and trends in applied microbiology, vol 2. FORMATEX, Badajoz, pp 586–593
Institute of Food Technologists (2016) Coloring foods & beverages. Available from: https://www.ift.org/news-and-publications/food-technology-magazine/issues/2016/january/columns/food-safety-quality-coloring-foods-and-beverages Accessed 2020 August 1
Jaeger H, Schulz M, Lu P, Knorr D (2012) Adjustment of milling, mash electroporation and pressing for the development of a PEF assisted juice production in industrial scale. Innov Food Sci Emerg 14:46–60
Jaeschke DP, Mercali GD, Marczak LDF et al (2019) Extraction of valuable compounds from Arthrospira platensis using pulsed electric field treatment. Bioresour Technol 283:207–212
Jaime L, Rodríguez-Meizoso I, Cifuente A et al (2010) Pressurized liquids as an alternative process to antioxidant carotenoids' extraction from Haematococcus pluvialis microalgae. LWT-Food Sci Techno 43(1):105–112
Jemai AB, Vorobiev E (2006) Pulsed electric field assisted pressing of sugar beet slices: towards a novel process of cold juice extraction. Biosyst Eng 93(1):57–68
Johnson EA (2003) Phaffia rhodozyma: colorful odyssey. Int Microbiol 6(3):169–174
Kempkes MA (2016) Pulsed electric fields for algal extraction and predator control. In: Handbook of electroporation. Springer International Publishing, pp 1–16
Lam MK, Lee KT (2015) Bioethanol production from microalgae. In: Handbook of marine microalgae. Academic Press, pp 197–208
Lamanauskas N, Bobinaitė R, Šatkauskas et al (2015) Pulsed electric field-assisted juice extraction of frozen/thawed blueberries. Zemdirbyste 102(1)
Lebovka NI, Praporscic I, Vorobiev E (2004) Effect of moderate thermal and pulsed electric field treatments on textural properties of carrots, potatoes and apples. Innov Food Sci Emerg 5(1):9–16
Lebovka NI, Shynkaryk MV, El-Belghiti K et al (2007) Plasmolysis of sugarbeet: pulsed electric fields and thermal treatment. J Food Eng 80(2):639–644
Leong HY, Show PL, Lim MH et al (2018) Natural red pigments from plants and their health benefits: a review. Food Rev Int 34(5):463–482
Li X, Xu J, Tang X et al (2016) Anthocyanins inhibit trastuzumab-resistant breast cancer in vitro and in vivo. Mol Med Rep 13(5):400–4013
Liu ZW, Zeng XA, Ngadi M (2018) Enhanced extraction of phenolic compounds from onion by pulsed electric field (PEF). J Food Process Pres 42(9):e13755
Liu Z, Esveld E, Vincken JP, Bruins ME (2019) Pulsed electric field as an alternative pre-treatment for drying to enhance polyphenol extraction from fresh tea leaves. Food Bioproc Tech 12(1):183–192
Loginova KV, Vorobiev E, Bals O, Lebovka NI (2011a) Pilot study of countercurrent cold and mild heat extraction of sugar from sugar beets, assisted by pulsed electric fields. J Food Eng 102(4):340–347
Loginova K, Loginov M, Vorobiev E, Lebovka NI (2011b) Quality and filtration characteristics of sugar beet juice obtained by "cold" extraction assisted by pulsed electric field. J Food Eng 106(2):144–151
López N, Puértolas E, Condón S, Raso J, Álvarez I (2009) Enhancement of the extraction of betanine from red beetroot by pulsed electric fields. J Food Eng 90(1):60–66
Luengo E, Álvarez I, Raso J (2013) Improving the pressing extraction of polyphenols of orange peel by pulsed electric fields. Innov Food sci Emerg 17:79–84
Luengo E, Álvarez I, Raso J (2014a) Improving carotenoid extraction from tomato waste by pulsed electric fields. Front Nutr 1:12
Luengo E, Condón-Abanto S, Álvarez I, Raso J (2014b) Effect of pulsed electric field treatments on permeabilization and extraction of pigments from Chlorella vulgaris. J Membr Biol 247(12):1269–1277

Luengo E, Martínez JM, Coustets M et al (2015a) A comparative study on the effects of millisecond- and microsecond-pulsed electric field treatments on the permeabilization and extraction of pigments from Chlorella vulgaris. J Membr Biol 248(5):883–891

Luengo E, Martínez JM, Bordetas A et al (2015b) Influence of the treatment medium temperature on lutein extraction assisted by pulsed electric fields from Chlorella vulgaris. Innov Food Sci Emerg 29:15–22

Luengo E, Martínez JM, Álvarez I, Raso J (2016) Effects of millisecond and microsecond pulsed electric fields on red beet cell disintegration and extraction of betanines. Ind Crop Prod 84:28–33

Macías-Sánchez MD, Serrano CM, Rodríguez MR, de la Ossa EM (2009) Kinetics of the supercritical fluid extraction of carotenoids from microalgae with CO2 and ethanol as cosolvent. Chem Eng J 150(1):104–113

Manikandan R, Thiagarajan R, Goutham G et al (2016) Zeaxanthin and ocular health, from bench to bedside. Fitoterapia 109:58–66

Martínez JM, Cebrián G, Álvarez I, Raso J (2016) Release of mannoproteins during Saccharomyces cerevisiae autolysis induced by pulsed electric field. Front Microbiol 7:1435

Martínez JM, Luengo E, Saldaña G et al (2017) C-phycocyanin extraction assisted by pulsed electric field from Artrosphira platensis. Food Res Int 99:1042–1047

Martínez JM, Delso C, Angulo J, Álvarez I, Raso J (2018) Pulsed electric field-assisted extraction of carotenoids from fresh biomass of Rhodotorula glutinis. Innov Food Sci Emerg 47:421–427

Martínez JM, Delso C, Álvarez I, Raso J (2019a) Pulsed electric field permeabilization and extraction of phycoerythrin from Porphyridium cruentum. Algal Res 37:51–56

Martínez JM, Gojkovic Z, Ferro L et al (2019b) Use of pulsed electric field permeabilization to extract astaxanthin from the Nordic microalga Haematococcus pluvialis. Bioresour Technol 289:121694

Martínez JM, Schottroff F, Haas K et al (2020a) Evaluation of pulsed electric fields technology for the improvement of subsequent carotenoid extraction from dried Rhodotorula glutinis yeast. Food Chem 323:126824

Martínez JM, Delso C, Álvarez I, Raso J (2020b) Pulsed electric field-assisted extraction of valuable compounds from microorganisms. Compr Rev Food Sci Food Saf 19(2):530–552

Maskooki A, Eshtiaghi MN (2012) Impact of pulsed electric field on cell disintegration and mass transfer in sugar beet. Food Bioprod Process 90(3):377–384

McMillan JR, Watson IA, Ali M, Jaafar W (2013) Evaluation and comparison of algal cell disruption methods: microwave, waterbath, blender, ultrasonic and laser treatment. Appl Energy 103:128–134

Mhemdi H, Bals O, Grimi N, Vorobiev E (2012) Filtration diffusivity and expression behaviour of thermally and electrically pretreated sugar beet tissue and press-cake. Sep Purif Technol 95:118–125

Mhemdi H, Bals O, Vorobiev E (2016) Combined pressing-diffusion technology for sugar beets pretreated by pulsed electric field. J Food Eng 168:166–172

Middelberg AP (1995) Process-scale disruption of microorganisms. Biotechnol Adv 13(3):49–551

Moliné M, Libkind D, van Broock M (2012) Production of torularhodin, torulene, and β-carotene by Rhodotorula yeasts. In: Microbial carotenoids from fungi. Humana Press, Totowa, pp 275–283

Parniakov O, Barba FJ, Grimi N et al (2014) Impact of pulsed electric fields and high voltage electrical discharges on extraction of high-added value compounds from papaya peels. Food Res Int 65:337–343

Parniakov O, Barba FJ, Grimi N et al (2015) Pulsed electric field assisted extraction of nutritionally valuable compounds from microalgae Nannochloropsis spp. using the binary mixture of organic solvents and water. Innov Food Sci Emerg 27:79–85

Parniakov O, Barba FJ, Grimi N et al (2016) Extraction assisted by pulsed electric energy as a potential tool for green and sustainable recovery of nutritionally valuable compounds from mango peels. Food Chem 192:842–848

Pashazadeh B, Elhamirad AH, Hajnajari H et al (2020) Optimization of the pulsed electric field-assisted extraction of functional compounds from cinnamon. Biocatal Agric Biotechnol:101461
Pasquet V, Chérouvrier JR, Farhat F et al (2011) Study on the microalgal pigments extraction process: performance of microwave assisted extraction. Process Biochem 46(1):59–67
Pataro G, Carullo D, Siddique MA et al (2018) Improved extractability of carotenoids from tomato peels as side benefits of PEF treatment of tomato fruit for more energy-efficient steam-assisted peeling. J Food Eng 233:65–73
Peiró S, Luengo E, Segovia F et al (2019) Improving polyphenol extraction from lemon residues by pulsed electric fields. Waste Biomass Valori 10(4):889–897
Plaza M, Santoyo S, Jaime L et al (2012) Comprehensive characterization of the functional activities of pressurized liquid and ultrasound-assisted extracts from Chlorella vulgaris. LWT-Food Sci Techno 46(1):245–253
Poojary M, Roohinejad S, Barba F et al (2016) Application of pulsed electric field treatment for food waste recovery operations. In: Handbook of electroporation. Springer International Publishing, pp 2573–2590
Praporscic I, Lebovka N, Vorobiev E, Mietton-Peuchot M (2007a) Pulsed electric field enhanced expression and juice quality of white grapes. Sep Purif Technol 52(3):520–526
Praporscic I, Shynkaryk MV, Lebovka NI, Vorobiev E (2007b) Analysis of juice colour and dry matter content during pulsed electric field enhanced expression of soft plant tissues. J Food Eng 79(2):662–670
Puértolas E, Barba FJ (2016) Electrotechnologies applied to valorization of by-products from food industry: main findings, energy and economic cost of their industrialization. Food Bioprod Process 100:172–184
Puértolas E, Cregenzán O, Luengo E, Álvarez I, Raso J (2013) Pulsed-electric-field-assisted extraction of anthocyanins from purple-fleshed potato. Food Chem 136(3–4):1330–1336
Qiu BS, Zhang AH, Liu ZL, Gao KS (2004) Studies on the photosynthesis of the terrestrial cyanobacterium Nostoc flagelliforme subjected to desiccation and subsequent rehydration. Phycologia 43(5):521–528
Rajaeifar MA, Hemayati SS, Tabatabaei M et al (2019) A review on beet sugar industry with a focus on implementation of waste-to-energy strategy for power supply. Renew Sustain Energy Rev 103:423–442
Rock C, Yang W, Goodrich-Schneider R, Feng H (2012) Conventional and alternative methods for tomato peeling. Food Eng Rev 4(1):1–15
Rodriguez-Amaya DB (2016) Natural food pigments and colorants. Curr Opin Food Sci 7:20–26
Román RB, Alvarez-Pez JM, Fernández FA, Grima EM (2002) Recovery of pure B-phycoerythrin from the microalga Porphyridium cruentum. J Biotechnol 93(1):73–85
Roohinejad S, Everett DW, Oey I (2014) Effect of pulsed electric field processing on carotenoid extractability of carrot purée. Int J Food Sci Technol 49(9):2120–2127
Roselló-Soto E, Barba FJ, Parniakov O et al (2015) High voltage electrical discharges, pulsed electric field, and ultrasound assisted extraction of protein and phenolic compounds from olive kernel. Food Bioproc Tech 8(4):885–894
Saini RK, Keum YS (2018) Carotenoid extraction methods: a review of recent developments. Food Chem 240:90–103
Saulis G (2010) Electroporation of cell membranes: the fundamental effects of pulsed electric fields in food processing. Food Eng Rev 2(2):52–73
Saulis G, Saulė R (2012) Size of the pores created by an electric pulse: microsecond vs millisecond pulses. BBA-Biomembranes 1818(12):3032–3039
Scherer D, Krust D, Frey W et al (2019) Pulsed electric field (PEF)-assisted protein recovery from Chlorella vulgaris is mediated by an enzymatic process after cell death. Algal Res 41:101536
Schmidt I, Schewe H, Gassel S et al (2011) Biotechnological production of astaxanthin with Phaffia rhodozyma/Xanthophyllomyces dendrorhous. Appl Microbiol Biotechnol 89(3):555–571

Segovia FJ, Luengo E, Corral-Pérez JJ et al (2015) Improvements in the aqueous extraction of polyphenols from borage (Borago officinalis L.) leaves by pulsed electric fields: pulsed electric fields (PEF) applications. Ind Crop Prod 65:390–396

Silve A, Brunet AG, Al-Sakere B et al (2014) Comparison of the effects of the repetition rate between microsecond and nanosecond pulses: Electropermeabilization-induced electro-desensitization? BBA-Gen Subjects 1840(7):2139–2151

Skorupskaite V, Makareviciene V, Ubartas M et al (2017) Green algae Ankistrodesmus fusiformis cell disruption using different modes. Biomass Bioenergy 107:311–316

Smeriglio A, Barreca D, Bellocco E, Trombetta D (2016) Chemistry, pharmacology and health benefits of anthocyanins. Phytother Res 30(8):1265–1286

Smith DS, Cash JN, Nip WK, Hui YH (eds) (1997) Processing vegetables: science and technology. CRC Press

Töepfl S (2006) Pulsed electric fields for permeabilization of cell membranes in food- and bioprocessing-applications, process and equipment design and cost analysis. PhD Thesis, University of Berlin, Germany

Van der Poel P, Schiweck H, Schwartz T (1998) Sugar technology beet and cane sugar manufacture. Verlag Dr. Bartens KG, Berlin, pp 479–563

Viskari PJ, Colyer CL (2003) Rapid extraction of phycobiliproteins from cultured cyanobacteria samples. Anal Biochem 319(2):263–271

Vorobiev E, Lebovka N (2010) Extraction from solid foods and biosuspensions enhanced by electrical pulsed energy (pulsed electric field, pulsed ohmic heating and high voltage electrical discharges). Food Eng Rev 2(2):95–108

Vorobiev E, Lebovka N (2017) Selective extraction of molecules from biomaterials by pulsed electric field treatment. In Handbook of electroporation, p 1–16

Walther C, Kellner M, Berkemeyer M et al (2017) A microscale bacterial cell disruption technique as first step for automated and miniaturized process development. Process Biochem 59:207–215

Xu F, Li Y, Ge X et al (2018) Anaerobic digestion of food waste–challenges and opportunities. Bioresour Technol 247:1047–1058

Yu X, Gouyo T, Grimi N et al (2016) Pulsed electric field pretreatment of rapeseed green biomass (stems) to enhance pressing and extractives recovery. Bioresour Technol 199:194–201

Zha J, Koffas MA (2017) Production of anthocyanins in metabolically engineered microorganisms: current status and perspectives. Synth Syst Biotechnol 2(4):259–266

Zhang Y, Liu Z, Sun J et al (2018) Biotechnological production of zeaxanthin by microorganisms. Trends Food Sci Technol 71:225–234

Zhou Y, Zhao X, Huang H (2015) Effects of pulsed electric fields on anthocyanin extraction yield of blueberry processing by-products. J Food Process Pres 39(6):1898–1904

Zoz L, Carvalho JC, Soccol VT et al (2015) Torularhodin and torulene: bioproduction, properties and prospective applications in food and cosmetics – a review. Braz Arch Biol Technol 58(2):278–288

Part IV
Pulsed Electric Fields Processing in the Food Industry

Pulsed Electric Field Process Performance Analysis

Henry Jaeger, Thomas Fauster, and Felix Schottroff

Introduction

Pulsed electric fields (PEF) applications in industrial scale are well established for the pre-treatment of potatoes in French fries and snack production as well as for fruit juice preservation. Further applications have been studied, and potential processing concepts have been developed but not yet transferred into industrial application. Certain requirements have to be met in order to successfully translate electroporation phenomena into improved product quality, process efficiency, or innovative processing concepts. In this regard, considerations regarding the viability of existing and new technical system have to be made. This viability depends on three fundamental categories of attributes: functionality, performance, and efficiency (Harvey 1986). Integrating a process performance analysis into these considerations may assure an efficient system design and may allow a more accurate

H. Jaeger (✉)
Institute of Food Technology, University of Natural Resources and Life Sciences (BOKU) Vienna, Vienna, Austria
e-mail: henry.jaeger@boku.ac.at

T. Fauster
Institute of Food Technology, University of Natural Resources and Life Sciences (BOKU) Vienna, Vienna, Austria

FFoQSI GmbH – Austrian Competence Centre for Feed and Food Quality, Safety and Innovation, Tulln, Austria
e-mail: thomas.fauster@boku.ac.at

F. Schottroff
Institute of Food Technology, University of Natural Resources and Life Sciences (BOKU) Vienna, Vienna, Austria

Core Facility Food & Bio Processing, University of Natural Resources and Life Sciences (BOKU) Vienna, Vienna, Austria
e-mail: felix.schottroff@boku.ac.at

J. Raso et al. (eds.), *Pulsed Electric Fields Technology for the Food Industry*, Food Engineering Series, https://doi.org/10.1007/978-3-030-70586-2_16

description of the systems behavior (Oakland 2008). Process performance in the meaning used in the following section is understood as the capability of a process to meet previously set requirements or to fulfill defined process goals. The conceptual idea is used as a tool for the systematic process analysis in order to reveal discrepancies between the process goal and the actual process outcome.

Taking into account the current applications of PEF in the food industry, the following key processing goals can be identified:

- Nonthermal inactivation of microorganisms resulting in a higher level of product quality retention
- Cell disintegration of plant tissue resulting in desired texture modification

From a process performance point of view, the following questions need to be answered in order to evaluate the capability of the PEF application to fulfill the abovementioned processing goals:

- What is the level of thermal and nonthermal microbial inactivation and what are other shelf life-limiting factors?
- Can the particularities resulting from PEF cell disintegration be converted into beneficial structural changes?

Answering these questions is challenging for the following reasons:

- A suitable experimental setup needs to be chosen, and experimental guidelines need to be fulfilled in order to gain data at standardized and comparable conditions.
- Suitable reference processes need to be selected in order to compare results to optimized processes established in the industry rather than comparing results to untreated control samples.
- The relevance and impact of results need to be quantified using data from industrial case studies.

In order to systematically evaluate the aforementioned aspects, a performance analysis should consider the multi-scale nature of a PEF process. Figuratively, phenomena are allocated to dimensions defined as microscale, mesoscale, and macroscale. For the aspects covered in the present chapter, Fig. 1 illustrates the different levels that are suggested for analysis. The phenomena can be further clustered into process (PRO) related, structure (S) related, or property (PRO) related as suggested by Windhab (2008) in the S-PRO2 scheme.

Table1 gives some examples for the different scale levels related to PEF processing of plant-based materials. Phenomena taking place on a cellular level such as the inactivation of microorganisms or permeabilization of plant cells as well as protective effects or the modification of food compounds are considered as the microscale. The mechanism of electroporation can also be described in the microscale, where a process-induced accumulation of charges at the cell membrane takes place leading to a property change by increasing the transmembrane potential. The result is a structure modification, namely, the pore formation in the membrane called electroporation (Weaver and Chizmadzhev 1996).

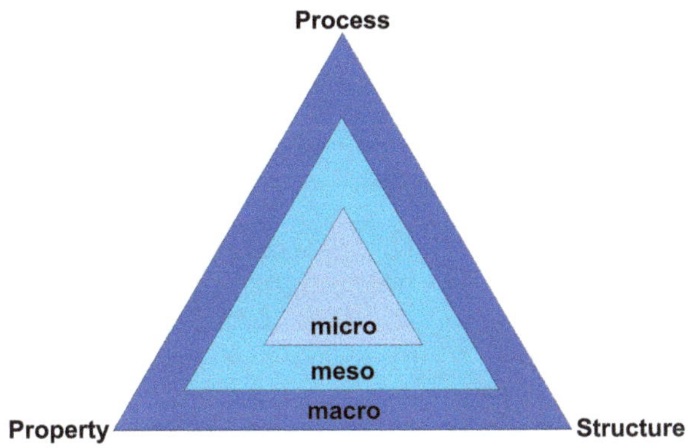

Fig. 1 Matrix including the hierarchic structure scheme for the multi-scale nature of PEF processing. For examples of phenomena related to the process, the structure, and the property of vegetables see table 1

Table 1 Effect of PEF on purge loss of meat

	Property	Process	Structure
micro	Transmembrane-potential	Charge accumulation by external electric field	Membrane pore
meso	Tissue softening	Cutting	Cutting surface
macro	Interaction → capability	Process design and integration	→ Physical integration

Phenomena related to the PEF treatment chamber and its performance such as electric field, flow velocity, or temperature distribution are considered as the mesoscale. The mesoscale is also applied to mass transfer processes taking place in plant raw materials such as the juice recovery from a fruit mash, considering mash structure and particle size distribution still as mesoscale parameters. Structural changes in the microscale such as pore formation lead to property changes in the mesoscale, e.g., tissue softening. The relevance of this change in property is related to cutting processes, where the structure of the cutting surface will be affected depending on tissue properties.

The macroscale, as defined in the present context, involves design and integration aspects of the PEF process considering a whole processing concept as well as connected processing steps. Hence, the definition of property and structure is now more connected to the processing line rather than to the product itself. Structure may be related to the physical implementation of the PEF system in an existing processing line, whereas the property may describe the capability of the different processing steps to interact and to be synchronized.

In addition to the different scales related to the size, the concept also applies to the timescale, which is another important aspect in dynamic processes. Whereas electroporation takes place in the microsecond range and may be temporary or permanent, affected processes such as extraction or drying have much higher time requirements. Hence, categories such as ultrashort time, short time, and long time can be defined in analogy to the micro-, meso-, and macroscale.

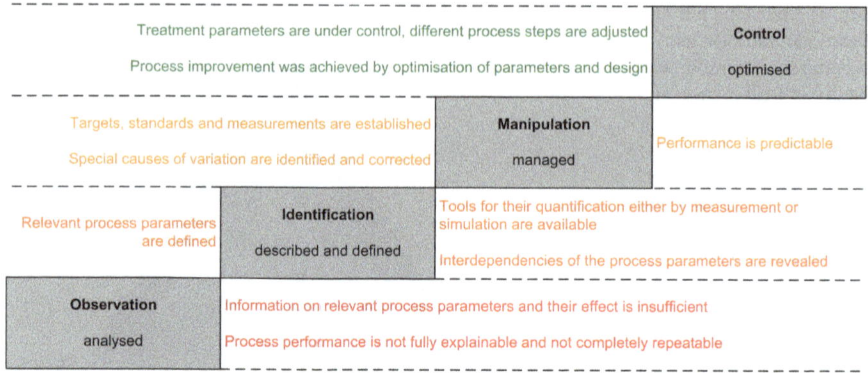

Fig. 2 Systematic approach suggested for a PEF process analysis and optimization. The model can be applied to different scales and levels of complexity as shown by the examples presented in this chapter

Regardless of the position of a specific aspect within the matrix shown in Fig. 1, a general approach is required and was used in the presented studies in order to perform a stepwise analysis of the underlying interactions. The procedure is illustrated in Fig. 2.

An optimized process as a basis for the successful implementation of new application concepts based on PEF into industrial scale needs to offer a maximum of process control with defined process standards as well as a set of measurements and key performance indicators in order to quantify relevant benefits.

Role of Experimental Setup

In order to evaluate the process performance of a PEF application, to enable comparability of results, and to conclude on benefits, it is necessary to use standardized protocols, treatment variables, and experimental setups. For this purpose, comprehensive guidelines should be considered (Raso et al. 2016). For the development of suitable application concepts and the optimization of existing PEF applications, it is essential to consider the generation of pulsed electric fields, their impact on biological cells, and the resulting benefit for food industry applications:

1. Particularities resulting from the different ways of the generation and technical application of PEF will determine the electric field characteristics including aspects such as the pulse shape or pulse duration as well as the distribution of the electric field in a treatment chamber. Exactly defined electric field conditions are a prerequisite for the subsequent evaluation of the electric field impact on the biological cell.
2. Direct and indirect methods are available in order to analyze the electric field effects on a cellular and tissue level. The degree of cell permeabilization needs to be quantified but also evaluated from a qualitative point of view (e.g., reversible or irreversible permeabilization).

3. As a third step, an analysis is required in order to evaluate the extent to which these basic effects can be converted into tangible process outcomes leading to beneficial applications. For example, a specific level of cell membrane permeabilization of fruit mashes determined in step two does not necessarily result in an increased juice yield since other impact factors such as mash structure or the solid-liquid separation method will affect the beneficial outcome of the cell disintegration. The same is true for the microbial inactivation by PEF. Pore formation detected in step two does not guarantee sufficient inactivation or might be related to treatment intensities that still compromise product quality.

Hence, it is essential to take all three steps into account and to consider the underlying basic principles in order to develop a comprehensive understanding of complex process applications. Please refer to previous chapters in this book in order to get detailed insights into the generation of pulsed electric fields and their impact on biological cells as well as examples for application concepts.

A particular point to be mentioned in this context is the interdependencies of the pulse generation system, the treatment chamber design, and the treatment medium properties. These interdependencies have a major impact on the performance of a PEF treatment and affect various process parameters as shown in Fig. 3.

The displayed relations (Fig. 3) of the different parameters illustrate the complex interdependencies. Taking into account the need of establishing a standard experimental procedure for the performance of PEF treatments in lab, pilot, and industrial scale, these parameters are essential in order to provide a detailed insight in PEF-related processing criteria. It is obvious that it is not possible to change one single parameter without affecting related treatment conditions. Hence, a systematic study of the impact of one single parameter by its variation while keeping all other parameters at a fixed level is impossible. Therefore, as a first step in the experimental design of the PEF treatment, it is necessary to choose the most suitable key parameter to describe the treatment intensity from parameters such as the electric field strength, the electric field exposure time, or the total specific energy input. As a second step, using the schematic shown in Fig. 3, it will be possible to evaluate the interdependencies of these with other parameters. These interdependencies need to be taken into account when interpreting obtained process results.

In addition, upscaling of experimental setups from lab to pilot or industrial scale is challenging. Even when using the same treatment chamber configuration, deviations in the dimension will have a major impact on electrical resistance and electric field strength distribution as well as on the parameters indirectly linked to the design such as the residence and treatment time or the total specific energy input (Jin et al. 2015). In this case, indirect measures are required in order to achieve comparable treatment intensities. A concept could entail the definition of a similar level of cell disintegration or microbial inactivation to be achieved by the different treatment systems. Based on such a comparable process result, the process parameters can be analyzed and compared in terms of the required treatment intensity or even in terms of the required energy consumption. A major research need can be defined regarding this aspect, since up to now, no suitable concepts for the process evaluation including indicators for a process performance analysis are available.

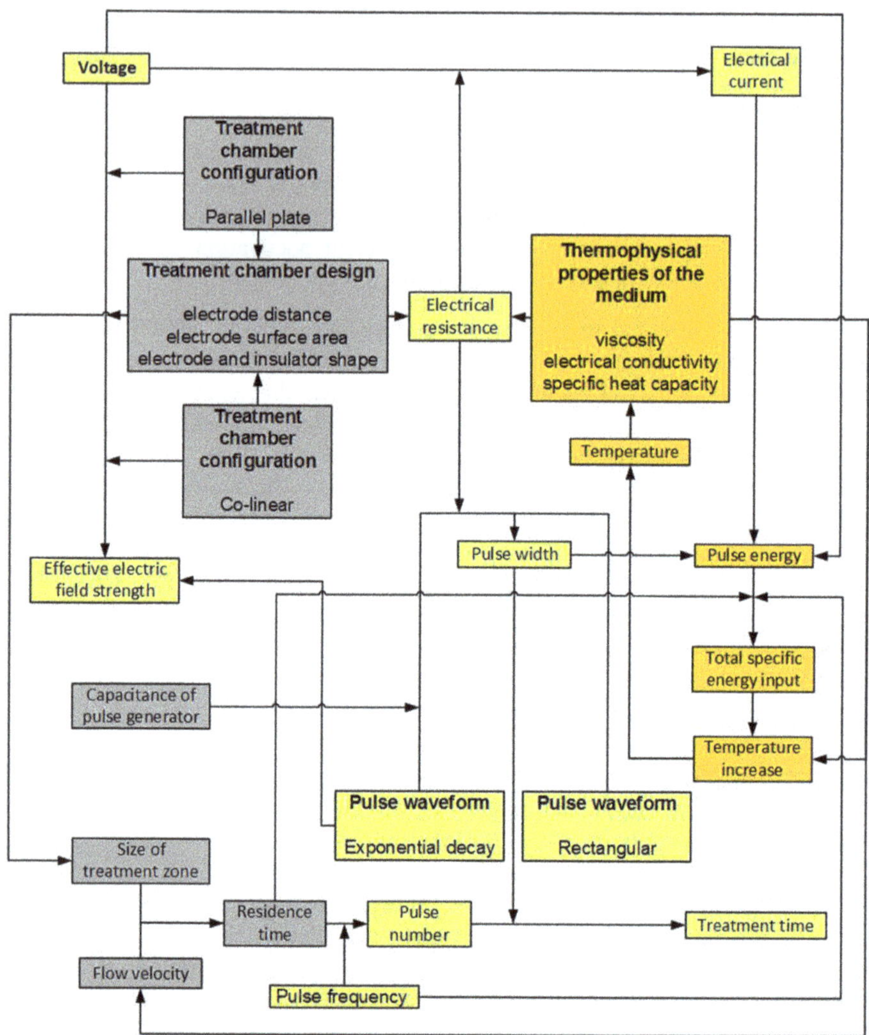

Fig. 3 Schematic drawing of the various interdependencies of PEF processing parameters related to the pulse generation system, the treatment chamber, as well as the treatment medium. Exemplarily considered are a parallel plate and colinear treatment chamber as well as exponential decay and rectangular pulses

Role of Thermal Effects and Complex Media in Microbial Inactivation

Thermal Effects

Although electroporation is a nonthermal mechanism for microbial inactivation, high-intensity PEF treatment as applied for shelf life extension purposes is linked to a significant temperature increase due to Joule heating (Salengke et al. 2012). In addition to this temperature increase, elevated preheating temperatures are used in order to facilitate electroporation effects, to improve the microbial inactivation efficiency, and to reach final product temperatures that also lead to partial inactivation of enzymes. The application of PEF in combination with mild heat is a promising technique for a gentle, multi-hurdle preservation process. The synergetic effect of temperature during pulsed electric field inactivation of microorganisms can be used to improve the inactivation results and/or to reduce the electrical energy costs and will contribute to an improvement of the process performance (Saldaña et al. 2014). Energy savings derive from the lower PEF treatment intensity (treatment time and total specific energy input) required for a certain level of microbial inactivation at increased temperature and from the possibility to recover the electrical energy dissipated during PEF treatment in form of thermal energy for preheating the incoming product (Heinz et al. 2003).

Apart from the overall liquid temperature measured at the outlet of the treatment chamber, treatment inhomogeneity and the occurrence of temperature peaks within the treatment chamber have to be considered as thermal impact factors. This is of particular importance when discussing PEF effects on functionality of heat-sensitive compounds, such as proteins, or when conducting kinetic inactivation studies (Jaeger et al. 2009a). Additional thermal effects may occur in case of application of high total energy inputs or insufficient temperature control or for unfavorable treatment chamber designs.

When applying treatment concepts considering the synergetic effect of PEF and temperature, it is of high relevance to perform an adequate temperature control in order to limit negative thermal effects on heat-sensitive compounds. In addition, analytical approaches are required in order to quantify the contribution of the electric field and thermal effects on the overall inactivation results. As the application of an electric field to a conductive matrix is always accompanied by an increase in temperature, the separation of the individual effects of heat and electric field is challenging. The validity of results is highly dependent on the experimental design, as well as the exact evaluation of process parameters, especially temperature. Despite experimental efforts, computational fluid dynamics (CFD) simulation coupled with kinetic data is an efficient tool for this purpose, as it is capable to also take into account electric field and temperature distributions within a treatment chamber. Approaches for the differentiation of thermal and electric field effects have been presented by Jaeger et al. (2010) and Schottroff and Jaeger (2020).

If thermal effects have to be enhanced in order to meet certain processing goals, then also other treatment options such as ohmic heating or conventional high-temperature short-time processing concepts need to be taken into account as alternative technologies (Jaeger et al. 2016; Ramaswamy et al. 2016).

Complex Media

The consideration of process-product interactions is crucial for the evaluation of the process performance of a specific PEF application. Each unit operation and almost each processing step within food production lead to desired or undesired changes of the food material and product properties (process-product impact). These modifications may be required to establish or improve the processability of a food or a raw material matrix by applying different types of pre-treatments without changing the major characteristics of the food. However, modifications may also be intended to significantly alter the appearance and quality attributes of the raw material matrix in order to produce a certain type of final product. Also, the permeabilization of microbial cells as part of the food product that is subjected to PEF preservation or the permeabilization of plant cells by PEF can be analyzed from the perspective of a process-product impact, and these phenomena are discussed in previous chapters.

In addition to the impact of food processing and particularly PEF processing on product properties, the product itself and its properties will affect the processing since material properties determine the performance of unit operations in food processing (product-process impact). The following section will focus on this product-process impact by exemplifying the role of the matrix for the PEF processing of food and its performance.

The microbial inactivation rates differ considerably between inactivation in simple media and inactivation in a complex matrix. This was partly attributed to the protective effect of some food compounds such as xanthan (Ho et al. 1995), proteins (Jaeger et al. 2009b; Sampedro et al. 2006), or fat (Grahl and Märkl 1996). Other studies did not reveal differences in the microbial inactivation conducted in buffers or complex media (Dutreux et al. 2000), or did not detect the occurrence of sublethally injured cells after PEF treatment of complex food systems (Walkling-Ribeiro et al. 2008). However, inactivation kinetics obtained from PEF treatment in buffer systems have only limited comparability with real food products, and the process performance of a PEF preservation might be much lower when processing concepts are transferred from model to real food products. In addition, model microorganisms used in most of the studies and the way of sample preparation differ significantly from the native state of the microbial population present in the real food system. The diverse and heterogeneous microbial flora present in real foods is hardly comparable to inoculated microorganisms in most cases, due to strong variability of microbial species and their physiological state. The consideration of microbial growth state, adaptation to the treatment media, and the existence of inhomogeneous microbial populations with less sensitive subpopulations seem to be the most challenging aspects when transferring inactivation results from lab scale to real products and industrial implementation (Morales-de la Peña et al. 2011).

Current industrial applications of PEF preservation are mostly dedicated to pasteurization of fruit juices (Timmermans et al. 2019), representing a matrix with comparably low complexity containing small amounts of protein and fat and mainly soluble compounds such as sugars and acids. In addition, fruit juices have a low pH that facilitates the microbial inactivation. On the other hand, most enzymes are not vigorously affected by electric fields, and consequently a certain process temperature is necessary, to obtain stability of the physiochemical properties of the juice through enzyme inactivation (Siemer et al. 2018).

Media complexity and the protective effect of food constituents become a limiting factor in microbial inactivation and for the transfer of PEF preservation concepts to products rich in protein or high in pH. Although the decontamination of liquids containing heat-sensitive bioactive compounds such as proteins or enzymes might be an emerging field for PEF processing, suitable process windows showing a sufficient process performance are narrow. As shown for the pasteurization of whey protein formulations (Schottroff et al. 2019; Schottroff et al. 2020a), the pH of the formulation as well as the protein content distinctly influenced the outcome of the inactivation, with a low pH, as an additional antimicrobial hurdle, enhancing the microbial log reduction, whereas higher protein contents reduced the efficacy of the treatment. However, the applicability of PEF treatment also shows limitations. Especially at neutral pH and reduced process temperatures, the effect of the electric field alone may not be sufficient to achieve the desired reduction of microbial counts as evaluated for the PEF decontamination of high solids (10–25%) whey protein formulations. For these matrices, an antimicrobial effect is only present when the process temperature is distinctly increased and the occurring effects are then mainly based on thermal mechanisms.

In addition to the media complexity that affects the microbial inactivation, media composition and media properties, such as electrical conductivity, thermal conductivity, viscosity, presence of suspended particles in liquids, as well as presence of air inclusions, determine the processability of a food by PEF and need to be taken into account for evaluating the process performance (Wölken et al. 2017).

The electric behavior of foods is not only depending on the chemical composition, such as water content, fat, protein, carbohydrates, and minerals, but also on the effects of dissolved and suspended solids or air. Foods containing different phases (i.e., solid particles, air bubbles, or large fat globules) with greatly different properties from those of the continuous phase will be considered as heterogeneous. It needs to be taken into account that electric behavior of one phase is essentially different from another phase, resulting in different electric field strength distribution, Joule heating rates, as well as dielectric strength. Information on the rheological properties of the food products to be PEF treated is required for the proper design of the treatment chamber geometry. Modifications of the treatment chamber geometry (Jaeger et al. 2009a; Schottroff et al. 2020b) can be used to affect the flow behavior and to induce mixing effects that may influence the treatment efficacy for heat-sensitive compounds and therefore improve the process performance toward a more gentle preservation process.

PEF Treatment of Plant Materials and Requirements for Process Integration

Role of Plant Tissue Structure for PEF Performance

PEF treatment of liquid or semisolid products for nonthermal pasteurization application aims at the permeabilization of microbial cells. Although variations regarding the sensitivity against PEF may occur within one single microbial population, the microbial cells in the population show an almost homogeneous distribution regarding structural characteristics. This is different in the case of PEF treatment of plant raw materials, since cells with distinctly different sizes and structural properties occur in the same raw material, forming different tissue segments as exemplified in Fig. 4. Hence, before applying PEF technology to plant raw materials, it is essential to analyze the specific particularities of the relevant tissue. Effective treatment conditions can only be determined by the consideration of raw material properties.

Cell size and cell size distribution within the tissue have a significant impact on the cell disintegration either by mechanical means or by PEF treatment. Grinding of tissues with small cells in comparison to large cells will result in a lower degree of cell disintegration since the resulting particles of a given size greater than the cell size contain a higher number of intact cells. An increase in cell disintegration can be achieved by increasing the grinding intensity leading to a further reduction in particle size. For larger cells, the same particle size distribution after grinding will result in a higher degree of mechanical disintegration already since a higher number of larger cells were destroyed. Differences in cell size of the raw material have to be taken into account for PEF processing as well. PEF applications such as the disintegration of fruit and vegetable mashes are mainly applied after mechanical disintegration (Abenoza et al. 2013; Mannozzi et al. 2018). Depending on the cell size, a different degree of mechanical cell disruption will occur and will affect the PEF performance, since a varying fraction of intact cells remains available for cell disintegration by PEF. Hence, the consideration of the cell disintegration index after grinding and mechanical disintegration is essential in order to evaluate the potential of an additional PEF application (Vorobiev and Lebovka 2020).

The cell size affects the electroporation since the transmembrane potential is proportional to the cell radius. Lower-intensity electric fields will be required to achieve electroporation in larger cells (Zimmermann et al. 1974). Hence, when dealing with a large variation in cell size, it would be required to optimize the magnitude of the applied electric field for each cell type. This is not possible for structured, complex raw materials. As a consequence, a selective electroporation of large cells will occur in heterogeneous samples, and the smallest cells need to be considered in order to develop processing concepts aiming a high level of cell disintegration.

PEF is affecting the cell membranes and thus can be expected to influence the texture of products in which the structure is largely dependent on the integrity of cells. The pre-treatment of potato tissue prior to French fries or snack production represents the example for the transfer of this phenomenon into industrial scale. PEF-induced changes occurring on a cellular level have an impact on the microstructures of the raw material or the food which is generated during further processing of the PEF-treated raw material. In case of the processing of PEF pre-treated potato, such changes in resulting microstructures during subsequent processing steps are the surface roughness and feathering after cutting or the crust formation and crust properties during frying (Fauster et al. 2018). Fundamental research on the modification of textural and structural properties of plant raw materials represents the basis for tailoring

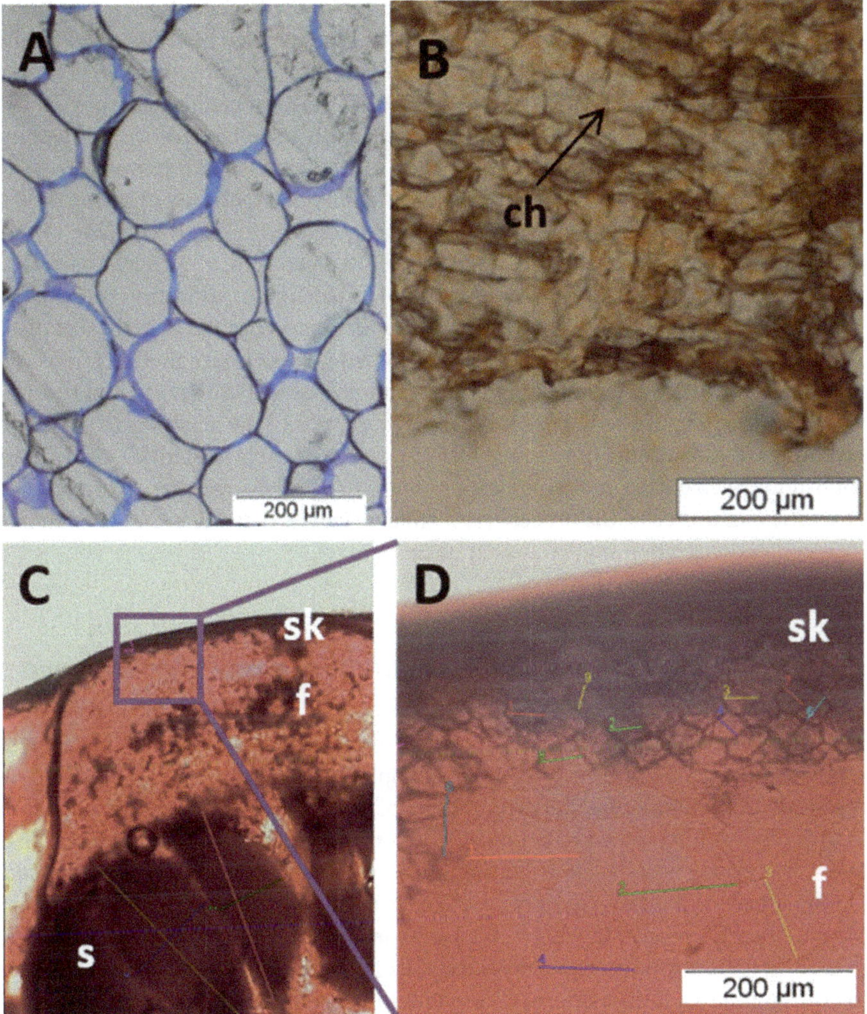

Fig. 4 Microscopic images of (**a**) apple tissue, (**b**) carrot tissue and (**c** and **d**) blueberry tissue. Substructures are shown such as chromoplasts (ch), the skin (sk), the flesh (f), and the seeds (s)

the applications with respect to the formation of targeted microstructures relevant for subsequent processing steps or specific product characteristics.

Destruction of the cell membranes and reduction of the cellular turgor are defined as the main impact factors for the modification of texture and viscoelastic properties of plant tissue after PEF treatment (Lebovka et al. 2004a, b; Pereira et al. 2009). However, application of PEF treatment will have limited effects on softening of the cell walls and reduction of their integrity. Hence, structural elements need to be identified and characterized in order to predict PEF effects and to assure a sufficient impact of PEF on structure-related properties of the raw material. The targeted PEF effect on the cell membrane is an advantage on the one hand but might be a limitation on the other hand, if additional effects of cell disintegration treatments such as heating or enzymatic maceration are required. These additional effects may include changes of the chemical structure of the cell wall, their breakdown or swelling, starch gelatinization, protein insolubilization, or the expulsion of trapped air.

The aspect of creating tailor-made structural characteristics by applying PEF to plant raw materials is also relevant for the treatment of fruit and vegetable mashes prior to juice recovery (Jaeger et al. 2012). Using mechanical grinding as well as PEF treatment, a process-specific induction of a structure modification is occurring. Whereas mechanical disintegration simultaneously modifies the particle size distribution, PEF disintegration occurs independent from particle size modifications. The impact of the different cell disintegration technologies and of the different resulting structural modifications on the juice release properties of the mash has to be considered for the subsequent juice recovery process since particularities of different solid-liquid separation technologies exist with regard to required optimal mash properties. In this case, the process performance can be defined as the translation of PEF cell disintegration into increased juice yield.

Role of Process Integration for Evaluation of PEF Performance

The use of PEF treatment aims at partially replacing or complementing existing food processing operations. When PEF is applied for the cell disintegration of raw materials from plant or animal origin, it is not the cell disintegration as such which is in the focus of the process, but it is the impact of the cell disintegration on subsequent processing steps such as cutting, drying, or extraction of target compounds. On the other hand, unit operations such as grinding of the raw material are applied prior to the PEF processing and will have an impact on PEF treatment performance as well as on resulting material properties. Hence, due to the various interactions of the different processing steps, it is essential to analyze not only the PEF process as such but also its integration into complex food processing systems.

Consideration of Connected Processing Steps

For the application of PEF for microbial inactivation, the process can be applied as a stand-alone system, and interactions with pre- as well as post-PEF processing operations are less pronounced. However, since the resistance of various types of microorganisms and enzymes covers an enormous range, a combination of several preservation methods based on different inactivation mechanisms will provide additional or synergetic inactivation results according to the hurdle concept (Leistner and Gorris 1995). Hence, the PEF process integration becomes an important aspect in the field of microbial inactivation as well since complex systems with different interacting steps will arise. The combination of PEF with other stress factors like mild heat, antimicrobial compounds, pH, or organic acids as well as the combination with other thermal or nonthermal decontamination techniques will determine further development.

The need for process integration is even more pronounced for PEF applications related to cell disintegration of plant raw materials. PEF-assisted juice recovery from fruit and vegetable mashes is used in order to exemplify the importance of considering pre- and post-PEF processing unit operations such as mechanical disintegration and solid-liquid separation in order to successfully transfer the cell disintegration provided by PEF into improved process results such as higher juice yields.

Figure 5 illustrates the different key steps in the juice production process, including milling, PEF treatment, and de-juicing as well as the related process results, such as particle size reduction and mechanical disintegration, cell disintegration by electroporation, and the solid-liquid separation.

Each of the processing steps can be performed using different pieces of equipment as well as variation of the processing parameters. The implementation of the PEF treatment in an existing juice processing line will be simple from a technical point of view, concerning the installation of the treatment chamber as well as the availability of industrial-scale pulse modulators.

Whereas mash characteristics related to particle size or cell disintegration can be mainly controlled by operational parameters, such as milling or PEF treatment intensities, the de-juicing process depends very much on the system that is used for the solid liquid separation. In addition to the operational parameters that can be adjusted for each of the de-juicing systems, the completely different design and working principle require a certain mash structure and pressing behavior on the one hand but offer also the selection of appropriate conditions for a given mash characteristic.

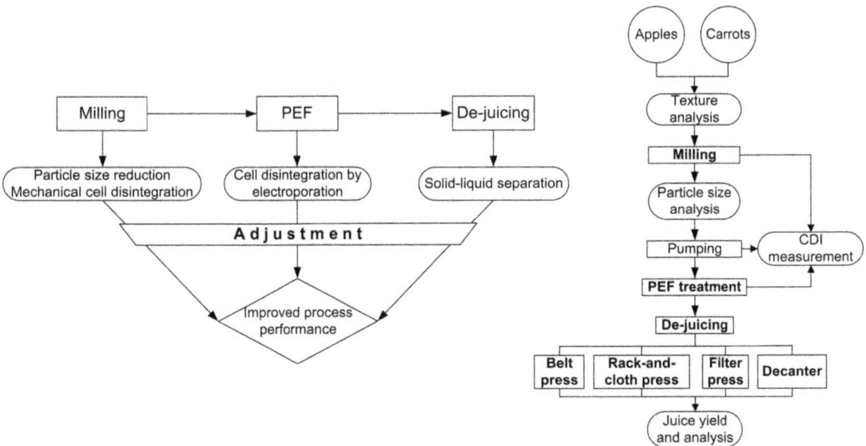

Fig. 5 Left: Concept for the improvement of the process performance of a PEF-assisted juice recovery by consideration of connected processing steps. Right: Juice recovery process from apples and carrots: processing steps and related process analysis. CDI refers to "cell disintegration index" as determined by impedance measurement

De-juicing systems are based on different separation principles as well as on design variations regarding the realization of the particular solid-liquid separation (Barrett et al. 2004). As discussed before, due to the process-product interactions, there will be an impact of the material structure on the separation process, i.e., juice release performance, as well as an impact of the process on the material, i.e., compaction of the mash. Phenomena are involved within each scale level as well as related to the process, the structure, and the property of the respective raw material. In order to evaluate juice yield results related to different mash structures and PEF treatment intensities, the operating principle of the solid-liquid separation system needs to be taken into account (Jaeger et al. 2012).

The consideration and combination of the various parameters related to material properties, disintegration technologies, as well as the performance properties of different de-juicing equipment leads to a complex multi-scale and multiparameter system. The analysis of the occurring interdependencies is the key factor in order to control the beneficial effect of a PEF treatment of the mash on juice yield (Jaeger et al. 2012). Favorable and unfavorable conditions can be identified regarding the mash structure as well as the de-juicing characteristics in order to improve the juice yield increase by PEF. The developed concept can be applied to different multistep processes that may include a potential PEF application. A process analysis and optimization provides the basis for the evaluation of the process performance, which in turn is a prerequisite for the decision regarding the use of the PEF treatment for a specific application or the redesign and adaptation of existing processing steps.

This is of particular importance when increasing the complexity and the scale of the PEF trials that require an appropriate experimental practice in order to allow the evaluation of the impact of the PEF treatment on the overall process performance. Integrating PEF into a multistage food processing line needs to focus less on the PEF treatment as such, rather than on the interdependencies of the different components in the system and on the evaluation of performance criteria of the traditional food processing equipment. Both aspects are of high relevance for the setup of an appropriate experimental design to study the PEF performance in the system and for the interpretation of obtained results. The consideration of connected processing steps is required from a technical point of view, since each component has a specific range of operating conditions that need to be adjusted. On the other hand, it is crucial to be aware of the properties of the raw material which is passing through the whole processing line and which is modified by each processing step (process-product impact). Depending on this modification, the performance of the subsequent processing step will be affected, since a product-process impact occurs. A successful integration of the PEF technology in a complex industrial processing environment as well as the evaluation of new application concepts of PEF in food processing will require such a multi-scale approach considering the process-structure-property interactions occurring in each single processing step and affecting connected processing steps.

Challenges for Industrial Implementation

In general, the implementation of new processes or methods into industrial scale is related to a number of challenges. To describe the impact of adaptations of processes or new technologies, a detailed validation of the whole processing line is needed. Further, the transfer of results from investigations in laboratory or pilot scale to industrial scale faces challenges for both sides, researchers and companies. Such challenges, among others, are also responsible for the limited implementation of PEF applications in the food industry. Although a few selected process parameters such as electric field strength, total specific energy input, or treatment time enable to characterize the PEF process from a treatment intensity perspective, the transfer of settings and findings from the laboratory to the industrial scale still reveals different issues regarding throughput, subsequent processing steps, and data acquisition possibilities in industrial scale. For trials in lab scale, raw materials can be accurately selected regarding their size, shape, and ripening stage. In contrast, the industrial production has to deal with different and continuously changing raw material properties. Moreover, due to the high throughput of industrial production lines, variation of processing conditions in industrial scale and the downscaling of industrial processes to the laboratory environment are challenging.

It is of major importance for the detection of PEF-related effects that are relevant for the industry to compare them with similar processes commonly used in the industry. In some cases, specific benefits resulting from PEF treatment might only occur due to comparison with reference processes that lack practical relevance or the use of non-optimized conventional processing conditions. As discussed above for the case study related to the integration of PEF into fruit and vegetable juice production, the mash pre-treatment by PEF may lead to a higher juice yield compared to untreated mash. However, the modern juice industry commonly applies thermal or enzymatic pre-treatments to facilitate and enhance the juice yield that need to be taken into account when comparing process efficiency and yield but also other parameters such as product quality and production costs. Hence, a process performance analysis needs to include suitable reference processes that are selected from an industry perspective and that are validated by reference data derived from optimized industrial practice.

Scientific investigations rarely describe the resulting economic benefits for specific PEF applications. However, main driver for the successful implementation of the technology in industrial-scale potato processing is the economic benefit resulting from the replacement of the thermal preheating step. Hence, process performance can not only be evaluated from technological perspective but needs to include economic profitability of a PEF application considering different scenarios already at an early stage.

Considering the implementation of industrial-scale PEF preservation, various limitations are present as well. Due to the fact that PEF is a relatively new technology, compared to heat treatment, an extensive database on the inactivation behavior of different foodborne pathogens of interest is missing. This is especially due to the fact that data from older publications are very difficult to compare and the fact that batch PEF treatment is often used for lab-scale studies with pathogens and the transferability of results to continuous systems is limited. Moreover, PEF treatment depends on distinctly more influence factors than just temperature and time, as it is the case for conventional thermal preservation processes. Therefore, the compilation of industrially realizable data and the selection of key processing variables is further complicated. Hence, this hampers the validation of PEF preservation processes.

In this context, there are also a variety of open research questions related to the determination of suitable microbial surrogate strains and indicator organisms, mimicking the behavior of pathogenic microorganisms in the presence of an electric field. Moreover, the limited scalability and occurrence of electric field and temperature inhomogeneities make the application of PEF preservation on an industrial scale more challenging. Lastly, an aseptic filling process may have to be implemented subsequent to PEF pasteurization, especially if it is supposed to replace a hot filling process. This would imply further investments as well as the implementation of more complex process technology, which not all companies will be willing to realize.

In summary, the major benefits for the food industry regarding the effects of PEF are improved food quality, increased profit by energy reduction or better raw material to product ratio, and the possibility to introduce new food products into the market. The outreach of PEF can be increased in different ways. In the case of the application of PEF for fruit juice preservation, only a single benefit – the quality improvement – is addressed. In the case of PEF application in potato processing, it is mainly economic benefits resulting from the replacement of thermal preheating and reduction of product losses. In addition, the modification of product texture allows the manufacturing of new products that satisfy consumer expectations. For these cases, a successful transfer of lab and pilot scale results to industrial scale, and the validation of the process performance criteria as well as expected benefits was possible in industrial scale.

Conclusion

PEF performance characteristics need to be studied based on a systematic process analysis approach including aspects related to different levels of scale as well as complexity. This approach represents a valuable tool in order to reveal potential aspects for a process improvement and a targeted process design. Process-product interactions need to be identified, and a bidirectional analysis needs to be performed considering PEF effects on food compounds and the raw material matrix as well as considering effects of the product to be treated on the PEF performance. The integration of the PEF treatment into complex processing concepts requires the consideration of connected processing steps. This is of particular relevance for the application of PEF for the disintegration of plant raw materials, since the PEF treatment serves as a pre-treatment for subsequent processing steps such as cutting, drying, extraction, or juice recovery. The adjustment and optimization of related processing steps will be essential in order to control and maximize beneficial PEF effects and to explore further application areas. The consideration of the technological process performance, economic benefits, as well as ecological criteria will determine the future implementation of the technology on an industrial scale.

References

Abenoza M, Benito M, Saldaña G, Álvarez I, Raso J, Sánchez-Gimeno A (2013) Effects of pulsed electric field on yield extraction and quality of olive oil. Food Bioprocess Technol 6:1367–1373

Barrett DM, Somogyi L, Ramaswamy HS (2004) Processing fruits: science and technology. CRC Press, 2nd

Dutreux N, Notermans S, Wijzes T, Gongora-Nieto MM, Barbosa-Canovas GV, Swanson BG (2000) Pulsed electric fields inactivation of attached and free-living *Escherichia coli* and *Listeria innocua* under several conditions. Int J Food Microbiol 54:91–98

Fauster T, Schlossnikl D, Rath F, Ostermeier R, Teufel F, Toepfl S, Jaeger H (2018) Impact of pulsed electric field (PEF) pretreatment on process performance of industrial French fries production. J Food Eng 235:16–22

Grahl T, Märkl H (1996) Killing of microorganisms by pulsed electric fields. Appl Microbiol Biotechnol 45:148–157

Harvey C (1986) Performance engineering as an integral part of system design. British Telecom Technol J 4:143–147

Heinz V, Toepfl S, Knorr D (2003) Impact of temperature on lethality and energy efficiency of apple juice pasteurization by pulsed electric fields treatment. Innov Food Sci Emerg Technol 4:167–175

Ho SY, Mittal GS, Cross JD, Griffith MW (1995) Inactivation of Pseudomonas fluorescens by high voltage electric pulses. J Food Sci 60:1337–1340

Jaeger H, Meneses N, Knorr D (2009a) Impact of PEF treatment inhomogeneity such as electric field distribution, flow characteristics and temperature effects on the inactivation of E. coli and milk alkaline phosphatase. Innovative Food Sci Emerg Technol 10:470–480

Jaeger H, Schulz A, Karapetkov N, Knorr D (2009b) Protective effect of milk constituents and sublethal injuries limiting process effectiveness during PEF inactivation of Lb. rhamnosus. Int J Food Microbiol 134:154–161

Jaeger H, Meneses N, Moritz J, Knorr D (2010) Model for the differentiation of temperature and electric field effects during thermal assisted PEF processing. J Food Eng 100:109–118

Jaeger H, Schulz M, Lu P, Knorr D (2012) Adjustment of milling, mash electroporation and pressing for the development of a PEF assisted juice production in industrial scale. Innov Food Sci Emerg Technol 14:46–60

Jaeger H, Roth A, Toepfl S, Holzhauser T, Engel K-H, Knorr D, Vogel RF, Bandick N, Kulling S, Heinz V, Steinberg P (2016) Opinion on the use of ohmic heating for the treatment of foods. Trends Food Sci Technol 55:84–97

Jin TZ, Guo M, Zhang HQ (2015) Upscaling from benchtop processing to industrial scale production: more factors to be considered for pulsed electric field food processing. J Food Eng 146:72–80

Lebovka NI, Praporscic I, Vorobiev E (2004a) Effect of moderate thermal and pulsed electric field treatments on textural properties of carrots, potatoes and apples. Innovative Food Sci Emerg Technol 5:9–16

Lebovka NI, Praporscic I, Vorobiev E (2004b) Combined treatment of apples by pulsed electric fields and by heating at moderate temperature. J Food Eng 65:211–217

Leistner L, Gorris LGM (1995) Food preservation by hurdle technology. Trends Food Sci Technol 6:41–46

Mannozzi C, Fauster T, Haas K, Tylewicz U, Romani S, DALLA ROSA, M. & JAEGER, H. (2018) Role of thermal and electric field effects during the pre-treatment of fruit and vegetable mash by pulsed electric fields (PEF) and ohmic heating (OH). Innovative Food Sci Emerg Technol 48:131–137

Morales-de la Peña M, Elez-Martínez P, Martín-Belloso O (2011) Food preservation by pulsed electric fields: an engineering perspective. Food Eng Rev 3:94–107

Oakland JS (2008) Statistical process control. Butterworth-Heinemann, Oxford

Pereira RN, Galindo FG, Vicente AA, Dejmek P (2009) Effects of pulsed electric field on the viscoelastic properties of potato tissue. Food Biophys 4:229–239

Ramaswamy S, Marcotte M, Sastry SK, Abdelrahim K (2016) Ohmic heating in food processing. CRC Press, Boca Raton

Raso J, Frey W, Ferrari G, Pataro G, Knorr D, Teissie J, Miklavčič D (2016) Recommendations guidelines on the key information to be reported in studies of application of PEF technology in food and biotechnological processes. Innovative Food Sci Emerg Technol 37:312–321

Saldaña G, Álvarez I, Condón S, Raso J (2014) Microbiological aspects related to the feasibility of PEF technology for food pasteurization. Crit Rev Food Sci Nutr 54:1415–1426

Salengke S, Sastry SK, Zhang HQ (2012) Pulsed electric field technology: modeling of electric field and temperature distributions within continuous flow PEF treatment chamber. Int Food Res J 3

Sampedro F, Rivas A, Rodrigo D, Martinez A, Rodrigo M (2006) Effect of temperature and substrate on PEF inactivation of Lactobacillus plantarum in an orange juice-milk beverage. Eur Food Res Technol 223:30–34

Schottroff F, Jaeger H (2020) Differentiation of electric field and thermal effects during pulsed electric field and ohmic heating treatments. In: Knoerzer K, Muthukumarappan K (eds) Innovative food processing technologies: a comprehensive review. Elsevier, Inc., Amsterdam

Schottroff F, Gratz M, Krottenthaler A, Johnson NB, Bédard MF, Jaeger H (2019) Pulsed electric field preservation of liquid whey protein formulations – influence of process parameters, pH, and protein content on the inactivation of *Listeria innocua* and the retention of bioactive ingredients. J Food Eng 243:142–152

Schottroff F, Johnson K, Johnson NB, Bédard MF, JAEGER H (2020a) Challenges and limitations for the decontamination of high solids protein solutions at neutral pH using pulsed electric fields. J Food Eng 268:109737

Schottroff F, Knappert J, Eppmann P, Krottenthaler A, Horneber T, Mchardy C, Rauh C, Jaeger H (2020b) Development of a continuous pulsed electric field (PEF) vortex-flow chamber for improved treatment homogeneity based on hydrodynamic optimization. Front Bioeng Biotechnol. in press

Siemer C, Toepfl S, Witt J, Ostermeier R (2018) Use of pulsed electric fields (PEF) in the Food Industry, DLG Expert Report, 5/2018, https://www.dlg.org/fileadmin/downloads/food/Expertenwissen/Lebensmitteltechnologie/e_2018_5_Expertenwissen_PEF.pdf

Timmermans RAH, Mastwijk HC, Berendsen LBJM, Nederhoff AL, Matser AM, Van Boekel MAJS, Nierop Groot MN (2019) Moderate intensity pulsed electric fields (PEF) as alternative mild preservation technology for fruit juice. Int J Food Microbiol 298:63–73

Vorobiev E, Lebovka N (2020) Techniques to detect electroporation. In: Vorobiev E, Lebovka N (eds) Processing of foods and biomass feedstocks by pulsed electric energy, Cham, Springer

Walkling-Ribeiro M, Noci F, Cronin DA, Lyng JG, Morgan DJ (2008) Inactivation of Escherichia coli in a tropical fruit smoothie by a combination of heat and pulsed electric fields. J Food Sci 73:M395–M399

Weaver JC, Chizmadzhev YA (1996) Theory of electroporation: a review. Bioelectrochem Bioenerg 41:135–160

Windhab E (2008) Structure-based quality and health related function tailoring in food systems - A process engineering approach. Dialogue on Food, Health and Society, 2008 Zurich

Wölken T, Sailer J, Maldonado-Parra FD, Horneber T, Rauh C (2017) Application of numerical simulation techniques for modeling pulsed electric field processing. In: MIKLAVCIC D (ed) Handbook of electroporation. Springer, Cham

Zimmermann U, Pilwat G, Riemann F (1974) Dielectric breakdown in cell membranes. Biophys J 14:881–899

Pulsed Electric Fields Industrial Equipment Design

Volker Heinz and Stefan Toepfl

Abbreviations

FDA	Food and Drug Administration
HACCP	Hazard Analysis and Critical Control Points
PEF	Pulsed electric field
R&D	Research and development

Introduction

The commercial viability of food processors is dependent on the use of optimum processes and process equipment throughout the production chain. Process performance criteria such as energy efficiency and robustness as well as equipment design considerations such as ease of use, reliability, operator as well as food safety play an important role in successful commercial exploitation of novel processes as PEF. Since first reports in the 1950s and 1960s (Amiali et al. 2007; Barbosa-Cánovas et al. 1998; Clark 2006; Doevenspeck 1961, 1984; Evrendilek et al. 2000; Heinz et al. 2003; Lebovka et al. 2005; Sitzmann and Münch 1988; Toepfl et al. 2006a; Zhang et al. 1995), substantial technical advancement in power electronics has been achieved, which facilitates the development of pulse generators with sufficient peak voltage, current and power level for large-scale food processing, but besides the pulsed power supply the treatment chamber or applicator design is of particular importance in the food industry and related applications. This chapter discusses important equipment design prerequisites and their impact on the selection of suitable equipment with regard to applicability, maintenance and process control.

V. Heinz
German Institute of Food Technologies (DIL), Quakenbrueck, Germany

S. Toepfl (✉)
Elea Technology GmbH, Quakenbrück, Germany
e-mail: s.toepfl@elea-technology.com

J. Raso et al. (eds.), *Pulsed Electric Fields Technology for the Food Industry*, Food Engineering Series, https://doi.org/10.1007/978-3-030-70586-2_17

Prerequisites of PEF Equipment Design

PEF systems basically are a setup of a pulsed power supply and a treatment chamber (Bluhm 2006; Gaudreau et al. 2010; Loeffler 2006). Up to the 1990s, mostly spark gaps and vacuum tubes have been applied for power switching, showing poor reliability and short lifetime (Sitzmann 2006; Sitzmann et al. 2016). Driven by medical, defence and energy conversion applications in particular for renewable energies, high-power semiconductor industry has made significant progress. Whereas high-performance thyristors or transistors are generally available today, the design of pulse generator concepts for food industry applications still can pose some challenges. A food processing factory is often characterized by fluctuating environmental conditions and temperature or relative humidity changes, a high load of dust and dirt as well lack of trained personnel with pulsed power background for service and maintenance. In comparison to other pulsed power applications, typically the design focus is less put on energy delivery accuracy than on reliability and service friendliness of the overall setup (Toepfl et al. 2006a). The use of custom-designed, single-stage pulsed power switches may result in equipment downtime and idle times until replacement. Recent developments, therefore, have focussed on modular designs to allow the use of standard components and to reduce maintenance efforts (Toepfl 2011b). To allow use in the food industry, the equipment has to be designed according to hygienic design guidelines, but also operator and machine safety are of importance.

Pulsed Power Supply

To design or select a suitable pulsed power supply, as a first step typically, the peak voltage and current and the average power needs are to be defined as follows:

- Peak voltage: The voltage requirement is dependent on the type of application, as cell disintegration requires much lower (average) electric field strength (~ 1 kV/cm) in comparison to microbial inactivation (~10–20 kV/cm). In addition, the desired throughput, as well as the product shape, will have an impact as they will define the electrode gap required.
- Peak current: The current is a result of voltage applied and dependent on product conductivity but also treatment chamber geometry due to electrode area and gap impact on the load resistance of the chamber.
- Average power: The average power is dependent on the desired processing capacity and specific energy input applied. The energy input level is dependent on the level of cell disintegration or microbial inactivation, ranging from less than 1 kJ/kg for plant tissue softening to 100 kJ/kg for microbial inactivation and exceeding 100 kJ/kg, e.g. for bacterial spore inactivation or biopolymer modification applications.

- Pulse waveform: Either exponential decay or rectangular pulses are applied; this is dependent on power switching circuit availability and cost framework. In recent years a shift to rectangular waveforms is observed due to higher energy efficiency. The use of monopolar as well as bipolar pulses has been reported. Monopolar pulses are more common as they allow a simpler design. For bipolar pulse generators, a positive and a negative stage will be required, so benefits regarding energy efficiency or electrode erosion are often overweighed by increasing investment costs.
- Pulse width: For most exponential decay pulse generators, the pulse width is not variable, but a result of the charging voltage and the capacity of the energy storage and the discharge circuit properties. Rectangular pulse shape generators often provide the option to vary pulse width in a range of some microseconds. Together with peak voltage and current, the pulse width defines the energy delivery per pulse; its variation, therefore, allows to control energy input.
- Pulse repetition rate: Pulse frequency has two major effects; on the one hand, it needs to be high enough to make sure every volume element passing the treatment zone is exposed to a sufficient number of pulses. As long a sufficient pulse number per volume element is applied pulse repetition in combination with energy delivered per pulse and product mass flow rate is used for specific energy input control.

To allow the use of standard switching devices, setups making use of pulse transformers or semiconductor-based Marx generators have been developed (Kern 2005; Kuthi et al. 2003; Toepfl et al. 2006a). In the first case, a low-voltage high-current thyristor or transistor is applied for power switching with a subsequent transformation to the desired voltage level. Pulse transformers allow compact equipment design with a low number of components but due to required matching of generator and load impedance have a more limited range of optimum operating conditions. Marx generators are based on using several stages of energy storage charged in parallel and discharged in series. Early designs have been based on untriggered or triggered spark gaps for switching; nowadays semiconductors are used which allow higher pulse repetition rates and low maintenance design.

As indicated above, the selection of a suitable pulse generator is based on a number of technical design criteria. Modern systems are based on semiconductor switching making use of standardized switching modules. Though from an application point of view the average power level defines the maximum treatment capacity, in most cases the peak power has shown to be the major challenge and cost driver. Lower average power systems are often done on the basis of pulse transformer setups, whereas larger-scale systems with high peak and average power requirement are mostly designed on Marx generator basis. The typical average power of today's PEF units is in a range of 30–400 kW (Frey and Sack 2021; Kempkes 2011; Loeffler 2006; Schultheiss et al. 2004; Witt et al. 2021). The peak power can range up to several hundred megawatts, in particular when using treatment chambers with large cross-section as applied for solid products or products with high electrical conductivity.

Treatment Chamber Design Considerations

In the treatment chamber, the electric energy is applied to the food product. Electrode configuration, area, gap and the flow pattern impact on treatment homogeneity, as well as electrical and hygienic properties. In addition, the type of materials used is of importance due to their food contact. Whereas most electrode materials are common in food contact areas, insulators used and sealings or gaskets with sufficient electric field resistance may be more challenging. For industrial processing in general, continuously operating equipment is desirable. Treatment chamber design is dependent on the type of application due to the field strength requirements but also product type (solid, semi-solid or liquid) as well as mode of transport. For liquid and semi-solid, pumpable products, tubular electrodes can be applied. Applications include cell disintegration of pumpable mashes or product water mixes; the typical configurations include co-linear, coaxial or concentric rift chambers (Huang and Wang 2009; Odriozola-Serrano et al. 2006). Treatment homogeneity is of particular importance for microbial inactivation applications (Witt et al. 2021). Whereas over-processed volume elements may be subject to quality deterioration and overheating, under-processed product may lead to insufficient microbial inactivation and product safety. Making use of fluid dynamics as well as electric field modelling, a number of optimized chamber designs are available (Fiala et al. 2001; Meneses et al. 2009), but in general co-linear chambers with a pinched insulator have shown sufficient treatment homogeneity and cleanability. Such chambers for extending shelf life of juices or dairy products are available with a diameter of 2 mm for lab-scale applications up to 50 mm for industrial-scale processing (Siemer et al. 2018; Toepfl 2011b) (see Fig. 1).

Fig. 1 Industrial-scale system for fruit juice preservation, up to 5.000 l/h

Dependent on product viscosity and the presence of particles, removable static or dynamic mixing elements can be applied to enhance processing homogeneity and allow a versatile chamber use. Electrode erosion has been a concern, but in a number of papers, it has been shown that electrochemical reactions and metal release to the product do not exceed acceptable limits of tap water directives when using stainless steel electrodes at optimized processing conditions (Arshad et al. 2020; IFT Nonthermal Processing Division 2011; Morren et al. 2003; Pataro and Ferrari 2020; Roodenburg et al. 2005a, b; Roodenburg 2011; Saulis et al. 2005; Toepfl et al. 2006b). As alternative titanium electrodes can be used for applications with high current density. The lower field strength requirements for cell disintegration use allow the design of larger electrode gap chambers. For example, for fruit, wine or olive mashes, treatment systems with nominal diameters of up to 150 mm are used with processing capacities between 1 and 100 t/h. To allow treatment of solid products, different chamber designs are required. For texture modification of fruits or tubers and to enhance cutting properties, belt or rotating wheel systems have been developed. Water is required as an energy-transferring media, and products are submerged and conveyed through an electrode setup by a belt or another mode of transport. As the product may sink or float, the raw material to be processed will impact on the design of in- and outfeed sections. In case of belt systems, the use of double belts has shown useful for treatment of floating products. The design of such systems has to incorporate PEF-related prerequisites such as allowing sufficient electric field strength as well as residence time and energy input, product-related parameters such as cross-section required as well as current food and operator safety directives. Operator safety could be best maintained by completely closed systems, but cleaning as well as service needs will require access to moving parts as well as treatment electrodes. Current designs make use of either removable covers or doors or a more open design using safety fencing. The choice between different concepts often is based on the availability of space in the respective environment as well as cleaning protocols and intervals applied. Continuous processing systems for solid products range from 1 t/h up to 80 t/h on single lines (see Fig. 2).

Current and Upcoming Industrial Applications and Systems Used

Though many potential applications have been identified (Barsotti et al. 1999; Huang and Wang 2009; Jeyamkondan et al. 1999; Knorr et al. 1994; Rastogi 2003), at present commercial applications are mostly found in fruit juice preservation, drying acceleration and vegetable and tuber processing (Arshad et al. 2020; Fauster et al. 2021; Jowitt 1997; Lammerskitten et al. 2019; Liu et al. 2020; Witt et al. 2021). In the area of juice processing, systems with a treatment capacity of up to 8.000 l/h have been installed, whereas for potato processing the capacity ranges up

Fig. 2 Two PEF treatment systems for processing of solid material in industrial scale

to 80 t/h. In the following section, commercial experience from existing as well as potential future applications will be discussed.

R&D and Smaller-Scale Systems

PEF can be applied in batch as well as continuous mode. Batch operation usually is applied for proof of principle testing as well as process optimization, e.g. in raw material testing associated to large-scale processing. For solid products such as roots, tubers or fruits, process scalability is typically given due to low energy input requirements and limited process-induced temperature increase. For higher energy input applications such as microbial inactivation, the applicability of PEF systems is often limited in comparison to continuous systems, due to poor scalability. In batch systems, the temperature increase induced by electric energy input is often less pronounced due to energy dissipation via the metal electrodes. Also, as usually no mixing is applied, treatment chamber geometry-related inhomogeneities play an important role. In continuous processing, steady-state conditions are achieved after some minutes of start-up, allowing more constant, reproducible processing. Product movement– even if flow pattern may differ from small to large scale – typically also benefits process homogeneity.

For early-stage testing, a number of different equipment concepts are available. While many research centres have created own experimental setups, such units due to lack of commercial availability and compliance to safety or food regulations often have limited use for pilot-scale testing or small-scale processing. Most of the available batch systems for processing solid materials are based on spark gap-driven discharge circuits, allowing flexible and versatile use but limited pulse repetition rates. With treatment chamber volumes of several litres, product amounts of some kg can be treated at time requirements of seconds to minutes. Such systems are available at a number of research centres in Europe, the United States as well as Asia. For microbial inactivation testing typically, semiconductor-based systems are used, as higher pulse repetition rates of a few hundred hertz are required. Such systems are available from Diversified Technologies (United States), ScandiNova (Sweden), Pulsemaster (the Netherlands) or Elea (Germany), exemplarily. Recently first multiple-use systems have been launched, allowing batch processing with treatment volumes up to 10 l as well as continuous operation for microbial inactivation as well as processing of mashed products (see Fig. 3).

Fig. 3 Multiple-use PEF system for batch and continuous operation

Fruit Juice and Smoothie Preservation

In recent years the European fruit juice market has shifted from ambient shelf-stable products towards fresh, often non-pasteurized products distributed in a chilled distribution chain (AIJN 2012). The consumer demand for fresh and natural taste has significantly changed the supply chain, as products are not produced from concentrate but manufactured from fresh fruits or frozen semi-fabricates close to the point of sale. Dependent on raw material quality as well as the production processes used, the shelf life of such fruit juices or smoothies is in a range of 7–10 days. To supply a high number of product types and allow sufficient time on the market shelf, significant distribution efforts as well as a high number of product changes result.

By PEF application, the microbial load can be reduced and the shelf life extended (Cserhalmi et al. 2006; Elez-Martínez and Martín-Belloso 2007; Schilling et al. 2008). A treatment with an initial temperature of 40 °C, a field strength of 20 kV/cm and an energy input of 120 kJ/kg results in a 4–5 log reduction of the total count of orange juice, exemplarily (Toepfl 2011b; Witt et al. 2021). In addition to inactivation of natural microbial flora, also the inactivation of inoculated *E. coli* 35218 has been shown. A 5 log inactivation of that target strain will be a key requirement for an FDA approval. As temperature has a synergetic effect on PEF efficacy, often combined approaches are used to reduce the amount of electrical energy required (Amiali et al. 2007; Evrendilek and Zhang 2003; Heinz et al. 2003). Making use of a setup of (often existing) preheaters, a high processing capacity can be achieved, while in comparison to a standard thermal processing, the product heat load is significantly reduced. When operating at maximum temperatures in a range of 45–50 °C, the product shelf life is extended up to 21 days while maintaining fresh-like taste and product quality. At present, such products are on market shelves in the Netherlands and the United Kingdom, where PEF processing equipment with a capacity of 1,500–2,000 l/h and 5,000–8,000 l/h is used (Irving 2012). The total processing costs including investment and operation are in a range of 0.01 Euro/l of product, justified by an extended distribution time and market range as well as significantly lower product return. With regard to the European Novel Food legislation, the impact on product quality parameters as well as the level of undesired substances has been evaluated (Aganovic; Cserhalmi et al. 2006; Schilling et al. 2008; Timmermans et al. 2011; Vervoort et al. 2011). After a PEF treatment, no significant process-induced changes have been observed; therefore products have not been considered as novel.

PEF equipment design for fruit juice processing requires systems with integrated process monitoring. Operation protocols will also have to cover start-up and shut-down as well as product change, where instead of product, water will have to be processed, which has a lower electrical conductivity and therefore energy input. To prevent (re)contamination of processing line parts after the treatment, area diversion valves are applied, discharging or collecting undertreated product in buffer tanks for reprocessing. To identify shift over from start-up liquid to product, a number of sensors are applied including conductivity, pH as well as turbidity measurement.

Modern PEF systems come with an integrated data logging or HACCP module for process monitoring and recording processing parameters. The units are implemented into standard CIP and SIP circuits, which require resistance against common detergents and disinfectants or the pressure and temperature ranges of a steam sterilization.

Besides fruit juices, the possibility to extend the shelf life of vegetable juices and protein-based products such as emulsions and dressings, dairy products or blood has been evaluated (Bendicho et al. 2002; Boulaaba et al. 2014; Buckow et al. 2014; Deeth and Datta 2018; Li et al. 2005; Reina et al. 1998; Sepulveda et al. 2005; Wiezorek et al. 2009). For fresh carrot juice, a 3–4 log inactivation was observed after a treatment at 30 °C and an energy input of 100 kJ/kg. The shelf life was extended from 2 to 6 days, though the close to neutral pH value. Vegetable juices are of high importance in Asia but also Eastern Europe. Due to their high content of nutrients and health-related substances, that product type can be of high interest for a further exploitation of the technique. Whereas for consumption milk, a thermal pasteurization is well feasible and accepted; other dairy-based products such as whey or milk protein concentrates or milk to manufacture raw milk cheese or formulations may be more relevant for PEF preservation. Past and ongoing research work has shown the general possibility to achieve a microbial inactivation in dairy products (Calderón-Miranda et al. 1999; Deeth and Datta 2018; Sepulveda et al. 2005; Shamsi et al. 2008; Sobrino-Lopez et al. 2009). Reduced protein denaturation and fouling and increased equipment operation time provide additional benefits of a reduced heat load in comparison to pasteurization.

A number of system manufacturers have started offering PEF systems for preservation applications with a focus on EU as well as US market. Equipment suppliers include Diversified Technologies (United States), Arc Aroma (Sweden), CoolWave Processing (the Netherlands), Pulsemaster (the Netherlands), Energy Pulse Systems (Portugal) as well as Elea (Germany).

Processing of Agricultural Products

Processing of vegetables and tubers results in a quick, energy-efficient cell disintegration (Angersbach et al. 2000; Fauster et al. 2021; Fincan and Dejmek 2003; Fincan et al. 2004; Lebovka et al. 2003, 2007). Processing of fruit and vegetable mashes such as apple or carrot causes a release of intracellular liquid and may be utilized to enhance juice yield. At present industrial systems to enhance the yield of clear or cloudy apple juice are operated in the European fruit juice industry in scales of 5–20 t/h. As the treatment significantly changes mash texture, a careful adaptation of the fruit mill, as well as the subsequent liquid-solid separation, is required to achieve a consistent yield increase (Schulz et al. 2011). Textural changes that occur after a treatment are caused by a loss of turgor pressure (Angersbach and Knorr 1997; Janositz and Knorr 2010; Lebovka et al. 2004; Rastogi et al. 1999). Such a tissue softening has been observed for potato, sugar beet and carrot, exemplarily. As a result, subsequent handling, pumping or cutting processes may be facilitated (Lindgren 2006). After treatment of potatoes with an energy input of 1–1.5 kJ/kg, an improved cutting was observed, causing less fracture and a smoother cut surface after industrial hydro-jet cutting. Due to tissue softening, less product breakage occurs in the following production stages. The process can be used to replace conventional preheating of potatoes (60 °C, 30 min) in French fries production in an industrial scale (Fauster et al. 2021; Siemer et al. 2018; Toepfl et al. 2006a). A higher cut quality and product length have been reported as major benefits. In potato as well as vegetable chips industry, improved slicing results in a yield increase due to reduced starch loss, a reduced oil uptake as well as the potential to reduce frying time and temperature. Processing scales in that industry sector usually are smaller than in French fries processing; single lines range from 0.5 to 3 t/h finished product corresponding to 2–12 t/h throughput at the stage of PEF applications. For that industry sector, as PEF is added as a new, additional treatment step, a compact equipment design is a key requirement, leading to the development of all in one, closed PEF systems (see Fig. 4).

These systems can also be applied for other applications such as fresh or frozen vegetable processing or drying acceleration, where current R&D work on basis of existing equipment availability is believed to lead to a fast commercial exploitation.

As PEF is applied as a pre-treatment step process, monitoring has a lower importance in such applications than for microbial inactivation use. The equipment design focus is usually put into equipment reliability, as processing lines operate 24/7 throughout most of the year with short shutdown and maintenance periods in between harvest seasons.

In the sugar industry, a pilot-scale equipment has been installed in 2002 (Kraus 2003, 2004). An enhanced sugar yield and purity and the possibility to reduce extraction temperature have been reported. In addition, softening of the cossettes has been observed, facilitating product transport through the extraction setup. Though processing capacities are higher than present equipment capacities for a bulk product like sugar, a scale-up is feasible based on existing concepts.

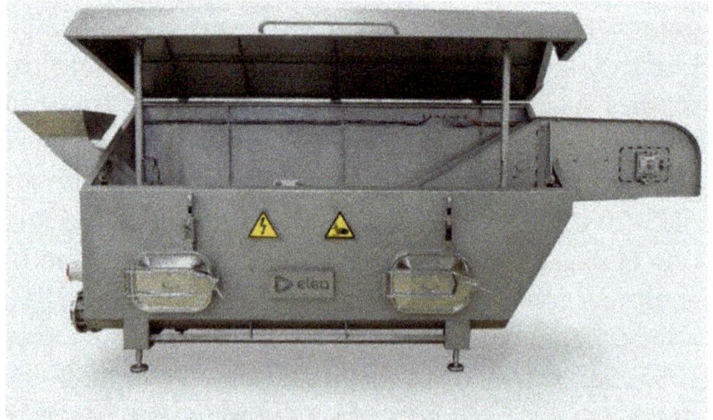

Fig. 4 PEF processing system for solid products with a capacity of 1 to 6 t/h

Also, the possibility to enhance peeling of some fruits or extraction of valuable plant oils has been shown (Toepfl 2011a). After an energy input of 1–7 kJ/kg, tomato or prune peel can be easily removed, similar than after a steam treatment.

An energy requirement of 1 kJ/kg is translated into 0.3 kWh energy requirement per metric ton of product or an installed power of 0.3 kW per 1 t/h treatment capacity. It needs to be kept in mind that in many applications not only the product is passing through the treatment chamber, but dependent on product geometry, water may be used as an energy-transferring medium. For example, when processing potatoes on belt systems, up to 3 parts of water per 1 part of the product is applied. The additional volume to be processed requires large cross-sections, often belts with a cross-section of 300–1000 mm, and electrode gaps in a range of 200 mm are applied, which results in peak current requirements of up to 5 kA. Other concepts are based on rotating wheel designs with the benefit of a narrower treatment chamber and smaller cross-section which results in lower peak power requirement (Schultheiss 2001). While the average power requirement will be at the same level, that approach may cause a higher probability of mechanical clogging of the treatment chamber. The total processing costs for a treatment of plant-based products are in a range of 0.5–3 Euro/t, dependent throughput, product type and desired level of tissue disintegration.

For solid products PEF systems are available from different equipment manufacturers as well; suppliers include ScandiNova (Sweden), Heat and Control (Australia), Pulsemaster (the Netherlands), Kea-Tec (Germany) as well as Elea (Germany).

Process Control Options

Whenever applied in food industry, the use of new processes also requires suitable process control options and the setup of a HACCP concept. Whereas PEF use for cell disintegration typically occurs at a noncritical process stage, any preservation application has to be considered as a critical control point. The consistent delivery of sufficient treatment intensity has to be maintained and recorded at all times (Nöhle 1994).

To monitor cell disintegration applications, an impedance measurement can be used to characterize the level of cell disintegration (Angersbach et al. 2000; Zhang and Willison 1991). Based on the frequency-dependent conductivity of untreated and treated tissue, the treatment impact can be detected.

For microbial inactivation, a similar concept than that used for thermal processing has been developed. Whereas for conventional thermal processing temperature and time are considered to sufficiently describe an inactivation process, for PEF a number of processing as well as product parameters have been used. The specific energy input and treatment temperature are generally considered as important process conditions. When describing treatment temperature, one needs to keep in mind that product temperature will change during PEF processing dependent on energy input. For most PEF setups, the energy input will not be a set value, but dependent value and result of other processing conditions set by the user, e.g. generator voltage, pulse frequency and product flow as well as product-related parameters such as electrical conductivity and specific heat capacity (see Fig. 5).

Process validation is based on a two-step approach. In a first step, the required energy delivery at the desired treatment temperature is elaborated in lab-scale challenge tests for each product group. During industrial-scale processes, the energy delivery is continuously monitored by the detection of pulse voltage and current. To validate power delivery to the product, the expected product temperature rise is calculated and compared to measured values pre- and post-treatment (see Fig. 6). Implementation of suitable processing protocols is key for successful commercial use. Making use of process monitoring system algorithms can be used to identify start-up and shutdown period and steady operation. In case of unforeseen situations, product can be diverted or disposed according to the manufacturer's operation protocols and a re-clean or sterilization protocol be triggered.

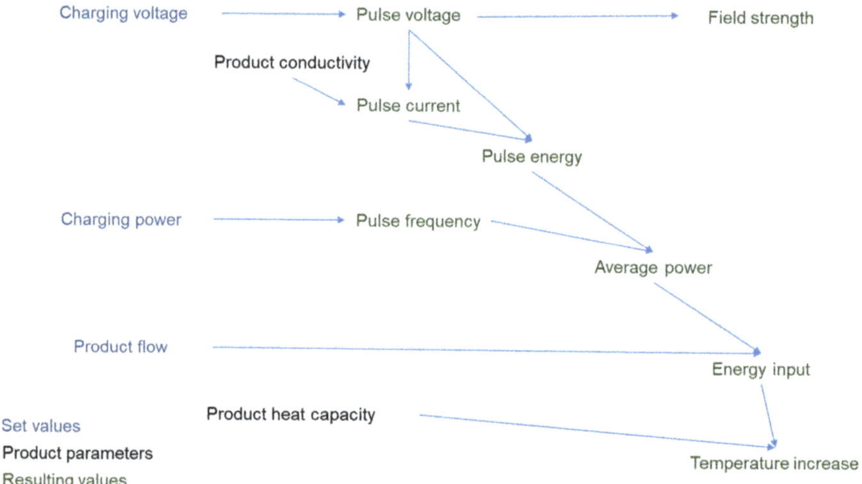

Fig. 5 Schematic view of the interaction of PEF processing and product parameters during microbial inactivation

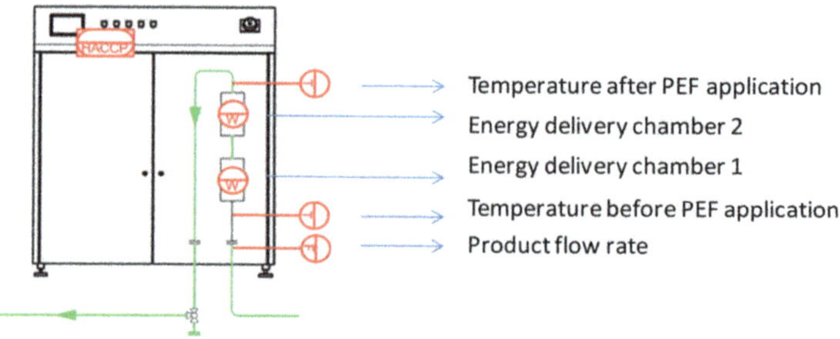

Fig. 6 Process monitoring options of PEF system

Conclusions and Outlook

PEF application allows achieving a cell disintegration and microbial decontamination of food products. The recent development of pulsed power systems suitable for food industry applications has allowed a successful transfer from lab scale to industrial applications. At present approx. 150 units are in commercial use worldwide, mainly in fruit juice and vegetable processing industry. Low energy requirements and processing costs, continuous operability and easy implementation are major benefits of the technique. Current equipment development work is focussing on system upscaling for products such as sugar beet as well as the design of dedicated application chambers for special products, e.g. for meat products or leafy material.

References

Aganovic K. Headspace fingerprinting and sensory evaluation to discriminate between traditional and alternative pasteurization of watermelon juice

AIJN (2012) Market report liquid fruit

Amiali M, Ngadi MO, Smith JP, Raghavan VGS (2007) Synergistic effect of temperature and pulsed electric field on inactivation of Escherichia coli O157:H7 and Salmonella enteritidis in liquid egg yolk. J Food Eng 79(2):689–694

Angersbach A, Knorr D (1997) Anwendung elektrischer Hochspannungsimpulse als Vorbehandlungsverfahren zur Beeinflussung der Trocknungscharakteristika und Rehydrationseigenschaften von Kartoffelwürfeln. Nahrung 41:194–200

Angersbach A, Heinz V, Knorr D (2000) Effects of pulsed electric fields on cell membranes in real food systems. Innov Food Sci Emerg Technol 1:135–149

Arshad RN, Abdul-Malek Z, Munir A, Buntat Z, Ahmad MH, Jusoh YMM et al (2020) Electrical systems for pulsed electric field applications in the food industry: an engineering perspective. Trends Food Sci Technol 104:1–13. https://doi.org/10.1016/J.TIFS.2020.07.008

Barbosa-Cánovas GV, Pothakamury UR, Palou E, Swanson BG (1998) Nonthermal preservation of foods. Marcel Dekker, New York

Barsotti L, Merle P, Cheftel JC (1999) Food processing by pulsed electric fields: 2. Biological effects. Food Rev Int 15(2):181–213

Bendicho S, Barbosa-Cánovas GV, Martin O (2002) Milk processing by high intensity pulsed electric fields. Trends Food Sci Technol 13:195–204

Bluhm HJ (2006) Pulsed power systems. Springer, Berlin

Boulaaba A, Kiessling M, Töpfl S, Heinz V, Klein G (2014) Effect of pulsed electric fields on microbial inactivation and gelling properties of porcine blood plasma. Innovative Food Sci Emerg Technol 23:87–93. https://doi.org/10.1016/j.ifset.2014.02.008

Buckow R, Chandry PS, Ng SY, McAuley CM, Swanson BG (2014) Opportunities and challenges in pulsed electric field processing of dairy products. Int Dairy J 34(2):199–212. https://doi.org/10.1016/j.idairyj.2013.09.002

Calderón-Miranda ML, Barbosa-Cánovas GV, Swanson BG (1999) Inactivation of Listeria innocua in skim milk by pulsed electric fields and nisin. Int J Food Microbiol 21:19–30

Clark P (2006) Pulsed electric field processing. Food Technol 60:66–67

Cserhalmi Z, Sass-Kiss A, Tóth-Markus M, Lechner N (2006) Study of pulsed electric field treated citrus juices. Innovative Food Sci Emerg Technol 7:49–54

Deeth HC, Datta N (2018) Non-thermal technologies: pulsed electric field technology and ultrasonication. In: Reference module in food science. Elsevier, Amsterdam

Doevenspeck H (1961) Influencing cells and cell walls by electrostatic impulses. Fleischwirtschaft 13(12):968–987
Doevenspeck H (1984) Elektroimpulsverfahren und Vorrichtung zur Behandlung von Stoffen
Elez-Martínez P, Martín-Belloso O (2007) Effects of high intensity pulsed electric field processing conditions on vitamin C and antioxidant capacity of orange juice and gazpacho, a cold vegetable soup. Food Chem 102(1):201–209
Evrendilek GA, Zhang QH (2003) Effects of pH, temperature, and pre-pulsed electric field treatment on pulsed electric field and heat inactivation of Escherichia coli O157:H7. J Food Prot 66(5):755–759
Evrendilek GA, Jin ZT, Ruhlman KT, Qiu X, Zhang QH, Richter ER (2000) Microbial safety and shelf-life of apple juice and cider processed by bench and pilot scale PEF systems. Innov Food Sci Emerg Technol 1:77–86
Fauster T, Ostermeier R, Scheibelberger R, Jäger H (2021) Pulsed Electric Field (PEF) application in the potato industry. In: Innovative food processing technologies. Elsevier, Amsterdam, pp 253–270
Fiala A, Wouters PC, van den Bosch E, Creyghton YLM (2001) Coupled electrical-fluid model of pulsed electric field treatment in a model food system. Innovative Food Sci Emerg Technol 2:229–238
Fincan M, Dejmek P (2003) Effect of osmotic pretreatment and pulsed electric field on the viscoelastic properties of potato tissue. J Food Eng 59:169–175
Fincan M, DeVito F, Dejmek P (2004) Pulsed electric field treatment for solid-liquid extraction of red beetroot pigment. J Food Eng 64:381–388
Frey W, Sack M (2021) Generator design and operation for pulsed electric field processing. In: Innovative food processing technologies. Elsevier, Amsterdam, pp 241–252
Gaudreau MPJ, Hawkey T, Petry J, Kempkes M (eds) (2010) Design considerations for pulsed electric field processing. Shaker, Aachen, Hamburg
Heinz V, Toepfl S, Knorr D (2003) Impact of temperature on lethality and energy efficiency of apple juice pasteurization by pulsed electric fields treatment. Innov Food Sci Emerg Technol 4(2):167–175
Huang K, Wang J (2009) Designs of pulsed electric fields treatment chambers for liquid foods pasteurization process: a review. J Food Eng 95(2):227–239
IFT Nonthermal Processing Division (ed) (2011) Corrosion Experiments in a PEF treatment chamber with stainless steel electrodes, Wageningen
Irving D (2012) We zijn nu al aan het opschalen. VMT Voedingsmi 16:11–13
Janositz A, Knorr D (2010) Microscopic visualization of Pulsed Electric Field induced changes on plant cellular level. Innovative Food Sci Emerg Technol 11:592–597
Jeyamkondan S, Jayas DS, Holley RA (1999) Pulsed electric field processing of foods: a review. J Food Prot 62(9):1088–1096
Jowitt R (ed) (1997) Engineering and food. Sheffield Academic Press, London
Kempkes M (2011) Pulsed electric field processing. In: Kempkes M (ed) iFood Conference 2011; October, 12th to 14th, 2011, Osnabrueck
Kern M (2005) Bipolarer Marxgenerator mit Schalterturm, direkter Triggerung und Gaskonditionierung für industriellen Dauerbetrieb
Knorr D, Geulen M, Grahl T, Sitzmann W (1994) Food application of high electric field pulses. Trends Food Sci Technol 5:71–75
Kraus W (2003) The 2002 beet campaign – VDZ Zweigverein Süd. Zuckerindustrie 128(5):344–354
Kraus W (2004) Reports on the 2003 campaign – VDZ Zweigverein Süd. Zuckerindustrie 129(5):349–363
Kuthi A, Alde R, Gundersen M, Neuber A (eds) (2003) Marx generator using pseudospark switches IEEE Xplore. Pulsed Power Conference, 2003. Digest of Technical Papers. PPC-2003. 14th IEEE International, 1. https://doi.org/10.1109/PPC.2003.1277701

Lammerskitten A, Mykhailyk V, Wiktor A, Toepfl S, Nowacka M, Bialik M et al (2019) Impact of pulsed electric fields on physical properties of freeze-dried apple tissue. Innovative Food Sci Emerg Technol 57:102211. https://doi.org/10.1016/j.ifset.2019.102211

Lebovka NI, Praporscic I, Vorobiev E (2003) Enhanced expression of juice from soft vegetable tissues by pulsed electric fields: consolidation stages analysis. J Food Eng 59:309–317

Lebovka NI, Praporscic I, Vorobiev E (2004) Effect of moderate thermal and pulsed electric field treatments on textural properties of carrots, potatoes and apples. Innov Food Sci Emerg Technol 5(1):9–16

Lebovka NI, Praporscic I, Ghnimi S, Vorobiev E (2005) Temperature enhanced electroporation under the pulsed electric treatment of food tissue. J Food Eng 69:177–184

Lebovka NI, Shynkaryk NV, Vorobiev E (2007) Pulsed electric field enhanced drying of potato tissue. J Food Eng 78(2):606–613

Li SQ, Zhang QH, Tony ZJ, Turek EJ, Lau MH (2005) Elimination of Lactobacillus plantarum and achievement of shelf stable model salad dressing by pilot scale pulsed electric fields combined with mild heat. Innovative Food Sci Emerg Technol 6:125–133

Lindgren M (2006) Root vegetable treatment: WO2006121397

Liu C, Pirozzi A, Ferrari G, Vorobiev E, Grimi N (2020) Impact of pulsed electric fields on vacuum drying kinetics and physicochemical properties of carrot. Food Res Int 137:109658. https://doi.org/10.1016/j.foodres.2020.109658

Loeffler M (2006) Generation and application of high intensity pulsed electric fields. In: Raso J, Heinz V (eds) Pulsed electric field technology for the food industry. Springer, Berlin

Meneses N, Jaeger H, Knorr D (2009) Coupling electric field, heat transfer and fluid dynamics modelling during PEF-treatment to predict impact on enzyme inactivation and retention. In: IFT annual meeting. IFT

Morren J, Roodenburg B, de Haan SWH (2003) Electrochemical reactions and electrode corrosion in pulsed electric field (PEF) treatment chambers. Innov Food Sci Emerg Technol 4(3):285–295

Nöhle U (1994) Präventives Qualitätsmanagement in der Lebensmittelindustrie Teil II: Risikoanalysen nach HACCP. Dtsch Lebensmitt Rundsch 90:351–354

Odriozola-Serrano I, Bendicho-Porta S, Martín-Belloso O (2006) Comparative study on shelf life of whole milk processed by high-intensity pulsed electric field or heat treatment. J Dairy Sci 89(3):905–911

Pataro G, Ferrari G (2020) Limitations of pulsed electric field utilization in food industry. In: Pulsed electric fields to obtain healthier and sustainable food for tomorrow. Elsevier, London, pp 283–310

Rastogi NK (2003) Application of high intensity pulsed electrical fields in food processing. Food Rev Int 19(3):229–251

Rastogi NK, Eshtiaghi MN, Knorr D (1999) Accelerated mass transfer during osmotic dehydration of high intensity electrical field pulse pretreated carrots. J Food Sci 64(6):1020–1023

Reina LD, Jin ZT, Zhang QH, Yousef AE (1998) Inactivation of *Listeria monocytogenes* in milk by pulsed electric field. J Food Prot 61:1203–1206

Roodenburg B (2011) Corrosion experiments in a PEF treatment chamber with stainless steel electrodes. In: IFT nonthermal processing division. Corrosion experiments in a PEF treatment chamber with stainless steel electrodes, Wageningen

Roodenburg B, Morren J, Berg HE, de Haan SWH (2005a) Metal release in a stainless steel pulsed electric field system. Part I. Effect of different pulse shapes; theory and experimental method. Innovative Food Sci Emerg Technol 6(3):327–336

Roodenburg B, Morren J, Berg HE, de Haan SWH (2005b) Metal release in a stainless steel pulsed electric field system. Part II. The treatment of orange juice; related to legislation and treatment chamber lifetime. Innov Food Sci Energ Technol 6(3):337–345

Saulis G, Lape R, Praneviciute R, Mikevicius D (2005) Changes of the solution pH due to exposure by high voltage electric pulses. Bioelectrochemistry 67:101–108

Schilling S, Schmid S, Jäger H, Ludwig M, Dietrich H, Toepfl S et al (2008) Comparative study of apple juice preservation by exposure to pulsed electric fields and thermal processing with respect to juice quality-related parameters and enzyme inactivation. J Agric Food Chem 56(12):4545–4554

Schultheiss C (2001) Verfahren zum kontinuierlichen nichtthermischen Aufschluss und Pasteurisieren industrieller Mengen organischen. Prozessgutes durch Elektroporation und Reaktor zur Durchführung des Verfahrens

Schultheiss C, Bluhm HJ, Sack M, Kern M (2004) Principle of electroporation and development of industrial devices. Zuckerindustrie 129(1):40–44

Schulz M, Jaeger H, Abstimmung von Zerkleinerungsgrad DK (2011) Elektroporation und Entsaftungssystem auf die Saftausbeute von Apfel- und Karottensaft. Fluss Obst 6:222–227

Sepulveda DD, Góngora-Nieto MM, Guerrero JA, Barbosa-Cánovas GV (2005) Production of extended shelf-life milk by processing pasteurized milk with pulsed electric fields. J Food Eng 67:81–86

Shamsi K, Versteeg C, Sherkat F, Wan J (2008) Alkaline phosphatase and microbial inactivation by pulsed electric field in bovine milk. Innovative Food Sci Emerg Technol 9(2):217–223

Siemer C, Toepfl S, Witt J, Ostermeier R (2018) Use of pulsed electric fields (PEF) in the food industry. https://www.dlg.org/en/food/topics/dlg-expert-reports/food-technology/dlg-expert-report-5-2018. Accessed 10 Sept 2020

Sitzmann W (2006) Technologieentwicklung der Elektroporation, Reinbek

Sitzmann W, Münch EW (1988) Das ELCRACK ® Verfahren: Ein neues Verfahren zur Verarbeitung tierischer Rohstoffe. Die Fleisch 40(2):22–28

Sitzmann W, Vorobiev E, Lebovka N (2016) Applications of electricity and specifically pulsed electric fields in food processing: historical backgrounds. Innovative Food Sci Emerg Technol 37:302–311. https://doi.org/10.1016/j.ifset.2016.09.021

Sobrino-Lopez A, Viedma-Martínez P, Abriouel H, Valdivia E, Gálvez A, Martin-Belloso O (2009) The effect of adding antimicrobial peptides to milk inoculated with Staphylococcus aureus and processed by high-intensity pulsed-electric field. J Dairy Sci 92(6):2514–2523

Timmermans RAH, Mastwijk H, Knol JJ, Quataert MCJ, Vervoort L, van der Plancken I et al (2011) Comparing equivalent thermal, high pressure and pulsed electric field processes for mild pasteurization of orange juice. Part I: Impact on overall quality attributes. Innov Food Sci Emerg Technol 12:235–243

Toepfl S (2011a) Pulsed Electric Field food treatment – scale up from lab to industrial scale. In: 11th ICEF: Athens, May 23rd to 26th

Toepfl S (2011b) Pulsed Electric Field food treatment – scale up from lab to industrial scale. Proced Food Sci 1:776–779

Toepfl S, Heinz V, Knorr D (2006a) Applications of pulsed electric field technology for the food industry. In: Raso J, Heinz V (eds) Pulsed electric field technology for the food industry. Springer, Berlin, pp 197–221

Toepfl S, Heinz V, Knorr D (2006b) High intensity pulsed electric fields applied for food preservation. Chem Eng Process. https://doi.org/10.1016/j.cep.2006.07.011

Vervoort L, van der Plancken I, Grauwet T, Timmermans RAH, Mastwijk H, Matser AM et al (2011) Comparing equivalent thermal, high pressure and pulsed electric field processes for mild pasteurization of orange juice. Part II: Impact on specific chemical and biochemical quality parameters. Innov Food Sci Emerg Technol 12:466–477

Wiezorek T, Heinz V, Toepfl S (2009) Ein Beitrag zur Nachhaltigkeit-Aufarbeitung von Schlachttierblut mittels gepulster elektrischer Felder (PEF). Rundsch Fleisch Lebensm 3

Witt J, Siemer C, Bostelmann A, Toepfl S (2021) Juice preservation by pulsed electric fields. In: Innovative food processing technologies. Elsevier, Amsterdam, pp 305–317

Zhang MIN, Willison JHM (1991) Electrical impedance analysis in plant tissues: a double shell model. J Exp Bot 42:1465–1475

Zhang Q, Barbosa-Cánovas GV, Swanson BG (1995) Engineering aspects of pulsed electric field pasteurization. J Food Eng 25:261–281

Process Validation and Hygienic Design for Pulsed Electric Field Processing

Claudia Siemer, Jim Gratzek, Volker Heinz, and Stefan Toepfl

Introduction

Pulsed electric field (PEF) technology is a mild food processing technology, suitable for various food applications. The technology can be used for the treatment of solid food material, such as potatoes or other vegetables, and liquids as well as semiliquid food. The purpose of the treatment is varied. Liquid food products, such as juices and purees, can be exposed to short electric pulses to achieve microbial inactivation for food safety and shelf life extension while preserving the food's fresh character. Short PEF pulses disrupt the osmotic balance of microorganisms, which results in permeation of the cell membranes. As a result of this, microorganisms can be inactivated by this method, and the microbial safety of the product is established. The application of electric pulses also affects plant cell behavior. The natural barrier separates the inner compounds of the cell from the environment. PEF induces an electroporation of this natural barrier resulting in an improved extraction of inner compounds, such as colorants or fatty acids. Moreover, the food product gets softer enabling easier processing. In the case of potato, the cutting requires much less force and the cut is more precise. This results in various advantages for the snack chip and French fry potato industries.

As PEF technology is and will be used in a growing number of food applications, specific PEF system requirements will be quite diverse. The criticality for example of delivering a "5-log kill" for food safety in a liquid food requires PEF apparatus,

C. Siemer (✉) · S. Toepfl
Elea Vertriebs- und Vermarktungsgesellschaft mbH, Quakenbrueck, Germany
e-mail: c.siemer@elea-technology.com

J. Gratzek
Food Physics, Boise, ID, USA

V. Heinz
German Institute of Food Technologies (DIL e.V.), Quakenbrueck, Germany

J. Raso et al. (eds.), *Pulsed Electric Fields Technology for the Food Industry*, Food Engineering Series, https://doi.org/10.1007/978-3-030-70586-2_18

which reduces residence time variability through turbulent flow and guarantees uniform and high-intensity electric field strength to ensure a minimum dose to deliver every food particle. For the treatment of potatoes for cutting improvement or olives for higher oil extraction, the importance of residence time, field strength, and pulse number is different and largely dependent on the operational needs of a producer or a specific downstream unit performance, not food safety. In every case, the PEF energy dose and variability for the specific application must be known and validated. Well-documented validation studies will lead to higher confidence in PEF and encourage a faster adoption. Equipment design differences for each application type will require different approaches to ensure sanitation and regulatory requirements. This chapter focuses on the implementation of PEF treatment in industrial processes with respect to the validation of the PEF process as well as the hygienic design.

Hygienic Design

The first reports of PEF use in foods were published in the 1950s in Ukraine, Moldova, and Germany (Sitzmann et al. 2016a). However, it took several decades to transfer the technology into industrial practice. From the 1990s on, more than 20 research groups have been identifying underlying mechanisms of action, influencing parameters and potential applications. For the first studies on PEF, lab-scale systems were used where hygienic design was of minor importance.

In general, PEF equipment requires a pulse modulator to supply high-voltage pulses and a treatment chamber, where the product is exposed to the electric field. At the beginning of PEF equipment design in 1990s, spark gaps and vacuum tubes with poor durability were applied for power switching (Sitzmann et al. 2016b). Considering the positive results of PEF technology for product quality increase and its impact on process efficiency, there was commercial demand for systems that offered improved durability and higher capacity. A solution was found in the development of high-power semiconductors. The current industrial-scale units are based on transformers, wherein a low-voltage switch is applied in combination with a pulse transformer, or Marx generator. Although industrial-scale PEF units are in operation, neither specific hygienic design guidelines nor specific processing standards exist. The industry currently relies on general hygienic design good manufacturing practices. As PEF is working with low temperatures for inactivation compared to standard thermal inactivation, good hygienic design is even more important. All aspects influencing the effectiveness of cleaning and the removal of product residues must be considered.

For the construction of PEF systems, cleanability and protection of the product from contamination with foreign bodies are the main criteria. In this respect, materials used should have a smooth, impermeable surfaces without cracks or crevices, process sensors must be cleanable, and there can be no allowance for uncleanable dead ends. As there are no specific hygienic guidelines for PEF systems, it is advised to use the standards 3-A Sanitary Standard (A Primer for 3-A Sanitary Standards and 3-A Accepted Practices; n.d.) and EHEDG Guidelines (EHEDG 2018).

As previously mentioned, the PEF system consists of two main parts, the pulse generator and the treatment chamber. For the construction of the power generator, safeguards to prevent food and moisture leakage into the generator should be in place; flashovers and power overloads can destroy equipment. As many food manufacturing sites are very humid, the generator needs to be isolated such that no water can condense in the cabinets and on system components, which would lead to electrical short circuits and equipment damage.

Treatment chamber hygiene and cleanability are most critical. The treatment chamber consists of two electrodes separated by insulating material. The flow pattern, treatment homogeneity, and electrical properties are influenced by the design of the treatment chamber. Material choice is complicated and critical due to extremely high voltage and the requirement of an insulator between the electrodes. As electrical energy is transformed into heat, crevices at the electrode insulator interface may arise due to differences in thermal expansion of the different materials. Good design will minimize this occurrence. Crevices promote accumulation of food deposits, which leads to chamber fouling and the possibility of inadequate cleaning leading to an inevitably system contamination.

As a PEF unit is part of a complete processing system and often added to existing lines, hygienic design for the whole system must be carefully reviewed and validated. It is recommended that Sanitation Standard Operating Procedures (SSOPs) including both cleaning in place (CIP) and manual cleaning be developed and validated for the entire installation prior to validation. SSOPs are typically developed by a team including machine vendors, sanitation experts, and plant quality assurance and operation management. Validation of SSOPs involves ensuring that the most difficult-to-clean system parts soiled with the most difficult-to-clean products are free from residues. SSOPs should be validated in triplicate and monitored as part of a company's quality system and life cycle management. Validation of a PEF system SSOPs should be done as a part of the overall PEF system validation. PEF system validation is explained in more detail in the next section.

Validation of PEF Process

Process validation is crucial for the industrial application of novel technologies. It is essential to validate the homogeneity of the process and to measure the actual process conditions at different positions in the process to make sure that each part of the product is subjected at least to the minimum PEF conditions. Validation is necessary in the R&D phase of the system development and also necessary upon installation and start-up of a new system. A validated system requires verification through (1) process measurements and record-keeping in daily operation and (2) ongoing maintenance and change control protocol including device recalibration.

The use of PEF for solid material focuses on the improvement of product quality and process efficiency. The validation of such processes aims at controlling a particular product attribute, like cutting or drying performance, after the PEF treatment. A certain PEF intensity causes disruption of cell membranes to achieve different effects, such as cutting improvement, extraction of valuable components, or process acceleration. For each application, the best and most efficient PEF settings for each product are defined and tested through trials, which vary in specific energy and electric field strength levels. Optimal operating levels depend on the product and the process aim. For the potato industry, the cutting force is measured as validation of the treatment. The potato is softened by PEF and this is studied by measuring the cutting force. The optimal treatment condition range is then used for industrial applications. In the case of potato processing, the PMD80 cut control device (Elea Vertriebs- und Vermarktungsgesellschaft mbH) can be used as a constant verification tool during processing (Fig. 1). Firmness of potatoes, as defined by its cellular structure, changes over time. A newly harvested potato is usually much firmer and harder to cut than a similar potato that had been stored for long period of time. As the purpose of PEF here is to soften the structure of the potato for consistent cutting performance, the PEF intensity required to achieve the cell structure for consistent processing decreases over time and with increased storage. The cut control provides an objective and repeatable measurement allowing an operator to adjust the PEF setting appropriately. If the force required to cut the potato is not in the desired range, the PEF settings can be adjusted by choosing a more or less intense PEF treatment. As it can be seen in Fig. 2, the force required to cut the potato is decreasing with increasing energy input.

For other purposes, such as the production of wine or olive oil, where a mash is treated, an electrical conductivity measurement can be used as a validation tool of the PEF process. When exposing biological cells to a PEF treatment, electrical conductivity changes. Angersbach et al. (1999) detected this change in electrical conductivity after PEF and introduced the concept of the Cell Disintegration Index (CDI). This index compares the conductivity of a PEF-treated product with an untreated product. A CDI value of 1 indicates maximum cell disruption; once the

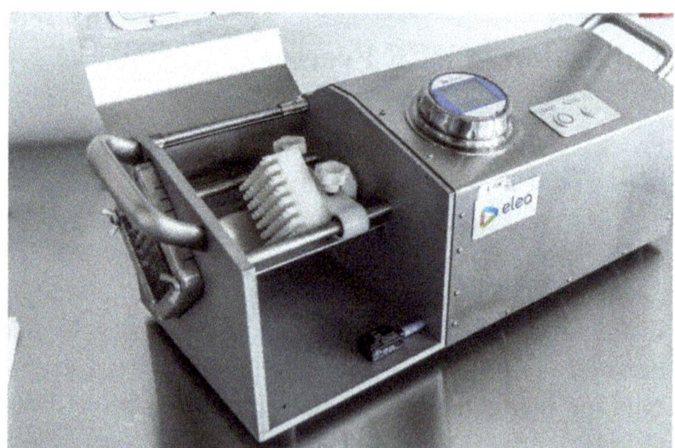

Fig. 1 Cut control to measure the cutting force of different solid food products, such as potato, before and after a PEF treatment

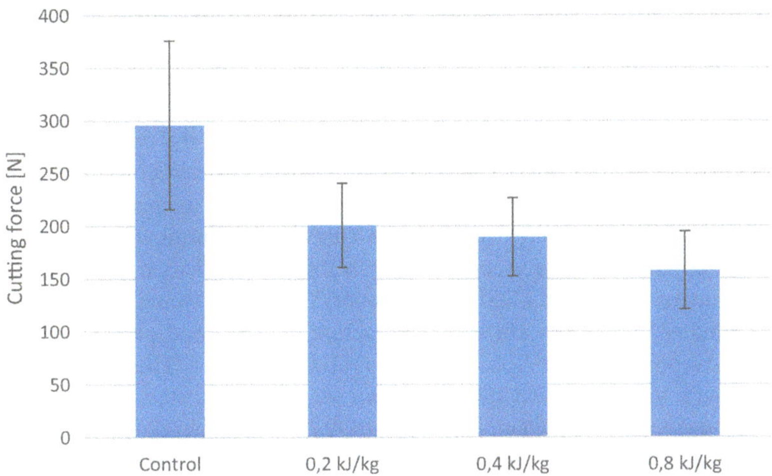

Fig. 2 Cutting force measurement of potato (Verdi) at an electric field strength of 1 kV/cm

CDI of 1 is reached, any additional energy applied by the PEF system will not change the CDI value. A CDI value of 0 represents the original tissue electrical conductivity, i.e., the conductivity value with no PEF treatment. The CDI measurement is an excellent measurement tool for a broad range of PEF applications. As PEF commercial know-how develops, the CDI will become an indispensable feedback tool for nearly all solid treatment applications (Angersbach et al. 1999).

For premium-quality liquid food products, like fruit juices and smoothies, PEF systems are still in the early stages of industrial application. Unique advantages have been demonstrated both for producers and retailers, especially due to the extended shelf life, high product quality, and much lower operational costs. Despite these advantages, the applications are limited to a group of innovative but relatively small producers. In contrast to this, conventional producers in the juice and soft drink industry are awaiting further technological development, as standardized process validation and control options are not yet concrete and available. In a survey of food professionals conducted by a collaboration between the Food Safety Working Group of CIGR (International Commission of Agricultural and Biosystems Engineering), Guelph Food Research Centre of Agriculture and Agri-Food Canada, and Campden BRI published in 2015, respondents indicated that while PEF is considered to be among the top three technologies of commercial significance now, as well as in the next 5–10 years in Europe, this was not the case in the USA (Jermann et al. 2015). In the same study, PEF was ranked fourth in the USA behind high pressure processing (HPP), microwave heating, and UV light. However, the survey authors noted that commercialized use of PEF may be more widely publicized in Europe compared to the USA, which may have impacted this result. Regardless, PEF was widely noted as having strong applicability for use in the drinks and beverages sector. Discussion with early adopters of the technology as well as those who are open to future adoption indicates a need for generally accepted process guidelines to ensure compliance with relevant food legislation. Such guidelines should speed up technology utilization.

Using PEF as the final pathogen control step for a ready to eat liquid production requires an identification of all relevant and likely food safety risks and their respective preventive control measures. Within a hazard analysis and critical control point (HACCP) system, food safety can be assured through process control and monitoring of critical factors during production, thereby lowering the need for end-product testing. The HACCP concept requires the study and identification of critical and hazardous conditions for a specific processing system and food that are important for the safety of the operator and the consumer. The HACCP concept was developed by the Pillsbury Company for the manufacture of safe food products for the US National Aeronautics and Space Administration (NASA). The aim was to have products with very low food safety risk in order to assure safe food for the astronauts in space (Barbosa-Cánovas et al. 1999a). The general background of the HACCP concept is the control of the processing line and product input reducing the need for sample testing of the finished product. Microbial, chemical, and physical hazards will be eliminated by anticipation rather than inspection. According to the National Advisory Committee on Microbiological Criteria for Foods (NACMCF), the HACCP system is based on seven principles (NACMCF, 1998):

1. Conduct a hazard analysis identifying all chemical, physical, and microbiological hazards.
2. Determine the critical control points (CCPs).
3. Establish critical limits.
4. Establish monitoring procedures.
5. Establish corrective actions.
6. Establish verification procedures.
7. Establish record-keeping and documentation procedures.

In order to establish a HACCP concept for a PEF process, the hazards have to be identified and assessed. More precisely, the critical control points (CCPs), process conditions, and monitoring tools must be evaluated and determined. A CCP is described as a "step at which control can be applied and is essential to prevent or eliminate a food safety hazard or reduce it to an acceptable level" (National Advisory Committee on Microbiological Criteria for Foods 1998). As PEF is a technology with multiple process and operation parameters (electric field strength, pulse waveform, energy input, and temperature), there is no general concept available so far to describe and monitor the treatment intensity and its distribution. Present users of the technology have identified their own, application specific concepts for process monitoring, which requires a high level of technical and scientific background as well as case by case discussions with food authorities. The development of a standardized, in-line process monitoring concept based on the specific energy delivered and the determination of treatment intensity distribution is crucial.

The development of a HACCP concept within a production line containing a PEF process as the last step for the preservation of the product is a preliminary step for getting the approval (or non-objection) by regulatory agencies, such as US Food and Drug Administration (FDA). The FDA requires processors of juice products to include in their HACCP plans control measures that will consistently produce at a minimum, a 5-log reduction, for a period at least as long as the shelf life of the product, when stored under normal and moderate abuse conditions, in the "pertinent microorganism" (Code of Federal Regulations CFR 21 n.d.). In this regulation, the "pertinent microorganism" is defined as "the most resistant microorganism of public health significance that is likely to occur in the juice," in which the term resistance relates to the resistance toward the preservation technology applied to a particular food matrix in order to achieve the required 5-log reduction. These requirements apply to all thermal and nonthermal methods for juice pasteurization. As the first step, a 5-log reduction of the most resistant pathogen must be demonstrated and documented through good scientific principles and practices. As a second step, the HACCP plan needs to be established containing the required CCPs identified in step one of the process. US processors and importers must have valid HACCP plans in place, which are subject to review by the US FDA. Not having a valid HACCP plan and supporting 5-log reduction data will likely lead to a product recall or other regulatory actions.

Validation of Microbial 5-Log Reduction

As with thermal processing, the purpose of using PEF as a last step in liquid food production is the inactivation of any pathogenic and spoilage microorganisms in the product. The microbial reduction depends on the specific PEF dose delivered, which is a product of electric field strength, accumulated PEF energy delivered in each pulse, pulse number, pulse type characteristics, and temperature management. Furthermore, the efficiency of microbial inactivation by PEF application also depends on the type of juice and microorganism, herein described as product factors and microbial factors, respectively. All these parameters must be considered for the validation of the PEF process. The FDA has summarized these critical factors, which should be considered when evaluating the effectiveness of a given PEF treatment to kill microorganisms in a food product (Table 1).

The process and product parameters selected for validation studies should consider the "worst-case scenario" conditions with respect to actual in-process parameters and microbiological survival. Monitoring approaches of in-process parameters should consider the inherent fluctuation in the parameter, such as temperature, electric field intensity, product conductivity, pH, or flow rate. Treatment time minimum, which is a factor of pulse width and pulse number, is determined by the fastest moving particle within the treatment cell. This is depending on whether the flow profile is laminar, transitional, or turbulent. Parameters selected for the study should reflect the "worst-case scenario" of the fluctuation observed, with consideration for what may promote the greatest survival of microorganisms. For example, if product pH ranges between 3.4 and 3.7 in each product, one should select a product with a pH

Table 1 Summary of critical factors and the related effect on microorganisms (adapted from FDA)

Category	Critical factor	Effect on microorganisms
Process factors	Electric field intensity	Increase results in higher inactivation
	Treatment time (pulse number × pulse width)	Increase results in higher inactivation
	Pulse shape	Order of increase resistance to PEF: Oscillatory < exponential decay < square wave
	Treatment temperature	Increase results in higher inactivation
Product critical factors	Conductivity	Decrease results in higher inactivation (strongly depends on the electrical characteristics of the PEF system)
	pH	Decrease results in higher inactivation
	Ionic strength	Decrease results in higher inactivation
Microbial factors	Type of microorganism	Order of increase resistance to PEF: Yeast < Gram-negative bacteria < Gram-positive bacteria
	Growth phase of the microorganism	Order of increase resistance to PEF: Logarithmic < stationary

of 3.7 or higher for the study, as this will be the least inhibitory to the survival of organisms.

Based on this the following measurements are required:

- PEF treatment parameters:
  - Electric field strength, length of the pulses, the number of pulses, specific energy: These parameters determine in combination with the product conductivity the intensity of the process. In addition to this, the field strength influences the rate of microbial inactivation.
- Product conductivity: The energy input into the product depends on the field strength, length of each pulse, and the number of pulses combined with the conductivity of the product. The corresponding temperature rise is based on a product's specific heat. As electrical conductivity increases at higher product temperature, the energy input will also vary with the product temperature. This is especially important when multiple treatment chambers are used in series.
- Flow conditions: Flow conditions determine the number of electrical pulses every product particle receives. Flow control is therefore necessary (product flow, distribution of flow, Reynolds number). Especially when more than one treatment chamber is used in parallel, it is necessary to control the flow in all treatment chambers. If one chamber is blocked, the higher flow rate in the other treatment chamber could result in under-processing.
- Temperature: The temperature of the product determines the efficiency of the PEF process. Moreover, the temperature also influences the electrical conductivity of the product.

Product Factors

Just as with any microbiological control process, food characteristics are important factors influencing the effectiveness of PEF application for the inactivation of microbial contaminants in fruit juices (Saldaña et al. 2009). The validation, done prior to commercial operation, should be done in a conservative way that takes product factors into account.

The chemical and physical composition of treated product represents a hurdle for microbial survival or growth in the product. Factors like conductivity, resistivity, dielectric properties, ionic strength, water activity, and pH can influence the treatment intensity required for a successful inactivation by PEF.

Particulate and pulp in the product may be an important factor to consider. Per studies by Mak et al. (2001), filterable pulp of apple cider increases the thermotolerance of *E. coli* O157:H7 compared to apple cider without pulp, while heat transfer was not affected (Mak et al. 2001). Similarly, orange pulp increases PEF tolerance of *Salmonella typhimurium* in orange juice compared to juice without pulp (Liang et al. 2002). Validation activities should be conducted on the real and final form of the product.

It is important that validation studies for juices reflect the worst-case conditions that likely to occur in industrial operation. Therefore, for validation studies pH and low organic acid content should be high and low, respectively, either by batch selection or chemical manipulation. If water activity is considered to be a significant factor for the product, higher Brix lots should be used for process and surrogate validation.

Microbial Factors

Considering the substantial differences in PEF sensitivity among different microorganisms, it is essential for a process validation to define the most pertinent microorganism for a particular fruit juice.

The likelihood of a pathogen to survive in a particular juice matrix depends on both, the type of microorganism and the ecological conditions prevailing during treatment and storage. The latter does not only comprise the characteristics of the juice (endogenous factors like pH) but also the conditions given by its treatments (exogenous factors, such as pasteurization and refrigerated storage conditions) (FDA 2004). With regard to varying product factors that might affect the efficacy of a treatment process to achieve a 5-log pathogen reduction, it is recommended to validate the process for each specific product. The concept of a safe harbor process does not yet exist in PEF processing; however, a change is expected, perhaps within a specific product category, i.e., citrus juice, with increased use and number of industrial validations.

Not all microorganisms are equally susceptible to PEF treatment at specific settings. This has also been considered during the search for a suitable surrogate. As each producer has to get approval on its own and with his own product, validation studies have to take place at the factory. Due to the fact that no factory wants to bring in pathogenic bacteria, so-called surrogates have to be identified which are nonpathogenic but show the same resistance toward the PEF treatment as the pathogenic bacteria. Within the search for these surrogates, intrinsic characteristics, like size, shape, species, or growth state, which play a major role in determining the level of susceptibility of a microbial strain to PEF treatment, have to be considered and evaluated. Generally, the larger the microbial cell, the more sensitive it is to PEF treatment. Yeasts, which are typically larger than bacteria, are generally more sensitive than bacteria (Heinz et al. 2001). Also, the higher sensitivity of microorganisms is noticed in the logarithmic phase than in the stationary phase (Álvarez et al. 2002).

Furthermore, the composition of the membrane plays an important role. Gram-negative bacteria, e.g., *E. coli*, are enclosed by a cell wall, which consists of only one murein layer. In contrast, the membrane of Gram-positive bacteria, e.g., lactobacillus strains, consists of several murein layers, which makes the membrane thicker and more resistant against the PEF treatment (Barbosa-Cánovas et al. 1999b).

Further to these intrinsic factors, also the cultivation of microorganisms has to be considered for the validation. It is important that standard microbiological analysis is used and that the bacteria are adapted to the product, which mostly means the bacteria need to be adapted to the acid environment. Different publications are available indicating the procedure of acid adaption for certain bacteria types and products (Leyer et al. 1995; Caggia et al. 2009; Gahan et al. 1996; Evrendilek and Zhang 2003).

These parameters should be considered when determining the relevant pathogen for the validation study. Also, of consideration are prior illness outbreaks associated with the type of juice product. Factors, which impact the ability of the organism to survive or grow in the product must also be considered, including the pH and storage condition of the final product. For defining the pertinent microorganisms (bacteria) of public health concern, which could be associated with different types of juice products, a general list of representative fruit and vegetable products, which may be used to manufacture cold-pressed juices currently in the marketplace, should be compiled, and a search of the scientific literature regarding outbreak information for each of the individual ingredients should be completed. The results of the investigation should be collected and evaluated. Based on information available, the vegetative organisms *Escherichia coli* O157:H7, *Salmonella*, and *Listeria monocytogenes*, as well as the spore-forming organism *Clostridium botulinum,* usually represent the pathogens of public health concern. They are known food safety risks and are reasonably likely to occur in juice products. It has to be taken into account that vegetative microorganisms are able to change the chemical composition of the product, which could induce spore germination. The pH and storage condition of the finished products must also be considered.

For refrigerated high-acid juice products, with pH of 4.6 or less, the enteric bacterial pathogens including *E. coli*, *Salmonella*, and *L. monocytogenes*, which are ubiquitous in the environment, are the microorganisms of public health concern. Even though these organisms may not always grow in high-acid juice products, they can survive for extended periods of time in the product, particularly at refrigeration (or frozen) temperatures (Miller and Kaspar 1994). As the infectious dose for these organisms is low (i.e., one to ten cells for *Salmonella* (D'Aous et al. 2001) and less than one cell per gram for *E. coli* O157:H7 (Meng et al. 2001)), the application of a process, which achieves a 5-log reduction of these organisms in accordance with 21 CFR 120 (Albala and Mentxaka 2015), is important to ensure safety. Although *C. botulinum* may be present in these products, it is not considered as a food safety risk as its growth and toxin production is required to cause illness, and this is prevented by the low pH of the product (pH < 4.6). Correspondingly, product factors or spoilage organisms, which might lead to an increase in product pH, may need to be considered in a HACCP risk analysis.

For low-acid, refrigerated juice with pH above 4.6, *C. botulinum* is considered to be a pertinent microorganism of public health concern. If these products are not processed to achieve commercial sterility, the product formulation and storage conditions must be adequate to prevent the growth of this organism. Even though these products are refrigerated, certain non-proteolytic strains of *C. botulinum* are capable of growth at refrigerated temperatures depending on the formulation and redox potential. Therefore, studies should be done on inactivation of non-proteolytic *C. botulinum.* It should be noted that toxin produced by *C. botulinum* was previously implicated in foodborne illness associated with the consumption of a refrigerated carrot juice product, which is a low-acid juice product with a natural pH of 6.8 (Sheth et al. 2008). Epidemiological results from the outbreak investigation indicated that as refrigeration was the only barrier to growth and toxin production by this organism in the carrot juice product, and as the implicated organism was *C. botulinum* type A, storage of the implicated product at elevated temperatures permitted the growth of this organism and its subsequent toxin production. The US FDA has demonstrated heightened concern over some nonthermally processed low-acid refrigerated juices ensuring their compliance with 21 CFR 120, specifically regarding the control of the *C. botulinum* food safety risk associated with these products. It requires consideration of the likelihood of temperature abuse (e.g., storage at 10 °C or higher for an extended period of time) and consideration of psychrotrophic *C. botulinum* in HACCP analyses. Thus, it is suggested that thorough scientific justification be provided to support the efficacy of these processes, products, and storage conditions to control the microbiological food safety risks associated with these types of juices.

PEF System Validation: Process Parameters

Beside the validation to demonstrate the accuracy and reliability of the process, the FDA is reviewing the equipment design, product specification, process design, and process validation (Larkin 1999). Table 2 gives a short overview about the required information.

In order to understand the critical process factors, one first need is to gain an understanding of the mechanisms by which PEF treatment inactivates microorganisms. Jeyamkondan et al. (1999) present two different theories regarding the impact

Table 2 Overview of required information for process validation (Larkin 1999)

Equipment design	Description of a system control mechanisms used and fail-safe procedures
Product specifications	Description of the product including information about physical and chemical characteristics as well as critical factors influencing the process
Process design	A complete description of the critical processing conditions Used in manufacturing the product including: Process records Pulse shape and pulse width Frequency Electric field strength Treatment time Records must be easy to access so that the FDA can verify if the required electric field and time are delivered according to operating procedures
Validation	A physical demonstration of the accuracy, reliability, and safety of the process assuring reproducibility Evidence of microorganism injury recovery studies as well as the resistance of certain pathogens concerning a given product must also be included

of an electrical field on the membrane of a microorganism. In the first theory, the authors state that there is a specific transmembrane potential, which a bacterial cell can tolerate. The transmembrane potential is influenced by the size of the bacterial cell and the applied electrical field strength. When the transmembrane potential is exceeded, pores are formed in the cell membrane, which disrupts the membrane's function resulting in cell death. In the second "electroporation theory," a combination of both thermal and electrical forces leads to disruption of the permeability of the cell membrane and denaturation of the transmembrane proteins, which results in cell death.

Regardless of the theory, it appears that the target of PEF within the bacterial cell, which results in its killing factor, is the cell membrane, and the effectiveness of the PEF treatment to kill the microorganisms is primarily the electrical field strength and accumulated pulse energy. PEF treatment is then characterized by initial temperature, PEF energy delivery, electrical field strength, and temperature rise after treatment.

The manufacturer of any given PEF system is obliged to ensure that the design provides effective and reliable treatment across the range of expected applications. To ensure good design, analysis of the design itself with respect to the spatial treatment difference itself is recommended. Depending on the configuration of the treatment chamber and the product viscosity, a three-dimensional spatial treatment intensity matrix can be compiled, both through numerical modelling and by careful measurement. While the overall energy delivery is detected by pulse monitoring indicating the voltage and current of each pulse, the spatial distribution of that energy can only be ensured by good design. Numerical methods are customarily used to estimate field spatial distribution; confirmation of this through physical temperature measurement should be done. The way of measuring the temperature of the product after PEF needs to be carefully evaluated. It is important to record the temperature as precisely as possible, e.g., by using fiber-optic sensors. Confirmation of the adequate cell design is done by measuring the temperature at suitable positions to confirm field strength calculations. While these measurements would not be used as direct supporting data for product HACCP, it is considered as an important prerequisite in machine design.

The main step in the PEF process is the application of pulses of an electrical field to a liquid stream. The frequency, shape, and amplitude of the pulses result in the intensity of the treatment. Together with the configuration of the treatment chamber, the efficiency is determined. It is important to measure and monitor that the pulses set on the PEF system are applied to the product. This monitoring system should establish a consistent and uniform treatment and that missing pulses trigger alarms. During system validation, it is routine to use a high-voltage oscilloscope and ammeter to confirm system measurement and software. During set-up and calibration, an oscilloscope can confirm and record each pulse, current, and voltage, respectively. They can also be used to ensure that software controls are functioning. For example, missing pulses would trigger a system error. Measurements with an oscilloscope will also confirm a calibrated PEF system attesting expected pulse width and pulse shape. From this, the PEF intensity, mainly specific energy and electric field strength, can be calculated, which can be compared to the PEF intensity set on the PEF system.

Temperature before and after PEF should be monitored as well as the flow rate (Fig. 3). The temperature increase depends on the accumulated pulse energy and specific heat of the product. The comparison of temperature measured at the pipe and calculated based on energy calculated by current and voltage measured by oscilloscope should be confirmed in the system set-up. While PEF temperature development within the treatment cell is practically adiabatic, the way of temperature measuring should be carefully chosen. Fiber-optic sensing systems as an example can provide a reliable signal that the energy absorption occurred. These measurements and correlations should be established during biological validation.

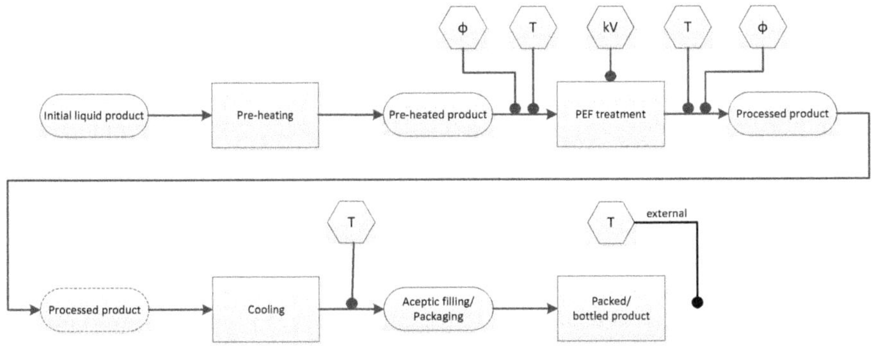

Fig. 3 Scheme for PEF processing of liquid food material for microbial decontamination

Considering the process factors explained, a HACCP plan can be set up and discussed with the food authorities. Once the established HACCP plan is accepted, the factory can start the production using PEF.

Conclusion

Pulsed electric field (PEF) technology first emerged in food processing in the early 1960s and had emerged relatively slowly until the recent acceleration in the potato sector. There has been great interest in academia resulting in thousands of publications and many patents. The industrial interest in this technology is growing across a growing number of applications, and as a result, companies are developing systems for new application types, reliability, and higher throughput. The hygienic design is of major importance for this development with respect to both electrical and microbiological safety. Higher durability and lower maintenance needs are also of focus. The choice of material is important, especially for the treatment chambers, to avoid any cracks and allow adequate CIP. Given that these systems are necessarily high voltage, construction must be electrically safe and suitable for wet factory environments.

To promote further industrial adoption, process validation standards should be adopted. The PEF treatment of solid material focusing on extraction or cutting improvement, improved PEF response measurement will enable the industry to better control and monitor PEF intensity and delivery. This will improve application exactness and aid in energy efficiency. For the PEF treatment of liquid food products, the process validation is more complex and critical as it is often the final safety step within a production line. Three categories must be evaluated and integrated in a HACCP plan to allow the use of PEF. The first category focuses on product factors, such as pH or conductivity. The effect of the different product characteristics has to be identified and evaluated. Microbiological factors must also be thoroughly evaluated. The microorganism of concern for a certain product type should be identified based on literature and networking in expert circles; the highly resistant and likely to occur microorganism should be used. Validation of the appropriate process parameters is of special attention given the emerging state of this technology. It must be ensured that every element of a product is effectively treated by PEF minimum dose, which is comprised of both field strength and accumulated pulse time. Each of these categories has to be carefully studied and implemented in a HACCP plan before using PEF.

In a recent survey of worldwide PEF installations and applications, systems at industrial scale have emerged, and hygienic design has improved over the past years. Process validation, especially for the treatment of solids, is well studied and tools for monitoring and controlling are available. In the case of applications, where PEF is the kill step, it is mandatory to establish a HACCP plan approved by food authorities.

References

A Primer for 3-A Sanitary Standards and 3-A Accepted Practices. (n.d.). https://www.3-a.org/Knowledge-Center/Resource-Papers/A-Primer-for-3-A-Standards-Practices. Accessed 18 May 2020

Albala K, Mentxaka AL (2015) Hazard analysis and critical control point (HACCP). In: The SAGE encyclopedia of food issues. SAGE, Washington DC

Álvarez I, Pagán R, Raso J, Condón S (2002) Environmental factors influencing the inactivation of Listeria monocytogenes by pulsed electric fields. Lett Appl Microbiol 35:489–493. https://doi.org/10.1046/j.1472-765X.2002.01221.x

Angersbach A, Heinz V, Knorr D (1999) Electrophysiological model of intact and processed plant tissues: cell disintegration criteria. Biotechnol Prog 15:753–762. https://doi.org/10.1021/bp990079f

Barbosa-Cánovas GV, Góngora-Nieto MM, Pothakamury UR, Swanson BG (1999a) Hazard analysis and critical control point (HACCP) in PEF processing. In: Preservation of foods with pulsed electric fields, pp 172–183. https://doi.org/10.1016/b978-012078149-2/50008-8

Barbosa-Cánovas GV, Góngora-Nieto MM, Pothakamury UR, Swanson BG (1999b) PEF inactivation of vegetative cells, spores, and enzymes in foods. In: Preservation of foods with pulsed Electric fields, pp 108–155. https://doi.org/10.1016/b978-012078149-2/50006-4

Caggia C, Scifò GO, Restuccia C, Randazzo CL (2009) Growth of acid-adapted Listeria monocytogenes in orange juice and in minimally processed orange slices. Food Control 20:59–66. https://doi.org/10.1016/j.foodcont.2008.02.003

Code of Federal Regulations CFR 21, Part 120, Subpart B, Sec. 120.24: Process controls

D'Aous J, Maurer J, Bailey J (2001) Chapter 8: Salmonella species. In: Doyle MP, Beuchat LR, Montville TJ (eds) Food microbiology: fundamentals and fundamentals, 2nd edn, pp 141–178

EHEDG (2018) Document 8: hygienic design principles. https://www.ehedg.org/guidelines/

Evrendilek GA, Zhang QH (2003) Effects of pH, temperature, and pre-pulsed electric field treatment on pulsed electric field and heat inactivation of Escherichia coli O157:H7. J Food Prot. https://doi.org/10.4315/0362-028X-66.5.755

FDA (2004) Guidance for industry: juice HACCP hazards and controls guidance, 1st edn. Final guidance. FDA, Silver Spring

Gahan CGM, O'Driscoll B, Hill C (1996) Acid adaptation of Listeria monocytogenes can enhance survival in acidic foods and during milk fermentation. Appl Environ Microbiol 62:3128–3132. https://doi.org/10.1128/aem.62.9.3128-3132.1996

Heinz V, Alvarez I, Angersbach A, Knorr D (2001) Preservation of liquid foods by high intensity pulsed electric fields – basic concepts for process design. Trends Food Sci Technol 12:103–111. https://doi.org/10.1016/S0924-2244(01)00064-4

Jermann C, Koutchma T, Margas E et al (2015) Mapping trends in novel and emerging food processing technologies around the world. Innov Food Sci Emerg Technol 31:14–27. https://doi.org/10.1016/j.ifset.2015.06.007

Jeyamkondan S, Jayas DS, Holley RA (1999) Pulsed electric field processing of foods: a review. J Food Prot 62:1088–1096. https://doi.org/10.4315/0362-028X-62.9.1088

Larkin J (1999) Regulatory issues and consumer acceptance in non-thermal food preservation. Inst Food Technol Annu Meet 30

Leyer GJ, Wang L-L, Johnson EA (1995) Acid adaptation of Escherichia coli O157:H7 increases survival in acidic foods. Appl Environ Microbiol 61(10):3752–3755

Liang Z, Mittal GS, Griffiths MW (2002) Inactivation of Salmonella Typhimurium in orange juice containing antimicrobial agents by pulsed electric field. J Food Prot 65:1081–1087. https://doi.org/10.4315/0362-028X-65.7.1081

Mak PP, Ingham BH, Ingham SC (2001) Validation of apple cider pasteurization treatments against Escherichia coli O157:H7, Salmonella, and Listeria monocytogenes. J Food Prot 64:1679–1689. https://doi.org/10.4315/0362-028X-64.11.1679

Meng J, Doyle M, Zhao T, Zhao S (2001) Enterohemorrhagic Escherichia coli. In: Doyle MP, Beuchat LR, Montville TJ (eds) Food and microbiology: fundamentals and frontiers, 2nd edn, pp 193–213

Miller LG, Kaspar CW (1994) Escherichia coli O157:H7 acid tolerance and survival in apple cider. J Food Prot. https://doi.org/10.4315/0362-028X-57.6.460

National Advisory Committee on Microbiological Criteria for Foods (1998) Hazard analysis and critical control point principles and application guidelines. J Food Prot 61:1246–1259. https://doi.org/10.4315/0362-028X-61.9.1246

Saldaña G, Puértolas E, López N et al (2009) Comparing the PEF resistance and occurrence of sublethal injury on different strains of Escherichia coli, Salmonella Typhimurium, Listeria monocytogenes and Staphylococcus aureus in media of pH 4 and 7. Innov Food Sci Emerg Technol 10:160–165. https://doi.org/10.1016/j.ifset.2008.11.003

Sheth AN, Wiersma P, Atrubin D et al (2008) International outbreak of severe botulism with prolonged toxemia caused by commercial carrot juice. Clin Infect Dis 47:1245–1251. https://doi.org/10.1086/592574

Sitzmann W, Vorobiev E, Lebovka N (2016a) Applications of electricity and specifically pulsed electric fields in food processing: historical backgrounds. Innov Food Sci Emerg Technol 37:302–311. https://doi.org/10.1016/j.ifset.2016.09.021

Sitzmann W, Vorobiev E, Lebovka N (2016b) Pulsed electric fields for food industry: historical overview. In: Miklavcic D (ed) Handbook of electroporation. Springer, Cham, pp 1–20

Environmental Impact Assessment of Pulsed Electric Fields Technology for Food Processing

Kemal Aganovic and Sergiy Smetana

Introduction

Food processing entails a series of steps and operations that turn raw biomass into final products suitable for human consumption while delivering energy and certain function in the human organism, as well as pleasure and joy. Efficiency and sustainability of the conversion can be directly influenced by the choice of a raw material and suitable processing technology. Besides raw material quality, selected technology and intensity of processing strongly influence important food attributes such as texture, flavor, appearance, and nutritional composition, and finally consumer acceptance. Food production involves numerous steps, mostly classifiable in mechanical, thermal, or special processing steps, e.g., sorting, washing, crushing, milling, extraction, separation, heating, conversion, etc. Accordingly, different processing steps have different demands in terms of intensity and energy required to achieve the desired goal (e.g., sorting or blanching vs. roasting, frying, or deep freezing). Often (in particular for energy intensive processes) methods and approaches are sought that can improve the efficiency and reduce the intensity, with the final goal of energy reduction and maintaining or improving product quality.

Owing to the perishable nature of fresh agricultural biomass, often a preservation step is required to extend shelf-life and provide safety. In many cases, preservation involves actions taken to maintain as much as possible fresh and natural attributes of the product while providing safety and extending shelf life. In general, preservation can be applied by (1) slowing down or inhibiting chemical deterioration and microbial growth (e.g., acidification, refrigerated storage, concentration, etc.), (2) directly inactivating spoilage microorganisms and enzymes, and (3)

K. Aganovic (✉) · S. Smetana
DIL Deutsches Institut für Lebensmitteltechnik e. V., Quakenbrück, Germany
e-mail: k.aganovic@dil-ev.de; s.smetana@dil-ev.de

J. Raso et al. (eds.), *Pulsed Electric Fields Technology for the Food Industry*, Food Engineering Series, https://doi.org/10.1007/978-3-030-70586-2_19

avoiding recontamination before and after processing (e.g., packaging, hygienic processing and storage, etc.) (Rahman 2012). Along with thermal processing, as the most often used preservation approach, often nonthermal food processing technologies are applied and described as energy and environmentally friendly. Owing to their different and specific working principle compared to heat, and driven by other forces than thermal energy, these technologies open new applications beyond bacterial inactivation and preservation. In case of PEF, first attempts focused on permeabilization of fruit and vegetable tissue (Flaumenbaum 1949), separation of phases (Doevenspeck 1960), or microbial inactivation (Sale and Hamilton 1967; Hamilton and Sale 1967). In the following years, the latter aspect was the most interesting and promising, resulting in a range of published papers and scientific reports for gentle preservation of liquid food products (Barbosa-Cánovas et al. 1999; Knorr et al. 1994). Despite early expectations and research and development, focusing mostly on microbial inactivation and gentle preservation of liquids, PEF was sooner established in targeted tissue softening and improving mass transfer, allowing for its scale-up and market maturity.

PEF is a technology relying on electrical energy accumulated in capacitors and delivered in form of high-intensity short-time pulses to a biomass contained in a treatment chamber with electrodes (Heinz et al. 2001). In this targeted energy delivery, minimal losses occur, compared to conventional thermal heating and unavoidable energy dissipation toward the environment (air, water). It should be noted that PEF is applicable to the foods with a certain degree on electrical conductivity and moisture, allowing for ion movement and treatment effects (Rodriguez-Gonzalez et al. 2015). PEF at the same time has a range of applications for improving mass transfer, microbial inactivation, or structure modification (Grahl and Märkl 1996; Ade-Omowaye et al. 2003; Lebovka et al. 2004). Despite a long history of research on PEF alone, in combination or in comparison with thermal and other preservation technologies, there are very few studies available, which would reveal the environmental benefits and drawbacks of PEF on a "fair" comparison basis, either preservation or mass transfer improvement level. In the following chapter, a brief review on environmental aspects of PEF in comparison with other conventional and emerging technologies is provided. The chapter will rely on life cycle thinking approach and relevant comparison basis to define the proper function of compared systems.

Environmental Assessment of a Food Processing Technology

Sustainability of products or technologies is currently assessed via a few standardized methods based on life cycle thinking. Life cycle thinking (Fava 1993) is a broad underlying concept, aimed at considering direct and indirect environmental impacts of complex systems, integrated into several related methods such as eco-efficiency, eco-design, and life cycle assessment (LCA). Special attention to futuristic estimates is devoted in anticipatory (Guinée et al. 2018), scenario-based (Fukushima and Hirao 2002), consequential (Zamagni et al. 2012), prospective, or ex ante LCA (Spielmann et al. 2005; Walser et al. 2011; Cucurachi et al. 2018;

Buyle et al. 2019). All of the mentioned methods have a certain application aim and level of accuracy.

Life cycle assessment (LCA), aiming at an environmental impact assessment, consists of a few main components, concepts, and guiding factors. ISO standards define four main stages of LCA: (1) goal and scope definition; (2) life cycle inventory (LCI); (3) life cycle impact assessment (LCIA); and (4) interpretation (ISO 14040 2006; ISO 14044 2006). Any LCA should include information on functional unit (FU), system boundaries, impact assessment methods, and timeframe (Guinée and Lindeijer 2002). Moreover, two main approaches in completing LCA should be outlined: attributional (information on the environmental burden associated with the specific product life cycle) and consequential (information on environmental burden appearing because of decision-making with consequences of the market changes) (UNEP/SETAC 2011). Such conceptualization allows for certain consistency between different studies and standardization and comparability of results.

When performing LCA for emerging food processing technologies, a practitioner should consider the comparability of a large-scale existing technology to the one being developed and assessed. In many cases, it is not possible to extrapolate the environmental impact on a small or a lab scale to a larger industrial setup. However, with the support of defined energy laws and modelling approaches, certain degree of transferability is possible (Caduff et al. 2011, 2012, 2014). Alternatively, downscaling the conventional technology to the lower scale for the fair comparability could be a viable strategy (Aganovic et al. 2017; Ites et al. 2020; Smetana et al. 2020).

Another essential point for any comparative LCA is the selection of a proper functional unit (FU), which should reflect not only physical nature of a product but also the functional properties of foods: nutrition, palatability, shelf life, etc. In the last decades, numerous LCA studies pointed to the need to include functional properties of food for the "fair" comparability of food production technologies. They included protein amount and digestibility, nutritional composition, and textural properties (Smetana et al. 2015; Weidema and Stylianou 2019; McAuliffe et al. 2020; Ulmer et al. 2020). In terms of PEF, recent digestibility studies indicated no significant differences and even improvements of PEF-treated products (Barba et al. 2017; Stübler et al. 2020; Chian et al. 2019). These functional aspects in LCA studies and comparisons are often not considered and currently represent a challenge. Furthermore, comparability of emerging processing technologies to conventional ones is complicated due to the different effects by different means on the biomass caused, e.g., by thermal treatment, electrical discharges, pressure changes, waves applied, etc. Such processes often result in new structure types forming a new product with distinguished properties.

Further, the intensities of the treatment have to be selected on a comparable basis or an effect. The aim of the treatment and the result thereof must be comparable at some level. Besides comparing different technologies at a comparable microbial inactivation, as usually used in many studies, comparison of these technologies can be performed also aiming on other parameters, e.g., comparable enzyme inactivation, energy and costs requirements, or retention level of a certain compound, among other parameters. Next, it is recommendable that comparisons are performed for the same or at least comparable technology readiness levels (TRL). More importantly, the comparison should be done at the same or at least comparable production scale (Aganovic et al. 2017). It is not viable and not fair to compare LCA based on data from prototypes, patent records, and theoretical calculations to analyses obtained from industrial systems (Gavankar et al. 2015). Therefore, the comparison of emerging technologies should be based on the data obtained from pilot or industrial technological scale (Aganovic et al. 2017), or ex ante approaches should be applied for upscaling modelling (Cucurachi et al. 2018). It is also necessary to indicate that in the food processing industry, such aspects as comparison of technologies in ideal conditions (e.g., specific energy consumed, comparison without loading-unloading timing) are quite common (Atuonwu et al. 2018, 2020; Atuonwu and Tassou 2018; Silva and Sanjuán 2019). While such approaches are viable for the first approximation of technological viability, they might not necessarily result into practically useful results.

Comparison of batch nature of processing versus continuous processing lines sets certain level of bias in favor of latter approaches. Therefore, such approaches should be avoided where possible. Moreover, different traditional and emerging processing technologies can be applied for a variety of objectives from microbial or enzyme inactivation, structure modification, or extraction (Perez and Pilosof 2004; Xiang et al. 2011; Zhao et al. 2007; Corrales et al. 2008); thus the fair comparison should include only relevant background of the comparing technologies.

Performances of empirical studies are quite specific, and it is sometimes difficult to draw conclusions based on assumptions and comparisons made in some studies or precisely export results to other studies. In this way, a sensitivity analysis with changes of the most impactful elements of the life cycle chains can be performed (Silva and Sanjuán 2019), to identify different scenario with more environmentally sustainable options. In the case where the use of energy is related to the highest environmental impacts, it is meaningful to conduct a scenario analysis by changing the energy mix for other types of fuels and sources. Use of natural gas compared to the energy sources that include gasoline or coal is recognized to be lower in general (Burnham et al. 2012). Similarly, the changes of transportation distances or transportation means, where the highest environmental impacts are related to transportation, might change the overall results in certain applications.

Sustainability Conditions for Different PEF Applications

As described in previous chapters of this book, PEF technology is based on application of high-intensity electric field pulses of short duration on the product contained between electrodes. This results in structural rearrangements, changes in conductivity, and increased cell membrane permeability to ions and macromolecules. As such increase in permeability is, presumably, related to the formation of nanoscale defects or pores on the cell membrane, the phenomenon is often called "electroporation" or "electropermeabilization" (Ivorra and Rubinsky 2010; Kotnik et al. 2012).

The electroporation effect can be utilized in different ways and for a range of applications in food, biotechnology, medicine or pharmacy, and similar areas. Depending on the desired effect to be achieved by the PEF treatment, different treatment intensities and treatment configurations can be applied. For many of these applications, energy requirements are related to electrical energy, directly used by a pulse generator and delivered to the product, and accompanying processes and auxiliary equipment. Unlike thermal or high hydrostatic pressure (HHP) treatment where process can be described by few parameters such as temperature, pressure, and time, the PEF treatment is characterized by several process parameters, such as peak voltage, electric field strength, treatment time, pulse form, pulse width, number of pulses, pulse specific energy, and pulse repetition rate. However, mainly electric field strength and specific energy input are considered as major treatment parameters, whereas the specific energy input can be considered as a dose parameter.

In the food processing, PEF can be applied for solid, liquid, and semiliquid goods (Siemer et al. 2014a). In PEF applications for solid foods, structural changes of biological material are desired, and specific energy levels are typically lower, compared, for example, to the energy needed for electroporation of bacterial cells. Typically, specific PEF energy of around 1–3 kJ/kg is required for softening roots and tubers, e.g., in applications for softening potatoes for easier processing, such as cutting and frying (Fauster et al. 2018), compared to conventional methods, such as mechanical processing (20–40 kJ/kg), enzymes (70–100 kJ/kg), or heat treatment (>150 kJ/kg) (Toepfl et al. 2006). For cell permeabilization aiming on extraction from plant-based materials, specific energy of around 3–10 kJ/kg is required (Yan et al. 2017). Significantly lower specific energy levels are required for applications, for example, in biotechnology, where nanosecond electrical pulses are used and recommended for inducing stress response in microorganisms and algae (Buchmann et al. 2019). For treatment of heat-sensitive liquid products with the aim of inactivation of vegetative bacteria and shelf life extension, applied energy levels are usually between 80 and 120 kJ/kg, depending on the product characteristics, process setup, and bacterial characteristics (Toepfl 2011). Aiming on inactivation of spores, the specific energy can reach up to 350 kJ/kg in combination with thermal energy (Siemer et al. 2014b). Conclusively, depending on the application of the PEF process and the desired outcome, specific energy requirements per unit of volume/weight can vary significantly. The abovementioned examples indicate a wide range of specific energy intensities applied to food biomass for a range of applications from plant cell permeabilization to spores inactivation. Therefore, the impact of PEF cannot be simply estimated without the background information. Moreover, the comparison of PEF to conventional treatment technologies should be performed only on the basis of the targeted effect. From the LCA perspective, not only the direct energy savings (or losses) should be considered but also the effects on the other chains of the production lines. For example, substitution of water preheating in French fries' production with PEF except for the direct energy savings effects has positive impacts on oil uptake, frying, water loss, etc. Moreover, the impacts could lead to even further extended consequences associated with the higher/lower acceptance of the new product and relevant effects on the market. Such indirect rebound effects should be considered in the LCA.

Life Cycle Impact Assessment and Comparison of Technologies

In comparison studies, very often a certain reference and benchmark is needed. In most of the comparison studies, PEF technology was compared to already existing and established treatments or other alternative processes, in terms of quality, process efficiency, or environmental impact. Environmental impact assessment of PEF processing should be performed at the scale similar to the one of the conventional processing. The assessment accounts for the electric energy directly delivered to the volume or weight unit of a product, but also for additional aspects and resources, depending on the application and type of processing, typically required for a complete operation. Thus, the energy needed for the production of material, its transportation, washing, sorting, crushing, heating up, cooling down, drying, cleaning, maintenance, possibility for energy recovery, and all other relevant processes in relation to the defined functional unit, should be taken into account.

Moreover, LCA practitioner would account for the potential differences in infrastructure (resources needed to build a processing line), water use and cleaning procedure (detergent and water), variations in input and output material flows, etc. The limits of consideration on the inclusion in the LCA are usually set through the system boundaries, which in most cases would include options from "cradle-to-gate," "gate-to-gate," or "cradle-to-grave." Such frame outlines the rough setup for the analysis, and indicating if raw material production is from upstream processes ("cradle"), processing of biomass into food ("gate-to-gate"), and end of life of the product through the consumption and waste treatment ("grave") are included in the scope of the study.

PEF for Preservation of Liquids

Depending on product characteristics (pH value, viscosity, electrical conductivity, etc.) and cell characteristics (size, shape, orientation in the electric field), processing conditions may vary, but usually, the electric field strength of at least 10 kV/cm is required to achieve microbial electroporation (Castro et al. 1993). There are numerous studies available describing inactivation of microorganisms by PEF, usually reporting electric field strength up to 60 kV/cm (Mosqueda-Melgar et al. 2008; Castro et al. 1993; Wouters et al. 2001), whereas in the large-scale applications, usually electric field strength of 10–25 kV/cm is used (Toepfl 2011). Required treatment intensity and related energy required depend on the different specific energy inputs needed to achieve the desired level of microbial inactivation in different products (e.g., conductivity, specific heat capacity, pH, etc.), different preheating temperatures, and possible different microbial population occurring in the product

(Aganovic et al. 2017). For preservation of liquid products (e.g., juices, smoothies, milk, etc.), synergistic effects between thermal and PEF treatment for achieving microbial inactivation have been reported (Heinz et al. 2003). PEF effect occurs due to the product's electrical conductivity and ability of electrical current to flow through it due to charged molecules. However, depending on the product conductivity and strength of an electric field (defined by the electrode geometry and electrical potential), a certain amount of electrical energy is dissipated as heat, based on the principle of Joule heating (Heinz et al. 2001). As illustrated in Fig. 1, in PEF applications, the liquid product is usually heated up to a certain temperature, typically between 20 and 40 °C. This heating is required at the starting phase, whereas on the course of the processing, up to 95% of heat recovery can be achieved depending on the processing setup and type of heat exchangers used, by preheating the cold product with warm product already treated by PEF. Based on yearlong experience and optimization, conventional thermal technologies in industrial conditions rely on almost 95% of thermal energy recovery rate making it hard to compete in terms of energy efficiency for PEF preservation (e.g., pasteurization) (Toepfl et al. 2006). Nevertheless, for heat-sensitive products or products where the application of plate heat exchangers for optimal energy recovery is not suitable, these emerging technologies may be economically and environmentally viable solution.

Consequently, depending on the amount of thermal energy that can be recovered, for example by using heat exchangers (Fig. 1), heat recovery will have a large impact on the total energy balance, regardless if for thermal or for PEF treatment

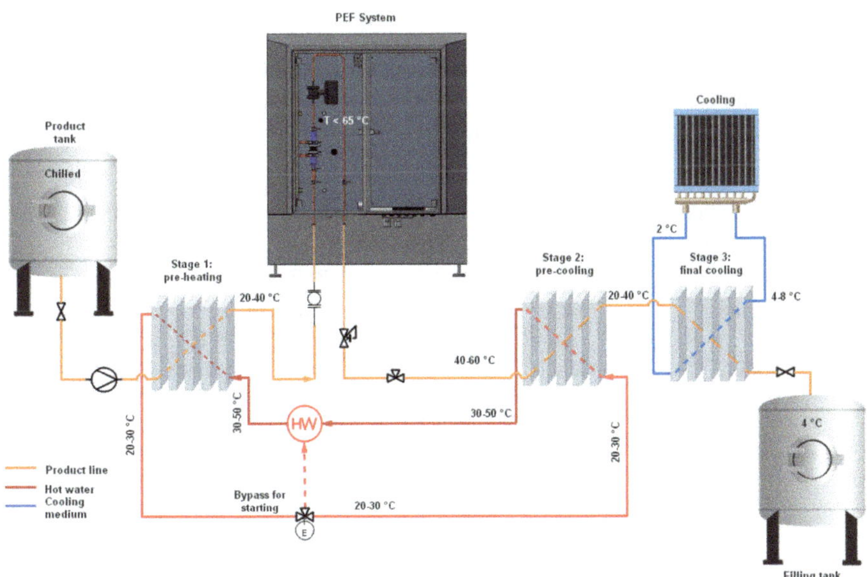

Fig. 1 An example flowchart of PEF treatment juice line, from product storage tank to filling tank, including system for preheating, PEF system, cooling, and heat recovery (Reproduced and adapted from Aganovic et al. 2017)

(Jouhara et al. 2018; Aganovic et al. 2017). In terms of energy balance, by increasing the inlet temperature of a product before PEF treatment, theoretically more energy could be recovered in the heat exchangers, and less energy would be required for cooling the final product. For example, preheating the product to moderate temperatures (e.g., 40–60 °C) and applying less energy from PEF (e.g., 60–80 kJ/kg), a more efficient process could probably be set up. Nevertheless, it is worth mentioning that this thermal-assisted setup with potentially improved efficient energy recovery would be suitable for less heat-sensitive products and products where relatively high specific PEF energy inputs are required to achieve the desired level of bacterial inactivation. From the quality perspective, it is an optimal or maximum temperature that the product may undergo while maintaining and reaching the desired quality characteristics that can serve as the upper processing limit.

In case of application of PEF and comparison to other technologies for preservation of juices, there are only two LCA studies published (Davis et al. 2010; Aganovic et al. 2017), including few more on other foods or other processes (Pardo and Zufía 2012; Cacace et al. 2020; Villamonte et al. 2014). Obviously, the comparison with available literature data has a very "shallow" character as multiple processing aspects were different in the studies, including the processing parameters, the scale of processing, options for heat recovery, etc., as well as a very limited number of studies, not only for PEF processing but for conventional and alternative processing of food in general. Even though different authors used different raw materials for the processing trials, the impact results were in close approximation to the ranges of the main midpoint results (Table 1). However, to make a comparison with literature data, it is necessary to account for the variations in systems and LCA methodologies.

A LCA performed on tomato juice and selected preservation technologies (thermal, PEF, and HHP) revealed an expected differentiation between the technologies, based on the differences in energy input required to achieve a comparable level of

Table 1 Main midpoint impact categories results for the complete assessment (from "farm to gate," FU 1 kg of preserved juice, bottled and ready for sale)

Impact category	Unit	Thermal	PEF	HPP
Land occupation	m^2 arable	TM: 0.14–0.50[a] WM: 0.19–0.27[a]	TM: 0.14–0.50[a] WM: 0.19–0.27[a]	TM: 0.14–0.50[a] WM: 0.19–0.27[a]
Global warming	kg CO_2 eq.	TM: 1.83–1.84[a] WM: 1.03–1.11[a] CJ: 0.75[b]	TM: 1.87–1.89[a] WM: 1.08–1.16[a] CJ: 0.75[b]	TM: 1.93–1.95[a] WM: 1.14–1.22[a] CJ: 0.75[b]
Nonrenewable energy use	MJ	TM: 22.56–25.89[a] WM: 16.49–19.58[a] CJ: 14.8[b]	TM: 23.14–26.76[a] WM: 17.07–20.46[a] CJ: 14.5[b]	TM: 23.78–27.71[a] WM: 17.71–21.41[a] CJ: 14.7[b]

Sources: [a]Aganovic et al. (2017); [b]Davis et al. (2010)
Note: Ranges of results are based on the use of IMPACT 2002+ and ReCiPe methodologies and are expressed as arithmetic mean values of two methodologies ± half deviation ranges; *TM* tomato juice, *WM* watermelon juice, *CJ* carrot juice

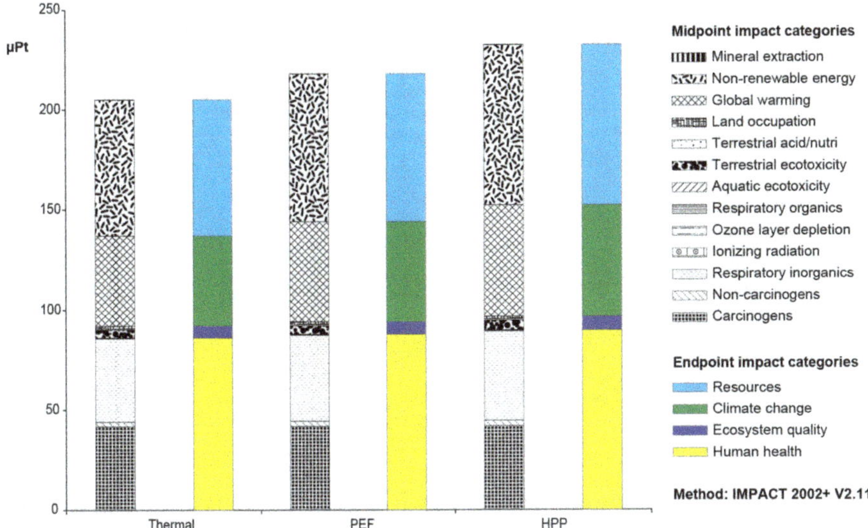

Fig. 2 Processing stage LCA technologies comparison (FU 1 kg of preserved tomato juice, bottled and ready for sale, raw material production excluded, scope – from "gate to gate"; zero values of aquatic acidification and eutrophication excluded; 1 kPt = impact of 1 European per year) (Reproduced from Aganovic et al. 2017)

microbial inactivation, analyzed in the scope of "processing stage LCA" (Aganovic et al. 2017). This outcome was expected considering that in "gate-to-gate" approach, the largest impact is usually associated with direct energy consumption (Fig. 2). Other inputs and outputs, like water consumed, bottle type, cleaning procedures, and similar, were intentionally levelled to obtain a "fair comparison" ground. Even the energy comparison was the highest for the HHP, around 85% of the environmental impact was associated with the production of the 250 mL PET bottles. In a scenario analysis, it is possible to identify alternative packaging options that potentially can decrease the overall environmental impact. Thus, high-density polyethylene (HDPE) or polypropylene (PP) bottles of the same weight as their polyethylene terephthalate (PET) counterpart may improve the overall environmental impact.

Further LCA in "cradle-to-gate" approach revealed the impact and potential changes in supply chains triggered by alterations of the different preservation technology. The results indicated only minor differences coming from the energy used for processing of the tomato juice, while differences were more pronounced when compared to another juice product, namely, watermelon juice (Fig. 3). Thus, besides the impact of packaging, the processing stage aiming at preservation will most likely play a minor role in the "cradle-to-gate" approach, regardless of the preservation technology. The highest impacts of a juice production from "farm to gate" will likely be associated with the energy and resources required for production, transportation, and storage of the raw materials. In another study where PEF was compared to HHP technology for preservation of carrot juice (Davis et al. 2010), an energy consumption of 14–15 MJ (3.89–4.17 kWh), ~0.75 kg CO2 eq. for climate change,

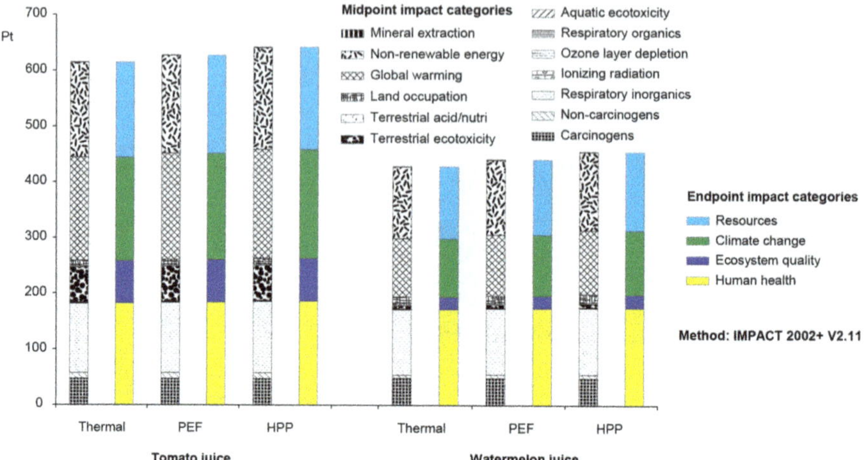

Fig. 3 Results of LCA comparison for different technologies (FU 1 kg of preserved juice, bottled and ready for sale, scope – from "farm to gate"; zero values of aquatic acidification and eutrophication excluded; 1 kPt = impact of 1 European per year) (Reproduced from Aganovic et al. 2017)

~0.85 g PO4 eq. for eutrophication, and ~ 2.2 g SO4 eq. for acidification were reported. Similarly, the conclusion was that selection of packaging should be carefully considered for a sustainable PEF process, described as a possible tool to design product with potential environmental advantage (Davis et al. 2010). Even when some results of these two studies are comparable, in general, due to different processing parameters, as well as due to variations in systems and LCA methodologies, deeper discussions and comparisons are very difficult.

Taking into account the different raw materials, treatment conditions, and other factors considered in other similar studies, it is difficult to compare obtained results with those already available in the literature. In some cases, the comparison of processing stage LCA would be possible if the full range of conditions (e.g., production capacity, holding time, temperature profile, energy recovery) are specified in literature sources. The production capacity of equipment is of special importance as it affects the product flow volume and efficiency of resource use. The lack of these parameters creates difficulties for the reproduction and comparison of the studies.

PEF in Potato Processing

By applying PEF on solid foods and agricultural goods, a loss of turgor pressure, followed by tissue softening and improved mass transfer, can be achieved. One of the most widespread applications of PEF so far based on these effects is found in processing of potatoes, in particular French fries and chips. By applying specific energy input of around 1–2 kJ/kg, numerous benefits in potato processing are achieved, such as smoother cut, less fracture, reduced starch and increased sugar release, less oil uptake during frying, less acrylamide formation, and several others (Janositz et al. 2011; Fauster et al. 2018; Ignat et al. 2015; Genovese et al. 2019). In addition, these systems replace typical conventional thermal systems for preheating the potatoes before cutting and frying (typical time-temperature combinations 30 min at around 60 °C), thus resulting in significant energy, time, and water savings (McHugh and Toepfl 2016). The first potato industrial application was installed in 2009, and although the range of benefits are multiple in terms of product quality and process efficiency, still, there is no single peer review study addressing LCA of this application. Most of the studies in this case report on energy and resources savings induced by the process. The study of Fauster et al. (2018) revealed a reduction of water usage by 90% by replacing traditional preheater of potatoes by PEF, while improving characteristics of potatoes and reducing energy consumption for preheating by up to 85% can be achieved. Industrial users of the technology reported an annual reduction of freshwater use by 8% and energy use by 5% while improving frozen product recovery (PotatoPro 2017).

PEF for Other Applications

Other applications of PEF process include pre-treatments and conditioning for different purposes, aiming mostly on improved mass transfer, obtaining a substance of interest, removing water from foods, or introducing a given substance into the food matrix (Puértolas et al. 2012; Donsì et al. 2010; Dymek et al. 2015; Gómez et al. 2019). While processing benefits in these applications are multiple and are reflected in improvement of, e.g., extraction yield, shortening of processing time, and improvement of product quality, the specific PEF energy inputs are rather low, compared to microbial inactivation using PEF (Barba et al. 2020). Due to small pores in the membranes caused by electroporation, different intracellular compounds of different molecular size and of fine purity can be extracted from different biomass (Yan et al. 2017). These include extraction of lipids, colorants (chlorophylls, carotenoids, betalains), sugars, polyphenols, proteins, polysaccharides, and many more, where often no use of solvents or chemicals is required (Chemat et al. 2015; Käferböck et al. 2020). Improvement of extraction yield or reduction of malaxation time in olive oil processing (Abenoza et al. 2013); improvement of color, anthocyanin content, and total polyphenolic index (López et al. 2008); or energy-efficient wine production with reduced days required for maceration (Ferreira et al. 2019) were also reported. In the range of scientific studies, combined approach of PEF with other means has been beneficially described, such as improved wine quality in a PEF and enzyme-combined applications (Fauster et al. 2020), improved quality of grapefruit juice after combined PEF and sonication treatment (Aadil et al. 2015), improved quality of red peppers in combination with PEF and osmotic dehydration (Ade-Omowaye et al. 2003), or synergistic effects with antimicrobials for bactericidal effects (Pol et al. 2000).

Not only for direct extraction from fruits and vegetables, PEF has been also reported as a promising and economically viable technology for recovery of high valuable compounds from side streams and waste of agricultural and food processing either alone (Golberg et al. 2016; Andreou et al. 2020; Luengo et al. 2014) or in combination with other mechanical, biotechnological, and conventional approaches toward development of greener and sustainable biorefineries (Puértolas and Barba 2016). In another PEF application for peeling, PEF was added as an addition to traditional steam peeling in the tomato processing, and a reduction of energy use by 20% for the thermophysical peeling stage was reported. In addition, PEF application also did not require air blowing during the washing stage. The resulting energy reductions improved environmental impact on several indicators, like climate change, ozone depletion, and terrestrial acidification among others (Arnal et al. 2018). Such applications allow for recovery and added value of usually underutilized resources and thus have beneficial environmental impact.

Due to lack of environmental studies involving PEF technology, which can be stated also for other emerging technologies, a good starting approach is evaluation of an energy balance and additional beneficial impacts of the treatment. An energy assessment for a processing technology can be obtained by calculation of the changes of internal energy, energy directly applied to a product, or a total energy consumed at a power source (Heinz et al. 2001, 2003). Internal energy is a characteristic of a system, in this case dependent on properties of food, such as specific heat capacity and thermal and electrical conductivity. These attributes will have direct impact on energy balance and effectiveness of a process. For the PEF treatment, the major parameter, beside moisture level, is electrical conductivity and dielectric constant of a food. For many goods and foods, these parameters are usually homogeneously distributed within the biomass. However, in some cases content and distribution of one phase, does not necessarily have to be the same and homogeneously distributed, and can have significantly different physical properties. For example in meat, content and distribution of fat and connective tissue can vary significantly, depending on many parameters (animal, feed, meat cut, etc.), but also in fiber orientation. Thus, these variables can have direct and significant impact on effectiveness of the treatment, as well as on the energy demand (Smetana et al. 2019). The applications of PEF for improving tenderness of meat and reduction of ageing time have been described as a result of increased permeabilization of cellular membranes due to PEF-caused electroporation (Warner et al. 2017; Faridnia et al. 2014; Alahakoon et al. 2019). Applied without significant heating when compared to conventional thermal processes, it has a potential to reduce needs for additional cooling, as well as to reduce needs for water consumption. Certain heating and cooking processes are associated with release of particulate matter and gases, which has not been reported for the PEF technology. In this way, substantial advantages in terms of energy and environmental aspects can be achieved for PEF technology compared to conventional processes.

Techno-economic Conditions

It is necessary to point out that, except for the environmental impact, there are many other factors which should be considered when considering PEF application on an industrial scale. Some of these factors include improvement in process efficiency through consideration of operational requirements and technical constraints, improvement of product quality, market and marketing opportunities, or provided production flexibility. Consideration of financial criteria of PEF application includes product throughput and output, installation cost, fixed capital cost, operating cost, maintenance costs, labor cost, etc. The capital investment costs are usually driven by the electrical capacity, whereas the power and operating costs are mostly related to required production capacity and applied processing conditions. For different applications, different and robust machines can be developed to allow flexible applications while lowering the investment costs and still providing advantages. Considering PEF investment costs and seasonal character of some agricultural commodities with varying production capacities (e.g., wine or olive oil production), there is a need for a different kind of solution. Here, different business models seem to be applicable, like machine leasing, farm cooperatives owning (or leasing) the machine or shared machines by a private entity.

Moreover, implementation of PEF on industrial scale should rely on the three key conditions to be considered: (i) the required process protocol (e.g., field strength, treatment time), (ii) product characteristics (e.g., electrical conductivity), and (iii) the desired throughput (Heinz et al. 2003). Parameters of direct importance considering electrical applications of PEF systems are electric field strength, the system's voltage, and size and configuration of treatment zone. These parameters play major role in reaching the desired electric field strength and depending on the design and size of the treatment zone, the distribution of the electric field, and the energy applied. In order to ensure correct estimation of costs and complexity of the PEF system, these processing parameters should be carefully assessed.

As in case of other emerging technologies, especially in their early application and life stage, PEF technology is considered to have high manufacturing and initial investment costs. Depending on the application and capacity of the machine, as well as level of collateral equipment (e.g., belt, heat exchanger), a PEF system purchase price is in the range from 250,000 Euros to more than 1 million Euros. However, an important measure for the commercial success of a technology is the return of investment, along with payback period, lifetime, and residual value. In the scenario of potato processing, return of investment for a 60 tons per hour PEF machine is considered to be around 2 years; the production costs are in the range of 0.5–0.7 Euro per ton and range of other benefits are provided. The costs for recovery of valuable compounds by PEF from different matrices are estimated around 0.1–0.5 EUR/ton, still much lower compared to costs of an enzymatic pre-treatment (7.5 Euros/ton) (Picart-Palmade et al. 2019). Prices for juices are estimated in the range of 0.01–0.2 EUR/L, compared to thermal treatment (e.g., high temperature short time, HTST) where costs are usually around 0.005–0.01 EUR/kg, depending on the throughput, annual operating hours, and energy recovery (Heinz and Buckow 2010). Similarly, for processing apple pulp, PEF costs were estimated around 2.69 EUR/ton compared to 8.50 EUR/ton for enzymatic maceration (Toepfl and Heinz 2011). With further process optimization and new developments, technologies such as PEF are rapidly maturing and experiencing improvements in machine design and manufacturing, allowing systems to be smaller and less expensive while still providing process and product improvements.

Conclusion

The current chapter summarizes a few major findings on environmental aspects of PEF processing based on information available in the literature. First of all, there is a lack of scientific information on the LCA of PEF technology, regardless of application. Interestingly, the LCA studies are mostly performed on the PEF application for preservation of juices or tomatoes peeling, where actually lower environmental benefits are expected. In other applications, e.g., for potato processing, LCA studies are currently completely missing, but benefits of PEF application in this field are known, and its industrial penetration is driven by energy savings, costs reduction, and product quality improvement.

"Gate-to-gate" scope sets specific challenges in the assessment of processing technologies (alternative technologies are not an exception). Thus, the biggest impact at the processing stage might be associated not with the change of processing parameters of application of alternative technology, but with the change of packaging (e.g., different plastic for bottles). Available studies where analysis were performed at the same scale and for the similar product, indicate that there are no significant differences in environmental impacts of the preservation technologies with "gate-to-gate" system boundaries. However, the energy balance and the outcome of environmental impact assessments of different processing technologies might change with increasing production capacity and consideration of large industrial scale.

One of the most important outcomes of the analysis in this chapter is connected with one more defined shortage in the literature. However, evaluation of new technologies significantly depends also on how consumers perceive environmental sustainability of a technology, or PEF process, and how it will impact their purchase decision. Some consumers are more open to innovation and believe that new technologies provide benefits and reduce risk (Bruhn 2007). Other aspects, such as price, health claims, and buying regional and local, will also have a significant role when deciding for product purchase. Emerging technologies, such as PEF, are needed to follow progress to more sustainable, more secure, and safe food (Siegrist and Hartmann 2020). Apart from the PEF, there are other food processing technologies like UV light, electron beam, gene technology, etc. that can provide several benefits in terms of processing and product quality, but are facing consumer and/or legal acceptance. The use of emerging food processing technologies, where suitable and beneficial, not only provides social benefits, but can also improve sustainability and environmental protection.

References

Aadil RM, Zeng X-A, Sun D-W, Wang M-S, Liu Z-W, Zhang Z-H (2015) Combined effects of sonication and pulsed electric field on selected quality parameters of grapefruit juice. LWT Food Sci Technol 62:890–893

Abenoza M, Benito M, Saldaña G, Álvarez I, Raso J, Sánchez-Gimeno AC (2013) Effects of pulsed electric field on yield extraction and quality of olive oil. Food Bioprocess Technol 6:1367–1373

Ade-Omowaye BIO, Rastogi NK, Angersbach A, Knorr D (2003) Combined effects of pulsed electric field pre-treatment and partial osmotic dehydration on air drying behaviour of red bell pepper. J Food Eng 60:89–98

Aganovic K, Smetana S, Grauwet T, Toepfl S, Mathys A, Van Loey A, Heinz V (2017) Pilot scale thermal and alternative pasteurization of tomato and watermelon juice: an energy comparison and life cycle assessment. J Clean Prod 141:514–525

Alahakoon AU, Oey I, Bremer P, Silcock P (2019) Quality and safety considerations of incorporating post-PEF ageing into the pulsed electric fields and sous vide processing chain. Food Bioprocess Technol 12:852–864

Andreou V, Dimopoulos G, Dermesonlouoglou E, Taoukis P (2020) Application of pulsed electric fields to improve product yield and waste valorization in industrial tomato processing. J Food Eng 270:109778

Arnal ÁJ, Royo P, Pataro G, Ferrari G, Ferreira VJ, López-Sabirón AM, Ferreira GA (2018) Implementation of PEF treatment at real-scale tomatoes processing considering LCA methodology as an innovation strategy in the agri-food sector. Sustainability 10:979

Atuonwu JC, Tassou SA (2018) Model-based energy performance analysis of high pressure processing systems. Innovative Food Sci Emerg Technol 47:214–224

Atuonwu JC, Leadley C, Bosman A, Tassou SA, Lopez-Quiroga E, Fryer PJ (2018) Comparative assessment of innovative and conventional food preservation technologies: process energy performance and greenhouse gas emissions. Innovative Food Sci Emerg Technol 50:174–187

Atuonwu JC, Leadley C, Bosman A, Tassou SA (2020) High-pressure processing, microwave, ohmic, and conventional thermal pasteurization: quality aspects and energy economics. J Food Process Eng 43:e13328

Barba F, Mariutti LRB, Bragagnolo N, Mercadante AZ, Barbosa-Cánovas GV, Orlien V (2017) Bioaccessibility of bioactive compounds from fruits and vegetables after thermal and nonthermal processing. Trends Food Sci Technol 67:195–206

Barba F, Parniakov O, Wiktor A (2020) Pulsed electric fields to obtain healthier and sustainable food for tomorrow. Academic Press, London

Barbosa-Cánovas GV, Pothakamury UR, Gongora-Nieto MM, Swanson BG (1999) Preservation of foods with pulsed electric fields. Elsevier, London

Bruhn CM (2007) Enhancing consumer acceptance of new processing technologies. *Innovative Food Science & Emerging Technologies, 8*(4), pp.555–558

Buchmann L, Frey W, Gusbeth C, Ravaynia PS, Mathys A (2019) Effect of nanosecond pulsed electric field treatment on cell proliferation of microalgae. Bioresour Technol 271:402–408

Burnham A, Han J, Clark CE, Wang M, Dunn JB, Palou-Rivera I (2012) Life-cycle greenhouse gas emissions of shale gas, natural gas, coal, and petroleum. Environ Sci Technol 46:619–627

Buyle, Audenaert, Billen, et al (2019) The Future of Ex-Ante LCA? Lessons Learned and Practical Recommendations. Sustainability 11:5456. https://doi.org/10.3390/su11195456

Cacace F, Bottani E, Rizzi A, Vignali G (2020) Evaluation of the economic and environmental sustainability of high pressure processing of foods. Innovative Food Sci Emerg Technol 60:102281

Caduff M, Huijbregts MA, Althaus H-J, Hendriks AJ (2011) Power-law relationships for estimating mass, fuel consumption and costs of energy conversion equipments. Environ Sci Technol 45:751–754
Caduff M, Huijbregts MA, Althaus H-J, Koehler A, Hellweg S (2012) Wind power electricity: the bigger the turbine, the greener the electricity? Environ Sci Technol 46:4725–4733
Caduff M, Huijbregts MA, Koehler A, Althaus HJ, Hellweg S (2014) Scaling relationships in life cycle assessment: the case of heat production from biomass and heat pumps. J Ind Ecol 18:393–406
Castro AJ, Barbosa-Cánovas GV, Swanson BG (1993) Microbial inactivation of foods by pulsed electric fields. J Food Process Preserv 17:47–73
Chemat F, Fabiano-Tixier AS, Vian MA, Allaf T, Vorobiev E (2015) Solvent-free extraction of food and natural products. TrAC Trends Anal Chem 71:157–168
Chian FM, Kaur L, Oey I, Astruc T, Hodgkinson S, Boland M (2019) Effect of pulsed electric fields (PEF) on the ultrastructure and in vitro protein digestibility of bovine longissimus thoracis. LWT Food Sci Technol 103:253–259
Corrales M, Toepfl S, Butz P, Knorr D, Tauscher B (2008) Extraction of anthocyanins from grape by-products assisted by ultrasonics, high hydrostatic pressure or pulsed electric fields: a comparison. Innovative Food Sci Emerg Technol 9:85–91
Cucurachi S, Van der Giesen C, Guinée J (2018) Ex-ante LCA of emerging technologies. Procedia CIRP 69:463–468. https://doi.org/10.1016/j.procir.2017.11.005
Davis J, Moates G, Waldron K (2010) The environmental impact of pulsed electric field treatment and high pressure processing: the example of carrot juice. In: Case studies in novel food processing technologies. Elsevier, London
Doevenspeck H (1960) Verfahren und Vorrichtung zur Gewinnung der einzelnen Phasen aus dispersen Systemen. DE 1:237–541
Donsì F, Ferrari G, Pataro G (2010) Applications of pulsed electric field treatments for the enhancement of mass transfer from vegetable tissue. Food Eng Rev 2:109–130
Dymek K, Dejmek P, Galindo FG, Wisniewski M (2015) Influence of vacuum impregnation and pulsed electric field on the freezing temperature and ice propagation rates of spinach leaves. LWT Food Sci Technol 64:497–502
Faridnia F, Bekhit AE-DA, Niven B, Oey I (2014) Impact of pulsed electric fields and post-mortem vacuum ageing on beef longissimus thoracis muscles. Int J Food Sci Technol 49:2339–2347
Fauster T, Schlossnikl D, Rath F, Ostermeier R, Teufel F, Toepfl S, Jaeger H (2018) Impact of pulsed electric field (PEF) pretreatment on process performance of industrial French fries production. J Food Eng 235:16–22
Fauster T, Philipp C, Hanz K, Scheibelberger R, Teufl T, Nauer S, Scheiblhofer H, Jaeger H (2020) Impact of a combined pulsed electric field (PEF) and enzymatic mash treatment on yield, fermentation behaviour and composition of white wine. Eur Food Res Technol 246:609–620
Fava JA (1993) Life cycle thinking: application to product design. In: Proceedings of the 1993 IEEE International Symposium on Electronics and the Environment. IEEE, pp 69–73
Ferreira VJ, Arnal ÁJ, Royo P, García-Armingol T, López-Sabirón AM, Ferreira G (2019) Energy and resource efficiency of electroporation-assisted extraction as an emerging technology towards a sustainable bio-economy in the agri-food sector. J Clean Prod 233:1123–1132
Flaumenbaum B (1949) Electrical treatment of fruits and vegetables before juice extraction. Trudy OTIKP 3:15–20
Fukushima Y, Hirao M (2002) A structured framework and language for scenario-based Life Cycle assessment. The International Journal of Life Cycle Assessment 7:317. https://doi.org/10.1007/BF02978679
Gavankar S, Suh S, Keller AA (2015) The role of scale and technology maturity in life cycle assessment of emerging technologies: a case study on carbon nanotubes. J Ind Ecol 19:51–60
Genovese J, Tappi S, Luo W, Tylewicz U, Marzocchi S, Marziali S, Romani S, Ragni L, Rocculi P (2019) Important factors to consider for acrylamide mitigation in potato crisps using pulsed electric fields. Innovative Food Sci Emerg Technol 55:18–26

Golberg A, Sack M, Teissie J, Pataro G, Pliquett U, Saulis G, Stefan T, Miklavcic D, Vorobiev E, Frey W (2016) Energy-efficient biomass processing with pulsed electric fields for bioeconomy and sustainable development. Biotechnol Biofuels 9:94

Gómez B, Munekata PES, Gavahian M, Barba FJ, Martí-Quijal FJ, Bolumar T, Campagnol PCB, Tomasevic I, Lorenzo JM (2019) Application of pulsed electric fields in meat and fish processing industries: an overview. Food Res Int 123:95–105

Grahl T, Märkl H (1996) Killing of microorganisms by pulsed electric fields. Appl Microbiol Biotechnol 45:148–157

Guinée JB, Lindeijer E (2002) Handbook on life cycle assessment: operational guide to the ISO standards. Springer, Dordrecht

Guinée JB, Cucurachi S, Henriksson PJG, Heijungs R (2018) Digesting the alphabet soup of LCA. The International Journal of Life Cycle Assessment 23:1507–1511. https://doi.org/10.1007/s11367-018-1478-0

Hamilton W, Sale A (1967) Effects of high electric fields on microorganisms: II. Mechanism of action of the lethal effect. Biochim Biophys Acta 148:789–800

Heinz V, Buckow R (2010) Food preservation by high pressure. J Verbr Lebensm 5:73–81

Heinz V, Álvarez I, Angersbach A, Knorr D (2001) Preservation of liquid foods by high intensity pulsed electric fields—basic concepts for process design. Trends Food Sci Technol 12:103–111

Heinz V, Toepfl S, Knorr D (2003) Impact of temperature on lethality and energy efficiency of apple juice pasteurization by pulsed electric fields treatment. Innovative Food Sci Emerg Technol 4:167–175

Ignat A, Manzocco L, Brunton NP, Nicoli MC, Lyng JG (2015) The effect of pulsed electric field pre-treatments prior to deep-fat frying on quality aspects of potato fries. Innovative Food Sci Emerg Technol 29:65–69

ISO 14040 (2006) Environmental management—life cycle assessment—principles and framework ISO 14040: 2006. ISO, Geneva

ISO 14044 (2006) 14044 (2006) NF EN ISO 14044: 2006–environmental management – life cycle assessment – requirements and guidelines. AFNOR, Paris

Ites S, Smetana S, Toepfl S, Heinz V (2020) Modularity of insect production and processing as a path to efficient and sustainable food waste treatment. J Clean Prod 248:119248

Ivorra A, Rubinsky B (2010) Historical review of irreversible electroporation in medicine. In: Rubinsky B (ed) Irreversible electroporation. Springer, Berlin/Heidelberg

Janositz A, Noack AK, Knorr D (2011) Pulsed electric fields and their impact on the diffusion characteristics of potato slices. LWT Food Sci Technol 44:1939–1945

Jouhara H, Khordehgah N, Almahmoud S, Delpech B, Chauhan A, Tassou SA (2018) Waste heat recovery technologies and applications. Therm Sci Eng Prog 6:268–289

Käferböck A, Smetana S, de Vos R, Schwarz C, Toepfl S, Parniakov O (2020) Sustainable extraction of valuable components from Spirulina assisted by pulsed electric fields technology. Algal Res 48:101914

Knorr D, Geulen M, Grahl T, Sitzmann W (1994) Food application of high electric field pulses. Trends Food Sci Technol 5:71–75

Kotnik T, Kramar P, Pucihar G, Miklavcic D, Tarek M (2012) Cell membrane electroporation – part 1: the phenomenon. IEEE Electr Insul Mag 28:14–23

Lebovka NI, Praporscic I, Vorobiev E (2004) Effect of moderate thermal and pulsed electric field treatments on textural properties of carrots, potatoes and apples. Innovative Food Sci Emerg Technol 5:9–16

López N, Puértolas E, Condón S, Álvarez I, Raso J (2008) Effects of pulsed electric fields on the extraction of phenolic compounds during the fermentation of must of Tempranillo grapes. Innovative Food Sci Emerg Technol 9:477–482

Luengo E, Álvarez I, Raso J (2014) Improving carotenoid extraction from tomato waste by pulsed electric fields. Front Nutr 1(1):12

McAuliffe GA, Takahashi T, Lee MR (2020) Applications of nutritional functional units in commodity-level life cycle assessment (LCA) of agri-food systems. Int J Life Cycle Assess 25:208–221
McHugh T, Toepfl S (2016) Pulsed electric field Processing for fruits and vegetables. *Food Technology Magazine*, 70:73–75
Mosqueda-Melgar J, Elez-Martínez P, Raybaudi-Massilia RM, Martín-Belloso O (2008) Effects of pulsed electric fields on pathogenic microorganisms of major concern in fluid foods: a review. Crit Rev Food Sci Nutr 48:747–759
Pardo G, Zufía J (2012) Life cycle assessment of food-preservation technologies. J Clean Prod 28:198–207
Perez OE, Pilosof AM (2004) Pulsed electric fields effects on the molecular structure and gelation of β-lactoglobulin concentrate and egg white. Food Res Int 37:102–110
Picart-Palmade L, Cunault C, Chevalier-Lucia D, Belleville M-P, Marchesseau S (2019) Potentialities and limits of some non-thermal technologies to improve sustainability of food processing. Front Nutr 5:20190117
Pol IE, Mastwijk HC, Bartels PV, Smid EJ (2000) Pulsed-electric field treatment enhances the bactericidal action of nisin against Bacillus cereus. Appl Environ Microbiol 66:428–430
POTATOPRO (2017) Sustainability report: potato processor Lamb Weston/Meijer continues to innovate. [Online]. PotatoPro.com. Available: https://www.potatopro.com/news/2017/sustainability-report-potato-processor-lamb-weston-meijer-continues-innovate. Access date: 23 Oct 2020
Puértolas E, Barba FJ (2016) Electrotechnologies applied to valorization of by-products from food industry: main findings, energy and economic cost of their industrialization. Food Bioprod Process 100:172–184
Puértolas E, Luengo E, Álvarez I, Raso J (2012) Improving mass transfer to soften tissues by pulsed electric fields: fundamentals and applications. Annu Rev Food Sci Technol 3:263–282
Rahman MS (2012) Food preservation and processing methods. In: Handbook of food process design. Blackwell, New York, pp 1–17
Rodriguez-Gonzalez O, Buckow R, Koutchma T, Balasubramaniam V (2015) Energy requirements for alternative food processing technologies—principles, assumptions, and evaluation of efficiency. Compr Rev Food Sci Food Saf 14:536–554
Sale A, Hamilton W (1967) Effects of high electric fields on microorganisms: I. Killing of bacteria and yeasts. Biochim Biophys Acta 148:781–788
Siegrist M, Hartmann C (2020) Consumer acceptance of novel food technologies. Nat Food 1:343–350
Siemer C, Aganovic K, Toepfl S, Heinz V (2014a) Application of pulsed electric fields in food. In: Conventional and advanced food processing technologies. Wiley, Chichester
Siemer C, Toepfl S, Heinz V (2014b) Inactivation of Bacillus subtilis spores by pulsed electric fields (PEF) in combination with thermal energy – I. Influence of process- and product parameters. Food Control 39:163–171
Silva VL, Sanjuán N (2019) Opening up the black box: a systematic literature review of life cycle assessment in alternative food processing technologies. J Food Eng 250:33–45
Smetana S, Mathys A, Knoch A, Heinz V (2015) Meat alternatives: life cycle assessment of most known meat substitutes. Int J Life Cycle Assess 20:1254–1267
Smetana S, Terjung N, Aganovic K, Alahakoon AU, Oey I, Heinz V (2019) Chapter 10 – Emerging technologies of meat processing. In: Galanakis CM (ed) Sustainable meat production and processing. Academic Press, London
Smetana S, Leonhardt L, Kauppi S-M, Pajic A, Heinz V (2020) Insect margarine: processing, sustainability and design. J Clean Prod 264:121670
Spielmann M, Scholz R, Tietje O, Haan P de (2005) Scenario Modelling in Prospective LCA of Transport Systems. Application of Formative Scenario Analysis (11 pp). The International Journal of Life Cycle Assessment 10:325–335. https://doi.org/10.1065/lca2004.10.188

Stübler A-S, Lesmes U, Juadjur A, Heinz V, Rauh C, Shpigelman A, Aganovic K (2020) Impact of pilot-scale processing (thermal, PEF, HPP) on the stability and bioaccessibility of polyphenols and proteins in mixed protein-and polyphenol-rich juice systems. Innovative Food Sci Emerg Technol 64:102426

Toepfl S (2011) Pulsed electric field food treatment-scale up from lab to industrial scale. Procedia Food Sci 1:776–779

Toepfl S, Heinz V (2011) Pulsed electric field assisted extraction—a case study. In: Nonthermal processing technologies for food. Section II: electromagnetic processes. Blackwell and Institute of Food Technologist, London, pp 190–200

Toepfl S, Mathys A, Heinz V, Knorr D (2006) Potential of high hydrostatic pressure and pulsed electric fields for energy efficient and environmentally friendly food processing. Food Rev Intl 22:405–423

Ulmer M, Smetana S, Heinz V (2020) Utilizing honeybee drone brood as a protein source for food products: life cycle assessment of apiculture in Germany. Resour Conserv Recycl 154:104576

UNEP/SETAC (2011) Life cycle initiative. Global guidance principles for life cycle assessment databases. A basis for greener processes and products. UNEP/SETAC, Nairobi

Villamonte G, Lamballerie MD, Jury V (2014) Consideration of the product quality in the life cycle assessment: case of a meat product treated by high pressure. Proceedings of the 9th international conference on life cycle assessment in the agri-food sector (LCA Food 2014), San Francisco, CA, USA, 8–10 October, 2014, American Center for Life Cycle Assessment, pp 1488–1496

Walser T, Demou E, Lang DJ, Hellweg S (2011) Prospective Environmental Life Cycle Assessment of Nanosilver T-Shirts. Environmental Science & Technology 45:4570–4578. https://doi.org/10.1021/es2001248

Warner R, McDonnell CK, Bekhit A, Claus J, Vaskoska R, Sikes A, Dunshea F, Ha, M. (2017) Systematic review of emerging and innovative technologies for meat tenderisation. Meat Sci 132:72–89

Weidema, BP, Stylianou KS (2019) Nutrition in the life cycle assessment of foods—function or impact? Int J Life Cycle Assess, July, pp 1–7

Wouters PC, Alvarez I, Raso J (2001) Critical factors determining inactivation kinetics by pulsed electric field food processing. Trends Food Sci Technol 12:112–121

Xiang BY, Ngadi MO, Ochoa-Martinez LA, Simpson MV (2011) Pulsed electric field-induced structural modification of whey protein isolate. Food Bioprocess Technol 4:1341–1348

Yan L-G, He L, Xi J (2017) High intensity pulsed electric field as an innovative technique for extraction of bioactive compounds—a review. Crit Rev Food Sci Nutr 57:2877–2888

Zamagni A, Guinée J, Heijungs R, et al (2012) Lights and shadows in consequential LCA. The International Journal of Life Cycle Assessment 17:904–918. https://doi.org/10.1007/s11367-012-0423-x

Zhao W, Yang R, Lu R, Tang Y, Zhang W (2007) Investigation of the mechanisms of pulsed electric fields on inactivation of enzyme: lysozyme. J Agric Food Chem 55:9850–9858

Regulation of Foods Processed by Pulsed or Moderate Electric Fields (PEF or MEF)

James G. Lyng, Fiona Lalor, Geraldine Quinn, and Selene Pedrós-Garrido

Abbreviations

AC	Alternating current
AHAW	Panel on Animal Health and Welfare
BIOHAZ	Panel on Biological Hazards
BSE	Bovine spongiform encephalopathy
CEP	Panel on Food Contact Materials, Enzymes and Processing Aids
CFR	Code of Federal Regulations
CFSAN	Center for Food Safety and Applied Nutrition
CFU	Colony-forming units
CJEU	Court of Justice of the European Union
CONTAM	Panel on Contaminants in the Food Chain
CSHPF	Conseil supérieur d'hygiène publique de France
CTA	Commission de technologie alimentaire
DG	Directorate General
DVFA	The Danish Veterinary and Food Administration
EC	European Commission
ECA	European Court of Auditors
ECB	European Central Bank
EFSA	European Food Safety Authority
EPA	Environmental Protection Agency
EU	European Union
FAF	Panel on Food Additives and Flavourings
FAO	Food and Agriculture Organization of the United Nations
FD&C	Federal Food, Drug, and Cosmetic (Act)
FDA	Food and Drug Administration (US)
FEEDAP	Panel on Additives and Products or Substances used in Animal Feed
FIN	Finland
FSA	Food Standards Agency

J. G. Lyng (✉) · F. Lalor · G. Quinn · S. Pedrós-Garrido
UCD Agriculture and Food Science Centre, School of Agriculture and Food Science,
University College Dublin, Dublin 4, Ireland
e-mail: james.lyng@ucd.ie; fiona.lalor@ucd.ie; geraldine.quinn@ucd.ie;
selene.pedros-garrido@ucd.ie

J. Raso et al. (eds.), *Pulsed Electric Fields Technology for the Food Industry*,
Food Engineering Series, https://doi.org/10.1007/978-3-030-70586-2_20

FSAI	Food Safety Authority of Ireland
FSIS	Food Safety and Inspection Service
FSMA	Food Safety Modernization Act
FSVP	Foreign Supplier Verification Programs
GMO	Panel on Genetically Modified Organisms
GRAS	Generally Recognized as Safe
HPP	High-pressure processing
MEF	Moderate electric fields
MEPs	Members of the European Parliament
MS	Member states
NDA	Panel on Nutrition, Novel Foods and Food Allergens
NFB	Novel Food Board
NFU	Novel Food Unit
NL	Netherlands
NMFS	National Marine Fisheries Service
PCBs	Polychlorinated biphenyls
PEF	Pulsed electric fields
PLH	Panel on Plant Health
PPR	Panel on Plant Protection Products and their Residues
QMV	Qualified majority vote
SNFA	Swedish National Food Agency
USA	United States of America
UK	United Kingdom
USDA	US Department of Agriculture
UV	Ultraviolet light
WHO	World Health Organization
WTO	World Trade Organization

Introduction

As we enter the third decade of the twenty-first century, our food system has never been so challenged and undermined to the extent we are currently experiencing. The world's population continues to grow at pace, and it is estimated that by 2050 the population of the globe will have reached 9.7 billion from today's figure of 7.5 billion (WRI 2016). This increase in population will result in an increased demand for more food and will have significant impacts on food production, leading to increased outputs, which in turn will accelerate the requirements for finite resources such as energy and water (Monforti-Ferrario et al. 2015). Allied to this is the ever increasing impact of the rise of chronic diseases associated with our diet, such as cardiovascular disease, cancer, diabetes and obesity (Eurostat 2014; Johnston et al. 2014), forcing consumers to reconsider their own behaviours while looking for foods that are healthier, more natural (less processed), more ethically produced, safe, and, in the long term, more sustainable (ETPs 2016). All these concerns with respect to our food system are even more acute as we embark on a period of increasing uncertainty post-Covid-19 and Brexit.

The application of emerging processing technologies can be used to address how we can make our food healthier, safer and more sustainable. PEF and MEF are two technologies with significant potential in this regard. These technologies differ from other forms of electro-processing (i.e. microwave or radio frequency waves) in that they both involve the direct application of electrical current to a foodstuff instead of the application of electromagnetic radiation at microwave or radio frequencies. PEF involves the application of higher field strengths (e.g. 1000–50,000 V/cm) in short pulses (e.g. 5–30 μs) at frequencies typically ranging from 1 to 1000 Hz with the primary function of inducing electroporation within the matrix. By contrast, Sastry (2008) defined MEF processing as the use of electric fields at levels ranging from about 1 to 1000 V/cm (i.e. lower than PEF) of arbitrary waveform and frequency (e.g. 50–25,000 Hz). Significant frictional heat generation (generally referred to as ohmic heating) can be induced in both PEF and MEF. Energy inputs encountered in PEF processing are generally too low for substantial heating to occur, and while MEF is more commonly used for heating applications, low-energy MEF, which avoids ohmic heating, is also encountered in terms of design. While PEF systems tend to be more complex and contain capacitors and pulse-forming networks, MEF systems tend to be simpler, in some cases just consisting of a main powered variable AC (alternating current) supply. Overall, the use and uptake of these technologies is not without some constraints, specifically in relation to their regulation, which can delay their commercial uptake and hinder them from achieving their full commercial potential. It is critical that the food processing industry, which hopes to avail of these technologies, is cognisant of the legislation pertaining to these processes and their application.

The aim of this chapter is to introduce legislation relevant to the uptake of MEF and PEF new processing technologies in the European Union and US contexts.

For our food system to function and to retain its integrity, regulation is essential. In its absence we can no longer have confidence in what we eat or drink or in those who supply our food from the farmer to the grocer. Not only is regulation vital to protect consumers' health, but it is also crucial to ensure that the foods and drink supply chain is robust and secure and that each element along the link remains viable. We have numerous examples of how the lack of regulation has had a detrimental effect on the food supply chain and undermined consumer confidence:

- In Italy in 1997, an outbreak of gastroenteritis caused by consuming corn salad contaminated with *Listeria monocytogenes* was recorded. More than 1500 people were affected, most of them children and staff at 2 primary schools whose cafeterias were served by the same caterer and where 292 had to be admitted to hospital (Twisselmann 2000).
- In 1998, the employees of a large company in Helsinki in Finland developed gastroenteritis after eating in their workplace cafeteria. It transpired that salad dressing made with frozen raspberries was contaminated with norovirus resulting in 509 people becoming ill (Tavoschi et al. 2015).

- One of the most notorious cases related to an *Escherichia coli* O157:H7 outbreak is in Wishaw in Lanarkshire, Scotland, in November 1996 where cross-contamination of raw and cooked meat caused the death of 21 people. This outbreak was a watershed in terms of how countries up to this point had dealt with food pathogens and outbreaks of food poisoning. National food regulatory agencies were established in the wake of this and similar outbreaks.
- Food poisoning outbreaks are not restricted to microbiological pathogens. In Belgium in January 1999, 500 tonnes of animal feed became contaminated with polychlorinated biphenyls (PCBs) and dioxins which lead to the Belgian dioxin crisis. Over 2500 poultry and pig farms were involved. The problem was identified by health inspectors in January, but the case was only revealed in May when it was reported by the media.

Collectively these and many other outbreaks, and the extensive associated publicity, increased consumer concern and awareness of the issue of food safety. As a result, where once enabling trade was the primary focus, the balance changed with consumer protection increasingly viewed as a priority for the entire food industry.

Regulation Within the European Union (EU)

Background to Regulation in the EU

Nationally, food safety and supply chains are regulated by agencies that operate at a local level. Thus, in global terms there are numerous such organisations in operation. To enable trade and ensure a free flow of food supply across jurisdictions, multinational agencies including the World Trade Organization (WTO), the World Health Organization (WHO) and the Food and Agriculture Organization of the United Nations (FAO) have roles to play in directing, coordinating and advising in relation to such regulations (Lalor 2016).

This section will overview regulation within the EU and will focus specifically on the issue of food processing technologies in this regard.

There are seven key institutions which are central to decision-making within the EU. These decision-making bodies are:

- European Court of Auditors (ECA)
- European Central Bank (ECB)
- Court of Justice of the European Union (CJEU)
- European Commission (EC)
- Council of the European Union
- European Council
- European Parliament

In terms of the regulation of food supply, only the latter four of the above institutions have a role to play. The European Parliament consists of 751 members (MEPs) who are elected every 5 years with the number of MEPs elected per country being proportional to its population.

The European Council is made up of the heads of state or heads of government of the EU member states and also includes the presidents of the European Commission and the European Council. It is responsible for defining 'the general political direction and priorities of the EU'.

The Council of the European Union comprises government ministers from EU member states with the ministerial portfolios of its members at a given time being relevant to the specific policy area being considered. This group is considered the 'voice of EU member governments, adopting EU laws and coordinating EU policies'. Presidency of the Council of the EU rotates among member states every 6 months.

The European Parliament, European Council and Council of the European Union have important roles to play in European regulation and legislation. However, in Europe, the most influential organisation in terms of food regulation is the EC. The EC comprises 27 commissioners (1 per member state) and has its headquarters in Brussels (Belgium). The EC is effectively the executive (civil service), and its remit includes the submission of proposed new legislation to the Parliament and Council. The EC is also responsible for administering the budget, implementing policy and negotiating international agreements while also ensuring compliance with European law ('guardian of the treaties'). The EC is structured into 53 departments and executive agencies each of which is responsible for a specific policy area. Prior to the bovine spongiform encephalopathy (BSE) crisis, the food safety fell under the remit of the DG Trade Commission Department but afterwards was moved to DG Santé. DG Santé has responsibility for EU policy on food safety and health and also is responsible for monitoring the implementation of related laws (EU 2019a, b). This department states that its main goal is 'to make Europe a healthier, safer place, where citizens can be confident that their interests are protected'. In order to achieve this goal, DG Santé aims to (a) protect and improve public health; (b) ensure Europe's food is safe and wholesome; (c) protect the health and welfare of farm animals; and (d) protect the health of crops and forests.

In the context of food processing and distribution, (a) and (b) above are the most relevant. In terms of food safety, the role of DG Santé is to monitor, listen and act, if necessary, so it can be considered as the risk manager for this area who takes on board scientific advice from other agencies as required (EC 2019a, b).

Within DG Santé, directorate E2 'Food processing technologies and novel foods' is the group of most relevance in relation to the processing of foods as it has responsibility for managing Novel Food Regulation.

A subsidiary organisation, which also has a vital role to play in this area, is the European Food Safety Authority (EFSA). EFSA is based in Parma (Italy) and was established following the publication of Regulation (EC) 178/2002 (EU 2002). EFSA's mission is to 'provide objective, science-based advice and clear and coherent communication, grounded in the most up to date scientific knowledge and data while also cooperating with EU member states, international bodies and other stakeholders'. While the EC can be viewed as occupying the role of risk manager, EFSA can be considered the risk assessor providing advice and scientific opinions, which are considered during the formation of European policy and legislation. Members of EFSA's scientific committees and panels are appointed for 3-year terms and consist of independent scientific experts in their fields. These committees and panels perform scientific assessments and also develop assessment methodologies (EFSA 2019). There are ten scientific panels covering the following areas:

(i) Animal health and welfare (AHAW)
(ii) Biological hazards (BIOHAZ)
(iii) Food contact materials, enzymes and processing aids (CEP)
(iv) Contaminants in the food chain (CONTAM)
(v) Food additives and flavourings (FAF)
(vi) Additives and products or substances used in animal feed (FEEDAP)
(vii) Genetically modified organisms (GMO)
(viii) Nutrition, novel foods and food allergens (NDA)
(ix) Plant health (PLH)
(x) Plant protection products and their residues (PPR)

In terms of novel processing and novel foods, the 'Nutrition, Novel Foods and Food Allergens' (NDA) Panel is the most relevant. Regulation (EU) 2015/2283 (EU 2015), which is the key legislation in this regard, defines a novel food as 'any food that was not used for human consumption to a significant degree within the Union before 15th May 1997, irrespective of the dates of accession of Member States to the Union'. This falls into any of the categories listed in the Regulation. Within the list of categories listed, a 'novel food' can be an innovative or newly developed food, a food produced using new technologies, or a food produced using a new production process. Foods that have been traditionally eaten outside of the EU but were not consumed within the EU to a significant extent prior to 15 May 1997 are also considered as novel foods. In order for a novel food to be sold for consumption within the EU, it must be evaluated to ensure it is safe and not nutritionally disadvantageous for consumers. If successful in this regard, it will be given a market authorisation, which generally also has a consumer labelling requirement (EU 2011, 2015). As an example, Dairy Crest (UK) was given novel food approval for UV-treated milk, and Article 2 of the implementing decision (2016/1189) specified that the designation for labelling shall be accompanied by 'contains vitamin D produced by UV-treatment' or 'milk containing vitamin D resulting from UV-treatment'.

In situations where little scientific evidence exists or the existing evidence is inconclusive or incomplete and as a result a thorough safety assessment cannot be performed, the 'precautionary principle' may be applied. The 'precautionary principle' is applied when there are reasonable grounds to suggest that an unacceptable level of risk to health exists and the available scientific evidence is not sufficient for an exhaustive and accurate risk assessment (Lalor 2016; EU 2012).

Novel Food Regulations 2283/2015 vs. 258/97

As previously mentioned, Regulation (EU) 2015/2283 (EU 2015), the 'novel foods' regulation, is the most relevant in terms of emerging processing technologies. The previous iteration of this regulation was Regulation (EC) 258/97/EC (EU 1997).

What Do These Two Regulations Have in Common?

Firstly, 15 May 1997 was considered a cut-off date. From this date, any food that had not been used for human consumption to a significant degree within the EU (regardless of the date of accession of member states to the Union) would be considered as 'novel food'.

Secondly, with slight differences, both Novel Food Regulations have several food categories to define what a novel food is. Probably the most relevant definition for emerging technologies is 'Food resulting from a production process [...] which gives rise to significant changes in the composition or structure of a food, affecting its nutritional value, metabolism or level of undesirable substances' (EU 2015). Then, if the emerging process does not give rise to significant changes in these areas, it does not have to be considered, i.e. just because a process technology is emerging, new or novel, it does not automatically mean it leads to the production of a novel food. If the application of any new process of a considered emerging technology (as PEF is) may impact on a food product, producing any of these changes, it is necessary to submit an application under the Novel Food Regulation. On 1 January 2018, Regulation (EU) 2015/2283 repealed Regulation (EC) 258/97. Incomplete applications which were submitted under Regulation 285/97 were transferred to Regulation 2015/2283.

What Are the Main Differences Between These Two Regulations?

Substantial equivalence was removed from Regulation 2283/2015 following the identification of an anomaly that was highlighted following a number of test cases. These cases demonstrated an unfair competitive advantage against companies who had invested in and developed new technologies, and/or innovations, and had applied for novel food approval under Regulation 258/97. Substantial equivalence allowed their competitors to apply for approval under this regulation but fast track their way through the process as subsequent toxicology and nutritional studies were not required. All Regulation 2283/2015 authorisations are now generic, which might appear to offer less protection. However, the new regulation actually provides better protection for applicants than the old Regulation 258/97 system. Under Regulation 2283/2015, companies can now choose to have proprietary data deemed confidential for a number of years during which time their competitors cannot use it without permission, forcing them to create their own data and allowing a period of time for the companies to recoup their investment costs.

Application Process

Regulation 258/97 also differs from Regulation 2283/2015 in terms of the application process. However, this chapter focuses on the current application process from Regulation 2283/2015, which has been improved and streamlined with respect to the previous one, but in terms of the time associated with processing an application, it is very similar.

This application process is outlined in Fig. 1. Under Regulation 2283/2015 the applicant prepares a dossier and submits it directly to DG Santé. Unlike the previous Regulation, the applicant must firstly take the decision of whether the product falls under the Novel Food Regulation. This decision can be made in consultation with the local competent authority (EC 2018; EU 2018a). However, as stated in 'Article 4: Procedure for determination of novel food status' (EU 2015), beyond the applicant or its local authority participating in this decision-making, in points 3 and 4, it is mentioned that this authority 'may consult' during the decision process with other member states (MS) and the Commission. Therefore, due to the fact that the authority is not obliged to consult with other MS, it may conclude that the product does not fall under the definition of a novel food, enabling MS to release a food product onto the market that they do not consider as novel. This could be a possible loophole in the wording of Regulation 2283/2015 which is something to consider.

Article 4:
Procedure for Determination of Novel Food Status

1. *Food business operators shall verify whether or not the food which they intend to place on the market within the Union falls within the scope of this Regulation.*
2. *Where they are unsure whether or not a food which they intend to place on the market within the Union falls within the scope of this Regulation, food business operators shall consult the Member State where they first intend to place the novel food. Food business operators shall provide the necessary information to the Member State to enable it to determine whether or not a food falls within the scope of this Regulation.*
3. *In order to determine whether or not a food falls within the scope of this Regulation, Member States may consult the other Member States and the Commission.*
4. *The Commission shall, by means of implementing acts, specify the procedural steps of the consultation process provided for in paragraphs 2 and 3 of this Article, including deadlines and the means to make the status publicly available. Those implementing acts shall be adopted in accordance with the examination procedure referred to in Article 30(3).*

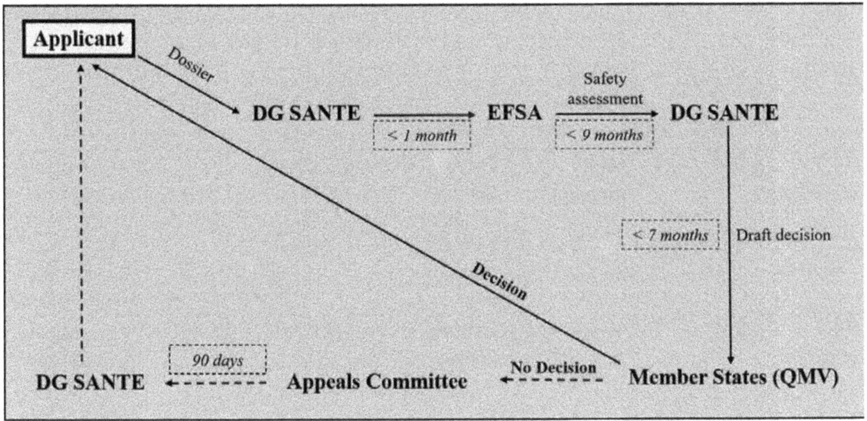

Fig. 1 Novel food authorisation process under Regulation (EU) 2283/2015

Once the decision is taken, if it is considered that the product falls under the scope of Regulation 2283/2015, the dossier submitted to DG Santé is passed on to EFSA within 1 month, after which time EFSA has 9 months to carry out a safety assessment.

Following this assessment, EFSA then provides feedback to DG Santé, which must draft a decision to the member states (in less than 7 months). At this point, if there are no objections and the application receives a positive decision, it can be placed on 'the Union list' (*Article 6: Union list of authorised novel foods*), and the applicant is allowed to place the product on the market. However, if a qualified majority vote (QMV) is not obtained, the application can pass through an appeals committee. A QMV can be defined as at least 55% of the members of the European Council representing the participating MS, comprising at least 65% of the population of these states. The appeals committee has 90 days to give feedback to DG Santé, which finally informs the applicant of the decision.

Novel Technology Applications Within the EU

All completed applications on the use of novel processing technologies under Regulations (EC) 258/97 and (EU) 2015/2283 are presented in Table 1. Only 3% of the total novel food applications relate to new processing technologies with new food ingredients representing the majority. Under both Regulations 258/97 and 2015/2283, only nine applications for new technologies have been made to date. These applications are based on the use of ultraviolet light (UV) or high-pressure processing (HPP) technologies only, which are still considered novel processing technologies, for specific food purposes. However, among the completed applications, none involving pulsed or moderate electric fields (PEF or MEF) has been

Table 1 Completed applications on the use of novel processing technologies under Regulations (EC) 258/97 and (EU) 2015/2283 of the European Parliament and the Council

Regulation	Technology	Food or food ingredient	Applicant	Application date (dd/mm/yyyy)	Initial assessment	Status
(EC) 258/97	HPP	Fruit preparations	Group Danone (France)	03/12/1998	(1) CSHPF, (2) CTA	Commission Decision (2001/424/EC). 23/05/2001 (EU 2001)
(EC) 258/97	UV	Yeast	Lallemand (France)	04/05/2012	FSA (UK)	Commission Implementing Decision (2014/396/EU). 26/06/2014 (EU 2014)
(EC) 258/97	UV	Bread	Viasolde AB (Sweden)	26/09/2012	NFB (FIN)	Commission Implementing Decision (2016/398/EU). 18/03/2016 (EU 2016a)
(EC) 258/97	UV	Milk	Dairy Crest Ltd. (United Kingdom)	26/09/2012	FSAI	Commission Implementing Decision (2016/1189/EU). 21/07/2016 (EU 2016b)

(continued)

Table 1 (continued)

Regulation	Technology	Food or food ingredient	Applicant	Application date (dd/mm/yyyy)	Initial assessment	Status
(EC) 258/97	UV	Mushrooms	Monaghan Mushrooms (Ireland)	04/02/2015	FSAI	No objections. Letter of Ireland 26/02/2016 (FSAI 2016)
			Ekoidé AB (Sweden)	29/11/2016	SNFA	Commission Implementing Decision (EU) 2017/2355. 14/12/2017 (EU 2017)
			Walsh Mushrooms (United Kingdom)	21/12/2016	FSAI	No objections. Letter of Food Safety Authority of Ireland 28/08/2017 (FSAI 2017)
(EU) 2015/2283	UV	Mushrooms (*Agaricus bisporus*)	Banken Champignons (Netherlands)	27/01/2016	NFU (NL)	Commission Implementing Regulation (EU) 2018/1011. 17/07/2018 (EU 2018b)
(EU) 2015/2283	UV	Baker's yeast	Lallemand Bio-Ingredients Division (Canada)	30/03/2017	DVFA	Commission Implementing Regulation (EU) 2018/1018. 18/07/2018 (EU 2018c)

Sources: EC (2019a, b)

CSHPF Conseil supérieur d'hygiène publique de France, *CTA* Commission de technologie alimentaire, *DVFA* the Danish Veterinary and Food Administration, *EC* European Commission, *EU* European Union, *FSA (UK)* Food Standards Agency (United Kingdom), *FSAI* Food Safety Authority of Ireland, *HPP* high-pressure processing, *NFB (FIN)* Novel Food Board (Finland), *NFU (NL)* Novel Food Unit (Netherlands), *SNFA* Swedish National Food Agency, *UV* ultraviolet light submitted for consideration to date. However, some novel processing technologies are already being applied to different foodstuffs for various processes, without having passed through the formal application process for novel foods. There are currently several PEF applications being used in different foods within the EU. For example, PEF is used to extend the shelf life of fruit juices from 7 to 21 days or to soften potatoes prior to cutting for chip manufacture. However, those applications have not been considered under the Novel Food Regulations 258/97 and 2015/2283, basically because these processes do not result in the production of a 'novel food'. As noted in Regulation 2015/2283, the process should lead to 'significant changes in the composition or structure of a food affecting its nutritional value, metabolism or level of undesirable substances' to be considered a 'novel food'.

Regulation Within the USA

Background

The responsibility for food safety and the protection against intentional and unintentional contamination and other emerging threats of the American food and agriculture supply chain is shared by state, local and territorial governments and the private sector. At least a dozen federal agencies implementing more than 35 statutes make up the federal part of the food safety system. Twenty-eight House and Senate committees provide oversight of these statutes.

In terms of the implementation of the regulations that govern food safety, there are four agencies that play major roles in carrying out these activities:

- Food and Drug Administration (FDA) of the Department of Health and Human Services

The FDA has jurisdiction over domestic and imported foods that are marketed in the USA except for meat and poultry products. FDA's Center for Food Safety and Applied Nutrition (CFSAN) has a role to ensure that these foods are safe, nutritious, wholesome and honestly and adequately labelled. CFSAN has jurisdiction over food processing plants and responsibility for approval and surveillance of food-animal drugs, feed additives and all food additives that can become part of food.

- Food Safety and Inspection Service (FSIS) of the US Department of Agriculture (USDA)

FSIS has the responsibility to ensure that meat and poultry products for human consumption are safe, wholesome and correctly marketed, labelled and packaged.

- Environmental Protection Agency (EPA)

The EPA registers and evaluates all pesticides distributed in the USA and establishes tolerances for pesticide residues in food commodities and animal feed. This becomes particularly important in terms of the role of the EPA in regulating for biotechnology in foods.

- National Marine Fisheries Service (NMFS)

NMFS conducts a voluntary seafood inspection and grading programme, which is primarily a food quality activity.

Each of these agencies' role is defined and clear. However, the principal two agencies involved in food safety, and those most frequently referred to when discussing food safety control in the USA, are the FDA and the USDA.

An example of the type of guidance provided by the FDA is a rule aimed at increasing the safety of fruit and vegetable juices published in 2001 (FDA 2001). This states that a preservation method or processing technology must be capable of inducing a 5 $\log_{10}$ CFU/mL reduction of the most resistant pathogen likely to be present in the product (Topalcengiz 2019). This rule is frequently cited in publications relating to the use of MEF, PEF or other novel technologies in the context of the preservation of fruit and vegetables (Gouma et al. 2020; Kim et al. 2018).

The agencies and some of their general principles as outlined above do provide a picture for how food safety is regulated in the USA. It is important, however, to be familiar with certain other terms and processes, e.g. Code of Federal Regulations (CFR), the Food Safety Modernization Act (FSMA) and Generally Recognized as Safe (GRAS). These form an integral part of legislation in the USA, and no food safety discussion is complete without their inclusion. These important concepts are briefly explained below:

(i) Code of Federal Regulations

The Code of Federal Regulations (CFR) is the official publication of all federal regulations. It is the 'codification of the general and permanent rules published in the Federal Register by the departments and agencies of the federal government'. The CFR is updated every year and is divided into 50 titles, each one representing a general subject area. Title 21 (Food and Drugs) of the CFR is reserved for rules of the Food and Drug Administration only. For example, Title 21, Chapter 1, Subchapter B, Part 179 provides regulations for irradiation in the production, processing and handling of food (FDA 2019a).

(ii) Food Safety Modernization Act

It would not be accurate to review any food safety system in the USA without referring to the Food Safety Modernization Act (FSMA). FSMA was signed in January 2011 by President Barak Obama and has changed the focus of how food safety is governed in the USA. The aim of this new Act in terms of food safety is to be proactive rather than reactive. Since its publication, the FDA has proposed and finalised critical regulations that established science and risk-based standards for the production and transportation of domestic and imported foods.

As the principal agency responsible for the enforcement of FSMA, the FDA identified seven major rules for its implementation (FDA 2017). These rules are:

1. Preventive control rules for human food
2. Preventive control rules for animal food
3. Produce safety rule
4. Foreign Supplier Verification (FSVP) rule
5. Accredited third party certification
6. Sanitary transportation rule
7. Intentional Adulteration Rule

Generally Recognized as Safe or GRAS is an acronym used, particularly in US regulatory context and that of the food industry, to refer to substances that, for their intended use, have been considered by the sponsor of the substance and appropriate experts to be safe. This concept of GRAS was added to the amendment of the Federal Food, Drug, and Cosmetic Act, of 1958, and substances that are considered GRAS, under the conditions of their proposed use, may be marketed for that use without FDA review or approval. The operation and administration of GRAS notifications and the petition process have evolved and changed many times since 1958. A summary is provided in this document: Substances Generally Recognized as Safe, rule issued by the FDA in July 2016 (FDA 2016).

Regulating 'New Technologies'

The most important criteria for novel technologies from a regulatory and legal standpoint will be for a processor to clearly ascertain that the process used is effective for its intended purpose and that food prepared using such a processing method is not misrepresented in a way that could lead to a consumer to think the food has properties that it does not possess.

In June 2014, the FDA published its new Guidance in the Federal Register (Vol. 79, No 124, Friday, June 27, 2014/Notices), 'Guidance for Industry: Assessing the Effects of Significant Manufacturing Process Changes, Including Emerging Technologies, on the Safety and Regulatory Status of food ingredients and Food Contact Substances, including Food Ingredients that are color Additives' (FDA 2014). This guidance represents the FDA's current thinking on the factors to be considered when determining whether changes in manufacturing processes result in changes in a food substance already in the market; impact the safety of the use of the food substance; change the regulatory status of the use of the food substance; or warrant a new regulatory submission to the (FDA 2014). While this document does pay specific heed to nanotechnology, it is a very useful reference for how the FDA is thinking in terms of other emerging technologies, and it is the most detailed information in this field available from the FDA. Examples of other technologies that have also completed the 'regulatory process' and are now widely commercially available are irradiation and biotechnology. It is worth reviewing how these are both dealt with by the FDA, bearing in mind that the examples given were addressed prior to the FDA guidance mentioned above.

Example A: Irradiation

Irradiation is a physical treatment where food is exposed to a defined dose of ionising radiation. It is designed to control insect infestation, reduce the number of pathogenic or spoilage microorganisms, and delay or eliminate natural biological processes such as ripening, germination or sprouting in fresh food (FSAI 2019).

In the USA, the Food Additives Amendment to the Federal Food, Drug, and Cosmetic Act (FD&C Act) of 1958 placed food irradiation under the food additive regulations, and so it is regulated as such, rather than as a food process. While the radiation itself is not an additive, the process of being irradiated is defined as one (Morehouse and Komolprasert 2004).

In general, a food additive regulation may be established or amended in one of two ways. The FDA proposes a regulation on its own initiative, where the public is given an opportunity to comment on the proposal and all substantive comments are considered. Alternatively, and more commonly, the food additive regulations are amended in response to petitions filed by proponents of an additive's use. The FDA has found irradiation of food to be safe under several conditions, and authorising regulations have been issued both in response to petitions and at the FDA's initiative (Morehouse 2002).

Heretofore, the FDA has approved food irradiation in the following foods:

- Beef and pork
- Crustaceans (e.g. lobster, shrimp and crab)
- Fresh fruit and vegetables
- Lettuce and spinach
- Poultry
- Seeds for sprouting
- Shell eggs
- Shellfish – molluscan
- Spices and seasoning

For completeness, it is important to know that the Code of Federal Regulations (Title 21: Part 179) (FDA 2019a) provides for the use of ionising radiation for treatment of foods. It also provides conditions for the use of other nonionising forms of electromagnetic radiation. All of these nonionising forms have lower energy, which is insufficient to knock electrons out of atoms (i.e. ionisation). These include radio frequencies, microwaves, ultraviolet light and pulsed light for the treatment of food. Also included are carbon dioxide lasers, which can be used for etching food surfaces allowing the replacement of physical labels on product surfaces with laser-etched 'natural' alternatives.

Example B: Biotechnology

The use of biotechnology in foods has enabled scientists to select and isolate specific genes with desirable traits from one plant and transfer these genes (and therefore this trait) to another plant. Examples include plants that confer pesticide resistance to certain pests or rice plants with enhanced nutritional properties.

In the USA, the regulation of biotechnology in food products is not different from the regulation of conventional products. The three agencies involved are the FDA, the USDA and the EPA. The FDA provides voluntary premarket consultations with food companies, seed companies and plant developers to ensure that biotech foods meet the same regulatory standards for safety. The USDA Animal and Plant Health Inspection Service (APHIS) licenses field testing of crops prior to their commercial release, and the EPA registers pesticides, including plants engineered to produce pesticides, and establishes levels at which pesticides in foods are permitted (Vogt and Parish 2001).

The Plant Biotechnology Consultation Program was established by the FDA to evaluate the safety of foods from new genetically modified organisms (GMO) before they enter the market. As part of this programme, the GMO plant developer meets with the FDA and submits a detailed food safety assessment to the FDA. The FDA evaluates the data and resolves any issues with the developer. Once satisfied that the new food is safe and the process is complete, the FDA makes this information public (FDA 2020). Examples of products that have completed the process include herbicide-tolerant soybean and insect-resistant cotton. As at April 2020, there are 168 completed consultations with the FDA (FDA 2019b).

Pulsed Electric Fields

The use of PEF technology as a preservation technique provides an alternative to thermal pasteurisation, and as a nonthermal technique, it can prevent thermal disruption of nutritional components in a food. It, therefore, has additional health benefits for consumers unlike other processes that might use chemical preservatives.

One of the first industrial applications of PEF technology was the ESTERIL process in Germany in the late 1980s for the electrical sterilisation and pasteurisation of pumpable electrically conductive media (Sitzmann 1995). Krupp Maschinentechnik GmbH, in association with the University of Hamburg, reported successful inactivation when PEF was applied to fluid foods such as orange juice and milk (Vega-Mercado et al. 2007). However, the widespread use of PEF in the food industry has proved to be slower than anticipated, because of the initial cost of the outlay required for equipment and also due to the uncertain regulatory process involved. Despite this uncertainty, in 1995 PurePulse Technologies developed CoolPure. This was a pulsed electric field processing system for antimicrobial treatment of liquids and pumpable foods. The same year, the FDA released a 'letter of no objection' for the use of pulsed electric fields, thus approving the industrial application of CoolPure. In 1996, PEF treatment of liquid eggs was also approved by the FDA with certain conditions.

In 2002, Gongora-Nieto et al. published a suggested procedural plan for how to get regulatory approval for the use of PEF. They identified that as well as complying with current regulations that might apply, PEF may also require premarket approval. This stems primarily from the use of different food contact materials throughout the process, e.g. the conditions inside the PEF chamber where electrodes could migrate into the food. The most sensible approach to this would be to submit a petition to the FDA for evaluation. After the evaluation, a ruling would be made by the FDA and published in the Code of Federal Regulations (Góngora-Nieto et al. 2002).

A key component of the process is the opening of dialogue with the FDA, and in so doing, users of PEF are recommended to submit a scientific review with necessary data to clearly characterise the product. When filing a new and emerging process, the following steps should be taken (Fig. 2):

1. Establish an active and continuous dialogue with the FDA during process development.
2. Meet with the FDA to describe the process.
3. Invite the FDA to a site visit.
4. Draft and provide the FDA with an outline of the proposed filing.
5. Identify the most resistant organism of public health concern, the most resistant organism for commercial viability and the least lethal treatment zone of the system (Larkin and Spinak 1997).

In terms of further detail required by the FDA, John Larkin (then an Associate Director of Research at the FDA) presented additional recommendations at the Institute of Food Technologists Annual Meeting in 1999. He recommended a further description of the equipment used, full specifications of the product and validation

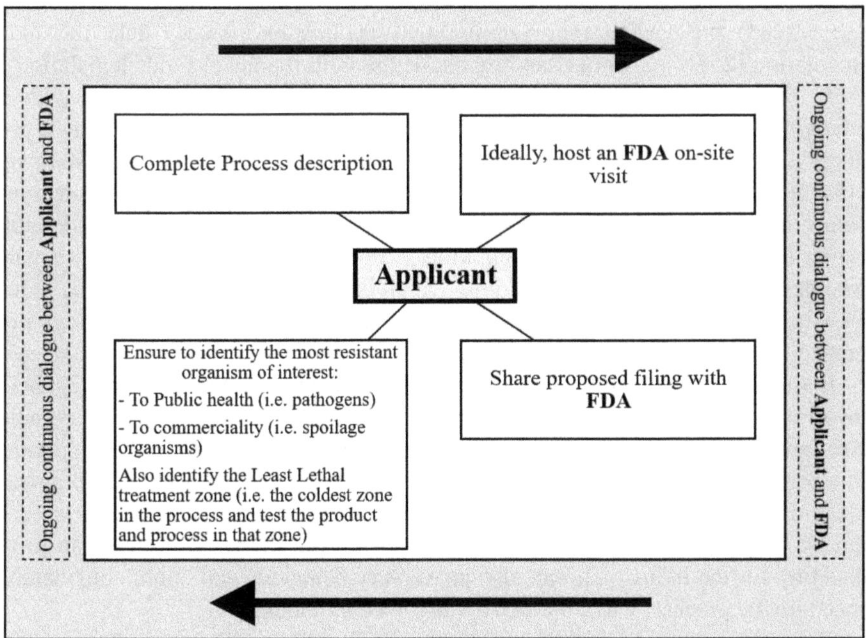

Fig. 2 Steps to be taken when filing a new and emerging process in the United States of America (Adapted from: Larkin and Spinak 1997)

(physical demonstration of the accuracy, reliability and safety of the process) (Larkin 1999).

In summary, the regulation for all 'new and emerging technologies' in the USA is not uniform. The procedures are not always clear for the industry. In some cases, a detailed petition submission appears to be required. In others, where the use of the technology is regulated as an 'additive', there may be scope to declare that its use results in foods which should be GRAS. In cases such as these, no further regulation is required. PEF technology has many potential benefits and commercial potential. A clear regulatory pathway would assist hugely and would give the industry a degree of certainty for future development.

Conclusions

In summary, while enabling trade and protecting the consumer have always featured to a certain extent in the development of food regulations, a number of high-profile food safety incidents occurring as far back as the 1980s have really raised the profile of consumer protection, in the drive for appropriate and suitable food regulation.

In the EU, responsibility for food regulation is divided: the European Commission is the ultimate risk manager, while EFSA provides scientific advice in the area of risk assessment. In terms of emerging processing technologies, the most relevant pieces of legislation are the Novel Food Regulations, and the updated Regulation (2015/2283) provides a more streamlined application process than the older Regulation (258/97). An outstanding challenge with the novel foods legislation in terms of novel processes is that just because a food may be processed with an emerging food processing technology, this does not necessarily mean that it produces a novel food, particularly if the production process does not give rise to significant changes in the composition or structure of the food to affect its nutritional value, metabolism or level of undesirable substances. While the procedure for application under the Novel Food Regulation is now relatively streamlined, knowing whether or not an application is required at all is still a huge dilemma for the industry. The guidance from the regulators is slow and the agencies do not always promote communication from industry.

In the USA, the procedure is considerably different. The two principal agencies involved in food safety are the FDA and the USDA. In these cases, both agencies conduct both risk assessment and risk management, and their portfolios are divided in terms of products rather than function. In terms of novel processes, the agency most involved is the FDA. The procedure for applying the use of a novel process in the USA involves extensive communication with the FDA. While this can prove daunting for the industry, it can also prove very beneficial and could considerably speed up the process, when the FDA is involved all along.

Given the potential role that emerging technologies may play in food processing in the future, having regulatory approval will play a large part in its development for commercial use. Whether the procedure adopted by the FDA or that of the European

Union is the most effective remains to be seen. Ultimately, the commercial uptake of emerging technologies will be driven by industry needs, but as food processing researchers, we should be mindful of the area of food regulation. As custodians of emerging technologies, we should behave responsibly and perform appropriate research to support the commercial uptake of these technologies to the betterment of society.

Acknowledgements This chapter is supported by the non-commissioned Food Research Measure, funded by the Irish Department of Agriculture, Food and the Marine and by European Union's Horizon 2020 under the ERA-NET for Sustainable Food Production and Consumption – 'SUSFOOD2', project MEFPROC (Improving Sustainability in Food Processing using Moderate Electric Fields (MEF) for Process Intensification and Smart Processing). The authors would also like to acknowledge the advice received from Dr. Patrick O'Mahony of the Food Safety Authority of Ireland and from Dr. Nathan Anderson, Center for Food Safety and Applied Nutrition, Office of Food Safety, Food and Drug Administration (USA).

References

EC – European Commission (2018) Commission implementing Regulation (EU) 2018/456 of 19 March 2018 on the procedural steps of the consultation process for the determination of novel food status in accordance with Regulation (EU) 2015/2283 of the European Parliament and of the Council on novel foods

EC – European Commission (2019a) Applications under Regulation (EC) No 258/97 of the European Parliament and of the Council. Available from: https://ec.europa.eu/food/sites/food/files/safety/docs/novel food_applications-status_en.pdf

EC – European Commission (2019b) List of authorisations under the former novel food regulation. Authorisation procedures. Available from: https://ec.europa.eu/food/safety/novel_food/authorisations/list_authorisations_en

EFSA – European Food Safety Authority (2019) Scientific Committee and panels. Available from: http://www.efsa.europa.eu/en/science/scientific-committee-and-panels

ETPs – European Technology Platforms (2016) Foods for tomorrow's consumers. Available from: http://etp.fooddrinkeurope.eu/

EU – European Union (1997) Regulation (EC) No 258/97 of the European Parliament and of the Council of 27 January 1997 concerning novel foods and novel food ingredients

EU – European Union (2001) Commission Decision of 23 May 2001 authorising the placing on the market of pasteurized fruit-based preparations produced using high-pressure pasteurisation under Regulation (EC) No 258/97 of the European Parliament and of the Council

EU – European Union (2002) Regulation (EC) No 178/2002 of the European Parliament and of the Council of 28 January 2002 laying down the general principles and requirements of food law, establishing the European Food Safety Authority and laying down procedures in matters of food safety

EU – European Union (2011) Regulation (EU) No 1169/2011 of the European Parliament and of the Council of 25 October 2011 on the provision of food information to consumers, amending Regulations (EC) No 1924/2006 and (EC) No 1925/2006 of the European Parliament and of the Council, and repealing Commission Directive 87/250/EEC, Council Directive 90/496/EEC, Commission Directive 1999/10/EC, Directive 2000/13/EC of the European Parliament and of the Council, Commission Directives 2002/67/EC and 2008/5/EC and Commission Regulation (EC) No 608/2004

EU – European Union (2012) Consolidated version of the Treaty on the Functioning of the European Union. Official Journal of the European Union, C 326, 26/10/2012 P.0001-0390
EU – European Union (2014) Commission Implementing Decision of 24 June 2014 authorising the placing on the market of UV-treated baker's yeast (*Saccharomyces cerevisiae*) as a novel food ingredient under Regulation (EC) No 258/97 of the European Parliament and of the Council
EU – European Union (2015) Regulation (EU) 2015/2283 of the European Parliament and of the Council of 25 November 2015 on novel foods, amending Regulation (EU) No 1169/2011 of the European Parliament and of the Council and repealing Regulation (EC) No 258/97 of the European Parliament and of the Council and Commission Regulation (EC) No 1852/2001
EU – European Union (2016a) Commission Implementing Decision (EU) 2016/398 of 16 March 2016 authorising the placing on the market of UV-treated bread as a novel food under Regulation (EC) No 258/97 of the European Parliament and of the Council
EU – European Union (2016b) Commission Implementing Decision (EU) 2016/1189 of 19 July 2016 authorising the placing on the market of UV-treated milk as a novel food under Regulation (EC) No 258/97 of the European Parliament and of the Council
EU – European Union (2017) Commission Implementing Decision (EU) 2017/2355 of 14 December 2017 authorising the placing on the market of UV-treated mushrooms as a novel food under Regulation (EC) No 258/97 of the European Parliament and of the Council
EU – European Union (2018a) List of national competent authorities responsible for the implementation of Commission implementing Regulation (EU) 2018/456. Available from: https://ec.europa.eu/food/sites/food/files/safety/docs/fs_novel-food_leg_list_comp_auth_reg_2018_en.pdf
EU – European Union (2018b) Commission Implementing Regulation (EU) 2018/1011 of 17 July 2018 authorising an extension of use levels of UV-treated mushrooms as a novel food under Regulation (EU) 2015/2283 of the European Parliament and of the Council, and amending Commission Implementing Regulation (EU) 2017/2470
EU – European Union (2018c) Commission Implementing Regulation (EU) 2018/1018 of 18 July 2018 authorising an extension of use of UV-treated baker's yeast (*Saccharomyces cerevisiae*) as a novel food under Regulation (EU) 2015/2283 of the European Parliament and of the Council and amending Commission Implementing Regulation (EU) 2017/2470
EU – European Union (2019a) Directorate-General Sante Available from: https://ec.europa.eu/info/departments/health-and-food-safety_en
EU – European Union (2019b) The European Union. What it is and what it does. Available from: https://europa.eu/european-union/about-eu/institutions-bodies_en
Eurostat (2014) Overweight and obesity – BMI statistics
FDA – Food and Drug Administration (2001) Hazard Analysis and Critical Control Point (HAACP); Procedures for the safe and sanitary processing and importing of Juice. F.D.A. (US) Federal Register: 6137–6202
FDA – Food and Drug Administration (2014) Guidance for industry: assessing the effects of significant manufacturing process changes, including emerging technologies, on the safety and regulatory status of food ingredients and food contact substances, including food ingredients that are color additives: 36533–36534
FDA – Food and Drug Administration (2016) Substance generally recognized as safe. Available from: https://www.federalregister.gov/documents/2016/08/17/2016-19164/substances-generally-recognized-as-safe#h-18
FDA – Food and Drug Administration (2017) Food Safety Modernization Act (FSMA). Available from: https://www.fda.gov/food/guidance-regulation-food-and-dietary-supplements/food-safety-modernization-act-fsma
FDA – Food and Drug Administration (2019a) Code of Federal Regulations. Title 21. Available from: https://www.accessdata.fda.gov/scripts/cdrh/cfdocs/cfcfr/cfrsearch.cfm
FDA – Food and Drug Administration (2019b) Consultations on food from new plant varieties. Available from: https://www.accessdata.fda.gov/scripts/fdcc/?set=Biocon
FDA – Food and Drug Administration (2020) How GMOs are regulated for food and plant safety in the United States. Available from: https://www.fda.gov/food/agricultural-biotechnology/how-gmos-are-regulated-food-and-plant-safety-united-states

FSAI – Food Safety Authority Ireland (2016) Authorisation letter of Food Safety Authority Ireland, 26 February 2016. Application for the approval of UV-treated mushrooms with increased vitamin D levels under the novel food Regulation (EC) No. 258/97

FSAI – Food Safety Authority Ireland (2017) Authorisation letter of Food Safety Authority Ireland, 28 August 2017. Application for the approval of UV-treated mushrooms (*Agaricus bisporus*) as a novel food under Regulation (EC) No. 258/97

FSAI – Food Safety Authority Ireland (2019) Irradiation of food. Available from: www.fsai.ie

Góngora-Nieto MM, Sepúlveda DR, Pedrow P et al (2002) Food processing by pulsed electric fields: treatment delivery, inactivation level, and regulatory aspects. Lwt-Food Sci Technol 35(5):375–388

Gouma M, Álvarez I, Condón S, Gayán E (2020) Pasteurization of carrot juice by combining UV-C and mild heat: impact on shelf-life and quality compared to conventional thermal treatment. Innov Food Sci Emerg 64:102362

Johnston JL, Fanzo JC, Cogill B (2014) Understanding sustainable diets: a descriptive analysis of the determinants and processes that influence diets and their impact on health, food security, and environmental sustainability. Adv Nutr: 2156–5376 (Electronic) doi:https://doi.org/10.3945/an.113.005553

Kim SS, Park SH, Kang DH (2018) Application of continuous-type pulsed ohmic heating system for inactivation of foodborne pathogens in buffered peptone water and tomato juice. Lwt-Food Sci Technol 93:316–322

Lalor F (2016) The regulatory environment for the food supply chain. In: Pataro G, Lyng JG (eds) High intensity pulsed light in processing and preservation of foods. Nova Science, New York, pp 251–265

Larkin JW (1999) Regulatory issues and consumer acceptance in non-thermal food preservation. Paper (30–6) presented at the Institute of Food Technologists Annual Meeting. Chicago, IL, USA

Larkin JW, Spinak SH (1997) Aspects of new technology implementation in the U. S. paper presented at the New Technologies Symposium NFPA Convention, Chicago, IL, USA

Monforti-Ferrario F, Dallemand JF, Pascua IP et al (2015) Energy use in the EU food sector: State of play and opportunities for improvement. JRC Science and policy report, European Commission

Morehouse KM (2002) Food irradiation—US regulatory considerations. Radiat Phys Chem 63(3):281–284

Morehouse KM, Komolprasert V (2004) Irradiation of food packaging. ACS Symp Ser 875:1–11

Sastry S (2008) Ohmic heating and moderate electric field processing. Food Sci Technol Int 14(5):419–422

Sitzmann W (1995) High voltage pulse techniques for food preservation. In: Gould GW (ed) New methods of food preservation. Blackie Academic & Professional, London, p 236

Tavoschi L, Severi E, Niskanen T et al (2015) Food-borne diseases associated with frozen berries consumption: a historical perspective, European Union, 1983 to 2013. Eur Secur 20(29):21193

Topalcengiz Z (2019) Assessment of recommended thermal inactivation parameters for fruit juices. Lwt-Food Sci Technol 115:108475

Twisselmann B (2000) Outbreak of listeria gastroenteritis in Italy caused by contaminated corn salad. Weekly releases (1997–2007) 4(18): 1610

Vega-Mercado H, Gongora-Nieto MM, Barbosa-Canovas GV, Swanson BG (2007) Pulsed electric fields in food preservation. In: Rahman MS (ed) Handbook of food preservation. Food science and technology. CRC Press, Taylor & Francis Group, Boca Raton, pp 783–813

Vogt DU, Parish M (2001) Food biotechnology in the United States: science, regulation, and issues. CRS Report for Congress. Received through the CRS Web. Available from: https://www.everycrsreport.com/reports/RL30198.html

WRI – World Resources Institute (2016) Shifting diets for a sustainable food future. Available from: http://www.wri.org/publication/shifting-diets

GPSR Compliance
The European Union's (EU) General Product Safety Regulation (GPSR) is a set of rules that requires consumer products to be safe and our obligations to ensure this.

If you have any concerns about our products, you can contact us on

ProductSafety@springernature.com

In case Publisher is established outside the EU, the EU authorized representative is:

Springer Nature Customer Service Center GmbH
Europaplatz 3
69115 Heidelberg, Germany

www.ingramcontent.com/pod-product-compliance
Ingram Content Group UK Ltd.
Pitfield, Milton Keynes, MK11 3LW, UK
UKHW021832270726
14058UKWH00001B/108
9783030705886